D0899179

Officially Withdrawn

The Foundations of Chemical Kinetics

The Foundations of
CHEMICAL KINETICS

Sidney W. Benson

PROFESSOR OF CHEMISTRY
UNIVERSITY OF SOUTHERN CALIFORNIA

28600

McGRAW-HILL BOOK COMPANY, INC.

New York   Toronto   London

1960

THE FOUNDATIONS OF CHEMICAL KINETICS

II
04778

This book is dedicated to the incomparable Pidge and to little Agoo and to Pidge's less substantial creatures—Cowardly Panther, Shabby Scales, and Midnight Rose—in whose delightful company many kinetic conceptions were simmered, distilled, and brought to more material form.

Preface

The Malthusian growth of the world's population, while more talked about, is not more striking than the concurrent expansion of the world's knowledge, particularly in the natural sciences, which include chemistry and chemical kinetics. One lamentable consequence of this explosive growth is illustrated by what we may describe as the "plight of the cultured kineticist." In 1920 this enthusiastic fellow could keep abreast of most of the developments in his field and find a common ground of discourse with any of his not too numerous colleagues. By 1960, chemists working in different areas of kinetics may find that they are not only unfamiliar with each other's work but that even their languages are strange to each other.

The present volume is an effort, in part, to present to the student at the senior or graduate level a broad survey of what the field of kinetics has become and acquaint him with the basic models and language in current use. To this end the first five chapters have been designed as a phenomenological description of kinetics and the last three chapters as a brief guided tour through the work on reactions in condensed states (solutions and solids). These, together with selected parts of Chaps. XI, XII, and XIII on gas-phase reactions, would comprise the content of an average graduate course in kinetics.

It was felt, however, that there is an important need on the part of the specialized student of kinetics for a more detailed and rigorous treatment of kinetic theory and practice which should be met within the confines of the same volume. It is for this second audience that a number of more detailed sections have found their way into the early chapters and that the very extensive discussion in Chaps. VI to XIV has been included.

Because the molecular theory of gases is so much more extensively developed than that of liquids or solids, the kinetics of gas reactions have been heavily emphasized relative to solutions and solids. The extensive discussion in Chap. XIII of the mechanism of complex gas-phase reactions is perhaps an exaggeration reflecting the peculiar tastes and interests of the author but may possibly be justified in terms of the importance of mechanism to the proper care and treatment of kinetic data. One of the very important aspects of the kinetic art, and it is still an art, is the dissection of an observed rate into the rate constants for elementary molecular steps. The lack of adequate consideration of mechanism was one of the principal reasons for a stalemate to theoretical progress in kinetics from the year 1929 until shortly after World War II.

In so large a book, one might expect to find some coverage of all topics in kinetics. This is, alas, not true of the present volume, and it is regretfully admitted that a number of topics, even some important ones, have either been slighted or altogether neglected. For these omissions the author makes the standard excuses of lack of space and conflicts in interest. He also begs the indulgence of his hard-working fellow kineticists whose efforts have been so lightly treated or ignored and hopes that what has been presented may compensate in part for what has been omitted.

The problems are arranged to follow the order of exposition of material in the text. The author has found that the last problem, labeled "special assignment" alias "operation referee," has been an extremely stimulating one to both students and instructor and very useful in judging how successful the course has been in rubbing off some kinetic sophistication and understanding on the student. He heartily commends it to other instructors.

The present book, which was begun more years ago than it is pleasant for the author to remember, has been nursed and encouraged in its tortoise-like progress by support to the author from a number of agencies. To these, which include the Guggenheim Foundation, the Fulbright Committee, the National Science Foundation, and the Office of Ordnance Research, I would like to acknowledge my grateful appreciation. This work would certainly have been impossible without their individual and joint assistance. I would also like to thank Prof. E. Bauer and Prof. M. Magat of the Laboratoire de Chimie Physique in Paris and Prof. Linus Pauling of the California Institute of Technology for the generous hospitality extended to me at those institutions during two separate sabbatical years.

In its various stages of metamorphosis this book has been inflicted upon captive students and unwary colleagues alike. To the former, some of whose names appear in the text at strategic points, I wish to express my deepest gratitude. Among the latter, I am specially indebted to Prof. Michel Magat, Prof. Richard A. Ogg, Jr., and Prof. Edward L. King for many stimulating discussions on kinetics and other topics. Dr. Alastair North and Prof. Martin Kilpatrick have been very helpful with portions of the text, and Professor King has been more than kind in criticizing large sections of it. Professor Richard M. Noyes has offered many valuable and detailed suggestions in the preparation of the final manuscript. Finally I should like to acknowledge my most profound indebtedness to the many kineticists whose work has served as guide and inspiration, in particular Prof. George B. Kistiakowsky, in whose classes and laboratory at Harvard many of the vistas of kinetics were first opened to the author. In a similar but less personal vein is the debt owed to Dr. Louis S. Kassel for his pioneering treatise on gas-phase kinetics.

Sidney W. Benson

Contents

PART II. STATISTICAL METHODS FOR TREATING SYSTEMS OF LARGE NUMBERS OF PARTICLES AT OR NEAR EQUILIBRIUM

PART III. THE KINETICS OF HOMOGENEOUS REACTION
IN GASES

PART ONE

The Phenomenological Description of Chemical Rate Processes

I

The Description of Reacting Systems

1. Description of Equilibrium Systems

In most general terms, physical chemistry has as its task the compact, quantitative description of the properties and behavior of matter. Within this framework we can make a useful, if sometimes arbitrary, distinction between systems whose properties do not change with time and those systems whose properties are time-dependent. Chemical kinetics is that branch of physical chemistry concerned with the study of the latter systems, and in particular with the subgroup of those systems whose chemical composition is changing with time, i.e., systems in which chemical reactions are occurring.

Precisely what is meant by the description of a chemical system? A description of a system may involve an almost infinite list of statements concerning system properties. We shall use "description" in the much more restricted sense indicated by the phrase "minimum description," namely, the minimum number of statements necessary and sufficient to describe the system such that from the information a similar system possessing identical properties may be constructed.

Thus for a "pure" substance in a state such that the properties of the substance are not changing with time, a minimum description would contain statements about the chemical composition, mass, pressure, temperature, volume, and state (gas, liquid, or solid) of the substance and the magnitude and position of external force fields. Of the first group of quantities one is extraneous, since there is a thermodynamic equation relating them, and from it, in principle, any one quantity may be calculated, given the value of the others.

From such a "minimum description," it is possible to reconstruct a similar system, identical in all properties. This implies that the gravimetric, optical, electrical, and other properties are uniquely determined by the thermodynamic state of the system and will be reproduced by reproducing the thermodynamic state.

In a similar fashion, thermodynamics provides us with a method for giving a minimum description of chemical systems (at equilibrium) of any degree of complexity. From this viewpoint, the program of thermodynamics can be formulated as the search for the relations which exist between two equilibrium states of a system and the external changes in energy (heat, work, etc.) that accompany the transformations from one state to another.

2. Description of Kinetic Systems

For the description of systems whose properties are changing with time, we may propose an extension of our previous definition. A minimum description of a kinetic system is the statement of the necessary and sufficient information which will permit us, *at each instant in time*, to construct (in principle, if not in practice) a similar system having identical properties.

A kinetic system is a system in unidirectional motion. It is not in a state of equilibrium, and although conforming to the first law of thermodynamics (conservation of energy), it escapes the complete restriction of the second law. Consequently, with fewer constraints on the system, and thus more freedom, the system becomes more difficult to describe. In fact, as we shall see later on, this difficulty in description becomes one of the real obstacles in the path of a satisfactory kinetic treatment. An even more formidable obstacle to description, however, lies in the multiplicity of essentially nonequilibrium factors which may under different conditions play a decisive role in determining the reaction path. There is, a priori, no simple compact statement of what constitutes an adequate description of a kinetic system. It is not difficult to see why in terms of a simple analogy.

A body of water on top of a hill may be described in terms of its equilibrium state. Similarly, the same body of water at some subsequent time may find its way to a lake at the bottom of the hill. There are perfectly definite descriptions of both states and of the energy differences between them. However, if we try to describe the transition, the water in process of flowing from the hilltop, we see that it may depend on almost innumerable factors: on the outlets, on the contour of the hillside, on the structural stability of the contour, on the numerous subterranean channels through the hillside that may exist and permit seepage. And finally, if someone has bored a hole under the hilltop, it will take careful experimental investigation to uncover this additional factor which will affect the flow.

The history of kinetic studies is replete with examples of important, hidden factors which may play a decisive role in the course of a reaction and which were not discovered for many years. In this sense, a good part of chemical kinetics must remain empirical for some time to come. This fact must not, however, detract from the formidable progress made in the quantitative understanding of the factors which have been found to influence the course of chemical reactions and which must provide the basis for future advances.

3. Phenomenological Approach

The most obvious goal of a kinetic study of a particular system is a purely empirical one: to discover methods for minimally describing the system. Stated differently, what are the quantities which will serve to uniquely define the system at each instant in time?

Implicit in such an investigation is the hope of subsequently finding some relation or relations between these quantities such that the knowledge of their magnitudes at some instant in time will be sufficient to permit a complete prediction of the future properties and states of the system. Thus in the case of one of the very simple reactions which seem to be understood, the decomposition of hydrogen iodide to give molecular hydrogen and iodine, it is possible, given the initial concentration of each species, the temperature, and some chemical constants, to predict the description of the system at all subsequent times.

An investigation such as the one described may be termed "phenomenological." It is concerned with describing the behavior and properties of the system in terms of macroscopically observable quantities such as pressure, temperature, composition, volume, and time. It satisfies the goal of providing an empirical description and prediction of reacting systems, and must of necessity precede other types of investigation. The results of such inquiries will be expressed in terms of general macroscopic variables (pressure, temperature, concentration, etc.), "chemical constants" which are specific for the particular system studied (e.g., specific rate constants, activation energies) and time-dependent relations between the constants.

4. Molecular Approach

A phenomenological description of nature is necessarily incomplete. With every new step that is made in our understanding of the microscopic or molecular structure of matter there is an added compulsion to review our macroscopic knowledge and relate or reduce it in terms of molecular structure. The goal of this molecular approach is thus to understand the macroscopic properties of systems in terms of their molecular structures and ultimately to reduce the chemical constants to molecular constants.

It is precisely such a program which has created a renewed interest in the phenomenological findings of kinetic studies in recent decades. The molecular interpretations of kinetic processes throw a new light on molecular structure, which adds to our understanding in this field. At the same time these interpretations must themselves conform to and be consistent with the knowledge gained from such diverse studies as measurement of dipole moments, X-ray and electron diffraction, heats of reaction, and stereochemistry.

In the present text we shall begin with a study of the phenomenological aspects of kinetic systems and then proceed to a critique of the bases for a molecular interpretation of the processes in these systems.

5. Macroscopic Variables of a Chemically Reacting System

As already stated, the macroscopic variables which influence the rate of a chemical reaction are not available from simple postulates, but on the contrary must be determined experimentally. For any particular system they may be large or small in number. Let us summarize some of the findings on this point.

5A. Temperature. Kinetic systems may be studied under a variety of experimental conditions. Within certain limits it is possible to impose on the system certain external constraints which simplify the study by reducing the number of variables. Thus it is general practice to enclose the reacting system in a thermostat to maintain it at constant temperature. This is a convenient method for isolating the temperature as a variable and studying its effect independently of the other variables. Reactions which proceed at constant temperature are called *isothermal*.

5B. Pressure and Volume. Two of the other important system variables which are usually at our disposal experimentally are volume and pressure. In the study of reactions between gases it is possible to keep either the pressure or the volume of the system fixed. The simplest procedure is to maintain a gas system in a vessel at fixed volume. For reactions in liquid and solid systems, the pressure is most conveniently controlled, volume control being either unimportant or unattainable owing to the small coefficient of compressibility of liquids.[a]

5C. Chemical Composition; Concentration of Reactants. The first important studies of reaction systems were begun with a study of the effect of the concentrations of the reacting species on the rate of a reaction. For gas reactions, concentrations are directly related through the equation of state to pressure, volume, and temperature. For liquid reactions the pressure assumes secondary importance as a variable, the volume of the system being very insensitive to either pressure or temperature changes.

Since the chemical stoichiometry of the reaction fixes relations between

---

[a] It is not always possible to isolate the experimental variables; and especially in rapid reactions, nature fixes its own conditions. In the cases of very rapid reactions, such as explosions, the reaction proceeds almost *adiabatically;* i.e., there is almost no heat exchange between the system and its surroundings during the course of the reaction. For obvious reasons such reactions are assuming an increasing practical importance.

the concentrations of the different species involved in the reaction, each particular concentration does not necessarily enter as an independent variable. Thus in the production of hydrogen iodide, $H_2 + I_2 \rightarrow 2HI$, the moles of hydrogen and iodine consumed must be equal to each other, while the moles of HI produced are twice as great as either.

5D. Chemical Composition; Concentration of Inert Substances. In a great many studies it has been found that an added substance which is not consumed in the chemical reaction and may thus be considered as "inert" has a pronounced effect on the rate of reaction. Thus in the case of the gas-phase thermal decomposition of certain hydrocarbons, ethers, and aldehydes[a] it has been found that added quantities of H_2, He, or N_2 may under certain conditions increase the rate of the reaction.[b] (H_2 is probably chemically active in these systems; see Chap. XIII.)

In liquid-phase reactions such effects may be much more pronounced. It is frequently found that a change in solvent may change the rate of a liquid reaction, in some cases by many orders of magnitude.[c to e] A most spectacular example of this is the reaction of ethyl iodide with trimethylamine to form the quaternary salt. The reaction rate is almost 1000 times greater in nitrobenzene than in cyclohexane under comparable conditions.

5E. Chemical Composition; Catalysis. Careful study has shown that certain substances present in a reaction system in only small quantities may have a considerable effect on the rate of the reaction. In the cases in which these substances are not consumed chemically the phenomenon is referred to as catalysis.

If the effect on the rate is to increase it, the substance is called a promoter (positive catalyst). If, on the other hand, the substance decreases the rate, it is called an inhibitor or retarder. Thus it has been found that the rate of decomposition of ClO^- ion, $2ClO^- \rightarrow 2Cl^- + O_2$, in aqueous solution is enormously increased by small concentrations of hydrogen ion.[f] Similarly it has been quite spectacularly demonstrated that small amounts of $HBr(g)$ can bring about the rapid oxidation of hydrocarbons at temperatures at which this process is otherwise infinitesimally slow.[g] Perhaps one of the most interesting examples investigated of the catalytic effect of trace impurities is the isomerization of normal to isobutane.[h]

---

[a] C. N. Hinshelwood and L. A. K. Staveley, *J. Chem. Soc.*, 818 (1936); *Proc. Roy. Soc.* (*London*), **A159**, 192 (1937).

[b] H. F. Cordes and H. S. Johnston, *J. Am. Chem. Soc.*, **76**, 4264 (1954). The decomposition of $NO_2Cl \rightarrow NO_2 + \frac{1}{2}Cl_2$ is actually first-order in Ar (Chap. XII) and first-order in NO_2Cl when Ar is present in large excess.

[c] H. G. Grimm et al., *Z. physik. Chem.*, **B13**, 301 (1931).

[d] N. Pickles and C. N. Hinshelwood, *Trans. Chem. Soc.* (*London*), **121**, 1353 (1936).

[e] N. Menschutkin, *Z. physik. Chem.*, **6**, 41 (1890).

[f] L. R. B. Yeatts and H. Taube, *J. Am. Chem. Soc.*, **71**, 4100 (1949).

[g] F. F. Rust and W. E. Vaughan, *Ind. Eng. Chem.*, **41**, 2595 (1949).

[h] P. A. Leighton and J. D. Heldman, *J. Am. Chem. Soc.*, **65**, 2276 (1943).

Aluminum chloride acts as a catalyst for this reaction, but only if hydrogen chloride and also trace amounts (0.001 mole per cent) of some olefin are present. Without the trace amount of olefin the reaction is extremely slow.[a]

An example of an inhibitor is the well-known poisoning of the contact catalyst platinum by As[b] and the inhibition of the explosion of $P_4(g)$ and O_2 by argon or nitrogen.[c]

In general, negative catalysts are rare. Most substances which appear to have the properties of negative catalysts turn out on closer analysis to be consumed during the reaction. A typical example is the inhibition of the thermal decomposition of diethyl ether by nitric oxide.[d]

5F. Effect of Surfaces; Heterogeneous Reactions. A system in which a reaction takes place more or less uniformly throughout the bulk of a single physical phase is said to be a *homogeneous reaction*. This is generally the case with most simple reactions occurring in solution. On the other hand, if the reaction occurs only at the interface between two phases, the reaction is said to be *heterogeneous*. There are a great many important examples of this type; among them may be noted the contact process for the reaction of SO_2 and O_2 at the surface of a platinum-asbestos catalyst and the familiar hydrogenation of unsaturated compounds in liquid suspensions of the Raney nickel catalyst, NiO_2.

Fitting neither of these extreme categories is another group of reactions, referred to as chain reactions, in which the rate of the reaction may be influenced by the chemical composition, total extent, and actual geometry of the surface enclosing the reacting system. Although these reactions have been classified as heterogeneous, the classification is not precise, because the reaction is not confined to the surface layers; rather, the surface contributes to or modifies processes occurring in the bulk of the gas phase. The chain oxidations of hydrogen, carbon monoxide, hydrocarbons, and phosphorus are typical examples of such reactions. Most of the gas-phase reactions which have been studied fall into this category.

6. Plotting a Course

It should be apparent from the preceding discussion that the science of chemical kinetics is as yet far from the stage at which we may state a few general laws which will suffice to explain the behavior of reacting systems. The present situation is one in which relatively few reactions have been extensively investigated over a truly broad range of experimental conditions. And of these, a majority appear to be extremely complex.

[a] H. Pines and R. C. Wackher, *J. Am. Chem. Soc.*, **68**, 595, 599, 1642 (1946).

[b] E. B. Maxted and R. G. Norrish, *J. Chem. Soc.*, 839 (1938); 252 (1940); 132 (1941).

[c] J. Tausz and H. Görlacher, *Z. anorg. Chem.*, **190**, 95 (1930).

[d] J. R. E. Smith and C. N. Hinshelwood, *Proc. Roy. Soc. (London)*, **A130**, 237 (1942).

However, from the mass of experimental data which has been accumulated there do appear certain common aspects which lend themselves to general classification and semiquantitative interpretation. It seems only proper, therefore, to begin our study with a consideration of these features, which have provided the most fruitful means for the investigation and description of reacting systems.

II

Mathematical Characterization
of Simple Kinetic Systems

1. Reaction Order

Conventional methods for investigating particular reaction systems usually begin with attempts to isolate the individual factors affecting the rate of the reaction so that each may be studied separately. Thus a vessel made in some particular size and shape of some "inert" material is chosen. The vessel is brought to constant temperature in a thermostat and the reaction materials (preheated if possible to the same temperature) are introduced as rapidly as possible with efforts made to ensure complete mixing.

Generally the chemical concentrations are known (from analysis) at the moment of mixing. The progress of the reaction is then studied as a function of time by any of a large number of available techniques. Similar experiments may then be performed by using different initial concentrations of the various materials and different temperatures.

In a typical experiment, the data usually consist of the concentrations of the various compounds present in the reaction mixture at different times during the course of the reaction. The first problem is to see if the data may be expressed in simple mathematical form. That is, we would like to see if we can find an equation for each species which represents the concentration of the species as a function of time.

Intuitively we would expect the data to show a monotonic decrease with time in the concentrations of each of the reacting species and a monotonic increase with time in the concentrations of the final products. Aside from this condition there is no a priori reason to expect the curves to be very regular or simple. It is therefore somewhat surprising to find that, for a preponderance of the reactions that have been studied, the curves of con-

10

centrations vs. time may be classified into a few very simple categories.

This finding is best represented in a language somewhat different from the direct language of our experimental data. If we talk for a moment, not of the dependence of the concentration C_A of a given species A on time, but of the dependence of the time derivative of the concentration dC_A/dt on the concentrations of the various species present, we find that usually the relation may be expressed as[a]

$$\frac{dC_A}{dt} = kC_A^\alpha C_B^\beta C_C^\gamma \cdots \qquad (II.1.1)$$

where k is some numerical constant; α, β, γ, . . . are usually integers; and C_A, C_B, C_C, . . . refer to the concentrations (in appropriate units) of the chemical species A, B, C, . . . present in the reacting system at a given time. Similar expressions will then be found for the time rate of change of the concentrations of the other species, dC_B/dt, dC_C/dt, and so on.

In terms of such an empirical description the particular reaction is said to have an order α with respect to species A, an order β with respect to species B, etc., and an over-all order given by the sum $\alpha + \beta + \gamma + \cdots$. The individual derivatives dC_A/dt, dC_B/dt, and dC_C/dt are referred to as the *rate of the reaction* with respect to species A, B, C, respectively, and are related by the stoichiometry of the chemical reaction.

In the reaction of $NO(g)$ with $O_2(g)$ given by $2NO + O_2 \rightarrow 2NO_2$

$$\frac{dC_{NO}}{dt} = \frac{2\,dC_{O_2}}{dt} = -\frac{dC_{NO_2}}{dt}$$

since O_2 is consumed at half the molar rate at which NO is consumed, the latter being equal to the molar rate of production of NO_2. In this particular reaction it is found experimentally that

$$\frac{dC_{NO}}{dt} = -k_{NO}C_{NO}^2 C_{O_2}$$

so that the reaction is experimentally third-order $(2 + 1)$, being second-order with respect to NO, first-order with respect to O_2, and zero-order with respect to NO_2.

When a chemical reaction may be described by a single stoichiometric equation, it is possible to define a rate of reaction for the system uniquely. The convention we shall use in this book is as follows: For the stoichiometric equation

$$aA + bB + \cdots \rightleftharpoons pP + qQ + \cdots$$

the rate of reaction R will be defined as:

$$R = -\frac{1}{a}\frac{dA}{dt} = -\frac{1}{b}\frac{dB}{dt} = \frac{1}{p}\frac{dP}{dt} = \frac{1}{q}\frac{dQ}{dt} = \cdots \qquad (II.1.2)$$

[a] For more complex systems, the rate law may be expressed as a sum of such terms.

where a, b, p, q, . . . are the coefficients in the balanced equation and for simplicity we replace C_A, C_B, . . . by A, B, ...

2. Nomenclature

The reader will observe that the discussion of the empirical description of kinetic systems in the preceding section was resolved by a verbal sleight of hand. Whereas the raw kinetic data are usually in the form of concentrations of chemical components at given times (and indeed our aim is to reproduce such data by mathematical equations that represent these concentrations as functions of the time), our very first statement involved an entirely different language, that of the concentration derivatives (the rates of reaction) and their relations to concentrations.

It is by no means accidental or arbitrary that the empirical and also, as we shall see later, the theoretical description of reaction systems should choose, as concepts providing a more convenient basis for discussion, the terms occurring in a differential equation rather than the integrated equation which this may yield.

The instantaneous rates at which concentrations are changing in time, represented by the derivatives dC/dt, are quantities which have an intuitive physical content which is missing in an integrated equation stating that the concentration itself is some particular function of time [for example, $C = C(t)$]. The time t in such equations is a parameter which does not convey much physical meaning. The time derivative dC/dt, however, represents a physically meaningful quantity: a velocity, a rate, which we can relate to other rate processes, the speed with which matter moves, with which molecules travel. And further we are interested not in the dependence of this rate upon time, but rather in the properties of the physical system itself, in this case the other concentrations. Thus the language of our investigation is directed toward the time derivatives of experimental quantities and the relation of these quantities to the properties of the system.[a]

The most meaningful units for expressing chemical concentrations are moles (or molecules) per unit of volume. The units for expressing specific reaction rates will, then, be concentration change per unit of time. Typical rate units are thus moles/liter-sec and molecules/cc-sec.

Since the empirical representation of reaction rates is in terms of the dependence of the reaction rates on other concentrations through a proportionality factor k, called the specific rate constant, the units for this specific rate constant k will depend on the order of the reaction. These

---

[a] This is a situation which has numerous analogies in other branches of science, for example, in Newton's laws of motion. In general, we find that time occurs in the equations of science in an implicit, rather than explicit, form.

units are, of course, inherent in the equation itself and are summarized in Table II.1.

3. Zero-order Reactions

According to our formulation, a reaction rate is said to be of zero order if the rate of the reaction is independent of the concentrations of the substances participating in the reaction. In that case the rate of change of the concentrations is a constant:

$$-\frac{dC}{dt} = k \qquad \text{(II.3.1)}$$

and on integration
$$C = -kt + \theta \qquad \text{(II.3.2)}$$

where θ is a constant of integration to be determined from the conditions of the reaction.

If the reaction may be designated symbolically by

$$aA + bB \rightarrow \text{products}$$

where a and b are integral coefficients in the balanced chemical equation, then

$$\frac{dC_A}{dt} = -k_A \qquad \frac{dC_B}{dt} = -k_B$$

TABLE II.1. UNITS OF SPECIFIC RATE CONSTANTS FOR REACTIONS
OF DIFFERENT ORDER

Total order of reaction	0	1	2	3	n
Units of specific rate constant k	$\dfrac{\text{moles}}{\text{liter-sec}}$	sec^{-1}	$\dfrac{\text{liters}}{\text{mole-sec}}$	$\dfrac{\text{liters}^2}{\text{mole}^2\text{-sec}}$	$\dfrac{\text{liters}^{n-1}}{\text{mole}^{n-1}\text{-sec}}$

Note: This table is based on the adoption of moles/liter for concentration units. More explicitly, the units should include the species involved. Thus for the reaction $H_2 + I_2 \rightarrow 2HI$, which is experimentally first-order with respect to H_2 and first-order with respect to I_2, the proper units of k_{HI} are

$$\frac{\text{moles HI} \times \text{liters}}{\text{moles } I_2 \times \text{moles } H_2 \times \text{sec}}$$

Since conventional usage has been to avoid identification of moles with species, this is written simply as liters/mole-sec. Such abbreviation can, however, lead to occasional confusion, and the nature of the abbreviation should be borne in mind.

Conversions to other units may be made by the application of the appropriate conversion factors to the above quantities.

where k_A and k_B are the specific rate constants for the disappearance of A and B, respectively.[a] From the balanced equation we see that

[a] The use of the minus sign in the equation is conventional so that the rate constants k will always be expressed as positive quantities. In the above case dC_A/dt and dC_B/dt are negative quantities because A and B are consumed as the reactions proceed.

$$R = -\frac{1}{a}\frac{dC_A}{dt} = -\frac{1}{b}\frac{dC_B}{dt} \qquad\qquad \text{(II.3.3)}$$

so that
$$\frac{k_A}{a} = \frac{k_B}{b} \qquad\qquad \text{(II.3.4)}$$

The integrated form of the above equations becomes

$$-C_A = k_A t + \theta_A \qquad -C_B = k_B t + \theta_B \qquad\qquad \text{(II.3.5)}$$

If at some time, which we can for convenience designate as t_0, the concentrations of A and B are respectively C_{A_0} and C_{B_0}, then we can eliminate the constants of integration θ_A and θ_B by substitution in the equations and write

$$C_A - C_{A_0} = -k_A(t - t_0) \qquad C_B - C_{B_0} = -k_B(t - t_0) \qquad \text{(II.3.6)}$$

There are very few cases of reactions of zero total order, and so these equations do not have wide application.[a] Most of the known cases involve heterogeneous reactions occurring on surfaces, such as the decomposition of nitrous oxide gas on hot platinum wire,[b] $2N_2O \rightarrow 2N_2 + O_2$, and the decomposition of ammonia gas on a hot platinum wire,[c] $2NH_3 \rightarrow N_2 + 3H_2$. The explanation seems to be that the reaction occurs only at the surface of the catalyst, and if the surface becomes saturated at a given gas or liquid concentration, further increase in the concentration in these phases cannot further change the surface concentration and so, beyond this point, the reaction seems to proceed at a rate independent of concentration in the gas phase.

4. First-order Reactions

A first-order reaction is one in which the rate of the reaction is proportional to the concentration of only one of the reacting substances. Algebraically,

$$\frac{dC_A}{dt} = -k_A C_A \qquad\qquad \text{(II.4.1)}$$

in which C_A is the concentration of species A being consumed (note minus sign) in the reaction and k_A is its specific rate constant. The reaction may be written as

$$aA + bB + \cdots \rightarrow \text{products}$$

Rearranging the terms, the differential equation becomes

$$\frac{dC_A}{C_A} = -k_A\, dt$$

---

[a] It is not at all uncommon for a rate to be zero-order with respect to one or more reactants.

[b] C. N. Hinshelwood, *J. Chem. Soc.*, **127**, 327 (1925).

[c] C. N. Hinshelwood and R. E. Burk, *J. Chem. Soc.*, **127**, 1105 (1925).

which can be immediately integrated to give

$$\ln C_A = -k_A t + \theta_A \tag{II.4.2}$$

where again θ_A is a constant of integration to be determined by the given experimental conditions. If the concentration of C_A is C_{A_0} at time $t = t_0$ (which we can set equal to zero for convenience), then θ_A can be eliminated by substitution and we find

$$\ln \frac{C_A}{C_{A_0}} = -k_A t \tag{II.4.3}$$

By using decadic instead of natural logarithms this may be written as

$$\log \frac{C_A}{C_{A_0}} = -0.434 k_A t \tag{II.4.4}$$

In exponential form

$$C_A = C_{A_0} e^{-k_A t} \quad \text{or} \quad C_A = C_{A_0} 10^{-0.434 k_A t} \tag{II.4.5}$$

The integrated form of the first-order equation (II.4.3) provides us with a simple graphical method of representation as shown in Fig. II.1a, in which

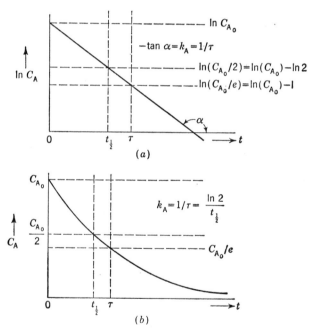

FIG. II.1. Graphical representation of first-order reactions. (a) Semilogarithmic plot; abscissa: time t (in arbitrary units); ordinate: natural logarithm of C_A (in arbitrary units). (b) Direct plot; abscissa: time t (in arbitrary units); ordinate: concentration of A (in arbitrary units).

$\ln C_A$ is plotted as a function of time t. For purposes of comparison Fig. II.1b shows the appearance of the data when the concentration C_A is plotted directly against the time. The logarithmic plot of Fig. II.1a is frequently used for experimental representation of data, because it is possible to present the data by a straight line. From Eq. (II.4.3) we see that minus the slope of this line, $-\tan \alpha$, is equal to the specific rate constant for the reaction k_A.

It is convenient to define two other quantities related to the specific rate constant. The first is called the *mean life* or *decay time* of the reaction and is represented by the symbol τ. It is defined as the time for the concentration of A to fall to $1/e$ of its initial value, where e is the natural number 2.718. By substitution in Eq. (II.4.3), we see that C_A will reach the value C_{A_0}/e after a time $\tau = 1/k_A$. That is, the mean life τ of the reaction is equal to the reciprocal of the rate constant.

An equally useful quantity is the half-life of the reaction designated by the symbol $t_{1/2}$. It is defined as the time required for the concentration C_A to fall to one-half of its initial value $C_{A_0}/2$. Again by substitution in Eq. (II.4.3) we see that C_A will have the value $C_{A_0}/2$ at the time

$$t = \frac{\ln 2}{k_A} = \frac{0.693}{k_A} = t_{1/2} \tag{II.4.6}$$

The useful property of the half-life $t_{1/2}$ is that, by virtue of the exponential decrease in concentration, the reaction will consume equal fractions of A in consecutive periods of time $t_{1/2}$. Thus the concentration will fall to $C_{A_0}/2$ in a time $t_{1/2}$. In an additional time $t_{1/2}$ it will fall to one-half of this value, or $C_{A_0}/4$, and so on.

It should be observed that both τ and $t_{1/2}$ are directly related to the specific rate constants and are independent of the initial concentrations.

There are many reactions which follow a first-order kinetic representation over a certain range of experimental conditions.[a] Some well-known cases are the gas-phase decomposition of N_2O_5, $N_2O_5 \rightarrow 2NO_2 + \frac{1}{2}O_2$;[b] the inversion of cane sugar, $C_{12}H_{22}O_{11} + H_2O \rightarrow 2C_6H_{12}O_6$;[c] the mutarotation of simple sugars;[d] the hydrogenation of ethylene on nickel catalysts;[e] and probably best known, the radioactive disintegration of unstable nuclei.

[a] At this point the reader must be cautioned against trying to extrapolate experimental data too far from the actual observed conditions. Most first-order reactions, especially in the gas phase, are quite complicated, and experience has shown that, although first-order kinetics may be obeyed in a limited experimental range, they may very well not be obeyed on extending the range.

[b] F. Daniels and E. H. Johnston, *J. Am. Chem. Soc.*, **43**, 53 (1921). See also R. A. Ogg, *J. Chem. Phys.*, **15**, 337 (1947); **15**, 613 (1947); **18**, 572 (1950).

[c] L. Whilhelmy, *Pogg. Annalen*, **81**, 413 (1850). E. A. Moelwyn-Hughes, *Z. physik. Chem.*, **B26**, 281 (1934).

[d] C. S. Hudson and J. K. Dale, *J. Am. Chem. Soc.*, **39**, 320 (1917). T. M. Lowry, *Trans. Chem. Soc. (London)*, **75**, 212 (1899). J. C. Kendrew and E. A. Moelwyn-Hughes, *Proc. Roy. Soc. (London)*, **A176**, 352 (1940).

[e] J. Horiuti and M. Polanyi, *Trans. Faraday Soc.*, **30**, 1164 (1934). A. Farkas and L. Farkas, *J. Am. Chem. Soc.*, **60**, 22 (1938). G. H. Twigg, *Trans. Faraday Soc.*, **46**, 152 (1950).

5. Second-order Reactions: Type I

From the definition of over-all order we see that there are two possible integral types of second-order reactions; in the first the rate is proportional to the square of a concentration of a single reacting species and in the second the rate is proportional to the first power of the product of the concentrations of two different species. The two types can be represented as follows:

$$aA + bB + \cdots \rightarrow \text{product}$$

Type I:
$$\frac{dC_A}{dt} = -k_A C_A^2 \qquad (II.5.1)$$

Type II:
$$\frac{dC_A}{dt} = -k_A C_A C_B \qquad (II.5.2)$$

Let us consider these separately, taking type I first. Here the rate of reaction depends experimentally only on the concentration of a single species as given by Eq. (II.5.1). This equation may be rearranged to permit simple integration, giving ($C_A = C_{A_0}$ at $t = 0$)

$$\frac{1}{C_A} - \frac{1}{C_{A_0}} = k_A t \qquad (II.5.3)$$

From Eq. (II.5.3) we can evaluate the half-life $t_{1/2}$. C_A will have the value $C_{A_0}/2$ when

$$t = t_{1/2} = \frac{1}{k_A C_{A_0}} \qquad (II.5.4)$$

We may observe that the half-life of a second-order reaction differs from that of a first-order reaction [Eq. (II.4.6)] in that the $t_{1/2}$ of the former depends on the initial concentration. As we shall see later, this provides a simple experimental test for distinguishing order.

Typical among the examples of second-order reactions of type I are the gas-phase thermal decomposition of hydrogen iodide, $2HI \rightarrow H_2 + I_2$;[a] the gas-phase thermal decomposition of NO_2, $2NO_2 \rightarrow 2NO + O_2$;[b] the liquid-phase decomposition of ClO^- ion, $2ClO^- \rightarrow 2Cl^- + O_2$;[c] and the dimerization of cyclopentadiene in either gas[d] or liquid phase,[e] $2C_5H_6 \rightarrow C_{10}H_{12}$. Actually, type I reactions are relatively rare in comparison with type II reactions.

[a] M. Bodenstein, Z. physik. Chem., **29**, 295 (1899). Also, G. B. Kistiakowsky, J. Am. Chem. Soc., **50**, 2315 (1928).

[b] M. Bodenstein and I. Ramstetter, Z. physik. Chem., **100**, 68, 106 (1928).

[c] F. Foerster, Z. Elektrochem., **23**, 137 (1917). Also, Giardini, Gazz. chim. ital., **54**, 844 (1924).

[d] G. B. Kistiakowsky and W. Mears, J. Chem. Phys., **5**, 687 (1937).

[e] H. Kaufmann and A. Wassermann, Trans. Chem. Soc. (London), 870 (1939).

6. Second-order Reactions: Type II

Here again the balanced equation may be represented by

$$aA + bB + \cdots \rightarrow \text{products} \tag{II.6.1}$$

while the rate of reaction is given by

$$\frac{dC_A}{dt} = -k_A C_A C_B \quad \text{or} \quad \frac{dC_B}{dt} = -k_B C_A C_B \tag{II.6.2}$$

where $k_A/a = k_B/b$ (Sec. II.3). Since both C_A and C_B are changing with time, we must write a relation connecting them in order to integrate the differential equation. This relation is provided by the balanced equation (II.6.1). If C_{A_0} and C_{B_0} represent the initial concentrations of A and B, respectively, at time $t = 0$ and x represents the concentration of A that has reacted, then the concentration of B that has reacted in the same time is given by bx/a. Consequently,

$$C_A = C_{A_0} - x \qquad C_B = C_{B_0} - \frac{bx}{a} \tag{II.6.3}$$

and from this

$$dC_A = -dx = \frac{a}{b} dC_B \tag{II.6.4}$$

On substitution in Eq. (II.6.2) we have

$$\frac{dC_A}{dt} = -\frac{dx}{dt} = -k_A(C_{A_0} - x)\left(C_{B_0} - \frac{bx}{a}\right) \tag{II.6.5}$$

On rearrangement

$$\frac{dx}{(C_{A_0} - x)(C_{B_0} - bx/a)} = k_A \, dt \tag{II.6.6}$$

This is a standard integral which may be found in tables.[a] It may be integrated after decomposing into partial fractions, and we obtain

$$\ln \frac{1 - x/C_{A_0}}{1 - bx/aC_{B_0}} = \left(\frac{bC_{A_0}}{a} - C_{B_0}\right) k_A t \tag{II.6.7}$$

By virtue of our initial definition [Eq. (II.6.3)] this may be written in terms of the instantaneous and initial concentrations:

$$\ln \frac{C_A/C_{A_0}}{C_B/C_{B_0}} = \frac{bC_{A_0} - aC_{B_0}}{a} k_A t \tag{II.6.8}$$

or

$$\ln \frac{C_A}{C_B} = \frac{bC_{A_0} - aC_{B_0}}{a} k_A t + \ln \frac{C_{A_0}}{C_{B_0}} \tag{II.6.9}$$

The final integrated forms of the second order as given by type I [Eq. (II.5.3)] and type II [Eq. (II.6.9)] permit us to plot the concentration-time relation in simple, graphical form. This is shown in Fig. II.2, where

[a] B. O. Peirce, "A Short Table of Integrals," p. 7 (no. 39), Ginn & Company, Boston, 1929.

in a (for type I), the relation takes the form of a straight line when the reciprocal of the concentration is plotted against time. The intercept at the ordinate has the value $1/C_{A_0}$, and the slope of the line equals the reaction-rate constant k_A. In b (type II) a logarithmic plot of the ratio

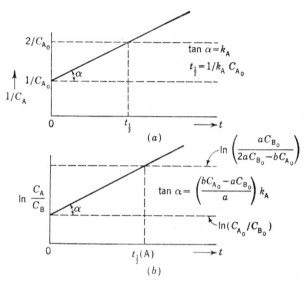

Fig. II.2. Graphical representation of second-order reactions. (a) Type I. $dC_A/dt = -k_A C_A^2$; $1/C_A = 1/C_{A_0} + k_A t$; abscissa: time t (in arbitrary units); ordinate: reciprocal of concentration, $1/C_A$ (in arbitrary units). (b) Type II. $dC_A/dt = -k_A C_A C_B$; abscissa: time t; ordinate: natural logarithm of the ratio of C_A/C_B.

C_A/C_B against time similarly gives a straight line whose slope has the value $[(bC_{A_0} - aC_{B_0})/a]k_A$ and from which the value of k_A may be calculated if the initial concentrations are known.[a] The intercept of this line on the ordinate ($t = 0$) has the value of $\ln(C_{A_0}/C_{B_0})$.

For purposes of graphical representation it is frequently advantageous to write Eq. (II.5.3) for type I in the form

$$C_A t = \frac{1}{k_A} - C_A \frac{1}{C_{A_0} k_A} \qquad (II.6.10)$$

A plot of the quantity $C_A t$ as ordinate against C_A as abscissa will give a straight line of intercept $1/k_A$ and slope $-1/k_A C_{A_0}$. This form has the double advantage that it confines the variable $C_A t$ to a shorter total range (C_A decreases as t increases) and represents ordinate and abscissa on a linear scale more closely commensurate with the errors involved in their measurement.

[a] Note that the slope may be either positive or negative, depending on whether the quantity $bC_{A_0} - aC_{B_0}$ is positive or negative.

It will be observed that the definition of the half-life for type II is ambiguous. We can define either of the two half-lives $t_{1/2}(A)$ and $t_{1/2}(B)$ with reference to the time required to use up half of the initial concentration of either species A or B, respectively. The half-life of species A is obtained by solving for $t_{1/2}(A)$ from Eq. (II.6.7) after substituting $x = C_{A_0}/2$. We obtain

$$t_{1/2}(A) = \frac{a}{k_A(bC_{A_0} - aC_{B_0})} \ln \frac{aC_{B_0}}{2aC_{B_0} - bC_{A_0}} \qquad \text{(II.6.11)}$$

Second-order reactions of type II are probably among the most common of all reactions studied. A few typical examples are the gas-phase formation of hydrogen iodide, $H_2 + I_2 \rightarrow 2HI$;[a] the reactions of free radicals with molecules, for example, $H + Br_2 \rightarrow HBr + H$;[b] the famous synthesis of urea from NH_4^+ and CNO^- ions;[c] the hydrolysis of organic esters in nonaqueous media;[d] and the reaction of tertiary alkyl amines with alkyl halides to produce quaternary ammonium salts, $R_3N + R'X \rightarrow R'R_3N^+ + X^-$.[e]

7. A Complication of Type II Second-order Reactions

When A and B are present in stoichiometric or nearly stoichiometric concentrations (that is, $aA_0 \cong bB_0$), then the solution given for type II second-order reactions [Eqs. (II.6.8) and (II.6.9)] becomes less and less useful as aA_0 approaches bB_0. The reason for this is that the term $\ln (AB_0/A_0B)$ will change very little with time. An alternative form of solution is possible under these conditions. The rate equation can be written as

$$-\frac{dA}{dt} = k_A AB \qquad \text{(II.7.1)}$$

where we are introducing the simplification of using A and B to represent concentration. For the instantaneous concentration of B we can write

$$B = B_0 - \frac{b}{a}(A_0 - A) = \frac{b}{a}\left(A + \frac{aB_0}{b} - A_0\right) = \frac{b}{a}(A + \Delta) \qquad \text{(II.7.2)}$$

The initial difference in stoichiometric concentrations, $\Delta = aB_0/b - A_0$,

[a] M. Bodenstein, *Z. physik. Chem.*, **13**, 56 (1894); **22**, 1 (1897); **29**, 295 (1898).

[b] See E. W. R. Steacie, "Atomic and Free Radical Reactions," A. C. S. Monograph No. 125, 2d ed., Reinhold Publishing Corporation, New York, 1954, for an excellent review of such reactions.

[c] J. Walker and R. J. Hambly, *Trans. Chem. Soc. (London)*, **67**, 746 (1895). J. C. Warner and F. B. Stitt, *J. Am. Chem. Soc.*, **55**, 4807 (1933).

[d] D. P. Evans et al., *Trans. Chem. Soc. (London)*, 1439 (1937). G. Davies and D. P. Evans, *ibid.*, 339 (1940). R. P. Bell, "Acid-Base Catalysis," Oxford Univ. Press, 1941.

[e] N. Menschutkin, *Z. physik. Chem.*, **6**, 41 (1890); R. G. W. Norrish and F. P. Smith, *Trans. Chem. Soc. (London)*, 129 (1928); E. A. Moelwyn-Hughes, *Chem. Revs.*, **10**, 241 (1932).

is now presumed to be a small fraction of these concentrations. The rate equation becomes

$$-\frac{dA}{dt} = \frac{bk_A}{a} A(A + \Delta) \tag{II.7.3}$$

which can also be written as

$$-\frac{dA'}{dt} = \frac{bk_A}{a}\left(A' + \frac{\Delta}{2}\right)\left(A' - \frac{\Delta}{2}\right) = \frac{bk_A}{a}\left(A'^2 - \frac{\Delta^2}{4}\right)$$
$$= \frac{bk_A}{a} A'^2\left[1 - \left(\frac{\Delta}{2A'}\right)^2\right] \cdots \tag{II.7.4}$$

where $A' = A + \Delta/2$. When $\Delta \leq A_0/4$, the term in brackets in Eq. (II.7.4) is initially ≥ 0.99 and changes to ≥ 0.96 when A is half-reacted. Under these conditions we can assume this term constant and replace it by its average value over the course of the reaction (that is, $1 - \Delta^2/4A_0'A_f'$). This permits direct integration, giving

$$\frac{1}{A'} - \frac{1}{A_0'} = \frac{bk_A t}{a}\left(1 - \frac{\Delta^2}{4A_0'A_f'}\right) \cdots \tag{II.7.5}$$

a result which resembles that for the type I equation (II.5.3).[a]

For graphical presentation one would plot $1/A' = 1/(A + \Delta/2)$ against time t to give a curve which should very closely approximate a straight line over the first 50 or 75 per cent of reaction. By dividing the slope by b/a and the term in brackets we can obtain k_A with good precision. Alternatively the equation can be thrown into the form

$$A't = \frac{a}{k_A b\,(1 - \Delta^2/4A_0'A_f')}\left(1 - \frac{A'}{A_0}\right) \cdots \tag{II.7.6}$$

so that a plot of $A't$ against A' can also be used to give a nearly straight line from which the slope can be calculated.

8. Third-order Reactions

Third-order reactions can be classified into three distinct types according to the general definition. If the stoichiometric equation is

$$aA + bB + cC + \cdots \rightarrow \text{products}$$

Type I:
$$\frac{dC_A}{dt} = -k_A C_A^3 \tag{II.8.1}$$

Type II:
$$\frac{dC_A}{dt} = -k_A C_A^2 C_B \tag{II.8.2}$$

[a] For a reaction of over-all order $n = \alpha + \beta + \gamma + \cdots$ with respect to reactants, Van't Hoff has shown that the integrated equation can be treated like one of the nth order in a single component when all the reactants are present in precisely stoichiometric concentrations. This can be useful in determining the individual orders. See footnote page 82.

Type III: $$\frac{dC_A}{dt} = -k_A C_A C_B C_C \qquad (II.8.3)$$

Since the mathematical principles for treating these equations are similar in kind to those previously described for first- and second-order reactions, we shall indicate the steps in the solutions with a minimum of text:

Type I: $$\frac{dC_A}{dt} = -k_A C_A^3$$

which gives an integration when $C_A = C_{A_0}$ at $t = 0$:

$$\frac{1}{C_A^2} - \frac{1}{C_{A_0}^2} = 2k_A t \qquad (II.8.4)$$

The half-life $t_{1/2}$ may be calculated by substituting $C_A = C_{A_0}/2$, whereupon we find

$$t_{1/2} = \frac{3}{2k_A C_{A_0}^2} \qquad (II.8.5)$$

Type II: $$\frac{dC_A}{dt} = -k_A C_A^2 C_B \qquad \text{or} \qquad \frac{dA}{dt} = -k_A A^2 B \qquad (II.8.6)$$

On making the substitutions

$$B = B_0 - \frac{b}{a}(A_0 - A) = \frac{b}{a}\left(A + \frac{aB_0}{b} - A_0\right) = \frac{b}{a}(A + \Delta)$$

where $\Delta = aB_0/b - A_0$, Eq. (II.8.6) becomes

$$\frac{dA}{dt} = -k_A \frac{b}{a} A^2(A + \Delta) \qquad (II.8.7)$$

which after it is rearranged and broken down into fraction form becomes

$$dA\left(\frac{\Delta}{A^2} + \frac{1}{A + \Delta} - \frac{1}{A}\right) = -\Delta^2 \frac{b}{a} k_A \, dt \qquad (II.8.8)$$

By integrating between the limits $A = A_0$ at $t = 0$ and $A = A$ at time t we obtain

$$\Delta\left(\frac{1}{A} - \frac{1}{A_0}\right) + \ln\left(\frac{A}{A_0}\frac{A_0 + \Delta}{A + \Delta}\right) = \frac{\Delta^2 b}{a} k_A t \qquad (II.8.9)$$

Or, by replacing $A + \Delta = aB/b$,

$$\Delta\left(\frac{1}{A} - \frac{1}{A_0}\right) + \ln\left(\frac{A}{A_0}\frac{B_0}{B}\right) = \frac{\Delta^2 b}{a} k_A t \qquad (II.8.10)$$

Either of these two last equations can be used to solve for a half-life for either A or B. When $\Delta/A_0 \leq 0.3$, we have the same difficulty discussed in the previous section for type II second-order reactions. Under these conditions we can recast Eq. (II.8.7) as $(A' = A + \Delta/3)$

$$\frac{dA'}{dt} = -k_A \frac{b}{a} A'^3 \left[1 - \frac{\Delta^2(A' - 2\Delta/9)}{3A'^3} \right] \qquad \text{(II.8.11)}$$

The term in brackets is close to unity, and by using its average value during the reaction it may be considered constant, permitting direct integration of the equation

$$\frac{1}{A'^2} - \frac{1}{A_0'^2} \cong \frac{2bk_A t}{a} \left[1 - \frac{\Delta^2(A_0' + A_f' - 4\Delta/9)}{6(A_0' A_f')^{3/2}} \right] \qquad \text{(II.8.12)}$$

Type III: $$\frac{dA}{dt} = -k_A ABC \qquad \text{(II.8.13)}$$

Substitute $$B = B_0 - \frac{b}{a}(A_0 - A) = \frac{b}{a}(A + \Delta_b)$$

$$C = C_0 - \frac{c}{a}(A_0 - A) = \frac{c}{a}(A + \Delta_c)$$

then $$\frac{dA}{dt} = \frac{-bc}{a^2} k_A A(A + \Delta_b)(A + \Delta_c) \qquad \text{(II.8.14)}$$

where $\Delta_b = aB_0/b - A_0$ and $\Delta_c = aC_0/C - A_0$, the difference in initial stoichiometric concentrations. Decomposition into fractions and integration gives

$$(\Delta_c - \Delta_b) \ln \frac{A_0}{A} - \Delta_c \ln \frac{A_0 + \Delta_b}{A + \Delta_b} + \Delta_b \ln \frac{A_0 + \Delta_c}{A + \Delta_c} = \frac{bc}{a^2} k_A \Delta_b \Delta_c (\Delta_c - \Delta_b) t$$

$$\text{(II.8.15)}$$

This expression is quite complex, as are the expressions of the half-lives, and it is doubtful if it has more than an academic interest, since there are very few examples of its actual application.

Third-order reactions are quite rare, the most famous examples being the reactions of NO with O_2, $2NO + O_2 \rightarrow 2NO_2$;[a] and with Cl_2, $2NO + Cl_2 \rightarrow 2NOCl$.[b]

In the liquid phase, part of the acid-catalyzed oxidation of I^- ion, $2H^+ + 3I^- + H_2O_2 \rightarrow 2H_2O + I_3^-$,[c] is caused by a third-order reaction of type III involving the species H_2O_2, I^-, and H^+.

It has also been shown that the gas-phase recombinations of atoms (for example, H, Cl, Br, O) and simple free radicals such as OH are third-order reactions.[d]

[a] M. Bodenstein, *Z. Elektrochem.*, **24**, 183 (1918). G. Kornfeld and E. Klinger, *Z. physik. Chem.*, **B4**, 37 (1929).

[b] M. Trautz, *Z. anorg. Chem.*, **88**, 285 (1914). W. Kraus and M. Saracini, *Z. physik. Chem.*, **A178**, 245 (1937). M. Trautz and V. P. Dalal, *Z. anorg. Chem.*, **102**, 149 (1918).

[c] H. A. Liebhafsky and A. Mohammad, *J. Am. Chem. Soc.*, **55**, 3977 (1933); *J. Phys. Chem.*, **38**, 857 (1934). F. Bell et al., *J. Phys. & Colloid Chem.*, **55**, 874 (1951).

[d] W. A. Noyes and P. A. Leighton, "The Photochemistry of Gases," Reinhold Publishing Corporation, New York, 1941.

TABLE II.2. SUMMARY OF IMPORTANT RELATIONS FOR DIFFERENT-ORDER REACTIONS

Reaction order	Differential equation	Rate constant k_A from integrated equation	Half-life, $t_{1/2}$
0	$-\dfrac{dA}{dt} = k_A$	$\dfrac{A_0 - A}{t}$	$\dfrac{A_0}{2k_A}$
1	$-\dfrac{dA}{dt} = k_A A$	$\dfrac{1}{t}\ln\dfrac{A_0}{A}$	$\dfrac{1}{k_A}\ln 2$
2 Type I	$-\dfrac{dA}{dt} = k_A A^2$	$\dfrac{1}{t}\left(\dfrac{1}{A} - \dfrac{1}{A_0}\right)$	$\dfrac{1}{k_A A_0}$
Type II	$-\dfrac{dA}{dt} = k_A AB$	$\dfrac{a\ln[(A/A_0)(B_0/B)]}{t(bA_0 - aB_0)}$	$\dfrac{a}{k_A(aB_0 - bA_0)}\ln\left(2 - \dfrac{bA_0}{aB_0}\right)$
3 Type I	$-\dfrac{dA}{dt} = k_A A^3$	$\dfrac{1}{2t}\left(\dfrac{1}{A^2} - \dfrac{1}{A_0^2}\right)$	$\dfrac{3}{2k_A A_0^2}$
Type II	$-\dfrac{dA}{dt} = k_A A^2 B$	$\dfrac{-a}{t(bA_0 - aB_0)}\left(\dfrac{1}{A} - \dfrac{1}{A_0}\right) + \dfrac{ab}{t(bA_0 - aB_0)^2}\ln\dfrac{A}{A_0}\dfrac{B_0}{B}$	$\dfrac{-a}{k_A A_0(bA_0 - aB_0)} - \dfrac{ab\ln(2 - bA_0/aB_0)}{k_A(bA_0 - aB_0)^2}$
Type III	$-\dfrac{dA}{dt} = k_A ABC$	See text	
n	$-\dfrac{dA}{dt} = k_A A^n$	$\dfrac{1}{(n-1)t}\left(\dfrac{1}{A^{n-1}} - \dfrac{1}{A_0^{n-1}}\right)$	$\dfrac{(2^{n-1} - 1)}{k_A A_0^{n-1}(n-1)}$

9. Fractional- and Higher-order Reactions

Because of the paucity of data there is no need to discuss reactions of any higher order than third. There are, however, a number of examples in which the order of a reaction has turned out to be fractional, so that it is worthwhile considering the mathematical representation of such relations. If we consider a reaction of arbitrary order n involving a single reactant, the equation may be written

$$\frac{dA}{dt} = -k_A A^n \qquad (II.9.1)$$

Except for the case that n is unity, this may be integrated by our usual methods to give

$$\frac{1}{A^{n-1}} - \frac{1}{A_0^{n-1}} = (n-1)k_A t \qquad (II.9.2)$$

where the symbols have their usual significance. For the half-life $t_{1/2}$ we can calculate

$$t_{1/2} = \frac{2^{n-1} - 1}{k_A A_0^{n-1}(n-1)} \qquad (II.9.3)$$

Some examples of fractional-order reactions are the interconversion of ortho- and para-H[a] [where the rate is $3/2$-order ($n = 3/2$)]; the gas-phase formation of phosgene,[b] $CO + Cl_2 \rightarrow COCl_2$ (which has an over-all order of $5/2$, being $3/2$-order with respect to Cl_2 and first-order with respect to CO); and the chlorine-catalyzed decomposition of ozone, $2O_3 \rightarrow 3O_2$ (which has a $3/2$ order with respect to ozone).[c] In addition, a great many heterogeneous reactions may follow fractional-order kinetics under different experimental conditions.

[a] K. F. Bonhoeffer and P. Harteck, *Z. physik. Chem.*, **B4**, 119 (1929). A. Farkas and L. Farkas, *Proc. Roy. Soc. (London)*, **A162**, 124 (1935).

[b] M. Bodenstein and H. Plaut, *Z. physik. Chem.*, **110**, 399 (1924).

[c] M. Bodenstein et al., *Z. physik. Chem.*, **B5**, 209 (1929). A. Hamann and H. J. Schumaker, *ibid.*, **B17**, 293 (1932).

III

Mathematical Characterization
of Complex Kinetic Systems

1. Classification of Complex Kinetic Systems

In Chap. II we discussed the reactions which lend themselves to a simple kinetic description in terms of reaction order. Unfortunately, there are a great many reactions which defy such simple description, and the experience of recent years has demonstrated that the bulk of gas-phase reactions, over an extended range of experimental conditions, are of the latter type. The reasons for this complexity are not difficult to understand. If a reaction proceeds by way of a single process, we might reasonably expect that a simple mathematical relation might characterize its behavior. If on the other hand there is a multiplicity of processes (not necessarily related) by which a reaction proceeds to completion, then we might well expect to find a concomitant difficulty in expressing the rate of the reaction in simple terms. The various complexities may be categorized as follows:

Opposing Reactions. If the products of a chemical reaction may themselves react to reproduce the original reactants, the apparent rate of the reaction will decrease as the reaction products accumulate. Eventually a state of dynamic equilibrium will be achieved; in it both of the reactions, forward and backward, will have equal rates. Such systems are subsumed under the category of *opposing reactions.* Their study is of great interest because the kinetic behavior of these systems can be related to the thermodynamic (equilibrium) properties of the final system.

Concurrent Reactions. If the reactants may combine with each other in two or more different ways to produce either the same or different products, the over-all rate of disappearance of reactants will be a composite of the individual reaction paths that are accessible. Such systems are termed systems of *concurrent* or *competing reactions,* and their kinetic behavior may *seem* to follow different simple kinetic laws under different experimental conditions.

Consecutive Reactions. It will frequently happen that the reactants in a system will not combine directly to give the final products but rather will first form intermediate substances which may in turn react either among themselves or with the reactants. This behavior may be restricted to two consecutive reactions or may be extended through a complex system of intermediates. Such reactions are termed *consecutive reactions,*[a] and the number of such types of reactions seems to be growing at an impressive rate. In a certain sense, almost all reactions may be considered complex systems of consecutive reactions.

It can be inferred from the above descriptions that chemical reactions may involve processes characteristic of one or all of these categories in such fashion as to become almost impossible of simple description or classification. Because of this near infinity of possible behaviors of reacting systems, we shall restrict our discussion in the present chapter to the most general methods for the mathematical description of such systems. At the present stage, this is all that can be done to provide a basis for their study. As the experimenter will easily discover, kinetic systems when investigated in detail display an anarchistic tendency to become unique laws unto themselves.

2. Opposing Reactions; First Order

The simplest case of opposing reactions is that of first-order reactions described in principle by the scheme

(1) $$A \xrightarrow{k_1} B$$

(2) $$B \xrightarrow{k_2} A$$

(III.2.1)

where k_1 and k_2 are the respective first-order rate constants. For the rate of reaction of A we can write[b]

$$\frac{dA}{dt} = -k_1 A + k_2 B \qquad (III.2.2)$$

where k_1 and k_2 are the specific reaction rate constants for the consumption

---

[a] When a set of steps reproduces the intermediate which started the set and the sum of the steps leads to a net reaction of starting material, the set is spoken of as a chain reaction.

[b] To avoid increasingly cumbersome notation, we shall adopt the convention of using the symbols A, B, C, ... to represent concentrations of species A, B, C, ... Although this risks an ambiguity in the text, in confusing concentrations and species, the distinction will generally be apparent from the discussion. Where this is not the case, the explicit distinction will be made.

and production of A by reactions 1 and 2, respectively. From these same equations (assuming the stoichiometry is $A \rightleftharpoons B$) we can write

$$B = B_0 + (A_0 - A) = (B_0 + A_0) - A \qquad \text{(III.2.3)}$$

where A and B are the instantaneous concentrations of A and B and A_0 and B_0 their initial concentrations, at time $t = 0$. On substituting these values in the initial equation we find

$$\frac{dA}{dt} = k_2(A_0 + B_0) - (k_2 + k_1)A \qquad \text{(III.2.4)}$$

which integrates to give

$$\ln \frac{k_1 A - k_2 B}{k_1 A_0 - k_2 B_0} = -(k_2 + k_1)t \qquad \text{(III.2.5)}$$

Equation (III.2.5), which expresses the concentration-time relation, is rather difficult to apply experimentally, since it requires a previous knowledge of k_1 and k_2, or at least their ratio. Although we shall say more about such problems in Chapter IV, it is well to consider here the relation of the above rate law to this eventual equilibrium reached in these systems.

At equilibrium, the rates of the forward and reverse processes are equal and we can write

$$k_1 A_{eq} = k_2 B_{eq} \qquad \text{(III.2.6)}$$

where A_{eq} and B_{eq} represent the concentrations of A and B present at equilibrium. From this we can obtain the equation

$$K = \frac{k_1}{k_2} = \frac{B_{eq}}{A_{eq}} \qquad \text{(III.2.7)}$$

which expresses the formal relation which must exist in this case between the thermodynamic equilibrium constant K and the individual specific rate constants.[a]

Equation (III.2.5) can now be written in terms of the equilibrium constant:

$$\ln \frac{AK - B}{A_0 K - B_0} = -(k_2 + k_1)t \qquad \text{(III.2.8)}$$

where K is now known from the value of the equilibrium concentrations [Eq. (III.2.7)]. This equation now permits a determination of the individual rate constants, since the sum $k_2 + k_1$ is measurable and the ratio $k_1/k_2 = K$ is known.

---

[a] The reader who is familiar with thermodynamics will recognize a difficulty at this point. Equation (III.2.7) is exact only if the activity coefficients of A and B are either equal to each other at all concentrations or both unity. Since this is not generally the case, we see that the empirical first-order rate laws cannot be exact but must themselves be modified to take into account the activity coefficients. This is a subject we shall consider later in some detail.

For the ratio $K = k_1/k_2$ to be very large or very small implies that one of the two reactions is slow compared to the other. In that case we can neglect the slower reaction, and the case reduces to one of a simple first-order reaction. There are a good many reactions that fit this description; among them are the gas-phase interconversion of cis-trans isomers such as isostilbene, $C_6H_5CH = CHC_6H_5$;[a] the catalytic interconversion of n-butane and isobutane,[b] C_4H_{10}, in solution; the racemization of α- and β-glucoses[c] and similar sugars;[d] and the interconversion of γ-hydroxy butyric acid into its lactone in water solutions.[e]

3. Opposing Reactions; Higher Orders

Two further cases of opposing reactions which can be easily resolved by present methods may be mentioned. They are opposing reactions of second order and of mixed order.

Case 1. The example of opposing reactions of second order may be represented by the equations

$$A + B \underset{k_2}{\overset{k_1}{\rightleftharpoons}} C + D \tag{III.3.1}$$

For simplicity we shall also assume that this represents the stoichiometry. The rate equation is

$$\frac{dA}{dt} = -k_1 AB + k_2 CD \tag{III.3.2}$$

or, letting $A = A_0 - x$,

$$\frac{dx}{dt} = k_1(A_0 - x)(B_0 - x) - k_2(C_0 + x)(D_0 + x) \tag{III.3.3}$$

or

$$\frac{dx}{dt} = (k_1 A_0 B_0 - k_2 C_0 D_0) - (k_1 A_0 + k_1 B_0 + k_2 C_0 + k_2 D_0)x$$

$$+ (k_1 - k_2)x^2 \tag{III.3.4}$$

which can be written as a standard equation of the form

$$\frac{dx}{\alpha + \beta x + \gamma x^2} = dt \tag{III.3.5}$$

and can be directly integrated.[f] The particular form of the integral is complicated and depends on the relation of the coefficients α, β, and γ.

[a] G. B. Kistiakowsky and W. R. Smith, *J. Am. Chem. Soc.*, **56**, 638 (1934).

[b] P. A. Leighton and J. Heldman, *J. Am. Chem. Soc.*, **65**, 2276 (1943).

[c] T. M. Lowry, *Trans. Chem. Soc. (London)*, **75**, 212 (1899).

[d] J. C. Kendrew and E. A. Moelwyn-Hughes, *Proc. Roy. Soc. (London)*, **A176**, 352 (1940).

[e] P. Henry, *Z. physik. Chem.*, **10**, 96, 98 (1892). F. A. Long et al., *J. Phys. & Colloid Chem.*, **55**, 814, 829 (1951).

[f] B. O. Peirce, "A Short Table of Integrals," p. 10 (nos. 67, 68) Ginn & Company, Boston, 1929.

For this particular case the quantity q, defined by

$$q = \beta^2 - 4\alpha\gamma$$

will always be positive, since on substitution we find

$$q = k_1{}^2(A_0 - B_0)^2 + k_2{}^2(C_0 - D_0)^2$$
$$+ 2k_1k_2[(A_0 + B_0)(C_0 + D_0) + 2A_0B_0 + 2C_0D_0] \qquad \text{(III.3.6)}$$

where each quantity on the right is positive.[a] The solution then takes the form

$$\ln \frac{x + (\beta - q^{\frac{1}{2}})/2\gamma}{x + (\beta + q^{\frac{1}{2}})/2\gamma} = tq^{\frac{1}{2}} + \theta \qquad \text{(III.3.7)}$$

where θ can be determined by the condition $x = 0$ at $t = 0$ or

$$\theta = \ln \frac{\beta - q^{\frac{1}{2}}}{\beta + q^{\frac{1}{2}}}$$

Once again the equation is difficult to use experimentally unless k_1 and k_2 are known, or at least unless their ratio is known. This knowledge may be obtained from a study of the equilibrium concentrations. Once known the data can be plotted and the individual specific rate constants determined.

Case 2. A case of mixed first- and second-order reaction can be represented by

$$A \underset{k_2}{\overset{k_1}{\rightleftharpoons}} B + C \qquad \text{(III.3.8)}$$

where again for simplicity we shall assume this represents the stoichiometry. The rate equation is

$$\frac{dA}{dt} = -k_1A + k_2BC \qquad \text{(III.3.9)}$$

which reduces on substitution to

$$\frac{dx}{dt} = k_1(A_0 - x) - k_2(B_0 + x)(C_0 + x) \qquad \text{(III.3.10)}$$

or $\qquad \dfrac{dx}{dt} = (k_1A_0 - k_2B_0C_0) - (k_1 + k_2C_0 + k_2B_0)x - k_2x^2 \qquad \text{(III.3.11)}$

It can be seen that this is of the same form as Eq. (III.3.4) and therefore admits of the same type of solution [Eq. (III.3.7)]. There are many examples of both of these types of system. Examples of opposing second-order reactions are represented by the gas-phase decomposition of HI,

---

[a] This implies that the quadratic expression $\alpha + \beta x + \gamma x^2$ has real roots. This is quite understandable, since the roots of this expression are precisely the values of x which make the quadratic zero. This in turn means that dx/dt will be zero when x has these values. However physically, when dx/dt is zero, the system is at equilibrium, and we know that there must be one real, positive value of x for which this is true.

$2HI \rightleftharpoons H_2 + I_2$;[a] the reactions of atoms with molecules;[b] and the displacement reactions between ions (such as I^-) and alkyl halides[c] in nonaqueous solution.

Examples of mixed order are provided by the ammonium cyanate–urea reaction, $NH_4^+ + CNO^- \rightleftharpoons (NH_2)_2CO$, in solution;[d] the dissociation of hexaphenyl ethane, $\phi_3C\!-\!C\phi_3$, into free radicals;[e] and the recombination of radicals at high pressures, $Br_2 \rightleftharpoons 2\ Br$.[f] There are other known examples of equilibrium involving competing reactions of higher order than second, but they are difficult to treat in any general way because the differential equation so obtained is cubic and does not admit of a simple solution. They are also rather scarce; a single, well-studied example is the equilibrium $2NO + O_2 \rightleftharpoons 2NO_2$[g] in the gas phase.

4. Concurrent Reactions

A case of concurrent reactions may be represented by the following stoichiometric scheme:

(1) $$A \xrightarrow{k_1} \text{products} \qquad \text{first order}$$

(2) $$A + B \xrightarrow{k_2} \text{products} \qquad \text{second order mixed} \qquad \text{(III.4.1)}$$

(3) $$A + A \xrightarrow{k_3} \text{products} \qquad \text{second order in A}$$

Here we have listed three alternate reaction paths whereby a substance A may disappear. It is clear that this scheme can be indefinitely extended to include other orders of reaction of A with other substances than B and finally the reactions of B itself with other substances as illustrated by

$$B + C \rightarrow \text{products} \qquad \text{second order mixed}$$
$$B \rightarrow \text{products} \qquad \text{first or higher order} \qquad \text{(III.4.2)}$$

If we take the first group of reactions [Eq. (III.4.1)], we can write an equation for the rate of disappearance of A:

$$-\frac{dA}{dt} = k_1A + k_2AB + 2k_3A^2 \qquad \text{(III.4.3)}$$

[a] M. Bodenstein, *Z. physik. Chem.*, **13**, 56 (1894); **22**, 1 (1897); **29**, 295 (1898).

[b] E. W. R. Steacie, "Atomic and Free Radical Reactions," A.C.S. Monograph No. 125, 2d ed., Reinhold Publishing Corporation, New York, 1954.

[c] M. Bodenstein, *Z. physik. Chem.*, **29**, 295 (1898). E. A. Moelwyn-Hughes, *Trans. Faraday Soc.*, **35**, 368 (1939).

[d] G. J. Burrows and E. W. Fawcett, *Trans. Chem. Soc.* (*London*), **105**, 609 (1914). J. Walker and F. J. Hambly, *ibid.*, **67**, 746 (1895).

[e] K. Ziegler et al., *Ann.*, **479**, 277 (1930); **504**, 131 (1933).

[f] E. Rabinowitch and W. C. Wood, *J. Chem. Phys.*, **4**, 497 (1936).

[g] M. Bodenstein, *Z. Elektrochem.*, **24**, 183 (1918). G. Kornfeld and E. Klinger, *Z. physik. Chem.*, **B4**, 37 (1929).

On replacing A and B by $A_0 - x$ and $B_0 - x$ we have, after collecting terms and factoring,

$$\frac{dx}{dt} = (A_0 - x)[(k_1 + k_2 B_0 + 2k_3 A_0) - (k_2 + 2k_3)x] \qquad \text{(III.4.4)}$$

This equation can be resolved by the method of partial fractions and integrated directly to give

$$-\ln (A_0 - x) + \ln [(k_1 + k_2 B_0 + 2k_3 A_0) - (k_2 + 2k_3)x]$$
$$= [k_1 + k_2(B_0 - A_0)]t + \theta_A \qquad \text{(III.4.5)}$$

or, by replacing $B_0 - x$ and $A_0 - x$ by B and A, respectively, and eliminating θ_A,

$$\ln \frac{A_0}{A} \frac{k_1 + k_2 B + 2k_3 A}{k_1 + k_2 B_0 + 2k_3 A_0} = [k_1 + k_2(B_0 - A_0)]t \qquad \text{(III.4.6)}$$

This equation is of an obviously complex form and does not admit of a simple experimental verification without some direct knowledge of the ratios of the three rate constants k_1, k_2, and k_3.[a] Similarly, the presence of the mixed second-order reaction A and B destroys the possibility of defining a half-life for such a system, since, for example, the amount of B that remains when A is half used up will clearly depend on the initial amounts of A and B present (that is, B_0 and A_0) as well as the relative values of the respective rate constants.

Special simpler cases of concurrent reactions may be obtained from the case treated here. Thus the case of any two of the three postulated reactions (III.4.1) is solved directly by setting the particular rate constant corresponding to the missing reaction equal to zero. If we want the case given by the reactions 1 and 3 of Eq. (III.4.1), the answer is obtained by setting $k_2 = 0$ in Eq. (III.4.6). It is then for the scheme

$$A \xrightarrow{k_1} \text{products} \qquad \text{first order}$$

$$A + A \xrightarrow{k_3} \text{products} \qquad \text{second order in A}$$

$$\ln \frac{A_0}{A} \frac{(k_1 + 2k_3 A)}{(k_1 + 2k_3 A_0)} = k_1 t \qquad \text{(III.4.7)}$$

Here a half-life can be defined for A.

(We shall consider in the next chapter special methods which can be employed for the experimental detection of concurrent reactions.)

There are a number of cases of such composite reactions. One example

---

[a] In practice such equations are usually verified in their differential form [Eq. (III.4.3)]. This could be done by plotting $\ln A$ against t, measuring the slope S, and replotting against A. The equation for S is $-S \equiv d \ln A/dt = k_1 + k_2(B_0 - A_0) + (2k_3 + k_2)A$.

which has been studied in some detail is the reduction of H_2O_2 by I^- ion. This reaction is well represented by the scheme[a]

$$\frac{d(I_3^-)}{dt} = k_1(H_2O_2)(I^-) + k_2(H_2O_2)(I^-)(H^+)$$

The organic chemist is only too well aware of the complexity represented by competing reactions, because many organic reactions involve precisely such schemes of concurrent reactions usually giving unwanted products.

5. Consecutive Reactions; First Order

The large number of systems which are now believed to consist of a series of consecutive reactions almost defies any simple description. The most important groupings may, however, be categorized as follows:

1. Consecutive reactions of first order
2. Consecutive reactions of second order
3. Consecutive reactions of mixed order

Of these, only the first case can be treated exactly. Let us consider a scheme of two consecutive first-order reactions with stoichiometry given by

$$\begin{aligned} A &\xrightarrow{k_A} B \\ B &\xrightarrow{k_B} C \end{aligned} \qquad \text{(III.5.1)}$$

The significance of this scheme is that B is produced as an unstable intermediate in the total over-all reaction represented by $A \rightarrow C$. By using our usual nomenclature we can write

$$\frac{dA}{dt} = -k_A A \qquad \text{(III.5.2)}$$

$$\frac{dB}{dt} = k_A A - k_B B \qquad \text{(III.5.3)}$$

$$\frac{dC}{dt} = k_B B \qquad \text{(III.5.4)}$$

Equation (III.5.2) is solved quite simply by our method for first-order equations [Eq. (II.4.5)]. The solution is

$$A = A_0 e^{-k_A t} \qquad \text{(III.5.5)}$$

If we now substitute this value for A in Eq. (III.5.3), we obtain on rearrangement

$$\frac{dB}{dt} + k_B B = k_A A_0 e^{-k_A t} \qquad \text{(III.5.6)}$$

[a] Liebhafsky and Mohammad, loc. cit. *Note:* Iodine is written I_3^- in the above case because it combines with the excess I^- ion present in the solution and forms the complex I_3^- ion.

This may be identified as a linear differential equation of first order.[a] The general solution of such an equation may be expressed as the sum of a particular solution determined by the right-hand side and a general solution determined by solving the equation obtained by setting the left-hand side equal to zero. The general solution is seen to be $e^{-k_B t}$, while the particular solution has the form $e^{-k_A t}$. The total solution may then be written as a linear combination of the two:

$$B = \alpha_1 e^{-k_B t} + \alpha_2 e^{-k_A t} \tag{III.5.7}$$

where the coefficients α_1 and α_2 are constants to be determined from the original equation and the initial conditions. If we substitute Eq. (III.5.7) back into the original differential equation (III.5.6), we obtain the relation

$$\alpha_2 = \frac{k_A A_0}{k_B - k_A} \tag{III.5.8}$$

On substituting this in our general solution [Eq. (III.5.7)], we have

$$B = \alpha_1 e^{-k_B t} + \frac{k_A A_0}{k_B - k_A} e^{-k_A t} \tag{III.5.9}$$

If we eliminate α_1 by using the initial conditions $B = B_0$ at $t = 0$:

$$\alpha_1 = B_0 - \frac{k_A A_0}{k_B - k_A} \tag{III.5.10}$$

which on substitution in the original equation and rearrangement yields

$$B = B_0 e^{-k_B t} + \frac{k_A A_0}{k_B - k_A} (e^{-k_A t} - e^{-k_B t}) \tag{III.5.11}$$

If $B_0 \neq 0$, we can write the equivalent form

$$\frac{B}{B_0} = e^{-k_B t} + \frac{A_0/B_0}{1 - k_B/k_A} (e^{-k_B t} - e^{-k_A t}) \tag{III.5.12}$$

The solution of Eq. (III.5.4) for C can be written very simply because there is the relation provided by the chemical equation (III.5.1):

$$A + B + C = \text{const} = A_0 + B_0 + C_0$$

Thus

$$C = C_0 + A_0(1 - e^{-k_A t}) + B_0 \left[1 - e^{-k_B t} - \frac{A_0/B_0}{1 - k_B/k_A} (e^{-k_B t} - e^{-k_A t}) \right] \tag{III.5.13}$$

It is interesting to study the solutions graphically. The behavior of A is

---

[a] A linear differential equation is one which may be written in the form

$$a_n D^n y + a_{n-1} D^{n-1} y + \cdots + a_1 D y + a_0 y = f(x)$$

where $D^n y$ represents the nth derivative of the dependent variable y with respect to x (that is, $D^n y = d^n y/dx^n$) and the coefficients a_n may be functions of x or constants (see, for example, H. Margenau and G. M. Murphy, "The Mathematics of Physics and Chemistry," 2d ed., D. Van Nostrand Company, Inc., Princeton, N.J., 1956.

represented in Fig. III.1 and is, of course, the usual exponential, first-order curve.

The behavior of B is more complicated and depends on the initial conditions A_0 and B_0 and the rate constants k_A and k_B. From Eq. (III.5.3) we

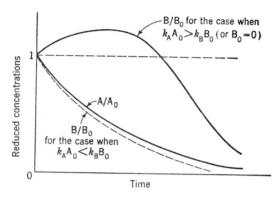

FIG. III.1. Variation of concentration with time for two consecutive first-order reactions $(A \rightarrow B \rightarrow C)$. *Note:* The lower curve for B/B_0 does not necessarily lie below that for A/A_0.

see that the slope $dB/dt = k_A A - k_B B$ will be negative when $k_A A < k_B B$, positive when $k_A A > k_B B$, and zero when $k_A A = k_B B$.

To investigate the occurrence of maxima or minima in the curve of B/B_0 vs. t, let us differentiate Eq. (III.5.12) with respect to t, set the derivative equal to zero, and solve for t. We obtain the following for the time t_{ex} at which an extremum will occur:

$$t_{ex} = \frac{1}{k_B - k_A} \ln \frac{k_B}{k_A} \left(1 + \frac{B_0}{A_0} - \frac{k_B}{k_A} \frac{B_0}{A_0} \right) \qquad \text{(III.5.14)}$$

We must first observe that there is at most a single extremum possible, since Eq. (III.5.14) has only a single solution. Next, this extremum must be a maximum, not a minimum, since the ratio B/B_0 is initially positive and reaches a final value, at $t = \infty$, of zero. Finally we can decide under what conditions the maximum will exist by observing the initial slope. From Eq. (III.5.3) we see that the initial slope is given by

$$\left(\frac{dB}{dt} \right)_0 = k_A A_0 - k_B B_0 \qquad \text{(III.5.15)}$$

This slope is negative when $k_A A_0 < k_B B_0$. But if this condition exists, then no maximum is possible, since in this case there would have to be a minimum preceding the maximum and we already know that only one extremum exists. In this case the curve for B/B_0 starts out at a value of 1 with a negative slope and decreases monotonically and uniformly to zero as shown in Fig. III.1.

If, on the other hand, $k_A A_0 > k_B B_0$, the initial slope is positive and the concentration of B grows to a maximum value at a time given by Eq. (III.5.14), whereupon it then decreases asymptotically to zero (Fig. III.1).

There are very few known chemical reactions which fit precisely into this scheme of two consecutive first-order reactions, but all of the well-known schemes of radioactive decay which involve two or more consecutive steps are characterized precisely by this type of behavior.

6. Infinite Sequence of First-order Reactions

The preceding scheme can be extended to the general case of an infinite sequence of first-order reactions. If we represent such an infinite sequence by the set of stoichiometric equations

$$C_1 \xrightarrow{k_1} C_2 \xrightarrow{k_2} C_3 \xrightarrow{k_3} C_4 \cdots \xrightarrow{k_{i-1}} C_i \xrightarrow{k_i} C_{i+1} \rightarrow$$

then we can write the following system of coupled linear differential equations:

$$\frac{dC_1}{dt} = -k_1 C_1$$

$$\frac{dC_2}{dt} = k_1 C_1 - k_2 C_2$$

$$\frac{dC_3}{dt} = k_2 C_2 - k_3 C_3 \qquad\qquad \text{(III.6.1)}$$

$$\frac{dC_i}{dt} = k_{i-1} C_{i-1} - k_i C_i$$

$$\begin{matrix} \cdot & \cdot & \cdot \\ \cdot & \cdot & \cdot \\ \cdot & \cdot & \cdot \end{matrix}$$

This set of equations can now be rewritten in more useful symbolic form if we let the symbol D represent differentiation with respect to time, that is, $D = d/dt$.

$$\begin{aligned}
(1) & & (D + k_1)C_1 &= 0 \\
(2) & & (D + k_2)C_2 &= k_1 C_1 \\
(3) & & (D + k_3)C_3 &= k_2 C_2 \\
(4) & & (D + k_i)C_i &= k_{i-1} C_{i-1}
\end{aligned} \qquad \text{(III.6.2)}$$

If now we differentiate equations 1 and 2 in this set, we obtain[a]

$$D(D + k_2)C_2 = D(k_1 C_1) = k_1 D C_1 \qquad\qquad \text{(III.6.3)}$$

If we add equations 1 and 2, we can eliminate $k_1 C_1$ and obtain a relation between DC_1 and $(D + k_2)C_2$, namely,

$$DC_1 = -(D + k_2)C_2 \qquad\qquad \text{(III.6.4)}$$

[a] Notice the identity $D(k_i C_i) = (d/dt)(k_i C_i) = k_i \, dC_i/dt = k_i D C_i$.

and on substituting this into Eq. (III.6.3) we obtain

$$D(D + k_2)C_2 = -k_1(D + k_2)C_2 \tag{III.6.5}$$

which on rearrangement and factoring becomes

$$(D + k_1)(D + k_2)C_2 = 0 \tag{III.6.6}$$

a second-order linear differential equation which has a known solution of the form

$$C_2 = \alpha_{21}e^{-k_1t} + \alpha_{22}e^{-k_2t} \tag{III.6.6a}$$

which we can recognize as the solution obtained for the case of two consecutive, first-order reactions [Eq. (III.5.7)].

By proceeding in this manner with successive differentiations and eliminations it can be shown that for the ith species we obtain an equation of the form

$$(D + k_1)(D + k_2)(D + k_3) \cdots (D + k_i)C_i = 0 \tag{III.6.7}$$

which is a linear, differential equation of the ith order. It has as its general solution (when the k_i are all different):

$$C_i = \alpha_{i1}e^{-k_1t} + \alpha_{i2}e^{-k_2t} + \cdots + \alpha_{ii}e^{-k_it} \tag{III.6.8}$$

where $(\alpha_{i1}, \alpha_{i2}, \ldots, \alpha_{ii})$ are constants of integration which may be determined from the initial conditions.[a]

[a] The set of i coefficients α_{ij} for each species C_i can be determined from the initial conditions. If for example we take as our initial conditions at $t = 0$ that $C_1 = C_1^\circ$ and all other $C_i^\circ = 0$ $(i \geqslant 2)$, then for C_1, $\alpha_{11} = C_1^\circ$. For C_2 there are two constants α_{21} and α_{22}. From the condition that $C_2 = 0$ at $t = 0$ we find on substitution in Eq. (III.6.6a) that $\alpha_{21} = -\alpha_{22}$. On substituting the solution for C_2 into the initial equation for DC_2 [Eq. (III.6.1)] we find that $\alpha_{21}k_1 + \alpha_{22}k_2 = k_1C_1^\circ$, which gives $\alpha_{21} = -\alpha_{22} = k_1C_1^\circ/(k_1 - k_2)$. For the ith species it can be shown from the initial set of equations [Eq. (III.6.1)] and the boundary conditions that dC_i/dt and all higher derivatives up to $d^{i-2}C_i/dt^{i-2} = D^{i-2}C_i$ vanish at $t = 0$. Inserting these values into Eq. (III.6.7) leads to the additional relation that $(D^iC_i)_0 = -(k_1 + k_2 + \cdots + k_i)(D^{i-1}C_i)_0$. Finally, the initial process of elimination which gave Eq. (III.6.7) yields the relation at $t = 0$ that $(D^{i-1}C_i)_0 = k_ik_{i-1}k_{i-2} \cdots k_2C_1^\circ$. On differentiating Eq. (III.6.8) for C_i i successive times and substituting $t = 0$ and utilizing the above relations for the derivatives we find the following set of i independent linear equations in the i quantities α_{i1} to α_{ii}:

$$\alpha_{i1} + \alpha_{i2} + \cdots + \alpha_{ii} = 0$$

$$k_1\alpha_{i1} + k_2\alpha_{i2} + \cdots + k_i\alpha_{ii} = 0$$

$$k_1^2\alpha_{i1} + k_2^2\alpha_{i2} + \cdots + k_i^2\alpha_{ii} = 0$$

$$k_1^{i-2}\alpha_{i1} + k_2^{i-2}\alpha_{i2} + \cdots + k_i^{i-2}\alpha_{ii} = 0$$

$$k_1^{i-1}\alpha_{i1} + k_2^{i-1}\alpha_{i2} + \cdots + k_i^{i-1}\alpha_{ii} = \left(\prod_{j=2}^{i} k_j\right)C_1^\circ$$

$$k_1^i\alpha_{i1} + k_2^i\alpha_{i2} + \cdots + k_i^i\alpha_{ii} = -\left(\prod_{j=2}^{i} k_j\right)C_1^\circ$$

If the k_i's are all different, these can be solved by standard methods for the α's in terms of the k_i's and C_1°.

If any two of the k's are equal, then the solution is slightly different. If, for example, $k_1 = k_2$, then the solution has the form

$$C_i = \alpha_{i1}e^{-k_1t} + \alpha_{i2}te^{-k_1t} + \cdots + \alpha_{ii}e^{-k_it} \qquad \text{(III.6.9)}$$

Or in general if all the k's are equal ($k_1 = k_2 = k_3 = \cdots = k$), the solution reduces to

$$C_i = e^{-k_1t}(\alpha_{i1} + \alpha_{i2}t + \alpha_{i3}t^2 + \cdots + \alpha_{ii}t^{i-1}) \qquad \text{(III.6.10)}$$

where again the α_{ij} are to be determined from the initial conditions. If the initial conditions are such that all of the concentrations $C_{i0} = 0$, except C_1 which $= C_{10}$, then the solution takes the very simple form

$$C_i = \alpha_{ii}t^{i-1}e^{-k_1t} \qquad \text{(III.6.11)}$$

and it can be shown[a] that

$$\alpha_{ii} = \frac{C_{10}k_1{}^{i-1}}{(i-1)!}$$

so that the solution takes the final form

$$\frac{C_i}{C_{10}} = \frac{(k_1t)^{i-1}}{(i-1)!} e^{-k_1t} \qquad \text{(III.6.12)}$$

This expression for the distribution of the instantaneous concentrations of product molecules C_i may be recognized as a Poisson distribution.[b] Each component C_i will reach a maximum concentration $C_{i,\max}$ under these conditions at a time

$$t_{\max}(i) = \frac{i-1}{k_1} \qquad \text{(III.6.13)}$$

after which its concentration slowly declines. While the treatment developed above for the extended set of consecutive first-order reactions has literal application only to radioactive decay series, it also approximates the behavior of certain polymerizations during the first few per cent of the reaction.[c] Thus if we have a sequence of steps in a chain polymerization represented by

$$M_1 + M_1 \rightarrow M_2$$
$$M_2 + M_1 \rightarrow M_3$$
$$M_3 + M_1 \rightarrow M_4$$
$$\begin{matrix} \cdot & \cdot & \cdot \\ \cdot & \cdot & \cdot \\ \cdot & \cdot & \cdot \end{matrix} \qquad \text{(III.6.14)}$$
$$M_i + M_1 \rightarrow M_{i+1}$$

[a] S. W. Benson, unpublished work.

[b] The mathematical formalism of the above scheme can be shown equivalent to a random walk or diffusion problem (Sec. VI.7).

[c] See Sec. XVI.10.

then although it turns out that the reactions of the above set are actually second-order, because of the large relative concentration of monomer M_1 present initially, we can make the approximation that M_1 is constant during the initial stages of the reaction, and the treatment then reduces to the above-outlined sequence of first-order reaction.[a]

6A. First-order Reactions; General Treatment. There has been a considerable amount of work done on the solution of particular and general systems of first-order reactions. All such systems are capable of exact, explicit mathematical solutions. If we consider the most general case of a system of s components C_1, C_2, ... , C_s in which first-order reactions of the following type may take place between any two components

$$C_j \underset{k_{mj}}{\overset{k_{jm}}{\rightleftharpoons}} C_m \qquad \text{(III.6A.1)}$$

Then we can write a set of s linear differential equations for the rates of reaction of the components. A typical such equation for the mth component C_m is

$$\frac{dC_m}{dt} = k_{1m}C_1 + k_{2m}C_2 + \cdots + k_{m-1,m}C_{m-1} + k_{m+1,m}C_{m+1} + \cdots$$
$$+ k_{sm}C_s - (k_{m1} + k_{m2} + \cdots + k_{m,m-1} + \cdots + k_{ms})C_m \qquad \text{(III.6A.2)}$$

or in abbreviated form
$$\frac{dC_m}{dt} = \sum_{i=1}^{s} k_{im}C_i \qquad \text{(III.6A.3)}$$

where we have defined the coefficient k_{mm} as

$$k_{mm} = -(k_{m1} + k_{m2} + \cdots + k_{m,m-1} + k_{m,m+1} + \cdots + k_{ms}) \qquad \text{(III.6A.4)}$$

It can be shown that the most general solution of a coupled set of linear, homogeneous first-order equations, represented by Eq. (III.6A.3), has the form

$$C_m = \sum_{j=1}^{s} a_{mj}e^{-\lambda_j t} + \theta_m \qquad \text{(III.6A.5)}$$

where θ_m is a constant of integration which may be determined from the initial conditions, namely, $C_m = C_m^\circ$ at $t = 0$, so that

$$\theta_m = C_m^\circ - \sum_{j=1}^{s} a_{mj} \qquad \text{(III.6A.6)}$$

---

[a] However, in this case we must change the second equation $(D + k_2)M_2 = k_1M_1^2 =$ const by setting $k_1M_1^2 =$ const. This introduces an additive constant in all the solutions.

The constants a_{mj} and λ_j may be determined[a] by substitution of the solutions for C_m [Eq. (III.6A.5)] into the original set of equations (III.6A.3). If we write

$$C_m = \sum_{j=1}^{s} a_{mj}e^{-\lambda_j t} + \theta_m \qquad C_i = \sum_{j=1}^{s} a_{ij}e^{-\lambda_j t} + \theta_i$$

Then

$$\frac{dC_m}{dt} = -\sum_{j=1}^{s} a_{mj}\lambda_j e^{-\lambda_j t}$$

and, by substituting these values in Eq. (III.6A.3), we have the identity

$$-\sum_{j=1}^{s} a_{mj}\lambda_j e^{-\lambda_j t} = \sum_{i=1}^{s}\sum_{j=1}^{s} k_{im}a_{ij}e^{-\lambda_j t} + \sum_{i=1}^{s} k_{im}\theta_i \qquad \text{(III.6A.7)}$$

Since the order of summation can be interchanged, we can rewrite this equation in the form

$$0 = \sum_{j=1}^{s} e^{-\lambda_j t}\left(a_{mj}\lambda_j + \sum_{i=1}^{s} k_{im}a_{ij}\right) + \sum_{i=1}^{s} k_{im}\theta_i \qquad \text{(III.6A.7a)}$$

For this last equation to hold true for all values of t, each coefficient of $e^{-\lambda_j t}$ in the equation must be zero and in addition the last term must be zero. For each value of m we will thus have s equations of the form

$$\lambda_j a_{mj} + \sum_{i=1}^{s} k_{im}a_{ij} = 0 \qquad j = 1, \ldots, s \qquad \text{(III.6A.8)}$$

and one of the form

$$\sum_{i=1}^{s} k_{im}\theta_i = 0 \qquad \text{(III.6A.9)}$$

We will have in all $s(s+1)$ equations, and they are generally just sufficient to solve for the s^2 coefficients a_{ij} and the s exponents λ_j. For example, if we keep j fixed and allow m to assume values from 1 to s, we obtain from Eq. (III.6A.8) the set of s equations

$$\begin{aligned}
(k_{11} + \lambda_j)a_{1j} + k_{21}a_{2j} \quad &+ \cdots + k_{s1}a_{sj} \quad = 0 \\
k_{12}a_{1j} \quad + (k_{22} + \lambda_j)a_{2j} + \cdots + k_{s2}a_{sj} \quad &= 0 \\
&\vdots \\
k_{1s}a_{1j} \quad + k_{2s}a_{2j} \quad &+ \cdots + (k_{ss} + \lambda_j)a_{sj} = 0
\end{aligned} \qquad \text{(III.6A.10)}$$

[a] The exponents λ_j must all be positive for the solutions to have physical meaning. A negative λ would imply a concentration going to infinity as $t \to \infty$. Imaginary values of λ are also excluded, since they would imply that equilibrium is never reached in the system but that instead the concentrations become periodic functions of time.

These are s linear equations with constant coefficients for the s quantities $a_{ij}, \ldots, a_{sj}$. Since there is no constant term in these equations, they will be compatible with each other only if the determinant of the coefficients vanishes:

$$
\begin{vmatrix}
k_{11} + \lambda & k_{21} & k_{31} & \cdots & k_{s1} \\
k_{12} & k_{22} + \lambda & k_{32} & \cdots & k_{s2} \\
k_{13} & k_{23} & k_{33} + \lambda & \cdots & k_{s3} \\
\cdot & \cdot & \cdot & & \cdot \\
\cdot & \cdot & \cdot & & \cdot \\
\cdot & \cdot & \cdot & & \cdot \\
k_{1s} & k_{2s} & k_{3s} & \cdots & k_{ss} + \lambda
\end{vmatrix} = 0
\qquad \text{(III.6A.11)}
$$

But the solution of this determinantal equation leads to an algebraic equation of the sth degree in λ.[a] The s roots of this equation give the values of $\lambda_1, \ldots, \lambda_s$ and so uniquely determine λ.

Having determined λ in this fashion,[b] we can return to Eq. (III.6A.10) and solve for the coefficients a_{ij}. The set (III.6A.10) determines uniquely not the a_{ij} but only their ratio. To determine the a_{ij} uniquely we must use the set of relations provided by Eqs. (III.6A.6) and (III.6A.9).

Let us take as a simple example the following set of first-order chemical reactions of simple stoichiometry:

$$
\begin{aligned}
C_1 &\underset{k_{21}}{\overset{k_{12}}{\rightleftharpoons}} C_2 \\
C_2 &\xrightarrow{k_{23}} C_3 \\
C_3 &\xrightarrow{k_{34}} C_4
\end{aligned}
\qquad \text{(III.6A.12)}
$$

From the definitions of k_{mm} [Eq. (III.6A.4)] we see that

$$
k_{11} = -k_{12} \qquad k_{22} = -k_{21} - k_{23} \qquad k_{33} = -k_{34} \qquad k_{44} = 0
$$

so that of the 16 possible rate constants for our system, 9 are zero.

The determinantal equation becomes on substitution:

$$
\begin{vmatrix}
\lambda - k_{12} & k_{21} & 0 & 0 \\
k_{12} & \lambda - k_{21} - k_{23} & 0 & 0 \\
0 & k_{23} & \lambda - k_{34} & 0 \\
0 & 0 & k_{34} & \lambda
\end{vmatrix} = 0
$$

[a] We have dropped the subscripts j on λ because every value of j leads to the same determinantal equation.

[b] If two or more of the roots had been identical, for example, $\lambda_1 = \lambda_2 = \lambda_3$, then the solution would have taken the form

$$
C_m = (a_{m1} + a_{m2}t + a_{m3}t^2)e^{-\lambda_1 t} + \sum_{j=4}^{s} a_{mj}e^{-\lambda_j t} + \theta_m
$$

See Eq. (III.6.10).

which has for its roots $\lambda_1 = 0$, $\lambda_2 = k_{34}$, λ_3, and λ_4, where λ_3 and λ_4 are the two roots (positive) of the quadratic equation

$$\lambda^2 - \lambda(k_{12} + k_{21} + k_{23}) + k_{12}k_{23} = 0$$

We can now determine the a_{ij} by substitution of these values of λ_j in Eqs. (III.6A.10) and (III.6A.9) and complete the solution. Since $\lambda_1 = 0$, we will find that one of the exponential terms disappears and $a_{11} = a_{21} = a_{31} = a_{41} = 0$. The remainder of the solution may be left as an exercise for the student.

Special cases of the above type of system have been worked out in detail by Korvezee,[a] Skrabal,[b] Balandin and Vaskevich,[c] and Erofeev,[d] while one of the first applications was made by Esson.[e] Swain[f] and Skrabal have given methods of applying the solution graphically to data for reactions involving three components. A general treatment, similar to the one presented here, has been given by Zwolinski and Eyring[g] and extended in a very useful fashion by Matsen and Franklin.[h] The latter authors have shown how it may be used to approximate the solution to sets of coupled second-order reactions when the total amount of reaction is not great. The chief difficulties with such complex reaction systems arise not so much from the mathematical solutions but from the application of the solutions to data when the experimental rate constants are unknown. No general methods have yet been devised for such applications, and the cases treated have been attacked more or less by trial and error and a judicious choice of experimental conditions. With the availability of analogue computers and high-speed calculators such difficulties may become unimportant.

7. Consecutive Reactions; Higher Order

In general there is no exact solution to any sequence of consecutive higher-order equations. The reason for this is that the differential equations are no longer linear equations (as they were in the case of first-order reactions), and nonlinear equations do not have exact solutions except in very particular cases. However, two exact methods are available for studying some aspects of these systems, and there is one more commonly used

[a] A. E. Korvezee, *Rec. trav. chim.*, **59**, 913 (1940).

[b] A. Skrabal, *Sitz. ber. Akad. Wiss. Wien*, **137**, 1045 (1928); *Z. physik. Chem.*, **B6**, 382 (1929).

[c] A. A. Balandin and D. N. Vaskevich, *J. Gen. Chem. U.S.S.R.*, **6**, 1870 (1936).

[d] B. V. Erofeev, *Zhur. Fiz. Khim.*, **24**, 721 (1950).

[e] W. Esson, *Phil. Trans. Roy. Soc. (London)*, **156**, 220 (1866).

[f] C. G. Swain, *J. Am. Chem. Soc.*, **66**, 1696 (1944).

[g] B. Zwolinski and H. Eyring, *J. Am. Chem. Soc.*, **69**, 2702 (1947).

[h] F. A. Matsen and J. L. Franklin, *J. Am. Chem. Soc.*, **72**, 3337 (1950).

method which is only approximate. The following will illustrate the use of these methods.

7A. Elimination of Time as an Independent Variable. Let us assume that we have the following scheme:

(1) $\qquad\qquad$ A $\rightarrow$ B $\qquad$ first-order k_1

(2) $\qquad\qquad$ A $+$ B $\rightarrow$ C $\qquad$ mixed second-order k_2 $\qquad$ (III.7A.1)

We can then write the following:

(1) $$\frac{dA}{dt} = -k_1 A - k_2 BA$$

(2) $$\frac{dB}{dt} = k_1 A - k_2 BA \qquad\qquad (III.7A.2)$$

(3) $$\frac{dC}{dt} = k_2 BA = -\frac{1}{2}\left(\frac{dA}{dt} + \frac{dB}{dt}\right)$$

Note that the third equation is not independent of the first two; it is related by the stoichiometry of the over-all reaction. We shall thus work with the first two equations. If we differentiate equation 2 of the set (III.7A.2) again and combine with equations 1 and 2 of the same set to eliminate A, we obtain a nonlinear, differential equation

$$(k_1 - k_2 B)\frac{d^2 B}{dt^2} + k_2\left(\frac{dB}{dt}\right)^2 + (k_1^2 - k_2^2 B^2)\frac{dB}{dt} = 0 \qquad (III.7A.3)$$

and although Eq. (III.7A.3) may be reduced further, it cannot be solved explicitly except by approximate methods. However, it may be observed that, if we divide equation 2 of set (III.7A.2) by equation 1, we can eliminate time as the independent variable and obtain an equation, involving only A and B, which is soluble:

$$\frac{dB/dt}{dA/dt} = \frac{dB}{dA} = \frac{k_1 A - k_2 BA}{-k_1 A - k_2 BA} = \frac{B - K}{B + K} \qquad (III.7A.4)$$

where we have replaced the ratio k_1/k_2 by K.

Equation (III.7A.4) may be rearranged to give

$$\frac{K + B}{K - B}\, dB = -dA \qquad\qquad (III.7A.5)$$

which may be integrated directly to give

$$B + 2K \ln (K - B) = A + \theta_A \qquad\qquad (III.7A.6)$$

By using the condition that B $= 0$ when A $=$ A$_0$ this is reduced to

$$B + 2K \ln \frac{K - B}{K} = A - A_0 \qquad\qquad (III.7A.7)$$

or, recasting this in more convenient form,

$$\frac{B}{K} + 2 \ln\left(1 - \frac{B}{K}\right) = -\frac{A_0}{K}\left(1 - \frac{A}{A_0}\right)$$
(III.7A.8)

The right-hand side of Eq. (III.7A.8) is always negative because $1 - A/A_0$, which represents the fraction of A used up, is always positive. Thus the left-hand side is negative also, which implies that B/K is always less than 1. Of the terms on the left-hand side, the logarithmic term is always negative and always the larger of the two terms in absolute magnitude. Further, we can show that, when A is used up (that is, $A = 0$), there will always be some B left. The amount of B left will depend both on the values of K and A_0.

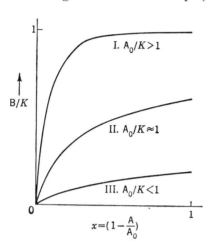

FIG. III.2. Two-step consecutive reaction. Variation of intermediate concentration. B is concentration of intermediate with maximum concentration K. x represents amount of reactant A used up.

If we plot B as a function of A, the curve will have the shape shown in Fig. III.2. It can be seen that B will approach its limiting value K only when A_0/K is very large compared to unity (curve I). If this condition does not hold, then the concentration of B will remain substantially smaller than K for the duration of the reaction (curves II and III). The values are shown in Table III.1.

TABLE III.1. CONCENTRATION OF INTERMEDIATES IN CONSECUTIVE REACTIONS

For reaction scheme: $A \rightarrow B$ First order: k_1
                       $A + B \rightarrow C$ Second order: k_2

Values given are for the ratio $B/K = Bk_2/k_1$ for different values of $A_0/K = A_0k_2/k_1$.

Extent of reaction $(1 - A/A_0)$:	0.1	0.2	0.3	0.4	0.5	0.7	0.9	1.0
Ratio A_0/K	Concentration							
100	0.99	1.00	1.00	1.00	1.00	1.00	1.00	1.00
10	0.53	0.74	0.85	0.91	0.95	0.98	1.00	1.00
1	0.09	0.17	0.23	0.30	0.35	0.43	0.50	0.54
0.1	0.010	0.020	0.028	0.037	0.046	0.064	0.082	0.091
0.01	0.002	0.004	0.006	0.008	0.009	0.010	0.012	0.014

This method is always applicable when the original equations are of such form that eliminating time as an independent variable leads to a differential equation in which the variables are separable. The following represent two further examples of its use and results.

Systems:

(1) $\qquad\qquad A + A \xrightarrow{k_1} B \qquad$ second order

(2) $\qquad\qquad B + B \xrightarrow{k_2} C \qquad$ second order

$$(III.7A.9)$$

Differential equations:

(1) $$\frac{dA}{dt} = -k_1 A^2$$

(2) $$\frac{dB}{dt} = \frac{k_1}{2} A^2 - k_2 B^2 \qquad\qquad (III.7A.10)$$

(3) $$\frac{dC}{dt} = \frac{k_2}{2} B^2 = -\frac{1}{4}\left(\frac{dA}{dt} + \frac{2dB}{dt}\right)$$

Combined equation:

$$\frac{dB}{dA} = -\frac{1}{2} + \frac{k_2}{k_1}\left(\frac{B}{A}\right)^2$$

or, by making the substitutions $B = Ay$ and $K = k_2/k_1$,

$$\frac{dy}{d\ln A} = Ky^2 - y - \frac{1}{2} \qquad\qquad (III.7A.11)$$

which is a separable equation that can be solved by standard methods. Solutions can be obtained for both B and A as functions of t.

Systems:

(1) $\qquad\qquad A + B \xrightarrow{k_1} C \qquad$ second order

(2) $\qquad\qquad B + C \xrightarrow{k_2} D \qquad$ second order

$$(III.7A.12)$$

Differential equations:

(1) $$\frac{dA}{dt} = -k_1 AB$$

(2) $$\frac{dB}{dt} = -k_1 AB - k_2 BC$$

$$(III.7A.13)$$

(3) $$\frac{dC}{dt} = k_1 AB - k_2 BC = \frac{dB}{dt} - \frac{2dA}{dt}$$

(4) $$\frac{dD}{dt} = k_2 BC = -\left(\frac{dB}{dt} - \frac{dA}{dt}\right)$$

Combined equation:

$$\frac{dC}{dA} = -1 + \frac{k_2}{k_1}\frac{C}{A}$$

By substituting $C = Ay$ and $K = k_2/k_1$,   (III.7A.14)

$$\frac{dy}{d \ln A} = -1 + (K - 1)y$$

which has for its solution

$$C = \frac{A}{K - 1}\left[1 - \left(\frac{A}{A_0}\right)^{K-1}\right] \qquad C = 0 \text{ when } A = A_0 \qquad \text{(III.7A.15)}$$

Similarly we can show that

$$B_0 - B = \left(\frac{2K - 1}{K - 1}\right)(A_0 - A) - \frac{A_0}{K - 1}\left[1 - \left(\frac{A}{A_0}\right)^{K}\right]$$

$$\text{(III.7A.16)}$$

$$D = (A_0 - A) - C = A_0 - \frac{A}{K - 1}\left[K - \left(\frac{A}{A_0}\right)^{K-1}\right]$$

Because of the interest in this particular system, the behavior of C/A is shown in Fig. III.3.[a]

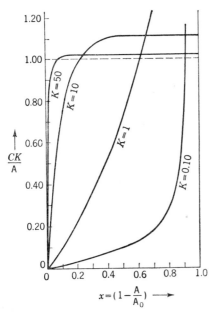

$$x = \left(1 - \frac{A}{A_0}\right) \longrightarrow$$

Fig. III.3. Concentration of intermediate in consecutive reactions, Eq. (III.7A.12).

7B. Reduction of Second-order to First-order Equation. If we have a system of consecutive second-order equations in which a common term appears throughout, then it is possible to make a change in variable which will reduce the system to a system of first-order equations which are linear and therefore susceptible to the exact treatment given previously.

Typical among such systems are the consecutive reactions which occur in growth problems such as the formation of precipitates in supersaturated gases or supercooled liquids, the onset of conditions leading to explosions, and the growth of polymers. In each of these cases the principal process responsible for the growth of large aggregates is the stepwise addition of the original species to the growing ensemble.

Thus a typical polymerization system in which we neglect cessation of

[a] For a more detailed consideration of these systems see S. W. Benson, *J. Chem. Phys.*, **20**, 1605 (1952).

growth and recombination of aggregates is the following series of second-order equations:[a]

$$M_1 + M_1 \xrightarrow{k_1} M_2$$
$$M_1 + M_2 \xrightarrow{k_2} M_3$$
$$M_1 + M_3 \xrightarrow{k_3} M_4$$

$$\cdot \quad \cdot \quad \cdot$$
$$\cdot \quad \cdot \quad \cdot \qquad \text{(III.7B.1)}$$
$$\cdot \quad \cdot \quad \cdot$$

$$M_1 + M_n \xrightarrow{k_n} M_{n+1}$$

in which M_n represents a polymer containing n monomeric units M_1. The differential equations for this system may be written as:

$$\frac{dM_1}{dt} = -k_1 M_1^2 - k_2 M_1 M_2 - k_3 M_1 M_3 - \cdots - k_n M_n M_1 - \cdots$$

$$\frac{dM_2}{dt} = k_1 M_1^2 - k_2 M_1 M_2$$

$$\cdot \quad \cdot \quad \cdot$$
$$\cdot \quad \cdot \quad \cdot \qquad \text{(III.7B.2)}$$
$$\cdot \quad \cdot \quad \cdot$$

$$\frac{dM_n}{dt} = k_{n-1} M_1 M_{n-1} - k_n M_1 M_n$$

The term M_1 occurs in each term, in each equation. If we divide each equation through by M_1, then all the right-hand terms become linear. If we now make the change in variable $dz = M_1\, dt$, we have a system of linear first-order equations:

$$\frac{dM_1}{dz} = -k_1 M_1 - \sum_2^\infty k_n M_n$$

$$\frac{dM_2}{dz} + k_2 M_2 = k_1 M_1$$

$$\cdot \quad \cdot \quad \cdot$$
$$\cdot \quad \cdot \quad \cdot \qquad \text{(III.7B.3)}$$
$$\cdot \quad \cdot \quad \cdot$$

$$\frac{dM_n}{dz} + k_n M_n = k_{n-1} M_{n-1}$$

If we neglect, in the first of these equations, the term $\sum_2^\infty k_n M_n$, the system reduces to the linear first-order sequence treated previously (Sec. III.6). This approximation will be valid only during the initial stages of the re-

[a] See Sec. XVI.10.

action, in which the concentrations of higher polymers are all small compared to the concentration of M_1.

If we make the assumption that $k_2 = k_3 = \cdots = k_n$, then we can get a simple equation for M_1. By adding all of the equations together, except for the first, we obtain

$$\sum_{2}^{\infty} \frac{dM_n}{dz} = k_1 M_1 \qquad \text{(III.7B.4)}$$

By differentiating the first equation with respect to z

$$\frac{d^2 M_1}{dz^2} = -k_1 \frac{dM_1}{dz} - k_2 \sum_{2}^{\infty} \frac{dM_n}{dz} \qquad \text{(III.7B.5)}$$

We can now eliminate the sum common to both of the last two equations, and we can then write

$$\frac{d^2 M_1}{dz^2} + k_1 \frac{dM_1}{dz} + k_1 k_2 M_1 = 0 \qquad \text{(III.7B.6)}$$

which is a linear second-order differential equation which can be solved. It has for a solution:

$$M_1 = a e^{w_1 z} + b e^{w_2 z} \qquad \text{(III.7B.7)}$$

where w_1 and w_2 are the negative or imaginary roots of the equation

$$w^2 + k_1 w + k_1 k_2 = 0$$

namely
$$w_{1,2} = -\frac{k_1}{2} \left[1 \pm \left(1 - \frac{4k_2}{k_1} \right)^{\frac{1}{2}} \right] \qquad \text{(III.7B.8)}$$

If $4k_2 > k_1$, signifying that the propagation rate is faster than the initiation rate, then the roots are imaginary and the real part of the solution may be written as

$$M_1 = C e^{-k_1 z/2} \sin \left[\frac{k_1}{2} \left(\frac{4k_2}{k_1} - 1 \right)^{\frac{1}{2}} z + \theta \right] \qquad \text{(III.7B.9)}$$

where C and θ are constants of integration.[a]

The variable z may be calculated graphically if the data for the rate of disappearance of M_1 with t are known. On integration we find

$$z = \int_0^t M_1 \, dt \qquad \text{(III.7B.10)}$$

so that $z = 0$ when $t = 0$ and $M_1 = M_{10}$.[b] We see that z is, in fact, the area

---

[a] These may be determined from the initial conditions. Thus if at $t = 0$, $M_1 = M_{10}$ and $M_2 = M_3 = \cdots = 0$, then $(dM_1/dz)_0 = -k_1 M_{10}$ and we find that $\theta = \pi/2 + \cot^{-1} (4k_2/k_1 - 1)^{\frac{1}{2}}$ and $C = M_{10} [4k_2/(4k_2 - k_1)]^{\frac{1}{2}}$.

[b] z will be finite when M_1 is used up (that is, $M_1 = 0$), because for the above case M_1 disappears at a rate faster than in an ordinary bimolecular reaction, and so the area under the curve M_1 versus t remains finite. In fact when $M_1 = 0$, $z_\infty = 2(\pi - \theta)/(4k_1 k_2 - k_1^2)^{\frac{1}{2}}$.

under the curve of M_1 as a function of time. It represents a change of time scale from the laboratory interval of time dt to a scale dz in which each interval is progressively shorter by an amount proportional to the diminution of M_1 as the reaction proceeds.

The rest of the solution may be obtained by the method outlined in Sec. III.6. By successive differentiation and elimination we arrive at the differential equation for M_n:

$$(D + k_n)(D + k_{n-1}) \cdots (D + k_2)M_n = k_{n-1}k_{n-2} \cdots k_1M_1$$

$$\text{(III.7B.11)}$$

or in more compact form

$$\prod_{j=2}^{n} (D + k_j)M_n = \prod_{j=1}^{n-1} k_jM_1 \qquad \text{(III.7B.12)}$$

This is a linear differential equation of order n and has a general solution obtained by setting the right-hand side equal to zero. The equation is solved in Sec. III.6. To this general solution we must add a particular solution. Since M_1 is a known function of z [Eq. (III.7B.7) or (III.7B.9)], the particular solution $\phi(z)$ can be found by known methods. In this case it has the form

$$\phi(z) = C_1e^{w_1z} + C_2e^{w_2z} \qquad \text{(III.7B.13)}$$

if M_1 is given by Eq. (III.7B.7). If M_1 has the form (III.7B.9), then

$$\phi(z) = C_1e^{-k_1/2}[\sin(az + \theta) + C_2 \cos(az + \theta)]$$

where $a = k_1/2(4k_2/k_1 - 1)^{1/2}$. In both cases C_1 and C_2 are not arbitrary constants but are determined by substituting this value of $\phi(z)$ for M_n in the differential equation. The total solution in this particular case is then (since we have assumed $k_2 = k_3 = \cdots = k_n = \cdots$)

$$M_n = e^{-k_2z}\left[\sum_{j=0}^{n-2} a_{nj}z^j\right] + \phi(z) \qquad \text{(III.7B.14)}$$

where once again the arbitrary constants a_{nj} are to be determined from the initial conditions that M_n and its first $n - 2$ derivations are all zero when $z = 0$.

8. Classification of Consecutive Reaction Systems

In dealing with systems of consecutive reactions we can classify the components of such a system into three groups.

Initiation Reactions. These are the first reactions responsible for the production of the intermediates which give rise to the subsequent reactions.

Propagation Reactions. These are all the reactions in which intermediate products react to produce further intermediates and usually result in removal of reactant.

Termination Reactions. These are the reactions in the sequence which result in the destruction of the intermediates.

Thus in a typical scheme of second-order reactions

$$
\begin{array}{lll}
(1) & \quad A + A \rightarrow B & \\
(2) & \quad B + A \rightarrow C & \qquad\qquad (\text{III.8.1}) \\
(3) & \quad C + A \rightarrow D \qquad \text{stable } D \equiv A_4 &
\end{array}
$$

reaction 1 is an initiation reaction for B; reaction 2 is a propagation reaction,[a] since it replaces the reactive intermediate B by the reactive intermediate C; and reaction 3 is a termination reaction, since it destroys a reactive intermediate C and produces in its place a nonreactive molecule D.

The results of the calculations in Sec. III.7 show that such systems of consecutive reactions that involve the production of reactive intermediates are susceptible of simple mathematical treatment under certain conditions. We shall treat this in the following section.

9. The Stationary-state Hypothesis; Chain Reactions

Let us consider the simplest possible case of a system of consecutive reactions involving two intermediates, namely

$$
\begin{array}{llll}
(1) & \quad A \xrightarrow{k_1} 2M_1 & \quad \text{first order} & \\
(2) & \quad M_1 + C \xrightarrow{k_2} P + M_2 & \quad \text{second order} & \\
(3) & \quad M_2 + A \xrightarrow{k_3} P + M_1 & \quad \text{second order} & \quad (\text{III.9.1}) \\
(4) & \quad M_1 + M_1 \xrightarrow{k_4} A & \quad \text{second order} &
\end{array}
$$

The over-all reaction is obtained by adding the 4 equations of set (III.9.1):

$$
A + C \rightarrow 2P
$$

where A and C are reactants, P is the product molecule, and M_1 and M_2 are the two reactive intermediates.[b] The individual rate equations for the stable molecules are then

[a] Reaction 2 is also an initiation reaction for the intermediate C. In the same sense all such propagation reactions in which an original "carrier" (in this case B) gives rise to a new carrier C are simultaneously initiation as well as propagation reactions. There are also propagation reactions in which the initial carrier produces a stable species and regenerates itself, e.g.,

$$
B + A \rightarrow E + B \qquad \text{E is stable}
$$

Such reactions are not important unless E, which has the same composition as A, is a distinguishable isomer. In such cases B is, in effect, a catalyst for the isomerization of A (e.g., Walden inversions and ortho- to para-H conversion).

[b] It can be seen that both A and C are symmetrical molecules with the formulas $A = (M_1)_2$ and $C = (M_2)_2$ and $P = (M_1 \cdot M_2)$ is the unsymmetrical product. The reaction $H_2 + Br_2 \rightarrow 2HBr$ can be looked on as an example of such a scheme. Notice that the propagation steps 2 and 3 constitute a chain.

$$-\frac{dA}{dt} = k_1 A + k_3 M_2 A - k_4 M_1^2$$

$$-\frac{dC}{dt} = k_2 M_1 C \qquad (III.9.2)$$

$$\frac{dP}{dt} = k_2 M_1 C + k_3 M_2 A$$

and for the unstable intermediates

$$\frac{dM_1}{dt} = 2k_1 A + k_3 M_2 A - k_2 M_1 C - 2k_4 M_1^2$$

$$\frac{dM_2}{dt} = k_2 M_1 C - k_3 M_2 A \qquad (III.9.3)$$

In addition we have the stoichiometric relations

$$\text{Moles of A used} = A_0 - A = \frac{P}{2} + \frac{M_1}{2}$$

$$\text{Moles of C used} = C_0 - C = \frac{P}{2} + \frac{M_2}{2} \qquad (III.9.4)$$

where it is assumed that P, M_1, and M_2 are zero at the start of the reaction. If now we can postulate that *the concentrations of intermediates* M_1 *and* M_2 *are at all times much less than the concentrations of reactants and products* A, C, and P, then we may neglect the terms involving intermediates in Eq. (III.9.4) which reduce to

$$A_0 - A = \frac{P}{2} \qquad C_0 - C = \frac{P}{2} \qquad (III.9.5)$$

and on differentiating with respect to time

$$-\frac{dA}{dt} = \frac{1}{2}\frac{dP}{dt} \qquad -\frac{dC}{dt} = \frac{1}{2}\frac{dP}{dt} \qquad (III.9.6)$$

Equations (III.9.6) imply that there is no time lag, or there is at least a negligible time lag, between the destruction of the individual reactants A and C and the appearance of the product P. By combining this condition with Eqs. (III.9.2) and (III.9.4) it leads to the further condition that dM_1/dt and dM_2/dt are also negligibly small and effectively zero.

If we make use of this result by setting these rates equal to zero in the chain equations (III.9.3), we can solve the equations as a simultaneous set of algebraic equations for the concentrations of M_1 and M_2. From the second of the two equations we find

$$k_2 M_1 C = k_3 M_2 A \qquad (III.9.7)$$

which on substitution into the first yields

$$M_1 = \left(\frac{k_1}{k_4}\right)^{1/2} A^{1/2}$$

(III.9.8)

and finally

$$M_2 = \frac{k_2}{k_3}\left(\frac{k_1}{k_4}\right)^{1/2} \frac{C}{A^{1/2}}$$

Our initial equations now reduce to

$$-\frac{dA}{dt} = k_3 M_2 A = k_2 \left(\frac{k_1}{k_4}\right)^{1/2} CA^{1/2}$$

$$-\frac{dC}{dt} = k_2 M_1 C = k_2 \left(\frac{k_1}{k_4}\right)^{1/2} CA^{1/2}$$

(III.9.9)

and these equations can now be solved by means of the usual substitutions

$$C = C_0 - x \qquad A = A_0 - x$$

(III.9.10)

giving

$$\frac{dx}{(C_0 - x)(A_0 - x)^{1/2}} = k_2 \left(\frac{k_1}{k_4}\right)^{1/2} dt$$

(III.9.11)

Equation (III.9.11) can be integrated directly by means of the substitution $(A_0 - x) = y^2$, and we finally obtain

Case 1. $C_0 - A_0 = \Delta > 0$:

$$\tan^{-1}\left(\frac{A}{\Delta}\right)^{1/2} = -k_2 \left(\frac{k_1}{k_4}\right)^{1/2} t + \tan^{-1}\left(\frac{A_0}{\Delta}\right)^{1/2}$$

(III.9.12)

Case 2. $A_0 - C_0 = \Delta > 0$:

$$\ln \frac{A^{1/2} + \Delta^{1/2}}{A^{1/2} - \Delta^{1/2}} = k_2 \left(\frac{k_1}{k_4}\right)^{1/2} t + \ln \frac{A_0^{1/2} + \Delta^{1/2}}{A_0^{1/2} - \Delta^{1/2}}$$

(III.9.13)

Case 3. $C_0 = A_0$:

$$\frac{1}{A^{1/2}} - \frac{1}{A_0^{1/2}} = k_2 \left(\frac{k_1}{k_4}\right)^{1/2} t$$

(III.9.14)

The necessary conditions for the application of the stationary-state hypothesis, that M_1 and M_2 be small compared to A or C, can be inspected in a quantitative fashion with Eq. (III.9.8). On rewriting these equations in terms of concentration ratios, we have

$$\frac{M_1}{A} = \left(\frac{k_1}{k_4 A}\right)^{1/2} \qquad \frac{M_2}{C} = \frac{k_2}{k_3} \frac{M_1}{A} = \frac{k_2}{k_3}\left(\frac{k_1}{k_4 A}\right)^{1/2}$$

(III.9.15)

Thus to make less than a 1 per cent error in reporting concentrations implies that $M_1/A \leq 0.01$ and $M_2/C \leq 0.01$. But this can be true for M_1/A only if $k_1 \leq 1 \times 10^{-4} k_4 A$ and for M_2/C if $M_1/A \leq 0.01$ and $k_2 \leq k_3$.

These inequalities often provide important checks on the self-consistency of an assumed mechanism when independent information about the individual rate constants is available or when reasonable guesses of their order of magnitude may be made.

10. The Induction Period

We have seen in Sec. III.9 that the kinetics of reactions which involve two or more consecutive intermediates is capable of a simplified treatment if the intermediate concentrations are small compared to both reactants and products. In such cases we can make the assumption that the rate of change of the concentrations of the intermediates with time is zero, a procedure which then permits us to solve the kinetic equations for the "stationary" concentrations of the intermediates and then eliminate them algebraically from the system of differential equations.[a]

But surely such a procedure does not apply to the initial stages of the reaction during which the concentration of intermediates is building up from zero to their stationary concentration. That is correct, and in applying our kinetic equations we must ignore this "induction period," to which our hypothesis does not apply. During the induction period, when all the intermediates have smaller concentrations than that attained in the stationary state, all rates, namely, the rate of disappearance of reactants and the rates of appearance of products,[b] will be smaller. How long a period will this be? We can give a simple relative answer to the question. If a_1 is the ratio of the stationary-state concentration of the first intermediate M_1 to reactant A_1, then we must wait at least long enough for this same fraction of A_0, namely, a_1A_0, to be used up before a stationary state is even possible. This provides a lower limit then to the induction period in terms not of time but of disappearance of reactant. An upper limit can be provided only by a solution of the original differential equations.[c]

However, the solutions given for the systems in Sec. III.7 indicate that the total induction time is roughly 4 to 10 times as long. In other words, if the stationary concentration of M_1 is 1 per cent of A_0, then about 4 to 10 per cent of A_0 will have disappeared before M_1 reaches 99 per cent of the stationary concentration (Figs. III.2 to III.4). A more detailed analy-

[a] This stationary-state method was first developed by the following: M. Bodenstein, *Z. physik. Chem.*, **85**, 329 (1913); *Ann. Physik*, **82**, 138 (1927). J. A. Christiansen, *Kgl. Danske Videnskab. Selskab., Mat.-fys. Medd.*, **1**, 14 (1919). K. F. Herzfeld, *Z. Elektrochem.*, **25**, 301 (1919); *Ann. Physik*, **59**, 635 (1919). M. Polanyi, *Z. Elektrochem.*, **26**, 50 (1920).

[b] This is not necessarily true for the initiation step involving a reactant.

[c] A very reliable answer can usually be obtained by integrating the equations for the intermediates on the assumption that the concentrations of reactants do not change appreciably during the induction period. The error made will usually be of the order of one-half the real change in the reactant concentrations in this interval. See Sec. XIII.4A.

sis of the induction period in chain reactions in which the above results were obtained has been given by the author.[a]

When the reaction involves a chain such as occurs in the reaction of H_2 and Cl_2

$$Cl + H_2 \xrightarrow{k_1} HCl + H$$

$$H + Cl_2 \xrightarrow{k_2} HCl + Cl$$

so that the initial intermediate is reproduced, it is possible for a large amount of product—here HCl—to be built up before the intermediates—H and Cl—reach their low stationary values. It can be shown that in such cases the induction period will be negligibly small only when the rate of the slowest step in the chain, k_1 or k_2, is very small compared to the rate of destruction of the intermediates by other processes.[b]

In analyzing all such reactions it is sometimes convenient to discuss the mean lifetime of an intermediate passing through a reaction step, which may be defined as the reciprocal of the rate constant (or its product with some concentration). Thus in the above scheme the mean time of Cl in step 1 is $\tau_1 = [k_1 \times (H_2)]^{-1}$ and for H in step 2, $\tau_2 = [k_2 \times (Cl_2)]^{-1}$. The mean time of chain cycle is then $\tau_c = \tau_1 + \tau_2$. If τ_d is the mean lifetime for the destruction of intermediates, then the above criteria imply that $\tau_d \ll \tau_c$ in order for the induction period to be small.[c]

In a very important paper, Semenov[d] has given a detailed analysis of these phenomena for both stationary and nonstationary (i.e., explosive) reactions. These will be discussed in greater detail in a later chapter. Frank-Kamenetskii[e] has also given a treatment of the problem, but in a more formal and general way.

11. Higher-order Reactions

While there are no exact, general methods for the solution of consecutive reaction schemes of higher order than the first, there has been a considerable amount of work done on special systems of this type. Thus exact solutions have been given by Jen-Yuan Chien[f] for the following reaction systems:

[a] S. W. Benson, *J. Chem. Phys.*, **20**, 1605 (1952). See also Sec. XIII.4A.

[b] The sum to be taken may include products of rate constants by concentrations if the rate constants compared are for reactions of different over-all order. Benson, *loc. cit.*

[c] See, W. H. Stockmayer, *J. Chem. Phys.*, **12**, 143 (1944), for a discussion of the steady state in polymerization reactions.

[d] N. N. Semenov, *J. Phys. Chem. U.S.S.R.*, **17**, 187 (1943); *J. Chem. Phys.*, **7**, 683 (1939).

[e] D. A. Frank-Kamenetskii, *J. Phys. Chem. U.S.S.R.*, **14**, 695 (1940).

[f] Jen-Yuan Chien, *J. Am. Chem. Soc.*, **70**, 2256 (1948).

(I) $A \rightarrow B$ $B + B \rightarrow C$
(II) $A \rightarrow B$ $B + D \rightarrow C$
(III) $A + A \rightarrow B$ $B \rightarrow C$
(IV) $A + A \rightarrow B$ $B + B \rightarrow C$
(V) $A + A \rightarrow B$ $B + D \rightarrow C$

Unfortunately, the explicit solutions to these systems usually involve fairly complicated functions such as Bessel, Haenkel, theta, and incomplete gamma functions in which the arguments depend on the rate constants.[a] The result is that these solutions are of use only if the ratios of the rate constants are known. Otherwise an inordinate amount of arithmetical labor is involved.

Adirovich[b] has given a detailed treatment of

(VI) $A + B \rightleftharpoons AB$ $AB \rightarrow D$

while less explicit solutions have been given for both the consecutive displacement reactions:[c]

(VII) $AX_4 + RY \rightarrow RAX_3 + XY$
                       $RAX_3 + RY \rightarrow R_2AX_2 + XY$

$$\cdot \qquad \cdot \qquad \cdot \qquad \cdot$$

$$\cdot \qquad \cdot \qquad \cdot \qquad \cdot$$

$$R_3AX + RY \rightarrow R_4A + XY$$

and the polymerization reactions discussed earlier[d] [Eq. (III.7B.1)].

12. Rate-determining Step

In systems of consecutive reactions it may sometimes occur that there is one step which is very much slower than all the subsequent steps leading to product. Then the rate of production of product may depend on the rates of all the steps preceding the last slow step but will not depend on any of the subsequent steps, all of which are rapid compared to the last slow step. Such a last slow step has been called, somewhat misleadingly, the *rate-determining step* of the reaction.

[a] Systems II and IV can be reduced to Riccati equations for which the solutions may be represented in terms of Bessel and Haenkel functions. System IV has been analyzed in some detail by A. D. Stepukhovich and L. M. Timonin, *Zhur. Fiz. Khim.*, **25**, 143 (1951). System III had been earlier reported on by A. A. Balandin and L. S. Leibenson, *Compt. rend. acad. sci. U.R.S.S.*, **39**, 22 (1943).

[b] E. I. Adirovich, *Doklady Akad. Nauk S.S.S.R.*, **61**, 467 (1948).

[c] R. M. Fuoss, *J. Am. Chem. Soc.*, **65**, 2406 (1943).

[d] S. Ya. Pshezhetskii and R. N. Rubinshtein, *J. Phys. Chem. U.S.S.R.*, **21**, 659 (1947). G. Natta, *Rend. inst. lombardo sci.*, **78**, p.I, 307 (1945). C. Potter and W. C. MacDonald, *Can. J. Research*, **25B**, 415 (1947). The latter authors use a method similar to the one developed here and outline graphical methods of application of the data to the solutions.

For illustration let us consider the following complex sequence of stoichiometric reactions as representing the reaction path for the over-all change $A + C \rightarrow P$:

$$A \underset{2}{\overset{1}{\rightleftharpoons}} B \qquad \text{both first order}$$

$$B + C \overset{3}{\rightarrow} D \qquad \text{second order mixed} \qquad \text{(III.12.1)}$$

$$D \overset{4}{\rightarrow} P \qquad \text{first order}$$

We can apply the stationary-state technique if both the intermediates B and D are small at all times compared to A and C. Their stationary-state concentrations will then be given by

$$B_{ss} = \frac{k_1 A}{k_2 + k_3 C} \qquad D_{ss} = \frac{k_3}{k_4} BC = \frac{k_1 k_3 AC}{k_4(k_2 + k_3 C)} \qquad \text{(III.12.2)}$$

This implies that $k_2 + k_3 C \gg k_1$ and $k_4 \geq k_3 C$. For the rate of appearance of products P we can write

$$\frac{dP}{dt} = k_4 D = \frac{k_1 k_3 AC}{k_2 + k_3 C} \qquad \text{(III.12.3)}$$

This simplifies in either of the two extreme cases:

Case 1. $k_2 \gg k_3 C$: $\dfrac{dP}{dt} \rightarrow \dfrac{k_1}{k_2} k_3 AC$

$$\text{(III.12.4)}$$

Case 2. $k_2 \ll k_3 C$: $\dfrac{dP}{dt} \rightarrow k_1 A$

In case 1 step 3 is the slow step and in case 2 step 1 is the slow step. It can be seen that in each of these extreme cases, the form of the over-all rate law is independent of the rates of all the much faster steps following the rate-determining steps. Note also that the steps leading up to the rate-determining step do appear.

The principle behind such terminology lies in the following method of analysis. In a system of consecutive reaction steps where reactant passes through a number of intermediate stages, the total time to produce a molecule of product is simply the sums of the discrete times necessary to pass through each consecutive stage of the reaction. The mean reaction time t_p is thus:

$$t_p = t_1 + t_2 + \cdots + t_n$$

But the over-all rate is simply the reciprocal of this mean time, $R_p = 1/t_p$, or[a]

---

[a] The rate of a step R_p can also be written in terms of rate constants and concentrations of reactants. Thus in the preceding example $R_3 = k_3 BC$. Where a reverse reaction is involved (e.g., an equilibrium), the net time for the step is increased by the "feedback" of intermediate.

$$R_p = \frac{1}{t_p} = \frac{1}{t_1 + t_2 + \cdots + t_n}$$

$$= \frac{1}{1/R_1 + 1/R_2 + \cdots + 1/R_n} \tag{III.12.5}$$

The philosophy of a rate-determining step is that one of these times is much larger than any of the others following it. Note that, for the stationary-state hypothesis to apply, all times involving formation of intermediates must be successively decreasing.

IV

Experimental Characterization
of Simple Kinetic Systems

1. Experimental Methods; Composition

Although a complete survey of the experimental methods which have been used for the study of reacting systems is outside the scope of this book, it is well to consider some of the more general methods which have been employed and some of the difficulties inherent in such studies.[a] The general problem involved in any experimental study of a kinetic system is to obtain a complete description of the state of the system over the duration of the reaction. Of the variables of the system, the temperature is generally kept constant (by employing a thermostat), and its effect on the rate is studied independently. Also, the volume is kept constant or nearly constant. The principal problem then resolves itself into devising methods for the chemical analysis of the system as a function of time.

The problem of chemical analysis is itself simplified by virtue of the stoichiometry of the given reaction. Consider the reaction of hydrogen and iodine to produce hydrogen iodide, $H_2 + I_2 \rightarrow 2HI$. If the number of moles of HI present at any time t is $x + x_0$, where $x = 0$ at $t = 0$, then $(H_2) = (H_2)_0 - x/2$ and $(I_2) = (I_2)_0 - x/2$, where the parentheses denote concentration.

If the composition of the system is known at any one time, then it will suffice to know the amount of any one of the species involved in the reaction as a function of time in order to completely establish the chemical composition as a function of time. This will be true of any system whose reaction can be specified by a simple stoichiometric equation. For such a

[a] An excellent treatise on the methods employed to study gas reactions will be found in A. Farkas and H. Melville, "Experimental Methods in Gas Reactions," Macmillan & Co., Ltd., London, 1939.

58

system it will suffice to analyze for only one of the changing chemical components. This is subject to the condition that there are no concurrent reactions and no appreciable concentrations of intermediates.[a]

In the more complex cases of concurrent or consecutive reactions the composition of the system will depend on a simultaneous knowledge of the amounts of two or more species present as a function of time. The number of composition variables to be studied will in general be equal to the number of independent chemical equations that describe the reaction.

Any one chemical or physical property or any combined set of chemical or physical properties that will uniquely define the composition of the system can be used to study the composition of the reaction system as a function of time. In the following subsections are described some of the more common methods employed.

1A. Total Pressure. If a reaction involves at least one gas and the stoichiometric equation for the reaction predicts a change in the number of moles of gas, then the total pressure will suffice to define the composition of the system at any time (if the gases are ideal).

Example

$$N_2(g) + 3H_2(g) \overset{cat.}{\rightleftharpoons} 2NH_3(g)$$

If $P^{(t)}$ represents the partial pressure at time t and $P^{(0)}$ the initial partial pressure, then

$$P^{(t)}_{total} = P^{(t)}_{N_2} + P^{(t)}_{H_2} + P^{(t)}_{NH_3}$$
$$P^{(0)}_{total} = P^{(0)}_{N_2} + P^{(0)}_{H_2} + P^{(0)}_{NH_3}$$

but $\qquad P^{(t)}_{N_2} - P^{(0)}_{N_2} = \tfrac{1}{3}(P^{(t)}_{H_2} - P^{(0)}_{H_2}) = -\tfrac{1}{2}(P^{(t)}_{NH_3} - P^{(0)}_{NH_3})$

from which we can deduce from the partial pressures at any time t in terms of the initial and total pressures:

$$P^{(t)}_{N_2} = P^{(0)}_{N_2} + \tfrac{1}{2}(P^{(t)}_{total} - P^{(0)}_{total}) = P^{(0)}_{N_2} + \tfrac{1}{2}\Delta P_{total}$$
$$P^{(t)}_{H_2} = P^{(0)}_{H_2} + \tfrac{3}{2}(P^{(t)}_{total} - P^{(0)}_{total}) = P^{(0)}_{H_2} + \tfrac{3}{2}\Delta P_{total}$$
$$P^{(t)}_{NH_3} = P^{(0)}_{NH_3} - (P^{(t)}_{total} - P^{(0)}_{total}) = P^{(0)}_{NH_3} - \Delta P_{total}$$

The procedure followed in studying such reactions is to fill a flask with known amounts of the reactants (or with a reaction mixture of known composition) and then follow the total pressure of the system as a function of time. Figure IV.1 illustrates such an apparatus. There is always the difficulty with such equipment of making corrections for the "dead space," i.e., the portions of connecting tubing which contain reacting gases but are outside the thermostat.[b]

---

[a] See Sec. III.9. Let us also note that in practice it is always wise to measure at least one other quantity as a check against unexpected errors, losses of material, or side reactions which may occur, even with the walls of the reaction vessel.

[b] The *effective volume* of this dead space is equal to its true volume multiplied by T_R/T_D, where T_R and T_D are the absolute temperatures of the reaction flask and dead space, respectively.

Because of its convenience this method is by far the most popular of all methods for studying gas reactions. It should be clear that the reactions studied must be slow enough that the time required to fill the vessel and reach temperature equilibrium is not a serious consideration. At high

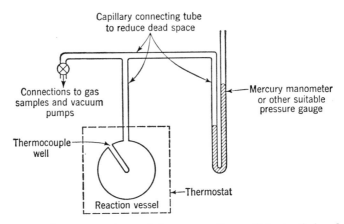

FIG. IV.1. Typical apparatus for studying gas reactions. *Note:* A stirring device operated magnetically may be used inside the reaction vessel. Also, the connecting tubing may sometimes be wound with nichrome heating elements to prevent condensation of less volatile vapors and reduce the effective volume of the dead space. The use of too small or too long capillary tubing is restricted by the difficulty of evacuating the reaction vessel, which increases with longer or smaller capillary leads.

relative densities, at which gas imperfections may become serious, this method, although still useful, requires tedious calculations in order to yield correct concentrations.

1B. Method of Sampling. Perhaps the simplest method of studying kinetic systems and one which is capable of yielding the most complete information is the method of sampling. This method consists in withdrawing small samples from the reaction mixture during the course of the reaction and then so treating the samples that the reaction is quenched. The samples are then analyzed for their composition.

The quenching may consist in sudden cooling to stop the reactions or in the addition of a chemical reagent which will remove (by reaction or neutralization) one of the reactive components. The difficulty with such a method is that it involves disturbing the reacting system. If the system is a gas system, then the sampling involves a sudden diminution in concentrations; if it is a liquid system, there need be no such disturbance.

In extreme form, the sampling procedure may be total, i.e., the entire reaction vessel containing the system may be suddenly quenched and a total analysis then made on the system. An identical system is prepared and quenched at a different time. By continued repetition of these steps the entire course of the reaction may be studied. While simple from the point of view of physical equipment, this procedure is costly in time and materials. It is, however, capable of yielding quite precise results.

1C. Flow Methods. One of the difficulties inherent in the study of complex kinetic systems is the variety of secondary reactions which may follow the initial reaction. It is possible in many cases to reduce the extent of these reactions to negligible proportions by limiting the kinetic study to the first few per cent of consumption of the initial species. This can be done in a static system by using the method of sampling. However, a very popular scheme which has been generally employed is to flow the reactants through the reaction vessel for periods of time which may be controlled. In this way a large amount of products may be accumulated (large in an absolute, not relative, sense, since it is small compared to the reactants used) without permitting secondary reactions to become important. It also permits the convenience of studying a reaction under conditions such that the concentrations of reactant species are kept almost constant. By studying the effects of changing initial concentrations, the results of a flow-method study permit a direct analysis of the differential form of the kinetic equation rather than its usually complicated integrated form.

Examples of this type of study are the pyrolysis of hydrocarbons and other organic compounds carried out by Szwarc and coworkers.[a] By confining the total extent of reaction to a few per cent, it was found possible to study the primary reactions occurring in what is normally a very complicated system. Thus in the gas-phase pyrolysis of toluene[b] the stages of the reaction may be represented by

$$\phi CH_3 \rightarrow \phi CH_2 + H \qquad \text{first order}$$
$$H + \phi CH_3 \rightarrow \phi CH_2 + H_2 \qquad \text{second order}$$
$$H + \phi CH_3 \rightarrow \phi H \quad + CH_3 \qquad \text{second order}$$
$$CH_3 + \phi CH_3 \rightarrow \phi CH_2 + CH_4 \qquad \text{second order}$$
$$\phi CH_2 + \phi CH_2 \rightarrow (\phi CH_2)_2 \qquad \text{second order}$$

If the reaction products are permitted to accumulate, then the reactive H atoms and CH_3 radicals will react further with them and the complexity of reactions and number of products increase.

There are many difficulties involved in the use of the flow method, the chief one being the difficulty of defining the reaction time and the reaction temperature. The reaction time is calculated from the rate of flow of the reactants through the system[c] and the length of the reactor zone. This assumes, however, that the gas comes to the reactor temperature immedi-

[a] M. Szwarc, *Chem. Revs.*, **47**, 75 (1950).

[b] M. Szwarc, *J. Chem. Phys.*, **16**, 128 (1948). M. Szwarc and J. S. Roberts, *ibid.*, **16**, 609, 981 (1948).

[c] Since a steady state of flow is usually established, the mass of material passing through any part of the flow system per unit time is constant.

ately on entering the reaction zone and that the reaction is immediately quenched on leaving the zone, neither of which assumptions can be true. These factors can be minimized by increasing the length of the reaction zone, decreasing the flow rate, or both. However, a decrease in the flow rate results in the back diffusion of products throughout the system and a loss of control of the range of reaction times. Because of these "end" effects, there is also a difficulty in defining the temperature of the reacting gas. At fast flow rates this will be a serious difficulty.

A further difficulty arises from the pressure drop through the reactor system,[a] which means that second- or higher-order processes will be going on at a faster rate near the reactor entrance (where pressures are higher) than at the reactor exit.[b] This becomes further complicated if there is a change in the number of moles of gas during the reaction. G. M. Harris[c] has given a treatment of these effects for first- and second-order reactions, and it has been a subject of investigation by chemical engineers,[d] who employ flow systems where possible.[e] In the case of reactions involving free radicals or unstable intermediates the difficulties are further increased, because the radicals or intermediates may continue chain reactions outside the reactor. If the amount of reaction is kept small, there arises an analytical problem of separating and measuring a small percentage of products in the presence of a large amount of reactant.

For the foregoing reasons flow methods are poor means of obtaining precise data in gas systems. On the other hand, the methods are very useful in the study of heterogeneous reactions in which the reaction zone is fairly well limited to the catalyst surface and the chief problem is one of diffusion from the gas phase (or solution) to the catalyst surface. In the case of liquid-phase reactions the method has been employed,[f] although it usually requires almost prohibitively large amounts of solvent. Finally

[a] This may be corrected for by using the equation for hydrodynamic flow of a gas. See, for example, F. Paneth and W. Lautsch, *Ber.*, **64**, 2708 (1931).

[b] R. Gomer, *J. Chem. Phys.*, **19**, 284 (1951).

[c] G. M. Harris, *J. Phys. & Colloid Chem.*, **51**, 505 (1947).

[d] See, for example, D. A. Hougen and K. M. Watson, "Chemical Process Principles," part III, John Wiley & Sons, Inc., New York, 1947.

[e] A further serious error discussed by M. Gilbert, *Combustion and Flame*, **2**, 137, 149 (1958), is due to the difference in residence times for sections of the gas near the walls as opposed to the center of the gas. This arises from the velocity gradient accompanying Poiseuille flow.

[f] A detailed study of a stirred liquid flow system using moderate volumes has been made by Hammett et al., *J. Am. Chem. Soc.*, **72**, 280, 283, 287 (1950), by using a design suggested by Denbigh, *Trans. Faraday Soc.*, **40**, 352 (1944); *ibid.*, **44**, 497 (1948). The system is quite successful for accurate measurement of rates for fast reactions. Piret and MacDonald, *Chem. Eng. Prog.*, **47**, 363 (1951), have made a detailed empirical analysis of mixing times, flow rates, and channeling in liquid reactors. The stirred flow reactor avoids some of the difficulties discussed above.

the method requires in many cases fairly elaborate experimental equipment for preheating the reactants, adjusting the flow of reactants precisely, and collecting and separating the products. A typical flow system is shown in Fig. IV.2.

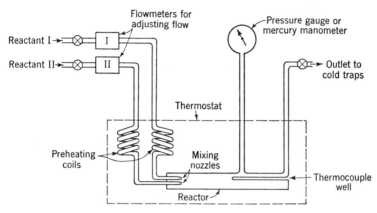

FIG. IV.2. Flow system for studying reaction systems. *Note:* Mixing nozzles are usually arranged to enter the vessel tangentially in order to avoid stagnation regions and also give some rotary mixing to the flow stream.

1D. Optical Methods of Analysis.   Optical methods of analysis of reaction systems are very convenient where they can be applied. The optical properties which characterize the system may be the absorption at one or more particular wavelengths (in the ultraviolet, visible infrared, or microwave region), the refractive index of the mixture, the optical rotation of one or more species, the light-scattering properties of large molecules, or the fluorescent emission of one or more of the substances present.

Any of these methods may be used in conjunction with a static or flow system or else a flow system in which the reaction mixture is cycled to the optical measuring system and returned to the original reaction vessel.

Each requires a very careful calibration of the system in order to obtain the composition from the measured optical property. If only one component is analyzed, then it is essential to show that its optical property is uniquely related to its concentration and is not dependent upon the other substances present. With the exception of optical rotation and refractive index, most of the optical methods do not give better than a 1 per cent accuracy under normal operating conditions, and generally the accuracy is much less.[a]

[a] A serious hazard encountered with optical methods arises from changes or shifts in absorption bands or in absorption coefficients with changes in temperature, medium, or pressure.

1E. Electrical Methods of Analysis. Among the various electrical properties which have been used for analysis and for which measuring apparatus may be directly incorporated in the reaction chamber are the dielectric constant; the electrical conductance; the pH by using glass, calomel, or H electrodes; the redox potential; and, in the case of gas reactions, the thermal conductivity. These properties are easily measured and lend themselves, as do the optical methods, to automatic recording devices. However, they also must be used only after careful calibration and do not give better than 1 per cent accuracy without unusual attention.

1F. Miscellaneous Methods of Analysis. Possible methods for studying systems are limited only by the ingenuity of the experimentalist, the available materials, and the nature of the reaction. Among the other methods which have found some wide adoption are:

1. *Dilatometric Methods.* Usually applied to liquid systems in which the volume of the system depends on the extent of the reaction. The reaction vessel is called a *dilatometer.*[a]

2. *Viscosity.* Applicable to both gas and liquid systems when there is a change in specific viscosity with extent of reaction. (Used frequently in studying polymerization.)

3. *Colorimetric Indicators.* Useful when one of the species is capable of being in rapid equilibrium with a dye substance such that the dye color is affected by the amount of the species present.

4. *Thermal Methods.* Applicable when the heat of the reaction is large enough to measure easily. It is a fairly elegant method when used in conjunction with an ice calorimeter for reactions that are run at 0°C. It requires more elaborate equipment at other temperatures, and it also requires good stirring equipment to maintain constancy of temperatures.

5. *Radioactive or Labeled Isotopes.* Aside from the difficulties of sample preparation these methods are sometimes the only ones available for certain types of reaction and are thus extremely valuable. However, the counting is seldom better than 1 per cent, so that it is not a very accurate method.

6. *Fast Reactions.* Where reactions are essentially completed in times of seconds or much less, ingenious methods have been devised to give measurements of the rate.[b] Such methods may involve static systems in which mixing is performed very rapidly. Also useful is excitation of the system by light for a specified period. A variant method involves a flow system in which reactants are rapidly mixed and flowed through a tube in which recording equipment can be employed to measure optical absorption, heat evolution (temperature), or electrical conductivity. The

---

[a] Dilatometers are extremely sensitive and permit precise measurement of very small amounts of reaction.

[b] For a review see the Symposium Volume, *Discussions Faraday Soc.,* **17** (1954).

earlier methods generally depended on flow systems, while the more recent methods have used static systems with very rapidly responding photocells to measure optical absorption, recording being done by means of a photocell or photomultiplier in conjunction with an oscilloscope. Such systems are, however, not isothermal, but closer to being adiabatic, and the rate constants must be corrected to constant temperature.

In this category mention should also be made of sound-dispersion techniques for investigating the rate of opposing reactions in a system in dynamic equilibrium. If a sound wave of frequency ν is passed through a system at equilibrium, then there will be an anomalously large dispersion of the sound energy at a frequency which corresponds to the frequency of one of the reactions going on in the system. At sound frequencies smaller than this reaction frequency, the system has time to adjust itself to the adiabatic disturbances caused by the sound wave. At much higher frequencies the system does not adjust itself rapidly enough. At some intermediate sound frequency there will be a resonance interaction, a high dispersion of the sound wave caused by this resonance, and a consequent rapid change in the apparent velocity of sound in the system.[a]

1G. Errors Due to Heterogeneity. All of the above methods are limited in their usefulness if the reactions studied are not strictly homogeneous. The problem of heterogeneity is usually serious only in connection with gas reactions, and the usual method of attack is to study the reaction in vessels which differ in their ratio of surface area to volume. This can be easily achieved by packing a vessel, whose area is known or can be measured, with pieces of glass or metal or, in extreme cases, with glass wool. If the rate of the reaction is independent of the packing (i.e., the surface/volume ratio) or nearly so, then the reaction may be assumed to be homogeneous.[b] If, however, there is an influence, as is generally the case for complex gas reactions, then the reaction is not purely homogeneous and methods must be devised for studying the heterogeneous contributions to the reaction. As we shall see later, this can become extremely complicated.

Another method which is employed is to change the vessel surface by using different vessel materials or by coating the walls with KCl, AgCl,

[a] Such a method has been of importance in measuring the rates of energy transfer from vibrations or rotations to translations. See Sec. VII.11. The method was first proposed by A. Einstein and was applied to the kinetic system $N_2O_4 \rightleftharpoons 2NO_2$ by W. T. Richards. For further treatment see S. H. Bauer and M. R. Gustavson, *Discussions Faraday Soc.*, **17**, 69 (1954).

[b] F. O. Rice and K. F. Herzfeld, *J. Phys. & Colloid Chem.*, **55**, 975 (1951). As pointed out in this article such a criterion may be deceptive in that it may simply mean that the reaction may still be heterogeneous but that the rate is not sensitive to the surface/volume ratio. This will happen in chain reactions if chains are both initiated and destroyed at the walls.

metal, paraffin, graphite, or some other presumably inert material. It is general experience in the pyrolytic decompositions of organic compounds that the kinetic results are very erratic during the first ten or so runs made in a new glass vessel and become reproducible only after the walls of the vessel have become coated with a heavy carbonaceous deposit formed during the reaction (see footnote [b] page 65).

2. Temperature Dependence of Rate Constants; Activation Energies

The rates of chemical reactions are very sensitive to temperature changes. For example, it is a very rough empirical rule that reactions double their speed with an increase in temperature of 10°C (at 300°K). This means that whereas for the first-order process an increase in the concentration of 100 per cent is required to double the specific rate, the same effect can be achieved by a 10°C increase in temperature.

The first quantitative, experimental formulation of the dependence of reaction rates on temperature was made by Hood and later extended by Arrhenius. The form of the dependence of the specific rate constant k on temperature is

$$\ln k = \ln A - \frac{E}{RT} \qquad \text{or} \qquad k = Ae^{-E/RT} \qquad \text{(IV.2.1)}$$

In Eq. (IV.2.1), which is known as the Arrhenius equation, k is the specific rate constant at some temperature, T is the absolute temperature, R is the gas constant, and A and E are empirical constants.

It is found experimentally that such an equation does indeed provide a reasonably accurate description of the temperature dependence of specific rate constants over a fairly wide temperature range and for constants representing reactions of different orders. Experimentally the constants A and E may be determined from a plot of the logarithm of the specific rate constant k against the reciprocal of the absolute temperature. Such a plot is shown in Fig. IV.3. (The use of these coordinates allows the data to be presented in the form of a straight line.)

As we shall see shortly, it is possible to attach some theoretical significance to the constants A and E in the Arrhenius equation. Out of this has grown a nomenclature in which E is called the *activation energy* for the reaction and A is called the *frequency factor*, or *preexponential factor*. It should be observed that, while E does indeed have the units of energy, A always has the same units as the specific rate constant k.[a] More accurate experiments have shown that the Arrhenius equation is only an approximate representation of the facts and a more accurate equation is

---

[a] A will have the units of a frequency (time^{-1}) only if k represents the rate constant for a first-order reaction. A more reasonable nomenclature for A would be to call it the Arrhenius factor.

$$\ln k = \ln A + \frac{C}{R} \ln T - \frac{E}{RT} \qquad \text{(IV.2.2)}$$

where once again it is possible to attempt a theoretical explanation[a] of the experimental constant C which in the above written form will have the units of a molar heat capacity.

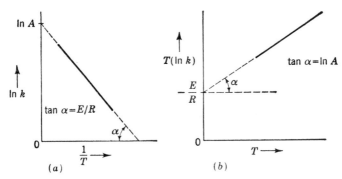

FIG. IV.3. Graphical methods of representing the temperature dependence of specific reaction rate constants. (a) For the Arrhenius equation in the form $\ln k = \ln A - E/RT$. (b) For the Arrhenius equation in the form $T \ln k = T \ln A - E/R$.

For the accuracy usually attained in kinetic measurements the Arrhenius equation (IV.2.1) is adequate, and, in fact, the data obtained for gaseous systems do not generally warrant the correction term present in Eq. (IV.2.2).[b] For the more accurate determinations and in particular for a number of reactions in solution,[c] for which C/R is rather large, the corrected equation (IV.2.3) is necessary. Some values of k, C, and A are listed in Table IV.1 for reactions in solution.

It should be emphasized that the above discussion applies only to individual, specific rate constants. It may well happen that a complex reaction may appear to follow a simple kinetic order over a limited range of experimental conditions. The apparent rate constant for such a reaction is not, however, a rate constant for a single process but rather is a perhaps complicated function of many rate constants. If such a rate constant is plotted against temperature it will not be surprising if it should turn out that the Arrhenius equation is not obeyed even crudely. On the contrary, such very marked disparity is very often a clue to the complexity of the reaction system.

[a] M. Trautz, Z. physik. Chem., 66, 496 (1909).

[b] See, however, L. S. Kassel, Proc. Nat'l. Acad. Sci., U.S., 16, 358 (1930), and N. F. H. Bright and R. P. Hagerty, Trans. Faraday Soc., 43, 697 (1947).

[c] E. A. Moelwyn-Hughes, Proc. Roy. Soc. (London), A164, 295 (1938). P. Johnson and E. A. Moelwyn-Hughes, ibid., A175, 118 (1940). J. C. Kendrew and E. A. Moelwyn-Hughes, ibid., A176, 352 (1940).

TABLE IV.1. SOME REACTIONS THAT INVOLVE THE CORRECTED FORM
OF THE ARRHENIUS EQUATION[a]

Reaction	Temp range, °C	$\ln A$	$-C/R$	E
[b] $Br\text{-}CH_2CO_2^- + S_2O_3^= \rightarrow S_2O_3\text{-}CH_2CO_2^= + Br^-$	6–31	—	−6	16.21
[c] $CH_3Br + H_2O \rightarrow CH_3OH + H^+ + Br^-$	17–100	259.40	34.3	46.72
[d] $CCl_3CO_2H \rightarrow CHCl_3 + CO_2 \uparrow$	50–100	110.82	10.07	42.91
[e] α-Glucose $\rightarrow \beta$-glucose	0–50	88.88	10.32	23.06

[a] $\ln k = -\dfrac{E}{RT} + (C/R) \ln T + \ln A$; A in sec^{-1}; E in Kcal/mole.

[b] Observed for an ionic strength of 0.2 M NaCl. The value of C is attributed to ion activities and vanishes at zero ionic strength. V. La Mer and M. E. Kamner, *J. Am. Chem. Soc.*, **57**, 2662 (1935).

[c] E. A. Moelwyn-Hughes, *Proc. Roy. Soc.* (*London*), **A164**, 295 (1938).

[d] P. Johnson and E. A. Moelwyn-Hughes, *ibid.*, **A175**, 118 (1940).

[e] J. C. Kendrew and E. A. Moelwyn-Hughes, *ibid.*, **A176**, 352 (1940).

3. Thermodynamic Interpretation of the Rate Equation

Let us consider a reaction that proceeds to a final state of dynamic equilibrium, for example,

$$a\text{A} + b\text{B} \rightleftharpoons c\text{C} + d\text{D} \qquad (\text{IV.3.1})$$

Without any further knowledge of the system or the mechanism we may write for R, the net rate of the reaction of any of the species A, B, C, or D, the stoichiometric relations

$$R = -\frac{1}{a}\frac{d\text{A}}{dt} = -\frac{1}{b}\frac{d\text{B}}{dt} = \frac{1}{c}\frac{d\text{C}}{dt} = \frac{1}{d}\frac{d\text{D}}{dt} \qquad (\text{IV.3.2})$$

$$= -aR_f + aR_b$$

In this last equation aR_f and aR_b are respectively the rates at which A is being consumed by the forward reaction and produced by the back reaction.[a] R_f and R_b may be simple or quite complicated functions of any or all of the concentrations and the temperature, so that in general $R_f = R_f(\text{A,B,C,D},T)$ and $R_b = R_b(\text{A,B,C,D},T)$. At equilibrium $R = 0$ and $R_f = R_b$, so that, if we designate the equilibrium concentrations at T_0 by subscripts e,

$$R_f(\text{A}_e,\text{B}_e,\text{C}_e,\text{D}_e,T_0) = R_b(\text{A}_e,\text{B}_e,\text{C}_e,\text{D}_e,T_0) \qquad (\text{IV.3.3})$$

and if we divide one side by the other we have

[a] More specifically, $R_b = 0$ when C or D $= 0$ and $R_f = 0$ when A or B $= 0$.

$$g(A_e,B_e,C_e,D_e,T_0) = \frac{R_f(A_e,B_e,C_e,D_e,T_0)}{R_b(A_e,B_e,C_e,D_e,T_0)} = 1 \qquad (IV.3.4)$$

where g is some function of the equilibrium concentrations and the temperature which is identically equal to unity.

But thermodynamically, if the system is ideal, so that we may ignore activity coefficients, the equilibrium constant K is given by

$$K = \frac{C_e^c D_e^d}{A_e^a B_e^b} = K(T_0) \qquad (IV.3.5)$$

and K can depend (at constant volume or pressure) only on the temperature. If again we divide one side of this last equation by the other we have

$$f(A_e,B_e,C_e,D_e,T_0) = \frac{1}{K(T_0)} \frac{C_e^c D_e^d}{A_e^a B_e^b} = 1 \qquad (IV.3.6)$$

where f is another function of the equilibrium concentrations and the temperature which is identically equal to unity. One particular set of conditions under which Eqs. (IV.3.6) and (IV.3.4) may both be true is that g is some function of f such that[a]

$$g = \phi(f) = 1 \qquad (IV.3.7)$$

Thus we might write

$$g = \phi(f) = \frac{\displaystyle\sum_{s=-M}^{N} x_s f^s}{\displaystyle\sum_{s=-M}^{N} x_s} \equiv 1 \qquad (IV.3.8)$$

which is identically equal to unity.[b] The coefficients x_s may depend on catalysts, extent of surface, etc., but not on the equilibrium concentrations.

However, it is found that R_f and R_b are usually not very complicated, so that it turns out that g can be written as

$$g = f^s = \frac{C_e^{cs} D_e^{ds}}{K^s A_e^{as} B_e^{bs}} = \frac{R_f}{R_b} \qquad (IV.3.9)$$

where s is usually equal to 1 or -1 and is generally an integer or ratio of small integers. If we follow this assumption, then it is seen that for such a system it is only necessary to measure either R_f or R_b in order to determine the other, since the equilibrium constant is in principle known. Further, if we assume that the mechanism of the reaction remains the

[a] See C. A. Hollingsworth, *J. Chem. Phys.*, **20**, 921 (1952), for the treatment given here. A more special solution is given by Manes et al., *ibid.*, **18**, 1355 (1950).

[b] Note that for every temperature there will be a new set of equilibrium concentrations but that the relation (IV.3.8) will automatically be satisfied for each such set of conditions, since $f \equiv 1$.

same whether or not we are close to equilibrium, then we need not measure the rates R_f or R_b at the equilibrium concentrations.

Example. For the equilibrium system

$$2NO(g) + O_2(g) \rightleftharpoons 2NO_2(g) \qquad (IV.3.10)$$

it is found that the initial rate of reaction of NO and O_2 in the absence of NO_2 is given quite accurately by

$$-\frac{d(NO)}{dt} = R_f = k_f(NO)^2(O_2) \qquad (IV.3.11)$$

We can now calculate R_b by use of Eq. (IV.3.9), and we find

$$R_b = R_f \frac{K^s(NO)^{2s}(O_2)^s}{(NO_2)^{2s}}$$

and on substituting for R_f from Eq. (IV.3.11)

$$R_b = k_f \frac{K^s(NO)^{2s+2}(O_2)^{s+1}}{(NO_2)^{2s}} \qquad (IV.3.12)$$

$$= k_b(NO)^{2s+2}(O_2)^{s+1}(NO_2)^{-2s} \qquad (IV.3.13)$$

where $k_b \equiv k_f K^s$ is the specific rate constant for the back reaction. Kinetic consideration of the actual reaction can now give us some information about the exponent s. Equation (IV.3.13) implies that the rate of decomposition of NO_2 will be zero if NO or O_2 is not present. It further implies that, unless s is negative, the decomposition of NO_2 is inversely proportional to the concentration of NO_2. This latter situation is physically unreasonable[a] and implies that s must be negative. The first situation is also improbable. To avoid its implications, we must make R_b independent of (NO) and (O_2), and that means setting $s = -1$. R_b then becomes

$$R_b = k_b(NO_2)^2 \qquad (IV.3.14)$$

which is in fact verified experimentally.

We shall find generally that such simplification of the relation between R_f/R_b and K is the case and that further, since both R_f and R_b can be expressed as simple algebraic products of specific rate constants times concentrations, the ratio of these constants will generally take the form

$$\frac{k_f}{k_b} = K \qquad (IV.3.15)$$

Thus if the equilibrium constant for a reaction is known and one of the specific rate constants has been measured, the other rate constants can usually be calculated from Eq. (IV.3.15).

[a] This is not to say that the experimental rate law may not have such a form over a restricted range of concentrations. This turns out to be the case for a number of chain reactions. In such cases s cannot be deduced.

3A. Activation Energies and Heats of Reaction. If we restrict our attention to the general relation between rate constants and the equilibrium constant implied by Eq. (IV.3.15), we can proceed to identify some other rate quantities with thermodynamic quantities. From Eq. (IV.3.15),

$$\ln K = \ln k_f - \ln k_b \qquad (IV.3A.1)$$

but thermodynamically

$$\ln K = -\frac{\Delta F}{RT} = -\frac{\Delta H}{RT} + \frac{\Delta S}{R} \qquad (IV.3A.2)$$

where ΔF, ΔH, and ΔS are respectively the standard changes in Gibbs free energy, enthalpy, and entropy accompanying the stoichiometric reaction[a] referred to some standard state of the substances involved. Now it is an empirical fact that both ΔH and ΔS can be represented reasonably accurately over restricted temperature ranges by equations of the form

$$\Delta H_T = \Delta H_{T_0} + \Delta C_p(T - T_0) = (\Delta H_{T_0} - T_0 \Delta C_p) + T \Delta C_p$$
$$\qquad (IV.3A.3)$$

$$\Delta S_T = \Delta S_{T_0} + \Delta C_p (\ln T - \ln T_0) = (\Delta S_{T_0} - \Delta C_p \ln T_0) + \Delta C_p \ln T$$

where ΔC_p is the mean difference (over the temperature range) in heat capacities for the reactants and products referred to their standard states.[b] If now we substitute Eq. (IV.3A.3) into Eq. (IV.3A.2), we have, on collecting terms,

$$\ln K_{eq} = \frac{\Delta S - \Delta C_p - \Delta C_p \ln T_0}{R} + \frac{\Delta C_p}{R} \ln T - \frac{(\Delta H - T_0 \Delta C_p)}{RT}$$
$$\qquad (IV.3A.4)$$

But, if we now use the corrected form of the Arrhenius equation (IV.2.2), we can write:

$$\ln k_f - \ln k_b = (\ln A_1 - \ln A_2) + \left[\frac{C_1 - C_2}{R}\right] \ln T - \frac{E_1 - E_2}{RT} \qquad (IV.3A.5)$$

and since by Eq. (IV.3.15) $\ln K_{eq} = \ln k_1 - \ln k_2$, we can make an immediate identification of the terms in Eqs. (IV.3A.4) and (IV.3A.5):

$$\ln A_1 - \ln A_2 = \frac{\Delta S - \Delta C_p - \Delta C_p \ln T_0}{R}$$
$$C_1 - C_2 = \Delta C_p \qquad\qquad (IV.3A.6)$$
$$E_1 - E_2 = \Delta H - T_0 \Delta C_p$$

[a] Note that, if the system is not ideal, then partial molal quantities $\Delta \bar{F}$, $\Delta \bar{H}$, and $\Delta \bar{S}$ must be used instead of the above.

[b] This follows at once from the equations:

$$H_T = H_{T_0} + \int_{T_0}^{T} \left(\frac{\partial H}{\partial T}\right)_P dT = H_{T_0} + \int_{T_0}^{T} C_p \, dT = H_{T_0} + \bar{C}_p(T - T_0)$$

$$S_T = S_{T_0} + \int_{T_0}^{T} \left(\frac{\partial S}{\partial T}\right)_p dT = S_{T_0} + \int_{T_0}^{T} \frac{C_p}{T} \, dT = S_{T_0} + \bar{C}_p (\ln T - \ln T_0)$$

We see that the difference in activation energies $E_1 - E_2$ is related to the standard heat of the reaction ΔH, that the ratio of the frequency factors A_1/A_2 is related to the standard entropy change for the reaction ΔS while the difference in the constants C_1 and C_2 is equal to the difference in the mean heat capacities ΔC_p of the reactants and products.[a] If the accuracy of the data or the temperature range is sufficiently restricted that the approximate form of the Arrhenius equation may be used to represent the rate constants, it implies that

$$C_1 - C_2 = \Delta C_p = 0$$

which then yields the identity

$$\ln A_1 - \ln A_2 = \frac{\Delta S}{R} \qquad E_1 - E_2 = \Delta H \qquad \text{(IV.3A.7)}$$

In such a case there is a direct equality between the activation-energy difference and the standard heat of the reaction as well as between the logarithm of the ratio of the frequency factors and the standard entropy change. This gives a rationale for the nomenclature *activation energy*, which is used to designate the constant E in the rate equation.

4. Entropy of Activation

The condition for thermodynamic equilibrium permits us to attach a thermodynamic significance to the constants in the Arrhenius equation, as we have just demonstrated in the case of reversible reaction systems. There is no real objection to extending this in a formal way to all reactions, so that we can rewrite the Arrhenius equation in the following form:

$$\begin{aligned} k &= A e^{-E/RT} \\ &= \nu e^{-F^*/RT} \\ &= \nu e^{S^*/R} e^{-H^*/RT} \end{aligned} \qquad \text{(IV.4.1)}$$

In these equations we have broken down the constant A into a product $\nu e^{S^*/R}$, where ν now has the dimensions of the rate constant k and in analogy to the nomenclature for A is referred to also as a frequency factor. S^* and H^* are referred to respectively as the entropy and enthalpy of activation, and as we shall see later they are capable of association with more conventional entropy and enthalpy changes.

Such a formulation of the temperature dependence of the rate constant is more satisfying in that it relates rate constants properly to thermodynamic equilibrium constants in the case of reversible reactions. In the case of reversible reactions we see that the frequency factor ν must be the same for the forward and back reactions, since

[a] When k_1 and k_2 differ in order, so that K_{eq} has units dependent on concentrations, the same standard states must be employed in using these relationships.

$$- \frac{\Delta F}{RT} = \ln K_{eq} = \ln \frac{k_f}{k_b} = \ln \frac{\nu_f}{\nu_b} - \frac{F_f^* - F_b^*}{RT} \qquad \text{(IV.4.2)}$$

and if $\Delta F = F_f^* - F_b^*$ [Eq. (IV.3A.7)], then ν_f must equal ν_b.

If we take the logarithmic partial derivative of k with respect to T [from Eq. (IV.4.1)], we have

$$\frac{\partial \ln k}{\partial T} = \frac{1}{\nu} \frac{\partial \nu}{\partial T} + \frac{1}{R} \frac{\partial S^*}{\partial T} - \frac{1}{RT} \frac{\partial H^*}{\partial T} + \frac{H^*}{RT^2} \qquad \text{(IV.4.3)}$$

or

$$\frac{\partial \ln k}{\partial T} = \frac{1}{\nu} \frac{\partial \nu}{\partial T} + \frac{H^*}{RT^2} \qquad \text{(IV.4.4)}$$

since $\partial S^*/\partial T = (1/T)(\partial H^*/\partial T)$ from thermodynamics. This suggests that the apparent constancy of E in the original Arrhenius equation (Fig. IV.3) may arise from a fortuitous cancellation of simultaneous changes in ν and E with temperature but in opposing directions.

The fact that very careful experimental results in solution show that $\partial \ln k/\partial(1/T)$ is not constant with temperature (see Table IV.1) but does change, indicates that such cancellation, while usual, is not exact. It should be observed that, from Eq. (IV.4.4), a variation of S^*, the entropy of activation, with temperature will have no influence on the variations of k with temperature, so that experimentally we can only hope to distinguish the quantities H^* and the product $A = \nu e^{S^*/R}$. Only some theoretical model can supply us with an additional relation whereby ν and S^* can be separately evaluated. Further, only if the quantity $\partial \ln \nu/\partial T = (1/\nu)(\partial \nu/\partial T)$ is small compared to H^*/RT^2 can we make the proper identification of H^* with the experimentally observed E. In fact from Eq. (IV.4.4) we see that

$$E = H^* + RT^2 \frac{\partial \ln \nu}{\partial T} \qquad \text{(IV.4.5)}$$

We shall return to this point in a later chapter.

4A. Relation between Reaction Rates and Free Energies; Rate Close to Equilibrium. The driving force of a chemical reaction under isothermal conditions is the difference in free energies of reactants and products. At equilibrium this difference is zero. While this does not tell us whether a reaction will or will not occur, it does tell us that, when reaction takes place, it must always go in the direction of lowering of the free energy of the system. It is thus useful to see if thermodynamics predicts any relation between rates and free energy differences. Such a study has been made by Prigogine et al.[a] and extended by Manes et al.[b] with the following results. (The derivation given here differs from those of the preceding authors, and the results are more general.)

[a] I. Prigogine et al., *J. Phys. & Colloid Chem.*, **52**, 321 (1948).

[b] M. Manes et al., *J. Chem. Phys.*, **18**, 1355 (1950).

From Eqs. (IV.3.4) and (IV.3.7) we see that at equilibrium the ratio of the rates of forward and reverse reactions is a function only of the equilibrium constant divided into the equilibrium-constant expression in concentration units:

$$\frac{R_f}{R_b} = g = \phi(f) = \phi\left[\frac{K(c)}{K(T)}\right] \tag{IV.4A.1}$$

where by $K(c)$ we mean the ratio of the concentrations as given in the equilibrium expression [Eq. (IV.3.6)]. If we assume that the mechanism of the reaction is not different at concentrations removed from equilibrium, the same equation (IV.4A.1) must hold when the system is not at equilibrium. But in this case the concentrations to use in $K(c)$ are not the equilibrium concentrations but the actual concentrations existing in the system, and $K(c)$ is no longer equal to $K(T)$. But the ratio $K(c)/K(T)$ is now related to the difference in free energy of the equilibrium system and the actual system ΔF, since by thermodynamics:

$$\begin{aligned}\Delta F &= \Delta F^\circ + RT \ln K(c) \\ &= -RT \ln K(T) + RT \ln K(c)\end{aligned} \tag{IV.4A.2}$$

or

$$\frac{\Delta F}{RT} = \ln \frac{K(c)}{K(T)} = \ln f \tag{IV.4A.3}$$

We can thus write in general

$$\frac{R_f}{R_b} = \phi(f) = \phi(e^{\Delta F/RT}) = \theta\left[\frac{\Delta F}{RT}\right] \tag{IV.4A.4}$$

for systems not at equilibrium and we see that the ratio of the rates of forward and reverse reaction are determined entirely by the difference in free energies of the actual system and its state of final equilibrium.

When the system is close to equilibrium (that is, $R_f/R_b \cong 1$) then $\Delta F/RT$ is close to zero and we can make a Taylor's series expansion of the function $\theta(\Delta F/RT)$ about the origin. We find

$$\frac{R_f}{R_b} = \theta_0 + \left[\frac{\partial\theta}{\partial(\Delta F/RT)}\right]_0 \frac{\Delta F}{RT} + \frac{1}{2!}\left[\frac{\partial^2\theta}{\partial(\Delta F/RT)^2}\right]_0 \left(\frac{\Delta F}{RT}\right)^2 + \cdots \tag{IV.4A.5}$$

and since $\theta_0 = 1$ (value at equilibrium) and ignoring higher powers of the small quantity $\Delta F/RT$,

$$\frac{R_f}{R_b} - 1 \cong \left[\frac{\partial\theta}{\partial(\Delta F/RT)}\right]_0 \frac{\Delta F}{RT}$$

or

$$R_f - R_b = R_b\left[\frac{\partial\theta}{\partial(\Delta F/RT)}\right]_0 \frac{\Delta F}{RT} \tag{IV.4A.6}$$

and the net rate of reaction $R_f - R_b$ near equilibrium is proportional to the excess free energy of the system.

If now we choose the usual case in which $\phi(f) = [K(c)/K(T)]^s$ [Eq. (IV.3.9)] then

$$\phi(f) = \left[\frac{K(c)}{K(T)}\right]^s = e^{s\,\Delta F/RT} = \theta\left(\frac{\Delta F}{RT}\right)$$

and

$$\left[\frac{\partial\theta}{\partial(\Delta F/RT)}\right]_0 = s\theta_0 = s$$

and our equation reduces to

$$R_f - R_b = sR_b\frac{\Delta F}{RT} \tag{IV.4A.7}$$

If one measures the net rate of reaction near equilibrium and plots the results against $\Delta F/RT$, the points should fall on a straight line whose slope is equal to sR_b. If one now repeats such measurements with different starting concentrations of the reacting species, one can deduce the dependence of R_b on concentrations and hence the form of the rate laws for the system. Such measurements have been made for the catalytic hydrogenation of benzene by Prigogine et al.[a] with confirmation of the relationships shown.

5. Experimental Determination of Reaction Order; Simple Systems

We are now equipped to consider in a more detailed fashion the interpretation of experimental results in terms of the various mathematical formalisms which we have developed for describing reaction orders. The practical problem is the converse of the problem we have considered. The problem presented by experimental data is to analyze it in such fashion as to fit it into one of the categories discussed. We shall consider in the present section a number of methods which have been generally employed for this task in the case of simple reaction systems.

5A. Method of Integration. If the rate of a given reaction depends directly on the concentration of starting reactants in a form

$$\text{Rate} = kA^xB^yC^z$$

it is possible from the stoichiometric relation between A, B, and C to integrate this equation and obtain an equation similar to one of those discussed in Chap. II. We then have the problem of comparing these different integrated expressions with the data to see which provides the most satisfactory fit. This can be done numerically or graphically. That is, we can take the experimental data and for each point calculate the value of k to be expected from an assignment of different orders for the

---

[a] *Loc. cit.* These results are of rather limited applicability because the series expansion [Eq. (IV.4A.5)] converges only very slowly and the condition $\Delta F/RT \ll 1$ is not usual for most reaction conditions.

reactants. The values of k so calculated will be the same for all points only if the individual orders have been correctly assigned. Since simple reactions will generally be zero-, first-, or at most second-order with respect to any particular reactant, it is not too difficult to make such a test at a number of selected points. The data at each of two times, sufficiently separated, generally suffice for such a test. Once such a calculation has been attempted for a number of trial orders it is not too difficult to make a guess at the proper choice of exponents.

Having finally decided on a reasonable selection of exponents, the data can be given the final test of calculating k at each point.

Example. The reaction of sodium ethoxide, $NaOC_2H_5$, with ethyl dimethyl sulfonium iodide, $(C_2H_5)(CH_3)_2SI$, in absolute ethanol solution has been shown to lead to the following products:

$$NaOC_2H_5 + C_2H_5\overset{\displaystyle CH_3}{\underset{\displaystyle CH_3}{S}}\!-\!I \rightarrow \begin{cases} NaI + C_2H_5OC_2H_5 + S(CH_3)_2 \\ or \\ NaI + C_2H_5OH + C_2H_4 + S(CH_3)_2 \end{cases}$$

where the rate of reaction seems to be independent of the ratio of the products formed (ether or alcohol plus olefin). Table IV.2 shows a typical set of data.

TABLE IV.2. REACTION OF $NaOC_2H_5 + C_2H_5(CH_3)_2SI$ IN C_2H_5OH[a]

Time, min	$NaOC_2H_5$[b]	$C_2H_5(CH_3)_2SI$[c]	$10^3 k$[d]
0	22.55	11.53	—
12	20.10	9.08	3.64
20	18.85	7.83	3.67
30	17.54	6.52	3.75
42	16.37	5.35	3.76
51	15.72	4.70	3.71
63	14.96	3.94	3.72
∞	11.02	—	—
			$k_{av} = 3.71 \times 10^{-3}$

[a] Data taken from E. D. Hughes et al., *J. Chem. Soc.*, 2072 (1948).

[b] The data were obtained by pipetting 5-cc samples of a cold reaction mixture into tubes; sealing the tubes, immersing them in a thermostat at 64.0°C (above data), and quenching them after times indicated; and titrating in cold acetone with sulfuric acid to measure residual $NaOC_2H_5$. The numbers given in columns 2 and 3 are in cc of $0.02134 N$ H_2SO_4 needed to titrate 5-cc sample, and they should be multiplied by $0.02134/5$ to get moles/liter.

[c] These are calculated from amount of $NaOC_2H_5$ measured and the stoichiometric equation.

[d] k in liters/mole-sec.

If we let $A = NaOC_2H_5$ and $B = C_2H_5(CH_3)_2SI$, then the kinetic equation to try is

$$-\frac{dA}{dt} = k_A A^x B^y = \cdots \tag{IV.5A.1}$$

where x and y are integers. In Table IV.3 are the rate constants k_A calculated from the integrated forms of Eq. (IV.5A.1) (see Table II.2) for different choices of x and y.

TABLE IV.3. TESTS OF DIFFERENT INTEGRATED RATE EQUATIONS (IV.5A.1)

			k_{calc} for					
t	A	B	$x = 0$ $y = 0$ $(\times 10^3)$	$x = 1$ $y = 0$ $(\times 10^3)$	$x = 0$ $y = 1$ $(\times 10^3)$	$x = 2$ $y = 0$ $(\times 10^4)$	$x = 0$ $y = 2$ $(\times 10^3)$	$x = 1$ $y = 1$ $(\times 10^3)$
0	22.55	11.53	—	—	—	—	—	—
30	17.54	6.52	0.1670	8.39	19.00	4.21	2.22	0.963
63	14.96	3.94	0.1205	6.51	17.05	3.57	2.65	0.956

It can be seen that none of the exponents except $x = 1$, $y = 1$ gives values of k that agree to within a few per cent. This choice does, and so the mixed second-order equation seems to be the correct choice for the rate law. The last column in Table IV.2 shows the values of the constant k calculated for each point in accordance with this law.

Alternatively, the data may be plotted by using as coordinates those quantities which will bring the data into the form of a straight line for the proper choice of reaction orders. While this is an excellent method for presenting the data, it is doubtful if it is any simpler or more accurate than the direct calculation.

The principal advantage of the method of integration is that it makes possible the estimation of reaction order from the data of a single experiment.

5B. Method of Half-lives. The time required to consume a given fraction, say one-half, of one of the starting materials will depend on the initial concentration of the reactants in a way which is fixed by the order of the reaction. If a number of experiments have been performed under conditions of very different initial concentrations of reactants, then it is possible from a comparison of the half-lives (or other fraction, whichever is more conveniently obtained from the data) in the different experiments to decide the proper order of the reaction.

Such a method has been used by Hinshelwood and Askey to determine the rate law for the decomposition of dimethyl ether, CH_3OCH_3.[a] Although the reaction is a quite complex chain reaction involving the for-

[a] C. N. Hinshelwood and P. J. Askey, *Proc. Roy. Soc. (London)*, **A115**, 215 (1927).

mation of formaldehyde, it is found that the principal products are

$$CH_3OCH_3 \rightarrow CH_4 + CO + H_2$$

and, since all products are gases at the temperatures employed, the reaction was followed manometrically, the CH_2O content having been established by independent chemical analysis.[a]

In Table IV.4 are listed the initial pressures of ether and the times required to reach 50 per cent pressure increase in the system, which would correspond roughly to 31 per cent of reaction. If the reaction were a first-order reaction, this column would be constant. However it has been shown[b] that the reaction is more nearly a $\frac{3}{2}$-order reaction. For a $\frac{3}{2}$-order reaction the time for completion of any given fraction of reaction will be inversely proportional to the square root of the initial pressure [Eq. (IV.5B.1)]. Column 3 is obtained by multiplying column 2 by the square root of the initial pressure. Over the 20-fold variation in P_0 this product is nearly constant, giving excellent evidence for the conclusion that the rate is closely $\frac{3}{2}$ order in ether.

TABLE IV.4. DECOMPOSITION OF DIMETHYL ETHER AT 504°C (777°K)[a]

P_0: initial press., mm Hg	$t_{0.31}$, sec	$P_0^{1/2}t_{0.31}$, sec × mm Hg$^{1/2}$	P_0: initial press., mm Hg	$t_{0.31}$, sec	$P_0^{1/2}t_{0.31}$, sec × mm Hg$^{1/2}$
28	1980	10.5	312	665	11.7
58	1500	11.4	321	625	11.2
91	1140	10.9	394	590	11.7
150	900	11.0	422	508	10.5
171	824	10.8	509	465	10.5
241	667	10.4	586	484	11.7
261	670	10.8			av = 11.0 ± 0.1

[a] Data from C. N. Hinshelwood and P. J. Askey, *Proc. Roy. Soc.* (*London*), **A115**, 215 (1927).

The method of half-lives offers a simple, direct approach to the estimation of reaction order and one which involves the least amount of computation when only one species is involved in the rate equation. It becomes, however, fairly complicated when applied to systems in which there is more than one reactant (see Table II.2).

The method may also be applied to the data from a single reaction if sufficient information is available. In such applications each point may be considered separately as the start of a new reaction for the subsequent

[a] CH_2O tends to approach a constant ratio to the residual ether of about 0.24.
[b] S. W. Benson, *J. Chem. Phys.*, **25**, 27 (1956); S. W. Benson and D. V. S. Jain, *ibid.*, in press

points. Thus in Table IV.5 we find a list of partial reaction times calculated for reactions of a single species from the rate equations for different orders. Also shown at the bottom of the table are a few partial time ratios as functions of the reaction order.

TABLE IV.5. RELATION BETWEEN PARTIAL REACTION TIMES
AND REACTION ORDERS[a]

Reaction order n:	0	1	2	3
Partial reaction time	Relations			
$t_{1/3}$	$\dfrac{A_0}{k} \times \dfrac{1}{3}$	$\dfrac{1}{k} \ln \dfrac{3}{2}$	$\dfrac{1}{kA_0} \times \dfrac{1}{2}$	$\dfrac{1}{2kA_0^2} \times \dfrac{5}{4}$
$t_{1/2}$	$\dfrac{A_0}{k} \times \dfrac{1}{2}$	$\dfrac{1}{k} \ln 2$	$\dfrac{1}{kA_0} \times 1$	$\dfrac{1}{2kA_0^2} \times 3$
$t_{2/3}$	$\dfrac{A_0}{k} \times \dfrac{2}{3}$	$\dfrac{1}{k} \ln 3$	$\dfrac{1}{kA_0} \times 2$	$\dfrac{1}{2kA_0^2} \times 8$
$t_{3/4}$	$\dfrac{A_0}{k} \times \dfrac{3}{4}$	$\dfrac{1}{k} \ln 4$	$\dfrac{1}{kA_0} \times 3$	$\dfrac{1}{2kA_0^2} \times 15$
t_f	$\dfrac{A_0}{k} \times f$	$-\dfrac{1}{k} \ln (1 - f)$	$\dfrac{1}{kA_0} \dfrac{f}{1 - f}$	$\dfrac{1}{2kA_0^2} \dfrac{f(2 - f)}{(1 - f)^2}$
Ratio $t_{1/2}/t_{1/3}$	1.500	1.701	2.000	2.400
Ratio $\dfrac{t_{3/4} - t_{1/2}}{t_{1/2}}$	0.500	1.000	2.000	4.000
Ratio $t_{f'}/t_{f''}$	f'/f''	$\dfrac{\ln (1 - f')}{\ln (1 - f'')}$	$\dfrac{f'(1 - f'')}{f''(1 - f')}$	$\dfrac{f'(2 - f')(1 - f'')^2}{f''(2 - f'')(1 - f')^2}$

[a] Calculated for the equation $-dA/dt = kA^n$; f = fraction of reaction completed at time t_f.

The table indicates, for example, that the difference in times between one-half and three-quarters completion divided by the half-life, $(t_{3/4} - t_{1/2})/t_{1/2}$, will be 0.500 for a zero-order reaction, 1.000 for a first-order reaction, etc. The method is extremely simple and involves a minimum of labor for quick estimation of reaction order. Its application becomes more involved for reactions of mixed order involving more than a single reactant. Depending on the data available, the student can calculate a similar list of more appropriate partial reaction times and their ratios.[a] Note that in general the time required to complete a given fraction f of reaction t_f will be proportional to the initial concentration to the $1 - n$ power

[a] For complex chain reactions which are not precisely describable by simple orders such a test may be misleading (see Benson, loc. cit.) and is not as good as lifetimes taken from different reactions.

$$t_f \propto A_0^{(1-n)} \qquad (IV.5B.1)$$

Thus for any two different initial concentrations A_0' and A_0'' and for the same fractions of reaction $f(A_0') = f(A_0'')$

$$\frac{t_{f'}}{t_{f''}} = \left(\frac{A_0'}{A_0''}\right)^{1-n}$$

or taking logarithms and rearranging

$$-(n-1) = \frac{\log (t_{f'}/t_{f''})}{\log (A_0'/A_0'')}$$

so that a plot of $\log t_f$ against $\log$ initial concentration should give a straight line of slope $= n - 1$.

Example. The decomposition of di-tertiary butyl peroxide (dtBP) has been found to follow this constant stoichiometry in the gas phase.[a]

$$(CH_3)_3C-O-O-C(CH_3)_3 \begin{cases} \xrightarrow{90\%} 2(CH_3)_2CO + C_2H_6 \\ \\ \xrightarrow{10\%} CH_3COC_2H_5 + (CH_3)_2CO + CH_4 \end{cases} \qquad (IV.5B.2)$$

In Table IV.6 are shown data for a typical run.

TABLE IV.6. DECOMPOSITION OF DI-TERTIARY BUTYL PEROXIDE AT 155°C[a]

Time, min	Total press., mm Hg	P_D,[b] mm Hg	$P_D^\circ - P_D$, mm Hg	Fraction reacted
0	173.5	169.3	0	0
3	193.4	159.3	10.0	0.0591
6	211.3	150.4	18.9	0.1116
9	228.6	141.7	27.6	0.1630
12	244.4	133.8	35.5	0.2098
15	259.2	126.4	42.9	0.2533
18	273.9	119.1	50.2	0.2965
21	286.8	112.6	56.7	0.3350

[a] Data of Raley et al., *J. Am. Chem. Soc.*, **70**, 88 (1948).

[b] P_D = partial pressure of the peroxide computed from the total pressure and the stoichiometry of the reaction [Eq. (IV.5B.2)] and corrected for added N_2 gas.

$$P_D = \tfrac{3}{2}P_{total}^\circ - \tfrac{1}{2}P_{total} - P_{N_2} = 256.0 - \tfrac{1}{2}P_{total}$$

If now we calculate the ratio $t_{f'}/t_{f''}$ for the points at 21 and 9 min, respectively, we find $t_{f'}/t_{f''} = 21/9 = 2.333$. If now from the formulas given in Table IV.5 and the values $f' = 0.3350$ and $f'' = 0.1630$ we calculate the expected value of $t_{f'}/t_{f''}$ for different reaction orders, we find that the ratio is 2.055, zero order; 2.292, first order; 2.588, second order; and 3.592, third order. From this trial calculation it is seen that the data fit

[a] J. R. Raley et al., *J. Am. Chem. Soc.*, **70**, 88 (1948).

best (within 2 per cent) a first-order law, and in fact further treatment of the data verifies this.[a]

5C. Method of Isolation; Pseudo-order Reactions. When the kinetic equation involves more than one reacting concentration (i.e., is of a mixed type), the two methods just discussed are still applicable but can involve tedious calculation, since the integrated kinetic equations are complicated (see Table II.2). In this case, if the experimental conditions lend themselves to it, it is possible to make an essential isolation of each of the reacting species by adjusting their concentrations so that one of them is present in considerable excess. In this case the concentration of the species present in great excess will remain almost constant during the course of the reaction, and the over-all order of the reaction will seem experimentally to be reduced.

Let us take the reaction whose stoichiometry is

$$A + B \rightarrow C + D \qquad (IV.5C.1)$$

where the true rate is given by

$$-\frac{dA}{dt} = k(A)(B) \qquad \text{mixed second order} \qquad (IV.5C.2)$$

If now we select A_0 to be much greater than B_0 (A_0 and B_0 represent initial concentrations), let us say $A_0 = 20B_0$, then we may write the approximate equation:

$$-\frac{dA}{dt} = -\frac{dB}{dt} \cong kA_0B = k'B \qquad (IV.5C.3)$$

where $k' = kA_0$. Equation (IV.5C.3) is, however, a first-order equation, whereas the true equation (IV.5C.2) is second-order mixed. The effect of choosing our concentrations in the manner specified was thus to lower the experimental order of the reaction. Equation (IV.5C.3) refers now to what is known as a pseudo-first-order reaction.

It can be seen that this method is equally valid for higher-mixed-order reactions.

True Rate Equation

$$-\frac{dA}{dt} = kAB^3C \qquad \text{mixed fourth order}$$

when $B_0 \gg A_0$ and $C_0 \gg A_0$, then

$$-\frac{dA}{dt} = k'A \qquad \text{pseudo first order}$$

where $k' = kB_0^2C_0$.

The order of the reaction with respect to the other components B and

---

[a] It is always wise to check calculations of order from fractional lifetimes experimentally by making runs under conditions of different initial concentrations to avoid fortuitous errors.

C can be determined by varying their excess concentrations one at a time and noting the dependence of k' so obtained upon their almost constant concentrations. Also the method can be applied by making either B_0 or C_0 small with respect to the other concentrations.

5D. Differential Method. It is sometimes convenient to deal not with the integrated rate equations but rather with their differential form directly. In this case it is necessary to have data not on the concentrations in the experimental system as a function of time, but rather of the rates of change of these concentrations as functions of the concentrations. Such data may be obtained graphically or algebraically from the normal data. Thus the rate of reaction of, let us say A, will be given by the tangent to the curve representing A as a function of time. Algebraically we can say, if A_1 and A_2 are the concentrations of A at 2 times t_1 and t_2, respectively, then the mean value of dA/dt is given by

$$\frac{\overline{dA}}{dt} = \frac{A_2 - A_1}{t_2 - t_1}$$

The rate so obtained will be an average rate during the period $t_2 - t_1$. The accuracy with which it is measured is, of course, dependent on the accuracy of measurement of the differences $A_2 - A_1$ and $t_2 - t_1$, which cannot be very great for small intervals.

We can now plot the quantities $\overline{dA/dt}$ against the concentrations of A or of more complicated functions of A and other reactants. Thus, if we consider the rate equation

$$-\frac{dA}{dt} = k_A A^n \qquad n\text{th order} \qquad \text{(IV.5D.1)}$$

it can be written as

$$\log\left(-\frac{dA}{dt}\right) = \log k_A + n \log A \qquad \text{(IV.5D.2)}$$

A plot of $\log\left(\overline{-dA/dt}\right)$ against $\log A$ should now yield a straight line whose slope is n, the order of the reaction.

If we take a more complicated reaction of mixed order[a]

$$-\frac{dA}{dt} = k_A A^x B^y \qquad \text{(IV.5D.3)}$$

then the same procedure gives

$$\log\left(-\frac{dA}{dt}\right) = \log k_A + x\left(\log A + \frac{y}{x} \log B\right) \qquad \text{(IV.5D.4)}$$

In this case a quick test will be to plot the experimental values of

---

[a] A method due to Van't Hoff for obtaining $x + y$ is to measure the rate under conditions such that A and B are present in stoichiometric concentrations. The equation then takes the form: $-dA/dt = k_A(A)^{x+y}$, and $x + y$ is readily obtained from the use of the integrated equation. See footnote page 21.

$\log(-dA/dt)$ against $\log A$. If the plot is a straight line, then $y = 0$; if it is not, then choice may be made of a few points at which the quantity $(y/x)\log B$ can be added to the abscissa $\log A$ to see if any simple choices of the coefficient y/x (which must be the ratio of small numbers $1/1$, $1/2$, $2/1$, etc.) will straighten the line. Three widely chosen points will suffice, say, the first, middle, and last points. If one of the choices does work, then the remainder of the data can be tested. In this way both x and y can be determined from a single experimental run without excessive calculation.

If sufficient data are available, then the most precise procedure is to plot A as a function of time t, draw a smooth line through the points, and read the slopes from the graph with a tangent meter or similar device.

In Table IV.7 this method is applied analytically to the data on the solvolysis of dimethyl ethyl sulfonium iodide by sodium ethoxide given originally in Table IV.2.

TABLE IV.7. DIFFERENTIAL METHOD OF RATE ANALYSIS APPLIED TO THE REACTION OF NaOC$_2$H$_5$ WITH C$_2$H$_5$(CH$_3$)$_2$SI IN C$_2$H$_5$OH[a]

Time, min	A, NaOC$_2$H$_5$	B, C$_2$H$_5$(CH$_3$)$_2$SI	$-\dfrac{\overline{dA}}{dt} \times 10^2$	$\overline{AB}$	$-\dfrac{10^4}{\overline{AB}}\dfrac{\overline{dA}}{dt}$
0	22.55	11.53			
			20.4	220	9.3
12	20.10	9.08			
			15.6	165	9.5
20	18.85	7.83			
			13.1	131	10.0
30	17.54	6.52			
			9.8	101	9.7
42	16.37	5.35			
			7.2	81	8.9
51	15.72	4.70			
			6.2	65	9.5
63	14.96	3.94			
∞	11.02	0			

[a] See Table IV.2 for discussion of analysis and units of concentration.

The analytical results of dividing the average rate for each interval by the product of the mean concentrations for the interval are shown in the last column, and it can be seen that the relative constancy of these differential rate constants indicates that a mixed-second-order law is correct for the system. Note that the intervals must be chosen to be much larger than the experimental errors involved in their measurement. Considerable improvement in this respect is obtained by first smoothing the data and then using tangents from the smoothed curve.

5E. Tracer Methods. With the availability of both stable and radioactive isotopes and equipment for their analytic determination, a new method presents itself for the study of kinetic systems, in particular for equilibrium systems in which equilibrium has already been established. To avoid ambiguity, let us consider a specific system in which equilibrium has been established:

$$2NO + Br_2 \underset{k_2}{\overset{k_1}{\rightleftharpoons}} 2NOBr \qquad (IV.5E.1)$$

Let R_{NOBr} = rate of disappearance of NOBr $\left(\dfrac{\text{moles NOBr}}{\text{liter-sec}}\right)$

back reaction

R_{Br_2} = rate of disappearance of Br_2 $\left(\dfrac{\text{moles } Br_2}{\text{liter-sec}}\right)$

forward reaction

The stoichiometry and equilibrium require

$$R_{NOBr} = 2R_{Br_2} = \text{const in time at equilibrium} \qquad (IV.5E.2)$$

Also we can write quite generally

$$\begin{aligned} R_{NOBr} &= k_2 f_2(NO, Br_2, NOBr) \\ R_{Br_2} &= k_1 f_1(NO, Br_2, NOBr) \end{aligned} \qquad (IV.5E.3)$$

where k_2 and k_1 are the specific rate constants and f_2 and f_1 are some unknown functions of the indicated concentrations. The problem is, of course, to find f_2 and f_1.

Suppose now that we introduce a small amount of a radioactive tracer into the system, for example, radioactive bromine, Br*, in the form of Br_2. By "small" we mean that the amount of added Br_2 is negligible with respect to the equilibrium amount of Br_2 already present in the system.[a]

Under these circumstances the addition of radioactive material will not disturb the equilibrium; at the same time the radioactive bromine will begin to react and distribute itself between the Br_2 and NOBr already present. If we assume that radioactive Br_2 molecules react at the same rate as normal Br_2, then we can say that the rate of production of radioactive NOBr*, R_{NOBr*}, is given by

$$\frac{d(NOBr^*)}{dt} = 2R_{Br_2}\frac{(Br_2^{**})}{(Br_2)} + R_{Br_2}\frac{(Br_2^*)}{(Br_2)} - R_{NOBr}\frac{(NOBr^*)}{(NOBr)} \qquad (IV.5E.4)$$

where $(Br_2^{**})/(Br_2)$ represents the fraction of Br_2 molecules that contain two Br* atoms, $(Br_2^*/(Br_2)$ is the fraction of Br_2 molecules with one Br* atom, and $(NOBr^*)/(NOBr)$ is the fraction of radioactive NOBr present.[b] Now the terms may be collected:

---

[a] It is possible to introduce quantities of radioactive material as small as 10^{-10} moles which are still active enough to be counted accurately.

[b] We have a factor of 2 before the first term because each molecule of Br_2 that disappears will produce 2NOBr*.

$$\frac{d(\text{NOBr}^*)}{dt} = R_{\text{Br}_2} \frac{2(\text{Br}_2^{**}) + (\text{Br}_2^*)}{(\text{Br}_2)} - R_{\text{NOBr}} \frac{(\text{NOBr}^*)}{(\text{NOBr})} \quad \text{(IV.5E.5)}$$

But $2(\text{Br}_2^{**}) + (\text{Br}_2^*)$ = number of Br* atoms present in the form of Br_2 molecules, which we can set equal to x. Similarly (NOBr^*) = number of Br* atoms present in the form of NOBr*, which we can set equal to $N - x$, where N = total number of Br* atoms added. If we now substitute

$$\frac{d(\text{NOBr}^*)}{dt} = \frac{d(N - x)}{dt} = -\frac{dx}{dt}$$

and, from Eq. (IV.5E.2),

$$R_{\text{Br}_2} = \tfrac{1}{2} R_{\text{NOBr}} = R \quad \text{(IV.5E.6)}$$

then

$$-\frac{dx}{dt} = R\left[\frac{1}{2}\frac{x}{(\text{Br}_2)} - \frac{N - x}{(\text{NOBr})}\right] \quad \text{(IV.5E.7)}$$

or

$$-\frac{dx}{dt} = R\left\{x\left[\frac{1}{2(\text{Br}_2)} + \frac{1}{(\text{NOBr})}\right] - \frac{N}{(\text{NOBr})}\right\} \quad \text{(IV.5E.8)}$$

which integrates quite readily (since R is constant) to give

$$\ln\left\{\frac{x}{N}\left[1 + \frac{2(\text{Br}_2)}{(\text{NOBr})}\right] - \frac{2(\text{Br}_2)}{(\text{NOBr})}\right\} = -Rt\left[\frac{1}{2(\text{Br}_2)} + \frac{1}{(\text{NOBr})}\right]$$

$$\text{(IV.5E.9)}$$

which can be rewritten as

$$\ln\left[\alpha_{\text{Br}_2} - \alpha_e\right] = \ln\left[1 - \alpha_e\right] - Rt\left[\frac{1}{2(\text{Br}_2)} + \frac{1}{(\text{NOBr})}\right] \quad \text{(IV.5E.10)}$$

where $\alpha_{\text{Br}_2} = x/N$ = fraction of specific activity present as Br_2 and α_e = equilibrium fraction of Br atoms in system present as Br_2.

By plotting $\ln (\alpha_{\text{Br}_2} - \alpha_e)$ against time we should obtain a straight line whose slope will give the coefficient of t in Eq. (IV.5E.10),[a] and since both (Br_2) and (NOBr) are known, we can thus calculate R. But by Eq. (IV.5E.6), $R = R_{\text{NOBr}} = k_2 f_2$, and if we now repeat the experiment with different equilibrium concentrations of NOBr present, we can observe the way in which f_2 varies with NOBr, Br_2, and NO and so obtain the order of the reaction. By virtue of the equilibrium condition, once $f_2(\text{NOBr})$ is known we can write (see page 70):

$$f_1(\text{NOBr}_2) = \frac{(\text{NO})^2(\text{Br}_2)}{(\text{NOBr})^2} f_2(\text{NOBr}) \quad \text{(IV.5E.11)}$$

While not too accurate because of the difficulty of precise counting over an extended range of concentrations, the method is excellent for investi-

---

[a] This implies that the rate at which a small amount of radioactive material introduced into an equilibrium mixture will distribute itself among the various species always follows a first-order law. This was first noted by H. A. C. McKay, *Nature*, **142**, 997 (1938), and proved quite generally by R. B. Duffield and M. Calvin, *J. Am. Chem. Soc.*, **68**, 557 (1946).

gating complex equilibria.[a] It rests, however, on a very important assumption: that the only available path for exchange of the isotopic species is through the reaction. Unless this is true the method cannot be applied.

6. Precision of Rate Measurements for Simple Systems

It is of some interest to investigate the nature of the errors in kinetic data so that we may obtain some idea of data reliability. There are two related quantities which we can calculate directly from an isothermal experiment: the specific rate constant or a partial reaction time (e.g., the half-life). These are not independent, and one may be calculated from the other. The quantity that completes the empirical description of a reaction system is the activation energy. We have seen that this may be calculated from a knowledge of the specific rate constants at different temperatures.

The errors inherent in any physical measurement are of two kinds. The first category, which is relatively simple to deal with, involves errors that are random. The second category, which is more difficult to detect and so also difficult to handle, includes systematic errors, i.e., errors which are not random but inherent in the reaction studied or the methods employed. A typical example of the latter would be the small contribution of a secondary reaction, the extent of which is determined by the concentrations and temperatures. It is thus inherent in the nature of the system observed, and the magnitude of the errors involved in neglecting this secondary reaction is not random but directly related to the state of the system. Errors due to small amounts of secondary reactions are the most frequent type of systematic error encountered in kinetic studies.[b]

6A. Precision of Specific Rate Constants. For a simple reaction system of a given order, the rate constant may be calculated from data giving the state of the system at two different times. This assumes that sufficient measurements have been made to establish the order of the reaction so that the functional relation is known.

Let us choose a simple second-order reaction as an example. The specific rate constant k_A will be given by (see Table II.2)

$$k_A = \frac{1}{t_2 - t_1} \frac{A_1 - A_2}{A_1 A_2} \qquad \text{(IV.6A.1)}$$

[a] C. I. Browne et al., *J. Am. Chem. Soc.*, **73**, 1946 (1951).

[b] Systematic errors sometimes show themselves in a uniform trend. Thus if there is a deviation of the experimental data from mean values which shows a consistent trend over a range of experimental conditions, rather than being randomly distributed, systematic error may be expected. For such a deviation to be significant it must exceed the estimated relative error in measurement of k_A. The a priori detection of systematic errors is an art rather than a science, since it implies a foreknowledge of the true behavior of the system.

where A_1 and A_2 are the concentrations of A present at time t_1 and t_2, respectively.

If we may assume that the errors made in measuring the four quantities A_2, A_1, t_2, and t_1 are independent of each other and completely random, then Δk_A, the expected resultant error in the dependent quantity k_A, is given by[a]

$$(\Delta k_A)^2 = \left(\frac{\partial k_A}{\partial t_1}\right)^2 (\Delta t_1)^2 + \left(\frac{\partial k_A}{\partial t_2}\right)^2 (\Delta t_2)^2 + \left(\frac{\partial k_A}{\partial A_1}\right)^2 (\Delta A_1)^2 + \left(\frac{\partial k_A}{\partial A_2}\right)^2 (\Delta A_2)^2$$
(IV.6A.2)

For most purposes it is convenient to rewrite Eq. (IV.6A.2) in terms of the relative errors $\Delta k_A/k_A$, $\Delta t_1/t_1$, etc., rather than in terms of the absolute errors in each of the variables. This can be done by dividing each term by k_A^2 and then in turn multiplying and dividing each term by the square of the independent variable:

$$\left(\frac{\Delta k_A}{k_A}\right)^2 = \left(\frac{t_1}{k_A}\frac{\partial k_A}{\partial t_1}\right)^2 \left(\frac{\Delta t_1}{t_1}\right)^2 + \left(\frac{t_2}{k_A}\frac{\partial k_A}{\partial t_2}\right)^2 \left(\frac{\Delta t_2}{t_2}\right)^2 + \left(\frac{A_1}{k_1}\frac{\partial k_1}{\partial A_1}\right)^2 \left(\frac{\Delta A_1}{A_1}\right)^2$$
$$+ \left(\frac{A_2}{k_2}\frac{\partial k_1}{\partial A_2}\right)^2 \left(\frac{\Delta A_2}{A_2}\right)^2 \quad \text{(IV.6A.3)}$$

or[b]

$$\left(\frac{\Delta k_A}{k_A}\right)^2 = \left(\frac{\partial \ln k_A}{\partial \ln t_1}\right)^2 \left(\frac{\Delta t_1}{t_1}\right)^2 + \left(\frac{\partial \ln k_A}{\partial \ln t_2}\right)^2 \left(\frac{\Delta t_2}{t_2}\right)^2 + \left(\frac{\partial \ln k_A}{\partial \ln A_1}\right)^2 \left(\frac{\Delta A_1}{A_1}\right)^2$$
$$+ \left(\frac{\partial \ln k_A}{\partial \ln A_2}\right)^2 \left(\frac{\Delta A_2}{A_2}\right)^2 \quad \text{(IV.6A.4)}$$

Equation (IV.6A.1) can be used to calculate the partial derivatives. On substituting these we find

$$\left(\frac{\Delta k_A}{k_A}\right)^2 = \left(\frac{t_1}{t_2 - t_1}\right)^2 \left(\frac{\Delta t_1}{t_1}\right)^2 + \left(\frac{t_2}{t_2 - t_1}\right)^2 \left(\frac{\Delta t_2}{t_2}\right)^2 + \left(\frac{A_2}{A_2 - A_1}\right)^2 \left(\frac{\Delta A_1}{A_1}\right)^2$$
$$+ \left(\frac{A_1}{A_2 - A_1}\right)^2 \left(\frac{\Delta A_2}{A_2}\right)^2 \quad \text{(IV.6A.5)}$$

To illustrate the use of this equation, let us select as an example a reaction in which $t_2 - t_1 = 1000$ sec, $\Delta t_1 = \Delta t_2 = 1$ sec, $A_2 = 0.90A_1$ (10 per

[a] This equation may be interpreted as the vector sum of the independent component vectors. The justification for taking such a vector sum as the resultant of the random errors will be found in the discussion of the random walk, Sec. VI.7.

[b] In the completely general case of a dependent variable $y = f(x_1, x_2, \ldots, x_n)$ which is a known function of the n independent variables $x_1, \ldots, x_n$, the expected relative error in y due to the relative errors in $x_1, \ldots, x_n$ is given by

$$\left(\frac{\Delta y}{y}\right)^2 = \sum_{i=1}^{n} \left(\frac{\partial \ln f}{\partial \ln x_i}\right)^2 \left(\frac{\Delta x_i}{x_i}\right)^2$$

cent reaction), and the relative error in analysis of A_1 and A_2 is ± 0.1 per cent. On substituting these values in the last equation we find

$$\frac{\Delta k_A}{k_A} = \pm(0.001^2 + 0.001^2 + 0.009^2 + 0.010^2)^{\frac{1}{2}} = \pm 0.014 = \pm 1.4\%$$

We see that the expected accuracy of k_A is much lower than the errors of measurement and that this arises principally because of the small reaction interval chosen.

Stating the problem conversely, what precision is required to measure k_A to within any given accuracy, say 0.1 per cent? We can say that the relative errors in each of the quantities in Eq. (IV.6A.2) must be less than ± 0.1 per cent.[a] This means in the above case that the interval $A_1 - A_2$ must represent at least 50 per cent reaction.

It can be seen that the principal random error in k_A is determined by the precision of analysis and the length of the interval chosen. In Table IV.8 are represented the expected percentage errors in k_A arising only from the errors of analysis. It turns out that this analytical error is about the same for all orders of reaction from zero to four and depends chiefly on the extent of reaction occurring between the two points selected.

It is clear from inspection of Table IV.8 what one of the dilemmas of

TABLE IV.8. ERRORS IN CALCULATED RATE CONSTANTS CAUSED
BY ANALYTICAL ERRORS

Per cent change in species analyzed:	1	5	10	20	30	40	50
Analytical precision, %	Error, per cent						
± 0.1	14	2.8	1.4	0.7	0.5	0.4	0.3
± 0.5	70	14	7	3.5	2.5	2	1.5
± 1.0	>100	28	14	7	5	4	3
± 2.0	>100	56	28	14	10	8	6

For the simple rate law, $dA/dt = -k_A A^n$ (for $n = 0, \frac{1}{2}, 1, \ldots, 4$).

experimental kinetics is. In order to measure k_A with 0.1 per cent accuracy, we are required to analyze the reactants with better than 0.1 per

[a] It should be noted that for completely random errors, the relative error in k_A may be improved by taking a great many measurements. For n measurements of equal estimated error (Table IV.8) the estimated error in the final mean will be decreased by a factor of $(n - 1)^{\frac{1}{2}}$. Thus 10 separate measurements of k_A each of equal estimated error Δk_A will yield a mean value of k_A in which the mean estimated error is $\Delta k_A/3$. To reduce Δk_A by a factor of 10, a prohibitive number of results is needed, namely $n = 100$.

cent accuracy at times separated by at least 1000 sec,[a] during which time the concentration must have changed by two- or threefold. If intermediate points are measured, the values of k_A obtained from them will have less than this accuracy. To obtain a second point with the same precision from the same experiment, the concentration must be allowed to decrease by another factor of 2, and so on for more points.

This situation is completely reversed if, instead of measuring the disappearance of reactants, we measure the appearance of products. In this latter case, if the products can be measured with precision, then the first points can be made to yield values k_A to within almost the same precision as the analytical data. Thus where possible it is desirable to follow the appearance of products during the initial stages of a reaction and the disappearance of reactants during the final stages.

6B. Weighting of Data. In principle there should be no difference between the procedure of obtaining n different analyses of a system during a single run and that of making n duplicate runs with but one analysis per run. If the analysis is simple, the first procedure is obviously less costly of time and equipment. Each pair of data provides a set of values from which k can be calculated. If all the data are taken from a single run, then the best value of k is presumably the average from all the points taken two at a time. If there are n points, then there will be $n(n-1)/2$ possible combinations[b] of which $n-1$ are independent.

These calculations are not, however, all of the same precision, and a simple arithmetic mean will give a disproportionately high weight to the less reliable points. If estimates can be made of the relative precision for each pair of points, then for completely random errors the calculated values of k should be weighted for averaging according to their expected errors. For example, a value having an estimated error of ± 2 per cent

[a] This assumes that our sampling times are smaller than ± 1 sec.

[b] The usual custom of using pairs of consecutive points to calculate a rate constant and then averaging these to obtain a final "best" rate constant puts almost all the weight on the first and last point [W. Roseveare, *J. Am. Chem. Soc.*, **53**, 1651 (1931)]. This can be seen from the example of a first-order rate where the successive time intervals are Δt_1, Δt_2, . . . corresponding to the successive concentrations A_0, A_1, ... For n such intervals the average constant is

$$k_{av} = \frac{k_1 + k_2 + \cdots + k_n}{n}$$

$$= \frac{1}{n}\left(\frac{1}{\Delta t_1}\ln\frac{A_0}{A_1} + \frac{1}{\Delta t_2}\ln\frac{A_1}{A_2} + \cdots + \frac{1}{\Delta t_n}\ln\frac{A_{n-1}}{A_n}\right)$$

In the extreme case that all the time intervals are equal this reduces to

$$k_{av} = \frac{1}{n\,\Delta t}\ln\frac{A_0}{A_n}$$

Similar results hold for reactions of other orders. This is not necessarily a bad result.

should be weighted $\frac{1}{2}$ (or less) with respect to a point having an error of ± 1 per cent.

Example. Let us suppose we have three experimentally calculated points together with their estimated errors as listed in the following table.

k_A (calc. exp.)	Estimated error, %	Relative assigned wt	Normalized assigned wt[a]
8.6	± 4	$\frac{1}{4}$	1.45
9.4	± 10	$\frac{1}{10}$	0.58
8.2	± 6	$\frac{1}{6}$	0.97

Unweighted mean, 8.73 ± 0.61; weighted mean, 8.60 ± 0.43.

[a] The relative weights can be taken as the reciprocal of the estimated error. The normalized weights which must be used in order to calculate the standard deviation of the final mean value are just the relative weights multiplied by a constant, so that the sum of the normalized weights $\Sigma \rho_i$ is exactly equal to n, the number of data. In the case cited $n = 3$. It should be observed from the example given that the process of averaging results of different estimated error does not necessarily improve the estimated error in the final mean over that of the best measurement; improvements of this kind are obtained only when many results of comparable error are averaged. (A single measurement with estimated error of ± 2 per cent is in principle better than four measurements each of ± 5 per cent error.) The value calculated from the usual unweighted mean would be 8.73 ± 0.61, while the weighted mean calculated by assigning weights which are inversely proportional to the estimated errors is 8.60 ± 0.43, a lower value and one with a smaller expected error. This latter value is a much better choice than the unweighted mean. Note that the expected error $\sigma = [\Sigma \rho_i^2 \Delta_i^2 / (n - 1)]^{\frac{1}{2}}$.

It is interesting to observe that the integral method of computing k_A generally places most weight on the initial concentration, since that quantity appears in the integrated rate equations (see Table II.2). Such a practice is wise only when the concentration is known with better precision than any of the others. If there is some other point for which the analytical data are better, then it is best to recast the equations in a form which refers the other concentrations to this point (i.e., a shift in coordinate axes).

A graphical representation of the data generally overcomes this difficulty because, with the exception of mixed-order reactions, the coordinates can be plotted independently of a knowledge of the initial concentrations.[a]

[a] Roseveare, *loc. cit.*, has given a detailed discussion of the errors involved in calculation of first- and second-order rate constants from data. E. A. Guggenheim, *Phil. Mag.*, **2**, 538 (1926), has discussed a method of treating first-order rate data which does not depend on a knowledge of the final or initial concentrations, and J. M. Sturtevant, *J. Am. Chem. Soc.*, **59**, 699 (1937), has devised a method for dealing with second-order rate data which similarly does not depend on a knowledge of final or initial concentrations. An analysis of methods of weighting data for first- and second-order reactions has also been made by J. A. Christiansen, *Z. physik. Chem.*, **A189**, 126 (1941).

6C. Precision in Measurements of Activation Energies. The value of the activation energy E may be calculated from a knowledge of the specific rate constants at two different temperatures. By rearranging the Arrhenius equation we have

$$\bar{E} = -\frac{RT_1T_2}{T_1 - T_2} \ln \frac{k_2}{k_1} \qquad \text{(IV.6C.1)}$$

where $\bar{E}$ is the average activation energy for the temperature range $T_1 - T_2$. Again if the errors in each of the experimental quantities are random, we may take the vector sum:

$$\left(\frac{\Delta\bar{E}}{\bar{E}}\right)^2 = \left(\frac{T_2}{T_1 - T_2}\right)^2 \left(\frac{\Delta T_1}{T_1}\right)^2 + \left(\frac{T_1}{T_1 - T_2}\right)^2 \left(\frac{\Delta T_2}{T_2}\right)^2$$

$$+ \left[\frac{1}{\ln(k_2/k_1)}\right]^2 \left[\left(\frac{\Delta k_1}{k_1}\right)^2 + \left(\frac{\Delta k_2}{k_2}\right)^2\right] \qquad \text{(IV.6C.2)}$$

It can be seen that the precision of measurement of $\bar{E}$ will depend critically on the size of the temperature interval chosen. Thus if $T_1 - T_2 = 10°C$ and the mean error in T_1 and T_2 is 0.2°C it will introduce an error of ±2 per cent in $\bar{E}$. In similar fashion, errors in measuring k_1 and k_2 will be magnified by the factor $1/\ln(k_2/k_1)$, so that if the reaction rate does not change appreciably over the temperature interval in question, this will cause enhanced errors in $\bar{E}$.

Increasing the interval to 20°C decreases this error by a factor of 2, but simultaneously it decreases the chances of observing with any accuracy a variation of E with temperature. Decreasing the interval, say, to 5°C, doubles this error and simultaneously increases the error in the last term, since this term contains a factor $\ln(k_2/k_1)$ in the denominator which becomes smaller as k_2 approaches k_1.

If k_2 and k_1 can each be estimated to ±1 per cent and their ratio is about 2 for a 10°C interval, then the error in E due to the last term in Eq. (IV.6C.2) will be about ±2.0 per cent.

To measure E over a 10°C interval to ±0.5 per cent will generally require a temperature error of less than ±0.03°C and a measurement of k_1 and k_2 to within ±0.3 per cent. This latter in turn will, as we have seen, require analytical precision of ±0.1 per cent over an extended range of concentration changes.

These calculations illustrate the difficulty of obtaining experimental values of the activation energy with precision and also indicate why it has been so difficult to observe temperature variations in E.

The precision in determining E within a 10°C interval at 300°K is generally about ±5 per cent for gas reactions and usually no better than ±3 per cent for reactions in solution. Over a range of 100°C, which is the usual temperature range accessible in the vicinity of room temperature, there would have to be a total change in E of at least twice the usual error of its determination (namely, 10 to 20 per cent) in order to detect a significant temperature variation. This is, however, very rarely the case.

It is interesting to understand the physical limitation imposed on the experimentally accessible temperature range for most reactions. Experimentally it is difficult to deal with reactions whose half-lives are less than 10 min ($\frac{1}{6}$ hr) or greater than 1 week (160 hr). But this represents only a 1000-fold range in rates.[a] If a given reaction doubles its speed for a 10°C rise in temperature (which is a fair average at 300°K), then a 1000-fold change in its rate is covered by a change in temperature of 100°C. If we try to extend this range by lowering the concentrations when the half-life is 10 min (i.e., at the high temperature), we can at best, for a second-order reaction, for example, decrease the concentration by a factor of about 100 to 1000.[b] That will bring up the half-life by a factor of 100 to 1000, which in turn will permit a further increase in temperature of 60 to 100°C. Thus the maximum over-all range will be 160 to 200°C, and generally it is much less. For the extreme ranges of half-life included in this range the precision of measurement falls off quite badly.

6D. Influence of Temperature on Precision. The dominant variable in a reaction is the temperature. Thus to measure k with any precision, the temperature must be controlled fairly accurately. We can calculate the influence of temperature from the differential form of the Arrhenius equation:

$$\frac{d \ln k}{dT} = \frac{E}{RT^2}$$
(IV.6D.1)

or rearranging terms:

$$\frac{dk}{k} = \frac{E}{RT}\frac{dT}{T} \quad \text{or} \quad \frac{\Delta k}{k} = \frac{E}{RT}\frac{\Delta T}{T}$$
(IV.6D.2)

For most reactions in their experimentally accessible range E/RT is found to be about 35. Thus to reduce the expected error in k to ± 0.1 per cent, T must be known to be about ± 0.003 per cent. For $T = 300°K$ (room temperature) this means a precision of $\pm 0.01°C$, and for $T = 600°K$ a control of $\pm 0.02°C$ (something well nigh impossible experimentally at 600°K).

[a] We might reduce the lower limit to 1 min, but that would only extend the range to 10,000 fold in rates.

[b] It is difficult to cover a concentration range for kinetic systems in excess of a factor of 100 to 1000 without elaborate techniques. In the case of solutions, solubility or molar density generally limits concentrations to a maximum of about 1 mole/liter. At concentrations below 0.01 or 0.001 M it is almost impossible to achieve accuracy in analysis of better than a few per cent. Note, however, the work of Yost and Johnston and others who, by the employment of spectroscopic methods of analysis in conjunction with oscilloscope recorders, have managed to measure reactions with half-lives of 0.1 sec and less [H. S. Johnston and D. Yost, *J. Chem. Phys.*, **17**, 386 (1949); R. L. Mills and H. S. Johnston, *J. Am. Chem. Soc.*, **73**, 938 (1951)].

To reduce the error in k to ± 1.0 per cent, T must be known to $\pm 0.1°C$ at 300°K (practical) and to $\pm 0.2°C$ at 600°K (difficult). This places a lower limit on the precision with which reaction rates may be measured at the higher temperatures. Above 600°K it is difficult to obtain temperature control to better than $\pm 1°C$ with a consequent error of ± 5 per cent in k, which in turn means about a ± 7 to ± 10 per cent error in E (for a 20°C interval).

An additional factor presents itself in the case of reactions which are strongly endothermic or exothermic. Unless the heat transfer to the thermostat is very good (i.e., much faster than the reaction rate), the internal temperature will not be that of the oven. If the heat of a gas reaction is, for example, 10 Kcal/mole, then during the course of the reaction, if the reaction vessel is perfectly insulating, there will be a temperature rise of $(10,000/C_p)°C$. Assuming a mean value of about 10 to 20 cal/mole-°C for C_p for most gases that would imply a temperature change of 1000 to 500°C in the course of the reaction! It can be shown[a] that the heat conductivity of gases is not sufficiently high to transport heat effectively, so that only convection or stirring can maintain a constant temperature. That convection is very rapid even in 1-liter glass vessels can be easily demonstrated by plunging such a vessel equilibrated to room temperature into an ice bath. If the pressure inside the vessel is measured as a function of time, it is found that temperature equilibrium is effectively reached in less than 2 min.[b]

This does not imply, however, that there may not exist gradients of the order of magnitude of a few degrees in gas-reaction vessels containing a fast reaction sufficiently endo- or exothermic. Only direct experiments can justify such assumptions.[c]

In the case of liquid systems the solvent is able to act as an effective thermostat because it usually exceeds the quantity of reactants in a mole ratio of from 100 to 1000 or more.[d]

6E. Summary. We can summarize the foregoing discussion as follows: To obtain k_A with an estimated error of $\pm x$ per cent, it is necessary to:

1. Measure *concentrations* with an accuracy of

$$\pm \frac{\text{Change in concentration}}{\text{Largest concentration}} \times \frac{x}{1.4} \qquad \text{per cent}$$

[a] See Sec. XIV.2.

[b] The author is indebted to Victoria Bordaz for carrying out a number of such experiments in his laboratory.

[c] From the foregoing it can be seen that gradients of even 1 to 2°C can cause serious errors in both k and E. See Sec. XIV.2.

[d] In liquid systems in which reactants are not at the thermostat temperature before mixing there may be considerable temperature lag and consequent error. W. S. Horton, *J. Phys. & Colloid Chem.*, **52**, 1129 (1948), has made an analysis of these errors for first-order reactions which shows that errors of up to 50 per cent may occur if the initial ΔT was 10°C.

2. Measure *time* with an accuracy of

$$\pm \frac{\text{Time interval}}{\text{Largest time}} \times \frac{x}{1.4} \qquad \text{per cent}$$

3. Measure *temperature* with an accuracy of

$$\pm \frac{x}{35} \text{ per cent} = \pm \frac{xT}{3500} \qquad °\text{K}$$

The resultant accuracy in E for two measurements of k_A each of accuracy $\pm x$ per cent will then be

$$\overline{\Delta^2(E)}^{1/2} = \pm \frac{\text{highest temperature}}{\text{temperature interval}} \times \frac{x}{18} \qquad \text{per cent}$$

These are rough but reliable rules. They are accurate for reactions for which E/RT is about 35.

V

Experimental Characterization of Complex Kinetic Systems

1. Characteristics of Complex Systems

By a complex reaction system we shall mean any reaction system in which we have more than one reaction process occurring. This may include systems in which we have two or more concurrent or consecutive reactions. Unfortunately such a definition in its subtler aspects may include all reaction systems, since, as we shall see in later chapters, even simple reaction systems involve both consecutive and concurrent reactions. We can, however, attempt to distinguish simple and complex systems pragmatically by saying that complex systems will include all those reactions whose rate expressions cannot be characterized experimentally by simple reaction orders over the accessible range of experimental conditions. That is, they will show systematic deviations from any of the simple rate laws by amounts which exceed the estimated experimental errors.

Such a distinction then provides us with at least an experimental method of differentiation. It is experimentally fairly simple to distinguish two types of these systems. The first type is the system which comprises a set of reversible reactions: the reaction does not proceed to completion but instead eventually reaches a state of dynamic equilibrium. Direct analysis of the final state of the system will generally reveal such equilibria.

The second class of complex systems is that in which concurrent reactions yield different sets of products. These are again simple to characterize in that at least two or more independent stoichiometric equations will be required to represent the reaction at any time. Thus, for the pyrolysis of toluene we need at least two such equations to account for the products[a]

[a] It is not intended to imply that this reaction is characterized only by two concurrent reactions. On the contrary, it is of a much more involved form.

$$2\phi CH_3 \rightarrow \phi CH_2\text{-}CH_2\phi + H_2$$
$$3\phi CH_3 \rightarrow \phi CH_2\text{-}CH_2\phi + \phi H + CH_4$$

Finally we have as our third category all reactions in which we have both of these complications or, speaking generally, those reactions which proceed through the formation of active intermediates. The experimental detection of such systems is sometimes extremely difficult, and the history of kinetics is replete with examples of reactions which have been mistakenly classified. Most notable has been the thermal decomposition of N_2O_5, which is now known to be quite complex,[a] although in 25 years at least 60 papers were written about it and all of them concluded it to be a simple first-order reaction.

2. Equilibrium Systems

The simplest case of opposing reactions (equilibrium system) is that of opposing first-order reactions:

$$A \underset{k_2}{\overset{k_1}{\rightleftharpoons}} B \qquad (V.2.1)$$

where

$$K = \frac{k_1}{k_2} = \frac{B_{eq}}{A_{eq}} \qquad (V.2.2)$$

When $K > 100$, $k_1 > 100k_2$, and over most of the reaction range we can neglect the reverse reaction k_2. This then resembles a simple first-order reaction. From the usual methods we can obtain k_1, and from a measurement of K we can then calculate k_2. When $K < 0.01$, the same simplification holds, except that we should then start experimentally with a system containing pure B rather than one containing pure A. When $100 > K > 0.01$, we must consider both reactions. The solution is given by Eq. (III.2.9):

$$\ln (AK - B) = \ln (A_0K - B_0) - (k_2 + k_1)t \qquad (V.2.3)$$

It is necessary to know K in order to apply this equation to data. K may be obtained from an analysis of the equilibrium concentrations. We can then plot $\ln (AK - B)$ against the time and obtain a straight line. The slope of the line will be the sum $k_2 + k_1$, while the intercept should be $\ln (A_0K - B_0)$. From the slope and the value of K we can now calculate both k_1 and k_2. For any pair of points separated by the time interval t, $k_2 + k_1$ will be given by

$$k_2 + k_1 = \frac{1}{t} \ln \frac{A_1K - B_1}{A_2K - B_2} \qquad (V.2.4)$$

By substituting $k_1 = k_2K$ and solving for k_2 we have

$$k_2 = \frac{k_1}{K} = \frac{1}{t(1 + K)} \ln \frac{A_1K - B_1}{A_2K - B_2} \qquad (V.2.5)$$

[a] R. A. Ogg, *J. Chem. Phys.*, **15**, 337 (1947).

From Eq. (V.2.5) it is now possible by the usual methods to estimate the errors in the calculation of k_1 or k_2. These errors will be found to be of the same order of magnitude as those found for a simple first-order reaction except in the region where $B/A \cong K$. In this latter region (near equilibrium) the accuracy falls off very badly because the numerator and denominator in the logarithm then approach zero very closely.[a]

As an example let us study the errors involved in a calculation of the constant k_2 in Eq. (V.2.5). By applying the methods of Sec. IV.6A we take the logarithmic differential of both sides of this equation.

$$d \ln k_2 = -d \ln t - d \ln (1 + K) + d \ln \ln \frac{A_1 K - B_1}{A_2 K - B_2} \quad (V.2.6)$$

or

$$\frac{dk_2}{k_2} = -\frac{dt}{t} - \frac{dK}{1 + K} + \frac{1}{\ln \left[(A_1 K - B_1)/(A_2 K - B_2)\right]}$$
$$[d \ln (A_1 K - B_1) - d \ln (A_2 K - B_2)]$$

$$= -\frac{dt}{t} - \frac{K}{1 + K} \frac{dK}{K} + \frac{1}{\ln \left[(A_1 K - B_1)/(A_2 K - B_2)\right]}$$
$$\left(\frac{K \, dA_1 + A_1 \, dK - dB_1}{A_1 K - B_1} - \frac{K \, dA_2 + A_2 \, dK - dB_2}{A_2 K - B_2} \right) \quad (V.2.7)$$

and, by substituting $\theta_1 = A_1 K - B_1$ and $\theta_2 = A_2 K - B_2$,

$$\frac{dk_2}{k_2} = -\frac{dt}{t} - \frac{K}{1 + K} \frac{dK}{K} + \frac{1}{\ln (\theta_1/\theta_2)} \left[\frac{KA_1}{\theta_1} \left(\frac{dA_1}{A_1} - \frac{B_1}{KA_1} \frac{dB_1}{B_1} + \frac{dK}{K} \right) - \right.$$
$$\left. \frac{KA_2}{\theta_2} \left(\frac{dA_2}{A_2} - \frac{B_2}{KA_2} \frac{dB_2}{B_2} + \frac{dK}{K} \right) \right] \quad (V.2.8)$$

If we now take a case in which $K = 1$, so that $A_{eq} = B_{eq}$, and consider two experimental points $A_1 = 0.8A_{eq}$ and $B_1 = 1.2B_{eq}$ and $A_2 = 0.9A_{eq}$ and $B_2 = 1.1B_{eq}$, then $\theta_1 = -0.4A_{eq}$, $\theta_2 = -0.2A_{eq}$, and the equation reduces to

$$\frac{dk_2}{k_2} = -\frac{dt}{t} - \frac{1}{2} \frac{dK}{K} - \frac{1}{\ln 2}$$

$$\left(2.0 \frac{dA_1}{A_1} - 3.0 \frac{dB_1}{B_1} - 4.5 \frac{dA_2}{A_2} + 5.5 \frac{dB_2}{B_2} - 2.5 \frac{dK}{K} \right)$$

[a] The same type of analysis will be found to hold for the systems of opposing reactions of higher order. So long as $K = k_1/k_2$ is very large or very small we need only consider either the forward or reverse reaction. In the range of values of K close to unity we must consider both reactions, and Eq. (V.2.5)—together with the estimated errors—becomes algebraically very complex. However, it is still true that the errors will be of the same order of magnitude as those for the simple higher-order reactions so long as we are less than 50 per cent from equilibrium. Past that point the errors are larger and grow with the closeness to equilibrium.

and we see that the principal errors in k_2 will arise from the bracketed term, i.e., the errors arising from the analysis. Thus if all the analyses are performed to an accuracy of ± 0.1 per cent, the bracketed term alone will contribute a mean error to k_2 of (neglecting the error in K):

$$\frac{\Delta k_2}{k_2} = 1.4[(2.0)^2 + (3.0)^2 + (4.5)^2 + (5.5)^2]^{1/2} \times 0.1 \text{ per cent}$$

$$= \pm 1.1 \text{ per cent}$$

while an analytical precision of ± 1 per cent will give rise to an error of ± 11 per cent in k_2.

3. Concurrent Reactions

The equations for concurrent reactions of different orders are discussed in Sec. III.4. Experimentally the concurrent reactions will be characterized by the fact that they will not fit any of the simple rate laws and in addition do not approach a final equilibrium. That is not, of course, sufficient to distinguish them from series of consecutive reactions, but in practice there are no general methods for such a distinction. The distinctions between the systems become almost an individual problem for each reaction studied and must be made on an experimental, rather than any formal, basis. Formal analysis can decide only whether or not any of the simpler reaction schemes can describe the experimental facts to within the estimated error of the measurements. It provides only a negative critique for the description of reactions.

Aside from the direct application of the rate equations to a set of concurrent reactions there are a few short-cut methods for analyzing such systems. One of these methods is applicable to the system in which a single reactant may react in two or more different ways to produce two distinct and independent sets of products. Consider the reactions:

$$\begin{aligned} \text{A} &\xrightarrow{k_1} \text{P}_1 + \cdots \qquad \text{order } \alpha \\ \text{A} &\xrightarrow{k_2} \text{P}_2 + \cdots \qquad \text{order } \beta \end{aligned} \qquad\qquad (\text{V.3.1})$$

We can divide the rate equations for P_1 and P_2 and obtain

$$\frac{d\text{P}_1}{d\text{P}_2} = \frac{k_1}{k_2} \text{A}^{\alpha - \beta} \qquad\qquad (\text{V.3.2})$$

If $\alpha = \beta$, we can integrate directly and obtain

$$\text{P}_1 = \frac{k_1}{k_2} \text{P}_2 + \phi \qquad\qquad (\text{V.3.3})$$

where ϕ is usually zero at zero time. If P_1 and P_2 are produced by mechanisms of the same order from identical reactants, they will thus be produced in a constant ratio. This ratio will be the ratio of the rate constants.

If $\alpha > \beta$, then the ratio of rates of production, dP_1/dP_2, will decline as the reaction proceeds. Conversely, if $\alpha < \beta$, the ratio will increase.

This method of comparative evaluation of two concurrent reactions is now one of the most powerful methods available for gaining information in complex reaction systems. In particular if one of the reactions is known from other work, then the comparative method is very convenient for investigating the second reaction.

In most systems of chain reactions there are frequently alternative modes of reaction for active intermediates. Consider, for example, a system in which the following reactions are taking place:

$$B + A \xrightarrow{k_1} P_1 + \cdots \qquad \text{second order}$$
$$B + C \xrightarrow{k_2} P_2 + \cdots \qquad \text{second order} \qquad (V.3.4)$$

Assume that these are the only reactions whereby A and C react. B may be another molecule or an active intermediate (e.g., a free radical). In that case the ratio of the two rate equations will be

$$\frac{dA}{dC} = \frac{k_1 A}{k_2 C} \qquad (V.3.5)$$

which integrates to

$$\ln A = \frac{k_1}{k_2} \ln C + \phi \qquad (V.3.6)$$

A logarithmic plot of A against C will now give a straight line with slope equal to k_1/k_2. If we perform the reaction at different temperatures and apply the Arrhenius equation to the ratio k_1/k_2, we can obtain the difference in activation energies $E_1 - E_2$.

Thus if Br_2 is allowed to react with a mixture of H_2 and D_2, it will be found in the initial stages of the reaction that a logarithmic plot of H_2 concentration against D_2 (or HBr vs. DBr) will be a straight line with slope equal to the ratio of k_1 to k_2 where

$$Br + H_2 \xrightarrow{k_1} HBr + H$$
$$Br + D_2 \xrightarrow{k_2} DBr + D \qquad (V.3.7)$$

and

$$\ln (H_2) = \frac{k_1}{k_2} \ln (D_2) + \text{const} \qquad (V.3.8)$$

In this way it has been found that the difference in activation energies for the two reactions[a] is 2.1 Kcal, being higher for deuterium.

4. Chain Reactions; Detection of Intermediates

The most fruitful progress in the field of experimental kinetics in recent years has come about through the development of experimental techniques

[a] F. Bach et al., Z. physik. Chem., **278**, 71 (1934).

for the study of the small amounts of active intermediates (mainly free radicals) produced in complex reaction systems. These techniques, together with the mathematics developed for their interpretation, are now providing an abundant body of literature for what promises to be a really sound basis for the understanding of the mechanisms of chemical kinetic processes. In the present section we shall describe briefly the most important of these techniques.

4A. Mirror Techniques. One of the first direct methods for demonstrating the existence of free radicals in chemical reactions consisted in the isolation of products of the radicals formed by allowing the radicals to react with metals deposited (by evaporation) on the walls of a glass tube in a flow reaction system.[a] A typical such system is shown in Fig. V.1. In

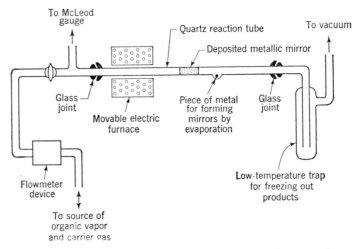

To McLeod gauge

Quartz reaction tube   To vacuum

Deposited metallic mirror

Glass joint

Piece of metal for forming mirrors by evaporation

Glass joint

Movable electric furnace

Low-temperature trap for freezing out products

Flowmeter device

To source of organic vapor and carrier gas

Fig. V.1. Apparatus for the chemical detection of free radicals.

principle the apparatus consists of a quartz tube a portion of which can be kept at high temperature (500 to 900°C) by means of a movable electric furnace. A volatile material (such as acetone or lead tetramethyl) is flowed through the tube past the furnace at low pressures (0.1 to 5.0 mm Hg) and high linear velocities (1 to 10 m/sec). This fast-moving stream, containing both reactants and products, then flows over a metal mirror which has been previously deposited on the walls of the tube by evaporation. Radicals formed by decomposition in the furnace can and do react with the metal mirror to yield volatile products which can be trapped in a tube im-

[a] For a résumé of this early work, see F. O. Rice and K. K. Rice, "The Aliphatic Free Radicals," The Williams & Wilkins Company, Baltimore, 1935. A more general treatise on free radicals will be found in E. W. R. Steacie, "Atomic and Free Radical Reactions," A.C.S. Monograph No. 125, 2d ed., Reinhold Publishing Corporation, New York, 1954.

mersed in a liquid nitrogen bath. There then remains the further problem of separating the products and reactants and identifying the former.

Table V.1 lists some of the metal alkyls which have been observed by such techniques.

TABLE V.1. METAL ALKYLS FOUND IN MIRROR REACTIONS

Reactive radical:	Methyl, CH_3	Methylene, CH_2	Ethyl, C_2H_5	Propyl, C_3H_7	Phenyl, C_6H_5	Benzyl, $C_6H_5CH_2$
Metal	Metal alkyl					
Pb	$Pb(CH_3)_4$	—	$PbEt_4$	—	—	—
Zn	$Zn(CH_3)_2$	—	$ZnEt_2$	—	—	—
Be	$Be(CH_3)_2$	—	$BeEt_2$	—	—	—
Hg	$Hg(CH_3)_2$	—	$HgEt_2$	$HgPr_2$	$Hg\phi_2$	$Hg(\phi CH_2)_2$
	$Hg_2(CH_3)_2$					
Sb	$Sb(CH_3)_3$	—	$SbEt_3$	—	—	—
	$Sb_2(CH_3)_4$		Sb_2Et_4			
As	$As(Me)_3$		$AsEt_3$			
	$As_2(Me)_4$	—	As_2Et_4	—	—	—
	$As(Me)_5$		$AsEt_5$			
Bi	$Bi(Me)_3$	—	$BiEt_3$	—	—	—
	$Bi_2(Me)_4$					
Te	$Te(Me)_2$	$(TeCH_2)_x$	$TeEt_2$	—	—	—
	$Te_2(Me)_2$					
Se	$Se(Me)_2$	$(SeCH_2)_x$	—	—	—	$Se(\phi CH_2)_2$
I	MeI	CH_2I_2	—	—	ϕI	—

In variations of the techniques, carrier gases such as He, H_2, and N_2[a] (which offer disadvantages owing to noncondensability) as well as condensable carrier gas, such as H_2O, CO_2, and heptane[b] have been employed. Also, it is possible to deposit mirrors not only by evaporation of metals directly but by decomposition of metal-organic compounds.[c]

It is possible to do semiquantitative work with this type of apparatus by studying the relative rates at which radicals can react with the mirror, compared to their rates of disappearance by recombination. Thus by studying the time taken by methyl radicals to remove a mirror placed at different distances from the oven zone, it was possible to show that recombination of the radicals is a slow process at low pressures.[d] The time for mirror removal can be estimated visually[e] or photometrically,[f] by direct

[a] F. Paneth and W. Hofeditz, *Ber.*, **62B**, 1335 (1929). F. Paneth and W. Lautsch, *Ber.*, **64B**, 2702 (1931).

[b] F. O. Rice et al., *J. Am. Chem. Soc.*, **54**, 3529 (1932).

[c] Paneth and Hofeditz, *loc. cit.*; Paneth and Lautsch, *loc. cit.*

[d] *Ibid.*

[e] *Ibid.*

[f] N. Prileshajeva and A. Terenin, *Trans. Faraday Soc.*, **31**, 1483 (1935).

weighing,[a] or by tracer techniques that make use of radioactive metals.[b] Another variation is to take advantage of the specificity of different radicals for different metals. Thus H atoms react with As, Sb, Se, Te, Ge, and Sn but not with Pb or Bi (which does, however, undergo slow distillation[c]).

By using a guard mirror of Pb to remove alkyl radicals, it is possible with a second mirror of Sb to study the H atoms present in a flow system independently of the presence of the alkyl radicals.[d]

The main difficulty with the mirror techniques as a quantitative tool lies in the extreme sensitivity of the mirrors to contamination by oxygen, nitrogen, parent compounds, or other radicals. Thus in the photolysis of ketones and fatty acids[e] the original reactants as well as acyl radicals are capable of desensitizing the mirror unless it is heated above 100°C. Another difficulty is that the mirror may not only react with the free radicals but also catalyze secondary reactions and recombination of radicals.

4B. Spectroscopic Techniques. While optical spectroscopy might be thought an ideal tool for the investigation of unstable intermediates, it has not been very useful in most cases. The difficulty lies in the low concentration of intermediates present, as well as in the difficulty of separating the spectra of the intermediates (emission or absorption) from those of other species present. There are, however, a growing number of cases in which such methods have been successfully applied. Thus emission spectra of excited radicals, atoms and ions have been observed in the case of glow and arc discharges and also for explosive reactions and flames. The radicals C_2, CH, HS, S_2, D, CN, NH, OH, PH, HgH are a few that have been identified in electrically excited emission.[f] Similarly, the radicals C_2, CH, OH, NH, SO, H, Cl, CHO are a few that have been observed in flames and explosions.[g] In both these instances, however, the observed emission spectra are capable of giving information only on the relative number of excited radicals and say nothing about the type or quantity of radicals present in unexcited states incapable of emission.

The difficulties inherent in the use of absorption spectra have been dis-

[a] F. Paneth and K. Hahnfield, unpublished work.

[b] P. A. Leighton and R. A. Mortensen, *J. Am. Chem. Soc.*, **58**, 488 (1936). Feldman et al., *ibid.*, **62**, 265 (1940). J. E. Ricci and T. W. Davis, *J. Chem. Phys.*, **13**, 440 (1945).

[c] T. G. Pearson et al., *Proc. Roy. Soc. (London)*, **A142**, 275 (1933).

[d] M. Burton, *J. Am. Chem. Soc.*, **58**, 692, 1645 (1936).

[e] T. G. Pearson et al., *J. Chem. Soc.*, 409 (1938).

[f] See G. Herzberg, "Molecular Spectra and Molecular Structure," 2d ed., D. Van Nostrand Company, Inc., Princeton, N.J., for an excellent review. Also A. G. Gaydon, "Dissociation Energies and Spectra of Diatomic Molecules," Chapman and Hall, Ltd., London, 1947.

[g] A. G. Gaydon, "Spectroscopy and Combustion Theory," Chapman and Hall, Ltd., London, 1948.

cussed in detail by Oldenburg,[a] who found the main problem to be the low concentrations. Oldenburg was, however, successful in using absorption techniques to study the OH radical produced in the $H_2 + O_2$ reaction. A more recent example of the use of absorption spectra has been demonstrated by Porter, who solved the problem by producing huge concentrations of free radicals with an ultra-high-energy light discharge.[b] At the high energies employed, it was found possible to obtain instantaneous radical concentrations as great as or greater than the concentration of parent compound.

Automatic recording infrared apparatus can also be used in studying the kinetics of both complex reactions and intermediates if their concentrations are not less than 1 per cent. In this way, ketene, CH_2CO, and acetaldehyde, CH_3CHO, have been identified as intermediates in the pyrolysis of ethylene oxide, C_2H_4O.[c]

4C. Mass Spectroscopy. The mass spectrograph seems ideally suited to a detailed study of complex kinetic systems, but progress in its utilization has been slow (owing to high costs) and the instances in which it has been used have not been strikingly successful with regard to the quantitative detection of intermediates. Rather, its principal application to date has been as an analytical tool for the determination of stable products.

The first extended study made with the aid of a mass spectrograph was that of Leifer and Urey,[d] who employed it in the study of the pyrolysis of dimethyl ether and acetaldehyde. While they were not successful in detecting radicals, they were able to demonstrate that formaldehyde is an intermediate in the decomposition of dimethyl ether and follow its concentration.

A more promising application was that of G. C. Eltenton,[e] who was able to construct a system capable of detecting free radicals in pyrolysis reactions and flames even at high pressures ($\sim$160 mm Hg). He was able to detect the presence of CH_3 radicals in the pyrolysis of hydrocarbons, the radicals CH_2 from CH_2N_2, and CHO and CH_3O in the combustion of CH_4 in O_2. The method depends in principle upon the fact that the electron energy required to ionize a radical is less than the electron energy required to produce the ionized radical from the original parent compound. This makes it possible to detect small amounts of radicals in the presence of

[a] Oldenburg, *J. Phys. Chem.*, **41**, 293 (1937).

[b] G. Porter, *Proc. Roy. Soc. (London)*, **A200**, 284 (1950).

[c] G. L. Simard et al., *J. Chem. Phys.*, **16**, 836 (1948). See also the study of the NO-N_2O_5 reaction by I. C. Hisatsune et al., *J. Am. Chem. Soc.*, **79**, 4648 (1957).

[d] E. Leifer and H. C. Urey, *J. Am. Chem. Soc.*, **64**, 994 (1942).

[e] G. C. Eltenton, *J. Chem. Phys.*, **15**, 455 (1947); *J. Phys. & Colloid Chem.*, **52**, 463 (1948). For extensions of this technique see F. P. Lossing et al., *Discussions Faraday Soc.*, **14**, 34 (1953).

large amounts of compounds whose own spectra would normally obscure the spectra of the radicals.

The method is not, however, simple in application and requires a kinetic system adjacent to the ionization chamber of a mass spectrograph. It also suffers from the danger that the radicals observed may be produced by secondary processes, either on the walls of the connecting tubing or from metastable radicals.

Even as an analytical instrument applied to flow systems and used for analysis of species present in abundance, the mass spectrograph may involve errors due to diffusive selection of lighter molecules and discrimination by the ion source of the more easily formed ions.[a]

4D. Chemical Methods. The mirror techniques already described illustrate but one of a more general system for detection of free radicals which is based on the chemical reactivity of the radicals. Thus if R represents a radical and Y is any stable chemical species capable of reacting with R, the introduction of Y into the kinetic system should result in a change in the original reaction and, depending on the specific nature of R and Y, the production of new products. From this point of view the substance Y acts as an inhibitor of the original reaction. An ideal inhibitor would react with radicals completely and as rapidly as they were formed and would thus give complete and unambiguous information on the first steps in a chain reaction from a study of the new products formed.

While there is no such substance as an ideal inhibitor, for particular cases there are certain compounds capable of combining with radicals to a sufficient extent to elucidate the nature of the primary reaction (i.e., the reaction initiating free radicals). The problem is much more difficult if the intermediates are unstable molecules.

Among the substances which have been used to inhibit chain reactions in the manner described are I_2,[b] Cl_2,[c] NO,[d] and propylene[e] vapors. Although the mode of reaction of each of these inhibitors is not well understood, it is clear in the case of the halogens that the reaction involved is

$$R + X_2 \rightarrow RX + X$$

The RX (halide) is isolated as a stable product and the X atoms presumably recombine to form X_2. In the case of NO, compounds of the type

[a] See F. P. Lossing and A. W. Tickner, *J. Chem. Phys.*, **20**, 907 (1952), for improvements in this technique.

[b] E. Gorin, *J. Chem. Phys.*, **7**, 256 (1939). J. N. Pitts and F. E. Blacet, *J. Am. Chem. Soc.*, **72**, 2810 (1950).

[c] E. Horn et al., *Trans. Faraday Soc.*, **30**, 189 (1934).

[d] L. A. K. Stavely and C. N. Hinshelwood, *J. Chem. Soc.*, 812, 818 (1936). F. J. Stubbs and C. N. Hinshelwood, *Proc. Roy. Soc.* (*London*), **A200**, 458 (1950); **A201**, 18 (1950).

[e] F. O. Rice and O. L. Polly, *J. Chem. Phys.*, **6**, 273 (1938). J. R. E. Smith and C. N. Hinshelwood, *Proc. Roy. Soc.* (*London*), **A180**, 237 (1942).

$R \cdot NO$ are formed.[a] The behavior of propylene is presumably to form the fairly stable radical R—CH_2—CH—CH_3, which may dimerize.[b] These methods are, for a number of reasons, only partially successful in elucidating the mechanism of the original reaction. In the first instance there is always the possibility that the added inhibitor may play a role in altering the original reaction. This is certainly evidenced in the case of NO; in some instances it may even accelerate the reaction.[c] A second difficulty is that the inhibition or capture of free radicals is incomplete, i.e., the radicals may react with other substances either more rapidly or rapidly enough to make the data ambiguous. Finally, there are always the problems of back reactions and of further decomposition of the radical-inhibitor products, found in the case of reactions of CH_3CO with I_2[d] and also for products RNO.[e] These same difficulties appear in the mirror techniques. In brief, while these methods are valuable in certain instances, their use must be circumscribed by a careful consideration of the reaction studied.

Some variants of these techniques have been described by Melville,[f] Benington,[g] and Bartlett.[h] The technique described by Melville involves the ability of H atoms and alkyl free radicals to reduce the blue oxide coat on a polished molybdenum surface. By photometrically studying the change in optical color of the surface it was shown possible to make calculations of the relative concentrations of radicals diffusing through a vapor stream to the metal oxide surface. The mathematical calculations are involved, and the difficulties already described for mirror techniques occur here too.

The method described by Benington involves the reaction of the free radicals produced in a gas stream with a thin film of organic solution containing triphenyl methyl radicals. The stream impinges upon the solution-covered surface, and the products of the reaction are measured spectroscopically. The work of Bartlett is more directly concerned with free radicals in solution. A fairly stable but nonetheless reactive free radical 1,1-diphenyl-2-picryl hydrazyl (DPPH), ϕ_2-N-N-$C_6H_2(NO_2)_3$, is capa-

[a] $CH_3 \cdot NO$ is stable at low temperatures and the isomer CH_2=NOH and dimer have been isolated from the reaction of NO with CH_3 radicals, C. S. Coe and T. F. Doumani, *J. Am. Chem. Soc.*, **70**, 1516 (1948).

[b] Alternatively, propylene can lose H atoms readily in abstraction by R to form RH plus the relatively stable allyl radical.

[c] Gorin, *loc. cit.*

[d] *Ibid.*

[e] F. H. Verhoek, *Trans. Faraday Soc.*, **31**, 1533 (1935). E. W. R. Steacie and R. D. MacDonald, *Can. J. Research*, **12**, 711 (1935). Smith and Hinshelwood, *loc. cit.*

[f] H. W. Melville, "Third Symposium on Combustion and Flame and Explosion Phenomena," The Williams & Wilkins Company, Baltimore, 1949.

[g] F. Benington, "Third Symposium on Combustion and Flame and Explosion Phenomena," *op. cit.*

[h] P. D. Bartlett and H. Kwart, *J. Am. Chem. Soc.*, **72**, 1051 (1950). See also Mueller et al., *Ann.*, **520**, 235 (1935).

ble of forming stable compounds with free radicals, and the products can be identified. It seems to be 100 per cent dissociated even at $-80°C$. This can be useful in the study of polymerization reactions when it can be shown that the DPPH does not initiate chains independently.

4E. Paramagnetic Properties. All free radicals are chemically unsaturated and possess an odd number of electrons. As a consequence they are paramagnetic. Any technique capable of detecting paramagnetism is therefore a potential tool for the detection of free radicals, if there are present no stable molecules such as O_2, NO, or NO_2, which are also paramagnetic.

The earliest attempt to detect the presence of free radicals in this way was through the catalytic conversion of ortho-para hydrogen mixtures. At equilibrium at room temperature, ordinary hydrogen consists of a mixture of 75 per cent ortho-H_2 (nuclear spins parallel) and 25 per cent para-H_2 (nuclear spins antiparallel). At low temperatures ($<90°K$) equilibrium mixtures may be prepared which contain up to 100 per cent pure para-H_2. The latter mixtures are metastable below $500°C$ and are slowly converted to the stable composition above that temperature. The thermal reaction has been well studied[a,b] and corresponds to a catalytic conversion by H atoms present at these temperatures.

$$H + o\text{-}H_2 \rightleftharpoons H + p\text{-}H_2$$

The reaction is catalyzed by any paramagnetic substance at much lower temperatures, and the rate of conversion of para-H_2 to ortho-H_2 can therefore be used as a measure of the concentration of such paramagnetic substances. A theoretical treatment which is in good qualitative agreement with the facts has been made by Wigner[c] and has been shown by Wilmarth et al.[d] to hold fairly well in solution.

The conversion has been used to demonstrate the existence of radicals in the photochemical decomposition of CH_3I,[e] HI,[f] C_2H_6,[g] and NH_3.[h,i] The chief difficulty with the method involves the inability to distinguish between different radicals and also the effects of paramagnetic impurities.

For long-lived free radicals such as metastable, electronically excited

[a] K. H. Geib and P. Harteck, *Z. physik. Chem.*, Bodenstein Festband, 849 (1931).

[b] L. Farkas, *ibid.*, **B10**, 419 (1930). A. Farkas, "Orthohydrogen, Parahydrogen and Heavy Hydrogen," Cambridge University Press, New York, 1935. G. Boati et al., *J. Chem. Phys.*, **24**, 783 (1956).

[c] E. P. Wigner, *Z. physik. Chem.*, **B23**, 28 (1933).

[d] Y. Claeys et al., *ibid.*, **16**, 425 (1948).

[e] W. West, *J. Am. Chem. Soc.*, **57**, 1931 (1935); *Ann. N.Y. Acad. Sci.*, **41**, 238 (1941).

[f] E. J. Rosenbaum and T. R. Hogness, *J. Chem. Phys.*, **2**, 267 (1934).

[g] F. Patat, *Z. physik. Chem.*, **B32**, 274, 294 (1936). F. Patat and H. Sachsse, *Z. Elektrochem.*, **41**, 493 (1935). H. Sachsse, *Z. physik. Chem.*, **B31**, 79, 87 (1935).

[h] L. Farkas and P. Harteck, *Z. physik. Chem.*, **B25**, 257 (1934).

[i] Geib and Harteck, *loc. cit.*

dye molecules[a] or triaryl methyl[b] more direct methods are available for the measurement of the concentrations by using sensitive Guoy balances or modifications thereof.

The recent developments of electron paramagnetic resonance (EPR) spectroscopy and nuclear magnetic resonance (NMR) spectroscopy have made available two new tools for the exploration of free radicals and metastable intermediates in chemical reactions. Atoms or radicals with unpaired electron species will absorb microwaves of the proper frequency when placed in a uniform magnetic field. Concentrations of radicals the order of 10^{-6} M can be detected in samples as small as 0.1 cc, and many radicals and paramagnetic species have been observed by this method.[c]

The NMR technique is restricted to observations of nuclei with magnetic moments and preferably of spin $\frac{1}{2}$. The absorption frequencies characteristic of such nuclei in strong magnetic fields are in the region of 1 to 100 megacycles/sec and are very sensitive to intramolecular and intermolecular environment. While not as sensitive as EPR in terms of concentrations (not less than 0.01 M), it is capable of measuring the presence of low concentrations of intermediates when they are in equilibrium with major species.[d]

Another instance of a reaction catalyzed by small concentrations of free radicals is the interconversion of geometrical isomers. The cis-trans isomerization of maleic to fumaric acids

$$
\begin{array}{cc}
\text{CH—COOH} & \text{HOOC—CH} \\
\| & \to \quad \| \\
\text{CH—COOH} & \text{CH—COOH}
\end{array}
$$

is catalyzed by small amounts of either Br_2 or I_2 and light,[e] as is the isomerization of allocinnamic acid[f]

$$
\begin{array}{cc}
\phi\text{—C—H} & \phi\text{—C—H} \\
\| & \to \quad \| \\
\text{H—C—COOH} & \text{HOOC—C—H}
\end{array}
$$

and the cis-trans isomerization of 1,2-dichloroethylene[g]

$$
\begin{array}{cc}
\text{Cl—C—H} & \text{Cl—C—H} \\
\| & \rightleftharpoons \quad \| \\
\text{Cl—C—H} & \text{H—C—Cl}
\end{array}
$$

[a] G. N. Lewis et al., *J. Chem. Phys.*, **17**, 804 (1949).

[b] C. S. Marvel et al., *J. Am. Chem. Soc.*, **62**, 1551 (1940); **63**, 1892 (1941); **64**, 1824 (1942).

[c] An excellent summary of recent work in these fields is found in the articles by H. McConnell, *Advances in Phys. Chem.*, **8** (1957), and by J. E. Wertz, *ibid.*, **9** (1958).

[d] Lewis et al., *loc. cit.*

[e] H. Eggert, *Z. Elektrochem.*, **33**, 542 (1927).

[f] F. Wachholtz, *Z. physik. Chem.*, **125**, 1 (1927).

[g] J. A. A. Ketelaar et al., *J. Phys. & Colloid Chem.*, **55**, 987 (1951).

While a small amount of addition to the double bond takes place, there is a very high quantum yield of isomerization that arises from the action of free halogen atoms. These isomerizations thus make possible a test for the existence of such radicals.

4F. Isotopic Exchange. An important subcategory of techniques which depend on chemical behavior is that involving the use of stable or radioactive isotopes. The utility of the methods is circumscribed principally by the availability of counting equipment, suitable isotopes, or apparatus for the quantitative detection of isotopic substitution. An interesting example of the methods is to be found in the work on the thermal and photochemical decomposition of acetaldehyde. The decomposition may be represented by

$$CH_3CHO \rightarrow CH_4 + CO$$

Smaller amounts of H_2 and $(CH_3CO)_2$ are found, along with C_2H_6 at higher temperatures. One of the important questions was whether the reaction occurred via free radicals

$$CH_3CHO \rightarrow CH_3 + CHO$$

followed by the secondary reaction chain

$$CH_3 + CH_3CHO \rightarrow CH_4 + CH_3CO$$
$$CH_3CO \quad \rightarrow CH_3 + CO$$

or occurred directly in one intramolecular rearrangement. The use of mixtures of CH_3CHO with CD_3CDO has finally shown that the major part of the reaction occurs via free radicals.[a-c] This conclusion was reached through the demonstration of the fact that the methane produced was the statistically expected mixture of CD_4, CH_4, CH_3D, and CD_3H. If the reaction did not involve free radicals, only CD_4 and CH_4 would be observed.

Similar methods have been employed in the photolysis of acetone in the presence of H_2,[d] the photolysis of light and heavy acetone mixtures,[e] the pyrolysis of ethane,[f] and the decomposition of C_2H_5Br in the presence of radioactive Br_2^*, HBr^*, and DBr^*.[g]

These methods are subject to the difficulties inherent in all such chemical methods and require careful experimental control to establish definitive results.[h] With adequate checks, however, they can be extremely valuable for establishing the nature of the intermediates in complex reactions and

[a] J. C. Morris, *J. Am. Chem. Soc.*, **63**, 2535 (1941); **66**, 584 (1944).

[b] P. O. Zemany and M. Burton, *J. Phys. & Colloid Chem.*, **55**, 949 (1951).

[c] L. A. Wall and W. J. Moore, *ibid.*, **55**, 965 (1951).

[d] H. S. Taylor and C. Rosenblum, *J. Chem. Phys.*, **6**, 119 (1938).

[e] S. W. Benson and C. W. Falterman, *ibid.*, **20**, 201 (1952).

[f] Wall and Moore, *loc. cit.*

[g] J. B. Peri and F. Daniels, *J. Am. Chem. Soc.*, **72**, 424 (1950).

[h] See, for example, recent work on acetaldehyde, Zemany and Burton, *loc. cit.*, and Wall and Moore, *loc. cit.*, in contrast to that of Morris, *loc. cit.*

also for obtaining quantitative data on the kinetic behavior of the intermediates.

4G. Photochemical Methods; Quantum Yield. The Einstein law of photoequivalence states that light is absorbed by molecules in discrete amounts as an individual molecular process; i.e., one molecule may absorb one photon at a time. Through measurements of light intensity and wavelength it is possible to determine quantitatively the number of photons of light absorbed during the course of a reaction. From the analysis of the products of such a reaction it is possible to calculate the ratio of the number of molecules decomposed to quanta absorbed, known as the quantum yield and usually designated by the symbol Φ. If this ratio exceeds unity, it provides unambiguous evidence for the existence of secondary processes and consequently indicates the presence of unstable intermediates.

Thus the photochemically induced reaction of H_2 and Cl_2 to form HCl has been shown to have quantum yields of up to 1,000,000.[a] This can be possible only if there is a chain reaction.

In a similar manner it has been shown that the photolysis of acetaldehyde at high temperatures is also a chain reaction.[b] The number of such reactions studied is by now too great to be adequately summarized here.

While a quantum yield in excess of unity is evidence of a chain reaction, it is not true that quantum yields lower than unity indicate the absence of chains. On the contrary, the photolysis of methyl iodide in solution may have a quantum yield of 0.008,[c] while it is known that the first step is to produce CH_3 radicals and I atoms.[d]

One of the outstanding virtues of photochemical studies is that in many cases there is independent evidence for the nature of the reaction following light absorption by a molecule, so that the primary process is understood. Under such conditions a comparison of the photochemical and thermal reactions is capable of yielding considerable information on the secondary reactions involved. Thus the absorption of light by Cl_2, Br_2, and I_2 is known to lead to the formation of free atoms, and the photochemical reactions of these halogens have been in many cases the key to understanding their thermal reactions.

4H. Photosensitization. Where photochemical reactions may not be directly initiated by light absorption because the substances do not absorb the wavelength available, it is possible to initiate the reaction by the use of a substance capable of absorbing the light and transferring the energy

---

[a] M. Bodenstein, *Z. physik. Chem.*, **85**, 329 (1913).

[b] J. A. Leermakers, *J. Am. Chem. Soc.*, **56**, 1537 (1934).

[c] W. West and B. Paul, *Trans. Faraday Soc.*, **28**, 688 (1932). W. West and L. Schlesinger, *J. Am. Chem. Soc.*, **60**, 961 (1938). R. D. Schultz and H. A. Taylor, *J. Chem. Phys.*, **18**, 194 (1950).

[d] The reason for the low quantum yields is generally the recombination of the primary products, e.g., $CH_3 + I \rightarrow CH_3I$ and reactions such as $CH_3 + I_2 \rightarrow CH_3I + I$.

to the reactants. Such a process is known as *photosensitization,* and a very popular photosensitizing agent is mercury. Mercury atoms show a strong absorption for the resonance radiations at 1849 and 2537 Å, both of which are easily produced in high intensities in mercury arcs. The excited Hg atoms so produced can transfer this energy and bring about a sensitized reaction. (1 photon at 2537 Å = 112 Kcal/mole; 1 photon at 1849 Å = 154 Kcal/mole.) In this way it is possible to produce H atoms from H_2[a,b] and from hydrocarbons[c] and start chain reactions at temperatures not normally accessible to such chains. Such studies have provided invaluable evidence on the kinetic behavior of radicals.

Other sensitizers which have been used are excited Xe, Cd, Zn, Na, and Cl atoms.[d]

4I. Chemical Sensitization. Equally valuable for the demonstration of the existence of free radicals and for their study are methods of chemical sensitization. Free radicals may be demonstrated to exist in a reaction by their ability to produce a sensitized decomposition of a material normally inert at the temperature employed. Thus, it has been demonstrated that, whereas acetaldehyde does not decompose at an appreciable rate at 300°C, a fast decomposition can be induced at that temperature by adding azomethane $(CH_3)_2N_2$ in small amounts.[e] The role of the azomethane is to produce methyl radicals which can then start a chain decomposition. Oxygen is similarly a chemical sensitizer for the decomposition of many hydrocarbons and aldehydes.

A very interesting example of a sensitizing agent which is finding wide application is dtBP, $(CH_3)C\text{-}O\text{-}O\text{-}C(CH_3)_3$. Through its thermal decomposition it is capable of forming large quantities of methyl radicals both in solutions and in the gas phase:[f]

$$[(CH_3)_3CO]_2 \rightarrow 2(CH_3)_3CO$$
$$(CH_3)_3CO \rightarrow CH_3COCH_3 + CH_3$$

Still another example, interesting for the light it throws on oxidation processes, is the sensitized oxidation of hydrocarbons by HBr.[g] The chain mechanism seems to be

[a] G. Cario and J. Frank, *Z. Physik,* **11,** 161 (1922). H. S. Taylor and J. R. Bates, *Proc. Nat'l Acad. Sci. U.S.,* **12,** 714 (1926).

[b] L. O. Olsen, *J. Chem. Phys.,* **6,** 307 (1938). F. F. Rieke, *ibid.,* **4,** 513 (1936).

[c] See E. W. R. Steacie, "Atomic and Free Radical Reactions," A.C.S. Monograph No. 125, 2d ed., Reinhold Publishing Corporation, New York, 1954.

[d] *Ibid.*

[e] A. O. Allen and D. V. Sickman, *J. Am. Chem. Soc.,* **56,** 1251, 2031 (1934).

[f] F. F. Rust et al., *J. Am. Chem. Soc.,* **70,** 88, 95, 1336, 3259 (1948). M. Szwarc and J. S. Roberts, *J. Chem. Phys.,* **18,** 561 (1950).

[g] F. F. Rust and W. E. Vaughan, *Ind. Eng. Chem.,* **41,** 2595 (1949). U.S. patent 2,403,771 (July, 1946); F. F. Rust et al., *Ind. Eng. Chem.,* **41,** 2612, 2597, 2609 (1949).

Initiation: $HBr + O_2 \overset{walls}{\rightleftharpoons} Br + HO_2$

Transfer: $Br + RH \rightleftharpoons HBr + R$

Chain: $R + O_2 \longrightarrow RO_2$

$$RO_2 + \begin{cases} HBr \longrightarrow RO_2H + Br \\ RH \longrightarrow RO_2H + R \end{cases}$$

with the net reaction for the chain steps

$$RH + O_2 \overset{HBr}{\longrightarrow} RO_2H$$

4J. The Induction Period. Many complex reactions are characterized by an induction period, i.e., a period at the beginning of the reaction during which the rate is smaller than for the remainder of the run. While it is sometimes possible that such an induction period can be accounted for in terms of slow approach to the proper temperature (i.e., insufficient preheating) or steady state, it is more generally a sign of a complex reaction. Such a period may also be caused by an inhibition arising from the presence of small amounts of impurities. Thus the $H_2 + Cl_2$ reaction may display anomalously long induction periods owing to the presence of small amounts of impurities, such as NH_3,[a] organic compounds,[b] ClO_2, O_3, or O_2,[c] which break the chains and are slowly consumed.

As we shall see later, almost all very fast reactions, including explosions, show such induction periods, and a theory of their character developed by both Semenoff[d] and Rice[e] is in good agreement with the facts.

4K. Inhibition. Since complex reactions are propagated through a series of reactions, it might be expected that they would show an unusual sensitivity to any substance or physical condition which might tend to interfere at any point with the propagation of the chain. It is thus not surprising to find that such reactions are capable of extreme inhibition by either chemical substances present in small amounts or by physical conditions that tend to change the nature of the vessel surfaces.

Most of the chain reactions show such chemical inhibition, and marked inhibition by trace impurities is excellent evidence for a chain reaction. Thus 0.01 mole per cent of O_2 is capable of reducing the quantum yield of the $H_2 + Cl_2$ reaction by 1000 fold.[f] In a similar fashion free radicals are easily adsorbed on vessel surfaces, and it is common experience to find that a change in vessel shape, in surface/volume ratio (e.g., that obtained by packing the vessel with glass), or the addition of inert gases such as

[a] C. H. Burgess and D. L. Chapman, *J. Chem. Soc.*, **89**, 1399 (1906).

[b] *Ibid.*

[c] J. G. A. Griffiths and R. G. W. Norrish, *Proc. Roy. Soc. (London)*, **A130**, 591 (1931); **A135**, 69 (1932); **A147**, 140 (1934).

[d] N. Semenoff, *Z. Physik*, **48**, 571 (1928).

[e] O. K. Rice et al., *J. Am. Chem. Soc.*, **57**, 2212 (1935).

[f] M. C. C. Chapman, *J. Chem. Soc.*, **123**, 3062 (1923). M. Bodenstein and W. Dux, *Z. physik. Chem.*, **85**, 297 (1913).

helium can have a pronounced effect on the reaction velocity. Similarly, changing the walls from glass to metal or coating them may also have a marked effect.[a] While these effects are not always definite evidence of a chain reaction, since they may only indicate a simple heterogeneous reaction on the vessel surface, they do constitute a strong suspicion of the former.

4L. Intermittent Activation. The activation energy required to initiate what are called thermal reactions is provided by molecular collisions. It is, however, possible to initiate such reactions at temperatures at which their "normal" (i.e., thermal) rates would be experimentally unobservable. This activation may be supplied by means of radiant energy (photochemical reactions), ionizing impacts by high-energy particles (e.g., alpha, beta, gamma, or X rays) or sensitization (see Sec. V.4H) by already excited molecules.

When the energy is supplied photochemically, the rate of the subsequent reaction will depend on the intensity of the light absorbed. In simple cases it may be directly proportional to some power of the absorbed radiation. Thus in the photochemical reaction of H_2 with Br_2 it is found that[b]

$$\frac{d(HBr)}{dt} = f(H_2, Br_2, HBr) \times (I_a)^{1/2} \qquad (V.4L.1)$$

where f is a complicated function of the concentrations indicated and I_a is the absorbed light intensity.

If the system is irradiated with the same intensity I_a through a rotating sector that has a slit that on the average admits a fraction α of the light (i.e., the slit subtends an angle $= \alpha \times 360°$) then the rate may, under certain conditions, depend on the speed of the sector. If the reaction proceeds in one step to the formation of final products, the rate of reaction cannot depend on the speed of rotation of the sector, and the total effect of the slit is simply to reduce the intensity I_a by the fraction α.

If, on the other hand, secondary processes are involved and the primary act of light absorption is the production of active intermediates, then the speed of rotation may become important.

Case 1. In case 1, *all* of the secondary steps are first-order with respect to active intermediates. Then the rate will not depend on the sector speed, since the rate of all processes depends in the same way on the concentrations of intermediates, namely, in a linear fashion, and the rate of reaction averaged over the light period and the dark period (when the radical concentrations are decreasing) will be the same as the rate obtained by simply using a light source of intensity αI_a. In a case of this type it will turn out that the over-all rate of reaction will be directly proportional to the first power of the amount of light absorbed.

[a] F. O. Rice and K. F. Herzfeld, *J. Phys. & Colloid Chem.*, **55**, 975 (1951).

[b] M. Bodenstein and H. Lutkemayer, *Z. physik. Chem.*, **114**, 208 (1925).

Case 2. In the second case the secondary steps include one or more processes which are second-order (mixed or pure) with respect to intermediates.

This is illustrated by the $H_2 + Br_2$ photolysis [Eq. (V.4L.1)], where the chain-termination steps are second-order in Br atoms:

$$Br + Br \rightarrow Br_2 \qquad (V.4L.2)$$

In polymerization reactions it is illustrated by chain-termination steps of the type

$$R_1 + R_2 \rightarrow R_1 \cdot R_2 \qquad (V.4L.3)$$

where R_1 and R_2 are two different free radicals and $R_1 \cdot R_2$ is a stable molecule.

In these instances the rates of over-all reaction will not be simply proportional to the light intensity. During the period of illumination, when the radicals are at their maximum concentration, the termination steps will be relatively faster with respect to other steps than they are during the dark periods, when the radical concentrations have begun to fall off. This arises from the fact that the rate of a second-order reaction falls off more rapidly with decreasing concentration than does the rate of a first-order reaction.

The result will be as follows: If the rate R may be written as some power n of the absorbed intensity

$$R = f(C)I_a^n \qquad (V.4L.4)$$

then at very low speeds, at which the radicals essentially all disappear before the next period of illumination begins, the effect of the sector is to give an average rate $\overline{R}_S$ (averaged over a revolution) which is equal to αR:[a]

$$\overline{R}_S = \alpha f(C)I_a^n = \alpha R \qquad (V.4L.5)$$

When the speed is very high (such that the period between successive illuminations is small compared with the mean time of propagation of the secondary reactions), the effect of the sector is to diminish the effective light intensity to αI_a and the fast rate $\overline{R}_f$ is

$$\overline{R}_f = f(C)(\alpha I_a^n) = \alpha^n R \qquad (V.4L.6)$$

In the case of the $H_2 + Br_2$ reaction ($n = \frac{1}{2}$),

$$\frac{\overline{R}_f}{\overline{R}_S} = \alpha^{n-1} = \frac{1}{\alpha^{\frac{1}{2}}} \qquad (V.4L.7)$$

It is clear that by using a sector with a slot of $30°(\alpha = 1/12)$ we can expect a difference in rates of $12^{\frac{1}{2}} = 3.46$ between very fast and very slow rotation. From the way in which the rate varies with frequency of rotation between these extremes it is possible to calculate the mean life of the chain

[a] This is equivalent to having the full intensity for a fraction of time α of the original time.

processes. The rate will be about midway between the limits when the half-lives of the processes responsible for the termination of radicals have the same order of magnitude as the intervals between successive illuminations.

The method should be generally applicable to a study of secondary processes (whatever the source of excitation) whenever the secondary processes proceed at rates exceeding the first order with respect to intermediates. It has been described in some detail by Noyes[a] and Dickinson[b] and has been applied to the study of polymerization by Melville[c] and Bartlett[d] and to the recombination of radicals by Noyes and Zimmermann,[e] Gomer,[f] and Rice.[g] The difficulties in the use of the method are not small and involve the general difficulties of photochemical techniques.[h]

4M. Thermal Methods for Detection of Free Radicals. The recombination of radicals and atoms liberates a considerable quantity of heat, one that is at least equal to the energy of the bond formed. Since, at low pressures, it has been found that recombination of radicals takes place heterogeneously, i.e., at surfaces, it is possible to measure the relative concentrations of radicals by measuring the heat liberated when they recombine on a surface.

The thermal element used may be a coated thermometer bulb,[i] a glowing wire,[j] or a thermocouple junction.[k,l]

The method requires careful calibration and a study of possible poisoning of the metal surface. It can, however, be used quite generally for all types of radicals and atoms. The atoms may be produced by the methods already discussed or by electrical excitation. One of the most useful devices

[a] W. A. Noyes, Jr., and P. A. Leighton, "Photochemistry of Gases," chap. 4, Reinhold Publishing Corporation, New York, 1941.

[b] R. G. Dickinson, J. Phys. Chem., **42**, 740 (1938).

[c] G. M. Burnett and H. W. Melville, Nature, **156**, 661 (1945). M. H. Mackay and H. W. Melville, Trans. Faraday Soc., **45**, 323 (1949).

[d] P. D. Bartlett and C. G. Swain, J. Am. Chem. Soc., **67**, 2273 (1945); **68**, 2381 (1946).

[e] J. Zimmermann and R. M. Noyes, J. Chem. Phys., **18**, 658 (1950). See also A. Berthoud and H. Bellenot, Helv. Chim. Acta, **7**, 307 (1923), and F. Brierly et al., J. Chem. Soc., 562 (1926), for early work.

[f] R. Gomer and G. B. Kistiakowsky, J. Chem. Phys., **19**, 85 (1951).

[g] W. L. Haden and O. K. Rice, ibid., **10**, 445 (1942). V. E. Lucas and O. K. Rice, ibid., **18**, 993 (1950).

[h] Noyes and Leighton, loc. cit. See also the article by H. W. Melville and G. M. Burnett in "Technique of Organic Chemistry," vol. 8, Interscience Publishers, Inc., New York, 1953, for a discussion of errors.

[i] K. F. Bonhoeffer, Z. physik. Chem., **113**, 119, 492 (1924).

[j] H. C. Urey and G. I. Lavin, J. Am. Chem. Soc., **51**, 3286 (1929).

[k] G. M. Schwab and H. Fries, Z. Elektrochem., **39**, 586 (1933). G. M. Schwab, Z. physik. Chem., **B27**, 452 (1935).

[l] For application to the study of molecular beams see C. A. Reilly and B. S. Rabinovitch, J. Chem. Phys., **19**, 248 (1951).

for studying the properties of H atoms is the Wood-Bonhoeffer discharge tube shown in Fig. V.2.

The tube was first used by Wood[a] and later modified by Bonhoeffer.[b] In operation, a fast stream of low-pressure H_2 (with or without carrier gas such as He or Ne) is drawn past a 3000- to 5000-volt discharge (up to

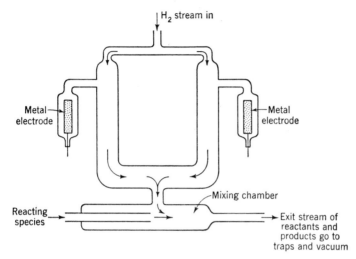

FIG. V.2. Discharge tube for the production of H atoms.

300 ma). Water, KCl, or syrupy H_3PO_4 may be used to poison the walls and prevent recombination of H atoms. The pressures are from 0.1 to 1.0 mm Hg (or higher with carrier gas). As high as 50 per cent H atoms can be produced, and many reactions can be studied, at room temperature. The tube has also been employed for studying N, O, Br, and Cl atoms. It is not easy to get an accurate measure of the H atom concentration, and consequently it is not simple to measure activation energies by a comparison of results at different temperatures. The activation energies obtained are generally calculated by means of an assumption about the value of the frequency factor for the reaction between H atoms and molecules.[c]

5. Résumé

It can be seen from the foregoing that, while simple reaction systems are amenable to a reasonable experimental and mathematical analysis, complex

[a] R. W. Wood, *Proc. Roy. Soc.* (*London*), **A97**, 455 (1920); **A102**, 1 (1922).

[b] Bonhoeffer, *loc. cit.; Z. physik. Chem.*, **116**, 391 (1925); *Z. Elektrochem.*, **31**, 521 (1925).

[c] See M. R. Berlie and D. J. Le Roy, *J. Chem. Phys.*, **20**, 201 (1951), for some improvements in the technique.

reaction systems require considerable, individual investigation. In particular there is no certain a priori method for predicting the course of a complex reaction or, even after careful study, of being sure that the mechanism assumed is the only one possible. Even in the case of many simple systems a unique mechanism can never be absolutely decided upon. One of the reasons for this ambiguity is that there are many alternate reaction paths which can lead to the same kinetic observations. The detailed techniques just described are generally necessary to verify the existence of postulated intermediates.

An even more difficult problem is that in the empirical treatment of reacting systems we have two simultaneous problems: (1) the selection of a rate law for the system and (2) the question of experimental accuracy. Regarding accuracy the best that can be done is to see whether or not the expected experimental errors are compatible with the observed deviations from a chosen rate law. While this is not a sufficient criterion for the adequacy of a rate law, it is certainly a necessary one. As we shall see shortly, there are a number of other criteria which can also be employed. Basically, only a very careful study of the system under varied experimental conditions can finally decide the reality of the postulated mechanisms.

6. The Molecular Approach

A scientific theory, like a mathematical system, never yields more than is built into it in the way of assumptions or postulates. The phenomenological approach presented in the preceding chapters could no more than characterize the kinetic behavior of systems in terms of the macroscopic variables used to describe them. We have obtained from this approach the kinetic quantities and rate constants or, in terms of the Arrhenius formulation, the frequency factors and the energies of activation. These quantities constitute our phenomenological category of kinetic language. If we are to relate them to the molecular properties of the reacting species, we must construct a new theory and a new nomenclature which starts with the molecule as the unit under consideration.

The results of such a theory will be to relate the macroscopic kinetic quantities to whatever new quantities we shall use to define our molecular unit. At this point we are faced with a dilemma. A molecule is in principle completely defined by the mass and atomic number of its representative atoms. A basic theory should therefore reduce our kinetic quantities to these more fundamental quantities and some universal constants such as the velocity of light c, Planck's constant h, and so on. While such a program is in principle possible, it is by no means practicable.

The answer to our dilemma is, then, to choose some less complete system of definition for the molecular unit, one that is capable of relating our macroscopic data to meaningful molecular quantities; i.e., we must select some reasonable model for the molecule, one that is simple enough to be dealt with mathematically and one whose properties can be related to other experimental molecular properties. In the following chapter we shall deal with some of the models that have been adopted and develop the mathematical tools for describing them.

PART TWO

Statistical Methods for Treating Systems of Large Numbers of Particles at or Near Equilibrium

VI

Representation of Equilibrium Properties

1. Description of a System of Discrete Particles

From a classical point of view the behavior of a system of discrete particles is uniquely determined by Newton's laws of motion and the laws of force acting between the particles. We can write for each particle in the system three second-order differential equations which determine the values of the three cartesian coordinates of the particle as functions of time.

$$m_i \frac{d^2 x_i}{dt^2} = F_{x_i} \qquad m_i \frac{d^2 y_i}{dt^2} = F_{y_i} \qquad m_i \frac{d^2 z_i}{dt^2} = F_{z_i} \qquad \text{(VI.1.1)}$$

or, in the more compact vector form,

$$m_i \frac{d^2 \mathbf{r}_i}{dt^2} = \mathbf{F}_i \qquad \text{(VI.1.2)}$$

where m_i refers to the mass of the ith particle; x_i, y_i, and z_i refer to the cartesian coordinates of the particle, and F_{x_i}, F_{y_i}, F_{z_i} refer to the instantaneous cartesian components of force acting on the particle.[a]

For a system of N particles there will be a total of $3N$ such equations. In principle we may solve these equations, and we should find for each equation two arbitrary constants of integration. For the entire set there will be $6N$ such constants of integration, and in order to eliminate these arbitrary constants, we would have to have $6N$ independent pieces of information. These might be the coordinate positions ($3N$) of each particle at two different times or the equivalent.[b] It is clear that the mechanical

---

[a] These components include the forces exerted on the ith particle by all the particles, the forces exerted by electromagnetic and other force fields, and the walls of the containing vessel.

[b] A more generally used set are the $3N$ values of the cartesian coordinates x_i, y_i, and z_i and the $3N$ values of the cartesian velocity components x_i, y_i, and z_i at a single instant. For many purposes the components of momentum ($m_i x_i, = p_{x_i}$, etc.) have a more fundamental significance and the $3N$ components of momentum may be used in place of the $3N$ velocity components.

behavior of the system is not uniquely defined until there is enough experimental information to determine the $6N$ constants. The state of such a system is completely determined when this internal information is given together with the external description of the system (i.e., the masses of the particles, the position and magnitude of force fields, and the positions of boundaries, walls, etc.). For the simplest example of physical interest, say a mole of helium,[a] we would have to have a total of about $6N = 6 \times 3 \times 6 \times 10^{23} = 1.08 \times 10^{25}$ pieces of independent information.[b]

There is clearly a broad gap between this impossible informational requirement and the handful of variables (P, V, T, mole fraction) needed to adequately describe the thermodynamic state of the system and so determine the macroscopic behavior of the system at equilibrium (see Secs. I.1 and I.2). Even the requirements for an empirical description of a kinetic system are nowhere so formidable.

The reason for this large discrepancy between the molecular and macroscopic requirements for a description is to be found in the fact that from the latter point of view we are not at all interested in the particular behavior of each molecule but are instead interested only in the average behavior of the system as a whole. If we can adopt a similar disinterest in individual molecules, we can perhaps hope to bridge the gap. Thus if we can somehow reduce our original system of $3N$ second-order differential equations to a small set that describes the average behavior of the whole, we shall have some chance of relating the macroscopic behavior of the system to the microscopic description.

This particular legerdemain is the function of the science known as statistical mechanics, some of whose aspects and findings we shall now proceed to discuss.

2. Distribution Functions

If we consider for simplicity a quantity of gas in a rigid container made of completely nonconducting walls, we know that it will be uniformly distributed[c] through the container and that the system will be characterized by a state of equilibrium, i.e., a fixed energy and a uniform pressure and temperature. From a molecular point of view the pressure arises from the random impacts of the molecules against the walls, and the energy of the

---

[a] This would include two electrons and one nucleus per atom. For a simpler case in which we treated the molecule as a unit it would be less by a factor of 3.

[b] From a quantum-mechanical viewpoint, the problem is not essentially different, but in fact more information is required. Since the nucleus and electrons cannot be treated as point particles, an additional set of three parameters must be specified for each such particle to describe the orientation of the axis of spin of the particle in space.

[c] We ignore here the variations in density caused by gravitational fields (usually insignificant) and the small variations in density near the walls that arise from forces of attraction or repulsion.

system is simply the sum of the energies of the individual molecules. If somehow we were provided with information not about the individual molecules, but rather about the number which had a given velocity,[a] then we could show with the use of a few simple assumptions that from such information it is possible to calculate the thermodynamic properties of the gas.

Information on the number of molecules that possess a given velocity is known as a *distribution function*, in this case a *velocity distribution function*. In the present instance such information quickly gives us the total kinetic energy of the system, since this is the arithmetic sum of all the individual kinetic energies. If $N(v_1)$ represents the number of molecules that have velocity v_1, the total kinetic energy (assuming identical masses) is

$$KE = \tfrac{1}{2}mv_1^2 N(v_1) + \tfrac{1}{2}mv_2^2 N(v_2) + \cdots$$

$$= \tfrac{1}{2}m \sum_i v_i^2 N(v_i) \qquad (VI.2.1)$$

or if $N(v_i)$ is given as a continuous function of v, we can replace the sum by the integral

$$KE = \int_0^\infty v^2 N(v)\, dv \qquad (VI.2.2)$$

where $N(v)\, dv$ is now the number of molecules with velocities between v and $v + dv$.

We shall show later that the pressure exerted by such an ideal gas is given by

$$P = \frac{1}{3}\frac{m}{V}\int_0^\infty v^2 N(v)\, dv = \frac{2}{3}\frac{KE}{V} \qquad (VI.2.3)$$

where V represents the volume occupied by the gas.

For the equilibrium properties of an ideal gas it is thus the distribution function of the velocities which is required. For nonideal gases or liquids, a position-distribution function is needed; for a system not at equilibrium but changing in time, distribution functions of velocity and position which were the proper functions of time would similarly serve to establish the properties of the system.

3. Some Simple Distribution Functions

Before proceeding further let us examine some very simple distribution functions. If we have a uniform rectangular rod of cross section α, density ρ, and length l, the matter in the rod is said to be uniformly distributed along its length. We can express this distribution of matter (per unit length) by

---

[a] We here ignore the internal energy of the molecules and the forces between them, which means that we treat the gas as ideal and monatomic. The inclusion of the internal energy does not complicate the problem, but the inclusion of the forces does (Sec. IX.5).

the constant $\alpha\rho$. If we construct a coordinate system whose x axis starts at one end of the rod and goes through the rod's center, then the rod spans the x axis from $x = 0$ to $x = l$ (Fig. VI.1). If $P(x)$ represents the mass distribution along the x axis such that the product $P(x)\,dx = $ the mass between x and $x + dx$, then

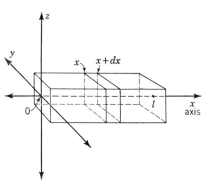

$$P(x) = \alpha\rho \qquad 0 \leqslant x \leqslant l \qquad (VI.3.1)$$

$P(x)$ is said to be a continuous, uniform, and bounded distribution. The total mass of the rod is obtained by integrating $P(x)\,dx$ between its bounds 0 and l:[a]

Fig. VI.1.

$$M = \int_0^l P(x)\,dx = \int_0^l \alpha\rho\,dx = \alpha\rho l$$
$$(VI.3.2)$$

A nonuniform distribution may be illustrated by assigning not a constant density ρ, but for example a density $\rho = ax$, that is, a density proportional to the distance from the origin. In this case $P(x) = \alpha\rho = a\alpha x$, and the total mass is

$$M = \int_0^l P(x)\,dx = \int_0^l a\alpha x\,dx = \frac{a\alpha l^2}{2} \qquad (VI.3.3)$$

4. Distributions in More Than One Variable

A distribution in more than one variable may be illustrated by the function describing the distribution of matter in a sphere with its center at the origin of a cartesian system. If the radius is c and density ρ, then the distribution of matter $P(x,y,z)$ is a function of three variables with the definition that $P(x,y,z)\,dx\,dy\,dz$ is the amount of matter in the element of volume bounded by the set of six cartesian planes through the points (x,y,z) and $(x + dx,\ y + dy,\ z + dz)$. The mass-distribution function $P(x,y,z)$ is then equal to the density ρ, for all points satisfying the condition $x^2 + y^2 + z^2 \leqslant c^2$. If we are interested in the distribution of matter in only two dimensions, we can obtain a function $P(x,y)$ from $P(x,y,z)$ by integrating over all z:

$$P(x,y) = \int P(x,y,z)\,dz \qquad (VI.4.1)$$

The limits of this integration must be between $z = \pm(c^2 - x^2 - y^2)^{\frac{1}{2}}$. On integration we find

[a] Formally we could integrate between $-\infty$ and ∞ if $P(x)$ had been defined for the entire range. This could be done by defining $P(x) = 0$ for $x < 0$ and for $x > l$. In this case $P(x)$ would no longer be continuous.

$$P(x,y) = \left[\rho z\right]_{-(c^2-x^2-y^2)^{1/2}}^{(c^2-x^2-y^2)^{1/2}} = 2\rho(c^2 - x^2 - y^2)^{1/2} \qquad (VI.4.2)$$

$P(x,y)$ has the significance that $P(x,y)\,dx\,dy$ represents the total amount of matter bounded by the four cartesian planes that go through the successive pairs of points (x,y), $(x,y + dy)$, $(x + dx,y)$, and $(x + dx, y + dy)$ and are parallel to the z axis.

In a similar fashion we can define $P(x)$, the distribution function along the x axis, by integrating $P(x,y)$ over y.

$$P(x) = \int P(x,y)\,dy$$

$$= 2\rho \int (c^2 - x^2 - y^2)^{1/2}\,dy$$

$$= \rho \left[y(c^2 - x^2 - y^2)^{1/2} + (c^2 - x^2)\sin^{-1}\frac{y}{(c^2 - x^2)^{1/2}} \right]_{-(c^2-x^2)^{1/2}}^{(c^2-x^2)^{1/2}}$$

$$= \pi\rho(c^2 - x^2) \qquad (VI.4.3)$$

We can now verify that $P(x)$ represents the correct distribution function of matter along the x axis by a final integration:[a]

$$\int_{-c}^{c} P(x)\,dx = \pi\rho \int_{-c}^{c} (c^2 - x^2)\,dx = \pi\rho \left[c^2 x - \frac{x^3}{3} \right]_{-c}^{c}$$

$$= \frac{4\pi}{3}\rho c^3 \qquad (VI.4.4)$$

which we recognize as the total mass M of the sphere.

5. Moments of a Distribution Function

The nth moment of a variable x which is distributed over a range of values according to a function $P(x)$ is defined by

[a] In similar fashion we could have defined the distribution functions $P(x,z)$, $P(y,z)$, $P(y)$, and $P(z)$. By generalizing for a distribution function of n variables $P(x_1, x_2, \ldots, x_n)$, we can define n different functions of $n - 1$ variables; $n(n - 1)/2$ of two variables, etc. Thus,

$$P(x_1,x_2,x_3) = \int_{x_4} \cdots \int_{x_n} P(x_1,x_2,x_3, \ldots, x_n)\,dx_4\,dx_5\,dx_6 \cdots dx_n$$

For any given problem we will select that function which involves the variables in which we are interested.

In dealing with distributions in more than one variable it is important to define the property of independence. The variables will be said to be independent if the distribution function may be written as a product of functions of each of one of the variables. Thus if we can write

$$P(x,y,z) = P(x)P(y)P(z)$$

then the variables and the distributions in the variables are independent of each other. This is true for the function $P(x,y,z)$ for the distribution of matter in a uniform sphere. However the functions $P(x,y)$, $P(y,z)$, $P(x,z)$ are not functions of independent variables.

$$\overline{x^n} = \int x^n P(x) \, dx \tag{VI.5.1}$$

where the limits of integration are the limits of the range of values accessible to x.

We can recognize that the first moment $\bar{x}$ is what is commonly known as the mean, or average, value of x. Similarly $\overline{x^n}$ would correspond to the average value of x^n. This assumes that the distribution function $P(x)$ is the "relative" or "normalized" distribution function; i.e., it satisfies the condition[a]

$$\int P(x) \, dx = 1 \tag{VI.5.2}$$

$\bar{x}$ is also known as the "center of gravity" of the distribution. If we consider the quantity $x - \bar{x}$ (i.e., the difference between any value of x and its average value), its average value will be zero:

$$\overline{(x - \bar{x})} = \int (x - \bar{x})P(x) \, dx = \int xP(x) \, dx - \int \bar{x}P(x) \, dx$$

$$= \bar{x} - \bar{x} \int P(x) \, dx = \bar{x} - \bar{x} = 0 \tag{VI.5.3}$$

Similarly, the average value of $(x - \bar{x})^2$ is given by:

$$\overline{(x - \bar{x})^2} = \int (x - \bar{x})^2 P(x) \, dx = \int (x^2 - 2x\bar{x} + \bar{x}^2)P(x) \, dx$$

$$= \int x^2 P(x) \, dx - 2\bar{x} \int xP(x) \, dx + \bar{x}^2 \int P(x) \, dx$$

$$= \overline{x^2} - 2\bar{x} \cdot \bar{x} + \bar{x}^2 = \overline{x^2} - \bar{x}^2 \tag{VI.5.4}$$

The average value $\overline{(x - \bar{x})^2}$ is thus equal to the difference between the second moment $\overline{x^2}$ and the square of the first moment $\bar{x}^2$. It can only be zero if all values of $x = \bar{x}$. Thus it is a measure of the spread of the distribution about the center of the distribution $\bar{x}$. It is usually designated by the symbol σ^2 and called the dispersion.[b]

The quantities $\overline{(x - \bar{x})^n} = \int (x - \bar{x})^n P(x) \, dx$ are known as the central moments of the distribution, since they are taken about the mean value or center $\bar{x}$. When $\bar{x} = 0$, they reduce to the regular moments previously defined. It should be observed that, if the distribution is symmetrical about its center [that is, $P(+x) = P(-x)$] then all odd moments are zero.[c]

[a] Any distribution function may be normalized by dividing it by the value of the total quantity. Thus if $\int P(x) \, dx = p_0 \neq 1$, then $P(x)$ divided by the constant p_0 [for example, $P(x)/p_0$] will be normalized.

[b] σ is known as the standard deviation of the distribution.

[c] This follows since

$$\int_{-b}^{b} x^{2n+1} P(x) \, dx = \int_{-b}^{0} x^{2n+1} P(x) \, dx + \int_{0}^{b} x^{2n+1} P(x) \, dx$$

and on substituting the new variable $-x$ for x in the last of these integrals we see that it just cancels the preceding one:

$$\int_{0}^{b} x^{2n+1} P(x) \, dx = \int_{0}^{-b} (-x)^{2n+1} P(-x) \, d(-x) = \int_{0}^{-b} x^{2n+1} P(x) \, dx$$

$$= -\int_{-b}^{0} x^{2n+1} P(x) \, dx$$

It has been shown[a] that, if all the moments of a distribution are known, then the distribution may under certain conditions be expressed as a series in terms of the variable and the moments. The expansion is given by

$$P(x) = \frac{1}{\sigma(2\pi)^{1/2}} e^{-x^2/2\sigma^2} \left[1 + \sum_{n=3}^{\infty} \frac{\phi(\overline{x^n})}{n!} H_n \left(\frac{x}{\sigma} \right) \right] \qquad \text{(VI.5.5)}$$

where $\phi(\overline{x^n})$ are known algebraic functions of the moments and $H_n(x/\sigma)$ are the Hermite polynomial functions.[b]

6. Some Important Theorems for Distribution Functions

Change of Variable. If $P(x)$ is a distribution function in the variable x and $x = f(y)$, then the corresponding distribution function in y will be

$$Q(y) = P[f(y)]f'(y)$$

where
$$f'(y) = df/dy.$$

Independent Distributions. If $x_1, \ldots, x_n$ constitute a set of n independent variables each with an independent distribution function $P_i(x_i)$, then the joint distribution function for the set of variables is the product of the individual distribution functions:

$$P(x_1, x_2, \ldots, x_n) = P_1(x_1)P_2(x_2) \cdots P_n(x_n) = \prod_{i=1}^{n} P_i(x_i)$$

Linear Combinations. If the quantity y is taken as some linear combination of the independently distributed variables $(x_1, x_2, \ldots, x_n)$, that is,

$$y = \sum_{i=1}^{n} a_i x_i, \text{ then}$$

$$\overline{y} = \sum_{i=1}^{n} a_i \overline{x}_i$$

$$\sigma_y^2 = \overline{(y - \overline{y})^2} = \sum_{i=1}^{n} a_i^2 \sigma_{x_i}^2$$

$$\mu_y^3 = \overline{(y - \overline{y})^3} = \overline{\left[\sum_{i=1}^{n} a_i(x_i - \overline{x}_i) \right]^3}$$

$$= \sum_{i=1}^{n} a_i^3 \mu_{x_i}^3$$

$$\mu_y^4 = \sum_{i=1}^{n} a_i^4 \mu_{x_i}^4 + 6 \sum_{i \neq j=1}^{n} a_i^2 a_j^2 \sigma_{x_i}^2 \sigma_{x_j}^2$$

[a] H. Cramer, "Mathematical Methods of Statistics," chap. 17, Princeton University Press, Princeton, N.J., 1946.

[b] H. Margenau and G. M. Murphy, "The Mathematics of Physics and Chemistry," chap. 3, D. Van Nostrand Company, Inc., Princeton, N.J., 1943.

7. Illustration: Random Walk in One Dimension

The use of distribution functions is well illustrated by an important and classical problem, the "random walk." Let us assume that a marker is placed at the origin ($x = 0$) of the x axis. We shall now allow it to move one step in the $+x$ direction with a probability p and one step in the negative direction of the x axis with a probability q, such that $p + q = 1$.[a]

We are now concerned with the position of the marker with respect to its starting position after n moves have been made. If $p = q$, we would intuitively guess that the marker would be near its starting position, there being equal probability of plus and minus moves.

We define $P(x,n)$ as the probability of finding the marker at the position x after n moves have been made. $P(x,n)$ is a bounded function, since $|x|$ cannot exceed $+n$. (Note that x and n are both integers.)

Now the chance that the marker will be at the position x after n moves is exactly the chance that the total number of positive moves n_+ exceeds the total number of negative moves n_- by the quantity x:

$$n_+ - n_- = x \qquad (VI.7.1)$$

where

$$n_+ + n_- = n \qquad (VI.7.2)$$

But the chance that there will be n_+ positive moves and n_- negative moves is given by the well-known binomial formula[b]

$$\frac{(n)!}{(n_+)!(n_-)!} p^{n_+} q^{n_-} \qquad (VI.7.3)$$

so that, since $n_+ = (n + x)/2$ and $n_- = (n - x)/2$, we have

$$P(x,n) = \frac{n!}{[(n + x)/2]![(n - x)/2]!} p^{(n+x)/2} q^{(n-x)/2} \qquad (VI.7.4)$$

$P(x,n)$ is the famous binomial or Bernouilli distribution, and it has a very useful asymptotic form when n is a very large number such that, as $n \to \infty$, $x^3/n^2 \to 0$.[c] We can then write

$$P(x,n) = \frac{1}{2(2\pi nqp)^{1/2}} e^{-(x-\bar{x})^2/8npq} \qquad (VI.7.5)$$

where it can be found from Eq. (VI.7.4) that $\bar{x} = n(p - q)$.

[a] This can be done, for example, if we move it in accord with the results of a random game. Thus if we have a box containing p black balls and q white balls and select a ball at random such that we move the marker one unit toward $+x$ every time a black ball is selected and one unit toward $-x$ when a white ball is selected. The ball is, of course, returned and the box is shaken after each choice. The distribution function obtained here is known as a *probability distribution*.

[b] The chance of obtaining n_+ independent events each of probability p and n_- independent events each of probability q is $p^{n_+} q^{n_-}$ multiplied by the number of different ways in which these events can occur, which is given by the coefficient above, $n!/[(n_+)!(n_-)!]$, also abbreviated $C_{n_+}^n$ or $\binom{n}{n_+}$.

[c] W. Feller, "An Introduction to Probability Theory and Its Application," chap. 7, John Wiley & Sons, Inc., New York, 1950.

If we make the substitution[a] $\sigma = 2(npq)^{1/2}$, this equation can be written as

$$P(n,x) = \frac{1}{\sigma(2\pi)^{1/2}} e^{-(x-\bar{x})^2/2\sigma^2} \qquad \text{(VI.7.6)}$$

In the special case that $p = q = \frac{1}{2}$ (i.e., the chances are equal for positive or negative moves) $\bar{x} = 0$ and $\sigma = n^{1/2}$, so that $P(n,x)$ may then be written as

$$P(n,x) = \frac{1}{n^{1/2}(2\pi)^{1/2}} e^{-x^2/2n} \qquad \text{(VI.7.7)}$$

The asymptotic expression for $P(n,x)$ given in Eqs. (VI.7.5) to (VI.7.7) may be recognized as the Gaussian distribution. It is very important for many physical problems, and we shall discuss it in some detail. The distribution has its center at $\bar{x} = n(p - q)$ and is in its approximate form symmetrical about $\bar{x}$. The exact form is not symmetrical about $\bar{x}$ but is nearly so for values of $x - \bar{x}$ that are small compared to n.

TABLE VI.1. COMPARISON OF THE BERNOUILLI AND GAUSSIAN DISTRIBUTIONS[b]

$x - \bar{x}$	Bernouilli distribution, Eq. (VI.7.4)	Gaussian distribution, Eq. (VI.7.7)
2	0.2301	0.2341
−2	0.2379	
4	0.1489	0.1445
−4	0.1409	
6	0.0589	0.0590
−6	0.0591	
8	0.0170	0.0160
−8	0.0143	
10	0.00343	0.00283
−10	0.00201	
12	0.00049	0.00033
−12	0.00015	
14	0.00005	0.00003
−14	0.000006	

Table VI.1 shows a comparison of the asymptotic and exact forms for the unfavorable case that $n = 100$ and $p = 0.7$. In this case x is even, $q = 0.3$, and $\bar{x} = 40$.

[a] There is a factor of 2 introduced here and also in Eq. (VI.7.5) for $P(x,n)$ (in the denominator of the coefficient) to take account of the fact that for any given n, x is restricted to half of the possible integral values between $+n$ and $-n$. If n is odd, then x must be odd; if n is even, then x must be even by the terms of its definition $x = n_+ - n_-$.

[b] Figures calculated from W. Feller, "An Introduction to Probability Theory and Its Application," chap. 7, John Wiley & Sons, Inc., New York, 1950.

It can be shown quite readily that, if the steps are of unequal length, the same form of asymptotic expression [Eq. (VI.7.6) or (VI.7.7)] is obtained except that in the place of n or σ we must write αn or $\alpha \sigma$, where α is the rms average of all the steps taken. That is[a]

$$\alpha^2 = \frac{x_1^2 + x_2^2 + \cdots + x_n^2}{n}$$

Finally it is possible to generalize the problem to two, three, or more dimensions under the provision that the motions are all made independently of each other. In this case the distributions for each dimension are independent and the total function is the product of the individual functions. The function for a random walk in three dimensions, with variable step, is then given by

$$P(x,y,z) = \frac{1}{(n_x n_y n_z)^{1/2}(2\pi)^{3/2}} e - (x^2/2n_x + y^2/2n_y + z^2/2n_z) \quad \text{(VI.7.8)}$$

where n_x is the total number of steps taken in the x direction multiplied by the rms average length of each step and n_y and n_z are similarly defined.

8. The Normal Law of Error

The distribution function for the random walk gives us an immediate approach to the theory of random errors. In any physical experiment there may be a number of factors disturbing an observation, with each one contributing an error of magnitude, let us say, ε_i (for the ith source) which may be positive or negative. The result of all of these individual errors is to produce a total error $x = \Sigma \varepsilon_i$ such that our observed measurement, call it m, differs from what we presume is the true value $\bar{m}$ by the amount x:[b]

$$m - \bar{m} = x \quad \text{(VI.8.1)}$$

If we assume that each of the individual errors may be assigned an equal probability of being positive or negative and all act independently, then the question of finding the distribution of possible errors is given precisely by the answer to the random walk problem with varied step. That is, the chance of making an error x is given by

$$P(x) = \frac{1}{\sigma(2\pi)^{1/2}} e^{-x^2/2\sigma^2} \quad \text{(VI.8.2)}$$

This Gaussian distribution function is symmetrical about the true value $\bar{m}$ (here choosen as the origin for x) and thus implies that positive errors are as probable as negative errors. It is normalized, since[c]

[a] See, for example, E. H. Kennard, "Kinetic Theory of Gases," chap. 7, McGraw-Hill Book Company, Inc., New York, 1938.

[b] This assumes, of course, that a true value $\bar{m}$ exists.

[c] Strictly speaking, we should integrate between $x = \pm \alpha n$ (α = rms error, n = number of factors contributing), but for large values of n—and it is only for these that the formula is valid—the infinite limits may be taken.

$$\frac{1}{\sigma(2\pi)^{\frac{1}{2}}}\int_{-\infty}^{+\infty} e^{-x^2/2\sigma^2}\,dx = 1 \qquad\qquad \text{(VI.8.3)}$$

If we calculate the moments, we find

$$\bar{x} = \overline{x^3} = \overline{x^{2n+1}} = 0 \qquad\qquad \text{(VI.8.4)}$$

i.e., all odd moments are zero, since $P(x)$ is symmetric about the origin. The second moment is[a]

$$\overline{x^2} = \int_{-\infty}^{\infty} \frac{x^2}{\sigma(2\pi)^{\frac{1}{2}}} e^{-x^2/2\sigma^2}\,dx = \sigma^2 \qquad\qquad \text{(VI.8.5)}$$

This Gaussian function is shown plotted in Fig. VI.2. The maximum value of $P(x)$ [obtained by differentiating Eq. (VI.8.2) and solving the

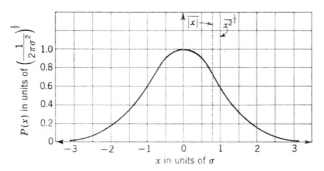

FIG. VI.2. Plot of the Gaussian distribution function

$$P(x) = \frac{1}{\sigma(2\pi)^{\frac{1}{2}}} e^{-x^2/2\sigma^2}$$

Ordinates: $P(x)$ in units of $\left(\dfrac{1}{2\pi\sigma^2}\right)^{\frac{1}{2}}$; abscissa: x in units of σ.

equation $dP/dx = 0$] occurs at $x = 0$. This is also then the most probable error; i.e., the most frequently occurring error is zero. Since $\bar{x} = 0$, the average error is also zero. The mean absolute error, defined as the average of $|x|$ (the absolute value of x) can be obtained from

$$\overline{|x|} = \int_{-\infty}^{\infty} |x| P(x)\,dx = 2\int_{0}^{\infty} xP(x)\,dx$$

$$= \sigma\left(\frac{2}{\pi}\right)^{\frac{1}{2}} = 0.798\sigma \qquad\qquad \text{(VI.8.6)}$$

The *half-width* of the distribution is defined as the value of x at which $P(x)$ has fallen to half its maximum value. We find $x_{\frac{1}{2}} = \sigma(2\ln2)^{\frac{1}{2}} = 1.175\sigma$. The *mean width* is given by the value of x at which $P(x)$ has fallen to $1/e$ of its maximum value. We find $x_{1/e} = \sigma\sqrt{2} = 1.414\sigma$. From a

[a] See Appendix B for formulas for the different moments and related integrals.

practical point of view we are interested in the area under the curve for different values of x; it is defined by the function

$$\text{Erf}(x) = \int_{-x}^{x} P(x)\, dx = 2 \int_{0}^{x} P(x)\, dx$$

$$= \frac{2}{\sigma(2\pi)^{\frac{1}{2}}} \int_{0}^{x} e^{-x^2/2\sigma^2}\, dx = \frac{2}{\pi^{\frac{1}{2}}} \int_{0}^{x} e^{-y^2}\, dy \qquad \text{(VI.8.7)}$$

The function $\text{Erf}(x)$ is called the probability integral or the error function, and standard tables for it have been prepared.[a] $\text{Erf}(x)$ represents the chance of finding an error of less than x in absolute magnitude. We have $\text{Erf}(0) = 0$, and $\text{Erf}(\infty) = 1$. $\text{Erf}(x)$ will have the value $\frac{1}{2}$ when $x = 0.675\sigma$, that is, half of the errors made will exceed 0.675σ in absolute magnitude.

The chance that an error which lies between x_1 and x_2 will be made is given by the area under the curve bounded by x_1 and x_2. If we call $E(x_1,x_2)$ the chances of such an error,

$$E(x_1,x_2) = \int_{x_1}^{x_2} P(x)\, dx = \frac{1}{2}[\text{Erf}(x_2) \pm \text{Erf}(x_1)] \qquad \text{(VI.8.8)}$$

The use of the plus or minus sign will depend on whether or not x_1 and x_2 have or do not have the same sign.

Some typical values of both $P(x)$ and $\text{Erf}(x)$ are shown in Table VI.2, from which we can see that the chance of having an error in excess of $\pm 2\sigma$ is only 46 in 1000 ($1 - 0.954$), for an error in excess of $\pm 4\sigma$, only 6.4 in 100,000. It can be seen why in laboratory practice it is customary to ignore results whose deviation from the mean exceeds the expected deviation σ (precision) by a factor of 4.

TABLE VI.2. SOME VALUES OF THE GAUSSIAN DISTRIBUTION AND ERROR FUNCTION

x/σ	$P(x)$	$\text{Erf}(x)$	x/σ	$P(x)$	$\text{Erf}(x)$
0.0	0.399	0.000	1.6	0.1109	0.890
0.2	0.391	0.158	1.8	0.0790	0.928
0.4	0.368	0.311	2.0	0.0540	0.954
0.6	0.333	0.451	2.5	0.0175	0.9876
0.8	0.290	0.576	3.0	0.00443	0.99730
1.0	0.242	0.683	3.5	0.000873	0.999534
1.2	0.194	0.770	4.0	0.000134	0.999936
1.4	0.1497	0.838	4.5	0.000016	0.999994

Note: $\qquad \text{Erf}(x) = \dfrac{2}{\pi^{\frac{1}{2}}} \int_{0}^{x} e^{-y^2}\, dy \qquad P(x) = \dfrac{\text{Erf}'(x)}{2} = \dfrac{e^{-x^2}}{\pi^{\frac{1}{2}}}$

Table VI.2 also shows that 15.8 per cent of all the errors will be less than 0.2σ in absolute magnitude and that 31.1 per cent of all random errors will be less than 0.4σ in absolute magnitude.

[a] See, for example, the Chemical Rubber "Handbook of Physics and Chemistry," or B. O. Peirce, "Standard Table of Integrals."

Any linear combination of quantities each of which is distributed according to the Gaussian curve is itself distributed according to the Gaussian curve (law of addition). Thus if $z = x_1 + x_2 + x_3 + \cdots + x_n$ where each quantity x_i is distributed according to

$$P(x_i) = \frac{1}{\sigma_i(2\pi)^{1/2}} \exp\left[-\frac{(x_i - \bar{x}_i)^2}{2\sigma_i^2} \right]$$

then z is distributed according to the law[a]

$$P(z) = \frac{1}{\sigma_z(2\pi)^{1/2}} \exp\left[-\frac{(z - \bar{z})^2}{2\sigma_z^2} \right] \tag{VI.8.9}$$

where
$$\bar{z} = \bar{x}_1 + \bar{x}_2 + \cdots + \bar{x}_n \tag{VI.8.10}$$
and
$$\sigma_z^2 = \sigma_1^2 + \sigma_2^2 + \cdots + \sigma_n^2 \tag{VI.8.11}$$

If $x_1, \ldots, x_n$ are a number of independent variables, each distributed in an arbitrary manner but with a common center, then the quantity $z = x_1 + x_2 + \cdots + x_n$ will be distributed about the same center with a distribution which approaches the Gaussian distribution as the number of variables x_n becomes infinite. This property has been proven and is known as the *central limit theorem*.[b] In practice even three or four variables will quickly combine to give the Gaussian form. The result is that in practice most symmetrical distributions are nearly indistinguishable from the Gaussian.

9. Note: Stirling's Approximation

We shall have need for an analytic representation of a factorial (for example, $x!$) in terms of known functions. For these purposes we can avail

[a] This can be proved for two variables, $z = x_1 + x_2$, as follows. The product $P(x_1)P(x_2)\, dx_1\, dx_2$ gives the chances that x_1 lies between x_1 and $x_1 + dx_1$ and that simultaneously x_2 lies between x_2 and $x_2 + dx_2$. The chances that z lies between z_1 and $z_1 + dz_1$ are then equal to this product integrated over all values of $x_1(\pm \infty)$ when x_2 is equal to $z - x_1$ and $x_2 + dx_2 = z - x_1 + dz$ (that is, $dx_2 = dz$).

$$P(z)\, dz = dz \int_{-\infty}^{\infty} P(x_1)P(z - x_1)\, dx_1$$

$$P(z) = \frac{1}{\sigma_1\sigma_2(2\pi)} \int_{-\infty}^{\infty} \exp\left[-\frac{(x_1 - \bar{x}_1)^2}{2\sigma_1^2} \right] \exp\left[-\frac{(z - x_1 - \bar{x}_2)^2}{2\sigma_2^2} \right] dx_1$$

which on integration reduces to

$$P(z) = \frac{1}{\sigma_z(2\pi)^{1/2}} \exp\left[-\frac{(z - \bar{z})^2}{2\sigma_z^2} \right]$$

where $\sigma_z^2 = \sigma_1^2 + \sigma_2^2$ and $\bar{z} = \bar{x}_1 + \bar{x}_2$. By induction this can now be extended to any number of variables with the results quoted.

[b] See, for example, H. Jeffrey, "Theory of Probability," p. 85, Oxford University Press, New York, 1948. See also Feller, *op. cit.*, chap. 10.

ourselves of the expansion known as Stirling's approximation. To a very high degree of accuracy the quantity $x! = x(x-1)(x-2) \cdots 1$ is given by

$$x! = (2\pi x)^{1/2} \left(\frac{x}{e}\right)^x e^{1/12x} \qquad \text{(VI.9.1)}$$

where e is the natural number. Another less accurate but more frequently used approximation is

$$x! = (2\pi x)^{1/2} \left(\frac{x}{e}\right)^x \qquad \text{(VI.9.2)}$$

or in logarithmic form[a]

$$\ln x! = \tfrac{1}{2} \ln (2\pi x) + x \ln x - x \qquad \text{(VI.9.3)}$$

The relative accuracy of these forms can be seen in Table VI.3.

TABLE VI.3

x	$x!$	$x!$ Eq. (VI.9.2)	$x!$ Eq. (VI.9.1)	Error, % Eq. (VI.9.2)	Error, % Eq. (VI.9.1)
1	1	0.922	1.0023	-8	0.2
2	2	1.919	2.0007	-4	0.04
5	120	118.02	120.01	-2	0.01
10	3.629×10^6	3.599×10^6	3.629×10^6	-0.8	0.000
100	9.333×10^{157}	9.325×10^{157}	9.333×10^{157}	-0.08	—

For chemical problems in which x is likely to be of the order of magnitude of Avogadro's number, a quite good approximation to $\ln(x!)$ is taken as

$$\ln(x!) \cong x \ln x - x \qquad \text{(VI.9.4)}$$

While the absolute error in $x!$ now increases with increasing x, the relative error decreases because $x!$ increases faster than $\ln (2\pi x)$.

[a] For very large values of x such as occur in molecular problems, the first term is negligible compared to x and we can write $\ln (x!) = x \ln x - x$ without appreciable error.

VII

The Kinetic Theory of Gases

1. Molecular Models

As is pointed out in Sec. VI.1, the exact treatment of an assembly of molecules is a mathematical impossibility. If we are to make any progress in the theoretical treatment of matter, we must somehow simplify the problem. This can be done (in a number of different ways) by replacing the exact description of a molecule (in terms of electrons, nuclei, etc.) by some mathematically more amenable substitute. Such a simplification, known as a model treatment, must then be justified, first, by showing that the model chosen resembles very closely the properties of a real molecule and, second, by showing (if possible) that under given conditions the relative errors made in the simplification are small. While such a procedure leaves much to be desired from a rigorous point of view, it has much to recommend it from the point of view of the insight to molecular behavior that it gives. Some of the more useful molecular models which have been employed for statistical studies are described in the following sections.

1A. Ideal Gas Model. This model pictures a molecule as a point particle (dimensionless) that has a mass equal to the molecular weight, exerts no forces on other molecules, and is capable of completely elastic collisions with the walls of a containing vessel. The ability of the model to represent the properties of matter depends on the property chosen and the experimental conditions. Thus the model will represent quite well the pressure-volume-temperature relation for gases under conditions that the average distance between molecules is large compared to their "diameters" and the temperature is not close to the condensation point. But it is obviously useless for giving any information about molecular collisions.

1B. Hard Sphere Model. Here the molecule is assumed to be the equivalent of a billiard ball. That is, the molecule is presented as a rigid sphere of diameter σ, mass m (the molecular weight), and the capability

135

of only perfectly elastic collisions with other molecules and the container walls.[a] This is a very popular model for investigating molecular collisions, but it cannot account for condensed phases (liquids and solids), because it assumes no force of attraction but only an infinite repulsive force when two molecules collide. Its advantage is that it characterizes the molecule by a single parameter σ, the molecular diameter.

1C. Hard Sphere Model with Central Attractive Forces. A great improvement over the hard sphere model physically, but one that is much more difficult to deal with mathematically, is the hard sphere molecule which is capable of exerting attractive forces, centrally directed (i.e., the force between two molecules depends only on the distance between them and is directed along their line of centers). The molecule is still spherical, and the closest distance of approach of two molecules is given by their mean diameter. This model can now account for the properties of condensed states.

1D. Symmetrical Molecule with Central Forces. If we discard the idea of a hard sphere and replace it by a molecule that is capable of exerting both attractive and repulsive forces but acts centrally, we have the closest approach yet to real molecules, and also the model that is most difficult to treat. Such a molecule is characterized completely by the function chosen to represent its force field. A function commonly used is the Lennard-Jones function

$$F(r) = -\frac{a}{r^7} + \frac{b}{r^n} \qquad (VII.1D.1)$$

where $F(r)$ is the force acting between two molecules, a and b are experimental constants, and n is a number usually between 9 and 13 whose choice depends on the substance. The first term (negative sign) represents an attraction between molecules, and the second term represents a repulsion. There will be some distance r_0 at which $F(r_0) = 0$, and the formula may be rewritten in terms of the dimensionless variable $\rho = r_0/r$:

$$F(\rho) = -\frac{a}{r_0^7} (\rho^7 - \rho^n) \qquad (VII.1D.2)$$

1E. Rectangular Well Model. A compromise between the oversimple hard sphere model, which is mathematically tractable, and the Lennard-Jones potential, which is very difficult to treat mathematically, is the rectangular well model. The force between two molecules is everywhere zero except at two distances of separation, σ_a, at which the force is infinitely large and attractive, and σ_r, at which the force is infinitely large and re-

---

[a] Since such a molecule has rotational energy, various authors have used the "completely smooth" billiard ball model which is incapable of changing its rotational energy on collisions (i.e., it slips), while others have used a "rough" billiard ball with a coefficient α_R ($0 < \alpha_R \leq 1$) to describe the extent to which its rotational energy is involved in a collision.

pulsive. In terms of $U(r)$, the energy of interaction of the two molecules, $F(r) = -dU/dr$, we can write

$$U(r) = 0 \qquad r > \sigma_a$$
$$U(r) = -U_0 \qquad \sigma_r < r < \sigma_a \qquad \text{(VII.1E.1)}$$
$$U(r) = \infty \qquad r < \sigma_r$$

This model is characterized by three parameters: the well depth U_0, the range of the attractive forces σ_a, and the hard sphere radius σ_r. Probably because of this it is able to represent many equilibrium and transport properties of real molecules with semiquantitative accuracy.

In our development of the kinetic theory we shall begin with the simplest of these models, that of the point particle.

2. Velocity Distribution of an Ideal Gas; Cartesian Coordinates

For an ideal gas we shall take the point particle with no forces between molecules. A real gas will approach this behavior when its density is such that the average distance between molecules is large compared to their diameter σ. This means that the volume of the vessel is very large compared to the total volume occupied by the molecules themselves, or $V \gg N\pi\sigma^3/6$, where N = number of molecules in the volume V.

If now we assume that the molecules move completely at random and independently of each other and that the system is at equilibrium and isolated (no exchange of energy between the gas and its environment) then the total energy of the gas is simply the kinetic energy attributable to the random motion of the molecules.[a] This total energy and the volume then fix completely the thermodynamic properties of the gas. If now we could know the probability distribution function for the molecular velocities (at equilibrium), that would determine uniquely the properties of the system.

Let us now select a three-dimensional system of cartesian axes fixed with respect to the vessel. Our assumption of complete randomness of molecular motion tells us that we shall expect to find as many molecules moving with velocity components of a given range along the x axis as along the y and z axes. That is, the motion is isotropic. If we define three distribution functions $P(v_x)$, $P(v_y)$, and $P(v_z)$ such that $P(v_x)\, dv_x$ represents the *fraction* of all the N molecules which have x velocity components between v_x and $v_x + dv_x$ and the other two functions have similar relations with v_y and v_z, the assumption of randomness tells us that the three functions are the same for the different components. The assumption of independent motion further tells us that the fraction of all molecules with

---

[a] We ignore here the internal energy which the molecules may have. This is justifiable if we assume that we have only perfectly elastic collisions between the molecules (no loss of kinetic energy on collision).

three velocity components between v_x and $v_x + dv_x$, v_y and $v_y + dv_y$, v_z and $v_z + dv_z$ is then the product

$$P(v_x,v_y,v_z)\, dv_x\, dv_y\, dv_z = P(v_x)P(v_y)P(v_z)\, dv_x\, dv_y\, dv_z \quad \text{(VII.2.1)}$$
or
$$P(v_x,v_y,v_z) = P(v_x)P(v_y)P(v_z)$$

since the motions along the three axes are assumed independent.

The scalar speed c of a molecule with vector velocity components v_x, v_y, v_z is given by $c^2 = v_x^2 + v_y^2 + v_z^2$. The assumption that space is isotropic for molecular velocities implies that the probability of finding a molecule of a given speed c will be independent of the direction of motion of the molecule. This in turn means that the joint distribution function $P(v_x,v_y,v_z)$ $= P(v_x)P(v_y)P(v_z)$ is constant for all combinations of the components that add up to a given speed c. Thus $P(v_x,v_y,v_z) = P(c)$, that is, it depends only on c and not on the distribution of c among its spatial components. This prescribes the functional form for $P(v_x)$, $P(v_y)$, and $P(v_z)$, and we can derive it as follows:

For any selected c we have the two simultaneous conditions:

$$P(v_x)P(v_y)P(v_z) = P(c) = \text{const} \quad \text{(VII.2.2)}$$
$$v_x^2 + v_y^2 + v_z^2 = c^2 = \text{const}$$

On taking the total differentials of these two equations we have

$$\frac{dP(v_x)}{dv_x} P(v_y)P(v_z)\, dv_x + \frac{dP(v_y)}{dv_y} P(v_x)P(v_z)\, dv_y + \frac{dP(v_z)}{dv_z} P(v_x)P(v_y)\, dv_z = 0 \quad \text{(VII.2.3)}$$

$$v_x\, dv_x + v_y\, dv_y + v_z\, dv_z = 0 \quad \text{(VII.2.4)}$$

The first of these two equations can be simplified by dividing through by the joint product $P(v_x)P(v_y)P(v_z)$, and it then becomes

$$\frac{d \ln P(v_x)}{dv_x} dv_x + \frac{d \ln P(v_y)}{dv_y} dv_y + \frac{d \ln P(v_z)}{dv_z} dv_z = 0 \quad \text{(VII.2.5)}$$

Equation (VII.2.4) can be solved jointly with the equation (VII.2.5) for the velocity components. The most convenient technique is Lagrange's method of undetermined multipliers. If we multiply Eq. (VII.2.4) by a fixed quantity λ and add it to Eq. (VII.2.5), we have

$$\left[\frac{d \ln P(v_x)}{dv_x} + \lambda v_x\right] dv_x + \left[\frac{d \ln P(v_y)}{dv_y} + \lambda v_y\right] dv_y$$
$$+ \left[\frac{d \ln P(v_z)}{dv_z} + \lambda v_z\right] dv_z = 0 \quad \text{(VII.2.6)}$$

In order for Eq. (VII.2.6) to hold true for all arbitrary variations in the quantities dv_x, dv_y, and dv_z, each term in brackets must vanish identically. Thus we have three similar equations in v_x, v_y, and v_z of which one is

$$\frac{d \ln P(v_x)}{dv_x} + \lambda v_x = 0 \quad \text{(VII.2.7)}$$

This can be integrated on sight to give

$$P(v_x) = Ae^{-\lambda v_x^2} \qquad \text{(VII.2.8)}$$

where A is a constant of integration to be determined by the normalization condition

$$\int_{-\infty}^{\infty} P(v_x)\, dv_x = 1 \qquad \text{(VII.2.9)}$$

On substitution and integration (Appendix B) we find that $A = (\lambda/\pi)^{1/2}$ or, writing for convenience $2\sigma^2 = 1/\lambda$,

$$P(v_x) = \frac{1}{\sigma(2\pi)^{1/2}} e^{-v_x^2/2\sigma^2} \qquad \text{(VII.2.10)}$$

which can be recognized as our Gaussian distribution function of dispersion $\overline{v_x^2} = \sigma^2$.

Similar equations hold for $P(v_y)$ and $P(v_z)$, so that we can write

$$P(v_x, v_y, v_z) = \frac{1}{\sigma^3(2\pi)^{3/2}} e^{-(v_x^2 + v_y^2 + v_z^2)/2\sigma^2} \qquad \text{(VII.2.11)}$$

It has become standard practice to write this last equation in terms of a quantity α, defined by $\alpha^2 = 2\sigma^2$:

$$P(v_x, v_y, v_z) = \frac{1}{\alpha^3 \pi^{3/2}} e^{-(v_x^2 + v_y^2 + v_z^2)/\alpha^2} \qquad \text{(VII.2.12)}$$

We see that for the individual components the most probable velocity is the same as the mean velocity, namely, zero ($\bar{v}_x = \bar{v}_y = \bar{v}_z = 0$). That means that the most frequently observed component in a trial sampling of the gas will be zero. Similarly, by drawing on the properties of the Gaussian distribution (Sec. VI.8), the rms components are $\alpha/2^{1/2}$:

$$\overline{v_x^2}^{1/2} = \overline{v_y^2}^{1/2} = \overline{v_z^2}^{1/2} = \frac{\alpha}{2^{1/2}}$$

The mean absolute velocity component is given by

$$\overline{|v_x|} = \overline{|v_y|} = \overline{|v_z|} = \frac{\alpha}{\pi^{1/2}} = 0.564\alpha \qquad \text{(VII.2.13)}$$

3. Velocity Distribution in Spherical Coordinates

$P(v_x, v_y, v_z)\, dv_x\, dv_y\, dv_z$ represents the fraction of all molecules that have velocities in the range v_x to $v_x + dv_x$, v_y to $v_y + dv_y$, and v_z to $v_z + dv_z$. For many purposes we are interested in expressing this in spherical coordinates. That is, we would like to know the fraction of molecules with velocity vectors in the range c to $c + dc$, θ to $\theta + d\theta$, and ϕ to $\phi + d\phi$ where c, θ, and ϕ are the spherical coordinates of the velocity vector. The relation between these systems is illustrated in Fig. VII.1. Algebraically we have

$$v_x = c \sin \phi \cos \theta \qquad v_y = c \sin \phi \sin \theta \qquad v_z = c \cos \phi \qquad \text{(VII.3.1)}$$

where θ has the range 2π to 0, ϕ has the range π to 0, and c the range 0 to ∞ (nonnegative). From these relations we can show that

$$c^2 = v_x^2 + v_y^2 + v_z^2$$
$$dv_x\, dv_y\, dv_z = c^2 \sin \phi \; d\phi \; d\theta \; dc \tag{VII.3.2}$$

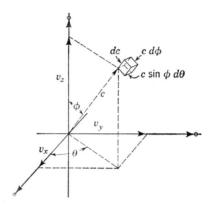

FIG. VII.1. Relation between cartesian coordinates and spherical coordinates.

so that we can write for the function $P(c,\theta,\phi)$ on substitution:

$$P(v_x,v_y,v_z)\, dv_x\, dv_y\, dv_z \rightarrow \frac{c^2 \sin \phi}{\alpha^3 \pi^{3/2}}\, e^{-c^2/\alpha^2}\, d\theta\, d\phi\, dc \tag{VII.3.3}$$

and

$$P(c,\theta,\phi) = \frac{c^2 \sin \phi}{\alpha^3 \pi^{3/2}}\, e^{-c^2/\alpha^2}$$

From this we can obtain the distribution function $P(c,\phi)$ by integrating over θ and the function $P(c)$ by integrating again over ϕ (Sec. VI.4):

$$P(c,\phi) = \int_0^{2\pi} P(c,\theta,\phi)\, d\theta = \frac{2c^2 \sin \phi}{\alpha^3 \pi^{1/2}}\, e^{-c^2/\alpha^2}$$
$$P(c) = \int_0^{\pi} P(c,\phi)\, d\phi = \frac{4c^2}{\alpha^3 \pi^{1/2}}\, e^{-c^2/\alpha^2} \tag{VII.3.4}$$

$P(c)$ will be of particular interest to us because it represents the fraction of molecules that have *speeds* in the range c to $c + dc$. It is no longer a Gaussian distribution but instead has the shape shown in Fig. VII.2, being zero at the origin. The function $P(v_x)$ is shown plotted for comparison.

The most probable speed c_p is obtained by solving

$$\frac{d\, P(c)}{dc} = 0 = \frac{4}{\alpha^3 \pi^{1/2}}\, e^{-c^2/\alpha^2} \left(2c - \frac{2c^3}{\alpha^2} \right)$$

from which we find $c_p = \alpha$ (negative value not allowed). The average speed $\bar{c}$ is given by

$$\bar{c} = \int_0^\infty cP(c)\, dc = \frac{4}{\alpha^3\pi^{1/2}} \int_0^\infty c^3 e^{-c^2/\alpha^2}\, dc \qquad \text{(VII.3.5)}$$

and by setting $y^2 = c^2/\alpha^2$ we find

$$\bar{c} = \frac{4\alpha}{\pi^{1/2}} \int_0^\infty y^3 e^{-y^2}\, dy \qquad \text{(VII.3.6)}$$

which can be integrated by parts to give $\bar{c} = 2\alpha/\pi^{1/2} = 1.1284\alpha$.

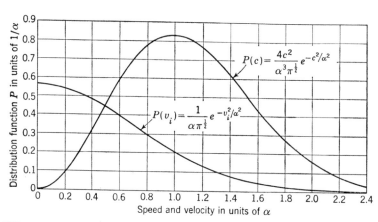

Fig. VII.2. Speed distribution function $P(c)$ and cartesian velocity component distribution function $P(v_i)$ for gases. Ordinate: distribution function P in units of $1/\alpha$; abscissa: speed and velocity in units of α.

The median speed c_m is that speed which is exceeded by half of the molecules. It can be obtained from the integral equation

$$\int_0^{c_m} P(c)\, dc = \frac{1}{2} = \frac{4}{\alpha^3\pi^{1/2}} \int_0^{c_m} c^2 e^{-c^2/\alpha^2}\, dc \qquad \text{(VII.3.7)}$$

On integrating by parts ($u = c$, $dv = ce^{-c^2/\alpha^2}\, dc$) and rearranging

$$\frac{1}{2} = \text{Erf}(c_m) - \frac{2}{\pi^{1/2}}\left(\frac{c_m}{\alpha}\right) e^{-(c_m/\alpha)^2} \qquad \text{(VII.3.8)}$$

From tables of the error function and exponentials this can be solved to give $c_m = 1.088\alpha$.

Finally the second moment $\overline{c^2}$ is given by

$$\overline{c^2} = \int_0^\infty c^2 P(c)\, dc = \frac{4}{\alpha^3\pi^{1/2}} \int_0^\infty c^4 e^{-c^2/\alpha^2}\, dc$$

and on substituting $y = c/\alpha$

$$\overline{c^2} = \frac{4\alpha^2}{\pi^{1/2}} \int_0^\infty y^4 e^{-y^2}\, dy \qquad \text{(VII.3.9)}$$

which can be reduced to standard integrals by integration by parts to give

$$\overline{c^2} = 3\alpha^2/2 = 1.5\alpha^2 \qquad\qquad (VII.3.10)$$

By taking square roots we find that the rms speed is given by $(\overline{c^2})^{1/2} = \alpha 1.5^{1/2} = 1.225\alpha$. In summary we have

$$c^2 = v_x^2 + v_y^2 + v_z^2$$

$$
\begin{aligned}
v_p &= 0 & c_p &= \alpha \\
\bar{v} &= 0 & \bar{c} &= 1.128\alpha \\
v_m &= \pm\,0.476\alpha & c_m &= 1.088\alpha \\
(\overline{v^2})^{1/2} &= 0.707\alpha & (\overline{c^2})^{1/2} &= 1.225\alpha \\
|\overline{v}| &= 0.564\alpha
\end{aligned}
\qquad (VII.3.11)
$$

4. Velocity Groups

For our subsequent work we will be very much interested in special groups of molecules that have speeds or velocity components in excess of a certain value. We may define special functions $f(v_x)$, $f(v_x,v_y)$, $f(c)$ as follows:

$$f(v_x) = 1 - \int_{-v_x}^{v_x} P(v_x)\,dv_x = \int_{-\infty}^{-v_x} P(v_x)\,dv_x + \int_{v_x}^{\infty} P(v_x)\,dv_x$$

$$f(v_x,v_y) = f(v_x)f(v_y) \qquad\qquad (VII.4.1)$$

$$f(c) = 1 - \int_0^c P(c)\,dc = \int_c^\infty P(c)\,dc$$

where $P(v_x)$, $P(v_x,v_y)$, and $P(c)$ are the relative distribution functions. It can be seen that $f(v_x)$ is then the fraction of all molecules that have an x velocity component in excess of the absolute value $|v_x|$. Similarly, $f(v_x,v_y)$ is the fraction of all molecules that simultaneously have x and y components in excess of $|v_x|$ and $|v_y|$. Finally, $f(c)$ represents that fraction of molecules with velocities in excess of the value c. On substituting the respective distribution functions, all of these integrals may then be expressed in terms of the Gaussian distribution function and the error integral. We find:

$$f(v_x) = 1 - \mathrm{Erf}\left(\frac{v_x}{\alpha}\right)$$

$$f(v_x,v_y) = 1 - \mathrm{Erf}\left(\frac{v_x}{\alpha}\right) - \mathrm{Erf}\left(\frac{v_y}{\alpha}\right) + \mathrm{Erf}\left(\frac{v_x}{\alpha}\right)\mathrm{Erf}\left(\frac{v_y}{\alpha}\right) \qquad (VII.4.2)$$

$$f(c) = 1 + \frac{2}{\pi^{1/2}}\left(\frac{c}{\alpha}\right)e^{-c^2/\alpha^2} - \mathrm{Erf}\left(\frac{c}{\alpha}\right)$$

The last of these can be expressed in terms of the second derivative of the error function:[a]

$$f(c) = 1 - \frac{1}{2} \text{Erf}'' \left(\frac{c}{\alpha}\right) - \text{Erf} \left(\frac{c}{\alpha}\right) \tag{VII.4.3}$$

When the ratio v_x/α or c/α exceeds unity, there is a convenient form which can be used to express these functions directly. On continued integration by parts we find

$$1 - \text{Erf}(x) = \frac{e^{-x^2}}{x\pi^{1/2}} \left(1 - \frac{1}{2x^2}\right) + \frac{3}{2\pi^{1/2}} \int_x^\infty \frac{e^{-x^2}}{x^4} dx \tag{VII.4.4}$$

and for values of $x > 5$ this is represented to within 1 per cent by

$$1 - \text{Erf}(x) \cong \frac{e^{-x^2}}{x\pi^{1/2}} \tag{VII.4.5}$$

In the range $x \geqslant 5$ we can then rewrite our formulas as

$$f(v_x) \cong \frac{\alpha e^{-v_x^2/\alpha^2}}{v_x \pi^{1/2}}$$

$$f(v_x, v_y) \cong \frac{\alpha^2}{\pi v_x v_y} e^{-(v_x^2 + v_y^2)/\alpha^2} \tag{VII.4.6}$$

$$f(c) \cong \frac{2c}{\alpha \pi^{1/2}} e^{-c^2/\alpha^2} \left(1 + \frac{\alpha^2}{2c^2}\right)$$

5. Translational Energy Distributions

If we define E_x, E_y, and E_z as the amounts of kinetic energy associated with the different cartesian components of velocity, then by definition we have

$$E_x = 1/2mv_x^2 \qquad E_y = 1/2mv_y^2 \qquad E_z = 1/2mv_z^2$$

and

$$E = E_x + E_y + E_z = 1/2mc^2 \tag{VII.5.1}$$

where E is the total translational energy of a molecule. From the definitions,

$$dE_x = mv_x \, dv_x = (2mE_x)^{1/2} \, dv_x \qquad dE = mc \, dc = (2mE)^{1/2} \, dc \tag{VII.5.2}$$

Now the chances that a molecule will have a translational energy com-

---

[a] If $\text{Erf}(x) = (2/\sqrt{\pi}) \int_0^x e^{-x^2} dx$, then the first derivative $\text{Erf}'(x) = (2/\sqrt{\pi}) e^{-x^2}$, the Gaussian function itself, and the second and third derivatives are given by

$$\text{Erf}''(x) = -\frac{4x}{\sqrt{\pi}} e^{-x^2} \qquad \text{Erf}'''(x) = \frac{4}{\sqrt{\pi}} e^{-x^2}(2x^2 - 1)$$

These notations are convenient, and tables of both the error function and its derivatives are available. See, for example, the Chemical Rubber "Handbook of Physics and Chemistry," or E. Jahnke and F. Emde, "Tables of Functions," 4th ed., Dover Publications, New York, 1945. Some values are listed in Table VI.2.

ponent E_x in the range E_x to $E_x + dE_x$ are precisely equal to the chances that it will have a velocity component whose absolute magnitude is in the range v_x to $v_x + dv_x$, where v_x and dv_x are given by Eqs. (VII.5.1) and (VII.5.2). Thus $P(E_x) dE_x = 2P(v_x) dv_x$, where a factor of 2 is introduced to take into account the fact that E_x is the same for v_x and $-v_x$ and E_x is only positive. On substitution

$$P(E_x) dE_x = \frac{1}{\alpha} \left(\frac{2}{\pi m E_x} \right)^{\frac{1}{2}} e^{-2E_x/m\alpha^2} dE_x \qquad (VII.5.3)$$

so that

$$P(E_x) = \frac{1}{\alpha} \left(\frac{2}{\pi m E_x} \right)^{\frac{1}{2}} e^{-2E_x/m\alpha^2} \qquad (VII.5.4)$$

with similar expressions for $P(E_y)$ and $P(E_z)$.

The expression for $P(E)$ becomes

$$P(E) = \frac{4}{\alpha^3} \left(\frac{2E}{\pi m^3} \right)^{\frac{1}{2}} e^{-2E/m\alpha^2} \qquad (VII.5.5)$$

The formula for an energy group, i.e., the fraction of molecules with energy components or speeds in excess of a certain quantity, is obtained by direct substitution of E from the definition [Eq. (VII.5.1)] into Eqs. (VII.4.2) and (VII.4.6). Thus the asymptotic expression for the number of molecules with total translational energy in excess of E is given by

$$f(E) \cong \frac{2}{\alpha} \left(\frac{2E}{\pi m} \right)^{\frac{1}{2}} e^{-2E/m\alpha^2} \left(1 + \frac{1}{4} \frac{m\alpha^2}{E} \right) \qquad (VII.5.6)$$

6. Pressure Exerted by an Ideal Gas; Collisions with a Wall

From the point of view of molecular motion the pressure exerted by a gas arises from the random collisions of the molecules with the container walls. At equilibrium these collisions must on the average be completely elastic, since the gas does not lose or gain energy from the container. Thus although each molecule which strikes the container wall with a normal component of momentum mv_z (z axis is selected as perpendicular to the wall) may not rebound elastically with a normal component $- mv_z$,[a] this condition must certainly be satisfied over a time interval of many molecular collisions. The result is that for a large number of collisions, the net transport of momentum to the walls by colliding molecules must be exactly compensated by an equivalent transport of momentum by molecules leaving the wall. Therefore if we can calculate the total amount of momentum carried to the walls in a given time, the net exchange of momentum between the gas and the surface will be exactly double it.

[a] It is quite possible to take as an extreme case that a molecule may stick to the walls on a collision, later to be evaporated. This will not violate the momentum conservation principle if the evaporating molecule carries away, on the average, as much momentum as the original molecule brought.

If we now consider an ideal gas made up of point particles, the particles cannot suffer collisions with each other and their velocities in the gas phase cannot change except after collisions with the walls of the container. The total transport of momentum in such a gas is by the individual molecules themselves. Let us calculate for such a gas the average pressure exerted on an element of wall surface dS.

By Newton's laws, the net perpendicular force exerted on a surface element dS will be equal to the rate of change of momentum perpendicular to the wall

$$\bar{F} = \frac{\Delta_\perp(\text{momentum})}{\Delta t} \qquad (VII.6.1)$$

The mean pressure is simply $\bar{P} = \bar{F}/dS$. But for our conservation principle to hold, $\Delta_\perp(\text{momentum})$ of the wall must be equal to twice the component of perpendicular momentum brought by all the molecules striking the surface in the time Δt. Let us set up coordinate axes at the surface element dS and investigate the collisions (Fig. VII.3).

Take a volume element in the gas $d\tau$ whose spherical coordinates are r, θ, ϕ with respect to the surface element dS. If N_g is the total number of molecules per unit volume in the gas, then $N_g\,d\tau$ will be the number of molecules in $d\tau$. Of this number a fraction $d\Omega/4\pi$ will have their velocity vectors lying in the solid angle $d\Omega$ subtended at $d\tau$ by the surface dS. Only these molecules are capable of striking dS.

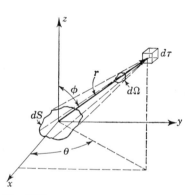

Fig. VII.3. Showing surface element dS subtending solid angle $d\Omega$ from position of volume element $d\tau$.

Now the solid angle $d\Omega = dS \cos \phi/r^2$, so that the total number of molecules in $d\tau$ capable of striking dS is equal to $N_g(dS \cos \phi/4\pi r^2)\,d\tau$. If we express $d\tau$ in spherical coordinates ($d\tau = r^2 \sin \phi\,d\phi\,d\theta\,dr$), then this becomes

$$\delta N(\tau) = dS\, N_g \frac{\sin \phi \cos \phi}{4\pi}\, d\theta\, d\phi\, dr \qquad (VII.6.2)$$

If we now integrate this expression over θ and ϕ, we obtain $\delta N(r)$, i.e., the total number of molecules lying in a shell of thickness dr at a distance r from the surface element dS, with velocity vectors pointing toward dS:

$$\delta N(r) = dS \frac{N_g}{4\pi}\, dr \int_0^{2\pi} d\theta \int_0^{\pi/2} \sin \phi \cos \phi\, d\phi$$
$$= dS \frac{N_g\, dr}{4} \qquad (VII.6.3)$$

Of the molecules δN, only a fraction given by $f(c)$ with speed in excess of r ($c \geqslant r$) will strike dS in 1 sec. The fraction of molecules with speeds in excess of r is given by $f(r)$ from Eq. (VII.4.2). $\delta N_w(r)$, the number of molecules in the shell which strike the element dS in 1 sec, is thus given by

$$\delta N_w(r) = \frac{dS\, N_g}{4} f(r)\, dr \qquad (VII.6.4)$$

If we now integrate this expression over all r, we obtain N_w, the total number of molecules striking the element dS in 1 sec.

$$N_w = \frac{dS\, N_g}{4} \int_0^\infty f(r)\, dr \qquad (VII.6.5)$$

On substituting the value of $f(r)$ and performing the integration,[a] we obtain

$$N_w = \frac{dS\, N_g}{4} \frac{2\alpha}{\pi^{1/2}}$$

but since the average speed $\bar{c} = 2\alpha/\pi^{1/2}$, this can be written

$$N_w = 1/4\bar{c}N_g\, dS \qquad (VII.6.6)$$

In order to derive an expression for the total perpendicular component of momentum transported per second, we take Eq. (VII.6.2) for $\delta N(r)$ the

[a] Since $f(r)$ [see Eq. (VII.4.2)] is given by

$$f(r) = 1 - \mathrm{Erf}\left(\frac{r}{\alpha}\right) + \frac{2}{\pi^{1/2}} \frac{r}{\alpha} e^{-r^2/\alpha^2} = 1 - \mathrm{Erf}\left(\frac{r}{\alpha}\right) - \frac{1}{2}\mathrm{Erf}''\left(\frac{r}{\alpha}\right)$$

we write

$$\int_0^\infty f(r)\, dr = \int_0^\infty \left[1 - \mathrm{Erf}\left(\frac{r}{\alpha}\right)\right] dr + \frac{2}{\pi^{1/2}} \int_0^\infty \frac{r}{\alpha} e^{-r^2/\alpha^2}\, dr$$

The first integral may be integrated by parts:

$$u = \left[1 - \mathrm{Erf}\left(\frac{r}{\alpha}\right)\right] \qquad dv = dr$$

then

$$du = -\mathrm{Erf}'\left(\frac{r}{\alpha}\right)\frac{dr}{\alpha} = -\frac{2}{\pi^{1/2}} e^{-r^2/\alpha^2}\frac{dr}{\alpha} \qquad \text{and} \qquad v = r$$

Thus

$$\int_0^\infty f(r)\, dr = \left\{r\left[1 - \mathrm{Erf}\left(\frac{r}{\alpha}\right)\right]\right\}_0^\infty + \frac{4}{\pi^{1/2}} \int_0^\infty \frac{r}{\alpha} e^{-r^2/\alpha^2}\, dr$$

The first term in brackets is zero at both limits $r = 0$ and $r = \infty$, and the second term integrates directly to give

$$\int_0^\infty f(r)\, dr = \frac{2\alpha}{\pi^{1/2}}$$

It can be seen that $r[1 - \mathrm{Erf}(r/\alpha)]$ is zero at $r = \infty$ by using the asymptotic expression for $\mathrm{Erf}(r/\alpha)$ as valid when r/α is large. From Eq. (VII.4.5) we see that $1 - \mathrm{Erf}(x) \to e^{-x^2}/x\pi^{1/2}$, so that $r[1 - \mathrm{Erf}(r/\alpha)] \to \alpha/\pi^{1/2} e^{-r^2/\alpha^2} \to 0$ as $r \to \infty$.

Although there are other more facile methods for evaluating the integral for N_w, the above approach has been given to illustrate a very general and powerful technique for tackling such problems.

number of molecules in $d\tau$ with velocity vectors directed toward dS. Of these molecules, the fraction $P(c)\,dc$ with speeds in the range c to $c + dc$ will carry a perpendicular component of momentum given by $mc \cos \phi$. The total perpendicular component of momentum transported by this group of molecules is thus

$$\perp \text{Mom}_r\,(c) = \frac{mN_g\,dS\,P(c)\,c}{4\pi}\,dc\,\sin\phi\,\cos^2\phi\,d\theta\,d\phi\,dr \quad \text{(VII.6.7)}$$

On integrating again over θ (from 2π to 0) and ϕ ($\pi/2$ to 0) we obtain the total perpendicular component of momentum carried by molecules of speed c, in the hemispherical shell dr, distant r from dS, with velocity vectors pointing toward dS:

$$\perp \text{Mom}_r\,(c) = \frac{mN_g\,dS\,cP(c)\,dc}{6}\,dr \quad \text{(VII.6.8)}$$

Now all molecules of speed c will reach dS in 1 sec if they lie in shells whose distance r from dS is less than c. If we integrate $\perp \text{Mom}_r\,(c)$ over all r from $r = 0$ to $r = c$, we will have the $\perp \text{Mom}\,(c)$ carried to dS in 1 sec by all molecules of speed c lying in the hemispherical shell, radius c.

$$\perp \text{Mom}\,(c) = \int_0^c \perp_r \text{Mom}\,(c)\,dr$$

$$= \frac{mN_g\,dS\,cP(c)\,dc}{6}\int_0^c dr \quad \text{(VII.6.9)}$$

$$= \frac{mN_g\,dS}{6}\,c^2 P(c)\,dc$$

By integrating over all c (0 to ∞) we find the total $\perp \text{Mom}$ carried to dS in 1 sec, namely

$$\perp \text{Mom} = \frac{mN_g\,dS}{6}\int_0^\infty c^2 P(c)\,dc \quad \text{(VII.6.10)}$$

where $\int_0^\infty c^2 P(c)\,dc = \overline{c^2}$, the mean square speed of the molecules.

From the discussion at the beginning of this chapter we know that the mean pressure is given by twice the total perpendicular component of momentum divided by the area. Thus

$$P = 2\left(\frac{\perp \text{Mom}}{dS}\right) = \frac{1}{3}\,mN_g\overline{c^2} \quad \text{(VII.6.11)}$$

or, since $\overline{c^2} = 3/2\alpha^2$,

$$P = 1/2mN_g\alpha^2 \quad \text{(VII.6.12)}$$

7. Absolute Temperature: Relation to Thermodynamics

The absolute temperature scale may be defined in terms of the properties of an ideal gas through its equation of state. Thus from the ideal gas law for 1 mole of gas

$$P = \frac{RT}{V} \tag{VII.7.1}$$

If we now compare Eqs. (VII.6.12) and (VII.7.1), we see that for one mole, $N_g = N_{Av}/V$, the concentration of gas molecules, so that we may equate

$$RT = 1/2 m N_{Av} \alpha^2 \tag{VII.7.2}$$

so that $\qquad \alpha^2 = \dfrac{2RT}{m N_{Av}} = \dfrac{2RT}{M} \qquad$ or $\qquad \alpha^2 = \dfrac{2kT}{m} \qquad$ (VII.7.3)

where M = the molecular weight of the gas and $k = R/N_{Av}$, or Boltzmann's constant.

This identification now makes for a completely consistent relation between the statistical properties of the gas and the thermodynamic variables. The total energy of the gas, which is simply its kinetic energy for ideal monatomic gases, becomes [Eq. (VI.2.2)] for 1 mole

$$
\begin{aligned}
E &= \tfrac{1}{2} m N_{Av} \int_0^\infty c^2 P(c) \, dc = \tfrac{1}{2} m N_{Av} \, \overline{c^2} \\
&= \tfrac{3}{4} m N_{Av} \alpha^2 = \tfrac{3}{2} N_{Av} kT = \tfrac{3}{2} RT
\end{aligned}
\tag{VII.7.4}
$$

and we can confirm the well-known properties of a monatomic ideal gas, $C_v = (\partial E/\partial T)_v = \tfrac{3}{2}R$, $C_p = C_v + R = \tfrac{5}{2}R$.

The following shows the appearance of some of our former quantities expressed in terms of T instead of the parameter α.

$$P(c) = \left(\frac{2}{\pi}\right)^{1/2} \left(\frac{m}{kT}\right)^{3/2} c^2 e^{-mc^2/2kT} \tag{VII.7.5}$$

$$c_p = \left(\frac{2kT}{m}\right)^{1/2} = 1.414 \left(\frac{kT}{m}\right)^{1/2} \qquad c_m = 1.540 \left(\frac{kT}{m}\right)^{1/2}$$

$$\bar{c} = \left(\frac{8kT}{\pi m}\right)^{1/2} = 1.596 \left(\frac{kT}{m}\right)^{1/2} \qquad \overline{c^2}^{1/2} = \left(\frac{3kT}{m}\right)^{1/2} = 1.732 \left(\frac{kT}{m}\right)^{1/2}$$

$$f(c) = 1 - \frac{1}{2} \operatorname{Erf}'' \left[\left(\frac{mc^2}{2kT}\right)^{1/2}\right] - \operatorname{Erf}\left[\left(\frac{mc^2}{2kT}\right)^{1/2}\right]$$

$$P(E) = \frac{2}{kT} \left(\frac{E}{kT}\right)^{1/2} e^{-E/kT}$$

$$f(E) \cong 2 \left(\frac{E}{\pi kT}\right)^{1/2} e^{-E/kT} \left(1 + \frac{1}{2}\frac{kT}{E}\right)$$

A table of values for the fundamental constants k, N_{Av}, R, etc., will be found in Appendix A.

8. Molecular Collisions

The point molecule, while adequate for representing the properties of an ideal gas, is incapable of accounting for molecular collisions. The simplest model that does have that property is the hard sphere model, and that is the model we shall use for our discussion.

8A. Mechanics of a Collision. The mechanical properties of a collision between two hard spheres of diameters σ_1 and σ_2 and masses m_1 and m_2 are obtained from the laws of conservation of energy and momentum.[a] A scheme of such a collision is shown in Fig. VII.4. At the moment of impact the two molecular centers will be separated by a distance $\sigma_{12} = (\sigma_1 + \sigma_2)/2$. If we take projections of their velocity vectors $\mathbf{v}_1$ and $\mathbf{v}_2$ perpendicular and parallel to this line of centers we will have two components for each: $(v_{1c}, v_{1\perp})$ and $(v_{2c}, v_{2\perp})$. If we assume no tangential forces acting during the collision, then the perpendicular components $(v_{1\perp}, v_{2\perp})$ are unchanged by the collision and we have only to consider the components (v_{1c}, v_{2c}) along the line of centers. From the law of conservation of momentum we have

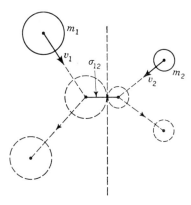

Fig. VII.4. Mechanics of a collision between two ideal spherical molecules.

$$m_1 v_{1c} + m_2 v_{2c} = m_1 v'_{1c} + m_2 v'_{2c} \qquad \text{(VII.8A.1)}$$

and from the conservation of energy the equation

$$\frac{m_1 v_{1c}^2}{2} + \frac{m_2 v_{2c}^2}{2} = \frac{m_1 v'^2_{1c}}{2} + \frac{m_2 v'^2_{2c}}{2} \qquad \text{(VII.8A.2)}$$

These can now be solved for the components (v'_{1c}, v'_{2c}) which the molecules will have after the collision.

$$v'_{2c} = \frac{2m_1 v_{1c} - v_{2c}(m_1 - m_2)}{m_1 + m_2} \qquad v'_{1c} = \frac{2m_2 v_{2c} - v_{1c}(m_2 - m_1)}{m_1 + m_2} \qquad \text{(VII.8A.3)}$$

When the masses are equal $(m_1 = m_2)$, we see that $v'_{1c} = v_{2c}$ and $v'_{2c} = v_{1c}$; that is, the net result of the collision is to exchange the head-on components of velocity.

When the masses are very different, for example, $m_1 \gg m_2$, we have approximately

$$v'_{2c} \cong 2v_{1c} - v_{2c} \qquad v'_{1c} \cong \frac{2m_2}{m_1} v_{2c} + v_{1c} \cong v_{1c} \qquad \text{(VII.8A.4)}$$

That is, the heavier molecule is virtually unchanged in velocity while the

---

[a] Note that for two spherical molecules with central force fields, the relative motion will take place in a plane. This plane will be defined by the centers of the two molecules and the direction of their relative velocity vector, which is parallel to the plane. No forces act perpendicularly to the plane.

lighter molecule rebounds almost as if it had struck a moving heavy wall.[a] That would be roughly the case in a collision between an H_2 molecule and an I_2 molecule or for an electron and an atom.

8B. Collision Frequency of a Single Molecule Moving through a Stationary Gas. Let us consider the random motions of a single molecule with velocity c that is moving through a gas composed of identical but stationary molecules. The molecule selected will then execute a series of zig-zag motions, with a constant speed c moving in straight lines between encounters.

If we imagine a typical path made by the molecule and surround it by a hypothetical cylinder of radius σ equal to the molecular diameter (Fig. VII.5), then any molecule with its center inside this cylinder will undergo

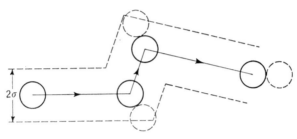

FIG. VII.5. Cylinder of exclusion swept out by molecule making consecutive collisions.

a collision with the moving molecule. If the average distance between molecules is large compared to the molecule diameters, then in 1 sec the moving molecule will sweep out an irregular cylindrical volume $= c\pi\sigma^2$. If the stationary molecules are randomly distributed and their density in the gas is N_g, there will be $N_g c\pi\sigma^2$ such molecular centers inside the cylinder. Thus the number of collisions made per second by our moving molecule is[b]

$$Z_c = N_g \pi\sigma^2 c \qquad (VII.8B.1)$$

We can define an average mean free path of our molecule L_c, which is equal to the average distance it traverses between collisions. L_c in the present case will be the ratio of c, the total path traversed in 1 sec, to the number of collisions Z_c.

$$L_c = \frac{c}{Z_c} = \frac{1}{\pi\sigma^2 N_g} \qquad (VII.8B.2)$$

[a] The velocity of the lighter particle relative to the heavy particle considered as fixed is $-v_{1c} + v_{2c}$. The particle will then rebound with this component reflected, namely $v_{1c} - v_{2c}$. If we now add to this latter the velocity of the heavy particle v_{1c}, we obtain $2v_{1c} - v_{2c}$, the component after reflection.

[b] An alternative approach is to use Eq. (VII.6.6) for the number of collisions of molecules with a surface S: $Z_c = N_w = \frac{1}{4}N_g \bar{c} S$. If we use for S the collisional surface of a molecule $4\pi\sigma^2$, then we obtain the above expression, Eq. (VII.8B.1).

8C. Collision Frequency of a Molecule Moving through Homogeneous Gas. If we change our assumptions to allow the molecule with velocity v_1 to move through not a stationary gas but rather one in which molecules are in random motion but with identical speeds c_2 ($c_2 = |v_2|$), then our collision frequency is different. Since the gas molecules are now assumed in motion, the net effect will be to have more collisions, since collisions will take place more frequently when the two velocity vectors v_1 and v_2 are directed toward each other. The result will be that the relative velocity of our molecule is effectively greater for collisions. To see this mathematically, let us consider the relative velocities of our probe molecule with

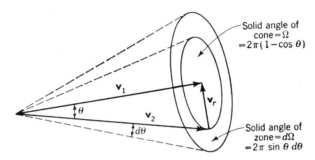

FIG. VII.6. Diagram of velocity vectors during molecular collision.

respect to a molecule v_2 with which it is about to collide (Fig. VII.6). If v_r is the relative velocity vector of the two molecules, then $v_r = v_2 - v_1$. The absolute value of v_r is given by

$$v_r = (v_1^2 + v_2^2 - 2v_1v_2 \cos \theta)^{\frac{1}{2}} \qquad \text{(VII.8C.1)}$$

The chance that the vector v_2 will lie within the cones which make a conical angle θ to $\theta + d\theta$ with respect to the vector v_1 is equal to $2\pi \sin \theta \, d\theta$ divided by the total solid angle, which is 4π. Thus $\frac{1}{2} \sin \theta \, d\theta$ is the fraction of molecules that have velocity vectors between θ and $\theta + d\theta$ with respect to v_2 at the moment of collision. The average relative velocity of colliding molecules $\bar{v}_r$ is then obtained by averaging over all θ (0 to π).

$$\bar{v}_r = \frac{1}{2} \int_0^\pi \sin \theta \, (v_1^2 + v_2^2 - 2v_1v_2 \cos \theta)^{\frac{1}{2}} \, d\theta \qquad \text{(VII.8C.2)}$$

On integration we find[a]

$$\bar{v}_r = \frac{1}{6v_1v_2} \left[\left(v_1^2 + v_2^2 - 2v_1v_2 \cos \theta \right)^{\frac{3}{2}} \right]_{\theta=0}^{\theta=\pi}$$

$$= \frac{1}{6v_1v_2} \left[(v_1 + v_2)^3 - |v_1 - v_2|^3 \right] \qquad \text{(VII.8C.3)}$$

[a] The absolute value $|v_1 - v_2|^3$ is written so that our signs will be consistent with the integrated quantity $[(v_1 - v_2)^2]^{\frac{3}{2}}$, which is always positive.

Depending on the relative magnitudes of $|v_1|$ and $|v_2|$, we have

$$v_r = v_1 + \frac{v_2^2}{3v_1} \quad |v_1| > |v_2| \quad v_r = v_2 + \frac{v_1^2}{3v_2} \quad |v_1| < |v_2| \quad \text{(VII.8C.4)}$$

When $|v_1| = |v_2|$ we have $v_r = \frac{4}{3}v_2$, so that the relative velocity for collision of moving molecules is, as earlier stated, greater than that for fixed molecules. Under these conditions,

$$Z_c = \frac{4}{3}N_g\pi\sigma^2 c \quad \text{and} \quad L_c = \frac{3}{4\pi\sigma^2 N_g} \quad \text{(VII.8C.5)}$$

8D. Collision Frequency of a Molecule in a Maxwellian Gas. We can now generalize our findings to the more interesting situation of a molecule of velocity $\mathbf{v}_1$ moving through a gas whose molecules have a Maxwellian distribution of velocities. By using our previous results we see that [Eq. (VII.8B.1)]:

$$Z_c = N_g\pi\sigma^2 v_r \quad \text{(VII.8D.1)}$$

where v_r is given by Eq. (VII.8C.4). In our Maxwellian gas, the fraction of molecules that have speeds in the range v_2 to $v_2 + dv_2$ is given by

$$P(v_2)\, dv_2 = \frac{4v_2^2}{\pi^{1/2}\alpha_2^3}\, e^{-v_2^2/\alpha_2^2}\, dv_2$$

The average number of encounters of molecule 1 is thus obtained by averaging Z_c over all velocities v_2:

$$Z_c = \frac{4N_g\pi\sigma^2}{\pi^{1/2}\alpha_2^3}\left[\int_0^{v_1}\left(v_1 + \frac{v_2^2}{3v_1}\right)v_2^2 e^{-v_2^2/\alpha_2^2}\, dv_2 + \int_{v_1}^{\infty}\left(v_2 + \frac{v_1^2}{3v_2}\right)v_2^2 e^{-v_2^2/\alpha_2^2}\, dv_2\right]$$
$$\text{(VII.8D.2)}$$

On reducing these integrals to standard forms we obtain

$$Z_c = \frac{N_g\pi\sigma^2\alpha_2}{\pi^{1/2}}\left[e^{-v_1^2/\alpha_2^2} + \left(\frac{2v_1}{\alpha_2} + \frac{\alpha_2}{v_1}\right)\int_0^{v_1/\alpha_2} e^{-x^2}\, dx\right] \quad \text{(VII.8D.3)}$$

which in terms of the error function and integrals is (here we set $y = v_1/\alpha_2$)

$$Z_c = \frac{N_g\pi\sigma^2\alpha_2}{2}\left[\text{Erf}'(y) + \left(2y + \frac{1}{y}\right)\text{Erf}(y)\right] \quad \text{(VII.8D.4)}$$

We note that as $v_1/\alpha_2 = y \to \infty$, this reduces to

$$Z_c \to \frac{N_g\pi\sigma^2\alpha_2}{2}\, 2y = N_g\pi\sigma^2 v_1$$

which is the limiting form obtained for a molecule moving with speed v_1 through a stationary gas [Eq. (VII.8B.1)]. If we let $\psi(y)$ represent the function in the brackets in Eq. (VII.8D.4), we can calculate various values of $\psi(y)$ as a function of y. They are shown in Table VII.1.

TABLE VII.1. RELATIVE COLLISION FREQUENCIES OF MOLECULES
WITH DIFFERENT VELOCITIES

$y = c/\alpha$	$\psi(y)$	$Z_c/\bar{Z}_c$	$L_c/\bar{L}_c$	$y = c/\alpha$	$\psi(y)$	$Z_c/\bar{Z}_c$	$L_c/\bar{L}_c$
0	2	0.632	0	4.0	8.25	2.60	1.370
0.5	2.440	0.770	0.579	5.0	10.20	3.22	1.385
1.0	2.941	0.929	0.961	10.0	20.10	6.35	1.406
1.5	3.66	1.155	1.160	20.0	40.05	12.64	1.414
2.0	4.50	1.420	1.256	40.0	80.0	25.25	1.414

Note: $\psi(y) = \mathrm{Erf}'(y) + \left(2y + \dfrac{1}{y}\right)\mathrm{Erf}(y)$ [Eq. (VII.8D.4)]

$$\frac{Z_c}{\bar{Z}_c} = \frac{\pi^{1/2}}{4\sqrt{2}}\psi(y) = 0.3158\psi(y) \qquad \frac{L_c}{\bar{L}_c} = \frac{v/\bar{Z}_c}{v/\bar{Z}_c} = \frac{y2^{3/2}}{\psi(y)}$$

8E. Collision Frequency between Maxwellian Molecules. Finally, we can calculate the average number of collisions made by a molecule going through a Maxwellian gas if the molecule does not have a fixed velocity v_1, but has instead a velocity distribution which is itself Maxwellian. This may be done by multiplying Z_c [Eq. (VII.8D.4)] by the Maxwellian distribution function and averaging over all values of v_1:

$$\bar{Z}_c = \int_0^\infty Z_c P(v_1)\, dv_1 \tag{VII.8E.1}$$

$$= \frac{2N_g\pi\sigma^2\alpha_2^4}{\pi^{1/2}\alpha_1^3} \int_0^\infty \left[y^2\, \mathrm{Erf}'(y)e^{-y^2\beta^2} + y^2\left(2y + \frac{1}{y}\right)e^{-y^2\beta^2}\, \mathrm{Erf}(y) \right] dy$$

where $\beta = \alpha_2/\alpha_1 = (m_1/m_2)^{1/2}$.

By using the type of reduction indicated in footnote (page 146) we find

$$\bar{Z}_c = \frac{2}{\pi^{1/2}} N_g\pi\sigma^2(\alpha_2^2 + \alpha_1^2)^{1/2} \tag{VII.8E.2}$$

or since $\alpha_2^2 = (\pi/4)\bar{v}_2^2$ and $\alpha_1^2 = (\pi/4)\bar{v}_1^2$ we have

$$\bar{Z}_c = N_g\pi\sigma^2(\bar{v}_1^2 + \bar{v}_2^2)^{1/2} \tag{VII.8E.3}$$

When $\alpha_1 = \alpha_2$, so that $\bar{v}_1 = \bar{v}_2$, this becomes

$$\bar{Z}_c = \sqrt{2}N_g\pi\sigma^2\bar{v} \tag{VII.8E.4}$$

Correspondingly, we have for different masses

$$\bar{L}_c = \frac{\bar{v}_1}{N_g\pi\sigma^2(\bar{v}_1^2 + \bar{v}_2^2)^{1/2}} \tag{VII.8E.5}$$

and for identical masses

$$\bar{L}_c = \frac{1}{N_g\pi\sigma^2\sqrt{2}} \tag{VII.8E.6}$$

8F. Bimolecular Collision Frequencies. We can immediately generalize our results to include the case of collisions between molecules that have different hard sphere diameters. If the diameters of the two species are σ_1 and σ_2, then collision occurs when their centers are a distance $\sigma_{12} = (\sigma_1 + \sigma_2)/2$ apart. This means that we would replace σ in the above formulas by the mean value σ_{12}.

If we are interested in the frequency of encounters of a selected molecule in a gas mixture containing N_1 molecules/cc of type 1, N_2 of type 2, etc., we can write

$$\bar{Z} = N_1\bar{Z}_1 + N_2\bar{Z}_2 + N_n\bar{Z}_n = \sum_{i=1}^{n} N_i\bar{Z}_i \qquad \text{(VII.8F.1)}$$

while for the mean free path we have

$$\bar{L} = \frac{\bar{c}}{Z} = \frac{\bar{c}}{\displaystyle\sum_{i=1}^{n} N_i Z_i} \qquad \text{(VII.8F.2)}$$

For many purposes we are interested not in the collisions made by a given molecule with others, but in the total number of collisions between two types, 1 and 2. If $\bar{Z}_{1,2}$ is the number of collisions made per second by a single molecule of type 1 with type 2, then the total number of collisions between these two types per unit volume per second is obtained by multiplying $\bar{Z}_{1,2}$ by the concentration N_1 of molecules of type 1. Thus

$$\bar{Z}(1,2) = N_1\bar{Z}_{1,2} = N_1 N_2 \pi \sigma_{12}^2 (\bar{v}_1^2 + \bar{v}_2^2)^{1/2} \qquad \text{(VII.8F.3)}$$

If the molecules are identical, we must divide this quantity by 2, since we would otherwise be counting each collision twice:

$$\bar{Z}(1,1) = \frac{1}{2} N_1\bar{Z}_{1,1} = \frac{N_1^2 \pi \sigma_1^2 \bar{v}_1}{\sqrt{2}} \qquad \text{(VII.8F.4)}$$

If we express these formulas in terms of temperatures, we have

$$\bar{Z}_{1,2} = N_2 \pi \sigma_{12}^2 \left(\frac{8kT}{\pi\mu}\right)^{1/2}$$

$$\bar{Z}(1,2) = N_1\bar{Z}_{1,2} = N_1 N_2 \pi \sigma_{12}^2 \left(\frac{8kT}{\pi\mu}\right)^{1/2}$$

$$\bar{Z}(1,1) = N_1^2 \pi \sigma_1^2 \left(\frac{4kT}{\pi m_1}\right)^{1/2}$$

$$L_c(1,2) = \frac{\bar{v}_1}{\bar{Z}_{1,2}} = \left(\frac{\mu}{m_1}\right)^{1/2} \frac{1}{N_2 \pi \sigma_{12}^2} \qquad \text{(VII.8F.5)}$$

$$L_c(2,1) = \frac{\bar{v}_2}{\bar{Z}_{2,1}} = \left(\frac{\mu}{m_2}\right)^{1/2} \frac{1}{N_1 \pi \sigma_{12}^2}$$

$$L(1) = \frac{\bar{v}_1}{\bar{Z}_{1,1}} = \frac{1}{\sqrt{2} N_1 \pi \sigma_1^2}$$

where μ is the reduced mass defined by $1/\mu = 1/m_1 + 1/m_2$.

Table VII.2 shows some values of collision frequencies and other molecular properties of different gases.

TABLE VII.2. SOME MOLECULAR PROPERTIES

Gas	M	σ, Å[a]	$\bar{v}$ at 25°C, km/sec	$\bar{L}_{STP}$, Å[b]	$\bar{Z} \times 10^{-28}$ [c]
H₂	2.016	2.74	1.772	1180	20.4
He	4.002	2.18	1.257	1765	9.13
N₂	28.02	3.75	0.475	596	10.22
C₂H₆	30.05	5.30	0.448	298	19.7
O₂	32.00	3.61	0.434	644	8.87
Ar	39.94	3.64	0.3975	633	8.07
CO₂	44.00	4.59	0.379	397	12.22
Kr	82.9	4.16	0.2760	485	7.32

[a] Calculated from viscosities.

[b] Equations (VII.8F.5), assuming 2.687×10^{19} molecules/cc at STP (ideal gas).

[c] $\bar{v} = 145.51(T/M)^{\frac{1}{2}}$ m/sec; $\bar{v}_{25°C} = 2.514/M^{\frac{1}{2}}$ km/sec; $\bar{Z} = 3.232 \times 10^4 N^2\sigma^2(T/M)^{\frac{1}{2}}$ collisions/cm³-sec; $\bar{Z}_{STP} = 3.852 \times 10^{44} \sigma^2/M^{\frac{1}{2}}$ collisions/cm³-sec; $\bar{L} = 2.331 \times 10^{-20} \left(\dfrac{T}{P_{mm} \text{ (Hg)}\sigma^2}\right)$ cm; $L_{STP} = \dfrac{8.39 \times 10^{-21}}{\sigma^2}$ cm.

Note: M = molecular weight, σ = diameter, $\bar{v}$ = average velocity, $\bar{L}_{STP}$ = mean free path, $\bar{Z}$ = frequency of collisions.

8G. Duration of a Collision. The hard sphere model is very useful because it permits us to describe molecular collisions in terms of a single, simple, molecular parameter, the collision diameter. It is, however, insufficient to permit a detailed description of a chemical reaction, which is an event that transpires *during* a collision between two molecules, because the duration in time of a hard sphere collision is precisely zero.

To extend the usefulness of the model to permit a description of chemical reactions, we must introduce another parameter, the effective duration of a collision. The rectangular well or central force models do this automatically by permitting molecular interaction over a range of distances. However, they are both more complex than the hard sphere model. We can "rescue" the hard sphere model by specifying a parameter σ_a, the effective diameter for chemical interaction, while keeping σ_r as the hard sphere core diameter. When the centers of two identical molecules are a distance $\leqslant \sigma_a$ apart, reaction may occur. The motion is reversed when they are σ_r apart, so that the "chemical collision" between two such molecules takes place over the path of maximum length $2(\sigma_a - \sigma_r)$. The effective reaction volume is $\pi(\sigma_a^3 - \sigma_r^3)/6$.

If the mean relative velocity of two molecules is v_r, then the maximum duration in time of a chemical collision is $t_{coll} = 2(\sigma_a - \sigma_r)/v_r$. Assuming that $\sigma_a - \sigma_r$ will be of the order of magnitude of 1 Å and the mean relative velocity v_r will be of the order of 4×10^4 cm/sec (Table VII.2), we see that the maximum collision time will be about 5×10^{-13} sec.[a]

8H. Termolecular Collisions; Collision Complexes. Having defined a collision radius σ_a for the hard sphere molecule M of core diameter σ_r, it is possible to define a collision complex $M \cdot M$ as a pair of molecules whose centers are a distance σ apart, where $\sigma_a \leqslant \sigma \leqslant \sigma_r$. For such a complex we can write the stoichiometric equation

$$M + M \underset{2}{\overset{1}{\rightleftharpoons}} M \cdot M \qquad \text{(VII.8H.1)}$$

At equilibrium the concentration of such complexes is given by

$$(M \cdot M) = K_{1.2}(M)^2 = \frac{k_1}{k_2}(M)^2 \qquad \text{(VII.8H.2)}$$

where k_1 is the collision frequency and $1/k_2$ is the mean lifetime of a collision pair. We can compute k_2 directly by suitable averaging, or we can calculate the equilibrium constant $K_{1.2}$ and then obtain k_2 as $k_1/K_{1.2}$.

The equilibrium concentration of collision complexes is just the number of molecules which lie within each other's collision range. This will be given by

$$(M \cdot M) = \frac{(M)^2}{2} \frac{4}{3} \pi(\sigma_a^3 - \sigma_r^3) \qquad \text{(VII.8H.3)}$$

where the factor of 2 is introduced to correct for double counting. Thus $K_{1.2} = \frac{2}{3}\pi(\sigma_a^3 - \sigma_r^3)$ in units of volume per molecule.

Since $k_1 = \bar{Z}_{1,1} = \pi\sigma_a^2\bar{v}/\sqrt{2}$ [Eq. (VII.8F.5)], we have for k_2

$$k_2 = \frac{k_1}{K_{1.2}} = \frac{\bar{v}}{(2\sqrt{2}/3)\sigma_a(1 - \sigma_r^3/\sigma_a^3)} \qquad \text{(VII.8H.4)}$$

The definition of the range or duration of a collision which makes possible the definition of a collision complex also makes it possible to discuss the rate of triple and higher-order collisions. The rate of triple collisions is just the rate at which binary collision complexes will be struck by another molecule. From Eq. (VII.8F.5),

$$Z_{MMM} = (M)(M \cdot M)\pi\left(\frac{\sigma_{MM} + \sigma_M}{2}\right)^2\left(\frac{8kT}{\pi\mu}\right)^{1/2} \qquad \text{(VII.8H.5)}$$

By substituting for $(M \cdot M)$ and μ we find

$$Z_{MMM} = (M)^3 \frac{\pi^2}{3}\sigma_a^5\left(1 - \frac{\sigma_r^3}{\sigma_a^3}\right)\left(1 + \frac{\sigma_{MM}}{\sigma_a}\right)^2\left(\frac{3kT}{\pi M}\right)^{1/2} \qquad \text{(VII.8H.6)}$$

for the rate of triple collisions in a hard sphere gas with extended collision range. If we use average values of $\sigma_a = 4.0$ Å, $\sigma_r/\sigma_a = \frac{2}{3}$, $\sigma_{MM}/\sigma_a = 1.5$,

---

[a] It is interesting to note that this is of the same order of magnitude as the period of a low-frequency molecular vibration. For real molecules v_r will be larger than average thermal velocities because of the increase in velocity due to attractive forces.

and $(3kT/\pi M)^{1/2} = 3 \times 10^4$ cm/sec, we find that Z_{MMM} is about 1.1×10^9 liters^2/mole^2-sec. Note that Z_{MMM} is a very sensitive function of the parameter σ_a.

9. Distribution of Free Paths

The mean free path already calculated gives the average over a great many collisions. We would like to find out how the individual paths between collisions vary. To do so, let us first calculate the probability that a molecule will go a distance x after its last collision without making a collision. If $f(x)$ represents this distribution function, then it is seen that it must satisfy the conditions $f(0) = 1$ and $f(\infty) = 0$.

Let us assume that the past history of a molecule has no influence on the molecule's chance of making a collision. That is, the chance of making a collision at any instant is completely independent of whether or not the molecule has just made a collision or has gone a long distance without any collisions. Thus if $f(x_1)$ represents the chance of the molecule moving a distance x_1 without collision and $f(x_2)$ the chance of moving a distance x_2 without collision, then the chance of the molecule having gone the distance $r = x_1 + x_2$ without a collision is the product

$$f(r) = f(x_1 + x_2) = f(x_1)f(x_2) \qquad \text{(VII.9.1)}$$

Equations of this type have a unique solution[a] of the form

$$f(x) = ke^{-\lambda x} \qquad \text{(VII.9.2)}$$

To satisfy our conditions $f(0) = 1$, we see that $k = 1$. Thus

$$f(x) = e^{-\lambda x} \qquad \text{(VII.9.3)}$$

Now we desire to find the chance that a molecule will go a distance x without making a collision but will subsequently make a collision in the next interval Δx. This is precisely the description of the distribution function $P(x)$ for mean free paths.

The quantity $f(\Delta x)$ is the chance of a molecule going the distance Δx without colliding. Then $1 - f(\Delta x)$ will be the chance of colliding in the interval Δx. $P(x)$ is then given by $f(x)$, the probability that the molecule will go the distance x without collision, times $1 - f(\Delta x)$, or

$$P(x) \, \Delta x = f(x)[1 - f(\Delta x)] = f(x) - f(x)f(\Delta x)$$
$$= f(x) - f(x + \Delta x)$$

or[b] $$P(x) = \frac{f(x) - f(x + \Delta x)}{\Delta x} \xrightarrow[\text{as } \Delta x \to 0]{\lim} -\frac{d\,f(x)}{dx} \qquad \text{(VII.9.4)}$$

[a] The solution can be obtained by application of the Lagrange method of undetermined multiplier (Sec. VII.2).

[b] Interpreted differently, $f(x)$ is the fraction of molecules that go a distance x without colliding; then $d\,f(x)/dx$ is the rate at which this changes, or the fraction of molecules that collide after x.

Thus on substituting $f(x) = e^{-\lambda x}$

$$P(x) = \lambda e^{-\lambda x} \qquad \text{(VII.9.5)}$$

The average mean free path $\bar{L}$ is given by

$$\bar{L} = \bar{x} = \int_0^\infty x P(x) \, dx$$

$$= \lambda \int_0^\infty x e^{-\lambda x} \, dx$$

which can be integrated directly to give

$$\bar{L} = \frac{1}{\lambda} \qquad \text{(VII.9.6)}$$

λ is the reciprocal of the mean free path $\bar{L}$, which we have already calculated. In terms of $\bar{L}$

$$P(x) = \frac{1}{\bar{L}} e^{-x/\bar{L}} \qquad \text{(VII.9.7)}$$

We see that $f(x)$ is the integrated form of $P(x)$ and corresponds to the fraction of molecules that have free path lengths in excess of x. Half of all molecules will have free paths in excess of $0.70\bar{L}$, whereas only 36.5 per cent of all molecules have free paths in excess of $\bar{L}$. Only 1.83 per cent of the molecules will have free paths in excess of $4\bar{L}$, and only 45 in every 1,000,000 will go distances greater than $10\bar{L}$.[a]

10. Molecular Effusion; Isotope Separation

If we have two vessels connected by a diaphragm with a hole in it (Fig. VII.7), the thermal motion of gas molecules in the vessels will bring about an interchange of matter between the two vessels. If one of the vessels is kept evacuated, there will be a one-way passage of molecules to the evacuated vessel (Fig. VII.7). If this hole is large compared with the mean free path $\bar{L}$ of the molecules, then the flow will obey the laws of hydrodynamics. If, however, the hole is small compared with the mean free path of the molecules, then the flow will take place in a completely random manner and there will be no significant disturbance in the properties of the molecules which remain in the flask.[b] This type of

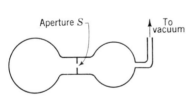

Aperture S

To vacuum

FIG. VII.7. Molecular effusion.

---

[a] It should be noted that the $\bar{L}$ to use in making calculations will depend on the properties of the molecule chosen. As we have seen (Sec. VII.8E), $\bar{L}$ will depend on the speed.

[b] In the case of hydrodynamic streaming, flow lines and velocity gradients will be set up in the vicinity of the aperture and the emergent stream will form a pronounced jet.

flow is known as molecular effusion and is predictable from the formula given in Eq. (VII.6.6) for the number of molecules striking a unit area per second, namely, $\frac{1}{4}\bar{c}N_g$. We can calculate from this formula the rate at which the density, and hence the pressure, changes in the flask:

$$\frac{dN}{dt} = -\frac{1}{4}\bar{c}N_g S = V\frac{dN_g}{dt} \qquad (VII.10.1)$$

where V is the volume of the vessel and S is the area of the aperture. This integrates directly to give

$$\ln\frac{N_g}{N_{g_0}} = -\frac{\bar{c}St}{4V} \qquad (VII.10.2)$$

or on substituting

$$\frac{N}{V} = N_g = \frac{P}{kT} \quad \text{and} \quad \bar{c} = \left(\frac{8kT}{\pi m}\right)^{\frac{1}{2}} = \left(\frac{8RT}{\pi M}\right)^{\frac{1}{2}}$$

$$\ln\frac{P}{P_0} = -\left(\frac{8RT}{\pi M}\right)^{\frac{1}{2}}\frac{St}{4V} \qquad (VII.10.3)$$

where it is assumed that the flow takes place at constant temperature.

We see that the only molecular constant involved is the molecular weight M of the gas, so that a measurement of the rate of flow of a gas through an orifice can be used to calculate M if S, the area of the orifice, is known. This formula was first verified by Knudsen[a] and has been applied since to the measurement of molecular weights of unknown gases.

If instead of a single gas we have a mixture, it can be seen from the formula that the lighter gases will diffuse through faster than the heavier gases. At equal concentrations the rates will be in the inverse ratios of the square roots of their masses. Such an effect, while small for isotopic molecules, is nevertheless sufficient, and if the separation process is repeated often enough, it is possible to make an almost quantitative separation of the species $U^{238}F_6$ (MW = 352) from $U^{235}F_6$ (MW = 349). The fractionation ratio is $(M_1/M_2)^{\frac{1}{2}} = 1.004$, and the separation was performed by the Manhattan Project.

The differential rates are such that the principle can also be employed for the microanalyses of gas mixtures. Thus Nash and Harris[b] were able to calculate composition from the effusive flows of binary and ternary gas mixtures.

It should be noted that the composition of the effluent mixture is not a random sample; rather, the effluent is richer in the high-speed molecules whose average energies can be shown to be $2RT$ as compared to $\frac{3}{2}RT$ for the gas inside. This results from the fact that high-speed molecules have a proportionately higher chance of escaping.

[a] M. Knudsen, Ann. Physik, 28, 75 (1909). Knudsen was able to measure the areas of irregular holes ($\sim$0.01 mm diameter) with the aid of a microscope.

[b] F. E. Harris and L. K. Nash, Anal. Chem., 22, 1552 (1950).

Another use of the same principle has been made by Langmuir[a] in calculating the vapor pressures of metals through measurement of the loss in weight of metal wires of known dimensions when heated in a vacuum to definite temperatures. The method depends on the principle that, at equilibrium, the number of molecules evaporating per second from the metal surface must be equal to the number striking from the gas phase. Since the latter is theoretically calculable and the former is experimentally observable (from loss in weight), the vapor pressure can be calculated. This method has been extended to organic liquids and irregular solids by Verhoek and Marshall[b] and forms the basis for the operation of molecular stills,[c] whereby high-boiling organic liquids can be successfully fractionated.

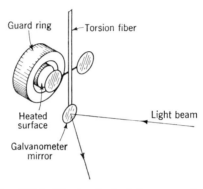

FIG. VII.8. Knudsen's absolute manometer.

When the thickness of the diaphragm is not negligibly small compared to the diameter of the hole, the molecules may be reflected back from the walls of the tube forming the aperture, and the flow is then reduced.[d]

A final example of the application of effusive flow is found in the so-called absolute manometer designed by Knudsen[e] for the measurement of very low pressures. If a disk is suspended near a heated surface at a distance which is small compared to the mean free path of gas molecules, then a molecular effusion will take place between the gas molecules in the space between the disk and surface and the rest of the gas (Fig. VII.8). The rate at which molecules enter this space will be proportional to $P_g T_g^{-1/2}$, where T_g refers to the temperature of the gas and P_g refers to the pressure. The rate at which molecules leave will be proportional to $P_s T_s^{-1/2}$, where T_s refers to the mean temperature of the space between disk and surface and

[a] I. Langmuir, *Phys. Rev.*, **2**, 329 (1913).

[b] F. H. Verhoek and A. L. Marshall, *J. Am. Chem. Soc.*, **61**, 2737 (1939).

[c] K. C. D. Hickman, *Chem. Revs.*, **34**, 51 (1944).

[d] See S. Dushman, "Scientific Foundations of Vacuum Technique," chap. 2.2, John Wiley & Sons, Inc., New York, 1949, for the various corrections which are applicable.

[e] M. Knudsen, *Ann. Physik*, **32**, 809 (1910).

P_s the pressure. This can be seen by rewriting Eq. (VII.10.1) in terms of pressures:

$$\frac{dN}{dt} = -\frac{PS}{(2\pi mkT)^{1/2}} \qquad (VII.10.4)$$

Thus at equilibrium, with equal numbers of molecules entering and leaving, we have

$$\frac{P_s}{P_g} = \left(\frac{T_s}{T_g}\right)^{1/2} \qquad (VII.10.5)$$

a result which indicates a difference in pressures and is known as "thermal transpiration," since the effect arises from the temperature gradient. Since P_s exceeds P_g, there will be a force acting on the disk proportional to $P_s - P_g$ and a resultant deflection of the fiber which can be measured optically by means of the galvanometer mirror. From this deflection, $P_s - P_g$ may be calculated and, since T_s and T_g are known, P_g and P_s may each be evaluated. The gauge may be used for pressures as low as 10^{-7} mm Hg.[a]

11. Energy Transfer in Collisions

At gas densities high enough that wall collisions are negligibly frequent compared to gas-phase collisions, changes in properties of a gas from point to point or at different times at a fixed point will occur at random as a consequence of the molecular collisions in the gas. Conversely, if by some process an unusual state that differs from the average equilibrium state is introduced into the gas, it will be quickly dissipated by the same process of collision. Molecules which can enter into chemical reactions are usually required to be energy-rich. They must acquire this energy by a series of "favorable" collisions with molecules of average energy, and if they do not react while they have this excess energy, they will lose the excess in subsequent "unfavorable" collisions with molecules of average energy. The hard sphere model is an instructive tool for examining these energy-exchange processes.

11A. Translational Exchange. In Sec. VII.8A we calculated the velocity components of two hard sphere molecules before and after a collision [Eq. (VII.8A.3)]. The change of kinetic energy ΔE of the two molecules is then equal and is given by

$$\Delta E_1 = -\Delta E_2 = \tfrac{1}{2}m_1(v_1^2 - v_1'^2) \qquad (VII.11A.1)$$

where we have dropped the subscript c (remembering that we are restricting our discussion to the components of velocity along the line of centers

---

[a] See Dushman, *op. cit.*, chap. 6, for an extended discussion of this gauge and several modifications.

at the moment of collision). By substituting from Eq. (VII.8A.3), we have
for the relative amount of kinetic energy transferred by species 1

$$\frac{\Delta E_1}{E_1} = \frac{4m_1m_2}{(m_1 + m_2)^2}\left(1 - \frac{v_2}{v_1}\right)\left(1 + \frac{m_2v_2}{m_1v_1}\right) \qquad (VII.11A.2)$$

$$= \frac{4\theta}{(1 + \theta)^2}\left(1 - \frac{v_2}{v_1}\right)\left(1 + \theta\frac{v_2}{v_1}\right) \qquad (VII.11A.3)$$

where $\theta = m_2/m_1$. Note that the quantity $4\theta/(1 + \theta)^2$ has a maximum
value of unity when $\theta = 1$ and falls to zero for both large and small θ. The
value of $\Delta E_1/E_1$ will have a maximum value of unity (for $\theta = 1$) when
$v_2/v_1 = 0$, corresponding to complete transfer of translational energy from
1 to 2.[a] One can conclude from this that translational energy is most
efficiently transferred between molecules of equal mass.[b]

11B. Inelastic Collisions. By an inelastic collision is meant one in
which the total translational energy of the system is not conserved. The
gain or deficit in translational energy must of course be balanced by other
changes in the colliding system, and these usually take the form of changes
in rotational or vibrational energy. In exceptional cases changes in electronic energy occur as well.

The smooth hard sphere model is not appropriate for describing inelastic collisions because the molecules it describes can exchange only
translational energy. However, a slight change in model makes it possible
to include rotational energy by simply introducing a coefficient of roughness a_r ($0 \leqslant a_r \leqslant 1$) which gives the amount of tangential force exerted
during a collision by two hard sphere molecules. To include vibrational
energy, we can assume that our hard sphere undergoes small oscillations
along some fixed axis in the molecule, or else if we wish to get away from
spheres, we can postulate a diatomic molecule made up of two hard spheres
m_A and m_B separated by the equilibrium distance $\sigma_{AB}^\circ \geqslant (\sigma_A^\circ + \sigma_B^\circ)/2$,
where σ_A° and σ_B° are the respective diameters of the two spheres. Figure
VII.9 shows a hypothetical collision between such a diatomic molecule of
mass $m_A + m_B$ and another hard sphere atom of mass m_C, the plane of the
paper being taken as the plane containing the centers of the three species
and the point of contact, which is coplanar with the three centers.

At the instant of collision there will be a force F exerted by the two colliding partners B and C on each other; it will have components $F_\perp$ and
$F_\|$ with respect to the axis σ_{AB} of the diatomic molecule. The perpendic-

---

[a] $\Delta E_1/E_1$ will of course become negative for values of $|v_2/v_1| > 1$, which simply means
that the head-on component of v_2 exceeds v_1 and energy is transferred to molecule 1 rather
than from it. Note that, on the average, the head-on component of translational energy
is one-third the total translational energy.

[b] These conclusions are not altered in going over to more complex molecular models,
since the utilization of the laws of conservation of momentum are not dependent on the
nature of the molecular forces.

ular component $F_\perp$ can only change the rotational energy of the diatom, while we shall assume that the component $F_\|$ is the one which is active in changing the vibrational energy of the diatom.[a]

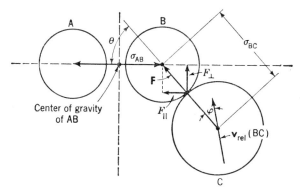

FIG. VII.9. Hypothetical collision between hard sphere atom and hard sphere molecule (diatomic).

Since we have assumed hard sphere molecules, a collision is instantaneous, and thus in the diagram the collision is essentially a hard sphere collision between B and C in which the two exchange components of velocity along their lines of centers. We can thus use Eq. (VII.8A.3) to give the components of velocity of B and C after the collision in terms of the components before the collision.

Equation (VII.11A.2) gives the fraction of the component of translational energy along the line of center which is transferred from C = 1 to B = 2. We see that the energy transfer in this case is determined exclusively by the masses of the contacting particles, in this case B and C, and that for most efficient transfer of energy the masses of B and C should be nearly equal.[b] However, of this energy, only part is available for inelastic exchange, since a considerable fraction must go into conserving the *total* momentum of the entire system, both linear momentum and angular momentum. To analyze this partition of energy in detail, let us consider two special cases in which the algebra is simplified.

11C. Vibrational Exchange. To consider the vibrational exchange, let us take a head-on collision between the atom and molecule in which the

[a] The components of these forces are of course determined by the magnitude of the instantaneous momenta of the species B and C at the moment of collision. F is proportional to the component of relative momentum of the pair B and C along the line of centers BC. A more complex case, in which we assume rough spheres, can be taken. In this case the components of momentum tangent to the spheres B and C at contact also contribute to changes in rotation and vibration.

[b] This is probably not too bad a result, since even for more realistic force fields, most of the interaction in such a collision is indeed between the contacting atoms, the forces between molecules falling very sharply with increasing distance.

centers of the three species are in a straight line as shown in Fig. VII.10. In this case no rotational energy exchange is possible. As in the previous

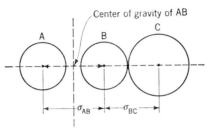

FIG. VII.10. Head-on collision between atom and diatomic molecule.

case (Fig. VII.9) the instantaneous exchange is between C and B and the fractional energy exchanged by C is given by Eq. (VII.11A.3):

$$\frac{\Delta E_C}{E_C} = \frac{4\theta_B}{(1 + \theta_B)^2}\left(1 - \frac{v_B}{v_C}\right)\left(1 + \theta_B \frac{v_B}{v_C}\right) \qquad (VII.11C.1)$$

with $\theta_B = m_B/m_C$. After the collision there must be conservation of momentum along the molecular axis, so that $m_{AB}(v'_{AB} - v_{AB}) = m_C(v_C - v'_C)$. The final component of velocity of the molecule along the line of centers v'_{AB} is given by

$$v'_{AB} = \frac{m_C}{m_{AB}}(v_C - v'_C) + v_{AB} \qquad (VII.11C.2)$$

On substituting for the final velocity component v'_C from Eq. (VII.8A.3) we have

$$v'_C = \frac{2m_B v_B}{m_{BC}} + \frac{(m_C - m_B)v_C}{m_{BC}}$$

and

$$\frac{v'_{AB}}{v_C} = \frac{2m_B m_C}{m_{AB} m_{BC}}\left(1 - \frac{v_B}{v_C}\right) + \frac{v_{AB}}{v_C} \qquad (VII.11C.3)$$

The fraction of energy transferred to vibration $\Delta E_v/E_C$ is then the difference between $\Delta E_C/E_C$ and that given to translation of AB:

$$\frac{\Delta E_v}{E_C} = \frac{\Delta E_C}{E_C} - \frac{1}{2}\frac{m_{AB}}{E_C}(v'^2_{AB} - v^2_{AB})$$

which on substitution becomes

$$\frac{\Delta E_v}{E_C} = \frac{4m_A m_B m_C}{m^2_{BC} m_{AB}}\left(1 - \frac{v_B}{v_C}\right)\left(1 + \frac{v_B}{v_C}\frac{m_{ABC}m_B}{m_A m_C} - \frac{v_{AB}}{v_C}\frac{m_{BC}m_{AB}}{m_A m_C}\right) \quad (VII.11C.4)$$

If we measure v_B, not with respect to fixed axes but relative to the center of gravity of the molecule AB, then we can write $v_B = v_{AB} + v_{BR}$, where v_{BR} is the relative velocity of B. This leads to

$$\frac{\Delta E_v}{E_C} = \frac{4m_A m_B m_C}{m^2_{BC} m_{AB}}\left(\frac{v_R}{v_C}\right)^2\left(1 - \frac{v_{BR}}{v_R}\right)\left(1 + \frac{v_{BR}m_B m_{ABC}}{v_R m_A m_C}\right) \quad (VII.11C.5)$$

where $v_R = v_C - v_{AB}$ = the relative velocity of C with respect to the center of gravity of the molecule AB. If now we break the total translational energy of the entire system into two parts E_G, the kinetic energy of the center of gravity of the entire system, and E_R, the relative kinetic energy of C with respect to AB, then

$$E_T = E_G + E_R \qquad (VII.11C.6)$$

with $E_G = \frac{1}{2}m_{ABC}v_G^2$, where $v_G = m_C v_C/m_{ABC} + m_{AB}v_{AB}/m_{ABC}$ and $E_T = \frac{1}{2}m_C v_C^2 + \frac{1}{2}m_{AB}v_{AB}^2$. Now E_G is a constant, since the total momentum of the system $m_{ABC}v_G$ is a constant, and we find on substitution that

$$E_R = \frac{1}{2}\frac{m_{AB}m_C}{m_{ABC}}(v_C - v_{AB})^2 = \frac{1}{2}\mu v_R^2 \qquad (VII.11C.7)$$

where μ is the reduced mass of the colliding system. Note that, since E_G is constant, only E_R is available for transfer to vibration.

On substitution in Eq. (VII.11C.5) and reduction, we find that

$$\frac{\Delta E_v}{E_R} = \frac{4m_A m_B m_C m_{ABC}}{m_{AB}^2 m_{BC}^2}\left[1 - \left(\frac{v_{BR}}{v_R}\right)^2\frac{m_{ABC}m_B}{m_A m_C} + \frac{v_{BR}}{v_R}\left(\frac{m_{ABC}m_B}{m_A m_C} - 1\right)\right]$$

$$(VII.11C.8)$$

If we consider the velocity-dependent terms in the brackets in Eq. (VII.11C.8), we see that the negative term involving v_{BR}^2 depends not on the algebraic sign but only on the absolute magnitude of v_{BR}. The last term in brackets will be positive when v_{BR} has the same sign as v_R and negative when the two have opposite signs.[a]

Now since $v_C - v_B = v_C - v_{AB} - v_{BR} = v_R - v_{BR}$, we see that, when v_R and v_{BR} have the same sign, the relative velocity of B and C is less than v_R, the relative velocity of AB and C (i.e., the bond AB is contracting as AB and C approach each other). Under these conditions the collision favors vibrational excitation if m_B is not too small compared to m_A and m_C. Conversely, a collision during the expansion cycle of the vibration works against vibrational excitation, or in the case that the bond energy is high compared to E_R, actually favors vibrational deactivation.

For real molecules, however, as opposed to hard sphere molecules, the collisions are not instantaneous; rather, they take place over a finite time[b] usually longer than vibrational times. The effect of such "slow" collisions is to make the momentum exchange between AB and C take place "smoothly" over many vibration cycles and thus relatively independently

[a] Unless of course the ratio $m_{ABC}m_B/m_A m_C < 1$ when the conclusions are reversed. This will happen usually if m_B is less than both m_A and m_C, or more exactly when $m_A/m_B \geqslant 1 + 2/(m_C/m_B - 1)$.

[b] Since relative velocities are of the order of about 5×10^4 cm/sec = 5×10^{12} Å/sec and molecular force fields extend over at least 1 to 2 Å, the time of a collision is about 2 to 4×10^{-13} sec. This is longer than vibrational times, which are in the region of 10^{-13} to 10^{-14} sec.

of the precise phase of the vibration. That is, we can average $(\Delta E_v / E_R)$ over many cycles, in which case $\langle v_{BR} \rangle = 0$ and $\langle v_{BR}^2 \rangle = E_0 m_A / m_{AB} m_B$,[a] so that Eq. (VII.11C.8) becomes

$$\frac{\langle \Delta E_v \rangle}{\langle E_R \rangle} = \frac{4 m_A m_B m_C m_{ABC}}{m_{AB}^2 m_{BC}^2} \left(1 - \frac{\langle E_0 \rangle}{2 \langle E_R \rangle} \right) \qquad \text{(VII.11C.9)}$$

where E_0 = the average energy of the oscillator measured from the ground state as zero.

The exchange of energy between an oscillator and a simple molecule was first analyzed from a classical viewpoint by Landau and Teller,[b] who showed that, for a "very slow" collision, the net inelastic transfer is zero. This can be seen intuitively by considering the behavior of an infinitesimal and nearly constant force applied to one atom of a vibrating molecule. On one half cycle when the force and motion are in phase there will be an increase in momentum and kinetic energy of this atom which will be almost precisely compensated in the next half cycle by a decrease in momentum and kinetic energy. Closer analysis shows that the net effect of such a force over a cycle is to slowly accelerate the entire oscillator but not to excite it.[c] The probability of inelastic transfer increases with the "hardness" of the collision. This latter is measured by the ratio of the time of a vibration to the collision time, $\tau_v / \tau_{coll} = v_R / 2\pi\nu\sigma$, where σ is the range of the intermolecular forces,[d] ν is the oscillator frequency, and v_R is the relative collision velocity.

The inelastic energy exchange thus depends, for not-too-slow collisions, only on the average relative translational energy[e] and the average oscillator energy. The mass-dependent factor in Eq. (VII.11C.9) is symmetrical in the mass ratios m_A / m_B and m_C / m_A and has a maximum value of unity

[a] During an unperturbed vibration, $m_A v_{AR} + m_B v_{BR} = 0$ from conservation of momentum, so that $m_B^2 \langle v_{BR}^2 \rangle = m_A^2 \langle v_{AR}^2 \rangle$. But the average kinetic energy of the oscillator, $F_{OK} = \frac{1}{2} [m_A \langle v_{AR}^2 \rangle + m_B \langle v_{BR}^2 \rangle] = \frac{1}{2} m_B \langle v_{BR}^2 \rangle m_{AB} / m_A$ is equal to one-half the total energy by the virial theorem.

[b] L. Landau and E. Teller, *Physik. Z. Sowjetunion*, **11**, 18 (1937). An excellent review of this and related work is to be found in the chapter on relaxation in gases by K. F. Herzfeld in "Thermodynamic and Physics of Matter," vol. I, Princeton University Press, Princeton, N.J., 1955.

[c] The quantum-mechanical problem was analyzed in semiempirical fashion by C. Zener, *Phys. Rev.*, **38**, 277 (1931), and *Proc. Cambridge Phil. Soc.*, **29**, 136 (1933), with qualitatively similar results.

[d] Actually σ is the range over which the forces change appreciably. For very small σ we have very "hard" collisions.

[e] This analysis is of course considerably oversimplified, because during a slow collision E_R changes from its initial value through a maximum (if there are attractive forces) to zero and then back again through another maximum and down to its final value. During all of this time the state of the oscillator is also undergoing a change from its initial to its final value. $\langle E_R \rangle$ is about one-half the initial value of E_R. For a quantum-mechancial system, the additional restriction of quantized vibrational levels makes the above method of analysis unusable.

when $m_A/m_B = m_C/m_B = 1 + \sqrt{2}$, although it does not change much when these ratios are in the range 1 to 4.[a] Thus we see again that inelastic collisions are most efficient when the masses in the colliding system are all comparable.

Despite the relatively high possible efficiencies of hard inelastic collisions, it turns out in practice that for real molecules these are usually quite small.[b] Part of the reason is apparent from consideration of the diagram in Fig. VII.9. For the hard sphere model used, only about one-third of all collisions will have velocity components along the molecular axis AB (i.e., about two-thirds of all collisions will take place laterally). Of this one-third, the average component of useful relative translational energy involved in the collision will be one-sixth of the total relative translational energy (averaging over θ and ϕ), so that in any given collision, only one-eighteenth of the average relative translational energy (which is $2kT$) is available for inelastic exchange with vibrations.[c]

For a quantum-mechanical oscillator, in which the energy levels are separated by energies large compared to kT, only those collisions in which the change in vibrational energy is very close to $h\nu$, where ν is the frequency of the oscillator, will be successful. This puts an even larger restriction on the possibility of an inelastic transfer of energy.

[a] In terms of E_C, E_R/E_C has the value 0.59 $v_R{}^2/v_C{}^2$ at the maximum. When $v_{AB} = 0$, so that $|v_R| = |v_C|$, we see that 40 per cent of the translational energy is unavailable for vibrational energy transfer.

[b] From sound-dispersion measurements in gases, it is possible to make estimates of the probability of deactivation of vibrations per collision, and these turn out to be of the order of 10^{-2} to 10^{-4}. See, for example, H. S. W. Massey and E. H. S. Burhop, "Electronic and Ionic Impact Phenomena," Oxford University Press, New York, 1952.

[c] This is of course for a classical oscillator which can have a continuous range of vibrational energies. Even in this case the probability falls as $E_0 \rightarrow 2E_R$[Eq. (VII.11C.9)].

VIII

Transport Properties of Gases

1. Mechanism of Transport

In Chap. VII we investigated the properties of a hard sphere gas molecule under conditions of low density and equilibrium. In the present chapter we shall extend this method to the calculation of the properties of gases under nonequilibrium conditions. As we shall see, this is a treatment which breaks down when the departure from equilibrium is great. However, it is the most fruitful method from the point of view of mathematical simplicity and the insight which it gives into the behavior of gases.

A fluid will be at equilibrium (in the absence of external fields) when there is a uniform distribution of matter and uniform pressure, composition, and temperature throughout the bulk of the fluid.[a] When any of these conditions is disturbed by the imposition of some external stress, there will be set up gradients within the fluid which create nonrandom flows that tend to bring about a new state of equilibrium. Thus when a gas is suddenly compressed, there is a momentarily greater density of matter near the moving piston and an increase in velocity of the molecules near the piston wall. These are manifested as density (or pressure) gradients and temperature gradients which lead to flows of matter and energy toward the other regions of the gas. These flows are referred to as transport processes, and they act to restore equilibrium.

We can define different types of transport in a gas as follows.

1A. Transport of Momentum. When there is a difference in pressure or momentum between two regions in a fluid, there will be a transport of momentum from one region to the other. From a molecular viewpoint this transport must be brought about by differences in the otherwise random motions of the molecules. Thus higher-velocity molecules in one region

---

[a] These conditions of equilibrium may be demonstrated thermodynamically.

will, through the mechanism of collisions, transfer this momentum to neighboring molecules, and ultimately the excess velocity is shared by the entire gas.

If we consider the transport of momentum between a gas and a wall which are in relative motion, there may be two types of effects. One of these is the transport of momentum normal to the surface, which is manifested as a force perpendicular to the surface or a pressure. The other is a tangential force which may be exerted on the wall and is referred to as a viscous drag.

1B. Transport of Energy. From a molecular viewpoint a difference in temperature between two regions of a fluid may be looked on as a difference in mean kinetic energy of the molecules in those regions. Collisions will again act to restore a uniform temperature through a sharing of the excess kinetic energy. While the transport mechanism is similar to that for momentum, the macroscopic effect is a transport of heat, or thermal conduction through the fluid.

1C. Transport of Matter. Differences in composition in a fluid will result in a flow of matter from regions of high concentration to regions of low concentration. This process is called diffusion, and it is brought about by a collision process and a process of unbalanced molecular flow.

We shall now examine, ad seriatum, the mechanisms of these different transport processes.

2. Viscosity of a Fluid

The phenomenon of friction may be described as the degradation of mechanical work (work performed by moving forces) into heat. From a molecular point of view such degradation occurs through the change in the uniform motion of an initially exerted force into an increased random motion of molecules. This latter is manifested as an increase in temperature. When such dissipation is observed in fluids, the process is referred to as internal friction, or viscosity.

We can formulate the phenomenon from a macroscopic point of view as

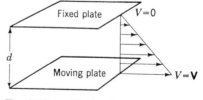

FIG. VIII.1. Illustrating viscous drag and Newtonian velocity gradient in fluid between two parallel moving plates.

follows: Let us suppose that we have two parallel, infinitely large plates that are separated by a distance d and contain between them a fluid. If now one of these plates is set into motion parallel to the other with a steady velocity V (see Fig. VIII.1), there will be a steady-state viscous force that opposes the motion of this plate and is equal and opposite to the force required to keep it in uniform motion. By symmetry there will be an equal viscous force that tends to bring the opposite plate into motion.[a] If this latter is kept fixed, then relative to it there will be a motion of the fluid parallel to the moving plate. The motion of the fluid will be different at different points between the two plates, and as a first approximation we shall assume that the velocity of the fluid varies linearly between the two plates (Newtonian flow).[b]

If **F** is the tangential force per unit area that acts on the plates and **V** is the difference in their velocity, then we can define the coefficient of viscosity η as the ratio[c]

$$\eta = \frac{\text{tangential force per unit area}}{\text{velocity gradient}}$$

$$= \frac{\mathbf{F}}{\mathbf{V}/d} = \frac{Fd}{V} \qquad \text{(VIII.2.1)}$$

From the definition we see that the coefficient of viscosity η will have the units (cgs system) of dynes-sec/cm^2. The unit 1 dyne-sec/cm^2 is called the poise, and a more frequently used unit is the centipoise (100 centipoises = 1 poise).

The mechanical power dissipated by such a system is equal to $FV = \eta V^2/d$ ergs per square centimeter of plate area per second.

3. Viscosity of a Dilute Gas

We can attempt to analyze, from the molecular viewpoint, the way in which viscous drag arises. If a surface is set in motion in a gas, the molecules striking the surface will, if the surface is rough,[d] rebound with a larger component of momentum in the direction of the motion. These molecules will in turn collide with other molecules and through a chain process transfer the added momentum received from the moving surface

---

[a] Note that these two forces, the one keeping the plate fixed and the other keeping the opposite plate in motion, act as a mechanical couple. For the system to remain in mechanical equilibrium there must be an equal and opposite couple to oppose this twisting couple. Note also that the shear imparts angular momentum and rotational energy to the gas.

[b] If the flow is non-Newtonian, it implies that there are inelastic processes in the fluid which are converting the energy imparted by the moving plate into thermal energy of the fluid. This in turn would mean that there are density and temperature gradients within the fluid. In all cases there will be such gradients and heat transfers at the walls, but normally for small velocity gradients they are small.

[c] Note that the definition does not depend on whether the fluid has a uniform gradient of velocity or not. For non-Newtonian flow, η will vary with V/d.

[d] A rough surface is defined as one for which each molecule on impact will rebound with the added momentum of the surface. A smooth surface will not be capable of exerting viscous drag. Obviously, real surfaces fall somewhere in between. The lack of accommodation is referred to as *slip*.

across the gas. It is this transport of momentum which constitutes the viscous drag.

To analyze the process, we must thus try to calculate the rate at which a gas can transport momentum. To do so, let us consider that the surface is rough and that the flow is Newtonian (constant velocity gradient). The viscous force acting will be the same throughout the system, and it will be given by the rate at which momentum is carried across a unit area.

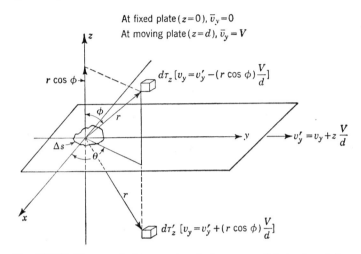

FIG. VIII.2. Transport of momentum between parallel moving plates.

Let us then consider an arbitrary plane parallel to our two plates (Fig. VIII.2). We will construct a set of cartesian axes with the y axis in the direction of the motion V and the z axis perpendicular to the motion.

If now we consider an element of area ΔS in this plane, there will be a flow of faster-moving molecules going through it from below and, under steady conditions, an equal flow of slower-moving molecules going through it from above. The difference between the two flows gives the net transport of momentum across ΔS.

In order to calculate the flow we must know something about the distribution of molecular velocities in the gas. Since the gas is not at equilibrium but only in a steady state, we cannot say that we have an equilibrium distribution. However we can make the approximation of assuming that the velocity distribution is "locally Maxwellian," i.e., that the molecules at any given point distant Z from the fixed plate have the normal distribution of velocities with respect to an average which is not zero but is given by the macroscopic stream velocity at that point. Thus at a point Z from the fixed plate the distribution is to be taken as

$$P(v_x,v_y,v_z) = \frac{1}{\pi^{3/2}\alpha^3} \exp\left(\frac{-\{v_x^2 + [v_y + (Z/d)V]^2 + v_z^2\}}{\alpha^2}\right) \quad \text{(VIII.3.1)}$$

so that though $\bar{v}_x = \bar{v}_z = 0$, $\bar{v}_y = ZV/d$, where V is the velocity of the moving plate and d is the distance between the plates.[a]

We see that the velocity and speed distribution functions are thus functions of position as well as velocity [that is, $P(v,z)$, $P(c,z)$].

We now compute as before the flow of molecules through the surface element ΔS. Let us consider a volume element in the gas $d\tau$, whose spherical coordinates with respect to ΔS are θ, ϕ, and $r(d\tau = r^2 \sin\phi \, d\theta \, d\phi \, dr)$. In this volume element there will be $\bar{Z} \, d\tau$ collisions per second, where $\bar{Z}$ is the average collision frequency previously calculated (Sec. VII.8E).

The number of molecules participating in these collisions will be $2\bar{Z} \, d\tau$, since two molecules take part in each collision. A fraction $P(c,z) \, dc$ of the molecules that have just suffered collision will have speeds in the range c to $c + dc$. A second fraction $\Delta S \cos\phi/4\pi r^2$ will have velocity directions subtended by the element ΔS. Finally, another fraction e^{-r/L_c} will have mean free paths greater than r. (L_c is the mean free path for molecules of speed c, see Sec. VII.8O.)

Thus the total number of molecules per second which leave $d\tau$ with speed c and are capable of arriving at ΔS without collision is given by the product

$$N_c = 2\bar{Z}_c \, d\tau \times P(c,z) \, dc \, \frac{\Delta S \cos\phi}{4\pi r^2} \, e^{-r/L_c}$$

$$= 2\bar{Z}_c P(c,z) \, dc \, \Delta S \, \frac{\sin\phi \cos\phi}{4\pi} \, e^{-r/L_c} \, d\theta \, d\phi \, dr \tag{VIII.3.2}$$

If the system is in a steady state, then this must also be the number of molecules that cross ΔS per second.[b] If we now assume that each molecule that suffers collision in $\Delta\tau$ has the distribution of velocities appropriate to that region, then each such molecule will have a y component of momentum which is lower than that of the molecules at the plane by an amount $(mV/d) \, r \cos\phi$. A similarly placed volume element below the plane will give a similar stream with an excess y component of momentum of the same amount. The net transport across the plane due to the two elements is thus the sum of these components or

$$\Delta\text{Mom } c = N_c \, \frac{2mV}{d} \, r \cos\phi$$

$$= \frac{4mV\bar{Z} \, \Delta S}{4\pi d} \, P(c,z) \, dc \sin\phi \cos^2\phi \, re^{-r/L_c} \, d\theta \, d\phi \, dr \tag{VIII.3.3}$$

[a] The mean square velocities are then

$$\overline{v_x^2} = \overline{v_z^2} = \frac{\alpha^2}{2} \qquad \overline{v_y^2}(Z) = \frac{\alpha^2}{2} + \left(\frac{ZV}{d}\right)^2$$

[b] Since none of these molecules suffers a collision before crossing the surface, N_c is actually the stream density of molecules of speed c that have made their last collision in $d\tau$ and are moving toward ΔS.

On integrating over the angles θ (2π to 0) and ϕ ($\pi/2$ to 0) we have for the net transport of momentum per unit area per second:[a]

$$\frac{\Delta \text{ Mom } c,z}{\Delta S} = \frac{2mV\bar{Z}}{3d} P(c) \, dc \, re^{-r/L_c} \, dr \qquad (\text{VIII.3.4})$$

By integrating over all r (0 to ∞) we find

$$\frac{\Delta \text{ Mom } c}{\Delta S} = \frac{2mV\bar{Z}}{3d} L_c^2 P(c) \, dc \qquad (\text{VIII.3.5})$$

If we make the approximation of replacing L_c by the Maxwell mean free path $\bar{L}$ averaged over all velocities, then we can integrate over all velocities and find[b]

$$\frac{\Delta \text{ Mom}}{\Delta S} = \frac{2mV}{3d} \bar{Z}\bar{L}^2 = \frac{mV}{3d} \bar{c}\bar{L}N_g \qquad (\text{VIII.3.6})$$

since $\bar{Z}(1,1) = \dfrac{1}{2}\dfrac{N_g\bar{c}}{L_c}$ (Sec. VII.8F), where N_g is the molecular density. We then find for the coefficient of viscosity:

$$\eta = \frac{F/\Delta S}{V/d} = \frac{\Delta \text{ Mom}/\Delta S}{V/d} = \frac{1}{3} mN_g\bar{c}\bar{L} \qquad (\text{VIII.3.7})$$

An exact integration of Eq. (VIII.3.5) can be made numerically[c] and leads to the value

$$\eta = \frac{1.051}{3} mN_g\bar{c}\bar{L} \qquad (\text{VIII.3.8})$$

A more important correction made by Jeans[d] has to do with the assumption that the molecule colliding in an element $d\tau$ has on the average the distribution of velocities appropriate to that element. It can be shown

---

[a] We can do this simply only by making the approximation that $P(c,z) \cong P(c)$. This will be valid only when the streaming velocity V is much less than α, the most probable speed. ($V \ll \alpha$.) When this condition is not fulfilled, then we must use the precise expression for $P(c,z)$, where $z' = r \cos \phi$, and the integrations become rather difficult. In such a case, however, when the plate is moving with molecular velocities, the use of a local Maxwellian velocity is probably very bad and the entire treatment breaks down.

[b] We have made the additional approximation of assuming that the number of collisions Z at any point is independent of d, the distance between plates. This is justifiable if the mean speed $\bar{c} \gg V L/d$, where $V L/d$ is the difference in velocity between two layers of gas separated by a mean free path. Under such conditions the molecular density in each layer is constant and most collisions then take place between molecules that have essentially the same relative Maxwellian distribution. When this condition is not satisfied, there will be important density gradients and thermal gradients, so that the entire analysis does not apply. This condition is the equivalent of saying that the velocity of the moving plate is small compared to the velocity of sound.

[c] P. G. Tait, *Trans. Roy. Soc. Edinburgh*, **33** (1887). See also E. H. Kennard, "Kinetic Theory of Gases," chaps. 3 and 4, McGraw-Hill Book Company, Inc., New York, 1938.

[d] J. H. Jeans, "Dynamical Theory of Gases," chap. 13, Cambridge University Press, New York, 1925.

that, in a collision of any given molecule, there will be a tendency for the molecule to retain some part of its component of velocity along its initial direction. For collisions between molecules of the same mass, averaged over all different velocities, this fraction will be 0.406.[a] That means that the average molecule leaving $d\tau$ after a collision has the distribution of velocities not of $d\tau$ but of some element farther away. Thus the transport is faster than calculated. The correction turns out to yield

$$\eta = 0.461 m N_0 \bar{c} \bar{L} \qquad \text{(VIII.3.9)}$$

A still more rigorous calculation, made by Chapman and Enskog,[b] in which the distribution function is no longer assumed Maxwellian gives

$$\eta = 0.499 m N_0 \bar{c} \bar{L} \qquad \text{(VIII.3.10)}$$

It is this last value which we shall use for experimental comparison. Expressed in terms of T, m, and σ it is

$$\eta = 0.499 m N_g \left(\frac{8kT}{\pi m}\right)^{1/2} \frac{1}{\sqrt{2}\pi\sigma^2 N_g}$$
$$= \left(\frac{mkT}{\pi}\right)^{1/2} \frac{1}{\pi\sigma^2} \qquad \text{(VIII.3.11)}$$

4. Comparison with Experiment

The last formula written for the coefficient of viscosity [Eq. (VIII.3.11)] indicates that η should be independent of the pressure and should vary as the square root of T. This rather surprising result with respect to pressure independence has been strikingly confirmed. Thus from 1×10^{-3} up to 20 atm pressure the change in the viscosity coefficient for most gases is less than 10 per cent. At very high pressures (above 100 atm) the viscosity becomes almost proportional to density, but here the mean free paths are of the same order of magnitude as the molecular diameters, and the whole treatment breaks down.

The result that the coefficient should increase as $T^{1/2}$ is equally surprising because it is contrary to our experience with liquids, the viscosity of which decreases as T increases. It is found experimentally that the increase in η is not given by any simple power of T but does increase faster than $T^{1/2}$. This may be qualitatively justified by considering not a spherical model for a molecule but one which is also capable of exerting attrac-

---

[a] This effect, called the persistence of velocities, arises from the fact that on the average the relative velocity during a collision will not be along the line of centers of the molecules but instead will be at some angle to it. Since only head-on components are exchanged in an encounter, this will result in part of the original forward component remaining unaffected.

[b] S. Chapman and T. G. Cowling, "The Mathematical Theory of Non-uniform Gases," chaps. 5, 10, and 12, Cambridge University Press, New York, 1939.

tive forces. A molecule capable of exerting attractive forces will have a mean collision diameter which depends on the range of the force field compared to the mean velocity of molecules. If we consider the path of a molecule near an attracting molecule, it will suffer a deflection which depends on the magnitude of the force and which will decrease as the relative velocity increases. Since the relative velocity is proportional to $T^{1/2}$, this means that the effective collision diameter $\pi\sigma^2$ should vary inversely with the absolute temperature.

Sutherland[a] has developed a formula for such a model which can be used to give a fair empirical fit over a temperature range of about 100 to 200°C for most gases. The formula may be written as a correction to the mean free path:

$$L_s = \bar{L}\,\frac{1}{1 + C/T} \qquad\qquad \text{(VIII.4.1)}$$

where C is a constant related to the attractive forces and diameter. The viscosity coefficient then becomes

$$\eta_s = \eta\,\frac{1}{1 + C/T} \qquad\qquad \text{(VIII.4.2)}$$

Experimental values of C vary from about 100 to 700 in degrees Kelvin.[b]

TABLE VIII.1. SOME EXPERIMENTAL VALUES OF VISCOSITY COEFFICIENTS

Gas	Temp range, °C	Sutherland[a] const C, °K	Coef of viscosity η(25°C), poise $\times 10^5$
Ammonia..........	−77–441	472 (401)	10.30
Argon.............	−183–827	133 (151)	22.67
Benzene...........	0–313	403 (562)	7.58
Carbon dioxide.....	−98–1052	233 (304)	15.03
Helium............	−258–817	97.6 (5)	19.66
Hydrogen..........	−258–825	70.6 (33)	9.04
Mercury...........	218–610	996 (950?)	25.03
Nitrogen..........	−191–825	102 (126)	17.80
Oxygen............	−191–829	110 (152)	20.78
Water vapor.......	0–407	659 (647)	9.84

Values taken from S. Dushman, "Scientific Foundations of Vacuum Technique," p. 37, John Wiley & Sons, Inc., New York, 1949.

[a] Values in parentheses are the critical temperatures of the gases in degrees Kelvin.

Table VIII.1 shows some of the experimental values that have been obtained for viscosity coefficients and Sutherland constants.[c] It should be

[a] W. Sutherland, *Phil. Mag.*, **36**, 507 (1893).

[b] See Landölt and Börnstein, "Physikalisch–Chemie Tabellen," Springer-Verlag, Berlin, 1923 and 1936.

[c] It might be expected that the Sutherland temperature and the critical temperature would be related, and as can be seen in Table VIII.1 that is qualitatively so.

pointed out that, because of the lack of independent data on molecular diameters, there is no good way of checking the dependence on cross section. Instead, viscosity data are usually used to calculate mean diameters and mean free paths.

5. Viscosity at Low Densities

If we consider a system in which the gas density is so low that the mean free path is of the same order of magnitude or greater than the distance between the plates, the entire mechanism of transport changes from transport through the collision between gas molecules to direct transport by collisions of molecules with the plates, since the average molecule now makes no collisions in going from one plate to the other. We can now calculate the momentum transport directly from our formula for the number of molecules striking a unit surface per second (Sec. VII.6). This rate is [Eq. (VII.6.6)]:

$$Z_w = \tfrac{1}{4} N_0 \bar{c} \tag{VIII.5.1}$$

Since each molecule thus striking has come from the opposite plate, it carries an excess of tangential component of momentum on striking the fixed plate equal to mV, where V is the difference in velocity. If on the average a fraction a of the molecules achieves the speed of the plate they collide with, it can be shown that, at a steady state, the excess velocity of molecules after striking the fixed plate is $(1 - a)V/(2 - a)$, while the excess velocity of molecules after striking the moving plate is $V/(2 - a)$.[a]

The net momentum transport per molecule is thus the difference between the two velocities times m, or $a/(2 - a) Vm$. By multiplying this by Z_w, we find the total viscous force per unit area:

$$F_s = \frac{1}{4} N_0 \bar{c} \frac{a}{2 - a} Vm \tag{VIII.5.2}$$

which on substituting for N_0 and $\bar{c}$ becomes

$$F_s = P \left(\frac{m}{2\pi kT} \right)^{\frac{1}{2}} \frac{Va}{2 - a} \tag{VIII.5.3}$$

Equation (VIII.5.3) shows that the viscous drag will be proportional to the pressure. A gauge embodying this principle has been designed by Langmuir[b] to measure very low pressures. It depends on the measure-

[a] There will be two different streams that have fixed excess components of velocity. When $a = 0$ (smooth surface), there is no momentum transport and the molecules have excess velocities $V/2$. For $a = 1$ (rough surface) the excess velocities become the excess velocities of the plates, namely 0 and V, respectively. The coefficient a is known as the accommodation coefficient for momentum and must be measured experimentally.

[b] S. Dushman, "Scientific Foundations of Vacuum Technique," chap. 6, John Wiley & Sons, Inc., New York, 1949.

ment of the damping force exerted on a suspended disk by a rarefied gas. However, it must be calibrated against other gases because of the accommodation coefficient a that appears in Eq. (VIII.5.3). A curious result of this low-pressure viscosity is that the viscous drag is independent of the separation of the two walls as long as the separation is much smaller than the mean free path.

6. Thermal Conductivity

The treatment of thermal conductivity parallels quite closely the treatment given for viscosity. Let us suppose that we have two parallel plates (Fig. VIII.3) at fixed temperatures T_1 and T_2, with a fluid between them.

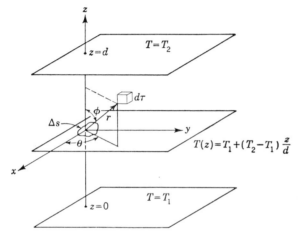

Fɪɢ. VIII.3. Thermal conduction between parallel plates.

There will be a transport of energy from the hotter plate T_2 to the colder plate T_1, again by a mechanism of molecular collision. We can define a coefficient of thermal conductivity Ω for the fluid as the ratio

$$\Omega = \frac{\text{heat carried per unit area per second}}{\text{temperature gradient}} \qquad \text{(VIII.6.1)}$$

The units of Ω are usually cal/deg-cm-sec.

Following the method already used, we select a plane at a distance Z from one of the two parallel surfaces ($Z = 0$ at T_1, $Z = d$ at T_2, $T_2 > T_1$) and choose an element of area ΔS in it. If we assume that the temperature varies linearly through the gas (Newtonian heat flow), the temperature gradient is a constant equal to $(T_2 - T_1)/d$ (Fig. VIII.3).

At a steady state, the flow of molecules through the area ΔS must be the same in both directions, while the net perpendicular component of momentum transported across ΔS must be independent of the location of

ΔS. These two conditions cannot be satisfied simultaneously without choosing distribution functions for molecular density and velocity which are functions of Z. To avoid such a choice, which would lead to an almost impossibly complicated calculation, let us rather make a crude but simpler calculation which leads to results of the correct order of magnitude and is in qualitative agreement with experiment.

Let us assume that the flow of molecules through ΔS from one side is given by the equilibrium value, $\frac{1}{4}N_0\bar{c}\,\Delta S$ [Eq. (VII.6.6)]. If now these molecules are assumed to have an average temperature characteristic of the gas at a distance proportional to the mean free path $\bar{L}$, the flow from the hot side will bring an excess of energy through ΔS given by

$$\Delta E_h = \frac{1}{4}N_0\bar{c}a\,\frac{\bar{L}}{d}\,(T_2 - T_1)C_v\,\Delta S \qquad \text{(VIII.6.2)}$$

where C_v is the molecular heat capacity of the gas and a is some constant of proportionality. The net energy flow per unit time across ΔS is twice this since there will be an equal flow of energy-poor molecules from the cold side:

$$\Delta E_t = 2\,\Delta E_h$$
$$= \frac{a}{2}N_0\bar{c}\bar{L}\,\frac{T_2 - T_1}{d}\,C_v\,\Delta S \qquad \text{(VIII.6.3)}$$

whereupon we have for the transport per unit area

$$\frac{\Delta E_t}{\Delta S} = \frac{a}{2}N_0\bar{c}\bar{L}\,\frac{T_2 - T_1}{d}\,C_v \qquad \text{(VIII.6.4)}$$

On dividing by the temperature gradient, $(T_2 - T_1)/d$, we obtain the coefficient of thermal conductivity

$$\Omega = \frac{\Delta E_t/\Delta S}{(T_2 - T_1)/d} = \frac{a}{2}N_0\bar{c}\bar{L}C_v \qquad \text{(VIII.6.5)}$$

But since the coefficient of viscosity η is equal to $\frac{1}{2}mN_0\bar{c}\bar{L}$ [Eq. (VIII.3.10)], we can write this as

$$\Omega = a\,\frac{C_v}{m}\,\eta \qquad \text{(VIII.6.6)}$$

It has been shown by Chapman,[a] by a very careful calculation in which proper approximations to the true distribution functions are used, that the constant $a = 2.522$ for hard sphere molecules and lies between 2.5 and 2.522 for molecules that repel each other as some inverse power of the distance. Thus the value of 2.5 may be taken without serious error.

It is found when rotational energy is considered[a] that the value of a falls to about 1.8 owing to the fact that the transfer of both translational and rotational energy is not as efficient as translational energy alone.

[a] Chapman and Cowling, op. cit., chap. 13.

For polyatomic gases that have internal vibrational energy, Eucken[a] has suggested a semiempirical formula for a which seems to fit the data fairly well:

$$a = \tfrac{1}{4}(9\gamma - 5) \qquad\qquad\qquad \text{(VIII.6.7)}$$

where γ = ratio of the specific heats C_p/C_v.

Table VIII.2 shows some of the measured values of Ω at 0°C and the calculated and experimental values of a.

TABLE VIII.2. THERMAL CONDUCTIVITIES OF GASES AT 0°C

Gas	$\Omega \times 10^{5,a}$	$a_{\exp}$	$a_{\text{calc}}{}^b$	Gas	$\Omega \times 10^{5,a}$	$a_{\exp}$	$a_{\text{calc}}{}^b$
H_2	41.6	2.02	1.92	NO	5.55	1.86	1.90
D_2	30.8			O_2	5.85	1.91	1.89
He	35.2	2.51	2.50	Ar	3.97	2.53	2.50
CH_4	7.21	1.73	1.70	CO_2	3.52	1.67	1.68
NH_3	5.22	1.41	1.70	N_2O	3.68	1.74	1.68
Ne	10.87	2.47	2.50	SO_2	2.06	1.49	1.64
CO	5.59	1.91	1.91	Cl_2	1.83	1.79	1.80
C_2H_4	4.07	1.44	1.56	Kr	2.12	2.54	2.50
N_2	5.80	1.97	1.91				

Values taken from S. Chapman and T. G. Cowling, "The Mathematical Theory of Non-uniform Gases," chap. 13, Cambridge University Press, New York, 1939.

[a] In cal/°C-cm-sec.

[b] See Eq. (VIII.6.7).

7. Thermal Conductivity at Low Pressures

When the gas density between two plates at different temperatures is such that the mean free path of the gas is much greater than the distance between the plates, the transport of thermal energy is directly by molecular impacts upon the plates. This process can be analyzed by following the procedure used for momentum transport at low densities (Sec. VIII.5).

If a is used to represent the accommodation coefficient of energy transfer between molecules and plates,[b] then the gas between the two plates may be assumed to consist of two independent Maxwellian distributions of densities N_1 and N_2 and temperatures T_1' and T_2', where the plate temperatures are T_1 and T_2, respectively, and the total gas density is $N_t = N_1 + N_2$.

[a] Eucken, *Physik. Z.*, **14**, 324 (1913).

[b] We define a as the fraction of the energy difference transferred upon impact between a molecule and a plate. Thus if a molecule has a mean temperature T' and strikes a plate at temperature T, it will rebound with a mean temperature $T' + a(T - T')$. Since the energy transfer of molecules is given by the product $C_v \Delta T$, the energy transfer will be $aC_v(T - T')$. In the above we will assume that T_2' is the temperature of the molecules after striking the plate T_2.

After a steady state has been reached, the total flow of molecules to the hot plate T_2 must be equal to the total flow of molecules to the cold plate. By using the formula for the number of impacts per unit area [Eq. (VII.6.6)] we have[a]

$$Z_w = \tfrac{1}{2}N_1\bar{c}_1 = \tfrac{1}{2}N_2\bar{c}_2 \qquad \text{(VIII.7.1)}$$

where $\bar{c}_1 = (8kT'_1/\pi m)^{1/2}$ and $\bar{c}_2 = (8kT'_2/\pi m)^{1/2}$ so that we can write

$$N_1(T'_1)^{1/2} = N_2(T'_2)^{1/2} \qquad \text{(VIII.7.2)}$$

The total net flow of energy[b] carried by these two streams per unit area per second is then given by the product

$$\begin{aligned}\Delta E_s &= Z_w(E_2 - E_1) \\ &= \tfrac{1}{2}N_1\bar{c}_1 C_v(T'_2 - T'_1) \qquad \text{(VIII.7.3)}\end{aligned}$$

As before, we can show that, after a steady state, $T'_2 = T_2 - [(1 - a)/(2 - a)] \Delta T$ and $T'_1 = T_1 + [(1 - a)/(2 - a)] \Delta T$, where $\Delta T = T_2 - T_1$. On substituting this in the above expression, we find

$$\Delta E_s = \frac{1}{2} N_1\bar{c}_1 C_v \Delta T \frac{a}{2 - a} \qquad \text{(VIII.7.4)}$$

which is the energy flow per unit area per second between the plates. But in terms of the total density N_t we have

$$N_1\bar{c}_1 = N_t \frac{\bar{c}_1\bar{c}_2}{\bar{c}_1 + \bar{c}_2} = N_t c_r \qquad \text{(VIII.7.5)}$$

so that we can write

$$\Delta E_s = \frac{1}{2} N_t C_v \Delta T \frac{a}{2 - a} c_r \qquad \text{(VIII.7.6)}$$

When ΔT is small compared to T_1, this may be reduced to an approximate form

$$\Delta E_s \cong \frac{1}{4} N_t\bar{c}_1 C_v \Delta T \frac{a}{2 - a} \qquad \text{(VIII.7.7)}$$

If the two plates are themselves immersed in a larger body of gas at temperature T_0, pressure P_0, and density $N_0 = P_0/kT_0$, then at steady-state conditions $N_0\bar{c}_0 = N_t\bar{c}_1$, so that we can write for the loss per unit temperature difference

---

[a] We use a factor of $\tfrac{1}{2}$ instead of $\tfrac{1}{4}$ because we have separated the gas into two distributions of density N_1 and N_2. The N_1 distribution have velocity components only toward plate 2 because they have all come from plate 1. This constitutes a reflection of all the reverse components and thus a doubling of the fraction of molecules moving in the forward direction.

[b] Note that there also will be a net transport of the perpendicular component of momentum because the component of velocity of the hot stream is greater than that of the cold stream. This results in a difference in pressure between the two plates that tends to push them apart. Because of the symmetry there is no tangential force acting.

$$\frac{\Delta E_0}{\Delta T_0} \cong \frac{P_0}{(2\pi mkT_0)^{1/2}} C_v \frac{a}{2-a} \tag{VIII.7.8}$$

This equation indicates that the heat loss at constant external temperature is proportional to the pressure P_0 (for a given gas). The principle of heat loss may thus be used in measuring very low pressures. The most common types of gauges that embody this idea are the thermocouple and Pirani gauges, in which the heat loss measured is from a thin, heated wire. These gauges are described in detail by Dushman;[a] they are widely used in high-vacuum work. Their principal disadvantage is that they are usually limited in use to the pressure range 0.1 to 0.001 mm Hg and must be calibrated for each different gas.

8. Mass Diffusion

If we have a container divided into two halves by a thin diaphragm with two different gases at the same temperature and pressure on opposite sides of the diaphragm, when the diaphragm is removed there will be a mixing of the two gases brought about by the random motion of the gas molecules. The process of pure gaseous diffusion is defined as this random mixing in the absence of convection currents or pressure gradients.[b]

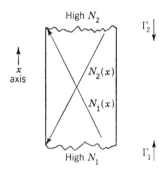

FIG. VIII.4. Illustrating mass diffusion in one dimension.

Under the conditions outlined there will be a net molecular transport of each species across the gas from the regions of high to those of low concentrations (Fig. VIII.4). If this diffusion takes place at constant pressure, then at all points $N_1 + N_2 = N_t =$ const. If we further assume that the flow is along the x axis of the tube (no horizontal gradient), then at each point

$$\frac{dN_1}{dx} = -\frac{dN_2}{dx} \tag{VIII.8.1}$$

If now we call Γ_1 and Γ_2 the net relative molecular flows of species 1 and 2 per unit area per unit time, we can define a coefficient of mass diffusion for each:

$$\Gamma_1 = -D_1 \frac{\partial N_1}{\partial x} \qquad \Gamma_2 = -D_2 \frac{\partial N_2}{\partial x} \tag{VIII.8.2}$$

---

[a] S. Dushman, *op. cit.*

[b] Normally when this experiment is performed an initial pressure gradient and temperature difference are established between the two gases. L. Miller, *Z. Naturforsch.*, **4A**, 262 (1949), has observed that under such conditions H_2 will be heated by 0.74°C on diffusing into argon which is itself cooled by 2.0°C.

where D_1 and D_2 are the coefficients of single diffusion for gases 1 and 2, respectively.[a] The units of D_1 and D_2 are generally unit area per unit time, or cm^2/sec if Γ_1 and N_1 are measured in similar units. We see by comparing Eqs. (VIII.8.1) and (VIII.8.2) that $\Gamma_1 + \Gamma_2 = 0$ only when $D_1 = D_2$. That is, there is uniform concentration throughout the system only when the diffusion coefficients are equal. Since in general there is no reason for assuming that $D_1 = D_2$, but quite to the contrary, we see that $\Gamma_1 + \Gamma_2 \neq 0$ and there will be an accumulation of molecules in one section of our column. Under steady-state conditions this is compensated for by a mass convection of the entire gas opposed to this motion. We can then write for the steady-state conditions

$$\Gamma_1 + \Gamma_2 + N_t v_c = 0 \quad \text{or} \quad (\Gamma_1 + N_1 v_c) + (\Gamma_2 + N_2 v_c) = 0 \quad (\text{VIII.8.3})$$

where v_c is the convection. From an experimental point of view we can only hope to measure the net molecular motions defined by

$$P_1 = \Gamma_1 + N_1 v_c \qquad P_2 = \Gamma_2 + N_2 v_c \qquad (\text{VIII.8.4})$$

where P_1 and P_2 are the observable permeation rates of species 1 and 2.

In terms of P_1 and P_2 we can define a new diffusion coefficient for the mixture:

$$P_1 = -D \frac{dN_1}{dx} \qquad P_2 = -D \frac{dN_2}{dx} \qquad (\text{VIII.8.5})$$

which now satisfies the condition $P_1 + P_2 = 0$.[b]

We can now employ our usual methods to calculate, at any point in the tube, the net flows of species 1 and 2 across a unit area. If we take a plane perpendicular to the direction of the concentration gradients (i.e., the x axis) then there will be a net flow of species 1 across this plane owing to the fact that there is a difference in the number of molecules of species 1 that strike it from opposite sides (Fig. VIII.5). By making the usual assumption of local equilibrium distributions we can use our formula [Eq. (VII.6.6)] for the number of collisions per unit area per second. We then have for the number of collisions of species 1 with a unit area per second, made on the high-concentration side,

$$Z_w(+) = \frac{1}{4} \bar{c}_1 \left(N_1 + b_1 \bar{L}_1 \frac{dN_1}{dx} \right) \qquad (\text{VIII.8.6})$$

and from the low-concentration side:

[a] By this we mean that under steady-state conditions such that there is no motion of the species 2, then Γ_1 will be the observed rate of diffusion of species 1. This is approximately the condition in a desiccator when a hydrate loses water by diffusion to a dehydrating agent through an essentially stagnant layer of air. It is also the permeation measured by radioisotope experiments.

[b] Under steady-state conditions we can also show that $N_2 D_1 + N_1 D_2 = N_t D$. Note that the driving force for the convection must be a pressure gradient in the gas opposing the diffusion of the more rapid species.

$$Z_w(-) = \frac{1}{4}\bar{c}_1\left(N_1 - b_1\bar{L}\frac{dN_1}{dx}\right) \tag{VIII.8.7}$$

where b_1 is some constant which will depend on the average distance traversed by a molecule before striking ΔS and dN_1/dx is the concentration gradient at the plane. We then find for the net flow:

$$\Gamma_1 = Z_w(+) - Z_w(-) = \frac{1}{2}b_1\bar{c}_1\bar{L}_1\frac{dN_1}{dx} \tag{VIII.8.8}$$

from which we have for the diffusion coefficient D_1

$$D_1 = \tfrac{1}{2}b_1\bar{c}_1\bar{L}_1 \tag{VIII.8.9}$$

By a similar calculation we find for D_2 the symmetrical value

$$D_2 = \tfrac{1}{2}b_2\bar{c}_2\bar{L}_2 \tag{VIII.8.10}$$

and finally for the coefficient for the mixture[a]

$$D = \frac{1}{N_t}(N_2D_1 + N_1D_2)$$

$$= \frac{\bar{b}}{2}\left(\frac{N_2}{N_t}\bar{c}_1\bar{L}_1 + \frac{N_1}{N_t}\bar{c}_2\bar{L}_2\right) \tag{VIII.8.11}$$

If we assume that concentration gradients are small and apply the same method for calculating $Z_w(\pm)$ that we used for viscosity, we find that

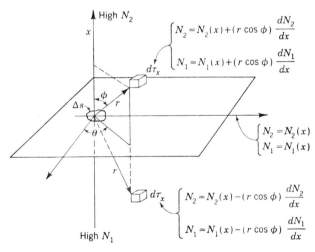

FIG. VIII.5. Illustrating mass diffusion in gases in one dimension.

$\bar{b} = \frac{2}{3}$. This formula, however, is not in good agreement with experiment, since it predicts for mixtures that have large differences in molecular

[a] This assumes $b_1 = b_2 = \bar{b}$.

weight a correspondingly large change in D with composition which is not observed experimentally.

A better agreement is obtained if we replace $\bar{L}_1$ by the quantity $\bar{L}(1,2)$ and $\bar{L}_2$ by the quantity $\bar{L}(2,1)$. $\bar{L}(1,2)$ is the mean free path of species 1 with respect only to species 2 and similarly for $\bar{L}(2,1)$.

$$\bar{L}(1,2) = \left(\frac{\mu}{m_1}\right)^{1/2} \frac{1}{\pi N_2 \sigma_{12}^2} \qquad \bar{L}(2,1) = \left(\frac{\mu}{m_2}\right)^{1/2} \frac{1}{\pi N_1 \sigma_{12}^2} \qquad \text{(VIII.8.12)}$$

where μ is the reduced mass, $1/\mu = 1/m_1 + 1/m_2$ [Eq. (VII.8F.5)]. Our equation then reduces to

$$D = \frac{\bar{b}}{2} \frac{1}{\pi N_t \sigma_{12}^2} \left(\frac{8kT}{\pi \mu}\right)^{1/2} \qquad \text{(VIII.8.13)}$$

A similar formula was obtained by Langevin in 1905 and later, on more rigorous grounds, by Chapman,[a] who calculated the value of the coefficient

$$\bar{b} = \frac{3\pi}{16} = 0.588 \qquad \text{(VIII.8.14)}$$

a value not too far from the $\frac{2}{3}$ above.

The use of the mean paths $\bar{L}(1,2)$ instead of $\bar{L}_1$ for the mixture implies that each gas is hindered in its diffusion only by the other. This is quite reasonable if we consider the actual collision process. When two identical molecules collide, they merely exchange head-on components of momentum and there is no net effect on the total component of momentum of the species along the direction of flow. Thus such collisions will to a first approximation not affect the molecular flows[b] or thus the diffusion. It is quite the other case with collisions between unlike pairs because such collisions do transfer components of momentum from one distinguishable species to the other.

9. Experimental Results for Diffusion

Chapman's approximate formula for the coefficient of diffusion may be rewritten as

$$D = \frac{3\pi}{32} \frac{1}{\pi N_t \sigma_{12}^2} \left(\frac{8kT}{\pi \mu}\right)^{1/2} \qquad \text{(VIII.9.1)}$$

This equation predicts a diffusion constant which is independent of composition, and that seems to be the experimental finding. The maxi-

[a] S. Chapman, *Phil. Trans. Roy. Soc. (London)*, **217A**, 115 (1918).

[b] The same result applies to the calculation of the number of molecules striking a wall. We obtain the result $\frac{1}{4} \bar{c} N_g$ for point molecules and the same result when we consider hard sphere molecules. The result is increased in the ratio of $1 + b/v$ when we consider dense gases, where b is the Van der Waals constant. At ordinary conditions, however, the correction b/v is small.

mum variation of D for mixtures is usually less than 10 per cent over the entire composition range even for extremely different gases such as H_2 and CO_2.

The inverse variation of D with pressure N_t has also been verified, while the temperature dependence (like that shown by viscosity) seems to be greater than that predicted.

An idea of the agreement of the formula with the facts may be obtained by a comparison of the mean diameters σ_{12} calculated from this equation and those computed from measurements of the viscosity of mixtures. A list of such values is given in Table VIII.3, which is taken from Chapman and Cowling.[a]

TABLE VIII.3. COEFFICIENTS OF DIFFUSION AND MEAN COLLISION DIAMETERS

Gas mixture	Mean coef of diffusion D, cm^2/sec^a	Mean collision diam σ_{12}, Å[b]	Mean collision diam from viscosity data $\frac{1}{2}(\sigma_1 + \sigma_2)$, Å
He-A	0.641	2.61	2.92
H_2-D_2	1.20	2.46	2.74
H_2-O_2	0.697	2.94	3.17
H_2-N_2	0.674	3.01	3.24
H_2-CO	0.651	3.05	3.25
H_2-CO_2	0.550	3.30	3.68
H_2-CH_4	0.625	3.14	3.44
H_2-SO_2	0.480	3.52	4.11
H_2-N_2O	0.535	3.35	3.70
H_2-C_2H_4	0.625	3.53	3.84
O_2-N_2	0.181	3.45 (3.49)[c]	3.69
O_2-CO	0.185	3.41 (3.53)	3.69
O_2-CO_2	0.139	3.73 (3.78)	4.12
CO-N_2	0.192	3.44 (3.60)	3.76
CO-CO_2	0.137	3.83 (3.89)	4.20
CO-C_2H_4	0.116	4.38 (4.37)	4.36
CO_2-N_2	0.144	3.74 (3.85)	4.19
CO_2-CH_4	0.153	4.08 (3.98)	4.39
CO_2-N_2O	0.096	4.30 (4.19)	4.65

[a] These are mean values over composition range at 0°C calculated from values given in the International Critical Tables.

[b] These hard sphere, mean diameters are calculated by Eq. (VIII.9.1).

[c] Values in parentheses obtained by assuming $\sigma(H_2) = \sigma(D_2) = 2.46$ Å.

We see that the viscosity diameters are about 10 per cent larger than the mean diameters calculated from Eq. (VIII.9.1). The difference probably arises from the inaccuracy in approximating the molecules by hard spheres. It probably also arises from the fact that quantum-mechanical

[a] Op. cit., p. 252.

considerations predict that interactions between identical molecules will have a larger cross section than will interactions between unlike molecules.[a] The viscosity data are of course for collisions of like particles. An idea of the self-consistency of the data may be obtained by comparing the figures in parentheses (Table VIII.3, column 3) with the adjacent figures. The values in parentheses were obtained by taking the diameter of $H_2 = D_2 = 2.46$ Å. It can be seen that these values agree within better than 4 per cent, which is fairly good considering the crudeness of the hard sphere model used.

10. Generalized Diffusion; Thermal Diffusion

Chapman, Enskog, and Cowling[b] have developed a general method for dealing with nonequilibrium states in not-too-dense gases which allows the approximate evaluation of the proper distribution functions to be used in place of the equilibrium distribution functions. While we shall not go into the details of the method, one of the results is of direct interest in the problem of diffusion in mixtures.

The authors have shown that the difference between the mean local velocities, evaluated over a small domain in a gas, may be written as

$$\mathbf{C}_1 - \mathbf{C}_2 = -\frac{N_t^2 D}{N_1 N_2}\left[\frac{\partial(N_1/N_t)}{\partial \mathbf{r}} + \frac{N_1 N_2(m_2 - m_1)}{N_t \rho_t}\frac{\partial \ln P}{\partial \mathbf{r}}\right.$$
$$\left. - \frac{\rho_1 \rho_2}{P \rho_t}(\mathbf{F}_1 - \mathbf{F}_2) + \frac{k_t}{T}\frac{\partial T}{\partial \mathbf{r}}\right] \qquad \text{(VIII.10.1)}$$

where $\mathbf{C}_1$ and $\mathbf{C}_2$ are the mean vector velocities[c] of species 1 and 2 at the point whose vector location is $\mathbf{r}$, pressure is P, and temperature is T. N_1 and N_2 are the molecular concentrations, and ρ_1 and ρ_2 are the densities of 1 and 2 such that $N_1 + N_2 = N_t$, $N_1 m_1 = \rho_1$, $N_2 m_2 = \rho_2$, and $\rho_1 + \rho_2 = \rho_t$. The molecular weights are m_1 and m_2, respectively, and $\mathbf{F}_1$ and $\mathbf{F}_2$ are the external forces (per unit mass) acting on species 1 and 2 at the point $\mathbf{r}$. D is the already discussed diffusion coefficient which as we have seen [Eq. (VIII.9.1)] depends on the variables T, N, m_1, and m_2 but not on composition (first approximation). The quantity k_t is dimensionless and is known as the coefficient of thermal diffusion. In general it is a complex function of all the variables, including composition.

[a] N. F. Mott and H. S. W. Massey, "The Theory of Atomic Collisions," 2d ed., chap. 12, Oxford University Press, New York, 1949. See also H. S. W. Massey and E. H. S. Burshop, "Electronic and Ionic Impact Phenomena," chaps. 7 and 8, Oxford University Press, New York, 1952.

[b] Chapman and Cowling, loc. cit.

[c] This means that we can write three independent equations for the x, y, z components of each vector quantity above. Each such equation will have the form of Eq. (VIII.10.1) and contain only quantities that involve the component chosen.

The significance of the above expression is that there will be a difference in the net molecular flows of two species in a gas mixture and hence a separation of species under conditions determined by the quantities in brackets.

The first term in the brackets $\partial(N_1/N_t)/\partial\mathbf{r}$ is simply the concentration gradient of species 1 and would give us the normal equation for mass diffusion at constant temperature and pressure.

The second term, involving the pressure gradient $\partial \ln P/\partial\mathbf{r}$, indicates that there will be a separation of molecules of different masses in a system in which there is a pressure gradient. Under such conditions (i.e., centrifugal fields) the heavier molecules tend to concentrate in the regions of high pressure.[a] The third term is different from zero only when there is a force field acting on the molecules which is different for different species. This would occur for example if one of the species were ionic and $\mathbf{F}$ represented an electric or magnetic field. The diffusion caused by such asymmetrically acting fields is spoken of as forced diffusion, and it can be observed in ionic migration.

The final term indicates that temperature gradients will cause a difference in diffusion of two species. In general the form of k_t is such that heavier molecules will migrate to regions of high temperature while lighter molecules will concentrate in regions of low temperature. This effect was first predicted by Chapman[b] in 1916 and has since been verified and extensively studied. Its sign is such that it opposes normal diffusion and will tend to produce experimentally smaller diffusion coefficients D if care is not taken to ensure uniform temperatures.

An exact calculation of k_t is extremely difficult, but approximate calculations have been made by Chapman.[c] The only case in which a reasonably simple formula is possible is that of a Lorentzian gas, i.e., one in which the difference in masses of the two atoms is large. For such a gas that has in addition a force field between two molecules separated by a distance r:

$$F(r) = \frac{a}{r^\nu} - \frac{c}{r^3} \qquad \text{(VIII.10.2)}$$

where a, c, and ν are constants (a/r^ν is the repulsive force), k_t is approximately given by

$$k_t = \frac{N_2}{N_t}\left[\frac{\nu - 5}{2(\nu - 1)} - ST^{-(\nu-3)/(\nu-1)}\right] \qquad \text{(VIII.10.3)}$$

[a] The earth's atmosphere would be expected to show such a concentration gradient were it not for the mixing caused by winds, etc.

[b] S. Chapman, *Phil. Trans. Roy. Soc. (London)*, **217A**, 115 (1918).

[c] E. Whalley and E. R. S. Winter, *Trans. Faraday Soc.*, **46**, 517 (1950), have succeeded in obtaining an equation similar to Chapman's and derived from the simple mean free path considerations developed here. The work is based on the earlier theory of Fürth, *Proc. Roy. Soc. (London)*, **A179**, 461 (1942), and permits evaluation of the coefficients k_t in excellent agreement with experimental data.

where S is a constant. It can be seen that k_t is thus a direct function of the exponent ν and so can be used to measure approximately the form of the force law between molecules.

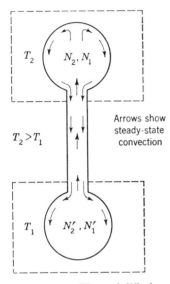

FIG. VIII.6. Thermal diffusion.

Experimentally k_t can be measured by connecting two vessels each of which contains a gas mixture and is at a different temperature than the other (Fig. VIII.6). When steady-state conditions have been established, there will be a difference in composition between the two vessels. Since there is no net flow at the steady state and no pressure gradients or asymmetric force field we see that [Eq. (VIII.10.1)]:

$$\frac{\partial(N_1/N_t)}{\partial \mathbf{r}} = -\frac{k_t}{T}\frac{\partial T}{\partial \mathbf{r}} \quad \text{(VIII.10.4)}$$

whereby we have at the steady state

$$k_t = \frac{N_1/N_t - N_1'/N_t'}{\ln (T_1/T_2)} \quad \text{(VIII.10.5)}$$

from which k_t can be calculated.

It is found that k_t is usually of the order of 0.1 or less. In a typical experiment with H_2-N_2 mixtures ($T_2 = 274°C$, $T_1 = 11°C$) the values of k_t reach a maximum of 0.10 for $\frac{2}{3}H_2$ and $\frac{1}{3}N_2$ and fall almost linearly to zero on both sides of this maximum. This would give a difference in mole fractions $H_2(T_1) - H_2(T_2)$ of only about 0.06.[a]

It is, however, possible by utilizing a cascade of such separations to multiply the separation. Clusius and Dickel[b] devised a column that consists of a long tube that is externally cooled and contains a heated wire along the axis. Convection causes a fractionation effect, since hot gas hitting the cold wall falls, cycles, and is reheated at the wire. The multiple effect of such convection cycles is such that it was possible to make an almost 100 per cent separation of H_2-CO_2 mixtures between the top and bottom of the column. ($T_{\text{wire}} = 600°C$, length of tube = 1 m.) The column has since been extensively developed for isotopic separation.

11. Criticism of the Nonequilibrium Kinetic Theory

The theory which we have developed for the kinetic behavior of nonequilibrium systems suffers at two important points. The most serious

[a] S. Chapman and F. W. Dootson, *Phil. Mag.*, **33**, 248 (1917). For detailed discussion of the theory see R. C. Jones and W. H. Furry, *Revs. Modern Phys.*, **18**, 151 (1946).

[b] K. Clusius and G. Dickel, *Naturwiss.*, **26**, 546 (1938); **27**, 148 (1939).

difficulty encountered is that we have had to make use of equilibrium distribution functions in order to simplify the mathematical calculations. This difficulty has to a considerable extent been rectified by the methods developed by Chapman, Enskog, and others in which a series of successive approximations makes it possible to obtain nonequilibrium distribution functions that more closely approach the physical system.

A more basic difficulty and one not yet adequately resolved is that encountered in the use of artificial models to represent molecules. From a rigorous point of view the entire behavior of a molecular encounter is determined by the force field surrounding each molecule. By representing molecular force fields by artificial models we avoid the impossible mathematical problem involved in the rigorous approach. The result, however, is to introduce an entirely new set of molecular parameters which remain as yet unpredictable from simpler molecular properties. In the case of the hard sphere model we have introduced the molecular diameter σ and the definition of a collision as two new parameters. But there exist two additional parameters which were somewhat concealed in the discussion, namely, the two accommodation coefficients, one for velocity transfers between molecules in collision and the other for collision between molecules and surfaces.

For the more complicated molecular models such as, for example, those that assume central forces, we replace the above set of parameters by a new set involved in defining the force field. If we add to this the problem of complex molecules (i.e., those with internal structure), then there is the additional set of parameters needed to define the interactions between the internal molecular motions and the external force fields. From the point of view of the hard sphere model this would involve the definition of still more accommodation coefficients to describe the efficiency of transfer of internal energy between colliding molecules.

Despite these difficulties, the kinetic theory in its simple equilibrium approximation and in its more accurate nonequilibrium representation is capable of reproducing physical behavior in a form which is mathematically simple, qualitatively correct in so far as it represents the interdependence of physical variables, and quantitatively correct to within better than an order of magnitude. As such it presents a valuable direct insight into the relations between molecular processes and macroscopic properties and, as we shall see, provides a valuable guide to understanding kinetic behavior.

However, before we proceed to an application of these findings to systems undergoing chemical change, we shall undertake in the next chapter the discussion of still more powerful methods for the analysis of the molecular behavior of equilibrium systems.

IX

Statistical Treatment
of Interacting Systems

1. The Canonical Distribution

While the equilibrium kinetic theory permits us to develop in fairly simple manner the properties of a dilute hard sphere gas, it becomes progressively more complicated and difficult to apply to both dense systems and systems in which there are forces acting between particles. To deal with such systems, we shall here outline briefly the very powerful statistical methods of Gibbs.[a]

From the classical mechanical point of view discussed in Sec. VI.1, any system of N particles is uniquely defined by a knowledge of $6N$ independent pieces of information together with the description of the components of the system (masses, force fields, etc.). These $6N$ quantities may be looked upon as the $6N$ constants of integration implicit in Newton's differential equations of motion.

We have seen that a statistical description of such a system is possible if we can express its average behavior in terms of a distribution function. Such a distribution function will in general be a function of all the individual variables needed to define the molecular system and also of the time. When the system is in a state of equilibrium, the distribution function which describes it is no longer a function of time but only of the molecular variables.

[a] J. W. Gibbs, "The Collected Works of J. Willard Gibbs," vol. II, "Statistical Mechanics," Yale University Press, New Haven, Conn., 1948. For a more detailed discussion of the statistics of both independent and interacting systems the student is referred to J. E. Mayer and M. G. Mayer, "Statistical Mechanics," John Wiley & Sons, Inc., New York, 1940, and also to R. H. Fowler and E. A. Guggenheim, "Statistical Thermodynamics," Cambridge University Press, New York, 1949. M. Dole, "Introduction to Statistical Thermodynamics," Prentice-Hall, Inc., Englewood Cliffs, N.J., 1954, is excellent for the beginning student.

In that case the distribution function cannot be any arbitrary function of the variables but only a function of the combinations of variables that allow it to be independent of time. Such combinations are called invariants of the system, and for any complex system only seven are known: the three components of total linear momentum, the three components of total angular momentum, and the total energy H. If we select a system at equilibrium and not in motion with respect to some set of fixed axes, only the total energy H remains as an invariant of interest.

If we select the position coordinates (total $3N$) and the components of momentum (total $3N$) of each particle in the system as our variables, we can write the total energy H as[a]

$$H = \sum_{i=1}^{3N} \frac{p_i^2}{2m_i} + U(x_1, y_1, z_1, \ldots, x_N, y_N, z_N) \qquad \text{(IX.1.1)}$$

Gibbs was able to show that, if we consider a very large number of identical systems distributed in such a way, the probability of finding one of them in a given state defined by the range $x_1, x_1 + dx_1, \ldots, z_N, z_N + dz_N, p_1, p_1 + dp_1, \ldots, p_{3N}, p_{3N} + dp_{3N}$ is given by the function

$$Ce^{-H/kT} dx\, dy_1 \cdots dz_N \cdot dp_1 \cdots dp_{3N} \qquad \text{(IX.1.2)}$$

then the mechanical properties of the system are in complete consistency with its thermodynamic behavior. C is here a constant which may depend on temperature, k is Boltzmann's constant, and T is the absolute temperature.

Such a distribution of systems is called a canonical ensemble, and the distribution function is $Ce^{-H/kT}$. If C is written in the form $e^{A/kT}$, then it can be shown that A may be identified with the thermodynamic Helmholtz free energy of the system.

2. The Phase Integral

Since $Ce^{-H/kT}$ is a distribution function, we obtain the value unity if we integrate it over all possible values of the variables. This normalization condition is[b]

[a] The x component of momentum of the ith particle is defined by $p_{x_i} = m_i v_{x_i} = m_i (dx_i/dt) = m_i \dot{x}_i$ (various abbreviations). Thus the first sum represents the kinetic energy. We assume implicitly that all the forces are conservative, so that the total potential energy $U(x_i, \ldots, z_n)$ is a function only of the positions of the particles. $-\partial U/\partial x_i$ is then the x component of force acting on the ith particle.

[b] C must have the units of $(xp_x)^{-3N}$ in order to make the probability a dimensionless quantity. The absolute magnitude of the dimensions are of little importance, since, as we shall see, it is only the relative probability of two states that is important. Quantum mechanics enables us to set this multiplicative constant equal to h^{-3N}, where h is Planck's constant. It should also be divided by $N!$ for a system of N indistinguishable molecules, since we are not able to distinguish configurations in which molecules have been interchanged.

$$\int \cdots \int C e^{-H/kT} \, dx_1 \cdots dz_N \, dp_1 \cdots dp_{3N} = 1 \qquad \text{(IX.2.1)}$$

where the integration is a $6N$-fold multiple integration. The value of the factor C is given by

$$\frac{1}{C} = e^{-A/kT} = \int \cdots \int e^{-H/kT} \, dx_1 \cdots dp_{3N} \qquad \text{(IX.2.2)}$$

This multiple integral (IX.2.2) is the famous phase integral of Gibbs and may be represented by the symbol Z. Thus

$$\frac{1}{C} = Z = e^{-A/kT} = \int \cdots \int e^{-H/kT} \, d\tau \qquad \text{(IX.2.3)}$$

where we abbreviate all the products $dx_1 \, dx_2 \cdots dp_{3N}$ by $d\tau$.

Z is a function not of the coordinates of any of the particles but only of the limits of integration (wall boundaries for $x_1 \cdots z_N$), the fields present, and the temperature.

The average value of any quantity may be obtained by the usual procedures of averaging, and we have the following formula for the average value R of any quantity which depends on the variables

$$\overline{R} = C \int \cdots \int R e^{-H/kT} \, d\tau$$

$$= \frac{\int \cdots \int R e^{-H/kT} \, d\tau}{\int \cdots \int e^{-H/kT} \, d\tau} \qquad \text{(IX.2.4)}$$

For the internal energy of the system E we have

$$E = \overline{H} = C \int \cdots \int H e^{-H/kT} \, d\tau$$

$$= C k T^2 \frac{\partial}{\partial T} \left(\int \cdots \int e^{-H/kT} \, d\tau \right)$$

$$= \frac{1}{Z} k T^2 \frac{\partial}{\partial T} Z \qquad \text{(IX.2.5)}$$

$$= k T^2 \frac{\partial \ln Z}{\partial T}$$

By similar methods we can obtain for the other thermodynamic functions:

$$A = -kT \ln Z$$

$$S = \frac{E - A}{T} = kT \left(\frac{\partial \ln Z}{\partial T} \right)_v + k \ln Z = k \left(\frac{\partial (T \ln Z)}{\partial T} \right)_v$$

$$P = -\left(\frac{\partial A}{\partial V} \right)_T = kT \left(\frac{\partial \ln Z}{\partial V} \right)_T \qquad \text{(IX.2.6)}$$

$$H = E + PV = kT^2 \left(\frac{\partial \ln Z}{\partial T}\right)_v + kT \left(\frac{\partial \ln Z}{\partial \ln V}\right)_T$$

$$F = A + PV = -kT \ln Z + kT \left(\frac{\partial \ln Z}{\partial \ln V}\right)_T$$

A knowledge of Z thus suffices for the prediction of all thermodynamic quantities.

3. Application to Gases

If we take a gas that consists of identical point particles that exert no mutual forces and rebound elastically at the walls of the container, then the potential energy $U = 0$ (except at the walls where $U = \infty$) and we can calculate Z directly. The integration over the position coordinates for each molecule simply gives the value V, the volume of the vessel, and since there are N particles, we obtain V^N. The rest of the integral is

$$Z = \frac{V^N}{N!} \int \cdots \int e^{-H/kT} \, dp_1 \cdots dp_{3N}$$

$$= \frac{V^N}{N!} \int \cdots \int \exp\left(-\frac{p_1^2 + p_2^2 + \cdots + p_{3N}^2}{2mkT}\right) dp_1 \cdots dp_{3N} \quad \text{(IX.3.1)}$$

but since each p_i ranges from $-\infty$ to ∞, this is equal to the product of $3N$ separate, independent integrals[a]

$$Z = \frac{V^N}{N!} \left(\int_{-\infty}^{\infty} e^{-p_i^2/2mkT} \, dp_i\right)^{3N}$$

$$= \frac{V^N}{N!} (2\pi mkT)^{3N/2} \quad \text{(IX.3.2)}$$

By substituting in the formula for E and P [Eqs. (IX.2.5) and (IX.2.6)], we find

$$E = \tfrac{3}{2} NkT \qquad P = \frac{NkT}{V} \quad \text{(IX.3.3)}$$

which are the relations for an ideal gas.

We see that in general, when the potential function U depends only on the position coordinates and not on the velocities, the integration of Z over the momenta can be obtained, and so in general we may write

$$Z = \frac{(2\pi mkT)^{3N/2}}{N!} \int \cdots \int e^{-U/kT} \, dx_1 \cdots dz_N$$

$$= \frac{(2\pi mkT)^{3N/2}}{N!} Q_c \quad \text{(IX.3.4)}$$

[a] The corrected function from quantum mechanics would be

$$Z = \frac{V^N}{N!} \left(\frac{2\pi mkT}{h^2}\right)^{3N/2}$$

where h is Planck's constant.

where the multiple integral Q_c is known as the configuration integral and depends on the potential energy of the molecules.

If we consider a gas made of hard sphere molecules of diameter σ, but exerting no forces, we can evaluate Q_c approximately as follows: Let us consider the first molecule. As its coordinates vary over the container, U is constant except at the walls or when its center is at a distance σ from the center of another molecule, where U becomes infinite. The result of the integration over the coordinates of this first molecule is then V (the volume of the container) minus $N - 1$ times $\frac{4}{3}\pi\sigma^3$ (the volume excluded by the other molecules).[a] Thus

$$Q_c = [V - (N - 1)\tfrac{4}{3}\pi\sigma^3] \int \cdots \int e^{-U/kT}\, dx_2 \cdots dz_{N-1} \quad (IX.3.5)$$

On repeating this for each molecule we have

$$Q_c = [V - (N - 1)\tfrac{4}{3}\pi\sigma^3][V - (N - 2)\tfrac{4}{3}\pi\sigma^3] \cdots [V - \tfrac{4}{3}\pi\sigma^3]V$$

$$= V^N\left[1 - \frac{(N - 1)8v}{V}\right]\left[1 - \frac{(N - 2)8v}{V}\right] \cdots \left[1 - \frac{8v}{V}\right]1 \quad (IX.3.6)$$

where $8v = \frac{4}{3}\pi\sigma^3 = 8$ times the volume of a single molecule.

If now the gas is dilute so that $V \gg 8Nv$, that is, the total volume is much greater than 8 times the volume of all molecules, we can take the logarithm of both sides and expand all the logarithmic terms in a series. We find

$$\ln Q_c - N \ln V = \ln\left[1 - \frac{(N - 1)8v}{V}\right] + \cdots + \ln\left(1 - \frac{8v}{V}\right)$$

$$\cong -\left[\frac{(N - 1)8v}{V} + \frac{(N - 2)8v}{V} + \cdots + \frac{1 \cdot 8v}{V}\right] \quad (IX.3.7)$$

which is an arithmetic series and yields

$$\ln Q_c - N \ln V \cong -\frac{N(N - 1)}{2}\frac{8v}{V}$$

or, since $N \gg 1$

$$\cong -N^2\frac{4v}{V} = -\frac{Nb}{V} \quad (IX.3.8)$$

where $b = 4Nv$, which may be recognized as the familiar Van der Waals constant. We then have

$$Q_c = V^N e^{-bN/V} = (Ve^{-b/V})^N$$

and on expanding the exponent, since b/V is small,

$$Q_c = \left[V\left(1 - \frac{b}{V} + \frac{b^2}{2V^2} - \frac{b^2}{3!V^3} + \cdots\right)\right]^N \sim \left(V - b\right)^N \quad (IX.3.9)$$

---

[a] This neglects configurations in which the second molecule is near a third molecule or the wall.

If we substitute this in the formula for Z and differentiate to obtain the thermodynamic pressure [Eq. (IX.2.6)], we have, since $\partial \ln Z / \partial V = \partial \ln Q_c / \partial V$,

$$P = \frac{NkT}{V - b} \tag{IX.3.10}$$

which is the first approximation to Van der Waals equation.[a]

4. Molecular Distribution Functions: Ideal Gas in a Force Field

The quantity $Ce^{-H/kT} dx_1 dy_1 \cdots dz_N dp_1 \cdots dp_{3N}$ may be written as $Ce^{-H/kT} d\tau_N d\phi_N$, where $d\tau_N$ represents the product of the $3N$ position differentials $dx_1 \cdots dz_N$ and $d\phi_N$ represents the product of the $3N$ momentum differentials $dp_1 \cdots dp_{3N}$. If we integrate this expression over all the momentum coordinates except those for particle 1 we obtain

$$C \int \cdots_{N-1} \int e^{-H/kT} d\tau_N d\phi_N$$

$$= C(2\pi mkT)^{\frac{3}{2}(N-1)} \exp - \left(\frac{p_{x_1}^2 + p_{y_1}^2 + p_{z_1}^2}{2mkT} + \frac{U}{kT} \right) d\tau_N dp_{x_1} dp_{y_1} dp_{z_1} \tag{IX.4.1}$$

If we now integrate over all the possible coordinate positions (over $d\tau_N$), we obtain

$$CQ_c(2\pi mkT)^{\frac{3}{2}(N-1)} \exp - \left(\frac{p_{x_1}^2 + p_{y_1}^2 + p_{z_1}^2}{2mkT} \right) dp_x dp_y dp_z \tag{IX.4.2}$$

where Q_c is the configuration integral [Eq. (IX.3.4)] over the position coordinates and we have dropped the subscripts. The expression obtained now represents the probability that particle 1 has momentum components in the range (p_x, p_y, p_z). From Eq. (IX.3.4) we see that $C = 1/Z = 1/Q_c(2\pi mkT)^{-\frac{3}{2}N}$, so that on substitution we have

$$\frac{1}{(2\pi mkT)^{\frac{3}{2}}} \exp - \left(\frac{p_{x_1}^2 + p_{y_1}^2 + p_{z_1}^2}{2mkT} \right) dp_x dp_y dp_z \tag{IX.4.3}$$

whence finally on replacing the momenta by the velocities ($p_x = mv_x$, etc.) we obtain

---

[a] In the integration in Eq. (IX.3.5) we neglected the possibility that on integrating over the volume for the first molecule, there might be two or more molecules in contact. In this case we should not subtract $(N - 1) \frac{4}{3}\pi\sigma^3$ but something less. This integration has been performed by both Boltzmann and Majumdar, *Bull. Calcutta Math. Soc.*, **21**, 107 (1929), and leads to the equation

$$\frac{PV}{NkT} = 1 + \frac{b}{V} + \frac{5}{8} \left(\frac{b}{V} \right)^2 + 0.2869 \left(\frac{b}{V} \right)^3 + \cdots$$

where the successive terms take into account the pressure of pairs of molecules $\frac{5}{8} b^2/v^2$, triple associations, etc. Only the first four terms shown have been evaluated.

$$\frac{1}{(2\pi kT/m)^{3/2}} \exp\left(- \frac{v_x^2 + v_y^2 + v_z^2}{2kT/m}\right) dv_x \, dv_y \, dv_z \qquad \text{(IX.4.4)}$$

which we can recognize as the Maxwell-Boltzmann distribution law for molecular velocities [Eq. (VII.2.11)].

We see then that, since we have made no assumptions about the nature of the potential energy U, the distribution of molecular velocities will be independent of the forces acting either between particles or through external fields.[a]

If now we select an ideal gas (no intermolecular forces) which is placed in some external force field, the potential energy is simply a sum of the individual potential energies of each molecule and the canonical distribution can be expressed as a product:

$$Ce^{-H/kT} \, d\tau_N \, d\phi_N = C \left(e^{-H_1/kT} e^{-H_2/kT} \cdots e^{-H_N/kT} \, d\tau_1 \, d\tau_2 \cdots d\phi_1 \cdots d\phi_N\right)$$
$$\text{(IX.4.5)}$$

H_1 is the total energy of particle 1 and depends only on its position in the force field, its momentum, etc.

$$H_1 = \frac{p_{x_1}^2 + p_{y_1}^2 + p_{z_1}^2}{2m_1} + U(x_1, y_1, z_1) \qquad H_2 = \cdots \qquad \text{(IX.4.6)}$$

Thus we can integrate over all the particles but one and obtain

$$P_1 \, dx_1 \, dy_1 \, dz_1 \, dp_{x_1} \, dp_{y_1} \, dp_{z_1} = \frac{1}{(2\pi mkT)^{3/2}} \frac{e^{-H_1/kT} \, dx_1 \, dy_1 \, dz_1 \, dp_{x_1} \, dp_{y_1} \, dp_{z_1}}{\iiint e^{-U_1/kT} \, dx_1 \, dy_1 \, dz_1}$$
$$\text{(IX.4.7)}$$

which represents the probability of particle 1 being in the volume element $dx \, dy \, dz$ and having momenta in the range (p_x, p_y, p_z). If we continue the integration over the momentum coordinates, we obtain

$$P(x, y, z) \, dx \, dy \, dz = \frac{e^{-U_1/kT} \, dx \, dy \, dz}{\iiint e^{-U_1/kT} \, dx \, dy \, dz} \qquad \text{(IX.4.8)}$$

which is the probability that particle 1 will be in the volume element $dx \, dy \, dz$.

If now U_1 is a constant, we see that there is equal probability for being in any volume element of our container. If U_1 depends on position, then the probability will be greatest for the largest values of the potential energy.[b]

Equation (IX.4.8) then gives us a molecular density distribution function for an ideal gas in the presence of an external field. In a gravitational

---

[a] That is, the velocity distribution law is valid subject only to the condition that the potential energy is independent of the velocities of the particles.

[b] This clearly depends on whether the force field is attractive or repulsive. For an attractive field (i.e., gravity) the density will be greatest nearer the source of the field.

field where $U = mgh$ (h being measured, say, from sea level) we see that the density distribution of molecules at two different heights h_1 and h_2 will be given by

$$\frac{P(h_1)}{P(h_2)} = e^{-mg(h_1 - h_2)/kT} \qquad \text{(IX.4.9)}$$

which we can recognize as the well-known exponential law for the density distribution of isothermal atmospheres.

5. Virial Theorem

There is another very powerful method for dealing with the statistics of interacting systems which may be derived directly from Newton's equations. For any system of particles we can write Newton's equations

$$m_1 \frac{d^2 x_1}{dt^2} = F_{x_1}(m) + F_{x_1}(e)$$

$$\vdots \qquad \qquad \vdots \qquad \qquad \vdots \qquad \qquad \text{(IX.5.1)}$$

$$m_N \frac{d^2 z_N}{dt^2} = F_{z_N}(m) + F_{z_N}(e)$$

where m_N is the mass of the Nth particle, $F_{x_N}(m)$ is the x component of instantaneous force acting on the Nth particle due to the other particles in the system, and $F_{x_N}(e)$ is the component due to external fields including the vessel walls.

If we multiply each equation by the quantity $x_1, \ldots, z_N$, respectively, and add all the equations, we obtain

$$\sum_{i=1}^{N} m_i \left(x_i \frac{d^2 x_i}{dt^2} + y_i \frac{d^2 y_i}{dt^2} + z_i \frac{d^2 z_i}{dt^2} \right) = \sum_{i=1}^{N} (x_i F_{x_i} + y_i F_{y_i} + z_i F_{z_i}) \quad \text{(IX.5.2)}$$

where $F_{x_i} = F_{x_i}(m) + F_{x_i}(e)$.

But now we can make use of the identity

$$x \frac{d^2 x}{dt^2} = \frac{d}{dt}\left(x \frac{dx}{dt} \right) - \left(\frac{dx}{dt} \right)^2 = \frac{1}{2} \frac{d^2}{dt^2}(x^2) - \left(\frac{dx}{dt} \right)^2 \qquad \text{(IX.5.3)}$$

and, since $v^2 = \left(\frac{dx}{dt} \right)^2 + \left(\frac{dy}{dt} \right)^2 + \left(\frac{dz}{dt} \right)^2$, we can write

$$\frac{1}{2} \sum_{i=1}^{N} m_i \frac{d^2}{dt^2}(x_i^2 + y_i^2 + z_i^2) - \sum_{i=1}^{N} m_i v_i^2 = \sum_{i=1}^{N} (x_i F_{x_i} + y_i F_{y_i} + z_i F_{z_i})$$

$$\text{(IX.5.4)}$$

But the first term can be rewritten, because $x_i^2 + y_i^2 + z_i^2 = r_i^2$, where r_i is the distance of the ith particle from the origin,

$$\frac{1}{2} \sum_{i=1}^{N} m_i \frac{d^2}{dt^2} (x_i^2 + y_i^2 + z_i^2) = \frac{1}{2} \frac{d^2}{dt^2} \left[\sum_{i=1}^{N} m_i r_i^2 \right] = \frac{1}{2} \frac{d^2}{dt^2} I \quad \text{(IX.5.5)}$$

where $2I$ is the sum of the component moments of inertia of the entire system.[a] If the entire system is at equilibrium, then its moment of inertia is constant and this first term in Eq. (IX.5.4) vanishes.[b] The second term is simply twice the total translational energy of the system, which for a system of N particles is equal to $2 \times \frac{3}{2}NkT = 3NkT$. We may thus write

$$3NkT = - \sum_{i=1}^{N} (x_i F_{x_i} + y_i F_{y_i} + z_i F_{z_i}) \quad \text{(IX.5.6)}$$

a result which is known as the Virial theorem.

If now we are dealing with a system of ideal molecules so that the internal forces are zero,[c] then F_{x_i}, F_{y_i}, F_{z_i} are the forces due only to external fields [i.e., all $F(m) = 0$] and the walls. If there are no external fields, then there is only the normal pressure exerted by the walls on the molecules and F is zero everywhere except at the walls, where it is equal to P per unit area. We can thus evaluate the sum. It is simply the line integral of the product of the pressure by the coordinates taken over the surface of the vessel. This can be shown quite generally to be equal to $-3PV$. In a particular instance if we consider a rectangular vessel with dimensions (a,b,c) (Fig. IX.1), then $\sum_{i=1}^{N} x_i F_{x_i}$ will be $aF_x = -aPbc = -Pabc = -PV$ at one wall and zero at the other, since x is taken as zero there. Similarly, we obtain the contributions $bF_y = -PV$ and $cF_z = -PV$ for the others.

For an ideal gas, the Virial theorem then leads to

$$3NkT = 3PV \quad \text{or} \quad PV = NkT \quad \text{(IX.5.7)}$$

which is the ideal gas law.

When we use a model that has forces between the molecules, the Virial theorem becomes

$$PV = NkT + \frac{1}{3} \sum_{i=1}^{N} (x_i F_{x_i} + y_i F_{y_i} + z_i F_{z_i}) \quad \text{(IX.5.8)}$$

---

[a] We can express

$$\sum m_i r_i^2 = \frac{1}{2} \sum m_i [(x_i^2 + y_i^2) + (y_i^2 + z_i^2) + (x_i^2 + z_i^2)]$$

as

$$= \frac{1}{2} [I_{zz} + I_{xx} + I_{yy}] = I$$

where I_{xx}, I_{yy}, I_{zz} are the moments of inertia about the x, y, and z axes.

[b] Note that it also vanishes if I is changing at a constant rate.

[c] If we consider a hard sphere molecule, the internal forces are zero except when molecules are in a state of collision.

where the forces F now refer *only* to the intermolecular forces exerted by the particles on each other.

If we assume that the forces between molecules are directed along their lines of centers and are additive,[a] we can further simplify this internal

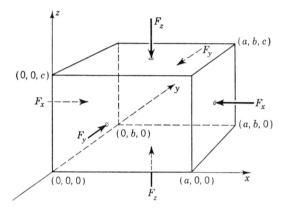

FIG. IX.1. Contribution of the pressure to the virial.

virial. If $F(r_{ij})$ represents the force between the ith and jth molecules, then $[(x_i - x_j)/r_{ij}]F(r_{ij})$ represents the x_i component (where r_{ij} is the distance between molecules i and j) and $x_i[(x_i - x_j)/r_{ij}]F(r_{ij})$ is the contribution of this component to the virial. However for the jth molecule there will be a similar term $-x_j[(x_i - x_j)/r_{ij}]F(r_{ij})$ with a minus sign, since $F(r_{ij}) = -F(r_{ji})$ by Newton's third law. The sum of these is then $[(x_i - x_j)^2/r_{ij}]F(r_{ij})$, and if we add over all pairs of molecules, we obtain

$$\frac{1}{3}\sum_{i=1}^{N} x_i F_{x_i} + \cdots = \frac{1}{3}\sum_{i,j}\left[\frac{(x_i - x_j)^2}{r_{ij}} + \frac{(y_i - y_j)^2}{r_{ij}} + \frac{(z_i - z_j)^2}{r_{ij}}\right]F(r_{ij})$$

$$= \frac{1}{3}\sum_{i,j} r_{ij}F(r_{ij}) \qquad\qquad (IX.5.9)$$

since $r_{ij}^2 = (x_i - x_j)^2 + (y_i - y_j)^2 + (z_i - z_j)^2$ and the sum is to be taken over all pairs of molecules. If we let $(4\pi r^2/V)P(r)\,dr$ be the probability of finding a molecule at a distance r to $r + dr$ from a central molecule, then the total number of pairs of molecules at a distance between r and $r + dr$ will be $\frac{1}{2}(N^2/V)4\pi r^2 P(r)\,dr$, since there are $N^2/2$ pairs of molecules. If further there is a potential energy function $U(r)$ such that $F(r) = -\partial U/\partial r$, we can write

$$\frac{1}{3}\sum_{i,j} r_{ij}F(r_{ij}) = \frac{-4\pi N^2}{6V}\int r^3\left(\frac{\partial U}{\partial r}\right)P(r)\,dr \qquad (IX.5.10)$$

[a] That is, the forces exerted by two molecules on a third is simply the sum of the forces that each would exert if the other were absent.

where the integration is over the entire volume of the container. If, however, the force between molecules is such that $\partial U/\partial r = -F(r)$ approaches zero very rapidly as r exceeds a few molecular diameters, then we can safely integrate from 0 to ∞ and write

$$PV = NkT - \frac{2\pi N^2}{3V} \int_0^\infty r^3 \left(\frac{\partial U}{\partial r}\right) P(r)\, dr \qquad (IX.5.11)$$

an equation which is known as the kinetic equation of state and is applicable to liquids as well as gases. From a knowledge of the distribution of molecules $P(r)$ and the potential energy $U(r)$ we can calculate a thermodynamic equation of state. Such calculations have been performed with fair success for liquid Hg,[a] liquid argon,[b] and other liquids[c] by using experimental data for the distribution function $P(r)$ obtained from X-ray scattering and hypothetical potential-energy functions $U(r)$.

If we select a gas that is so dilute that we need only consider that two molecules at a time will be close enough to each other to give a considerable force between them, we can use the Gibbs distribution $e^{-U/kT}$ for $P(r)$ and rewrite Eq. (IX.5.11) as

$$PV - NkT = -\frac{2\pi N^2}{3V} \int_0^\infty r^3 \frac{dU}{dr} e^{-U/kT}\, dr$$

$$= -\frac{2\pi N^2}{3V} \int_0^\infty r^3 kT \frac{d}{dr}(1 - e^{-U/kT})\, dr \qquad (IX.5.12)$$

where we have chosen to write $1 - e^{-U/kT}$ in place of $-e^{-U/kT}$ in order to obtain a function that vanishes rapidly at both $r = 0$ and $r = \infty$. On integrating by parts

$$PV - NkT = -\frac{2\pi N^2 kT}{3V} \left\{ \left[r^3(1 - e^{-U/kT}) \right]_0^\infty - \int_0^\infty 3r^2(1 - e^{-U/kT})\, dr \right\}$$

$$= \frac{2\pi N^2 kT}{V} \int_0^\infty r^2(1 - e^{-U/kT})\, dr \qquad (IX.5.13)$$

since the first term in brackets vanishes at $r = 0$ and also at $r = \infty$ if U approaches 0 faster than $1/r^3$. For the hard sphere model ($U = \infty$ at $r < \sigma$, $U = 0$ at $r > \sigma$) this can be integrated immediately to give Van der Waals first approximation

$$PV = NkT + \frac{NkT}{V} \tfrac{2}{3}\pi N \sigma^3$$

$$= NkT \left(1 + \frac{b}{V}\right) \qquad (IX.5.14)$$

[a] J. H. Hildebrand and R. L. Scott, "Solubility of Non-electrolytes," chap. 5, Reinhold Publishing Corporation, New York, 1950.

[b] G. W. Jura, *J. Phys. & Colloid Chem.*, **52**, 40 (1948). See also G. W. Jura and C. W. Garland, *J. Am. Chem. Soc.*, **74**, 6033 (1952).

[c] J. G. Kirkwood et al., *J. Chem. Phys.*, **18**, 1040 (1950), calculate the theoretical value of $P(r)$ for hard sphere molecules.

6. Quantum Statistics

The Gibbs formulation of an equilibrium statistics is in error when applied to the internal motions of molecules and also to the external motions of light molecules such as helium or hydrogen at low temperatures. The difficulty arises from the restrictions placed upon the mechanical behavior of such systems by quantum mechanics. Whereas a system that follows the laws of classical mechanics may assume any given mechanical configuration and any given energy, the laws of quantum mechanics restrict the energy of many systems to a discrete number of possible values.

To take a very simple example, an ideal gas molecule in a cubic box of side l may only take on translational energies given by $E_T = (h^2/8ml^2)(n_x^2 + n_y^2 + n_z^2)$, where h is Planck's constant, m is the mass, and n_x, n_y, n_z are numbers which may take on only integral values 1, 2, 3, etc. The translational energy is thus said to be quantized. Similar kinds of restrictions are placed on the rotational energy and vibrational energy of complex molecules.

It is, however, possible to replace the phase integral of Gibbs by a sum-over-states or partition function, which in quantum statistics plays the same role for the calculation of thermodynamic properties that Gibbs' phase integral plays in classical statistics. The partition function Q is defined as

$$Q = \sum_i g_i e^{-E_i/kT} \qquad (IX.6.1)$$

where E_i refers to the total energy of a molecule in the ith energy level and g_i is a specific weight that is equal to the number of different states of the molecule that have the same energy E_i.[a]

The individual products $g_i e^{-E_i/kT}$ then represent the relative probability of finding a molecule in the ith energy level with energy E_i, and so $g_i e^{-E_i/kT} = P(E_i)$ represents a discrete energy distribution function for molecules restricted to quantized energy states.

The Gibbs phase integral may be employed in place of the partition function without serious error whenever the temperature of the system is such that the energy kT is much larger than the difference in energy of two adjacent quantum states.

7. Partition Functions

For many purposes it is possible to separate the total energy of a molecule into almost independent terms. We may thus write:

---

[a] To take the case above for translational energy, there is only one state ($n_x = n_y = n_z = 1$) for which the total $E_T = 3h^2/8ml^2$. There are however three states (n_x or n_y or $n_z = 2$; the others $= 1$) for which $E_T = 6h^2/8ml^2$ and for large energies, $g_E \cong (4\pi ml^2/3h^2)(2mE)^{1/2}$, where $g_E \, dE$ is now the number of states with energies in the range E to $E + dE$.

$$E_{\text{total}} = E_{\text{elec}} + E_{\text{vib}} + E_{\text{rot}} + E_{\text{tr}} \qquad \text{(IX.7.1)}$$

where the consecutive terms refer to the electronic energy, the vibrational energy arising from the internal motions of the nuclei, the rotational energy of the molecule as a whole, and finally the translational energy. For diatomic molecules we have the following approximate formulas for the last three terms:

$$E_{\text{vib}} \cong (v + \tfrac{1}{2})h\nu_0 \qquad g = 1$$

$$E_{\text{rot}} \cong \frac{h^2 J(J+1)}{8\pi^2 I} \qquad g = 2J + 1 \qquad \text{(IX.7.2)}$$

$$E_{\text{tr}} \cong \frac{h^2}{8ml^2}(n_x^2 + n_y^2 + n_z^2) \qquad g = \frac{4\pi ml^2}{3h^2}(2mE_T)^{1/2}$$

where v, J, n_x, n_y, n_z are quantum numbers which may take on only integral values, ν_0 is the vibrational frequency of the molecule, and I is the moment of inertia of the molecule.

By following the Boltzmann formulation for the distribution function of a set of independent molecules among various energy states we have

$$P_E \propto e^{-E/kT} \qquad \text{(IX.7.3)}$$

where the proportionality factor can be determined. If the energies are assumed separable, then

$$P_E \propto e^{-E_{\text{elec}}/kT} e^{-E_{\text{rot}}/kT} e^{-E_{\text{vib}}/kT} e^{-E_{\text{tr}}/kT}$$

and for the partition function we can write

$$Q = (\Sigma g_{\text{elec}} e^{-E_{\text{elec}}/kT})(\Sigma g_{\text{vib}} e^{-E_{\text{vib}}/kT})(\Sigma g_{\text{rot}} e^{-E_{\text{rot}}/kT})(\Sigma g_{\text{tr}} e^{-E_{\text{tr}}/kT})$$
$$= Q_{\text{elec}} Q_{\text{vib}} Q_{\text{rot}} Q_{\text{tr}} \qquad \text{(IX.7.4)}$$

since the individual sums are independent. Thus for energies which are separable in this manner we can say that the partition function Q will be a product of individual partition functions for the different kinds of energy.

8. Translational Partition Function

The translational energy of a particle is always separable from the internal energy, and in the absence of fields we can write for a perfect gas of N particles

$$Q_{\text{tr}} = \left(\frac{2\pi mkT}{h^2}\right)^{3N/2} \frac{V^N}{N!} \qquad \text{(IX.8.1)}$$

This will be recognized as the classical result with the additional normalizing factor h^2 in the denominator. It is valid only when $kT \gg h^2/8ml^2$, where the volume of the gas $V = l^3$. Since the presence of constants in Q does not affect the relative values of thermodynamic quantities, the quantized molecules will have the same thermodynamic formula for the translational energy as the classical particles, except for an additive constant in the entropy term.

9. Vibrational Partition Function

For N diatomic molecules that have only a single internal vibration the partition function is easily determined from the approximate formula for the energy levels

$$E_v = (v + \tfrac{1}{2})h\nu_0$$
$$Q_v = e^{-Nh\nu_0/2kT}(1 - e^{-h\nu_0/kT})^{-N}$$
$$= e^{-E_0/kT}(1 - e^{-h\nu_0/kT})^{-N} \qquad \text{(IX.9.1)}$$

where $E_0 = \tfrac{1}{2}h\nu_0$ represents the lowest vibrational energy possible for the N molecules and is referred to as the zero-point energy, because it represents the residual energy of the molecules at $0°K$. While this introduces an additive constant E_0 in the formulas for A, F, E, and H, it leaves C_v and C_p unaffected.

When $h\nu_0 \ll kT$, it is possible to expand the exponential, and by using only the first term we find

$$Q_v \xrightarrow{h\nu_0 \ll kT} e^{-E_0/kT} \left(\frac{kT}{h\nu}\right)^N \cong \left(\frac{kT}{h\nu}\right)^N \qquad \text{(IX.9.2)}$$

which again corresponds, except for the constant h, to the classical result. When $h\nu_0 \gg kT$, the exponential is close to zero and the vibrational partition function becomes

$$Q_v \xrightarrow{h\nu_0 \gg kT} e^{-E_0/kT} \qquad \text{(IX.9.3)}$$

The significance of this latter case is that the molecules are effectively "frozen" into their lowest vibrational states ($v = 0$) at the temperature in question.

For more complex molecules it is possible to analyze the vibrational motions of the nuclei into a set of so-called normal coordinates, so that, to a first approximation,[a] we can represent the total vibrational energy as a sum of independent terms each one associated with one of the normal coordinates:

$$E = \sum_{i=1}^{n} (v_i + \tfrac{1}{2})h\nu_i \qquad \text{(IX.9.4)}$$

where n represents the total number of such normal coordinates.[b]

[a] This approximation becomes increasingly poor as the molecular excitation increases and atomic vibrations attain larger amplitudes.

[b] In general n $= 3N - 6$ for a complex molecule, where N is the total number of atoms in the molecule. For a linear molecule n $= 3N - 5$. It must be observed that the normal coordinates are generally linear combinations of the position coordinates and there is no necessarily simple relation between a normal coordinate and the change in length of a single bond.

Under these conditions the vibrational partition function for N independent molecules is given by

$$Q_v = (q_v)^N$$

$$= \exp\left(-\frac{Nh\Sigma\nu_i}{2kT}\right)(1 - e^{-h\nu_1/kT})^{-N}(1 - e^{-h\nu_2/kT})^{-N} \cdots (1 - e^{-h\nu_n/kT})^{-N}$$

$$= e^{-E_0/kT}\prod_{i=1}^{n}(1 - e^{-h\nu_i/kT})^{-N} \qquad (IX.9.5)$$

where E_0 is now the total zero-point energy of the N molecules, $\sum_i \frac{1}{2}h\nu_i$.

10. Rotational Partition Function

For a diatomic molecule the allowed rotational states will depend on whether or not the two nuclei are identical. At temperatures at which the energy difference of adjacent rotational states is small compared to kT we can write the approximate partition function for N molecules:

$$Q_{\text{rot}} = \left(\frac{8\pi^2 IkT}{\sigma h^2}\right)^N \qquad (IX.10.1)$$

where I is the moment of inertia perpendicular to the axis $[I = \mu r^2; \mu = $ reduced mass $= m_1 m_2/(m_1 + m_2)]$ and σ, the symmetry number, equals 2 if the nuclei are identical and is otherwise equal to 1.

An additional constant factor enters into these expressions if the nuclei have nonzero spins. If i_1 and i_2 are the two nuclear spins of a diatomic molecule, the complete rotational function is written as

$$Q_{\text{rot}} = \left[(2i_1 + 1)(2i_2 + 1)\left(\frac{8\pi^2 IkT}{\sigma h^2}\right)\right]^N \qquad (IX.10.2)$$

With the exception of the constant terms for nuclear spin and h, this is again identical with the classical result.

For complex, nonlinear molecules the rotational partition function may be written as

$$Q_{\text{rot}} = \left[\frac{g_N \pi^{1/2}}{\sigma}\left(\frac{8\pi^2 kT}{h^2}\right)^{3/2}(I_{xx}I_{yy}I_{zz})^{1/2}\right]^N \qquad (IX.10.3)$$

where $g_N = \prod_{k=1}^{N}(2i_k + 1)$ is the statistical spin factor for the N nuclei and σ represents the number of discrete positions which the molecule can assume about its symmetry axes.[a]

[a] CO_2 (linear) has $\sigma = 2$, NH_3 and PH_3 have $\sigma = 3$, C_2H_4 has $\sigma = 4$, and CH_4 and C_6H_6 have $\sigma = 12$. If there is no symmetry in the molecule, $\sigma = 1$.

The quantities I_{xx}, I_{yy}, I_{zz} represent the moments of inertia of the molecule about three principal axes through its center of gravity.[a]

11. Equilibrium Constant

By definition, the equilibrium constant for any chemical equilibrium is

$$-RT \ln K_{eq} = \Delta F°$$ (IX.11.1)

where $\Delta F°$ represents the difference in free energies per mole of reactants and products referred to some standard state.

If we consider the equilibrium

$$a\text{A} + b\text{B} \rightleftharpoons c\text{C} + d\text{D}$$ (IX.11.2)

then $$-RT \ln K_{eq} = (cF_C° + dF_D°) - (aF_A° + bF_B°)$$ (IX.11.3)

But if Q'_N is the partition function for an ideal gas of N identical molecules, we have the relation

$$Q'_N = \frac{1}{N!} (Q)^N$$ (IX.11.4)

where Q is now the molecular partition function and we have divided by $N!$ to take into account the indistinguishability of the molecules. By using the relation $F = A + PV = -kT \ln Q'_N + PV = -kT \ln Q'_N + NkT$ (for an ideal gas), and the value of Q_N from Eq. (IX.11.4) we have, on using Stirling's approximation ($\ln N! = N \ln N - N$),

$$\frac{-F}{NkT} = \ln \frac{Q}{N}$$ (IX.11.5)

so that the equilibrium constant may be written as

$$K_{eq} = \frac{(Q_C/N)^c(Q_D/N)^d}{(Q_A/N)^a(Q_B/N)^b}$$ (IX.11.6)

If we separate out the zero-point energies of the species and write the residual partition functions as $Q°$, where $Q = Q°e^{-E°/RT}$, then this can be rewritten as

$$K_{eq} = \frac{(Q_C°/N)^c(Q_D°/N)^d}{(Q_A°/N)^a(Q_B°/N)^b} e^{-\Delta E_0°/RT}$$ (IX.11.7)

where $\Delta E_0°$ is the difference in internal energies (including zero-point

[a] In general for any three cartesian axes (x',y',z') which are not the principal axes,

$$I_{xx} I_{yy} I_{zz} = \begin{vmatrix} I_{x'x'} & -I_{y'x'} & -I_{z'x'} \\ -I_{x'y'} & I_{y'y'} & -I_{z'y'} \\ -I_{x'z'} & -I_{y'z'} & I_{z'z'} \end{vmatrix}$$

where $I_{x'y'}$ is defined as $I_{x'y'} = \sum_i m_i x'_i y'_i$ and $I_{x'x'} = \sum_i m_i(y_i'^2 + z_i'^2) =$ moment of inertia about the x' axis.

energies) of the reactants and products at standard concentrations and $0°K$ (that is, ΔE_0° is the heat of reaction at $0°K$).

When independent data are available (as from spectroscopy) for the vibration frequencies and values of ΔE_0°, Eq. (IX.11.7) yields excellent results for K_{eq}.[a]

12. An Example: The Dissociation of Iodine

We can illustrate the use of Eq. (IX.11.7) by a calculation of the gas-phase equilibria at high temperatures of

$$I_2 \rightleftharpoons 2I \qquad (IX.12.1)$$

We can write

$$K_{eq} = \frac{(Q_1^\circ/N)^2}{(Q_2^\circ/N)} e^{-\Delta E_0^\circ/RT} \qquad (IX.12.2)$$

where the subscript 1 refers to I atoms and 2 refers to I_2 molecules. On substituting the individual partition function,

For I: $Q_1^\circ = g_1 \left(\dfrac{2\pi m_1 kT}{h^2}\right)^{3/2} \dfrac{V}{N}$

For I_2: $Q_2^\circ = g_2 \left(\dfrac{2\pi m_2 kT}{h^2}\right)^{3/2} \dfrac{V}{N} \left(\dfrac{8\pi^2 IkT}{\sigma h^2}\right) (1 - e^{-h\nu_0/kT})^{-1}$ (IX.12.3)

where g_1 and g_2 are statistical factors for the electronic spins.[b] For I atoms $g_1 = 4$ ($S_1 = \frac{3}{2}$); for I_2, $g_2 = 1$ ($S_2 = 0$). The nuclear spins have been omitted because they will cancel out in the ratio. The symmetry number $\sigma = 2$, and $m_2 = 2m_1 = 2m_I$, so that we can substitute and write finally

$$K_{eq} = \frac{8}{r_0^2} \left(\frac{kTm_1}{\pi h^2}\right)^{1/2} \left(\frac{V}{N}\right) (1 - e^{-h\nu_0/kT}) e^{-\Delta E_0^\circ/RT} \qquad (IX.12.4)$$

If now we use the spectroscopic data $\Delta E_0^\circ = 35.55$ Kcal/mole, $\nu_0 = \omega_0 c$ (c = velocity of light) and $\omega_0 = 213.67$ cm^{-1}, and $r_0 = 2.660$ Å, all the other numbers are known constants and we can calculate K_{eq}. Such a

---

[a] See, for example, S. Glasstone, "Theoretical Chemistry," D. Van Nostrand Company, Inc., Princeton, N.J., 1944, for a more detailed discussion of these calculations and of quantum statistics.

[b] We have all along neglected the electronic partition function. The reason has been that the energy separation of electronic states is usually so great that only one electronic state, the lowest, is ever occupied at most temperatures. The partition function for electronic states is simply $\Sigma g_1 e^{-\epsilon i/kT} \cong g_1$ if only the lowest state is occupied. For molecules that have zero magnetic moment, which is usual, $g_1 = 1$. For molecules that possess a magnetic moment and total spin (that is, O_2, NO, atoms, etc.) $g_1 = 2j + 1$, where j = the total spin.

calculation has been performed by Gibson and Heitler[a] and leads to results in excellent agreement with experiment.[b]

13. Application to Rate Processes

It is interesting to see how the statistical treatment of equilibrium systems may be applied to the calculation of kinetic data, at least to the same accuracy as was obtained when the assumption of a Maxwellian distribution was employed. Let us assume that we have a mixed gas of hard sphere molecules A and B, capable of forming a weakly bound complex AB. Let us further assume that the molecules A and B possess no internal energy. We can then write the molecular partition functions for A, B, and AB:

$$Q_A = \left(\frac{2\pi m_A kT}{h^2}\right)^{3/2} V e^{-E_A^\circ/RT}$$

$$Q_B = \left(\frac{2\pi m_B kT}{h^2}\right)^{3/2} V e^{-E_B^\circ/RT} \qquad \text{(IX.13.1)}$$

$$Q_{AB} = \left(\frac{2\pi m_{AB} kT}{h^2}\right)^{3/2} V \left(\frac{8\pi^2 IkT}{h^2}\right) (1 - e^{-h\nu/kT})^{-1} e^{-E_{AB}^\circ/kT}$$

For the chemical equilibrium

$$AB \rightleftharpoons A + B \qquad \text{(IX.13.2)}$$

we can write

$$K_{eq} = \frac{(Q_A/N)(Q_B/N)}{(Q_{AB}/N)} e^{-E_0^\circ/RT}$$

$$= \frac{(2\pi m_A kT/h^2)^{3/2}(2\pi m_B kT/h^2)^{3/2}(V/N)^2(1 - e^{-h\nu/kT})}{(2\pi m_{AB} kT/h^2)^{3/2}(V/N)(8\pi^2 IkT/h^2)} e^{-E_0^\circ/RT}$$

$$\text{(IX.13.3)}$$

If now we substitute $m_{AB} = m_A + m_B$, $I = \mu\sigma_{AB}^2$, and $\mu = m_A m_B/(m_A + m_B)$ and cancel appropriate terms, we have

$$K_{eq} = \frac{V}{N}\left(\frac{\mu kT}{8\pi}\right)^{1/2} \frac{1}{h\sigma_{AB}^2}(1 - e^{-h\nu/kT}) e^{-E_0^\circ/RT} \qquad \text{(IX.13.4)}$$

If now we admit that the forces between A and B even at small distances are small, so that $h\nu \ll kT$, and further that $E_0^\circ \ll RT$, we can expand the two exponential terms and find

[a] G. E. Gibson and W. Heitler, Z. Physik, **49**, 465 (1928).

[b] In the units used here, the standard state is taken as 1 molecule/cc and $V/N = 1$, and concentrations are in these units. That is, we should set $K_{eq} = N_I^2/N_{I_2}$, where N_I and N_{I_2} are both in units of molecules/cc. Changes to other units may be made by use of the gas laws. Thus if we use C (moles per liter) then $C_I^2/C_{I_2} = N_I^2/N_{I_2} \times (10^3/N_{Av})$ where N_{Av} = Avogadro's number. Similarly, if we wish to use partial pressures (atmospheres) then, since $P = CRT$, $P_I^2/P_{I_2} = (C_I^2/C_{I_2})(RT)$.

$$K_{eq} = \frac{V}{N} \left(\frac{\mu}{8\pi kT} \right)^{1/2} \frac{\nu}{\sigma_{AB}^2} \qquad (IX.13.5)$$

In this form the units of K are molecules/cc and $V/N = 1$ in the standard state. Thus

$$\frac{N_A N_B}{N_{AB}} = K_{eq} = \left(\frac{\mu}{8\pi kT} \right)^{1/2} \frac{\nu}{\sigma_{AB}^2} \qquad (IX.13.6)$$

where the N's are molecular concentrations. On rearrangement we have:

$$N_{AB}\nu = N_A N_B \sigma_{AB}^2 \left(\frac{8\pi kT}{\mu} \right)^{1/2} \qquad (IX.13.7)$$

But this can be interpreted quite readily because ν is now the limiting value (as the forces between A and B approach zero) of the vibration frequency of A and B in the molecule AB. Since in the limiting case each such vibration leads to a dissociation into free species A and B, the quantity νN_{AB} must represent the rate at which the species AB is decomposing to produce A and B. Finally since there is an equilibrium, this must also equal the rate at which A and B are combining to form AB. We can recognize that the right-hand side of Eq. (IX.13.7) is precisely the rate at which N_A and N_B make collisions [Eq. (VII.8F.5)], so that we see that our collision formula is also contained in the equilibrium constant. However, this is not strange, since both were derived from the same postulates. We shall see later how the partition function may be used to deduce formulas for various collision processes.

PART THREE

The Kinetics of Homogeneous
Reactions in Gases

X

Spontaneous Decomposition of Excited Molecules

1. Chemical Stability and Temperature

The first part of this text was concerned with an empirical analysis of reaction systems. There we were principally concerned with providing an as simple as possible self-consistent description of reaction systems. We shall now set ourselves the problem of trying to understand these reaction systems and the empirical description we have made of them from the point of view of the molecular structure of matter. In simplest terms we now ask, why do molecules undergo reactions?

Let us first note that a given set of molecules does not always react, but rather that there are specific conditions which must be met in order that reactions proceed at a measurable rate. If we take temperature for consideration as a variable, we know that at low enough temperatures almost any chemical reaction may be stopped. But the molecular theory of matter has shown that temperature is simply an average measure of the mean energy of a molecule in thermal equilibrium. An increase in temperature implies an increase in molecular energy.

Thus we see that the reactivity of a system is directly related to the energy content of the system. This relation may be more strikingly demonstrated if we consider the magnitude of the temperature dependence of chemical reaction systems. It is a crude but useful rule that the rate of most chemical reactions doubles when the temperature increases by 10°C. How may we interpret this rule? At 300°K a change in temperature of 10°C will result in about a 3 per cent (10/300) increase in the absolute temperature and hence (since the mean energy of a molecule is roughly proportional to the absolute temperature) a 3 per cent increase in molecular energy. We thus observe that, although the mean energy of a molecule has increased by only 3 per cent, the molecule's chances for decomposition have increased by 100 per cent! We must conclude that, if one requirement for chemical reaction is the possession of molecular energy, it cannot be the average molecule which is undergoing reaction but rather some special group of molecules.

That conclusion is further evidenced if we consider two typical equilibrium energy-distribution curves for identical systems at two different temperatures (Fig. X.1). Here $P(E)\, dE$ represents the fraction of mole-

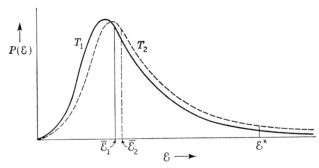

FIG. X.1. Energy-distribution function $P(E)$ as a function of E at two temperatures, $T_2 > T_1$.

cules that have energies between E and $E + dE$. We see that at the higher temperature T_2 there has been very little shift in the fraction of molecules in the low-energy groups around the mean energy $\overline{E}$. If, however, we consider the change in population of some high-energy states as indicated by the respective areas under the curves beyond E^*, we see that the relative increase in these types of energy-rich molecules is much greater than a factor proportional to T. In fact since the population of high-energy states near E^* is roughly proportional to $e^{-E^*/kT}$, we see that, when $E^* \gg kT$, a small change in T can produce a large relative change in this Boltzmann factor.[a]

2. Mechanical Stability of Molecules

For a molecule to be considered stable in a mechanical sense means that when its component parts, nuclei or electrons, are displaced from their equilibrium positions, there are internal attractive forces in the molecule that tend to restore the particles to their initial positions. If that were true for all displacements of the particles no matter how large in magni-

[a] The percentage change in population of the energy group E to $E + dE$ per percentage change in temperature is given by $\partial \ln P(E)/\partial \ln T$, which is approximately equal to E/kT when E/kT is large. Thus for $E/kT = 35$, a 1 per cent change in T produces a 35 per cent change in $P(E)$.

tude, then molecules would be infinitely stable and incapable of decomposition. The fact that decomposition can occur means that there do exist positions of the nuclei at which the restoring forces become zero and beyond which they either remain zero or become repulsive rather than attractive. This is illustrated in Fig. X.2 for the decomposition of the

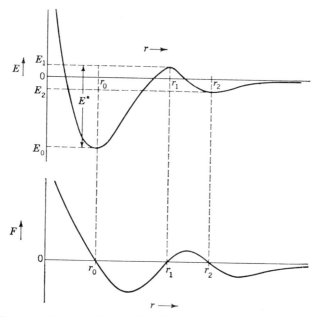

Fig. X.2. A potential-energy diagram that illustrates a hypothetical relation between the potential energy E as a function of the distance of separation r of two atoms A and B. The lower diagram is the corresponding force F acting between the nuclei, obtained from the relation $F = -\partial E/\partial r$.

hypothetical diatomic molecule AB. The upper curve shows a hypothetical potential energy E as a function of the distance of separation r of the two atoms. This energy will be a minimum $-E_0$ for the equilibrium distance r_0 and a maximum E_1 at the distance of instability r_1 and will have another minimum $-E_2$ at the distance r_2.[a] Correspondingly, the force F acting between the two atoms is given by the relation $F = -dE/dr$ and is positive (repulsive) for $r < r_0$, negative (attractive) when r is between r_0 and r_1, positive again between r_1 and r_2, and zero at r_0, r_1, and r_2.

If the molecule AB is in the energy state corresponding to E_0 on our diagram, then the critical energy $E^* = E_1 - E_0$ will have to be supplied

[a] This curve has been deliberately chosen with a maximum for illustrative purposes. There is no reason, in general, to believe that simple atoms will have the maximum shown. There is reason to believe, however, that more complex fragments such as free radicals and molecules will have curves similar to the one drawn here. The region beyond r_2 would then correspond to attractive Van der Waals forces.

to it before it can be dissociated into A and B. If that is done, the atoms will separate and will have left a translational energy E_1 when they are far apart.[a] No molecules AB with internal energy content less than E^* can dissociate, while all molecules with energy in excess of E^* must dissociate in one vibration unless this energy is lost. These latter molecules we shall refer to as being *critically energized*, or energized complexes. Energized complexes which have an internuclear separation r_1 are referred to as transition complexes because that distance marks a borderline between reactants and products.

3. Spontaneous Decomposition of an Energized Molecule; Potential Energy

A polyatomic molecule will be unstable when there is some nonequilibrium configuration of the nuclei for which the force acting to restore the equilibrium configuration becomes zero and becomes repulsive for more distorted configurations. As expressed in terms of the total potential energy of the molecule, we can say that a molecule is unstable at any configuration of the nuclei that corresponds to a maximum in the potential-energy curve. Since the potential energy of a molecule can always be expressed as a function of the internuclear distances,[b] we can write for the potential energy U of a molecule with N atoms:

$$U = U(r_1, r_2, \ldots, r_m) \qquad (X.3.1)$$

where $m = 3N - 6$ is the number of independent internuclear distances measured from some arbitrary origin (usually the center of gravity of the molecule).[c]

The condition for instability is that there exist a set of values $(r_1', r_2', \ldots, r_m')$ for which

$$(\delta U)_{r_i'} = 0 \qquad \text{and} \qquad (\delta^2 U)_{r_i'} < 0 \qquad (X.3.2)$$

We can further identify the energy of the molecule at this configuration, $U(r_1', r_2', \ldots, r_m')$, with the critical energy for decomposition E^*.

For a triatomic molecule U will be a function of three internuclear distances, and a potential-energy diagram analogous to Fig. X.2 for a diatomic molecule would have to be constructed in four dimensions. For such a case we can construct a three-dimensional model in which we can

---

[a] Since the potential energy E has fallen to zero at large distances, the excess energy E_1 must be transformed into kinetic energy. If atoms A and B are brought together, they will not be able to form a stable molecule AB unless they are provided initially with energy E_1. Even should this happen, the molecule then formed will not be stable unless enough energy is lost by it to bring the resultant AB below E_1 in total energy.

[b] This assumes that the electronic energy of the molecule can be expressed as a continuous function of the internuclear distances.

[c] For linear molecules $m = 3N - 5$.

plot surfaces of constant potential energy as a function of the internuclear distances. Figure X.3 shows two such surfaces for a hypothetical molecule ABC in which our coordinate axes are r_{AB}, r_{BC}, and r_{AC}. The inner ellipsoidal surface I would represent a configuration of the molecule with potential energy only slightly greater than the minimum potential energy of the molecule at its equilibrium configuration r_{AB}°, r_{BC}°, r_{AC}°.

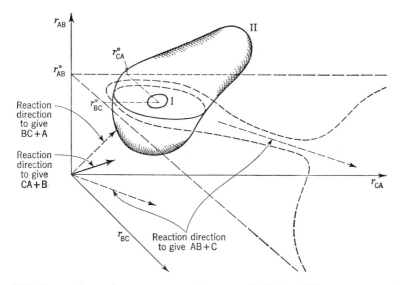

FIG. X.3. Two surfaces of constant potential energy (I, II) for ABC, a triatomic molecule, as a function of the internuclear distances.

The outer distorted surface II corresponds to a molecule that has much greater internal energy but less than the critical energy.[a] The diagonal arrows indicate the directions along which dissociation into an atom and a diatomic molecule would take place.

If we were to draw a plane perpendicular to the r_{AB} axis, it would intersect surface II in the half-solid curve shown. If we were now to draw a third surface that corresponded to an energy greater than the critical energy needed to rupture the bond between atom C and the species AB, it might make an intersection with the above plane such as that indicated by the dotted curve which is not closed. An atom of C approaching the molecule AB (in which the distance r_{AB} is kept fixed) will move in between the arms of this curve and then back out unless in the interim the excess energy of the system is removed and the molecule ABC is deactivated.

The general mathematical solution to this problem is given by Eq.

---

[a] The distortion represents a weakening of the restoring forces with large extensions of the three bonds.

(X.3.2). The condition $(\delta U)_{r'} = 0$ can be expressed in terms of m algebraic equations

$$\left(\frac{\partial U}{\partial r_1}\right)_{r_i'} = 0 \quad \left(\frac{\partial U}{\partial r_2}\right)_{r_i'} = 0 \quad \cdots \quad \left(\frac{\partial U}{\partial r_m}\right)_{r_i'} = 0 \quad (X.3.3)$$

which can now be solved for the values of $r_1, \ldots, r_m$. There will be not a single set of solutions, but an entire family of solutions each corresponding to a different possible decomposition or rearrangement of the molecule. And for each reaction in which we may be interested there may be a number of different configurations of the molecule each corresponding to the same critical energy required for that reaction (i.e., splitting of any of three equivalent H atoms in CH_3F).

While this particular problem of mechanical stability can be solved in principle by means of the Schroedinger equation, exact solutions have never been obtained for any polyatomic molecules except H_2 and H_2^+. Later on we shall consider some of the approximate treatments which have been made.

4. Rate of Spontaneous Decomposition

Though we cannot hope to give a precise picture of the mode of decomposition of a polyatomic molecule, we can represent schematically, as in Fig. X.4, the general behavior of the potential energy of a molecule as a function of configuration. In Fig. X.4 is shown a schematic hyper-

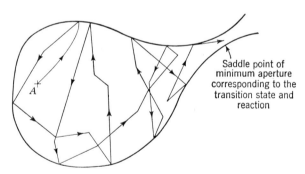

Saddle point of minimum aperture corresponding to the transition state and reaction

$A^{\ddagger}$

Fig. X.4. Schematic representation of the motion in phase space of the internal coordinates of a critically energized molecule with a single possible mode of decomposition. Bounding surface is one of constant potential energy.

surface of constant potential energy for a critically energized molecule of N atoms that have one mode of decomposition. (The reader must supply imagination at this point and pretend that this surface is m-dimensional, where $m = 3N - 6$.)

The erratic curve (enclosed within the surface) represents a possible internal motion of a molecule which, being suddenly energized and finding itself in the configuration corresponding to point A, begins at once to execute random internal vibrations which successively change its internal coordinates in accordance with the curve shown. The molecule eventually decomposes where it starts moving through the narrow pass shown.

Any isolated molecule so energized is forced to move within the hypervolume bounded by the surface until it either passes through the decomposition pass or loses its energy by deactivation. We can compare such a motion to a random walk or diffusion. The mean rate of reaction will then be inversely proportional to the time taken for the molecule, starting from any point within the surface to reach the configuration that corresponds to the decomposition. If we call $\overline{t(E)}$ the mean time of diffusion from any point (within the hypervolume of internal energy E) to the pass, then $\overline{k(E)} \cong 1/\overline{t(E)}$ will be the mean rate of decomposition of such a molecule.[a]

The "motion" of the internal coordinates of the particle within the energy hypersurface is analogous to the random spatial motion of a molecule in a box with rough walls which has a small hole in it. Thus we might guess intuitively that $k(E)$ will be given approximately by the product ν of the average "frequency of motion" of the molecule within the hypervolume times the ratio of the area of the pass at its minimum cross section to the total area of the bounding hypersurface. The area of minimum cross section must increase as the energy is increased above the critical energy E^*, while for energies less than E^* the area is zero. We should further expect that, as the number of atoms in the molecule increases, the chances of decomposition at any critical energy are decreased (i.e., the hypersurface increases in area, meaning that the number of configurations corresponding to no reaction is greater). Thus for a diatomic molecule with only one internal distance, a critically energized molecule will decompose in the time of one vibration or less. A triatomic molecule, however, may assume many configurations in which the critical

---

[a] We can define for such a random walk a function $P(r,E,t)\,dt$, which is the probability that a molecule of energy E starting from the configuration r will reach the pass in the time t to $t + dt$. Averaging over all configurations r, we find the average probability for any initial configuration:

$$\overline{P(E,t)} = \int_{\text{all } r} P(E,r,t)\,dr$$

which has the dimensions of probability per unit time. If all configurations are not equally probable, we must take this into account in the averaging. Then

$$\overline{t(E)} = \int_0^\infty t\,\overline{P(E,t)}\,dt \qquad \text{and rigorously} \qquad \overline{k(E)} = \int_0^\infty \frac{1}{t}\,\overline{P(E,t)}\,dt$$

We note that $\overline{P(E,t)}$ must have the property that $\overline{P(E,0)} = 0$ and $\overline{P(E,t)} \to 1$ as $t \to \infty$. H. A. Kramers, *Physics*, **7**, 284 (1940), has given an approximate solution to this problem based on a simplified analogy to a model for brownian motion in one dimension.

energy is distributed between its different bonds before the energy finally finds its way into the bond which will be broken.

5. Quantitative Theories of Spontaneous Decomposition

The first quantitative formulation of the decomposition process was made by Marcellin,[a] who did not attempt a solution. An interesting analysis was later made by Polanyi and Wigner,[b] who treated the molecule as an elastic medium in which decomposition occurred when certain elastic waves (equivalent to the atomic vibrations) reinforced each other sufficiently to break a bond. The form of the law they then deduced was the same as that derived from discrete models and has been of considerable use in discussing the decomposition of unstable atomic nuclei, which is a related problem.

The first usable results were obtained by Rice and Ramsperger[c] and Kassel,[d] who were able to deduce from a simplified model of a molecule, consisting of a set of harmonic oscillators,[e] that

$$\overline{k(E)} = 0 \qquad \text{for } E < E^*$$

$$\overline{k(E)} = A \left(1 - \frac{E^*}{E}\right)^{n-1} \qquad \text{for } E \geq E^* \qquad (X.5.1)$$

where n is the number of internal coordinates at the molecule and A is some molecular constant proportional to the frequency with which energy is able to pass from one degree of freedom of the molecule to another. This is equivalent to the result obtained by Polanyi and Wigner.

The essential correctness of Eq. (V.5.1) for reactions that give the Arrhenius equation as a "high-pressure" limit has been demonstrated by Slater.[f] The proof is accomplished by a Laplace transform of the equation

---

[a] M. R. Marcellin, Thesis, University of Paris, 1914, Gauthiers-Villars, Paris, 1914; *Ann. Physik*, **3**, 158 (1915).

[b] M. Polanyi and E. Wigner, *Z. physik. Chem.*, Haber Festband, 439 (1928).

[c] O. K. Rice and H. C. Ramsperger, *J. Am. Chem. Soc.*, **49**, 1617 (1927); **50**, 617 (1928). H. C. Ramsperger, *Chem. Revs.*, **10**, 27 (1932). O. K. Rice, *Phys. Rev.*, **32**, 142 (1928).

[d] L. S. Kassel, *J. Phys. Chem.*, **32**, 225, 1065 (1928). See also L. S. Kassel, "Kinetics of Homogeneous Gas Reactions," chap. 5, ACS Monograph, Reinhold Publishing Corporation, New York, 1932, for a fuller account of these developments.

[e] The deduction is based upon assuming that one of the oscillators is a weak bond which will break when energy E^* is present in it. For a molecule that consists of n weakly coupled harmonic oscillators the chance that, when the molecule has energy E, at least E^* of it will be localized in one oscillator is given by $(1 - E^*/E)^{n-1}$. The rate at which such an event happens, $\overline{k(E)}$, is then presumed proportional to this ratio, the constant of proportionality being A, the mean rate of internal energy transfer in the molecule. This derivation may be justified for a classical and for a quantized molecule.

[f] N. B. Slater, *Proc. Leeds Phil. Lit. Soc., Sci. Sect.*, **6**, 259 (1955); *Phil. Trans. Roy. Soc.* (*London*), **A246**, 57 (1953).

$$Ae^{-E/RT} = \int_0^\infty k(E)P(E)\, dE \qquad (X.5.2)$$

where $P(E)$ is the equilibrium distribution of energy states for a reactant molecule [Eq. (X.7.3)]. An alternative hypothesis was proposed by Hinshelwood and others[a] based on ad hoc speculations which are now known to be completely unjustified. They assume that $\overline{k(E)}$ has the form

$$\begin{aligned} k(E) &= 0 && \text{for } E < E^* \\ k(E) &= \text{const} && \text{for } E \geq E^* \end{aligned} \qquad (X.5.3)$$

6. Detailed Theory of Spontaneous Decomposition

In the Rice-Ramsperger-Kassel model (RRK) for spontaneous decomposition it is assumed that the total energy distributed among the n weakly coupled harmonic oscillators that make up the molecule[b] is freely available for redistribution. In this sense the $n - 1$ oscillators coupled to the weak one act as an energy reservoir for it. Slater[c] has criticized this model for not being sufficiently detailed and has proposed that, over short periods of time, the rate of energy transfer between oscillators may be slow enough to be negligible. As an extreme case, he points out that oscillators that belong to different symmetry classes of the molecular vibrations cannot exchange energy.[d] A further restriction on energy exchange results from the discreteness of the energy levels of quantized systems. That is, a molecule can change its internal energy distribution only between states whose total energies are practically identical. Small differences between such vibrational states can be absorbed by changes in rotational energy, but this provides a buffer of unknown efficiency.

Slater proposes instead that the detailed distribution of energy among the harmonic oscillators that make up the molecule be the starting point for the theory. Instead of assuming that one of these oscillators corresponds to the mode of rupture of the molecule, he assumes that there is a "critical bond distance" q_0 in the molecule such that, if the atoms attain the distance corresponding to q_0, reaction will occur. The distance q_0 must of course correspond to the configuration of what we have called the "transition state" for the molecule. The potential energy corresponding to q_0 is the critical energy E^* for reaction.

The distances q_i between the atoms in a molecule can be related by

---

[a] G. N. Lewis and D. F. Smith, J. Am. Chem. Soc., **47**, 1508 (1925). G. N. Hinshelwood, Proc. Roy. Soc. (London), **A113**, 230 (1927). R. H. Fowler and E. K. Rideal, ibid., 570 (1927).

[b] $n \leq 3N - 6$ for N-atom nonlinear molecules. When there are oscillators that do not exchange energy in the molecule, then these are not counted in n.

[c] Slater, loc. cit.

[d] This reasoning is based on an assumption that the symmetry of the molecule does not change in passing to the transition state, and it may not hold rigorously.

linear algebraic equations to the coordinates Q_i used to represent the normal modes of vibration of the molecule. Thus the critical bond distance q can be written as

$$q = \sum_{i=1}^{s} \alpha_1 Q_1 = \sum_{i=1}^{s} \alpha_i Q_i^\circ \cos (2\pi \nu_i t + \theta_i) \qquad (X.6.1)$$

where Q_i° is the maximum displacement[a] from its equilibrium value for the ith normal coordinate Q_i of frequency ν_i and phase θ_i. Slater now assumes that the rate of decomposition can be equated to one-half the rate[b] at which the trigonometric sum representing q reaches the critical extension q_0.

This rate $L(E)$ will be different for each detailed assignment of the energies E_i to the oscillators and is a very complex function of the values E_i. For sufficiently large molecules with values of E^*/kT not too small, Slater has estimated that this detailed model behaves like a Kassel model [Eq. (X.5.1)] in which the number of Kassel oscillators $n = (s + 1)/2$.

On the basis of the RRK model, the over-all rate of unimolecular reaction can be written as

$$\text{Rate} = (B) \int_0^\infty P(E)k(E)\, dE \qquad (X.6.2)$$

where $P(E)$ is the fraction of molecules B which have energies in the range E to $E + dE$ and $k(E)$ is their specific rate of decomposition. The Slater formulation gives

$$\text{Rate} = (B) \int_0^\infty \cdots \int P(E_1, \ldots, E_s)L(E_1, \ldots, E_s)\, dE_1 \cdots dE_s \qquad (X.6.3)$$

where $P(E_1, \ldots, E_s)$ is the fraction of molecules with the internal energy distribution $E_1, \ldots, E_s$ [Eq. (X.7.2)].

When $P(E)$ is not perturbed by the reaction, so that the distribution of critically energized molecules is that characteristic of equilibrium, the RRK model leads to a specific first-order rate constant of the form $k = A \exp (-E^*/RT)$, where A is the frequency of internal energy transfer between oscillators. The Slater formulation in these circumstances gives $k = \bar{\nu} \exp (-E^*/RT)$, both results being similar in form to the Arrhenius equation. The A factor in the RRK model represents the frequency of energy transfer between oscillators, which for weakly coupled oscillators would be of the order of their beat frequencies, or about 10^{12} to 10^{14} sec^{-1}. In the Slater model, $\bar{\nu}$ represents a weighted rms frequency of the normal frequencies which describe the decomposition [Eq. (X.6.1)]

[a] The motion of each normal coordinate is given by a single cosine term with maximum amplitude $Q_i^\circ = (2E_i/k_i)^{1/2}$, where E_i is the energy and k_i is the force constant of the ith oscillator. Not all s coordinates may be needed to describe q.

[b] The value of one-half is used because reactant molecules can approach q_0 only from values less than q_0.

and which thus also is expected to lie in the range of 10^{12} to 10^{14} sec^{-1}. In this high-pressure or equilibrium limit these models cannot be distinguished and the justification of a choice between them has to be made on other grounds. They will differ under conditions in which the internal energy distributions $P(E)$ and $P(E_1, \ldots, E_s)$ are no longer the equilibrium values. But then the differences will be not in form but in the effective number of oscillators assignable to the decomposition.

As we shall see later, kinetic studies of unimolecular reactions are a very insensitive and indirect method of gaining information about the rate of spontaneous decomposition. For this purpose, what is needed are direct observations of the function $k(E)$ as a function of E. Photochemical experiments provide one of the few methods available for putting definite amounts of energy into an absorbing molecule. The excited state produced by light absorption can engage in a number of competing acts.[a] If two of them are collisional deactivation and chemical decomposition, then experiments on the quantum yield as a function of pressure at several wavelengths can give values proportional to $k(E)$.

The effect of wavelength on the quantum yield of CO production from the photolysis of CH_2CO can be satisfactorily interpreted on the basis of the RRK model,[b] but unfortunately the lack of data on the precise bond dissociation energy prevents a unique assignment of parameters. The wavelength dependence of the fluorescence of photoexcited β-naphthylamine[c] has also been reasonably well interpreted in terms of the rate of spontaneous isomerization to a metastable state incapable of fluorescence. A model for $k(E)$ equivalent to the RRK model was used.

While these results are only tentative in view of the need to clarify the nature of the electronic states involved in the collisional deactivation and reaction, they do indicate the possibilities that such studies afford for measurements of $k(E)$.[d]

Because of the difficulty of distinguishing the RRK and Slater models experimentally and the unsettled question of whether or not internal energy transfer may take place between loosely coupled oscillators, we

---

[a] This subject is extensively reviewed by W. A. Noyes, Jr., et al., *Chem. Revs.*, **56**, 49 (1956).

[b] Gerald B. Porter, personal communication, finds that the data can be fitted with values of n in the range of 8, $\bar{\nu} \cong 10^{13}$ sec^{-1}, and $E^* = 72$ Kcal/mole. However, higher values of n (that is, 10) and $\bar{\nu}$ (10^{14} sec^{-1}) are also consistent with the data.

[c] M. Boudart and J. T. Dubois, *J. Chem. Phys.*, **23**, 223 (1955).

[d] Similar studies in a mass spectrometer, by using the electron beam to provide an initial distribution of excited ions, have been interpreted semiquantitatively in terms of the RRK model (D. P. Stevenson, private communication). However, the data are much less precise, and the assumptions needed to use them are open to considerable question. A similar theory has been used with some success to account for the mass spectral patterns of hydrocarbons, H. M. Rosenstock et al., *Proc. Nat'l Acad. Sci. U.S.*, **38**, 667 (1952).

shall use the simpler RRK form [Eq. (X.5.1)] in our applications, while recognizing that the effective number of oscillators may be much less than the number present in the molecule.

The chief difficulty with both the RRK and Slater models arises from the feature that, for complex molecules, the passage from configurations that correspond to energized complexes to the configuration of the transition state may produce gross changes in weak frequencies such as hindered rotations and bending vibrations. Such changes may alter enormously the values of $k(E)$. Phenomenologically this would correspond to large entropy charges associated with the change from energized complex to transition state. That such changes do in fact occur is indicated by the very high values of the frequency factors which have been found for the first-order, high-pressure, limiting rate constants for the decompositions of $C_2H_6 \rightarrow 2CH_3$, $N_2O_5 \rightarrow NO_2 + NO_3$, $N_2O_4 \rightarrow 2NO_2$, etc. (see Tables XII.4 and XII.5). For these reactions $\bar{\nu}$ is in the range 10^{16} to 10^{18} sec^{-1} and the region of "pressure dependence" of the rate constants is in an anomalously high range. R. A. Marcus[a] has reformulated the RRK theory to take such changes into account and has been able to predict the rate of the C_2H_6 reaction. No such attempt has been made to do so for the Slater model; for the algebra would present formidable difficulties.

7. Distribution Function for Energized Molecules

At equilibrium the distribution of molecules among the various energy states E is given by the Maxwell-Boltzmann expression. Thus for a molecule with n classical internal, harmonic oscillators, the fraction of molecules with energy $E_1, E_2, \ldots, E_n$ present in these oscillators is

$$P(E_1, \ldots, E_n) = \frac{e^{-(E_1 + E_2 + \cdots + E_n)/RT}}{RT} = \frac{e^{-E/RT}}{RT} \quad (X.7.1)$$

The fraction of molecules $P(E)$ with total internal energy in the range E to $E + dE$, with $E = E_1 + E_2 + \cdots + E_n$, is obtained by integrating this expression over all values of E_i subject to their sum being E:

$$P(E) = \int \cdots \int P(E_1, \ldots, E_n) \, dE_1 \cdots dE_n$$
$$= \frac{e^{-E/RT}}{RT} \int \cdots \int dE_1 \cdots dE_n \quad (X.7.2)$$

On performing the term-by-term integration between limits $E - E_2 - $

[a] R. A. Marcus, J. Chem. Phys., **20**, 364 (1952).

$E_3 - \cdots - E_n$ and 0 for E_1, and similarly for the successive E_i, we find[a]

$$P(E) = \left(\frac{E}{RT}\right)^{n-1} \frac{e^{-E/RT}}{(n-1)!RT} \qquad (X.7.3)$$

When a chemical reaction is occurring, these distribution functions are perturbed by the reaction, and in general we may expect to find the stationary-state concentration of energized molecules lower than that at equilibrium. Lindemann[b] was the first to propose a scheme for computing the influence of the reaction on these distributions. His scheme involved the competition between destruction of energized molecules by collisional deactivation and by chemical reaction. If B represents a molecule which ultimately reacts spontaneously to produce products, then

$$B + B \underset{d}{\overset{a}{\rightleftharpoons}} B^* + B$$

$$B^* \overset{r}{\rightarrow} products \qquad (X.7.4)$$

where B^* represents an energized molecule of B and we have neglected any back reaction. The stationary-state concentration of B^* is then

$$(B^*)_{ss} = \frac{k_a(B)^2}{k_d(B) + k_r} \qquad (X.7.5)$$

and the stationary rate of appearance of products is:

$$\frac{d\,(products)}{dt} = \frac{k_a k_r(B)^2}{k_d(B) + k_r} \qquad (X.7.6)$$

When $k_d(B) \gg k_r$, that is, collisional deactivation is much more probable than spontaneous decomposition, we have what is referred to as the high-pressure limit:

$$\frac{d\,(products)}{dt} \xrightarrow{k_d(B) \gg k_r} \frac{k_a}{k_d} k_r(B) \qquad (X.7.7)$$

which represents a first-order dependence on (B).

At the other, low-pressure extreme, where the time between collisions is so long that every energized molecule decomposes before it can be collisionally deactivated, we find

$$\frac{d(products)}{dt} \xrightarrow{k_d(B) \ll k_r} k_a(B)^2 \qquad (X.7.8)$$

[a] R. A. Marcus and O. K. Rice, *J. Phys. & Colloid Chem.*, **55**, 894 (1951), have proposed the function

$$P(E) = \frac{E^{n-1}}{(n-1)!\prod_i (h\nu_i)}$$

as a semiclassical approximation to the more real case of quantized molecules.
[b] F. A. Lindemann, *Trans. Faraday Soc.*, **17**, 598 (1922).

This represents a second-order reaction whose rate constant is the frequency of occurrence of activating collision between B molecules.

If we use the RRK model then we find for the stationary concentration of energized molecules:

$$B^*(E) = \frac{k_a(B)^2}{k_d(B) + k_r} = \frac{k_a(B)/k_d}{1 + k_r/k_d(B)}$$

$$= \frac{P(E)(B)}{1 + k_r(E)/(B)k_d(E)} \qquad (X.7.9)$$

where all the rate constants will in principle depend on the internal energy E. For the Slater model, a similar expression can be found for $B^*(E_1, \ldots, E_s)$. In the next chapter we shall consider the application of these expressions to models for unimolecular processes.

The important feature to note about the Lindemann model is that it predicts that any spontaneous molecule reaction process can exhibit a first-order rate law at sufficiently high pressures and a second-order rate law at sufficiently low pressures. These features have now been well confirmed experimentally.

XI

Unimolecular Reactions

1. Detailed Scheme for a Unimolecular Reaction: Isomerization

With the results thus far obtained we can now make a detailed formulation of a simple chemical reaction, isomerization. Some examples of chemical isomerizations which have been studied in the gas phase and found to follow homogeneous, first-order rates are certain cis-trans isomerizations of olefins[a] and some allylic rearrangements, such as the rearrangement of vinyl allyl ether, $CH_2\!\!=\!\!CH\!-\!O\!-\!CH_2\!-\!CH\!\!=\!\!CH_2$,[b] to allyl acetaldehyde, $CH_2\!\!=\!\!CH\!-\!CH_2\!-\!CH_2\!-\!CHO$.

For the general isomerization reaction in which we admit the possibility of equilibrium we can write the chemical equation

$$A \rightleftharpoons B \tag{XI.1.1}$$

and the detailed reaction scheme[c]

$$\tag{XI.1.2}$$

By A* or B* we mean any energized molecule of A or B with energy in excess of the critical energy E^* for the reaction. A* is differentiated from B* rather arbitrarily in terms of the operational criterion that A* is any form of active species which can be formed by activation of A and on de-

[a] See Table XI.1.
[b] F. W. Schuler and G. W. Murphy, *J. Am. Chem. Soc.*, **72**, 3155 (1950).
[c] S. W. Benson and A. E. Axworthy, Jr., *J. Chem. Phys.*, **21**, 428 (1953).

activation can only go back to A.[a] We can represent the scheme in terms of a schematic potential-energy diagram as shown in Fig. XI.1, where a cut through the potential-energy surface is shown corresponding to the minimum values of U for different values of L. All states to the left of L_0 are A or A* and to the right are B or B*. The reaction as shown is endothermic, since the minimum energy of B is shown higher than that for A. The difference in these two energies corresponds to the energy of the reaction ΔE.

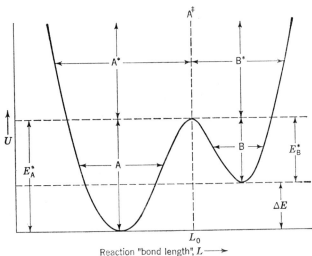

Reaction "bond length", $L \longrightarrow$

Fig. XI.1. Schematic representation of the potential-energy profile for an isomerization A $\rightleftharpoons$ B.

In the above scheme we have represented the reaction processes responsible for the production of A* and B* from A and B, respectively, as collisional processes in which M represents any molecule in the system. The differential equations for the system are

$$\frac{d\text{A}}{dt} = -k_1\text{MA} + k_2\text{MA*} \qquad \frac{d\text{B}}{dt} = -k_1\text{MB} + k_2\text{MB*}$$

$$\text{(XI.1.3)}$$

for the products and reactants, and for the intermediates they are

$$\frac{d\text{A*}}{dt} = k_1\text{MA} - (k_2\text{M} + k_\text{A})\text{A*} + k_\text{B}\text{B*}$$

$$\frac{d\text{B*}}{dt} = k_1\text{MB} - (k_2\text{M} + k_\text{B})\text{B*} + k_\text{A}\text{A*} \qquad \text{(XI.1.4)}$$

[a] It is necessary to introduce such a distinction between A* and B* in order to avoid real paradoxes in describing the system. Physically we might say that the configuration of A* corresponds more closely to A and B* to B, or in terms of the Slater model that in all A* the critical length L is less than L_0, while in B*, $L > L_0$. The transition state A‡ corresponds to $L = L_0$ and is the same for A* and B*.

By applying the steady-state hypothesis $(A^*,B^*) \ll (A,B)$ and setting $dA^*/dt = dB^*/dt = 0$ we can solve for the stationary-state concentrations A_{ss}^* and B_{ss}^*. For A_{ss}^* (the expression for B_{ss}^* is symmetrical in form) we find

$$A_{ss}^* = \frac{k_{1'}k_B MB + k_1 MA(k_{2'}M + k_B)}{(k_2 M + k_A)(k_{2'}M + k_B) - k_A k_B} \tag{XI.1.5}$$

On substitution into the equation for dA/dt and further simplification we have

$$-\frac{dA}{dt} = \frac{k_A k_1 k_{2'} MA - k_B k_{1'} k_2 MB}{k_A k_{2'} + k_2 k_{2'} M + k_2 k_B} \tag{XI.1.6}$$

Of the two terms in the numerator the second refers to the reverse reaction, $B \rightarrow A$, while the first represents the contribution of the forward reaction. At equilibrium $dA/dt = 0$, the two terms are equal, and we see that

$$\frac{B_{eq}}{A_{eq}} = \frac{k_A}{k_B} \frac{k_1}{k_2} \frac{k_{2'}}{k_{1'}} = K_{eq} = \frac{k_A}{k_B} K_{A^*} K_{B^*}^{-1} \tag{XI.1.7}$$

where we have set $K_{A^*} = k_1/k_2$ and $K_{B^*} = k_{1'}/k_{2'}$, which are respectively the equilibrium constants for A^* and B^*. By rearranging Eq. (XI.1.6) we can write for the rate of consumption of A (when $k_{1'} = 0$ or $B = 0$):

$$-\frac{1}{A}\frac{dA}{dt} = \frac{k_A K_{A^*}}{1 + k_A/k_2 M + k_B/k_{2'}M} \tag{XI.1.8}$$

while a completely symmetrical expression could be written for the rate of production of A from B.

In the case that $B = 0$ or $k_{1'} = 0$ (i.e., back reaction is too slow to be observable) the over-all rate of the reaction will be given by integrating Eq. (XI.1.8) over all energies from 0 to ∞.[a] We find

---

[a] The treatment presented here has been oversimplified to the extent that we should have started with an infinite set of coupled equations that represent the kinetics of not a single species A^*, but of an infinite set of species $A^*(E)$ differing in their internal energies E. A typical rate equation in such a scheme would have been:

$$\frac{dA(E_i)}{dt} = \sum_k \sum_{j=0}^{\infty}{}' k_{ji}\, M(E_k)A(E_j) - \sum_k \sum_{j=0}^{\infty} k_{ij}M(E_k)A(E_i) - k_A A(E_i) + k_B B(E_i)$$

where $A(E_i)$ is a molecule of A with energy E_i and k_{ji} represents the rate at which molecules of A with energy E_j collide with molecules M and are brought into the class $A(E_i)$, while k_{ij} represents the reverse process, namely, conversion of $A(E_i)$ to $A(E_j)$. The other two terms are unchanged, since $A(E_i)$ can pass unimolecularly into only one energy state of B, namely, $B(E_i)$. In such a scheme we should also have to take into account the energies of $M(E_k)$ and sum over all energies E_k compatible with the restriction that the total energies of A and M are conserved before and after the collision and that momentum is also conserved. It is difficult to estimate the errors of the present treatment without a complete solution to the above rigorous problem. Such an attempt has been made by I. Prigogine et al. [J. Phys. & Colloid Chem., **55**, 765 (1951); see article for other, earlier references and R. Willaert, Thesis, Université Libre de Bruxelles, 1951]

$$-\frac{1}{A}\frac{dA}{dt} = \int_0^\infty \frac{k_A(E)P_A(E)\,dE}{1 + k_A(E)/k_2(E)M + k_B(E)/k_{2'}(E)M} \quad (XI.1.9)$$

$P_A(E)$ is that fraction of A molecules with internal energy E in s degrees of freedom. The ratio $K_A(E)/Mk_2(E)$ is the ratio of the rates of spontaneous isomerization of A* to the rate of deactivation of A* by collision. Similarly, $k_B(E)/Mk_{2'}(E)$ is the ratio of the rate of isomerization of B* (to give A*) to the rate of deactivation of B* to give B.[a]

The term $d\ln A/dt$, which we would expect to be equal to a rate constant and be a function only of temperatures for a first-order reaction, has now become a function of temperature and composition, since the term M in the integral represents the total concentration of all molecules in the system.[b] Thus in the simplest system in which we might have expected to find a first-order rate law we find instead that the rate of the reaction has a rather complicated dependence on both concentration and composition. In order to proceed with the analysis of this system, we shall have to make some further investigation of the quantities that appear in the integrand of Eq. (XI.1.9).

2. Pressure Dependence of Unimolecular Reactions; The High-pressure Limit

Equation (XI.1.9) is of such form that it becomes independent of the concentration of the gas as $M \to \infty$. Under these conditions (which we shall discuss later) the terms in the denominator of the integrand approach unity and the limiting rate at high concentrations is given by

$$-\frac{1}{A}\left(\frac{dA}{dt}\right)_\infty = \int_{E^*}^\infty k_A(E)P_A(E)\,dE = k_\infty \quad (XI.2.1)$$

where the quantity k_∞ is a first-order rate constant which is independent of concentration and may be identified with the experimentally observed value of k for the reaction [Eqs. (X.6.2) and (X.6.3)].

---

for a reasonable but somewhat artificial model of energy exchange. His results indicate that the perturbation in the Maxwellian distribution of energies is seriously affected by the reaction only when the reaction is strongly exothermic. However, his results cannot be applied directly to chemical reactions because he did not use a reasonable model for the constants k_A and k_B.

[a] It is important to note that even when the back reaction B → A is negligibly small, the rate of deactivation of B* is still important, since every B* will re-form A* unless it becomes deactivated by a collision.

[b] M = A + B plus any foreign gases present. This further implies that the expressions $Mk_2(E)$ and $Mk_{2'}(E)$ are not simple functions but rather complex functions. In the above case we should replace $Mk_2(E)$ by a sum $Mk_2(E) = Ak_{2A}(E) + Bk_{2B}(E) + \cdots$, where the individual rate constants $k_{2A}, k_{2B}, \ldots$ represent the varying efficiencies of A, B, ... for deactivation of A.

The significance of neglecting the terms in the denominator of Eq. (XI.1.9) at high concentrations is that collisional deactivation is so much more rapid than reaction that all A* are in virtual thermodynamic equilibrium with A, while of all the B* that are formed, only an insignificant fraction are reconverted to A* before they are collisionally deactivated to give B.

If now we use the value of k_A given by the RRK theory and are willing to make some approximation for $P_A(E)$, we can evaluate the integral in Eq. (XI.2.1). The true value of $P_A(E)$ is of course given by an exact solution of the quantum-mechanical equation for the molecule. A first simple approximation is to assume that $P_A(E)$ is given by the classical approximation in which we assume that the molecule is made up of s classical harmonic oscillators of frequencies ν_i. In that case $P(E)$, the probability of finding energy E divided among these oscillators, is obtained by a direct integration of the Gibbs' phase integral[a] and gives the value

$$P_A(E) = \left(\frac{E}{kT}\right)^{s-1} \frac{1}{(s-1)!} \frac{e^{-E/kT}}{kT} \qquad (XI.2.2)$$

The value of $k_A(E)$ can be taken as $\bar{\nu}(1 - E^*/E)^{s-1}$ for values of $E > E^*$ and $k_A(E) = 0$ for $E < E^*$. Then k_∞ is

$$k_\infty = \frac{\bar{\nu}}{(s-1)!} \int_{E^*}^\infty \left(1 - \frac{E^*}{E}\right)^{s-1} \left(\frac{E}{kT}\right)^{s-1} e^{-E/kT} \frac{dE}{kT} \qquad (XI.2.3)$$

which on substitution of $x = E - E^*$ leads to

$$k = \frac{\bar{\nu}e^{-E^*/kT}}{(s-1)!} \int_0^\infty \left(\frac{x}{kT}\right)^{s-1} \frac{e^{-x/kT}}{kT} dx \qquad (XI.2.4)$$

where the integrand may be recognized as the gamma function $\Gamma(s) = (s-1)!$, and so

$$k_\infty = \bar{\nu}e^{-E^*/kT} \qquad (XI.2.5)$$

If we compare this result with the Arrhenius[b] expression for a first-order constant, $k = Ae^{-E/RT}$, we see that we can make the following identifications:

$$A = \bar{\nu} \qquad E = N_{Av}E^* \qquad (XI.2.6)$$

This means that the so-called frequency factor for unimolecular reaction rate constants should be of the order of magnitude 10^{12} to 10^{14} sec^{-1} (Sec. X.6). When the observed values are either larger or smaller, then we should look for some marked structural changes in the change from normal A to A‡. As we shall see later, the high-pressure rate constants for unimolecular reactions are generally (Sec. X.7) of the order of magnitude of 10^{13} sec^{-1}, so that this simplified theory is in good agreement at this point

[a] See Eq. (X.7.3).
[b] See Eq. (X.5.2).

with the observed facts. It should finally be noted that E^* does not depend on temperature, nor does $\bar{\nu}$.

3. Pressure Dependence of Unimolecular Reactions; Moderate Pressures

In order to see under what conditions the high-pressure limit for unimolecular reactions will be established, let us return to Eq. (XI.1.9) and examine the terms more closely. The equation is

$$-\frac{1}{A}\frac{dA}{dt} = \int_{E^*}^{\infty} \frac{k_A(E)P_A(E)\,dE}{1 + k_A(E)/Mk_2(E) + k_B(E)/Mk_{2'}(E)} \quad \text{(XI.3.1)}$$

It is of interest to look at the numerator $k(E)P(E)$ (omitting subscripts) of the integrand, which determines the high-pressure rate. In Fig. XI.2

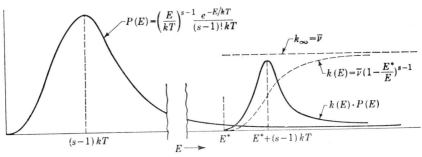

FIG. XI.2. Graphical representation of the reaction rate function $k(E)$, the classical energy-distribution function $P(E)$ for a group of s oscillators, and their product $k(E)P(E)$.

are plotted curves for $k(E)$, $P(E)$, and the product $k(E)P(E)$. This product has the form

$$k(E)P(E) = \bar{\nu}\frac{(E - E^*)^{s-1}e^{-E/kT}}{(kT)^s(s-1)!} \quad \text{(XI.3.2)}$$

when the RRK function $k(E)$ [Eq. (X.5.1)] and the classical energy-distribution function [Eq. (XI.2.2)] are employed. Since $k(E)$ is zero when $0 \leqslant E \leqslant E^*$, we can introduce the variable $x = E - E^*$, which allows us to write the limits in Eq. (XI.3.1) as 0 to ∞ and gives Eq. (XI.3.2) the form

$$k(x)P(x) = \frac{\bar{\nu}e^{-E^*/kT}}{(s-1)!(kT)}\left(\frac{x}{kT}\right)^{s-1}e^{-x/kT} \quad \text{(XI.3.3)}$$

where the first term on the right-hand side is independent of x.

It will be seen that, where $P(E)$ has a maximum at $E = (s-1)kT$ and is exponentially peaked[a] about that maximum, the product $k(E)P(E)$ has a similar type of maximum near the value $E = E^* + (s - 1)kT$. Thus most of the contribution to the area under this curve comes from molecules with energies near this maximum.[a] The average energy of the molecules under the product curve $k(E)P(E)$ is readily shown[b] to be $E^* + skT$, for which the mean rate of decomposition is

$$k(E_m) = \bar{\nu}\left(1 - \frac{E^*}{E_m}\right)^{s-1} = \bar{\nu}\left(1 + \frac{E^*}{skT}\right)^{1-s} \qquad (XI.3.4)$$

or in logarithmic form

$$\log\frac{k(E_m)}{\bar{\nu}} = (1 - s)\log\left(1 + \frac{E^*}{skT}\right) \qquad (XI.3.5)$$

Because of its importance in determining the integrand, this function is tabulated in Table XI.1 for different values of the ratio E^*/kT and s, the number of oscillators in the molecule. It is to be noted that, as $s \to \infty$, $k(E_m) \to \bar{\nu}e^{-E^*/kT}$, which is the rate of decomposition of a single oscillator that is always in thermodynamic equilibrium.

We can observe from the table that in the high-pressure limit, where the denominator in the integrand [Eq. (XI.3.1)] is unity, the mean lifetime of the average reacting molecule will be $\tau_m = 1/\bar{\nu} \times 10^{2.967}$ (for $s = 4$, $E^*/kT = 35$). If we choose as a mean value for $\bar{\nu}$ 10^{13} sec^{-1}, this becomes 10^{-10} sec. On the other hand, keeping $E^*/kT = 35$ but letting $s = 12$, we find $\tau_m \cong 3 \times 10^{-7}$ sec, so that the more complicated the molecule the greater will be its mean lifetime before decomposition. ($E^*/kT = 35$ is also about the average value for most unimolecular reactions.)

If we now turn to the terms in the denominator of Eq. (XI.3.1), we see that they will not be important unless $Mk_2(E)$ and $Mk_{2'}(E)$ are comparable to or greater than $k_A(E)$ and $k_B(E)$, respectively.

From the RRK theory we have

$$k_A(E) = \bar{\nu}_A\left(1 - \frac{E_A^*}{E_A}\right)^{s_A-1} \quad \text{while} \quad k_B(E) = \bar{\nu}_B\left(1 - \frac{E_B^*}{E_B}\right)^{s_B-1}$$

$$(XI.3.6)$$

We can set $s_A = s_B = s$ and $\bar{\nu}_A = \bar{\nu}_B = \bar{\nu}$, and then $E_A = E_B + E_R$ and

[a] The half-width of the peak [i.e., the value of x at which $k(x)P(x)$ has fallen to one-half its maximum value] is given by $2|x_{1/2}| \cong 2.4(s - 1)^{1/2}$, while the tenth width is $2|x_{0.1}| \cong 5.8(s - 1)^{1/2}$. For $s = 5$, $k(x)P(x)$ has fallen to half its maximum value at $x_{1/2} = (s - 1) + 2.4 = 1.6$ to 6.4, while for $s = 10$, the range is $x_{1/2} = (s - 1) + 3.6 = 5.4$ to 12.6, where x is in units of kT.

[b] This is done by evaluating: $\langle E \rangle_{av} = \langle x + E^* \rangle_{av} = E^* + \langle x \rangle_{av}$ and

$$\langle x \rangle_{av} = \frac{\int_0^\infty xP(x)k(x)\,dx}{\int_0^\infty P(x)k(x)\,dx} = \frac{\int_0^\infty (x/kT)^s \cdot e^{-x/kT}\,dx}{\int_0^\infty (x/kT)^{s-1}\,e^{-x/kT}\,dx} = \frac{kT\Gamma(s + 1)}{\Gamma(s)} = skT$$

TABLE XI.1. VARIATION OF THE MEAN REACTION RATE $k(E_m)$ WITH
CRITICAL ENERGY E^* AND NUMBER OF DEGREES OF FREEDOM s

$\dfrac{E^*}{kT}$	20	25	30	35	40
s	Mean reaction rate[a]				
2	1.041	1.130	1.204	1.267	1.322
3	1.770	1.940	2.083	2.205	2.312
4	2.334	2.580	2.787	2.967	3.123
5	2.796	3.111	3.380	3.612	3.816
6	3.185	3.565	3.890	4.175	4.425
7	3.520	3.960	4.340	4.668	4.960
8	3.808	4.305	4.670	5.110	5.446
9	4.603	4.620	5.100	5.515	5.885
10	4.291	4.895	5.418	5.878	6.291
11	4.500	5.150	5.710	6.210	6.660
12	4.686	5.379	5.984	6.523	7.007
15	5.154	5.965	6.680	7.350	7.895
18	5.518	6.425	7.240	7.975	8.64
21	5.82	6.62	7.70	8.52	9.26
∞	6.68	10.85	13.02	15.19	17.36

[a] Values given are for $-\log [k(E_m)/\bar{\nu}]$ from Eq. (XI.3.5).

$E_A^* = E_B^* + E_R$, where E_R is the energy change of the over-all reaction (for exothermic reactions $E_B > E_A$). On substitution

$$k_A(E) = \bar{\nu} \left(\frac{E - E_A^*}{E_A} \right)^{s-1} \quad \text{and} \quad k_B(E) = \bar{\nu} \left(\frac{E - E_A^*}{E_A - E_R} \right)^{s-1}$$

$$\text{(XI.3.6a)}$$

so that, if the reaction is exothermic, $k_A > k_B$ by the factor $(1 - E_R/E_A)^{s-1}$, while $k_B > k_A$ by the same factor for an endothermic reaction. Only for thermoneutral reactions will $k_A = k_B$. We can thus write

$$k_B(E) = \theta(E) k_A(E) \qquad \text{(XI.3.6b)}$$

where $\theta(E) = (1 - E_R/E_A)^{1-s}$.[a]

As a first approximation to $k_2(E)$ and $k_{2'}(E)$ we can write

$$k_2(E) = k_{2'}(E) = \lambda_D Z' \qquad \text{(XI.3.7)}$$

where, in view of the similarity of A* and B*, we have equated the rate constants for collisional deactivation, Z' is the collision frequency of the complex, and λ_D is some constant accommodation coefficient equal to that

[a] The significance of this relation between $k_A(E)$ and $k_B(E)$ arises from the fact that the factors $(1 - E^*/E)^{s-1}$ represent the ratio of the number of molecular quantum states at the saddle point L_0 to the total number of states on the hypersurface of energy $E \geqslant E^*$. This ratio is strongly dependent on the value of E^*, as indicated, and decreases greatly for any given value of E as E^* increases.

fraction of collisions of the complex which result in a change in status to inactive species.[a]

We can thus write, to this degree of approximation, for the denominator in the integrand of Eq. (XI.3.1)

$$1 + \frac{k_A(E)}{Mk_2(E)} + \frac{k_B(E)}{Mk_{2'}(E)} \sim 1 + \frac{\beta\bar{v}(1 - E^*/E_A)^{s-1}}{M\lambda_D Z_A'} \qquad (XI.3.8)$$

where $\beta = \beta(E) = 1 + \theta(E)$ [Eq. (XI.3.6b)] and where $Z_A' = \pi\sigma_{AM}^2$ $(8kT/\pi\mu)^{1/2}$ [Eq. (VII.8F.5)], the collision frequency for A^* (σ_{AM} = mean collision diameter, μ = reduced mass of collision complex). For most molecules at STP with diameters of about 3.5 Å, $Z'M$ is about 3×10^9 sec^{-1}, and for lack of any other information, if we set $\lambda_D = 1$ we can evaluate the integrand of Eq. (XI.3.1) numerically for any given concentration. This has been done by graphical integration by Kassel[b] and by a rather neat analytic method by Marcus.[c] By substituting $y = x/kT = E_A - E^*/kT$ into the integrand we can now write Eq. (XI.3.1) as

$$-\frac{1}{A}\frac{dA}{dt} = \frac{k_\infty}{(s-1)!} \int_0^\infty \frac{y^{s-1}e^{-y}\, dy}{1 + (\beta\bar{v}/M\lambda_D Z_A')[y/(y+y^*)]^{s-1}} = k_{exp} \qquad (XI.3.9)$$

where k_{exp} is the experimentally observed first-order rate constant for the isomerization. Without going through this laborious integration we can calculate for simple cases the value of the characteristic concentration (or pressure at given T) for which the denominator will reach a value of 2 when y has that value which gives the numerator the value near its peak (Fig. XI.2). Such a value will indicate the range of pressures in which the rate given by Eq. (XI.3.8) will begin to fall off from the high-pressure limiting rate. This value will be reached when

$$\beta k_A(E_m) = Mk_2 = M\lambda_D Z_A' \qquad (XI.3.10)$$

---

[a] This is certainly very crude in that it assumes that the rate of deactivation is independent of the internal energy. There is some experimental evidence that, for I_2 molecules, the rate of loss of vibrational energy on collision is about 100 fold greater for the highly excited states than for the lower states. Eliashevich [Phys. Rev., **39**, 532 (1932); J. Exptl. Theoret. Phys. (U.S.S.R.), **2**, 59 (1932)] and Mott and Massey ("The Theory of Atomic Collisions," 2d ed., chap. 12, Oxford University Press, New York, 1949) have made some crude quantum-mechanical calculations which indicate that the probability of losing or gaining 1 quantum of vibrational energy by a harmonic oscillator to an atom is proportional to the energy of the oscillator. Other work on this problem has been confined to an experimental study of the dispersion of sound in gases. These measurements show [W. Griffith, J. Appl. Phys., **21**, 1319 (1950)] that for the lowest vibrational states the value of λ_D is about 10^{-3} but may vary considerably from gas to gas and is strongly dependent on the chemical nature of the colliding gases.

[b] L. Kassel, J. Phys. Chem., **32**, 225, 1065 (1928).

[c] R. A. Marcus, J. Chem. Phys., **20**, 364 (1952).

If we use the values of $k(E_m)$ from Table XI.1, we obtain the values $\lambda_D P(35)$ shown in Table XI.2 that satisfy Eq. (XI.3.10) for the choice of values $E^*/kT = 35$, $\bar{\nu} = 10^{13}$ sec^{-1}, $Z'_A = 2.5 \times 10^9$ sec^{-1} atm(STP)$^{-1}$, and $\beta = 1$. The values shown in Table XI.2, while only roughly representative

TABLE XI.2. PROBABLE VALUES OF CHARACTERISTIC PRESSURES AT WHICH
UNIMOLECULAR RATE CONSTANTS SHOW PRONOUNCED FALLING
OFF FROM HIGH-PRESSURE LIMITS $[k_A(E_m) = Mk_2]$

Minimum no. of atoms N^a	Number of degrees of freedom S	Mean rate of decomposition $k(E_m)$, sec^{-1}	$\lambda_D P(35)$, mm Hg (STP)
2	1	$\bar{\nu} = 1 \times 10^{13}$	3.1×10^6
2	2	5.4×10^{11}	1.7×10^5
3	3	6.2×10^{10}	1.9×10^4
3	4	1.1×10^{10}	3300
3	5	2.4×10^9	730
4	6	6.7×10^8	210
4	7	2.2×10^8	67
4	8	7.8×10^7	24
5	9	3.1×10^7	8.7
5	10	1.3×10^7	4.0
5	11	6.2×10^6	1.9
6	12	3.0×10^6	0.91
7	15	4.5×10^5	0.14
8	18	9.5×10^4	0.029
9	21	3.0×10^4	0.0091
∞	∞	6.5×10^{-3}	2.0×10^{-11}

[a] The number of internal degrees of freedom for an N-atom molecule is $3N-6$, non-linear, and $3N-5$, linear, with the possibility of increasing this by 1 if two degrees of rotational freedom can contribute their energy to the chemical reaction (i.e., rotational energy is not conserved; only angular momentum is).

(since they may depend so much on λ_D, σ_{AM}, etc.), give a general idea of the pressure region in which the high-pressure rate may be expected to show a serious drop.[a]

4. Pressure Dependence of Unimolecular Reactions; Low-pressure Limit

In Sec. XI.3 we investigated the range of concentrations in which a unimolecular isomerization will show a pronounced dependence of its experimentally observed specific rate constant on total concentration. In

[a] Note that the characteristic pressure will be higher if λ_D is less than 1 or if $\bar{\nu} > 10^{13}$ sec^{-1}. The values in Table XI.2 are thus only probable values. It is actually very unlikely that $\lambda_D = 1$, since we should not expect a single collision to be effective in deactivating a molecule which has energy very far in excess of E^*.

this range the concentration dependence of k_{exp} is not simple and is given by the rather formidable integral of Eq. (XI.3.9). We shall now investigate the behavior of k_{exp} at extremely low concentrations and also at extremely high concentrations to see if a more tractable result can be obtained.

By rewriting k_{exp} from Eq. (XI.3.9)

$$k_{exp} = \frac{k_\infty}{(s-1)!} \int_0^\infty \frac{y^{s-1}e^{-y}\, dy}{1 + (C/M)[y/(y+y^*)]^{s-1}} \qquad (XI.4.1)$$

where the constant $C = \beta\bar{v}/\lambda_D Z'_A$ and for a typical set of values (see page 155) will lie in the range 6.7×10^3 [when M is in atm(STP)].

As already observed (Fig. XI.2), the numerator in this integral $G(s) = y^{s-1}e^{-y}$ is exponentially peaked about the value $y_{max} = s - 1$ and has a half-width of $2.4(s-1)^{1/2}$ (footnote page 231). Its shape is reproduced in Fig. XI.3 (upper curve). The lower curve in Fig. XI.3 shows the be-

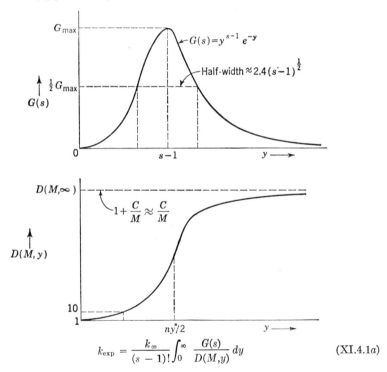

$$k_{exp} = \frac{k_\infty}{(s-1)!} \int_0^\infty \frac{G(s)}{D(M,y)}\, dy \qquad (XI.4.1a)$$

FIG. XI.3. Pressure dependence of a unimolecular isomerization rate constant.

havior of the bracketed terms in the denominator, $D(M,y) = 1 + C/M$ $[y/(y+y^*)]^{s-1}$. The integral is obtained by dividing $G(s)$ by $D(M,y)$ and integrating under the curve obtained. Now most of the value of this curve will still come from the region under the exponential peak bounded by $y = s - 1 \pm 1.2(s-1)^{1/2}$. If for this range of values of y the denominator $D(M,y)$ is close to its upper limit $D(M,\infty)$, we can neglect unity in $D(M,y)$ and write[a]

$$k_{\text{exp}}(M \to 0) = \frac{k_{\infty}M}{C(s-1)!} \int_0^{\infty} (y + y^*)^{s-1} e^{-y} \, dy \qquad (XI.4.2)$$

which on integrating by parts becomes

$$k_{\text{exp}}(M \to 0) = \frac{k_{\infty}M}{C} \left[\frac{y^{*(s-1)}}{(s-1)!} + \frac{y^{*(s-2)}}{(s-2)!} + \cdots + 1 \right] \qquad (XI.4.3)$$

Equation (XI.4.3) may be evaluated exactly or approximated (when $y^* \gg s - 1$) by taking the first term

$$k_{\text{exp}}(M \to 0) \cong \frac{k_{\infty}M}{C} \frac{y^{*(s-1)}}{(s-1)!} \qquad (XI.4.4)$$

and making the substitutions $k_{\infty} = \bar{\nu}e^{-E^*/kT}$, $C = \bar{\nu}/\lambda_D Z'_A$, and $y^* = E^*/kT$. We then find

$$k_{\text{exp}}(M \to 0) = \lambda_D Z'_A \frac{Me^{-E^*/kT}}{(s-1)!} \left(\frac{E^*}{kT} \right)^{s-1} \qquad (XI.4.5)$$

for the limiting value of the experimental rate constant at very low pressures. This may be placed in a more conventional form by the substitution

$$P(E^*) = \frac{k_1(E^*)M}{\lambda_D Z'_A M} = \frac{e^{-E^*/kT}}{(s-1)!} \left(\frac{E^*}{kT} \right)^{s-1}$$

where $P(E^*)$ is the equilibrium fraction of species A^* with energy E^*, $k_1(E^*)M$ is the rate at which such species are formed by collisions, and $\lambda_D Z'_A M$ is the rate at which they are destroyed by deactivating collisions.[b] In this case we have

$$k_{\text{exp}}(M \to 0) = k_1(E^*)M \qquad (XI.4.6)$$

The significance of these results is that the time between deactivating collisions is so long that every energized species reacts before it is deactivated. We see further that under these conditions most of the reaction is accounted for by just those species with energies E^* and also that the

---

[a] This implies that (C/M), $[y/(y + y^*)]^{s-1} \gg 1$ for values of $y = (s-1) \pm 1.2(s-1)^{1/2}$, and before applying Eq. (XI.4.3) this condition should be confirmed.

[b] We have eliminated the quantity β which appears in $D(M,y)$ by setting it equal to unity. If the over-all reaction is thermoneutral, then $\beta = 2$ and a factor of 2 should appear in the denominators of Eqs. (XI.4.5) and (XI.4.6). This arises from the fact that this equation was derived for the unimolecular isomerization equation (XI.1.9) and we have set the rates of isomerization $k_A = k_B$ and also the rates of deactivation $k_2 = k'_2$. At these very low pressures we have the result that A^* and B^* are in equilibrium with each other but not with processes of activation or deactivation, which are much slower. Thus only half of the active species reaches the final state B. If k_A were much greater or much smaller than k_B (which occurs if the reaction is very energetic), then the factor of 2 disappears and $\beta = 1$ as implied above.

experimental rate constant is under these conditions directly proportional to the total pressure M.

We can find the experimental activation energy for the reaction under these limiting conditions by taking the logarithmic derivative with respect to T of both sides of Eq. (XI.4.5):

$$\ln [k_{\exp}(M \to 0)] = \ln \lambda_D + \ln Z'_A - \frac{E^*}{kT}$$

$$+ (s - 1)[\ln E^* - \ln (kT)] - \ln (s - 1)! \qquad \text{(XI.4.7)}$$

$$\frac{E_{\exp}}{kT^2} = \frac{\partial \ln k^\circ_{\exp}}{\partial T} = \frac{1}{2T} + \frac{E^*}{kT^2} - \frac{s - 1}{T} \qquad \text{(XI.4.8)}$$

or
$$E_{\exp} = E^* - (s - \tfrac{3}{2})kT \qquad \text{(XI.4.9)}$$

since we have assumed λ_D independent of temperature and Z'_A is proportional to $T^{1/2}$ [Eq. (VII.5F.8)]. We see that the observed activation energy has fallen off from its high-pressure limit of E^* by an amount equal to $(s - \tfrac{3}{2})kT$.[a]

These same results may be obtained in a more formal way by differentiating $k_{\exp}$ [Eq. (XI.4.1)] with respect to M. We find for k and the first and second derivatives as $M \to 0$:

$$k_{\exp} = \frac{k_\infty}{(s - 1)!} \int_0^\infty \frac{y^{s-1} e^{-y}\, dy}{\{1 + (C/M)[y/(y + y^*)]^{s-1}\}}$$

$$\xrightarrow{M \to 0} \frac{M k_\infty}{C(s - 1)!} \int_0^\infty (y + y^*)^{s-1} e^{-y}\, dy \qquad \text{(XI.4.10)}$$

$$\left(\frac{\partial k_{\exp}}{\partial M}\right) = \frac{C k_\infty}{(s - 1)!} \int_0^\infty \frac{y^{2s-2} e^{-y}\, dy}{(y + y^*)^{s-1}\{M + C[y/(y + y^*)]^{s-1}\}^2}$$

$$\xrightarrow{M \to 0} \frac{k_\infty}{C(s - 1)!} \int_0^\infty (y + y^*)^{s-1} e^{-y}\, dy \qquad \text{(XI.4.11)}$$

which is the result already obtained [Eq. (XI.4.2)], while [b]

---

[a] The significance of this result is not that molecules with energies less than E^* may react but rather that the experimental activation energy will be the difference in average energy between reacting molecules and normal molecules. At these low pressures the specific rate of reaction of molecules with energies in excess of E^* is so great compared to their rate of production that the stationary-state concentration of such molecules is negligible compared to that of molecules with energy very close to E^*. Since the average molecule in the system has energy skT, we see that at these extremely low pressures the average reacting molecule has energy $E^* + kT$. The difference is $E^* - (s - 1)kT$, and this differs from $E^* - (s - \tfrac{3}{2})kT$ by the quantity $\tfrac{1}{2}kT$ contributed by the temperature dependence of the collision frequency.

[b] Note that this last integrand becomes infinite at the origin, so that the second derivative becomes infinite as $M \to 0$. This implies that we cannot expand $k_{\exp}$ in a powers series in M around the point $M = 0$ and further that the pressure dependence of $k_{\exp}$ close to $M = 0$ is greater than the first power of M but less than M^2.

$$\frac{\partial^2 k_{exp}}{\partial M^2} = \frac{-2Ck_\infty}{(s-1)!} \int_0^\infty \frac{y^{2s-2}e^{-y} \, dy}{(y+y^*)^{s-1}\{M + C[y/(y+y^*)]^{s-1}\}^3}$$

$$\xrightarrow{M\to 0} -\frac{2k_\infty}{C^2(s-1)!} \int_0^\infty \frac{(y+y^*)^{2s-2}e^{-y} \, dy}{y^{s-1}} \qquad \text{(XI.4.12)}$$

To obtain the shape of the curve for k_{exp} as a function of M at very high

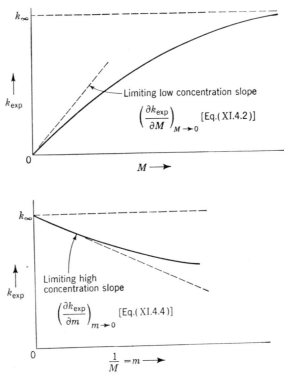

Fig. XI.4. The dependence of the unimolecular specific rate constant on concentration of M.

concentrations, we differentiate k_{exp} with respect to the variable $m = 1/M$. Then from Eq. (XI.4.10) we have

$$k_{exp}(m \to 0) = k_\infty \qquad \text{(XI.4.13)}$$

$$\left(\frac{\partial k_{exp}}{\partial m}\right) = -\frac{k_\infty C}{(s-1)!} \int_0^\infty \frac{y^{2s-2}e^{-y} \, dy}{\{1 + mC[y/(y+y^*)]^{s-1}\}^2 (y+y^*)^{s-1}}$$

$$\xrightarrow[m\to 0]{M\to\infty} -\frac{k_\infty C}{(s-1)!} \int_0^\infty \frac{y^{2s-2}e^{-y} \, dy}{(y+y^*)^{s-1}} \qquad \text{(XI.4.14)}$$

while for the second derivative we have

$$\left(\frac{\partial^2 k_{exp}}{\partial m^2}\right) = \frac{2k_\infty C^2}{(s-1)!} \int_0^\infty \frac{y^{3s-3}e^{-y}\,dy}{\{1 + mC[y/(y + y^*)]^{s-1}\}^3(y + y^*)^{2s-2}}$$

$$\xrightarrow{m\to 0} \frac{2k_\infty C^2}{(s-1)!} \int_0^\infty \frac{y^{3s-3}e^{-y}\,dy}{(y + y^*)^{2s-2}} \qquad (XI.4.15)$$

so that k_{exp} and all its derivatives with respect to $1/M = m$ will be well defined and finite as $M \to \infty\,(m \to 0)$. Despite this, the Taylor's series expansion of $k_{exp}(m)$ about $m = 0$ is at best only semiconvergent, so that k_{exp} approaches its high-pressure limit only very slowly as $m = 1/M \to 0$.[a] The behavior of k_{exp} with concentration is shown in Fig. XI.4.

5. Detailed Scheme for Unimolecular Decompositions

If we consider the reversible chemical reaction[b]

$$D \rightleftharpoons A + B \qquad (XI.5.1)$$

we can represent the detailed scheme for the reaction by

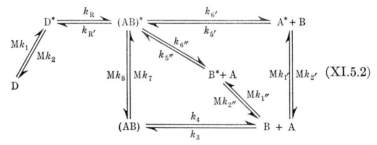

$$(XI.5.2)$$

where AB is some loose association complex of A and B with not enough energy to react to form D*, [AB]* is any complex of A and B with internal energy sufficient to permit reactions to form D* (the critically energized complex of D), and A* and B* are of necessity somewhat arbitrarily defined as species having sufficiently high total energy that a properly oriented collision with the complementary reactant (reactions 5' and 5'') may lead to formation of [AB]* (see footnote page 226).

By again applying the steady-state treatment to the intermediates D*, AB, A*, B*, and [AB]* we arrive at an expression for the net rate of the reaction $R = -dD/dt = dA/dt = dB/dt$:

$$R = \frac{(D)(M)k_r k_1 - (A)(B)(M)K_{AB}k_{r'}k_2}{k_r + (M)k_2} \cdot \frac{Z}{Z + (M)k_{r'}k_2/[k_r + (M)k_2]} \qquad (XI.5.3)$$

[a] The integrals (XI.4.14) and (XI.4.15) can be evaluated with good accuracy by the method appropriate to exponentially peaked functions, namely the saddle-point method (Marcus, loc. cit.).

[b] S. W. Benson and A. E. Axworthy, Jr., J. Chem. Phys., **21**, 428 (1953).

where Z is defined by [a]

$$Z = \frac{k_{2''}k_{6''}}{k_{2''} + k_{5''}a} + \frac{k_{2'}k_{6'}}{k_{2'} + k_{5'}b} + \frac{k_4 k_8 (M)}{k_4 + (M)k_7} \qquad (XI.5.4)$$

where $a = A/M$, $b = B/M$ (mole fractions), and K_{AB} is the equilibrium constant for the equilibrium between $[AB]^*$ and A and B.[b]

By defining R_f as the rate of the reaction when A or B $\ll$ D and R_b as the rate of the back reaction when D $\ll$ A or B, we can set $R = R_f - R_b$ and write for R_f

$$R_f = \frac{DK_D k_r}{1 + k_r/Mk_2} \frac{Z}{Z + k_{r'}(1 + k_r/Mk_2)^{-1}} \qquad (XI.5.5)$$

where $K_D = k_1/k_2$ and the first term in brackets represents the expected form for a unimolecular reaction [Eq. (XI.1.8)]. The second bracketed term, which we will designate by $F(Z)$, reflects the effect of the complex $[AB]^*$ on the forward reaction R_f.

At pressures sufficiently high that $Mk_2 \gg k_r(E_m)$, where $k_r(E_m)$ represents the mean rate constant, we have the high concentration limit for R_f

$$R_f = DK_D k_r \left(\frac{Z}{Z + k_{r'}}\right)_{M \to \infty} = DK_D k_r F_\infty(Z) \qquad (XI.5.6)$$

and we must now investigate the term $F_\infty(Z)$ to assess the importance of the remaining terms in M and mole fractions.

From Eq. (XI.5.4) we can rewrite Z as

$$Z = \frac{k_{6'}}{1 + k_5'b/k_{2'}} + \frac{k_{6''}}{1 + k_{5''}a/k_{2''}} + Mk_8 \qquad (XI.5.7)$$

where we have made the assumption that $k_4 \gg Mk_7$.[c] For convenience we may further write Z as

---

[a] The three terms in Eq. (XI.5.4) for Z can be related to the three processes whereby $[AB]^*$ can be formed from A and B, while the term in the last denominator of Eq. (XI.5.3) is related to the additional processes of formation and destruction of [AB] through the species D*.

[b] In the absence of reaction we would have the following equilibrium relations:

$$([AB]^*)_{eq} = \frac{k_5''}{k_6''}(A)(B^*) \quad = \frac{k_5'}{k_6'}(A^*)(B) \quad = \frac{k_7}{k_8}(AB)$$

$$\| \qquad\qquad \| \qquad\qquad \|$$

$$\frac{k_1''k_5''}{k_2''k_6''}(A)(B) = \frac{k_1'k_5'}{k_2'k_6'}(A)(B) = \frac{k_3k_7}{k_4k_8} = K_{AB}(A)(B)$$

which relates the various rate constants to the equilibrium constant K_{AB} for the equilibrium between $[AB]^*$ and A + B.

[c] k_4 is a unimolecular rate constant for a reaction that requires little or no activation energy. From the RRK theory it will have the value $k_4 \cong \bar{\nu}_4 \cong 10^{13}$ sec^{-1}. On the other hand, k_7 is a bimolecular rate constant for a reaction that requires activation energy. As we shall see later, Mk_7 will have a mean value of about $10^{11} e^{-E/RT}$ sec^{-1} when M is 1 mole/liter. Even at high concentrations we may thus certainly expect $k_4 \gg Mk_7$.

$$Z = \frac{k_6}{1 + k_5 m/k_2} + Mk_8 = \frac{k_6}{1 + K_{5\cdot2}m} + Mk_8 \qquad (XI.5.8)$$

where the unprimed values of k_2, k_5, and k_6 represent some respective weighted mean values of the primed values for deactivation of A* or B*, "sticky" collision to form [AB]*, and decomposition of [AB]* into A and B with distribution of energy between them. The mole fractions a and b are lumped together into the quantity m.

If we substitute this form of Z in the expression for $F(Z)$ [Eq. (XI.5.5)], we find

$$F(Z) = \frac{k_6/(1 + mK_{5\cdot2}) + Mk_8}{k_6/(1 + mK_{5\cdot2}) + Mk_8 + k_{r'}/(1 + k_r/Mk_2)} \qquad (XI.5.9)$$

and $F(Z)$ will differ from unity only if $k_{r'}/(1 + k_r/Mk_2)$ is of order of magnitude or greater than both Mk_8 and $k_6(1 + mK_{5\cdot2})^{-1}$.

By using the RRK theory we see that $k_6 \cong \bar{\nu}_6 \cong 10^{13} \text{sec}^{-1}$, since it represents a unimolecular process that requires no activation energy. On the other hand, $k_{r'} = \nu_{r'}(1 - E^{*'}/E')^{s-1}$, so that, setting $\bar{\nu}_6 = \nu_{r'}$ [since the complex [AB]* is the same for both processes], we see that

$$\frac{k_6}{k_{r'}} = \left(1 - \frac{E^{*'}}{E'}\right)^{1-s} \qquad (XI.5.10)$$

which will be very large if $E^{*'} > 0$ and close to unity only if little or no activation energy is required for the formation of [AB]* from A and B.[a]

In the usual case when $E^{*'}$ is greater than zero we can expect $k_r \ll k_6$ and $F(Z) = 1$.[b] In the event that $E^{*'}$ is close to zero, as might be expected when A and B are free radicals, we can neglect the deactivation step for A* and B* and set $K_{5\cdot2} = 0$.

In addition, Mk_8, the rate of collisional deactivation of [AB]*, will be very much less than either k_6 or $k_{r'}$ (see footnotes on page 240), so that $F(Z)$ is

$$F(Z) \cong \frac{k_6}{k_6 + k_{r'}(1 + k_R/Mk_2)^{-1}} = \frac{1}{1 + (k_{r'}/k_6)(1 + k_r/Mk_2)^{-1}} \qquad (XI.5.11)$$

At relatively high total concentration ($Mk_2 \gg k_r$) this has the value $(1 + k_{r'}/k_6)^{-1} \cong \frac{1}{2}$, since in this case $k_6 \cong k_{r'}$, and at very low total concentrations ($Mk_2 \ll k_r$) the value unity.

If we substitute the value of $F(Z)$ [Eq. (XI.5.11)] into R_f

$$R_f (E^{*'} = 0) = DK_D k_r \frac{1}{1 + k_r/Mk_2 + k_{r'}/k_6} \qquad (XI.5.12)$$

---

[a] This can very well be the case (that is, $E^{*'} = 0$) when A and B are free radicals. In such a case, however, AB, B*, and A* lose their significance and every collision of A and B may lead to [AB]*. This implies formally that $k_2 \gg k_5$ and $K_{5\cdot2} = 0$.

[b] This assumes that $mK_{5\cdot2}$ is of order of magnitude of unity or less. While there is no direct evidence on this point, it seems a reasonable assumption.

or since $k_6 \cong k_{r'}$ when $E^{*\prime} = 0$

$$R_f (E^{*\prime} = 0) \cong DK_D k_r \frac{1}{2 + k_r/Mk_2} \qquad \text{(XI.5.13)}$$

which is to be contrasted with the value of R_f when $E^{*\prime} > 0$ (i.e., when $F(Z) = 1$):

$$R_f (E^{*\prime} > 0) = DK_D k_r \frac{1}{1 + k_r/Mk_2} \qquad \text{(XI.5.14)}$$

At the high-pressure limit ($Mk_2 \gg k_R$) we have

$$R_{f\infty} (E^{*\prime} = 0) = \frac{DK_D k_r}{2} \qquad R_{f\infty} (E^{*\prime} > 0) = DK_D k_r$$

$$\text{(XI.5.15)}$$

We see that the unimolecular decomposition reactions differ in their concentration dependence from isomerization reactions (Sec. XI.3). In the latter case the complex corresponding to products will influence the rate of reaction [Eq. (XI.3.1)]. In the former case treated above, we see that the complex corresponding to products [AB]* will influence the rate only when the activation energy $E^{*\prime}$ of the reverse reaction (A + B → D) is zero [Eq. (X.5.12)]. Otherwise, when $E^{*\prime} > 0$, there is no effect of [AB]* in the rate. The reason for this physically is that, once [AB]* is formed, the probability that it will split into fragments (k_6) is so much greater than all competing processes that the latter have no effect on the rate. The only exception to this occurs when $E^{*\prime} = 0$, in which case [AB]* has equal chance of splitting or reforming D*, and this effectively reduces the rate of reaction by a factor of 2 [this holds even at the high-pressure limit, Eq. (XI.5.15)].

6. Quantum Effects in Unimolecular Reactions

One difficulty with the Slater and RRK theories as we have employed them lies in the use of a classical mechanical model of a molecule.[a] The first formulation of the problem of a chemical reaction in quantum-mechanical terms was made by London,[b] who used a very crude approximation to resolve the problem mathematically.[c] A study by Golden et al.[d]

[a] The Kassel theory is of course quantized and in essential agreement with the classical model. N. B. Slater, *Proc. Roy. Soc. Edinburgh*, **64**, 161 (1955), has given a quantum version with results somewhat different from his classical model.

[b] F. London, *Z. Physik*, Sommerfeld Festband, 104 (1928); *Z. Elektrochem.*, **35**, 552 (1929).

[c] A. Heitler and F. London, *Z. Physik*, **44**, 455 (1927). London, *ibid.*, **46**, 455; **50**, 24 (1928).

[d] S. Golden, *J. Chem. Phys.*, **17**, 620 (1949). S. Golden and A. M. Peiser, *ibid.*, **17**, 630 (1949); *J. Phys. & Colloid Chem.*, **55**, 789 (1951).

has refined the calculation to some extent but hardly to the point of quantitative usefulness. A crude quantum-mechanical model capable of yielding results in a direct manner was first proposed by Rice and Rams-perger[a] and Kassel.[b] They proposed a molecule composed of a group of s weakly coupled harmonic oscillators that have identical frequencies. It was then postulated that such a molecule decomposed when one oscillator contained a critical energy equal to m quanta or more.

If such a molecule has s oscillators (s degrees of freedom) and there are j quanta distributed among them, the total number of ways of distributing the j quanta[c] among the s oscillators is $q(s,j)$

$$q(s,j) = \frac{(j + s - 1)!}{j!(s - 1)!}$$ (XI.6.1)

The number of ways of distributing the j quanta if we require that m of them be present in one of the oscillators is

$$q_m(s,j) = \frac{(j - m + s - 1)!}{(j - m)!(s - 1)!}$$ (XI.6.2)

so that the probability $P_m(s,j)$, of finding m or more quanta in the "reactive" oscillator when the molecule has a total of j quanta ($j \geq m$) is[d]

$$P_m(s,j) = \frac{q_m(s,j)}{q(s,j)} = \frac{(j - m + s - 1)!j!}{(j - m)!(j + s - 1)!} = \prod_{k=1}^{s-1} \left(1 - \frac{m}{j + k}\right)$$ (XI.6.3)

If now we assume that the rate at which the quanta become redistributed is equal to some constant $\bar{\nu}$, we can write for the mean rate of decomposition $\overline{k(E_j)}$:

$$\overline{k(E_j)} = \bar{\nu} P_m(s,j)$$

or in the classical mechanical limit:

[a] O. K. Rice and H. C. Ramsperger, *J. Am. Chem. Soc.*, **49**, 1617 (1927); **50**, 617 (1928).

[b] L. S. Kassel, *J. Phys. Chem.*, **32**, 225, 1065 (1928).

[c] This is equivalent to the problem of distributing j balls into s boxes. If we arrange the j balls and s boxes in a line starting with one of the s boxes at left end of the line and adopt the convention that the number of balls in any box is given by the number that fall to the right of it on the line, then there will be $(j + s - 1)!$ ways of arranging these objects in a line (the first box is fixed). Then there are $j!$ permutations of the indistinguishable balls and $(s - 1)!$ permutations of the boxes leading to the value above for the number of distinguishable configurations.

[d] Note that in passing to the classical mechanical limit by allowing $m \to \infty$, $j \to \infty$, and $\nu \to 0$ such that $mh\nu = E^*$ and $jh\nu = E$, where E^* and E are kept constant, both j and $m \gg (s - 1)$ and we can apply Stirling's approximation to $P_m(s,j)$, giving

$$P_m(s,j) \to (1 - m/j)^{s-1} = (1 - E_m/E_j)^{s-1}$$ (XI.6.4)

which is the classical result for the chance that energy E_m or greater will be found in one of s identical, simple, harmonic oscillators containing a total energy E_j.

$$k(E_j) \to \bar{\nu}(1 - E_m/E_j)^{s-1} \qquad (XI.6.5)$$

which is identical in form to the mean rate given by the classical model [Eq. (X.5.1)].

While the Slater model does not lend itself to a simple solution in terms of the quantum theory, the fact that it agrees in form with the simple quantum model of Rice-Ramsperger and Kassel suggests that we can write the following expression for the mean rate of decomposition of a critically energized molecule of energy $E \geqslant E^*$;

$$\overline{k(E)} = \frac{\nu g_s(E - E^*)}{g_s(E)} \qquad (XI.6.6)$$

in which $g_s(E)$ is the number of quantum states of a molecule that has s degrees of internal freedom and total energy E, $g_s(E - E^*)$ is the total number of quantum states of the same molecule in which energy E^* is localized in some set of the normal coordinates such that, if left there, the molecule will decompose in one vibration, and ν is the mean rate at which energy is transferred from one normal coordinate to another.

In terms of the potential-energy diagram (Fig. X.4), $g_s(E)$ represents the total number of possible states bounded by the hypersurface of energy E, while $g_s(E - E^*)$ represents the total number of states within the same hypersurface that satisfy the condition that there is energy at least equal to E^* in the proper coordinates.[a]

The total rate of reaction is then given by multiplying $\overline{k(E)}$ by the probability $P(E)$ of finding a molecule with total energy E and summing over all energies $E \geqslant E^*$.

$$\text{Rate} = \sum_{E=E^*}^{\infty} k(E)P(E) \qquad (XI.6.7)$$

In the high-pressure limit, at which the molecules have the equilibrium thermal distribution of energies, we can use the quantum-mechanical distribution function for $P(E)$ [Eq. (IX.6.1)].

$$P(E) = \frac{g_s(E)e^{-E/kT}}{Q} \qquad (XI.6.8)$$

where Q is the partition function for the molecule. On substituting for $\overline{k(E)}$ from Eq. (XI.6.6) and for $P(E)$ above into Eq. (XI.6.7) we have

$$R = \frac{\nu}{Q} \sum_{E=E^*}^{\infty} g_s(E - E^*)e^{-E/kT} \qquad (XI.6.9)$$

$$= \nu \frac{Q^{\ddagger}}{Q} = \nu K^{\ddagger} \qquad (XI.6.10)$$

[a] In terms of our model, ν then represents the mean frequency of motion within the hypersurface and $k(E)$ the mean probability of diffusion per unit of time from any nonreactive state to a reactive state.

where $Q^{\ddagger}$ is defined by the sum and can be interpreted as the partition function for species that have at least energy E^* in the reactive degrees of freedom. $K^{\ddagger}$ is then the equilibrium constant for species A^* having the configuration at the saddle point, or more commonly, the transition state.

If we use E^* as our zero of energy[a] for the complex, we can substitute $E = E - E^*$, and then

$$Q^{\ddagger} = e^{-E^*/kT} \sum_{E=0}^{\infty} g_s(E)e^{-E/kT}$$

so that
$$R_{\infty} = \nu e^{-E/kT} \frac{Q^{\ddagger}(E)}{Q}$$

where
$$Q^{\ddagger}(E) = \sum g_s(E)e^{-E/kT} \qquad (XI.6.11)$$

If we compare this with the classical form for R_{∞}, namely, $\bar{\nu}e^{-E^*/kT}$, we see that the quantum-mechanical rate differs by a factor $Q^{\ddagger}(E)/Q(E)$, the ratio of the internal partition functions[b] of $A^{\ddagger}$, the transition complex, and normal A. This ratio in the form given, with the difference in zero-point energy of the reaction factored out, can be written as $e^{S^*/R}$, so that the difference between the classical and quantum-mechanical rates is related basically to a difference in entropy between normal and transition-state molecules.

If we use the approximation of factoring these internal partition functions into electronic, vibrational, and rotational partition functions, Q(internal) $= Q_{elec}Q_{vib}Q_{rot}$ and we can write (Sec. IX):

$$\frac{R_{\infty}(\text{qm})}{R_{\infty}(\text{class})} = \frac{Q^{\ddagger}(E)}{Q} = e^{S^{\ddagger}/R}$$

$$= \frac{Q_{elec}^{\ddagger}Q_{vib}^{\ddagger}Q_{rot}^{\ddagger}}{Q_{elec}Q_{vib}Q_{rot}} \qquad (XI.6.12)$$

For many reactions, there will be no change in electronic state or multiplicity associated with the transition state, nor is there any change in nuclear multiplicity so that we can set $Q_{elec}^{\ddagger}/Q_{elec} = 1$.[c] Further, the ratio of rotational functions can be written (Sec. IX. 10) as $(A^{\ddagger}B^{\ddagger}C^{\ddagger}/ABC)^{3/2}$ for a nonlinear molecule and as $I^{\ddagger}/I$ for a linear molecule, where I refers to the moment of inertia and A, B, and C are the principal moments of

[a] In the quantum-mechanical problem, E^* is the difference in energies between the lowest quantum states of A and A^* and hence is measured from the zero vibrational levels.

[b] Note that the difference in zero-point energies E^* has been factored out.

[c] There is of course a large group of reactions, those involving free radicals or atoms, for which this is not true. Here the electronic contributions can be considerable. See O. K. Rice, *J. Chem. Phys.*, **9**, 258 (1941).

inertia for the respective species and the ratio $Q_R^{\ddagger}/Q_R$ will be close to unity.[a]

It is thus possible to write as a further approximation for nonradical reactions

$$\frac{R_{\infty(qm)}}{R_{\infty(class)}} \approx \frac{Q_{vib}^{\ddagger}}{Q_{vib}} \qquad (XI.6.13)$$

and since the vibrational partition functions Q_{vib} can generally be factored as a product of contributions from each normal coordinate (Sec. IX.9)

$$Q_{vib} = \prod_i q_i = \prod_i (1 - e^{-h\nu_i/kT})^{-1} \qquad (XI.6.14)$$

it is only when the frequency $\nu_i \lesssim kT/h$ that q_i differs appreciably from unity (for $\nu_i \gg kT/h$, $q_i = 1$), so that the q_i and hence Q_{vib} will be unity except for the contributions of low-frequency vibrations. If some of the frequencies of A are lowered in changing to A‡, then $q_i^{\ddagger}/q_i > 1$ for those particular frequencies and we should expect an enhancement of the rate. If on the other hand some originally low frequencies of A are increased (as in cyclization), then we may expect to find $q_i^{\ddagger}/q_i < 1$ and a consequent falling off in the rate.[b]

In summary we can say that we may expect unimolecular reaction rate constants to have the high-pressure limit

$$k_{\infty} = \nu e^{S^{\ddagger}/R} e^{-E^*/RT} \qquad (XI.6.15)$$

where the frequency factor ν will be about 10^{13} sec^{-1} and the contributions of $S^{\ddagger}$ will be significant only when the activation process is accompanied by a change in the number of low-frequency vibrations or rotations in the molecule. As we shall see, most of the unimolecular reactions which have been studied have values of the preexponential factor (i.e., the product $\nu e^{S^{\ddagger}/R}$) of about 10^{13} sec^{-1}, so that for these reactions $S^{\ddagger} \cong 0$.

For the region in which the rate of these reactions is dependent on

---

[a] Unless the dimensions of the molecule change appreciably in transition, these ratios will be of order of magnitude of 1, being slightly greater than 1 for bond-breaking reactions and perhaps slightly smaller than 1 for elimination or cyclization reactions.

[b] At room temperatures, 300°K, $kT/h \cong 6 \times 10^{12}$ sec^{-1} = 200 cm^{-1}. Bond-stretching frequencies are generally in the range 500 to 2500 cm^{-1} and so contribute little at this temperature to Q_{vib}. Bending vibrations are lower and may contribute appreciably, while the greatest contributions may be expected from the very low frequencies arising from almost free internal rotations (e.g., rotation about single bonds in hydrocarbons). Undoubtedly the greatest effect on the ratio $Q_{vib}^{\ddagger}/Q_{vib}$ will come from the freeing or freezing of these low-frequency vibrations in the activation process. In the cyclization of a hydrocarbon or other chain molecule we might thus expect a decrease in rate and for decyclization reactions the opposite effect. The isomerizations of cyclopropane to propylene and cyclobutane to ethylene both seem to have abnormally high entropies of activations. The dissociation of $N_2O_4 \rightarrow 2NO_2$ seems also to have a high entropy of activation, possibly arising from the two degenerate bending modes of the NO_2 groups changing to free rotations about the O_2N—NO_2 bond (Table XI.4). Similarly, the dissociation of $C_2H_6 \rightarrow 2CH_3$ has an abnormally high frequency factor ($\sim 10^{18}$ sec^{-1}).

concentration the rate constants may be evaluated by substituting the quantum-mechanical forms in the appropriate expression, so that for a unimolecular decomposition we may write

$$k_{\text{exp}} = \int_{E^*}^{\infty} \frac{k_r K_D \, dE}{1 + k_r/Mk_2} \qquad \text{classical} \qquad (XI.5.14)$$

$$k_{\text{exp}} = \frac{1}{Q} \sum_{E=E^*}^{\infty} \left[\frac{\nu g_s(E - E^*)e^{-E/kT}}{1 + \nu g_s(E - E^*)/Mk_2 g_s(E)} \right] \qquad \text{quantum-mechanical}$$

$$(XI.6.16)$$

Marcus[a] has shown how to make reasonably good approximate calculations of this latter function and has applied it to the reactions of hydrocarbons and alkyl halides.

7. Equilibrium Theory of Reaction Rates; The Transition-state Method

In the preceding sections we have shown that at sufficiently high concentrations or for molecules sufficiently complex, unimolecular reactions may proceed in such a manner that the disturbance of the equilibrium distribution of reactive molecules does not significantly affect the rate of reaction. Under such conditions it is permissible to approximate the rate of a unimolecular reaction by its high-pressure rate. But such a pseudo-equilibrium invites still another approach to the formulation of the theory of reactions. This approach, known as the *transition-state method*, was first outlined by Marcellin,[b] was developed further by Rodebush[c] and Rice and Gershinowitz,[d] and was put in its present form by Eyring[e] and Evans and Polanyi.[f,g]

The method proposes that every molecule A that reacts does so by being excited to some state A[+] of energy $E \geqslant E^*$, the critical energy for the reaction, and having the configuration that corresponds to a maximum in its

[a] R. A. Marcus, *ibid.*, **20**, 352 (1952). R. A. Marcus and O. K. Rice, *J. Phys. & Colloid Chem.*, **55**, 894 (1951).

[b] M. R. Marcellin, Thesis, University of Paris, 1914; Gauthiers-Villars, Paris, 1914; *Ann. Physik*, **3**, 158 (1915).

[c] W. H. Rodebush, *J. Am. Chem. Soc.*, **45**, 606 (1923); *J. Chem. Phys.*, **1**, 440 (1933); **3**, 242 (1935); **4**, 744 (1936).

[d] O. K. Rice and H. Gershinowitz, *J. Chem. Phys.*, **2**, 853 (1934); **3**, 479 (1935).

[e] H. Eyring, *ibid.*, **3**, 107 (1935). W. F. K. Wynne-Jones and H. Eyring, *ibid.*, **3**, 492 (1935). See also S. Glasstone et al., "The Theory of Rate Processes," McGraw-Hill Book Company, Inc., New York, 1941.

[f] M. G. Evans and M. Polanyi, *Trans. Faraday Soc.*, **31**, 875 (1935); **33**, 448 (1937). M. Polanyi, *J. Chem. Soc.*, 629 (1937).

[g] For a critical discussion of this approach, see E. A. Guggenheim and J. Weiss, *Trans. Faraday Soc.*, **34**, 57 (1938), and especially the discussion in this symposium.

potential energy. This configuration is referred to as the transition state. In terms of the energy diagram (Fig. XI.1) this state would lie at the saddle point of the energy hypersurface. It is then proposed that species $A^{\ddagger}$ is in thermal equilibrium with the normal species A, so that for the chemical reaction path

$$A \underset{}{\overset{K^{\ddagger}}{\rightleftharpoons}} A^{\ddagger} \overset{\nu^{\ddagger}}{\rightarrow} \text{products} \qquad (XI.7.1)$$

we can write for the concentration of $A^{\ddagger}$

$$(A^{\ddagger}) = K^{\ddagger}(E)A \qquad (XI.7.2)$$

where $K^{\ddagger}(E)$ is the thermodynamic equilibrium constant. If it is now assumed that the frequency with which $A^{\ddagger}$ passes over the top of the barrier is given by the frequency of vibration of some normal coordinate $\nu^{\ddagger}$ in the system that describes the decomposition,[a] then the rate of reaction R (assuming there is no back reaction) is given by

$$R = \nu^{\ddagger}\kappa(A^{\ddagger}) = \nu^{\ddagger}\kappa K^{\ddagger}(E)A \qquad (XI.7.3)$$

where $\kappa^{\ddagger}$ is some fraction between 0 and 1, called the transmission coefficient for the reaction.[b] Since the experimental high-pressure rate for a unimolecular reaction is first-order, we can set $R = k_{exp}A$ and thus write for the experimentally determined, first-order, specific rate constant

$$k_{exp} = \sum_{E=E^*}^{\infty} \nu^{\ddagger}\kappa K^{\ddagger}(E)$$

$$= \nu\kappa K^{\ddagger} \qquad (XI.7.4)$$

where the sum is taken over all energy states above E^*, the minimum critical energy for the reaction, and it is assumed that $\nu^{\ddagger} = \nu$ and κ are independent of the energy of the reacting species.[c]

We see that the results of this approach are identical in form with the result obtained by an application of the detailed theory to a quantum-mechanical model for the reaction [Eq. (XI.6.10)].

If now we write $K^{\ddagger}$ in terms of the product of partition functions (Sec. IX.7),

$$K^{\ddagger} = \frac{Q^{\ddagger}e^{-E^*/RT}}{Q} = \frac{Q_{elec}^{\ddagger}Q_{rot}^{\ddagger}Q_{vib}^{\ddagger}}{Q_{elec}Q_{rot}Q_{vib}} e^{-E_0/RT} \qquad (XI.7.5)$$

[a] Because at the saddle point the potential energy decreases with displacement in the reaction direction, the "force constant" is negative and the "frequency" imaginary in a formal sense.

[b] The transmission coefficient corrects for the number of complexes that pass over the barrier and are reflected back again before being deactivated to final products. Generally it is assumed that $\kappa = 1$. The RRK and Slater models for spontaneous decompositions need similar correction.

[c] C. Fréjacques, in a private communication, has calculated the sum when it is assumed that the hypersurface in the vicinity of the saddle point is parabolic in shape, so that $\nu^{\ddagger}$ is a function of the energy of the complex. The correction amounts to a factor of $\sqrt{3}$ for the simple system $H + H_2$ (exchange).

where $Q^{\ddagger}$ refers to the transition state and Q to normal species and E_0 is the difference in energy between the ground states of $A^{\ddagger}$ and A. As before we can set $Q_{\text{elec}}^{\ddagger} = Q_{\text{elec}}$, and on expanding $Q_{\text{vib}} = \prod q_i$ (product of partition functions for normal vibrations), we have

$$K^{\ddagger} = \frac{Q_{\text{rot}}^{\ddagger}}{Q_{\text{rot}}} \frac{\prod_i q_i^{\ddagger}}{\prod_i q_i} e^{-E_0/RT} \tag{XI.7.6}$$

or for the specific rate constant

$$k_{\text{exp}} = \nu \kappa e^{-E_0/RT} \left(\frac{Q_{\text{rot}}^{\ddagger}}{Q_{\text{rot}}}\right) \frac{\prod_i q_i^{\ddagger}}{\prod_i q_i} \tag{XI.7.7}$$

If now it is assumed that the rupture of the molecule that leads to reaction corresponds to the motion of one of the normal coordinates of the complex of frequency $\nu = \nu_c$ and further that $\nu_c \ll kT/h$, then the individual vibrational partition function for this coordinate is $q_c^{\ddagger} = (1 - e^{-h\nu_c/kT})^{-1} \cong kT/h\nu_c$. On substitution and cancellation we have

$$k_{\text{exp}} = \kappa \frac{kT}{h} e^{-E_0/RT} \left(\frac{Q_{\text{rot}}^{\ddagger}}{Q_{\text{rot}}}\right) \frac{\prod_i^{s-1} q_i^{\ddagger}}{\prod_i^{s} q_i} \tag{XI.7.8}$$

where the product is over only the $s - 1$ internal coordinates of the transition-state complex because we have factored one out.[a]

In the case that the vibrations of complex and normal species are all high (ν_i, $\nu_i^{\ddagger} > kT/h$) or identical and that $\kappa = 1$ and the dimensions of the molecule are not changed by activation so that $Q_{\text{rot}}^{\ddagger} = Q_{\text{rot}}$ we have

$$k_{\text{exp}} = \frac{kT}{h} e^{-E_0/RT} \tag{XI.7.9}$$

The more general result corresponding to Eq. (XI.7.8) is usually written as

[a] An alternative derivation usually given sets the over-all rate of reaction equal to $\kappa \bar{\nu}^{\ddagger} C^{\ddagger}/2 = (\kappa/2) v^{\ddagger} A K^{\ddagger}/\delta$, where $C^{\ddagger}/2$ is the concentration of transition-state complexes per unit length of the reaction coordinate δ moving toward product and $v^{\ddagger}$ is the mean velocity of $A^{\ddagger}$ along $\delta(v^{\ddagger}/\delta \cong \bar{\nu}^{\ddagger})$. If we assume that the potential energy of the system is constant at the saddle point over the distance δ, then we may factor the partition function of $A^{\ddagger}$ into a product of a partition function for all the other normal modes of the molecule $Q^{\ddagger}$ times a translational partition function for the motion along the normal mode which corresponds to the reaction coordinate δ. This latter has the form $(2\pi\mu kT/h^2)^{1/2}\delta$, and the average velocity $v^{\ddagger}$ from kinetic theory is $(2kT/\pi\mu)^{1/2}$, so that we can write $\kappa \bar{\nu}^{\ddagger} C^{\ddagger}/2 = \kappa(kT/h)AQ^{\ddagger}/Qe^{-E_0/RT}$, which is equivalent to the result already given.

$$k_{exp} = \kappa \frac{kT}{h} \frac{Q^{\ddagger}}{Q} e^{-E_0/RT} \qquad (XI.7.10)$$

where it is understood that $Q^{\ddagger}$ is the incomplete partition function for the activated complex, one of the vibrational coordinates having been factored out.[a]

8. Summary of the Various Theories of Unimolecular Reactions

We may at this point profitably summarize the conclusions of the various theories which have been presented for the specific rate constant of a unimolecular reaction. For the rate constant in the high-pressure or equilibrium limit we have the following results

Arrhenius Equation:
$$k = Ae^{-E/RT} \qquad (XI.8.1)$$

Detailed Theory (Classical):
$$k = \nu e^{-E^*/RT} \qquad (XI.8.2)$$

Detailed Theory (Quantum Mechanical):
$$k = \nu^* K^{\ddagger}$$
$$= \nu^* e^{-F^{\ddagger}/RT} = \nu^* e^{S^{\ddagger}/R} e^{-H^{\ddagger}/RT} \qquad (XI.8.3)$$
$$= \nu^* \frac{Q^{\ddagger}}{Q} e^{-E_0^{\ddagger}/RT} \qquad (XI.8.3a)$$

Transition-state Theory:
$$k = \frac{kT}{h} K^{\ddagger}$$
$$= \frac{kT}{h} e^{-F^{\ddagger}/RT} = \frac{kT}{h} e^{S^{\ddagger}/R} e^{-H^{\ddagger}/RT} \qquad (XI.8.4)$$
$$= \frac{kT}{h} \frac{Q^{\ddagger}}{Q} e^{-E_0/RT} \qquad (XI.8.4a)$$

The detailed theory in its classical form [Eq. (XI.8.2)] presents a frequency factor ν which is in principle calculable from spectroscopic frequencies of the normal molecule and which should lie in the range 10^{12} to 10^{14} sec^{-1}. It is independent of temperature, so that the quantity E^*, which is the difference in energy between a critically activated species and a normal molecule (both in their lowest energy states) may be identified with the experimental activation energy. The theory represents a decided advance over the collision theory in presenting chemical reactions from

---

[a] A more detailed argument has been made by M. Szwarc, *Chem. Revs.*, **47**, 75 (1950), for the general validity of Eq. (XI.7.10) for all unimolecular decompositions in which the reaction occurs via the breaking of a single bond. This argument has been advanced to take account of the special case that the hypersurface does not have a maximum (i.e., saddle point) and the potential energy of the two fragments decreases monotonically to zero as the bond between them is increased over its equilibrium distance.

the point of view of molecular structure. It suffers seriously, however, in utilizing a classical model for the structure. One of the consequences of the latter is that all internal modes of normal and active species are fully excited, frequencies are identical, and the classical differences in entropy between the states neither affect the over-all rate constant nor appear explicitly in the rate equation.[a]

By modifying the classical theory and using a quantum-mechanical model for the critically activated complex it is possible to extend the detailed theory [Eqs. (XI.8.3) and (XI.8.3a)] in such a way as to include consideration of structural changes on the internal frequencies. In Eq. (XI.8.3) ν^* represents a weighted average of the internal frequencies of a species that has the configuration of the transition state and $K^{\ddagger}$ represents the equilibrium constant for the equilibrium between this transition state and normal molecules. $F^{\ddagger}$, $H^{\ddagger}$, and $S^{\ddagger}$ respectively represent the standard change in free energy, enthalpy, and entropy in producing the transition state. In Eq. (XI.8.3a) the rate expression has been placed in a form suitable for statistical calculation. $Q^{\ddagger}$ and Q are respectively the partition functions for the transition state and normal species measured from their lowest energy states, while $E_0^{\ddagger}$ is the difference in zero-point energies of the two species (i.e., heat of activation at absolute zero).

While this improvement in model introduces an entropy of activation and allows proper consideration to be given to structural changes, it suffers in introducing a transition-state complex whose properties cannot be observed or verified independently of the kinetic data. Thus ν^* is now an average frequency for the transition state, and while we may expect it to lie in the same range as ν for the normal molecule, such an equivalence is purely *ad hoc*. The net gain afforded by such a model lies in the degree to which it permits of a self-consistent scheme of speculation on the properties of the transition-state complex and the way in which such speculation may give us valid insights to the relation between molecular structure and chemical reactivity. In practice we can identify $H^{\ddagger}$ with the experimental activation energy, but there is no obvious way of dividing the experimental frequency factor between ν^* and $S^{\ddagger}$. The best that we can do is explore experimental frequency factors that differ markedly from 10^{13} sec^{-1} in the hope of understanding how some hypothetical structure of the activated complex can explain such differences. As we shall see, such speculations can be quite fruitful, and additional evidence with which to analyze them is available from the effects of third bodies—in particular, solvents—on the course of the reaction. Until the Schroedinger equation can itself be solved explicitly, we shall have to be content with such advances as are represented by the transition-state theories.

[a] The major difficulty arises from trying to adapt the theory of normal coordinates, which was designed for small vibrations, to chemical reactions that involve large displacements.

The transition-state theory itself as given in Eqs. (XI.8.4) and (XI.8.4a) differs from the previous theories in its attempt to eliminate the uncertainty inherent in the frequency factor ν^* by the replacement with the universal frequency factor kT/h. We can, for example, derive Eq. (XI.8.4a) from Eq. (XI.8.3a) if we are willing to say that the partition function $Q^{\ddagger}$ for the transition-state complex can be factored into a product of partition functions (including the internal vibrations) and that one of the internal frequencies, let us call it ν_e, corresponds to the motion over the top of the barrier. If then this frequency $\nu_e \ll kT/h$ so that the corresponding vibrational partition function q_e can be expanded and approximated as $q_e \cong kT/h\nu_e$ and finally that $\nu^* = \nu_e$, we obtain Eq. (XI.8.4a), where $Q^{\ddagger}$ refers to the partition function for the transition-state molecule from which one degree of vibration has been factored. $K^{\ddagger}$, $F^{\ddagger}$, $S^{\ddagger}$, and $H^{\ddagger}$ are then the thermodynamic quantities for this amputated species, while $E_0^{\ddagger}$ is identical with E_0^*.

While such an elimination of ν^* is certainly desirable, it is doubtful if the methods employed in eliminating it can be justified in any degree rigorous enough to permit of certainty in then assigning values to $S^{\ddagger}$ from the experimental frequency factor.[a] At the present stage of development of the theory it is just as reasonable to arbitrarily assign to ν^* a value of 10^{13} sec^{-1} and calculate S^* from experiment as it is to replace it by kT/h and then calculate $S^{\ddagger}$. In point of fact we should be deluding ourselves if entropy calculations made on such a basis were given precise significance. It is only on the basis of very large deviations of observed frequency factors from 10^{13} sec^{-1} that we have any justification for looking to the molecular structure for explanations in terms of an entropy factor.[b]

9. Some Unimolecular Isomerizations

In Table XI.3 are summarized most of the available data on homogeneous gas-phase isomerizations. In the high-pressure regions the rate constants have been observed to be first-order and could be fitted to the

---

[a] The value of kT/h is 6×10^{12} sec^{-1} at 300°K, while very weak, internal vibrations are usually not less than about 300 cm^{-1} (33μ) or about 9×10^{12} sec^{-1}. The assumption $\nu_e \ll kT/h$ can thus be justified only if we are willing to assign the reaction coordinate a frequency lower than even the frequency of internal rotation of very heavy groups. Alternative methods of deriving the transition-state formula (Glasstone et al., *op. cit.*, chap. 4) in which the reaction coordinate is made equivalent to a free translation are equally qualitative. Equally dubious is the identification of the reacting mode with a single one of the normal modes of the transition complex.

[b] In either of the quantum-mechanical formulations we may expect the experimental frequency factor to depend on the nature of the interactions governing the transition across the barrier. Thus reactions involving the interactions of relatively high vibration frequencies (for example, R—H bending or stretching) may show higher frequency factors than those that involve low vibration frequencies (for example, R—X where X is a heavy group). See the next section for some illustrations.

Arrhenius equation $k = Ae^{-E/RT}$ to within experimental error. Only one of these reactions, the isomerization of cyclopropane, has been observed to show a large falling off in rate constant with decrease in pressure. This occurs in the region of 10 mm Hg, and the pressure dependence of the apparent first-order rate constant can be fitted by a Kassel-type equation (XI.6.16) with $n = 13$ and a collision diameter σ of 3.9 Å. Somewhat larger (or smaller) values of n and smaller (or larger) values of σ are also compatible with the data.[a] Slater[b] has shown that his model also has the proper pressure dependence by using $s = 14$. However, neither treatment can predict the high frequency factor of 1.5×10^{15} sec^{-1}, which is about 100 fold higher than might be anticipated from purely mechanical considerations. A factor of 12 is accountable on consideration of the possibility of any of six equivalent H atoms moving to either of two equivalent C atoms. However, the remainder must surely reflect a real entropy increase in going from normal cyclopropane to the transition state, and only the modified quantum-mechanical model or a transition-state model which makes allowances for such entropy changes can be expected to provide a true explanation.

It is interesting to observe that the isomerization of cis or trans 1,2-dideuteriocyclopropane (Table XI.3) is about three times faster than the isomerization to cyclopropane and almost surely proceeds via ring cleavage to form the intermediate trimethylene diradical. This is probably the path for the propylene isomerization as well.[c]

Pinene   Limonene

The reactions of d-pinene to give dl-limonene or l-pinene which have been shown to be complicated experimentally (see footnote b', Table XI.3)

[a] This would seem to imply that the Kassel model for energy transfer is not too bad for this complex molecule. If we correct the probable values of Table XI.2 to the high frequency factor and the lower gas density at 450°C, we obtain a value of about 13 oscillators, also in agreement.

[b] N. B. Slater, *Proc. Roy. Soc. (London)*, **A218**, 224 (1953).

[c] The failure to detect the short-lived radical (lifetime $\cong 10^{-6}$ sec) by competing second-order processes (i.e., radical traps such as NO) cannot be taken as definitive.

TABLE XI.3. UNIMOLECULAR ISOMERIZATIONS IN THE GAS PHASE

Reactant	Product	$\log A$ (A in sec^{-1})	E_{act}, Kcal/mole	
Cyclopropane[a]	Propylene	15.17	65.0	
Cis-isostilbene[c]	Trans-isostilbene	12.78	42.8	
Methyl cis-cinnamate[d]	(Trans form)	10.54	41.6	
Cis-cyanostyrene[e]	(Trans form)	11.60	46	
Trans-cyanostyrene[e]	(Cis form)	11.8	46	
2,2-diamino-6,6-dimethyl diphenyl[f]	(Racemization)	10.37	45.1	
Vinyl allyl ether[g]	n-Pentaldehyde-ene-4	11.27	30.6	
Dimethyl maleate[h]	Dimethyl fumarate	5.11	26.5	
Dimethyl citraconate (trans)[c]	(Cis form)	5.0	25	
Cis-2-butene[i]	(Trans form)	0.30[i']	18	(52 ± 8)
d-Pinene[b]	dl-Limonene (dipentene)	(14.72)[b']	(43.7)[b']	
	Allocimene	13.8	42.7	
Cis-1,1'-dideuterioethylene[j] . . .	Trans isomer	12.48	61.3	(65)
Isopropenyl allyl ether[k]	Allyl acetone	11.27	29.3	
Trans-1,2-dichloroethylene[l] . . .	Cis form	12.7	42	
α-1,2-Diphenyl chloroethylene[m]	β Form	11.2	37	
α-1,2-Diphenyl-1',2'-dichloroethylene[m]	β Form	10.0	34	
Cis-cyclopropane-d_2[n]	Trans-cyclopropane-d_2	16.0	64.2	
Cyclobutene[o]	1,3-Butadiene	13.08	32.5	

[a] T. C. Chambers and G. B. Kistiakowsky, *J. Am. Chem. Soc.*, **56**, 399 (1934), and E. S. Corner and R. N. Pease, *ibid.*, **67**, 2067 (1945), assume free radical intermediate. H. O. Pritchard et al., *Proc. Roy. Soc.* (*London*), **A217**, 563 (1953).

would appear also to involve the breaking of a four-membered ring, and in this case the complexity of products shown to exist in the liquid-phase reaction as well as the much lower activation energies (compared to other rings) would appear to indicate a free radical intermediate.[a]

In contrast to these cases, the isomerization of vinyl allyl ether to n-pentaldehyde-ene-4 (Table XI.3), which may be pictured as going through the hypothetical six-membered ring complex shown in brackets,[b]

[a] See R. L. Burwell, Jr., *J. Am. Chem. Soc.*, **73**, 4461 (1952) for a discussion of the biradical intermediate.

[b] A four-membered intermediate ring can also be drawn.

has a negative entropy of activation of about 8 cal/mole-$°K$. The very similar isomerization of isopropenyl allyl ether to allyl acetone has almost identical rate parameters and can also be pictured as going through either a four- or six-membered ring complex.

b D. F. Smith, *J. Am. Chem. Soc.*, **49**, 43 (1927). See also F. H. Thurber and C. H. Johnson, *ibid.*, **52**, 786 (1930). J. E. Hawkins and J. W. Vogh, *J. Phys. Chem.*, **57**, 902 (1955).

b' More recent work by Hawkins and coworkers [R. E. Fuguitt and J. E. Hawkins, *J. Am. Chem. Soc.*, **69**, 319 (1947); **67**, 242 (1945); H. G. Hunt and J. E. Hawkins, *ibid.*, **72**, 5618 (1950)] on the liquid-phase isomerization shows that there are three simultaneous reactions of first order occurring, the racemization to β-pinene with $\log A = 14.0$, $E = 44.1$ Kcal/mole, and the two listed above. In addition, the allocimene undergoes reversible dimerization and irreversible first-order decomposition to α,β-pyrenones. The values of Smith in parentheses cannot therefore be taken too seriously.

c G. B. Kistiakowsky and W. R. Smith, *J. Am. Chem. Soc.*, **56**, 638 (1934).

d *Ibid.*, **57**, 269 (1935).

e *Ibid.*, **58**, 2428 (1936).

f *Ibid.*, **58**, 1043 (1936).

g F. W. Schuler and G. W. Murphy, *ibid.*, **72**, 3155 (1950).

h G. B. Kistiakowsky and M. Nelles, *Z. physik. Chem.*, Bodenstein Festband, 369 (1931).

i G. B. Kistiakowsky and W. R. Smith, *J. Am. Chem. Soc.*, **58**, 766 (1936). Values in parentheses are W. F. Anderson et al., *ibid.*, **80**, 2384 (1958), and are the more reliable. The latter authors found evidence for chain reaction.

i' W. MacF. Smith, *J. Chem. Phys.*, **20**, 1808 (1952), has shown that *cis*-butene-2 when activated by 48.3 Kcal from quenching of Na (^2P) atoms at 170°C does not isomerize. This throws some doubt on the homogeneity of the reaction reported by Kistiakowsky and Smith

j B. S. Rabinowitch et al., *J. Chem. Phys.*, **20**, 1807 (1952); **23**, 315, 2439 (1955). Latter result gives the high E_{act}.

k L. Stein and G. W. Murphy, *J. Am. Chem. Soc.*, **74**, 1041 (1952).

l J. L. Jones and R. L. Taylor, *J. Am. Chem. Soc.*, **62**, 3480 (1940).

m T. W. Taylor and A. R. Murray, *J. Chem. Soc.*, 2078 (1938).

n B. S. Rabinowitch et al., *J. Chem. Phys.*, **28**, 504 (1958).

o W. Cooper and W. D. Walters, *J. Am. Chem. Soc.*, **80**, 4220 (1958).

Of the remaining reactions, those that involve cis-trans isomerizations of olefins show a rather surprising range of kinetic parameters with frequency factors of 2 sec^{-1} for cis-2-butene, 10^5 sec^{-1} for dimethyl maleate, and about 10^{13} sec^{-1} for cis-isostilbene and dideuterioethylene. It is to be noted that the activation energies vary in an inverse manner (to the frequencies) for all of these reactions, so that the rate constants of isomerization are about the same. All of these isomerizations are unfortunately complicated by the simultaneous occurrence of small amounts of side reaction and a small amount of sensitivity to the reaction-vessel surface, and they also are catalyzed by free radicals. For these reasons those reactions with strikingly small frequency factors (such as 10^5 sec^{-1} or smaller) are either heterogeneous or possibly chain reactions, and a more extensive investigation of them would be needed before any explanation should be made.[a]

It has been proposed[b,c] that cis-trans isomerizations should be capable of proceeding via two different paths. One of the paths would involve a torsional motion around the double bond. That path would require large activation energies but would have normal frequency factors. The second path would involve an excitation of the double bond corresponding to the formation of a biradical with two unpaired electrons and thus free rotation about the resultant single bond. If a normal ethylenic molecule could somehow make the transition from its normal (singlet) state to the biradical (triplet) state, the activation energy might be much less; in some cases it has been calculated as low as 25 Kcal/mole.[d] However, such transitions are "forbidden" by quantum-mechanical rules because they involve a change in multiplicity of the total electron spin[e] of the molecule. As a result it is expected that such transitions occur with a frequency which is about 10^5 sec^{-1}.[f] While the isomerizations of dimethyl maleate and di-

[a] In the case of the cis-2-butene isomerization the authors suggest that the low-frequency factor (2 sec^{-1}) and low activation energy are evidence for a chain reaction. If that is true, the reaction is complex and no meaning can be ascribed to the frequency factor or activation energy without a detailed analysis of the mechanism. Recent work indicates that it is a heterogeneous reaction.

[b] J. L. Magee et al., J. Am. Chem. Soc., **63**, 677 (1941).

[c] H. M. Hulbert et al., Ann. N.Y. Acad. Sci., **44**, 371 (1943).

[d] R. S. Mulliken, Phys. Rev., **41**, 75 (1932). M. Kasha, Chem. Revs., **41**, 401 (1947).

[e] H. Eyring et al., "Quantum Chemistry," chap. 16, John Wiley & Sons, Inc., New York, 1944. This rule, known as the Wigner spin-conservation rule, is presumed to hold well only for relatively light atoms, in which interactions between electronic orbital motions and electronic spin are not strong. However, the experimental evidence is thus far very conflicting. Thus the inelastic collisions of a normal with an excited He atom, such as He (n^1P) + He (1^1S) → He (1^1S) + He (n^3D) have been shown [Lees and Skinner, Proc. Roy. Soc. (London), **A137**, 186 (1932)] to have a normal frequency factor. For further discussion of this point see H. S. W. Massey and E. H. S. Burhop, "Electronic and Ionic Impact Phenomena," chap. 7, Oxford University Press, New York, 1952.

[f] Magee, loc. cit.

methyl citraconate might seem to represent an experimental verification of such an effect, the experimental facts are much too meager to justify such a conclusion and seem incompatible[a] with any proposals yet made.[b]

10. Some Unimolecular Fission Reactions; Decomposition into Stable Molecules

First-order unimolecular fission reactions in which the products are stable molecules have been compiled in Table XI.4. As can be seen, the frequency factors all fall in the range between 3×10^9 to about 10^{16} sec^{-1}, with most of them very close to the value of 10^{13} sec^{-1}.

The two groups of reactions with low frequency factors, namely, the decompositions of the acetals and the decompositions of the chloroformate esters, would both have to go through the formation of four-membered-ring transition complexes, although the acetals can also decompose via a six-membered ring. While it might be expected that the formation of a

Acetal complex (4 ring) | Acetal complex (6 ring) | Chloroformate ester (4 ring) | Trichloromethyl chloroformate (4 ring)

four-ring complex would be accompanied by a decrease in entropy, the minus 10 to 15 cal/mole-°K indicated by these frequency factors and the normal frequency factor observed for trichloro methyl chloroformate are difficult to reconcile with this point of view (see footnote r, Table XI.4). In addition, the data of M. Szwarc and J. Murawski on the decomposition of acetic anhydride[c] indicate that the data on the acetals must be in error because the anhydrides produced from the acetals are decomposing at a faster rate than the acetals themselves (Table XI.4).

[a] In the case of dimethyl maleate the rate was observed to fall off with pressures near 100 mm Hg, and a side reaction in which CO_2 was evolved also took place. The region of pressure fall-off together with the low frequency factor is in direct contradiction to the Slater theory or any reasonable variant. It is the feeling of the author that these are chain sensitized reactions.

[b] M. Calvin and H. W. Alter, *J. Chem. Phys.*, **19**, 768 (1952), have attempted to investigate cis-trans isomerizations in the liquid phase. They found experimental difficulties in side reactions, and while they could represent their data by means of first-order plots, they did not investigate any catalytic effects of the presence of free radicals. Their frequency factors fell in the range 10^5 sec^{-1} to 10^{10} sec^{-1}, for which they found no reasonable explanation.

[c] *Trans. Faraday Soc.*, **47**, 269 (1951). Note that acetic anhydride can decompose via a six-membered ring complex. The loss of entropy of about five units seems reasonable for such a ring.

TABLE XI.4. UNIMOLECULAR GAS-PHASE DECOMPOSITIONS
THAT GIVE STABLE MOLECULES

Reactant	Products	$\log A$, sec^{-1}	E_{act}, Kcal/mole
$CH_3CH_2Cl^a$...............	$C_2H_4 + HCl$	14.6	60.8
$CCl_3-CH_3^b$...............	$CCl_2=CH_2 + HCl$	12.5	47.9
$CH_3CH_2Br^c$...............	$C_2H_4 + HBr$	12.86	52.3
		$(13.42)^{ee}$	$(53.9)^{ee}$
$CH_3-CHCl_2^a$...............	$CH_2=CHCl + HCl$	12.1	49.5
$CH_3CH_2CH_2Cl^d$............	$CH_3CH=CH_2 + HCl$	13.45	55.0
$CH_3CHClCH_3^e$.............	$CH_3CH=CH_2 + HCl$	13.4	50.5
$CH_3CHClCH_2Cl^f$............	$CH_3CH=CHCl + HCl$	13.8	54.9
$CH_3CH_2CH_2Br^c$............	$CH_3CH=CH_2 + HBr$	13.60^{ee}	47.7
		$(12.90)^{ee}$	$(50.7)^{ee}$
$CH_3CHBrCH_3^{c,g}$............	$CH_3CH=CH_2 + HBr$	13.00	50.7
		$(13.61)^{g,ee}$	$(47.8)^{g,ee}$
Isobutyl bromide^{ee}...........	Isobutene + HBr	13.05	50.4
t-Butyl chloride^f............	Isobutene + HCl	12.4	41.4
t-Butyl bromide^h............	Isobutene + HBr	13.3	40.5
		14.0^{ee}	42.2^{ee}
n-Butyl bromide^{ee}...........	Butene-1 + HBr	13.18	50.9
n-Butyl chloride^d............	Butene-1 + HCl	14.0	57.0
Sec-butyl bromide^{ee}.........	Butene + HBr	12.63	43.8
t-Butyl alcohol^i............	Isobutene + H_2O	14.68	65.5 ± 7.0
		$(11.5)^{i'}$	$(54.5)^{i'}$
t-Butyl acetate^j............	Isobutene + CH_3COOH	13.34	40.5
t-Butyl propionate^k.........	Isobutene + C_2H_5COOH	12.79	39.16
Cyclohexyl bromide^{ee}........	Cyclohexene + HBr	13.38	46.1
t-Amyl alcohol^i............	Pentenes + H_2O	13.51	60.0 ± 3.7
(−)Menthyl chloride^l.......	(+)2-p-Menthene, (+)3-p-		
	Menthene + HCl	12.6	45.0
$CH_3CH(OOCCH_3)_2^m$.........	$CH_3CHO + (CH_3CO)_2O$	10.3	32.9^o
$CH_3CH(OOCC_2H_5)_2^n$.........	$CH_3CHO + (C_2H_5CO)_2O$	10.4	32.9^o
$C_2H_5CH(OOCCH_3)_2^n$.........	$C_2H_5CHO + (CH_3CO)_2O$	10.5	32.9^o
$(CH_3CO)_2O^o$...............	$CH_2=C=O + CH_3COOH$	12.0	34.5^o
$ClCOOC_2H_5^p$...............	$C_2H_5Cl + CO_2$	10.7	29.4^r
$ClCOOCH(CH_3)_2^q$...........	$(CH_3)_2CHCl + CO_2$		
	(also some		
	$CH_3CH=CH_2 + HCl$)	9.5	26.4^r
$ClCOOCCl_3^r$...............	$COCl_2$	13.15	41.5
$CH_2=CH-O-C_2H_5^s$........	$C_2H_4 + CH_3CHO$	11.43	43.8
Cyclobutane^t...............	C_2H_4	15.6	62.5
Perfluoro-cyclobutane^u.......	C_2F_4	15.95	74.1
Cyclohexene^v...............	$C_2H_4 + $ butadiene	12.9	57.5^w
Vinylcyclohexene-3^x........	Butadiene	15.7	61.8
Cyclopentene^y.............	Cyclopentadiene + H_2	13.04	58.8
Dicyclopentadiene^z.........	Cyclopentadiene	13.0	33.7
Endomethylene-tetrahydro- benzaldehyde^{aa}...........	Acrolein + cyclopentadiene	12.34	33.6
Ring trimer of CH_3CHO^{bb}....	CH_3CHO	15.1	44.2
Ring trimer of $C_3H_7CHO^{bb}$...	C_3H_7CHO	14.4	42.0

TABLE XI.4. UNIMOLECULAR GAS-PHASE DECOMPOSITIONS
THAT GIVE STABLE MOLECULES (Continued)

Reactant	Products	$\log A$, $\sec^{-1}$	E_{act}, Kcal/mole
Ring trimer of			
$(CH_3)_2CHCHO$[bb]	$(CH_3)_2CHCHO$	14.5	42.0
F_2O_2[cc]	$F_2 + O_2$	12.77	17.3 (est)
N_2O_4[dd]	NO_2	16 (est)	13.0
Ethyl acetate[ee]	$CH_3COOH + C_2H_4$	12.5	47.8
Isopropyl acetate[ff]	$CH_3COOH + C_3H_6$	13.0	45.0
Acetyl cyclobutane[gg]	$C_2H_4 + CH_3COCH{=}CH_2$	14.53	54.5

[a] D. H. R. Barton and K. E. Howlett, J. Chem. Soc., 155, 165 (1949).

[b] D. H. R. Barton and P. F. Onyan, J. Am. Chem. Soc., 72, 988 (1950).

[c] A. T. Blades and G. W. Murphy, ibid., 74, 6219 (1952). In presence of excess toluene carrier gas in flow system they found no evidence of free radicals. For contrast see J. B. Peri and F. Daniels, ibid., 72, 424 (1950), who found surface-catalyzed radical reaction in static systems.

[d] D. H. R. Barton et al., J. Chem. Soc., 2033 (1951).

[e] D. H. R. Barton and A. J. Head, Trans. Faraday Soc., 46, 114 (1950).

[f] D. H. R. Barton and P. F. Onyan, ibid., 45, 725 (1949).

[g] A. MacColl and P. T. Thomas, J. Chem. Phys., 19, 977 (1951).

[h] G. B. Kistiakowsky and C. H. Stauffer, J. Am. Chem. Soc., 59, 165 (1937).

[i] R. F. Schultz and G. B. Kistiakowsky, ibid., 56, 395 (1934).

[i'] J. A. Barnard, Trans. Faraday Soc., 55, 947 (1959).

[j] C. E. Rudy, Jr., and P. Fugassi, J. Phys. & Colloid Chem., 52, 357 (1948).

[k] E. Warwick and P. Fugassi, ibid., 52, 1314 (1948).

[l] D. H. R. Barton et al., J. Chem. Soc., 453 (1952).

[m] C. C. Coffin, Can. J. Research, 5, 636 (1931).

[n] Ibid., 6, 417 (1932). J. C. Cornell et al., ibid., 18, 410 (1940).

[o] M. Szwarc and J. Murawski, Trans. Faraday Soc., 47, 269 (1951). There seems to be an inconsistency between these reaction rates and those reported by Coffin for the decomposition of the acetals. In the latter reactions the anhydrides which are produced should themselves be decomposing at a rate 10 times faster than the acetals themselves.

[p] A. R. Choppin et al., J. Am. Chem. Soc., 61, 3176 (1939); A. R. Choppin and G. F. Kirby, ibid., 62, 1592 (1940).

[q] A. R. Choppin and E. L. Compere, ibid., 70, 3797 (1948).

[r] H. C. Ramsperger and G. Waddington, ibid., 55, 214 (1933). It seems rather peculiar that this reaction which is so similar to [p] and [q] should give such different products and have so much smaller rate constant. One would suspect the former of being heterogeneous.

[s] A. T. Blades and G. W. Murphy, ibid., 74, 1039 (1952). S. Wang and C. A. Winkler, Can. J. Research, 21B, 97 (1943). The latter found a free radical reaction; the former claim to have suppressed it with toluene.

[t] C. T. Genaux and W. D. Walters, J. Am. Chem. Soc., 73, 4497 (1951). F. Kern and W. D. Walters, Proc. Nat'l Acad. Sci. U.S., 38, 937 (1952); J. Am. Chem. Soc., 75, 6196 (1953).

[u] B. Atkinson and A. B. Trenwith, J. Chem. Phys., 20, 754 (1952).

[v] L. Küchler, Nachr. Ges. Wiss. Göttingen, 1, 231 (1939); Trans. Faraday Soc., 35, 874 (1939).

[w] The equilibrium calculations of D. Rowley and H. Steiner, Discussions Faraday Soc., 10, 198 (1951), together with their data for the forward reaction indicate that the activa-

Of the reactions with large frequency factors, the decompositions of the aldehyde trimers, vinylcyclohexene-3, cyclobutane, and perfluorocyclobutane fall into a class, since they all involve the opening up of rather tight rings. If the transition complexes for these reactions are loose complexes, the high frequency factors are not unreasonable. The dissociation of N_2O_4 has a frequency factor in excess of 10^{15} sec^{-1}, and this seems reasonable if in the transition complex the originally stiff vibrations of the two NO_2 groups change over to only slightly hindered rotations.[a] This dissociation is of further interest in that it is one of the few unimolecular reactions for which the experimental first-order rate

[a] The dissociation has been studied photometrically by observing the increase in NO_2 following the passage of an adiabatic shock wave through a mixture of N_2O_4 in carrier N_2 gas. The method is admittedly crude and the activation energy (and thus frequency factor) is difficult to measure in this system, but there seems to be little doubt that the frequency factor exceeds 10^{14} sec^{-1}. The reaction also shows the typical pressure dependence of a nonequilibrium, unimolecular decomposition including an apparent activation energy which is smaller than the enthalpy change.

The entropy of activation would amount to about 10 cal/mole-°C, which is quite reasonable when compared with the over-all entropy change in the reaction of about 45 cal/mole-°C (standard state 25°C, 1 atm pressure). The standard entropy change due to translation is 32.4 cal/mole-°C, which leaves 12.6 cal/mole-°C for rotation and vibrational changes, a figure that should be compared with the 10 cal/mole-°C entropy of activation. This would indicate that the transition-state complex is much more like two loosely bound NO_2 molecules than it is like N_2O_4.

---

tion energy should be 6 Kcal higher. In that case the frequency factor would correspond to log A (sec^{-1}) = 14.5, which would be more compatible with their data on the vinylcyclohexene-3 decomposition.[x] The argument is not completely convincing, and the situation appears at present rather confusing.

[x] N. E. Duncan and G. J. Janz, *J. Chem. Phys.*, **20**, 1644 (1952).

[y] D. W. Vanas and W. D. Walters, *J. Am. Chem. Soc.*, **70**, 4035 (1948).

[z] J. B. Harkness et al., *J. Chem. Phys.*, **5**, 682 (1937).

[aa] G. B. Kistiakowsky and J. R. Lacher, *J. Am. Chem. Soc.*, **58**, 123 (1936).

[bb] C. C. Coffin, *Can. J. Research*, **7**, 75 (1932); **9**, 603 (1933). N. A. D. Parlee et al., *ibid.*, **18**, 223 (1940).

[cc] H. J. Schumacher and P. Frisch, *Z. physik. Chem.*, **B37**, 1, 18 (1937).

[dd] T. Carrington and N. Davidson, *J. Chem. Phys.*, **19**, 1313 (1951); *J. Phys. Chem.*, **57**, 418 (1953).

[ee] J. H. S. Green et al., *J. Chem. Phys.*, **21**, 178 (1953). These authors claimed to suppress chain reactions in EtBr, n-PrBr, n-BuBr, and i-BuBr by use of cyclohexene. It is not clear whether or not such suppression is completely effective or whether heterogeneous reaction on the glass surface persists. See also *J. Chem. Soc.*, 2455 (1955). A. E. Goldberg and F. Daniels, *J. Am. Chem. Soc.*, **79**, 1314 (1957), have shown that HBr accelerates rate, while C_2H_4 and n-hexane inhibit rate. They interpret reaction as combination of unimolecular split into C_2H_4 + HBr plus wall-catalyzed chain sensitized by HBr.

[ff] A. T. Blades, *Can. J. Chem.*, **32**, 366 (1954).

[gg] L. G. Daignault and W. D. Walters, *J. Am. Chem. Soc.*, **80**, 541 (1958).

constant shows an unmistakable decrease with decreasing total pressure.[a] The activation energy in this region also shows the pressure dependence predicted by the RRK theory.

Of the remaining reactions with high frequency factors, the decompositions of ethyl chloride, n-butyl chloride, and t-butyl alcohol seem anomalous, both with respect to the decompositions of similar compounds and in regard to the types of transition complexes which would have to be postulated. Thus a splitting of HCl from C_2H_5Cl would involve the formation of a four-membered ring complex, and it would seem that a decrease rather than an increase in entropy should be expected. (The same thing is true of the others.) In view of the fact that these reactions are all enormously sensitive to the state of the walls of the reaction vessel it is probably premature to accept these values as final.[b]

11. Some Unimolecular Fissions; Production of Free Radicals

The data on the unimolecular rates of decomposition of stable molecules into free radicals which are summarized in Table XI.5 are probably quantitatively the least reliable of all the data given on unimolecular reactions. The reason for this is that the rate constants given have had to be inferred from an over-all observed rate for a set of complex reactions. This is an inherent difficulty with reactions that involve the production of free radicals, which, being active intermediates, of necessity disappear by secondary reactions.

In the decomposition of toluene, for example,[c] the products are found to be principally H_2; CH_4; bibenzyl,[d] $(\phi CH_2)_2$; and presumably benzene, ϕH. The mechanism proposed is

[a] The high frequency factor and low activation energy both tend to bring this pressure dependence into the range of about 4 atm despite the fact that a six-atom molecule would normally not be expected to show such an effect except at pressures of about 1 mm Hg.

[b] Daniels and coworkers have shown [$J. Am. Chem. Soc.$, **72**, 424 (1950); $J. Chem. Phys.$, **7**, 756 (1939)] that the decomposition of C_2H_5Br proceeds via chains starting and stopping at the walls, so that, although the reaction is not terribly sensitive to surface/volume ratio, the walls always play a role in it. These results were demonstrated by means of radioactive tracers. This and results of other workers (to be discussed) show the difficulty of demonstrating experimentally the presence of catalyzed reactions or radical chains.

[c] M. Szwarc, $Chem. Revs.$, **47**, 75 (1950), has written an excellent review summarizing the difficulties involved in calculating individual rate constants from kinetic data for systems of consecutive reactions.

[d] C. Horrex et al., $Discussions Faraday Soc.$, **10**, 232 (1951), have reported that small amounts of stilbene and even smaller amounts of anthracene are also formed. The amounts of those increase as the temperature is raised and also as the pressure of inert carrier gas is increased.

TABLE XI.5. UNIMOLECULAR GAS-PHASE REACTIONS THAT PRODUCE RADICALS

Reactant	Products	log A, sec^{-1}	E_{act}, Kcal/mole
$\phi CH_3{}^a$	$\phi CH_2 + H$	12.32	77.5
$m\text{-}CH_3\text{-}\phi\text{-}CH_3{}^a$	$CH_3\!-\!\phi\!-\!CH_2 + H$	12.62	77.5
$p\text{-}CH_3\text{-}\phi\text{-}CH_3{}^a$	$CH_3\!-\!\phi\!-\!CH_2 + H$	12.68	76
$o\text{-}CH_3\text{-}\phi\text{-}CH_3{}^a$	$CH_3\!-\!\phi\!-\!CH_2 + H$	12.68	75
$\phi CH_2\text{-}CH_3{}^c$	$\phi CH_2 + CH_3$	13.0	63
$\phi CH_2\text{-}C_2H_5{}^{e'}$	$\phi CH_2 + C_2H_5$	13.5	57.5
$\phi_2 CH_2{}^w$	$\phi_2 CH + H$	13	73
$\alpha\text{-Me-naphthalene}^t$	$\alpha\text{-Me-naphthyl} + H$	13.18	73.5
$\beta\text{-Me-naphthalene}^t$	$\beta\text{-Me-naphthyl} + H$	13.18	73.5
ϕBr^t	$\phi + Br$	13.3	70.9
$\phi CH_2\text{-}Br^b$	$\phi CH_2 + Br$	13.0	50.5
$\phi CH_2\text{-}NH_2{}^d$	$\phi CH_2 + NH_2$	12.8	59
$\phi CH_2\text{-}COCH_3{}^e$	$\phi CH_2 + COCH_3$	13.9	63
$N_2H_4{}^f$	$2NH_2$	12.6	60
$[(CH_3)_3CO]_2{}^g$	$2(CH_3)_3CO$	13.3$^{g'}$	34
$CH_3COCOCH_3{}^e$	$2CH_3CO$	13.8	60
$CH_3CO\text{-}O\text{-}O\text{-}COCH_3{}^j$	$2CH_3COO$	14.9	31
$CH_3OOCH_3{}^y$	$2CH_3O$	15.4	36.1
$C_2H_5\text{-}O\text{-}O\text{-}C_2H_5{}^k$	$2C_2H_5O$	13.3 '14.8)^y	31.7 (34.1)^y
$CH_2\text{-}CHCH_2Br^{b,h,i}$	$CH_2\!-\!CHCH_2 + Br$	12.7 (12.32)$^{h'}$	47.5 (45.5)$^{h'}$
$Cl\text{-}CH_2\text{-}CH_2Cl^v$	$ClCH_2\!-\!CH_2 + Cl$	13	70
$CH_3O\text{-}NO^m$	$CH_3O + NO$	13.25	36.4
$C_2H_5O\text{-}NO^n$	$C_2H_5O + NO$	14.15	37.7
$n\text{-}C_3H_7\text{-}O\text{-}NO^o$	$n\text{-}C_3H_7O + NO$	14.43	37.6
$Iso\text{-}C_3H_7\text{-}O\text{-}NO^p$	$Iso\text{-}C_3H_7O + NO$	14.11	37.0
$n\text{-}C_4H_9O\text{-}NO^q$	$C_4H_9O + NO$	14.50	37.0
$CH_3\text{-}NO_2{}^u$	$CH_3 + NO_2$	13.6	50.6
$C_2H_5ONO_2{}^r$	$C_2H_5O + NO_2$	15.8(?)	39.9
$N_2O_5{}^s$	$NO_2 + NO_3$	14.8 $\pm$ 1.5, 12.8$^{s'}$	21 ($\pm$2), 19$^{s'}$
$C_2H_6{}^x$	$CH_3 + CH_3$	17.3 (est)	84

^{a} M. Szwarc, *Nature*, **160**, 403 (1947); *J. Chem. Phys.*, **16**, 128 (1948).

^{b} M. Szwarc and B. N. Ghosh, *J. Chem. Phys.*, **17**, 744 (1949).

^{c} M. Szwarc, *ibid.*, 431.

^{d} *Ibid.*, 505; *Proc. Roy. Soc.* (*London*), **A198**, 285 (1949).

^{e} M. Szwarc and J. Murawski, private communication.

$^{e'}$ C. H. Leigh and M. Szwarc, *J. Chem. Phys.*, **20**, 403 (1952).

^{f} M. Szwarc, *Proc. Roy. Soc.* (*London*), **A198**, 267 (1949).

^{g} J. H. Raley et al., *J. Am. Chem. Soc.*, **70**, 88 (1948). M. Szwarc and J. S. Roberts, *J. Chem. Phys.*, **18**, 561 (1949).

$^{g'}$ Other values have been reported [J. Murawski et al., *J. Chem. Phys.*, **19**, 698 (1951)], with log A = 14.7 and E = 36 Kcal/mole.

^{h} A. Maccoll, *ibid.*, **17**, 1350 (1949).

^{i} M. Szwarc et al., *ibid.*, **18**, 1142 (1950).

^{j} O. J. Walker and G. L. E. Wild, *J. Chem. Soc.*, 1132 (1937).

$$
\left.
\begin{array}{l}
\phi CH_3 \xrightarrow{1} \phi CH_2 + H \\
H + \phi CH_3 \rightarrow \phi CH_2 + H_2 \\
H + \phi CH_3 \rightarrow \phi H + CH_3 \\
CH_3 + \phi CH_3 \rightarrow \phi CH_2 + CH_4 \\
2\,\phi CH_2 \rightarrow \phi CH_2 - CH_2 \phi
\end{array}
\right\}
= \phi CH_3 \rightarrow \frac{x}{2} H_2 + \frac{2+x}{6} (\phi CH_2)_2
$$

$$
+ \frac{1-x}{3} CH_4 + \frac{1-x}{3} \phi H
$$

If it is then assumed that each mole of permanent gas produced (H_2 or CH_4) corresponds to the initial decomposition (*via reaction*, 1) of 1 mole of toluene, the total rate of production of gas ($H_2 + CH_4$) is then a direct measure of the amount of toluene initially decomposed and the rate of gas production may be equated to the rate of the reaction. In this way the rates of such fission reactions have been obtained.[a]

It is an unwritten law of chemical kinetics that all proposed mechanisms are inherently suspect, so that detailed data derived from proposed mechanisms always have somewhat of a tentative quality.[b] Despite these rather pessimistic observations the body of data concerning free radicals is rapidly reaching such a point that the results appear to have a great deal of self-consistency.

[a] See S. W. Benson and J. H. Buss, *J. Phys. Chem.*, **61**, 104 (1957), for a criticism of some of these reactions.

[b] Or to quote a popular aphorism, "A mechanism may be disproven but never proven." This is perhaps equally applicable to all generalizations.

---

[k] E. J. Harris, *Proc. Roy. Soc. (London)*, **A173**, 126 (1939). E. J. Harris and A. Egerton, *ibid.*, **A168**, 1 (1938). For most recent work see R. E. Rebbert and K. J. Laidler, *J. Chem. Phys.*, **20**, 574 (1952).

[l] M. Szwarc and A. Shaw, private communication.

[m] E. W. R. Steacie and G. T. Shaw, *Proc. Roy. Soc. (London)*, **A146**, 388 (1934).

[n] *Ibid.*, *J. Chem. Phys.*, **2**, 345 (1934).

[o] *Ibid.*, **3**, 344 (1935).

[p] *Ibid.*, *Proc. Roy. Soc. (London)*, **A151**, 685 (1935).

[q] E. W. R. Steacie and W. McF. Smith, *J. Chem. Phys.*, **4**, 504 (1936).

[r] G. K. Adams and C. E. H. Bawn, *Trans. Faraday Soc.*, **45**, 494 (1949). See also L. Phillips, *Nature*, **165**, 564 (1950), and F. H. Pollard et al., *ibid.*, 564 (1950), for work on alkyl nitrates.

[s] R. L. Mills and H. S. Johnston, *J. Am. Chem. Soc.*, **73**, 938 (1951). Johnston, *ibid.*, 4542.

[s'] A. R. Arnell and F. Daniels, *J. Am. Chem. Soc.*, **74**, 6209 (1952), give these values, which are admittedly crude in so far as the precise values of E are concerned.

[t] M. Szwarc and D. Williams, *J. Chem. Phys.*, **20**, 1171 (1952).

[u] L. J. Hillenbrand and M. L. Kilpatrick, *ibid.*, **19**, 381 (1951).

[v] K. E. Howlett, *Trans. Faraday Soc.*, **48**, 25 (1952).

[w] C. Horrex and J. O. McCrae, *Discussions Faraday Soc.*, **10**, 234 (1951).

[x] Estimated from data on recombination of CH_3 radicals and the ΔS° for reaction. See Table XII.8.

[y] P. L. Hanst and J. G. Calvert, *J. Phys. Chem.*, **63**, 104 (1959).

It will be noted that, as distinct from the unimolecular reactions previously considered, the frequency factors reported here all lie in the range of 2×10^{12} to about 10^{16} sec^{-1}, no very low values having been reported. In the fission reactions in which radicals are formed there is the possibility that excited electronic states may participate and thus contribute to the entropy of the energized state, transition state, or both.[a] In addition, the stretching of a bond between two initially bound radicals may permit hindered rotations to become free rotations in the transition state. This is probably the explanation for the high frequency factor, as in the C_2H_6 and N_2O_4 dissociations and also in the case of N_2O_5. As we shall see later, high frequency factors for reactions in which large radicals are formed seem to be the rule.

12. Unimolecular Reactions; Résumé

In concluding our discussion of unimolecular reactions we may observe that in general the experimental work seems to be in fair agreement with the theoretical expectations. Most frequency factors lie in the expected range of 10^{12} to 10^{14} sec^{-1}. Of the few values that lie below this range most seem explicable in terms of a reasonable decrease in entropy accompanying the formation of a transition complex.[b] Similarly, most of the higher values seem explicable in terms of a reasonable entropy increase accompanying formation of the transition complex. There seem to be exceptions to both of these cases, and they have been noted in the preceding sections. They certainly warrant further study.

The various detailed theories which have been proposed for unimolecular reactions predict that, at sufficiently low total concentrations, all unimolecular processes should show a falling off of their apparent first-order constants with decreasing pressure. Convincing experimental evidence on this point is at present rather meager. The reasons for the sparsity of data arise from at least three sources. First and most difficult to deal with is the fact that the decompositions of most molecules are not simple but involve complex intermediate steps. Secondly, the over-all major reaction is seldom quantitative but usually involves varying amounts of side reac-

---

[a] When these entropy effects appear in both the transition state and the energized molecule, the pressure range for falling off of the first-order rate constant is that calculated in Table XI.2 with a normal (that is, 10^{13} sec^{-1}) frequency factor. When, however, the increased entropy appears only in the transition state, the high frequency factor is to be used in estimating the region of pressure fall-off. When an upper excited electronic state is contributing to the reaction, the value for E^* must be measured from its lowest energy, not that for the ground state.

[b] There have been reported a number of cases for the decomposition of radicals (for example, $CH_3CO \rightarrow CH_3 + CO$) with frequency factors of 10^{10} sec^{-1}. However, there is strong evidence for believing that they are really bimolecular energy-transfer reactions, and not first-order.

tions. Finally, the walls of the reaction vessel are not inert but contribute a heterogeneous component to the reaction.

Rice and Herzfeld[a] have considered these difficulties in some detail and pointed out some of the problems involved in trying to obtain unequivocal evidence concerning their importance. As an illustration we may consider the decomposition of t-butyl chloride to give isobutene and HCl. It has been observed [b,c] that carbonaceous deposits are produced on the walls of the reaction vessel during the decomposition. (This has been generally observed in the pyrolysis reactions of organic compounds.) It is further found that reproducible results are not obtained until some 20 or more consecutive runs have been made in the vessel, and even more striking has been the observation that exposing the "reproducible" surface to a pressure of 4 mm Hg of O_2 for 15 min "deconditioned" the walls again so that a further reconditioning was necessary.

Peri and Daniels[d] in a very careful study of the C_2H_5Br decompositions have found even more striking effects and by use of radiobromine were able to demonstrate the decisive role of the walls in the reaction.[e]

A final example of the effectiveness of surfaces is the finding of Libby and Johnston[f] that the gas-phase exchange of radio Cl_2 with HCl has a half-life of 3 min[g] (or less) at 25°C in a 500-cc pyrex bulb, while coating the vessel with fluorocarbon (C_nF_{2n+2}) raised the half-life to 11 to 16 hr.

Thus before any given set of data may be reliably used in applying the details of a theory of unimolecular reactions, the chemical complexity of the reaction and the effects of the walls must first be completely established. There are at present only a few cases for which an exhaustive and convincing study has been presented; they are the decompositions of N_2O_5,[h] cyclopropane,[i] and N_2O_4.[j] In all of these cases the frequency factors reported have been abnormally high (10^{15} to 10^{16} sec^{-1}), which has had as a consequence that the characteristic pressure at which the first-order rates could be observed to fall was higher than would have been predicted for molecules of such complexity (see Table XI.2). In each case the reactions show the qualitative features which are to be expected in the intermediate

[a] F. O. Rice and K. F. Herzfeld, *J. Phys. & Colloid Chem.*, **55**, 975 (1951).

[b] D. Brearly et al., *J. Am. Chem. Soc.*, **58**, 44 (1936).

[c] D. H. R. Barton and P. F. Onyan, *Trans. Faraday Soc.*, **45**, 725 (1949).

[d] J. B. Peri and F. Daniels, *J. Am. Chem. Soc.*, **72**, 424 (1950).

[e] See footnote [c], Table XI.4.

[f] W. H. Johnston and W. F. Libby, *ibid.*, **73**, 854 (1951). See also H. Dodgen and W. F. Libby, *J. Chem. Phys.*, **17**, 951 (1949).

[g] This is the order of magnitude of the time of diffusion of a molecule from the center of the flask to the walls.

[h] H. S. Johnston, *J. Chem. Phys.*, **20**, 1103 (1952).

[i] A. F. Trotman-Dickenson et al., *Proc. Roy. Soc. (London)*, **A217**, 563 (1953).

[j] T. Carrington and N. Davidson, *J. Chem. Phys.*, **19**, 1313 (1951); *J. Phys. Chem.*, **57**, 418 (1953).

concentration region; i.e., both the rate constant and activation energy decrease with decreasing pressure (at constant temperature).[a]

A final point of interest is that in each of the above cases the rate constants in the intermediate region also responded to inert gases such as He, N_2, CO, and CO_2, so that the falling off in the observed rates could be prevented by the addition of one of those gases. This is a rather striking verification of the hypothesis of collisional activation and one that has been long sought.[b]

[a] For further discussion of N_2O_5 see Sec. XIII.19.

[b] Most of the work on such effects, prior to about 1940, has been shown to be inconclusive in the sense that the reactions studied were later demonstrated to be rather complex chain reactions or heterogeneous. R. N. Pease, *J. Chem. Phys.*, **7**, 749 (1939), has discussed this work in some detail.

XII

Bimolecular and Termolecular Reactions

1. Association Reactions

Bimolecular reactions can be divided into two important categories, associations and exchanges, which can be characterized as follows:

Association Reactions:

$$A + B \rightarrow D \qquad \text{both three and four center}$$

Exchange Reactions:

$$(1) \quad A + BC \rightarrow AB + C \qquad \text{three center}$$
$$(2) \quad AB + CD \rightarrow AC + BD \qquad \text{four center}$$

Whereas exchange reactions, both 1 and 2, are of the same order with respect to the reverse processes, the association reactions are not, because the reverse reaction represents a unimolecular decomposition. We can apply the methods developed in Sec. XI.5 for studying these systems.

The case of association reactions has already been studied in terms of the inverse process (unimolecular decomposition), and the model for that case, represented by Eq. (XI.5.2), and the results obtained from the model can be used here. The net rate of the over-all reaction R' is given by $-R$ of Eq. (XI.5.3), namely,

$$R' = \frac{(A)(B)(M)K_{AB}k_{r'}k_2 - (D)(M)k_rk_1}{k_r + (M)k_2} \cdot \frac{Z}{Z + (M)k_{r'}k_2/[k_r + (M)k_2]} \tag{XII.1.1}$$

where the terms are as defined in Sec. XII.5. The initial rate of reaction that obtains when $(D) = 0$ or k_1 is negligibly small is given by R'_f (note $R' = R'_f - R'_b$):

$$R'_f = \frac{(A)(B)(M)K_{AB}k_{r'}k_2}{k_r + (M)k_2} \frac{Z}{Z + (M)k_{r'}k_2/[k_r + (M)k_2]} \quad \text{(XII.1.2)}$$

or in form similar to Eq. (XI.5.5)

$$R'_f = \frac{(A)(B)K_{AB}k_{r'}}{1 + k_r/Mk_2} \frac{Z}{Z + k_{r'}(1 + k_r/Mk_2)^{-1}} \quad \text{(XII.1.3)}$$

The high concentration limit $R'_{f\infty}$ is attained when $Mk_2 \gg \overline{k_r(E)}$, that is, when the probability of deactivation of the complexes associated with the product D* is much greater than that of the competing process D* → [AB]*:

$$R'_{f\infty} = (A)(B)K_{AB}k_{r'} \frac{Z}{Z + k_{r'}} = (A)(B)K_{AB}k_{r'}F_\infty(Z) \quad \text{(XII.1.4)}$$

which is similar in form to the expression obtained for $R_{f\infty}$ in Eq. (XI.5.6).

The analysis of $R'_{f\infty}$ now follows the analysis of $F(Z)$ which was made in Sec. XI.5, and by applying these results we have two possible extremes for the values of $R'_{f\infty}$. When the critical energy $E^{*'}$ for the association is zero, we have

$$R'_{f\infty} (E^{*'} \leq 0) = \tfrac{1}{2}(A)(B)K_{AB}k_{r'} \quad \text{(XII.1.5)}$$

while for $E^{*'} > 0$ (that is, $E^{*'}$ of order of kT):

$$R'_{f\infty} (E^{*'} > 0) = (A)(B)K_{AB}k_{r'} \quad \text{(XII.1.6)}$$

For the intermediate concentration regions[a] we find

$$R'_f (E^{*'} \leq 0) = (A)(B)K_{AB}k_{r'} \left[2 + \frac{k_r}{Mk_2} \right]^{-1} \quad \text{(XII.1.7)}$$

$$R'_f (E^{*'} > 0) = (A)(B)K_{AB}k_{r'} \left[1 + \frac{k_r}{Mk_2} \right]^{-1} \quad \text{(XII.1.8)}$$

The interesting result in this case is that the concentration dependence of association reactions is identical in form to that for unimolecular decomposition. This is of course what we might have expected from the principle of microscopic reversibility. The significance of the results has already been discussed in Sec. XI.5.

2. Exchange Reactions

Ignoring the subclasses of exchange reactions, the general class may be represented by the chemical equation

$$A + B \rightleftharpoons C + D \quad \text{(XII.2.1)}$$

and we can write a detailed scheme as follows:

[a] The over-all rates are obtained by integration of R'_f over all energies.

$$(XII.2.2)$$

By again carrying out the approximate steady-state treatment utilized for unimolecular reactions (Sec. XI.5) we find for the net rate of reaction

$$R = R_f - R_b$$

$$= [(A)(B)K_{AB}k_r - (C)(D)K_{CD}k_{r'}] \frac{ZZ'}{ZZ' + k_r Z' + k_{r'} Z} \quad (XII.2.3)$$

where Z is defined as before [Eq. (XI.5.7)]:

$$Z \cong \frac{k_{6'}}{1 + k_{5'}b/k_{2'}} + \frac{k_{6''}}{1 + k_{5''}a/k_{2''}} + Mk_8 \quad (XII.2.4)$$

and Z' is symmetrically defined in terms of the appropriate rate constants and mole fractions. But if we now carry through the same analysis of the bracketed expression in Z and Z' as in Sec. XI.5,[a] we find that under all conditions it never deviates significantly from unity. As before, the significance of this result is that bimolecular fission of the critical complex will always be so much faster than the mean rates of the reaction steps k_r and $k_{r'}$ that the critical complexes $[AB]^*$ or $[CD]^*$ are effectively maintained at their equilibrium levels.[b] Thus bimolecular exchange reactions are qualitatively different from all the categories of reactions previously discussed and can be treated as though the energized complexes were in thermodynamic equilibrium with the initial reactants.

[a] This term may be written as $(1 + k_r/Z' + k_{r'}/Z)^{-1}$. The analysis of Sec. XI.5 shows that it is essentially pressure-independent under all conditions, i.e., the dissociation reactions k_6 are always faster than Mk_8, since the former require no activation energy and have inherently larger frequency factors. Thus $Z \cong k_{6'} + k_{6''} = k_6$ and $Z' \cong k_{6'''} + k_{6''''} = k_{6'}$. This term will further always be approximately unity, since $k_6 < k_{r'}$ and $k_{6'} < k_r$ unless one of the processes has zero activation energy. In that case the term will have the value $\frac{1}{2}$.

The significance of this case is that, when, for example, the reaction C + D has no activation energy, then each $[CD]^*$ has equal chance of splitting into C + D or reforming $[AB]^*$, and so only half of $[AB]^*$ which form $[CD]^*$ will go on to give C + D.

[b] The same results can be obtained for any reactions of higher order in which forward and back reactions are at least bimolecular. Conversely, only those higher-order reactions for which the reverse step is unimolecular can show a nonequilibrium dependence of the rate constant, e.g., $A + B + C \rightleftharpoons D$ [S. W. Benson and A. E. Axworthy, Jr., J. Chem. Phys., **21**, 428 (1953)].

Under conditions that the reverse reaction can be neglected, we can write for the rate of the forward reaction, $R = R_f - R_b$

$$R_f = (A)(B)K_{AB}k_r \qquad (XII.2.5)$$

where K_{AB} is the equilibrium constant for the reaction $A + B \rightleftharpoons [AB]^*$ and k_r is the specific rate constant for the change of $[AB]^*$ to $[CD]^*$, both of these being now functions of the energy of $[AB]^*$. The total rate is then obtained by integrating (or summing) over all energy states $[AB]^*$:

$$\text{Total rate} = \frac{-d(A)}{dt} = \int_0^\infty R_f \, dE = (A)(B) \int_0^\infty K_{AB}(E)k_r(E) \, dE \qquad (XII.2.6)$$

But $K_{AB}(E)$ can be written as a quantity $K_{AB}^\circ(E^*)$ independent of E and a factor $P(E)$, the probability that $[AB]^*$ have energy E in excess of E^*. Then

$$\frac{-d(A)}{dt} = (A)(B)K_{AB}^\circ(E^*) \int_0^\infty P(E)k_r(E) \, dE \qquad (XII.2.7)$$

If we now use the RRK form for $k_r(E)$ and the classical distribution function for $P(E)$ (Chap. X), we can perform the integration directly and we find[a]

$$\frac{-d(A)}{dt} = (A)(B)K_{AB}^\circ \bar{\nu} = (A)(B)\bar{\nu}e^{S^*_{AB}/R}e^{-E^*/RT} \qquad (XII.2.8)$$

where $\bar{\nu}$ is a mean weighted average of the normal frequencies of the complex $[AB]^*$ contributing to the description of the reaction path and E^* is the critical energy for the reaction.

By setting $-d(A)/dt = k_B(A)(B)$, we find for the experimental bimolecular rate constant k_B:

$$k_B = \bar{\nu}K_{AB}^\circ = \bar{\nu}e^{S^*_{AB}/R}e^{-E^*/RT} \qquad (XII.2.9)$$

If, however, we make the assumption that $k_r(E)$ can be approximated by the quantum-mechanical relation (Sec. XI.6):

$$k_r(E) = \frac{\nu g_s(E - E^*)}{g_s(E)} \qquad (XII.2.10)$$

where ν is some average of the normal frequencies of the critical complex $[AB]^*$ with energy E, $g_s(E)$ is the total number of quantum states of $[AB]^*$, and $g_s(E - E^*)$ is the total number of quantum states of $[AB]^*$ when it has the configuration corresponding to the transition state (see page 245), then we can follow the procedure used in Sec. XI.6, and we find for this approximate quantum-mechanical treatment

$$k_B = \nu K_{AB}^\ddagger$$

$$= \nu \frac{Q^\ddagger(AB)}{Q(A)Q(B)} e^{-E_0^*/kT} \qquad (XII.2.11)$$

[a] Benson, ibid., **20**, 1064 (1952).

where $K_{AB}^{\ddagger}$ is the true equilibrium constant for the transition-state complex, $Q^{\ddagger}(AB)$ is the partition function for this species, and $Q(A)$ and $Q(B)$ are the partition functions for A and B.

In thermodynamic language[a] we can write

$$k_B = \nu e^{-\Delta F^{\ddagger}/RT} = \nu e^{\Delta S^{\ddagger}/R} e^{-\Delta H^{\ddagger}/RT} \qquad (XII.2.12)$$

where $\Delta S^{\ddagger}$, $\Delta H^{\ddagger}$, and $\Delta F^{\ddagger}$ represent respectively the standard entropy, enthalpy, and free energy of formation of the transition-state complexes.

3. Collision Model for Bimolecular Reactions

It is interesting to compare the preceding results with the result obtained from a simple collision treatment of bimolecular reactions. For the system A + B → C + D we can write for the rate of reaction

$$-\frac{d(A)}{dt} = k_B(A)(B) = Z(A,B)P(A,B) \qquad (XII.3.1)$$

where $Z(A,B)$ is the total number of collisions per unit volume per unit time of A and B and $P(A,B)$ is the probability that any given collision of A and B will lead to reaction. By using the value of $Z(A,B)$ obtained for hard sphere molecules we may write for k_B, the experimental bimolecular specific rate constant

$$k_B = \bar{Z}'_{AB}P(A,B) \qquad (XII.3.2)$$

where $\bar{Z}'_{AB}$ is the specific collision frequency [Eq. (VII.8F.5)].

If now we assume that reaction cannot occur unless the collision takes place in such a manner that the relative kinetic energy of the colliding molecules (along their line of centers) is in excess of a minimum energy E, which we will call the critical energy, then we must use instead of $\bar{Z}'_{AB}$, $\bar{Z}'_{v'}$

[a] Note again the thermodynamic convention. The free energy change in a reaction $n_1 X_1 + n_2 X_2 \rightleftharpoons m_1 Y_1 + m_2 Y_2 + \cdots$ is given by $\Delta F = \Sigma n_i \mu_i$, where n_i refers to the number of moles of the ith species whose chemical potential $\bar{F}_i = \mu_i$. In terms of mole fractions x_i and for ideal mixtures of gases we can write

$$\Delta F = \Sigma n_i \mu_i = \Sigma n_i \overset{\circ}{\mu}_1 + RT\Sigma n_i \ln x_i = \overline{\Delta F^{\circ}} + RT \Sigma \ln x_i^{n_i}$$

which is zero at equilibrium so that $\overline{\Delta F^{\circ}} = -RT\Sigma \ln x_i^{n_i} = -RT \ln K(x_i)$ is the standard free energy change in the reaction referred to the pure substances ($\mu_i = \overset{\circ}{\mu}_i$ when $x_i = 1$) as the standard states. $K(x_i)$ is the equilibrium constant defined in terms of mole fractions. For an ideal gas we can now relate $K(c_i)$ and $K(p_i)$ to $K(x_i)$ by the definition of $K(x_i)$ and the ideal gas law relation, $p_i = x_i P$ (P = total pressure), $c_i = n_i/V = P_i/RT = x_i P/RT = x_i V$, where V = total volume. Then $K(p_i) = K(x_i) P^{\Delta n}$ and $K(c_i) = K(x_i) V^{\Delta n}$, where Δn is the change in number of moles in the reaction. Thus $\overline{\Delta F^{\circ}}(x_i) = \overline{\Delta F^{\circ}}(c_i) + \Delta n\, RT \ln V = \overline{\Delta F^{\circ}}(P_i) + \Delta n\, RT \ln P$, where P are the pressure units used and V the specific volume units. For the bimolecular reaction we have been discussing, $\Delta n = -1$, since the formation of the activated complex is an association.

the collision frequency for encounters in which the relative velocity is $v_r(E = \frac{1}{2}\mu v_r^2)$ and integrate over-all values of v_r from E to ∞.

Such a treatment has been carried out[a] and leads to the result

$$Z_{v_r} = \bar{Z}'_{AB} e^{-E/RT} \qquad (XII.3.3)$$

with

$$\bar{Z}'_{AB} = \pi\sigma_{AB}^2 \left(\frac{8kT}{\pi\mu}\right)^{\frac{1}{2}} \qquad (VII.8F.5)$$

If we assume that the critical energy need not be localized in the two degrees of freedom of the relative velocity components but may be dispersed in internal degrees of freedom of the colliding molecules, then we find[b] for the total frequency of collision in which there is at least energy E distributed in n chemical degrees of freedom[c]

$$Z'(n) = \bar{Z}'_{AB} \sum_{k=0}^{n-1} \frac{1}{k!} \left(\frac{E}{RT}\right)^k e^{-E/RT} \qquad (XII.3.4)$$

If $E \gg (n-1)RT$, we can approximate this by taking only the first term in the series and we have

$$Z'(n) \cong \bar{Z}'_{AB} \frac{1}{(n-1)!} \left(\frac{E}{RT}\right)^{n-1} e^{-E/RT} \qquad (XII.3.5)$$

To obtain this last equation, however, we have made the very unreasonable assumption that $P(A,B)$ is a constant, so that k_B is given by the first term:

$$k_B \cong \bar{Z}'_{AB} P \frac{1}{(n-1)!} \left(\frac{E}{RT}\right)^{n-1} e^{-E/RT} \qquad (XII.3.6)$$

while the usual collisional treatment, which assumes $n = 1$, gives an exact value:

$$k_B = \bar{Z}'_{AB} P e^{-E/RT} \qquad (XII.3.7)$$

in which P is known as the steric factor.

A further form can be obtained if we use a detailed set of values for Z and P. If we allow both Z and P to be functions of the critical energy E, we can rewrite Eq. (XII.3.2) as

$$k_B = \int_0^\infty \bar{Z}'_{A,B}(E) P_{AB}(E) \, dE \qquad (XII.3.8)$$

[a] R. H. Fowler, "Statistical Mechanics," Cambridge University Press, New York, 1936. See also L. Kassel, "Kinetics of Homogeneous Gas Reactions," chaps. 2 and 5, Reinhold Publishing Corporation, New York, 1932.

[b] Ibid.

[c] By this we mean a degree of freedom whose total energy can be written as a sum of two terms each of which is a perfect square. Thus the vibrational energy of a simple, one-dimensional harmonic oscillator represents one classical degree (two square terms), while three-dimensional translational energy has three components (three square terms) and is thus $\frac{3}{2}$ classical degrees of freedom.

Now assuming that the internal energies of A and B can be represented by the energy of a set of classical harmonic oscillators, we can calculate $\bar{Z}'_{AB}(E)$, the frequency of collisions in which the total internal, rotational, and translational energies of the colliding molecules are between E, and $E + dE$:

$$\bar{Z}'_{A,B}(E) = \bar{Z}'_{AB} \frac{1}{(n-1)!} \left(\frac{E}{kT}\right)^{n-1} \frac{e^{-E/kT}}{kT} \qquad (XII.3.9)$$

in which $2n$ = the total number of square terms in the internal energy expression.

If now we further assume that the probability of reaction or collision $P_{AB}(E)$ is given by a function of the form

$$P_{AB}(E) = 0 \qquad \text{when } E < E^*$$

$$P_{AB}(E) = P\left(1 - \frac{E^*}{E}\right)^{n-1} \qquad \text{when } E \geqslant E^* \qquad (XII.3.10)$$

Then $\quad k_B = \bar{Z}'_{AB} \dfrac{P}{(n-1)!} \displaystyle\int_{E^*}^{\infty} \left(\frac{E}{kT}\right)^{n-1} \left(\frac{E - E^*}{E}\right)^{n-1} e^{-E/kT} \dfrac{dE}{kT}$

which on substitution of the variable $x = (E - E^*)/kT$ reduces to the gamma function and

$$k_B = \bar{Z}'_{AB} P e^{-E^*/kT} \qquad (XII.3.11)$$

which is identical with the simple model above for critical energy in two translational degrees of freedom [Eq. (XII.3.7)] of the collision pair.

If we write the collision frequency $\bar{Z}'_{AB}$ in explicit form

$$\bar{Z}'_{AB} = \pi\sigma_{AB}^2 \left(\frac{8kT}{\pi\mu}\right)^{1/2} \qquad (XII.3.12)$$

it can be rewritten as

$$\bar{Z}'_{AB} = \frac{kT}{h} \frac{(2\pi m_{AB} kT/h^2)^{3/2}}{(2\pi m_A kT/h^2)^{3/2}(2\pi m_B kT/h^2)^{3/2}} \frac{8\pi I kT}{h^2} \qquad (XII.3.13)$$

in which $I = \mu\sigma_{AB}^2$ and the various terms in parentheses can now be identified as the translational molecular partition functions (Sec. IX.8) $q_{AB(tr)}$, $q_{A(tr)}$, $q_{B(tr)}$ of the complex AB and the species A and B, respectively. The last term in Eq. (XII.3.13) can be interpreted as the rotational partition function of a complex AB that consists of two rigid groups A and B. Thus we can write

$$Z'_{AB} = \frac{kT}{h} \frac{q_{AB(tr)}}{q_{A(tr)} q_{B(tr)}} q_{AB(rot)} \qquad (XII.3.14)$$

and finally if we further assume that the groups A and B are connected by a bond whose frequency $\nu \ll kT/h$, then the vibrational partition function for such a weak bond is approximately given by $q_{AB(vib)} \cong kT/h\nu$ so that

$$\bar{Z}'_{AB} = \nu \frac{q_{AB(tr)}}{q_{A(tr)}q_{B(tr)}} \, q_{AB(rot)}q_{AB(vib)} = \nu \frac{Q_{AB}}{Q_A Q_B} \qquad (XII.3.15)$$

$$= \nu K_{AB} \qquad (XII.3.16)$$

where K_{AB} is the equilibrium constant for the formation of a weakly bound complex AB from rigid groups A and B.

By using $\bar{Z}'_{AB}$ in this form we can now rewrite k_B as [Eq. (XII.3.11)]

$$k_B = \nu K_{AB} P e^{-E^*/kT} \qquad (XII.3.17)$$

which now has the same form as the equilibrium rate constant predicted by the detailed theory [Eq. (XII.2.11)].[a]

This simple collisional theory is thus in qualitative agreement with the more sophisticated treatments. The current usage is to refer to P as the "steric factor" and to the product $\bar{Z}'_{AB}P$ as the frequency factor or, more properly, the preexponential factor. In a later section we shall consider these in greater detail.

4. Transition-state Model

In view of the results derived for second-order exchange reactions (Sec. XII.2), namely, that the critical complex is always in effective equilibrium with the reacting species, we can feel justified in applying a quasi-equilibrium theory such as the transition-state theory to these reactions. The reaction can be represented by

$$\begin{aligned} A + B &\rightleftharpoons AB^{\pm} & K^{\pm}_{AB} \\ AB^{\pm} &\rightarrow \text{products} & \nu^{\pm} \end{aligned} \qquad (XII.4.1)$$

in which $AB^{\pm}$ represents the activated complex having the configuration that corresponds to the saddle point of our potential-energy diagram and $\nu^{\pm}$ is its rate of decomposition. The concentration of $AB^{\pm}$ is given by $K^{\pm}_{AB}(A)(B)$, so that the over-all rate of reaction is given by

$$\begin{aligned} \text{Rate} = \frac{-d(A)}{dt} &= \kappa\nu^{\pm}(AB)^{\pm} \\ &= \kappa\nu^{\pm}K^{\pm}_{AB}(A)(B) \end{aligned} \qquad (XII.4.2)$$

For the experimentally observed specific rate constant $k_B = -1/(A)(B) \, dA/dt$:

$$k_B = \kappa\nu^{\pm}K^{\pm}_{AB} \qquad (XII.4.3)$$

where the transmission coefficient κ represents the fraction of effective decompositions.[b]

[a] We would have to equate P with the ratio of the vibrational partition functions, $P = q_{AB(vib)}/q_{A(vib)} q_{B(vib)}$ and interpret E^* as the energy of the bond A—B.

[b] κ is called the transmission coefficient and takes into account the rate of reverse reaction. If we make comparisons with Sec. XII.1, κ may be equated to the expression in Z and Z' [Eq. (XII.2.3)]. Based on the discussion there we may expect κ to vary between $\frac{1}{2}$ and 1 and generally to equal 1 unless the reverse reaction has zero activation energy.

And once more repeating the steps of Sec. XI.7, we replace $K_{AB}^{\ddagger}$ by the quantity $kT/h\nu^{\ddagger}K_{AB}^{\ddagger}e^{-E_0°/RT}$, in which $\nu^{\ddagger}$ is the vibration frequency of the normal mode leading to decomposition of $AB^{\ddagger}$, it being assumed that $kT \gg h\nu^{\ddagger}$. $E_0°$ is the activation energy at $0°K$, and $K_{AB}^{\ddagger}$ is now the equilibrium constant $K_{AB}^{\ddagger}$ with the indicated terms factored out.[a] Then we can write, neglecting κ,

$$k_B = \frac{kT}{h} K_{AB}^{\ddagger} \tag{XII.4.4}$$

5. Comparison of Different Theories of Bimolecular Rate Constants; Experimental Activation Energies

We have seen in Sec. IV.4 that for thermodynamic consistency a specific rate constant should be capable of being represented in the form

$$k = \nu K^* = \nu e^{-F^*/RT} = \nu K^{*\prime} e^{-E_0^*/RT} \tag{XII.5.1}$$

where ν is a frequency and K^* is a quantity with the dimensions and units of an equilibrium constant to which we relate an empirical free energy of activation F^*.[b] $K^{*\prime}$ is defined by Eq. (XII.5.1), and E_0^* can be interpreted as an energy of activation at $0°K$. The factored equilibrium constant $K^{*\prime}$ can also be represented in statistical form as a ratio of partition functions if we can associate some definite chemical process with the various thermodynamic terms.

If we now inspect the results of the various theories we have just considered, we will see that they can all be cast into the form demanded by Eq. (XII.5.1), and the thermodynamic significance of the terms can now be interpreted in terms of the equilibria proposed by the various models. From laboratory experiments we can calculate specific rate constants at different temperatures, and following the Arrhenius method we can define an experimental energy of activation by the equation

$$E_{exp} = RT^2 \frac{\partial (\ln k_{exp})}{\partial T} = -R \frac{\partial (\ln k_{exp})}{\partial (1/T)} \tag{XII.5.2}$$

Variations of E_{exp} with temperature are in general small, and for the

---

[a] $K_{AB}^{\ddagger}$ is the equilibrium constant for an equilibrium in which the number of moles change, and so will have the units of $(\text{concentration})^{-1}$. It is thus dependent on our choice of units. If for example $K_{AB}^{\ddagger}$ is expressed in units of concentration, for example, $K_{AB}^{\ddagger}(c)$, then $K_{AB}^{\ddagger}(P)$, where P is pressure units, is obtained (for ideal gases) by using the conversion factor from the ideal gas law, $P = cRT$, $K_{AB}^{\ddagger}(P) = K_{AB}^{\ddagger}(c) (RT)^{-1}$. Similarly, to obtain $K_{AB}^{\ddagger}(N)$ where the units are mole fractions, we have $K_{AB}^{\ddagger}(N) = K_{AB}^{\ddagger}(c) C_{total} = K_{AB}^{\ddagger}(P) P_{total}$. These relations will be of importance when we consider standard states and various thermodynamic relations (see footnote pg. 271).

[b] These requirements are needed to represent the case of an equilibrium system in which the ratio of the k's for the forward and reverse reactions must usually reduce to the equilibrium constant. See Sec. IV.3.

limited temperature range generally accessible experimentally we lose little by setting $E_{exp} = E + CT$, where C and E are constants.[a] This allows us to integrate Eq. (XII.5.2) and obtain the modified Arrhenius form

$$\ln k_{exp} = \frac{E}{RT} + \frac{C}{R} \ln T + \ln k_0 \qquad (XII.5.3)$$

or $$k_{exp} = k_0 T^{C/R} e^{-E/RT} \qquad (XII.5.4)$$

where k_0 is a constant of integration.[a]

When the rate constant can be decomposed into a product of two terms, one of them being of the form $e^{-E/RT}$, where E is a constant, we refer to them as the exponential temperature term and the preexponential factor (frequency factor), respectively. In Eq. (XII.5.4) the preexponential factor is $k_0 T^{C/R}$.

If we apply the same treatment to Eq. (XII.5.1), we find

$$E_{exp} = RT^2 \frac{\partial \ln k}{\partial T} = -R \frac{\partial \ln \nu}{\partial(1/T)} - R \frac{\partial \ln K^*}{\partial(1/T)} = \frac{-R \, \partial \ln \nu K^{*\prime}}{\partial(1/T)} + E_0^*$$

$$= -R \frac{\partial \ln \nu}{D(1/T)} + H^* \qquad (XII.5.5)$$

but there is no further way of experimentally identifying ν and K^* individually, because the temperature dependence of ν and H^* cannot be separated. This dilemma is frequently resolved by arbitrarily assuming that ν is temperature-independent.

In Table XII.1 we list the values of the specific rate constants for bimolecular reactions and their experimental activation energies and preexponential factors as defined in the foregoing.

We see in Table XII.1 that we cannot separately identify the terms in the rate-constant expression for the thermodynamics equation or the collision theories without special assumptions. A complete identification of all the terms, frequencies, energies of activation and entropies of activation from experimental data is possible only for the Arrhenius equation and the transition-state theory.[b]

6. The Preexponential Factors of Bimolecular Reactions

The preexponential factor predicted by the hard sphere collision theory is given by (Table XII.1) $\bar{Z}'_{AB} P$ in which we can evaluate $\bar{Z}'_{AB}$ from kinetic theory (Table VII.2): $\bar{Z}'_{AB} = 4.57 \times 10^4 \, \sigma^2_{AB}(T/\mu)^{1/2}$ cc/molecule-sec =

---

[a] Usually the experimental data do not justify the use of the correction term C. When they do, the constants of Eq. (XII.5.3) should be fitted to the data by a least-squares fit.

[b] As pointed out in Sec. XI.8, the accurate assignment of entropies of activation on the basis of a universal frequency factor kT/h has rather dubious validity and may be misleading if interpreted too literally.

$2.74 \times 10^{28} \sigma^2_{AB} (T/\mu)^{1/2}$ cc/mole-sec. There is no a priori knowledge of σ_{AB}, but we can assume the viscosity diameters and place the uncertainty in the factor P. For a typical set of values at $T = 300°$K such as $\sigma_{AB} = 4 \times 10^{-8}$ cm and $\mu_{AB} = 30$ we have $\bar{Z}'_{AB} = 2.32 \times 10^{-10}$ cc/molecule-sec $= 1.39 \times 10^{14}$ cc/mole-sec.

This simple collision theory thus predicts preexponential factors of about 10^{14} cc/mole-sec, since we expect $P \leq 1$. Values of $P < 1$ are interpreted kinetically as due to improperly oriented collisions ("steric" hindrance) or thermodynamically as a negative entropy of activation, i.e., a loss of freedom of A and B in forming the collision complex. As we shall see, these results are in good qualitative agreement with observations and $\bar{Z}'_{AB}$ does indeed seem to be an upper limit for bimolecular frequency factors.[a]

Now it was pointed out in Sec. XII.3 that $\bar{Z}'_{AB}$ can also be written [Eq. (XII.3.14)] as

$$\bar{Z}'_{AB} = \frac{kT}{h} \frac{q_{AB(tr)}}{q_{A(tr)} q_{B(tr)}} q_{AB(rot)} = \frac{kT}{h} K_T q_{AB(rot)} \qquad (XII.6.1)$$

where K_T is defined as the ratio of the translational partition function and $q_{AB(rot)}$ is the rotational partition function of the complex AB considered as a hypothetical diatomic molecule AB.[b]

Let us compare this with the preexponential factors Z, predicted by the other theories, writing them in terms of partition functions:[c]

Detailed Theory, Quantum Mechanics:

$$Z_D = \nu K_T \frac{q^{\ddagger}_{AB(rot)}}{q_{A(rot)} q_{B(rot)}} \frac{q^{\ddagger}_{AB(vib)}}{q_{A(vib)} q_{B(vib)}} \qquad (XII.6.2)$$

Transition-state Theory:

$$Z_T = \frac{kT}{h} K_T \frac{q^{\ddagger}_{AB(rot)}}{q_{A(rot)} q_{B(rot)}} \frac{q'^{\ddagger}_{AB(vib)}}{q_{A(vib)} q_{B(vib)}}$$

The detailed theory differs from the transition-state theory in replacing kT/h by ν, a specific molecular constant, and in using $q^{\ddagger}_{AB(vib)}$, the true vibrational partition function of AB‡ rather than $q'^{\ddagger}_{AB(vib)}$, which has one less vibrational degree of freedom. However, since ν is expected to be about 10^{13} sec^{-1} and $kT/h = 6 \times 10^{12}$ sec^{-1} at $300°$K and the extra vibrational term in $q^{\ddagger}_{AB}$ contributes a factor of less than 10, it can be seen

---

[a] In those cases in which higher values have been reported, they have usually been traced to errors in the supposed mechanism or to errors in determining the activation energy.

[b] $\bar{Z}'_{AB}$ thus lends itself to a ready statistical interpretation, namely, the frequency of occurrence of species A and B within a collision diameter of each other, it being assumed that they exert no forces on each other.

[c] For the purposes of the present comparison we are comparing not the true preexponential factors (Table XII.1) but rather the factors of the term $e^{-E°/RT}$, where $E°$ is the activation energy at $0°$K.

TABLE XII.1. COMPARISON OF BIMOLECULAR SPECIFIC RATE CONSTANTS, ACTIVATION ENERGIES, AND PREEXPONENTIAL FACTORS PREDICTED BY VARIOUS THEORIES

Theory	Specific rate const k_B	Exp. activation energy (E^*_{exp}) Eq. (XII.5.2)	Preexponential factor
Arrhenius [modified Eq. (XII.5.3)]	$k_0 T^{C/R} e^{-E/RT}$	$E + CT$	$k_0 T^{C/R}$
Thermodynamic [Eq. (XII.5.1)]	$\nu K^* = \nu e^{-F^*/RT} = \nu K^{*\prime} e^{-E_0^*/RT}$	$H^* - R \dfrac{\partial \ln \nu}{\partial(1/T)} = E^*_0 - R \dfrac{\partial \ln (\nu K^{*\prime})}{\partial(1/T)}$	$\nu K^{*\prime} = \nu e^{S^*/R}$ (not separable)[a]
Collision model (hard sphere)	$\bar{Z}'_{AB} P e^{-E^*/RT}$	$E^* + \dfrac{RT}{2} - R \dfrac{\partial \ln P}{\partial(1/T)}$	$\bar{Z}'_{AB} P$ (not separable)[b]
Collision model[c] (Sutherland diameter)	$Z_{AB} \left(1 + \dfrac{C}{T}\right) P e^{-E^*/RT}$	$E^* + \dfrac{RT}{2} - RT\left(\dfrac{C/T}{1 + C/T}\right) - R \dfrac{\partial \ln P^b}{\partial(1/T)}$	$\bar{Z}'_{AB} P \left(1 + \dfrac{C}{T}\right)^b$ (not separable)
Detailed theory (quantum mechanical)	$\nu e^{-F^*/RT}$	H^*	$\nu e^{S^*/R}$
Transition-state theory	$\dfrac{kT}{h} e^{-F^*/RT}$	$H^{\ddagger} + RT$	$\dfrac{kT}{h} e^{S^*/R}$

[a] If we assume $\partial \ln \nu/\partial(1/T) = 0$, then the product $\nu K^{*\prime}$ can be determined.

[b] In practice it is assumed (arbitrarily) that $\partial \ln P/\partial(1/T) = 0$ to enable an identification of the steric factor P to be made.

[c] Note that the term $RT[(C/T)/(1 + C/T)]$ varies between 0 ($C/T \ll 1$) and RT ($C/T \gg 1$).

[d] Since ν and $\bar{\nu}$ are not known a priori except that they are constants and in the range 10^{13} sec^{-1}, it is not possible to separately determine S^* and the values of ν or $\bar{\nu}$.

that the values of the factors predicted by the two theories will be the same within an order of magnitude, the values predicted by the detailed theory being slightly higher.

In comparing both these theories with the simple collision theory we see that what has happened is that the rather vague[a] steric factor P of the latter has been replaced by the—in principle more meaningful—ratios of partition functions of the species A, B, and AB‡. In the case that A and B are atoms then $P = 1$, and all theories have the same factors.[b] The same is true if we assume that the groups A and B interact so weakly in forming the transition state that their rotational and vibrational modes are unaltered.

The detailed and transition-state theories thus allow us to interpret the steric factor P in molecular terms. Each of the species A and B has three degrees of translational freedom plus rotational and vibrational degrees of freedom, depending on their complexity. In forming the complex AB, there are the same total number of degrees of freedom but they are differently distributed, since the complex has only three degrees of translational freedom and a maximum of three degrees of rotational freedom. Thus at least three translational degrees of freedom and possibly three rotational degrees of freedom have been transformed into vibrational degrees of freedom in forming the complex. This represents a considerable loss of freedom (and therefore entropy), since rotational motion is much more restrictive than free translation and vibrations are more restricted than both.

To obtain a more quantitative idea of these changes, let us consider individual partition functions for single-component, translational, rotational, and vibrational motions. They are given by (Secs. IX.8, 9, 10):

$$f_{\text{tr}} = \left(\frac{2\pi m k T}{h^2}\right)^{\!\frac{1}{2}} = 5.75 \times 10^6 (MT)^{\frac{1}{2}} \text{ cm}^{-1} \qquad \text{(XII.6.3)}$$

$$f_{\text{rot}} = \left(\frac{8\pi^2 I k T}{h^2}\right)^{\!\frac{1}{2}} = 2.03 \times 10^7 (IT)^{\frac{1}{2}}$$

or, by setting $I = \mu\sigma^2$, where μ is in molecular-weight units,

$$f_{\text{rot}} = 2.03 \times 10^7 (\mu T)^{\frac{1}{2}}\sigma \qquad \text{(XII.6.4)}$$

$$f_{\text{vib}} = (1 - e^{-h\nu/kT})^{-1} \qquad \text{(XII.6.5)}$$

By picking typical values at $T = 300°\text{K}$, $M = 50$, $\mu = 30$, $\sigma = 3 \times 10^{-8}$ cm, $\nu = 10^{13}$ sec^{-1} (333 cm^{-1}), we find $f_{\text{tr}} = 7.05 \times 10^8$ cm^{-1}, $f_{\text{rot}} = 43$, and $f_{\text{vib}} = 1.23$. (In general we find that f_{tr} is between 10^8 and 10^9 cm^{-1}, f_{rot} between 10 and 100, and f_{vib} between 1 and 10.)

---

[a] This is a somewhat unfair statement since the collision theory can be given equally satisfactory status by specifying the details of the collision.

[b] That is, A and B have no rotation or vibrations, so we can neglect these terms in Eq. (XII.6.2).

TABLE XII.2. APPROXIMATE STERIC FACTORS FOR BIMOLECULAR REACTIONS FROM EQUILIBRIUM THEORIES

Type of reaction	Type of complex[a]	Detailed theory		Transition-state theory	
		Steric factor $P_{D^*} \times \dfrac{kT}{h\nu}$ Eq. (XII.6.6)	Probable[b] value of $P_{D^*} \times \dfrac{kT}{h\nu}$	Steric factor P_T Eq. (XII.6.7)	Probable value
Atom + atom[c]	Linear	f_{vib}	3	1	1
Atom + linear molecule	Linear	f_{vib}^3/f_{rot}^2	3×10^{-2}	f_{vib}^2/f_{rot}^2	10^{-2}
	Nonlinear	f_{vib}^2/f_{rot}	3×10^{-1}	f_{vib}/f_{rot}	10^{-1}
Atom + polyatomic molecule (nonlinear)	Nonlinear	f_{vib}^3/f_{rot}	3×10^{-2}	f_{vib}^2/f_{rot}^2	10^{-2}
Two linear molecules	Linear	f_{vib}^5/f_{rot}^4	3×10^{-4}	f_{vib}^4/f_{rot}^4	10^{-4}
	Nonlinear	f_{vib}^4/f_{rot}^3	3×10^{-3}	f_{vib}^3/f_{rot}^3	10^{-3}
Linear molecule + nonlinear molecule	Nonlinear	f_{vib}^5/f_{rot}^4	3×10^{-4}	f_{vib}^4/f_{rot}^4	10^{-4}
Two nonlinear molecules	Nonlinear	f_{vib}^6/f_{rot}^5	3×10^{-5}	f_{vib}^5/f_{rot}^5	10^{-5}

[a] A linear molecule (N atoms) has three translational degrees of freedom, two rotational degrees, and $3N-5$ vibrational degrees. A nonlinear molecule (N atoms) has one more rotational degree and one less translational degree. Both have $3N$ degrees.

[b] In calculating the probable values it is assumed that for each rotational degree of freedom, the corresponding partition function may be assigned a value of 30 per degree (f_{rot}), while each vibrational degree (f_{vib}) may be assigned a probable value of about 3. Actual values may of course show considerable deviation [Eqs. (XII.6.4) and (XII.6.5)].

[c] In reactions of free radicals or atoms to form molecules the electronic partition function may not be negligible, since atoms or radicals generally have odd numbers of electrons and hence a multiplicity of electronic states, while the molecules will not.

Table XII.2 lists some steric factors predicted by the detailed and transition-state theories for different structures of A, B, and $AB^{\ddagger}$. For purposes of comparison with the hard sphere collision theory we can define a steric factor for the detailed (P_D) and transition-state theories (P_T) in terms of their preexponential factors Z_D and Z_T as follows:[a]

$$P_D = \frac{Z_D}{\bar{Z}'_{AB}} = \frac{h\nu}{kT} \frac{f^*_{AB(rot)}}{q_{A(rot)}q_{B(rot)}} \frac{q_{AB*(vib)}}{q_{A(vib)}q_{B(vib)}} \qquad \text{(XII.6.6)}$$

$$P_T = \frac{Z_T}{\bar{Z}'_{AB}} = \frac{f^{\ddagger}_{AB(rot)}}{q_{A(rot)}q_{B(rot)}} \frac{q^{\ddagger}_{AB(vib)}}{q_{A(vib)}q_{B(vib)}} \qquad \text{(XII.6.7)}$$

Since $h\nu/kT$ will generally be greater than unity, we see that the steric factors computed by the detailed theory will generally be about an order of magnitude greater than those computed from the transition-state theory and may in fact be somewhat greater than unity. Unfortunately, collision diameters are not well enough known to check such calculations with any reliability.

The probable values quoted in Table XII.2 must be considered as only suggestive, since the ratio f_{vib}/f_{rot} may vary from a value of 1 to 10^{-2} and a mean value of 10^{-1} has been used. However, the qualitative trend is clear, and we should expect that steric factors of bimolecular reactions should show a decrease as the complexity of the reacting species increases. This qualitative rule is well borne out, as will be evident from the data presented in the following sections.

7. A Priori Calculation of Preexponential Factors of Bimolecular Reactions

There have been many attempts made to calculate the preexponential factors of bimolecular reactions from molecular constants based on the considerations of the transition-state theory.[b] Such efforts depend on a number of "educated guesses" as to the vibrational properties and structure of the transition-state complex, an assumption about the "transmission coefficient" for the reaction, and the assumption of the validity of the normal coordinate treatment for computing the thermodynamic properties of polyatomic molecules.

Of these assumptions the last is probably the least cause of error. The

[a] In utilizing the form of $\bar{Z}'_{AB}$ given by Eq. (XII.6.1) the rotational partition function for the complex q_{AB} is, strictly speaking, that for a hypothetical diatomic molecule made from A and B. It thus contains two rotational degrees of freedom. The quantities $f^*_{AB\ (rot)}$ appearing in Eqs. (XII.6.6) and (XII.6.7) are actually ratios of the true rotational partition functions for $AB^{\ddagger}$, $q^{\ddagger}_{AB(rot)}$ to that for the hypothetical diatomic species AB, that is, $f^*_{AB\ (rot)} = h^2 q^{\ddagger}_{AB(rot)}/8\pi^2\mu\sigma^2_{AB}kT$.

[b] S. Glasstone et al., "Theory of Rate Processes," McGraw-Hill Book Company, Inc., New York, 1941, present a summary of all of this work up to 1941.

usual assumption of unity for the transmission coefficient is also not likely to be a serious source of error. However the assignment of structural parameters and vibrational frequencies to the complex is at best pure guesswork that becomes more questionable as the number of atoms in the complex is increased. As we shall see, however, since it is the relatively small contributions of the vibrations and rotations to the partition function which are involved, the errors are probably within a factor of 10, or at most 100. Finally the assumption of a universal frequency factor kT/h for the decomposition of the complex may also be in error by a factor of about 10.

Despite these difficulties, this type of calculation is of some interest in providing a good guess of the order of magnitude to be expected for the preexponential factor, and we shall present some examples of its use.

An additional and much more reliable method for the calculation of preexponential factors presents itself when one of the rate constants for a reversible reaction is known and it is desired to compute the factor for the reverse reaction. Such a calculation depends on a knowledge or calculation of the entropy change in the reaction. An illustration of this type of calculation will also be presented.

8. The Reaction $H + H_2 \rightleftharpoons H_2 + H$ (Exchange)

The rate of exchange of H atoms with H_2 molecules (which can be measured experimentally by the H atom catalyzed exchange of ortho- and para-H) can be pictured as going through the transition state H_3.

$$H + H_2 \rightleftharpoons H_3 \rightarrow H_2 + H$$

If we use the transition-state theory, we can write for the specific rate constant, assuming a transmission coefficient of unity,

$$k = \frac{kT}{h} K^{\ddagger}$$

$$= \frac{kT}{h}\left(\frac{g_{H_3}}{g_H g_{H_2}}\right) \frac{N^{3/2} h^3 (3/2)^{3/2}}{(2\pi kT)^{3/2}} \frac{q_r(H_3)\sigma_{H_2} h^2}{8\pi^2 I_{H_2} kT} \frac{1 - e^{-h\nu/kT}}{\prod\limits_i {}^{\ddagger}(1 - e^{-h\nu_i^{\ddagger}/kT})} e^{-E_0/RT}$$

$$(\text{XII.8.1})$$

in which the g factors are the electronic statistical weights of the three species, $q_r(H_3)$ is the rotational partition function of H_3, $\sigma_{H_2} = 2$ is the symmetry number of H_2, ν is the vibrational frequency of H_2, I_{H_2} is its moment of inertia, and the product $\Pi^{\ddagger}$ is taken over all but one of the vibrational frequencies $\nu_i^{\ddagger}$ of the transition-state complex. N is Avogadro's number, 6.02×10^{23}, and E_0 is the activation energy of the reaction at $0°K$.

If we assume a symmetric linear configuration for the complex H_3, (symmetry number = 2), moment of inertia $I^{\ddagger}$, it will have four vibration frequencies (a doubly degenerate bending frequency and two stretching frequencies). We can then set $q_r(H_3) = 4\pi^2 I^{\ddagger} kT/h^2$. For H_2, $g_{H_2} = 1$, while for H atoms $g = 2$. If we further assume that $g_{H_3} = 2$,[a] then the expression for k can be written as

$$k = \frac{h^2 N^{3/2}}{2(kT)^{1/2}} \left(\frac{3}{\pi}\right)^{3/2} \frac{I^{\ddagger}}{I_{H_2}} \frac{1 - e^{-h\nu/kT}}{\prod_i {}^{\ddagger}(1 - e^{-h\nu_i^{\ddagger}/kT})} e^{-E^{\circ}/RT} \qquad \frac{cc}{\text{molecule-sec}}$$

$$\text{(XII.8.2)}$$

The ratio of the moments of inertia may be explicitly evaluated[b] as

$$\frac{I^{\ddagger}}{I_{H_2}} = \frac{4r_1^2}{r_0^2} \qquad \text{(XII.8.3)}$$

where r_0 is the equilibrium distance in H_2 and r_1 the distance from the central atom in the H_3 complex. The ground-state vibrational frequency of H_2 is known from spectroscopic data, $\nu = 4395.2$ cm^{-1} and from the same sources $r_0 = 0.7416$ Å. For H_3 the distance r_1 has to be guessed as well as the doubly degenerate bending frequency ν_b and the symmetric stretching frequency ν_s.[c]

An empirical quantum-mechanical analysis of H_3 has been made by Eyring and Polanyi[d] and others, and they arrive at $r_1 = 1.354$ Å, $r_2 = 0.753$ Å, $\nu_s = 3650$ cm^{-1}, and $\nu_b = 670$ cm^{-1}.[e] Of the vibrations ν_s and ν_b for H_3 and ν for H_2 only ν_b will contribute appreciably to the partition function at temperatures below 1500°K, so that we can neglect ν and ν_s and write for k

$$k = \frac{h^2 N^{3/2}}{(kT)^{1/2}} \left(\frac{3}{\pi}\right)^{3/2} \frac{r_1^2}{r_0^2} (1 - e^{-h\nu_b''/kT})^{-2} e^{-E_0/RT} \qquad \text{(XII.8.4)}$$

which on substitution of the above numbers and constants becomes

$$k = 2.03 \times 10^{12} T^{-1/2} (1 - e^{-966/T})^{-2} e^{-E_0/RT} \qquad \text{liters/mole-sec} \qquad \text{(XII.8.5)}$$

In order to compare this with experimental values we must know the

---

[a] H_3 will have at least one unpaired electron and so must be at least a doublet state.
[b] For a linear, triatomic molecule ABC the moment of inertia is

$$I_{ABC} = \frac{1}{m_{ABC}} (m_A m_{BC} r_{AB}^2 + 2m_A m_B r_{AB} r_{BC} + m_C m_{AB} r_{BC}^2)$$

where $m_{ABC} = m_A + m_B + m_C$; $m_{AB} = m_A + m_B$; etc.
[c] The asymmetric stretching frequency in this treatment is assumed to correspond to the reaction coordinate which leads to exchange and is thus left out of the sum.
[d] H. Eyring and M. Polanyi, Z. physik. Chem., B12, 279 (1931). For a complete discussion see Glasstone et al., loc. cit.
[e] Their data are based on less accurate values of H_2 than are now known from spectroscopic data. The more accurate data have been given above for H_2. They also use an unsymmetrical transition state for H_3, which is probably not correct.

value of E_0 (i.e., the activation energy at 0°K). In the absence of such information[a] we can only compare the frequency factors.

The experimental activation energy is given by $RT^2(\partial \ln k/\partial T) = E_{\exp}$, and on this basis we obtain from Eq. (XII.8.5)

$$RT^2 \frac{\partial \ln k}{\partial T} = E_0 - \frac{1}{2} RT + \frac{1932R}{e^{966/T} - 1} = E_{\exp} \qquad (XII.8.6)$$

which we can equate to $E_{\exp}$ and solve for E_0. On substituting for E_0 back in Eq. (XII.8.5) we have

$$k = 2.03 \times 10^{12}\ T^{-\frac{1}{2}}\ e^{\theta - \frac{1}{2}}\ e^{-E_{\exp}/RT} \qquad \text{liters/mole-sec} \qquad (XII.8.7)$$

where $\theta = 1932T^{-1}/(e^{966/T} - 1)$ and the term in brackets can now be compared with experimental frequency factors A from equations in the Arrhenius form $Ae^{-E/RT}$.[b] For comparisons with the collisional form of the bimolecular rate equation we must compare $2.03 \times 10^{12}\ T^{-1}e^{\theta-1}$ with the preexponential term A_c.[c] The observed data (Table XII.5) which are in the collisional form give $\log A_c = 9.07$.[d] Calculated values are 9.52 (300°K), 9.43 (700°K), and 9.39 (1000°K), which can be considered as rather striking agreement, the difference being about a factor of 2 in A_c.

The largest uncertainties in this calculation of the preexponential factors are the internuclear distance r_1 in the transition complex, the bending frequency of the complex, and the transmission coefficient. The error in the last is totally uncertain, but a crude guess is that it is not in error by more than a factor of 2. The distance r_1 is not likely to be in error by more than a factor of 2, which could introduce an additional error of 4 in the rate constant. If the frequency of the low bending mode was in

---

[a] There have been many attempts, some completely empirical, some less so, to calculate activation energies, but none of them has been able to give results reliable to better than a *roughly* estimated 10 Kcal/mole, which is useless for quantitative work. Glasstone et al., *loc. cit.*, outline a method which is the same as that used to get the molecular constants of the transition complex given above.

[b] Note that from Eq. (XII.8.6) we obtain $E_{\exp} = E_0 + (\theta - \frac{1}{2})RT$, which predicts that the experimentally observed activation energies should increase with temperature, since θ increases with T. Thus $E_{\exp} - E_0$ is -0.14 Kcal/mole at 300°K, 0.60 Kcal/mole at 700°K, and 1.4 Kcal/mole at 1000°K. The data (Table XII.5) actually do show such a trend, although they are not sufficiently accurate to make quantitative comparisons possible.

[c] If we make comparisons with the collision equation $k_c = A_c T^{\frac{1}{2}} e^{-E_{coll}/RT}$, then $E_0 = E_{coll} + RT - \theta RT$ and $k = [2.03 \times 10^{12}\ T^{-1}e^{\theta-1}]\ T^{\frac{1}{2}} e^{-E_{coll}/RT}$ and we should then compare A_c with this bracketed term.

[d] The data in Table XII.5 are for the conversion of para- to ortho-H. This rate must be multiplied by $\frac{2}{3}$ to obtain the total exchange for the above comparison, since the rate of para → ortho conversion is $3 \times$ the reverse rate and the total rate is the sum of the two.

error by even a factor of 3 (lower), this would introduce an error of 2 at 300°K and an error of 5 at 1000°K.[a]

In summing up we see that the assumption of a much looser complex might have raised k by a factor of 8 to 20, depending on the temperature, while the uncertainty in the transmission coefficient could introduce another factor of 2. Considering the over-all defects of the theory we may thus estimate the uncertainties in the preexponential factors calculated in this way at about an average factor of 10. The agreement obtained here for H_3, a factor of 2, is thus well within these expected uncertainties.

9. The Reaction $H_2 + I_2 \rightarrow 2HI$

On the assumption that the complex is a planar molecule H_2I_2 with an axis of symmetry, the reaction of H_2 with I_2 has been analyzed.[b]

The rate constant k_r can be written, neglecting κ (the transmission coefficient) as

$$k_r = \frac{kT}{h} \frac{h^3}{(2\pi\mu^* kT)^{3/2}}$$

$$\left[\frac{8\pi^2 (8\pi^3 ABC)^{1/2}(kT)^{3/2}/h^3\sigma^*}{(8\pi^2 I_1 kT/h^2\sigma_1)(8\pi^2 I_2 kT/h^2_2\sigma_2)} \right] \frac{\prod\limits^{5}(1 - e^{-h\nu_i^*/kT})^{-1}e^{-E_0/RT}}{(1 - e^{-h\nu_1/kT})^{-1}(1 - e^{-h\nu_2/kT})^{-1}} \quad (XII.9.1)$$

where a subscript 1 refers to H_2, a subscript 2 refers to I_2, * refers to the complex, μ^* is the reduced collision mass $(m_1 m_2/m^*)$, the term in brackets represents the ratio of the rotational partition functions, and the symmetry numbers are respectively $\sigma_1 = \sigma_2 = \sigma^* = 2$. On reduction we can write

$$k_r = \frac{h^3}{2\pi^2 kT2^{1/2}} \left[\frac{(ABC)^{1/2}}{\mu^{*3/2}I_1I_2} \right] \frac{(1 - e^{-h\nu_1/kT})(1 - e^{-h\nu_2/kT})e^{-E_0/RT}}{\prod\limits^{5}(1 - e^{-h\nu_i^*/kT})} \quad (XII.9.2)$$

The bond distances in the planar complex have been estimated at

---

[a] The additional uncertainty arising from the configuration of the complex, i.e., linear or bent, is within the same limits of error as that arising from the uncertainty in distances. The error arising from additional electronic states is perhaps not serious for H_3 but can be much more so for heavier atoms with p electrons. This can in the cases of more complicated species introduce another factor of 5.

[b] A. Wheeler et al., *J. Chem. Phys.*, **4**, 178 (1936). W. Altar and H. Eyring, *ibid.*, **4**, 661 (1936). See also Glasstone et al., *loc. cit.*

H—H $= 0.97$ Å, I—I $= 2.95$ Å, and H—I $= 1.75$ Å, the known distances in H_2 and I_2 being 0.7416 Å and 2.667 Å, respectively. From these the values of $A = 921.5$, $B = 6.9$, $C = 928.4$, $I_1 = 0.456$, and $I_2 = 748.5$ have been calculated (in units of 10^{-40} g-cm^2).[a]

For the six vibrations of the transition complex, an empirical method of evaluation has given the following (divided according to symmetry groups): I, 994 cm^{-1}; II, 86, 1280, and 965* cm^{-1}; III, 1400 and 1730 cm^{-1}; while for H_2, $\nu_1 = 4395.2$ cm^{-1} and for I_2, $\nu_2 = 214.57$ cm^{-1}. The errors made in assigning the frequencies of the complex are not known, and in the partition function the starred frequency of 965 cm^{-1} is omitted, because this is assumed to be the frequency corresponding to the reaction coordinate. On substitution of numbers we find

$$k_r = \frac{530}{T} \times 10^{10} Q_v e^{-E_0/RT} \qquad \text{liters/mole-sec} \qquad \text{(XII.9.3)}$$

where Q_v is the ratio of the vibrational partition functions. At 600°K, $kT/hc = 416$ cm^{-1} and Q_v (600°K) $= 2.60$, the only significant contributions coming from the 86 cm^{-1} vibration of the complex and ν_2 for I_2. At 600°K we thus have k_r (600°K) $= 2.3 \times 10^{10} e^{-E_0/RT}$ liters/mole-sec. By setting $E_0 = E_c + \tfrac{3}{2}RT - \partial \ln Q_v/\partial T$, where E_c is the activation energy determined from the experimental fit to the collision equation, we find k_r (600°K) $= 1.1 \times 10^{10} e^{-E_c/RT}$ liters/mole-sec, while the experimental value at 600°K turns out to be (Table XII.4) $8.1 \times 10^{10} e^{-E_c/RT}$, higher by a factor of about 8, which can be considered reasonable agreement.

10. Reactions of CH_3 radicals and H atoms

A number of calculations have been made for reactions of the type $CH_3 + HR \rightleftharpoons [CH_3 \cdot HR] \rightarrow CH_4 + R$ and also for similar H abstraction reactions by H atoms and halogen atoms.

Bywater and Roberts[b] have made a detailed comparison of such reactions for a series of compounds from which the changes in entropy of activation can be considered in some detail. Table XII.3 lists the calculated standard (300°K, 1 atm) molar entropies of translation and rotation of complexes made of H or CH_3 and RH and the standard entropy changes in the reactions $H + HR \rightleftharpoons H_2R^*$, $CH_3 + HR \rightleftharpoons CH_4R^*$ (neglecting vibration) calculated by these authors.

[a] *Ibid.*

[b] S. Bywater and R. Roberts, *Can. J. Chem.*, **30**, 773 (1952). See also D. R. Herschbach et al., *J. Chem. Phys.*, **25**, 736 (1956), for analysis of more complex reactions. Also H. S. Johnston et al., *ibid.*, in press (1959); H. S. Johnston and D. J. Wilson, *J. Am. Chem. Soc.*, **79**, 29 (1957); and K. S. Pitzer, *ibid.*, **79**, 1804 (1957).

TABLE XII.3. TRANSLATIONAL AND ROTATIONAL ENTROPIES AT 300°K
AND 1 ATM OF SOME TRANSITION COMPLEXES

RH:		H_2	CH_4	C_2H_6	Isobutane
Nature of entropy	Compound	Molar entropy			
S°_{tr}	RH	28.2	34.3	36.1	38.1
	H··HR	29.3	34.4	36.2	38.2
	CH_3··HR	34.4	36.2	37.4	38.8
S°_{rot}	RH	2.1	10.0	16.4	22.4
	H··HR	8.6	15.1	19.9	22.5
	CH_3··HR	15.1	18.0	23.7	24.3
ΔS°_{tr}	$H + HR \rightarrow H_2R*$	−25.1	−25.9	−26.0	−26.0
	$CH_3 + HR \rightarrow CH_4R*$	−27.9	−32.2	−33.0	−33.4
ΔS°_{rot}	$H + HR \rightarrow H_2R*$	5.1	2.3	−0.1	0.0
	$CH_3 + HR \rightarrow CH_4R*$	+3.1	−4.1	−7.0	−8.8
$\Delta S^\circ_{int\ rot}$	$H + HR \rightarrow H_2R*$	0.0	0.0	+0.1	0.3
	$CH_3 + HR \rightarrow CH_4R*$	0.0	3.0	4.0	4.0
$\Delta S^\circ_{tr+rot+int\ rot}$	H_2R*	−20.0	−23.6	−26.0	−25.7
	CH_4R*	−24.8	−33.3	−36.0	−38.2

Note: Symmetry contributions are not included in S°_{rot}. For isobutane it is assumed that tertiary H atom is attacked. Units are cal/mole-°K.

In these calculations the electronic contributions have been assumed to cancel, and vibrational assignments and internal rotation barriers were calculated according to Pitzer.[a] Other assumptions have been discussed by the authors. The vibrational contributions (not shown in table) nearly cancel each other near 300°K and can be neglected below 500°K in the calculation of ΔS° for the reactions, so that the ΔS° shown in the last two lines of Table XII.3, aside from symmetry changes, can be equated to the standard entropy of activation.

It can be seen that the largest contributions to the activation entropy, as expected, come from the loss in translational degrees of freedom.[b] This increases with increasing disparity in the masses of the reacting species, as can be seen by the change for $CH_3 + HR$ in going from $RH = H_2$ to RH = isobutane. A similar effect is seen in the ΔS°_{rot}, which is positive (3.1) for $CH_3 + H_2$ and negative for CH_3 + isobutane (−8.8). As the molecule becomes very large compared to the attacking radical or atom, the entropy of activation reaches a limiting large value, so that we would expect the values for larger hydrocarbons to be only slightly greater than the values recorded for isobutane. The estimated error is probably not greater than 2 to 3 cal/mole-°K for the entropy changes, which would

[a] K. S. Pitzer, J. Chem. Phys., **5**, 469, 473 (1937).

[b] To convert to standard state of 1 mole/cc at 300°K, subtract $R \ln 24,600 = 20.2$ cal/mol-°K from S°_{tr} and add this to ΔS°_{tr}.

represent a factor of about 3 in the preexponential factors for these reactions, from these sources above. From the known experimental data on the abstraction reactions, Bywater and Roberts found agreement for the calculated and observed preexponential of within a factor of 10, which can be considered quite good.

11. Estimations of Preexponential Factors from Calculated Equilibrium Constants and Experimental Rate Constants

One of the chief difficulties with the preceding methods of calculating preexponential factors, aside from the uncertainties of the transition-state theory, lies in the necessarily speculative character of the parameters used to describe the vibrations and rotations of the hypothetical transition-state complex. These latter speculations are not needed in the calculation of equilibrium constants, and with the growing body of data from spectroscopic and other sources on the molecular properties of atoms and free radicals it is now possible to make good estimates of the molar entropies of these species, and from the latter, fairly reliable estimates of the entropy changes in reactions involving atoms and free radicals. If rate data and activation energies are available for one of a pair of reversible reactions and the entropy change for the reaction can be estimated, then the preexponential factor for the reverse reaction can be calculated. Thus for the pair of opposing bimolecular reactions

$$A + B \underset{k_2}{\overset{k_1}{\rightleftharpoons}} C + D \qquad (XII.11.1)$$

the equilibrium constant $K = k_1/k_2$, and if we represent the rate constants by either the Arrhenius equation $Ae^{-E/RT}$ or the collision equation $A'T^{\frac{1}{2}}e^{-E/RT}$, the ratio of the preexponential factors is in either case equal to $e^{\Delta S/R}$, where ΔS is the entropy change in the reaction. If A_2 is known and ΔS is estimated, then

$$\ln A_1 = \ln A_2 + \frac{\Delta S}{R} \qquad (XII.11.2)$$

A. F. Trotman-Dickenson[a] has employed this method to calculate the preexponential factors of a number of free radical reactions. For checks on the calculated entropies of methyl radicals he compares his values with the entropies of similar molecules. Thus the standard entropies (25°C, 1 atm) of CH_4, 44.5 cal/mol-°K, and NH_3, 45.9 cal/mol-°K, are compared to a calculated value for CH_3 of 45.5 (neglecting electronic degeneracy),[b]

[a] A. F. Trotman-Dickenson, ibid., **21**, 211 (1953). For more refined methods of calculation see references in footnote to page 286. Also see Appendix D.

[b] For a planar CH_3 radical (symmetry number, $\sigma = 6$) $S° = 44.0$; for a nonplanar radical ($\sigma = 3$) with high inversion frequency $S° = 44.1$; and for one with very low inversion frequency $S° = 45.5$. This can be compared with $S° = 46.1$ calculated by Bywater and Roberts, loc. cit.

with an estimated uncertainty of not more than ± 1 cal/mole-°K. In similar fashion $S°(C_2H_5)$ is calculated as 56.3 (again neglecting electron degeneracy), to be compared with $S°(C_2H_6) = 54.9$, $S°(C_2H_4) = 52.5$, and $S°(CH_3NH_2) = 57.7$.

For metathetical reactions of the type

$$CH_3 + HR \rightleftharpoons CH_4 + R$$

or more generally for alkyl radicals R,

$$R + HR' \rightleftharpoons RH + R'$$

Trotman-Dickenson has shown that $\Delta S° \cong 0$, so that the preexponential factors should be the same for the forward and reverse reactions. In a quite similar fashion, the reactions of H atoms with hydrocarbons of the type

$$H + HR \underset{k_2}{\overset{k_1}{\rightleftharpoons}} H_2 + R$$

have $\Delta S° \cong 5$ to 7 cal/mole-°K, so that $A_1/A_2 \geqslant 10$ is in good agreement with the few experimental data available for comparison.

In decomposition reactions of the type

$$R\text{—}R' \underset{k_2}{\overset{k_1}{\rightleftharpoons}} R + R'$$

the rates of the reverse reaction have been measured directly for a number of systems. Thus for $C_2H_6 \rightleftharpoons 2CH_3$, Trotman-Dickenson's data give $\Delta S° = 35.1 + 2S°_{elec}(CH_3)$. This last figure is at least $2R \ln 2$ and may be higher, giving $\Delta S° \geqslant 37.8$ cal/mole-°K for a standard rate of ideal gas at 1 mole/cc at 25°C this becomes $\Delta S_3 \geqslant 17.8$. From the data of Kistiakowsky and Roberts,[a] the recombination rate constant for methyl radicals at 25°C is 3.1×10^{13} cc/mole-sec, so that if we assume no activation energy for the recombination, the minimum value of the preexponential factor for the decomposition of C_2H_6 is 2.4×10^{17} sec^{-1}, an extraordinarily high frequency factor for a unimolecular decomposition. This value is undoubtedly correct to within a factor of 3 and is probably better than that. Trotman-Dickenson has calculated a similarly high value for the frequency factor for $n\text{-}C_4H_{10} \rightarrow 2C_2H_5$, and Carrington and Davidson (loc. cit.) have found a high value for $N_2O_4 \rightarrow 2NO_2$. For the decomposition of propyl radicals to give $C_2H_4 + CH_3$, $n\text{-}C_3H_9 \rightarrow C_3H_6 + CH_3$, $CH_2OCH_3 \rightarrow CH_2O + CH_3$, and $CH_2OH \rightarrow CH_2O + H$, Trotman-Dickenson has estimated frequency factors (from experimental data) of the order of 10^9 to 10^{10} sec^{-1}.[b]

[a] G. B. Kistiakowsky and E. K. Roberts, J. Chem. Phys., **21**, 1637 (1953).

[b] These reactions are probably not first-order. S. W. Benson and D. V. S. Jain, J. Chem. Phys., **31**, 1008 (1959), have shown that CH_2OCH_3 decomposition is second order at 500°C.

12. Bimolecular Exchange Reactions between Molecules

Reliable data on gas-phase bimolecular exchange reactions between molecules are rather rare. Most of these data are presented in Table XII.4, where k is given in terms of the simple collision equation, $\log k = -E/2.3RT + 0.5 \log T + A'$. The temperature-independent term A', which is equal to the preexponential factor divided by $T^{1/2}$, is shown in column 3, and a collisional steric factor P is calculated in the last column on the arbitrary basis of a uniform collision diameter for all reactions of 3.5 Å.[a]

In general it can be seen that the steric factors are less than unity and tend toward smaller values for the more complicated species. With the exception of the HI and NOCl decompositions they are of the orders of magnitude predicted by the equilibrium theories (Table XII.2). The value for NOCl is not subject to literal interpretation because there is such a marked change in the energy of activation with temperature. Thus at 200°C the experimental activation energy is 22.7 Kcal/mole, which would lead to a steric factor of about 0.01. This rather large temperature dependence of the activation energy for this reaction has been interpreted as arising from the complex path of the reaction.[b]

The HI reactions are worthy of some interest, since the forward and reverse reactions are both bimolecular. If we write the rate constants for the forward and reverse reactions in the form of the equilibrium theories, we have:

$$H_2 + I_2 \underset{2}{\overset{1}{\rightleftharpoons}} 2HI$$

$$k_1 = \nu_1 e^{S_1^*/R} e^{-H^*/RT} \qquad k_2 = \nu_2 e^{S_2^*/R} e^{-H_2^*/RT}$$

while
$$K_{eq} = \frac{k_1}{k_2} = e^{\Delta S/R} e^{-\Delta H/RT}$$

so that for consistency we must have $\nu_1 = \nu_2$, $S_1^* - S_2^* = \Delta S$, and $H_1^* - H_2^* = \Delta H$. But since ΔS, the entropy change in the reaction, is known to be positive, this requires that $S_1^* > S_2^*$, that is, the entropy of activation for the forward reaction shall exceed that for the reverse reaction. But this in turn requires that the preexponential factor of the

[a] We could of course use viscosity diameters for these calculations, but they are of rather dubious significance for such purposes and in any case would not change any of the above values by more than a factor of 2.

[b] Thus we could have competing reactions (1) NOCl + NOCl → (NO)₂ + Cl₂ followed by (NO)₂ ⇌ 2NO (rapid) or (2) NOCl + NOCl ⇌ NOCl₂ + NO followed by the rapid decomposition NOCl₂ → NO + Cl₂. If both of these reactions occurred at once, we might expect a composite rate constant with different activation energies and frequency factors which could account for the above behavior. Other explanations offered have been some heterogeneity of the reaction and the formation of a very loose critical complex. Chain reactions are also known to be important at the higher temperatures.

TABLE XII.4. BIMOLECULAR EXCHANGE REACTIONS BETWEEN MOLECULES

Reactants and products	Preexponential factor log $(A/T^{1/2})$, liters/mole-sec	Activation energy Kcal/mole	log $(\overline{Z}'_{AB}/T^{1/2})$,[a] liters/mole-sec	Steric[a] factor P
[b]$H_2 + I_2 \rightarrow 2HI$	9.518 (9.78)[b']	38.9 (40.7)	10.38	0.14 (0.33)
[c]$D_2 + I_2 \rightarrow 2DI$	9.79	39.5	10.23	0.36
[b,d]$HI + HI \rightarrow H_2 + I_2$	9.31 (10.001)[c']	42.5 (45.9)[c']	9.33	0.10
[c,d]$DI + DI \rightarrow D_2 + I_2$	9.22 (9.914)[c']	43.0 (46.4)[c']	9.33	0.78
[e]$NO_2 + NO_2 \rightarrow 2NO + O_2$	8.419	26.6	9.85	0.038
[f]$NOCl + NOCl \rightarrow 2NO + Cl_2$	9.514	25.8	9.47	1.1
[g]$NO + ClNO_2 \rightarrow NOCl + NO_2$	7.73	6.6 (± 0.3)	9.76	0.01
[g,h]$NO + O_3 \rightarrow NO_2 + O_2$	7.80	2.3 (± 0.3)	9.90	0.008
[g,i]$NO_2 + O_3 \rightarrow NO_3 + O_2$	8.60	6.7 (± 0.6)	9.84	0.06
[j]$NO + Cl_2 \rightarrow NOCl + Cl$	8.00	19.6	9.87	0.014
[k]$CO + Cl_2 \rightarrow COCl + Cl$	8.5	51.3	9.87	0.04

[a] Calculated from $\overline{Z}'_{AB}/T^{1/2} = 2.74 \times 10^{25}\sigma^2\mu^{-1/2}$ or (assuming $\sigma = 3.5$ Å) $= 3.36 \times 10^{10}\mu^{-1/2}$ liters/mole-sec. This must be divided by 2 when A = B (identical reactants). We define $P = A/\overline{Z}'_{AB}$. Use of viscosity diameters would not change any P by more than a factor of 2.

[b] M. Bodenstein, Z. physik. Chem., **13**, 56 (1894); **22**, 1 (1897); **29**, 295 (1898). L. Kassel, Proc. Nat'l Acad. Sci. U.S., **16**, 358 (1930), has shown that a better fit is obtained by using a term 10.5 log T in place of the 0.5 log T of the collision equation. Similarly, a 16 log T term gives a better fit for the HI decomposition. This would be theoretically absurd.

[b'] Values in parentheses are from J. H. Sullivan, J. Chem. Phys. **30**, 1292, 1577 (1959), and have been corrected for the atomic chain. They are to be preferred.

[c] A. H. Taylor, Jr., and R. H. Crist, J. Am. Chem. Soc., **63**, 1377 (1941). J. C. L. Blagg and G. M. Murphy, J. Chem. Phys., **4**, 631 (1936).

[c'] The values given in parentheses are those of Bright and Hagerty, and they do not agree with the adjacent values, which are those of Bodenstein. Kassel, loc. cit., has shown that the simple collisional form of the rate equation will not fit the experimental data adequately over any extended temperature range but that the activation energy and preexponential factor show a marked temperature dependence. On the other hand both Taylor and Crist and Bright and Hagerty have shown that the Bodenstein values for both k and K_{eq} are probably in error at the higher temperatures. In particular they disagree with spectroscopic determinations of K_{eq}. S. W. Benson and R. Srinivasan, J. Chem. Phys., **23**, 200 (1955), have shown that the high-temperature data are complicated by an atomic chain. The best value is that of Sullivan, 8.97.

[d] N. F. H. Bright and R. P. Hagerty, Trans. Faraday Soc., **43**, 697 (1947).

[e] M. Bodenstein and Ramstetter, Z. physik. Chem., **100**, 106 (1922). L. Kassel, "Kinetics of Homogeneous Gas Reactions," Reinhold Publishing Corp., New York, 1932, has shown that an equation which fits the data for the equilibrium constant and reverse rate over a broader range is log $k = -5.480/T + 0.5$ log $T +$ 7.604 (liters/mole sec), which gives $E_{act} = 25.07$ Kcal/mole and reduces P to 7×10^{-4}. This latter also agrees better with the thermal data. Note: Kassel has erroneously divided rate constants by factor of 2.

[f] T. Welinsky and H. A. Taylor, J. Chem. Phys., **6**, 466 (1938). The activation energy quoted is for 250°C. It shows considerable variation with temperature, and the steric factor quoted above is thus subject to broad temperature changes. There is still a considerable question whether this reaction proceeds by a composite path. The same is true of the NO₂ decomposition. P. G. Ashmore and J. Chanmugam, Trans. Faraday Soc., **49**, 254, 265 (1953), have shown that the chain decomposition of NOCl is important at 300°C.

[g] E. C. Freiling et al., J. Chem. Phys., **20**, 327 (1952). The data given by the authors for this reaction, and also [h] and [i], have been recalculated to the form of the collision equation.

[h] H. S. Johnston and H. J. Crosby, ibid., **19**, 799 (1951).

[i] H. S. Johnston and D. M. Yost, ibid., **17**, 386 (1949). The ultimate products are N₂O₅, presumed to arise from the fast association NO₂ + NO₃ → N₂O₅. The reaction is extremely exothermic and as measured is almost adiabatic. This introduces some uncertainty into the activation energy, since NO₂, which is measured photometrically, is in mobile equilibrium with N₂O₄.

[j] P. G. Ashmore and J. Chanmugam, Trans. Faraday Soc., **49**, 270 (1953).

[k] Calculated from known rate of COCl + Cl → CO + Cl₂ and thermodynamic data for the equilibrium constant. See Table XII.5, footnote h.

forward reaction k_1 exceed that of the reverse reaction. If we compare the preexponential factors for these reactions (column 3, Table XII.4), we see that such is indeed the case.

13. Bimolecular Exchange Reactions That Involve Free Radicals or Atoms

The wealth of data on bimolecular reactions that involve free radicals or atoms is more testimony to the growing awareness of the importance of these intermediates in kinetic systems and the frequency of their occurrence than to the great accuracy of the results. With few exceptions such data must be inferred from postulated mechanisms concerning the steps involved in a complex system, and the validity of any given mechanism

TABLE XII.5. SOME BIMOLECULAR EXCHANGE REACTIONS THAT INVOLVE ATOMS

Reactants	Products	Preexponential factor $\log (A/T^{1/2})$, liters/mole-sec	Activation energy, Kcal/mole	Steric factor $P = A/Z'$ (σ, Å)
a,bH + p-H$_2$	H + o-H$_2$	8.94	5.5	0.035 (2.74)
a,bH + D$_2$	HD + H	9.00	6.5	0.044 (2.74)
aD + H$_2$	HD + H	8.99 (9.33)	4.9 (5.4)	0.047 (2.74)
aD + o-D$_2$	D + p-D$_2$	8.71	6.0	0.029 (2.74)
cH + Br$_2$	HBr + Br	9.83	0.9	0.15 (4.0)
cH + HBr	H$_2$ + Br	8.912	0.9	0.033 (3.0)
iH + O$_2$	OH + O	6.96	17.0	2.7×10^{-4} (3.5)
lH + HCl	H$_2$ + Cl	9.40	4.0	0.11 (3.0)
fH + C$_2$H$_6$	H$_2$ + C$_2$H$_5$	8.01	6.4	3.1×10^{-3} (3.5)
eH + CH$_4$	H$_2$ + CH$_3$	8.55	8.5	0.019 (3.0)
gD + D$_2$S	D$_2$ + DS	9.99	5.0	0.32 (4.0)
lCl + H$_2$	HCl + H	9.69	5.0	0.27 (3.0)
hCl + COCl$_2$	Cl$_2$ + COCl	10.5 (9.94)	20.5 (23.0)	6 (1.0–4.0)
hCl + COCl	Cl$_2$ + CO	10.1 (9.13)	0.5 (1.94)	2 (0.2–4.0)
kCl + NOCl	Cl$_2$ + NO	8.59	0.75	0.042 (4.0)
cBr + H$_2$	HBr + H	9.308	17.6	0.12 (3.0)
jBr + CH$_4$	HBr + CH$_3$	9.170	17.8	0.20 (3.1)
jBr + CH$_3$Br	HBr + CH$_2$Br	9.130	15.6	0.23 (3.8)
dBr + CHCl$_3$	HBr + CCl$_3$	7.82	8.86	0.0010 (5.0)
dBr + CCl$_3$Br	Br$_2$ + CCl$_3$	9.37	9.76	0.035 (5.0)
cBr + HBr	Br$_2$ + H	9.13	41.8	0.20 (4.0)
iO + OH	O$_2$ + H	5.65	0	3.7×10^{-5} (3.5)
mO + O$_3$	O$_2$ + O$_2$	8.98	5.6	0.011 (4.0)
nI + H$_2$	HI + H	9.75	33.4	0.35 (3.5)

a K. H. Geib and P. Harteck, *Z. physik. Chem.*, Bodenstein Festband, 849 (1931). A. Farkas and L. Farkas, *Proc. Roy. Soc. (London),* **A152**, 124, 152 (1935). The activation energies are not known to better than ± 1.5 Kcal, and so the differences above may not be real. The mean diameter for σ used above (2.74 Å) is larger than that used by

can rest only on the consistency with which it fits into a growing body of interrelated reactions and the perseverance and ingenuity of the kineticist in devising still more stringent tests for it. A consequence of this situation is that the qualitative data on free radical reactions are necessarily less precise than data for simple kinetic systems and the calculated acti-

---

Farkas and Farkas (2.3 Å). There is also a discrepancy in the data in that the activation energy appears to increase with increasing temperature. See also L. Farkas and E. Wagner, *Trans. Faraday Soc.*, **32**, 708 (1936), and R. Klein et al., *J. Chem. Phys.*, **30**, 58 (1959).

[b] M. van Meersche, *Bull. soc. chim. Belges*, **60**, 99 (1951), and G. Boati et al., *Nuovo cimenti*, **10**, 993 (1953).

[c] M. Bodenstein and W. Muller, *Z. Elektrochem.*, **30**, 416 (1924). W. Jost, *Z. physik. Chem.*, **B3**, 95 (1929). M. Bodenstein and G. Jung, *ibid.*, **121**, 127 (1926). $k(Br + H_2)$ is observed, as is the ratio $k(H + Br_2)/k(H + HBr)$. The constants $k(H + HBr)$, $k(H + Br_2)$, and $k(Br + HBr)$ have been computed from these values and the statistically calculated equilibrium constants.

[d] J. H. Sullivan and N. Davidson, *J. Chem. Phys.*, **19**, 143 (1951); **17**, 176 (1949). A. A. Miller and J. E. Willard, *ibid.*, **17**, 168 (1949).

[e] Estimated from reaction rate of $CH_3 + H_2 \rightarrow CH_4 + H$ [T. G. Majury and E. W. R. Steacie, *Discussions Faraday Soc.*, **14**, 45 (1953)] and the equilibrium constant for this system [A. F. Trotman-Dickenson, *J. Chem. Phys.*, **21**, 211 (1953)]. Data on the first rate constant are known only relative to the constant for CH_3 radical recombination. For this latter reaction, data of Gomer and Kistiakowsky are used. The activation energies are uncertain by ±1.5 Kcal. Discharge-tube work of LeRoy and Berlie gives 6.6 ± 1 Kcal for $H + CH_4$ with steric factor of 1×10^{-4} [*Discussions Faraday Soc.*, **14**, 121 (1953)].

[f] M. R. Berlie and D. J. LeRoy, *Discussions Faraday Soc.*, **14**, 50 (1953), used Wood's tube to generate H atoms. B. deB. Darwent and R. Roberts, *ibid.*, 55, find $E = 9.0$ Kcal/mole for $D + C_2H_6$ with steric factor of 0.6. This seems too high for both.

[g] Darwent and Roberts, *loc. cit.*, obtained values from photolysis of D_2S in presence of H_2. These results seem unreasonably high, and it is possible that the photolysis mechanism is not correct. Thus the quantum yield can rise above unity, and the fate of the DS radicals is uncertain. Also, at the wavelengths used, the D atoms have excess energies of >40 Kcal/mole, so that there should be very important "hot radical" effects in the system.

[h] W. G. Burns and F. S. Dainton, *Trans. Faraday Soc.*, **48**, 39 (1952). The values in parentheses are from M. Bodenstein et al., *Z. physik. Chem.*, **40B**, 121 (1938); and W. Brenschede, *ibid.*, 237; **41B**, 254 (1938). The latter are not as reliable, although both sets of steric factors seem anomalously high.

[i] B. Lewis and G. von Elbe, "Combustions, Flames and Explosions," p. 59, Academic Press, Inc., New York, 1951. The value for $H + O_2$ is estimated from chain reactions of $H_2 + O_2$. The value for $O + OH$ is estimated from the first reaction and the equilibrium constant. It is likely that the activation energies are too low. See Sec. XIV.7.

[j] G. B. Kistiakowsky and E. R. van Artsdalen, *J. Chem. Phys.*, **10**, 305 (1942); **12**, 469 (1944).

[k] W. G. Burns and F. S. Dainton, *Trans. Faraday Soc.*, **48**, 39 (1952).

[l] P. G. Ashmore and J. Chanmugam, *ibid.*, **49**, 254 (1953). The reaction $H + HCl$ is calculated from $Cl + H_2$ and the equilibrium constant.

[m] S. W. Benson and A. E. Axworthy, Jr., *J. Chem. Phys.*, **26**, 1718 (1957).

[n] J. H. Sullivan, *ibid.*, **30**, 1292, 1577 (1959).

vation energies and frequency factors cannot be treated with the reliability given those from the simple systems.

Table XII.5 is a compilation of some of the data on reactions that involve the attack of atoms on molecules or radicals. In most of the cases the data have been taken directly from the original authors and transformed into the form of the collision equation, when not already in that form. In some cases the rate constant for a reverse process has been calculated from that of the forward reaction and thermodynamic data for the equilibrium. Where that has been done the footnotes will so indicate. The steric factors shown in the last columns have been calculated on the basis of an arbitrary collision diameter σ which is indicated in parentheses. It is quite evident that the steric factors so calculated cannot be literally interpreted, since variations within a factor of 3 could equally well have been obtained by using different but equally justifiable cross sections.

It will be noted that on the whole the steric factors are close to 0.1 (within a factor of 3). For a few reactions they seem to be much smaller; generally for the reactions of atoms with more complex molecules and for a few other reactions they are of the order of unity. It is probable that the very low values such as 2.7×10^{-4} for $H + O_2 \rightarrow HO + O$ and 3.7×10^{-5} for the reverse case arise from large experimental errors. This is also very likely the case for the reactions $H + CH_4 \rightarrow H_2 + CH_3$ and $H + C_2H_6 \rightarrow H_2 + C_2H_5$. One of the principal difficulties with the data in such systems is that the effects of surface reactions on the observed rates are at present difficult, if not impossible, to calculate. This is particularly true for the high values found for $Cl + COCl_2 \rightarrow Cl_2 + COCl$ and $Cl + COCl \rightarrow Cl_2 + CO$; for Noyes[a] has recently shown that wall reactions can be important in that system.

We have omitted from Table XII.5 the large body of the literature on the reactions of the alkali metal atoms with halogens and alkali halides[b] obtained by the "diffusion flame technique" of Polanyi. The reason is that the data from these reactions cannot be used to obtain absolute rate constants or activation energies directly, but instead, assumptions must be introduced concerning the kinetic cross reactions and diffusion constants. Reed and Rabinowitch[c] have given an excellent analysis of some of the more troublesome features of the technique.[d] It is of interest, however,

[a] R. M. Noyes, *J. Am. Chem. Soc.*, **73**, 3039 (1951). R. M. Noyes and L. Fowler, *ibid.*, **73**, 3043 (1951).

[b] M. Polanyi, "Atomic Reactions," Williams & Norgate, Ltd., London, 1932. H. V. Hartel and M. Polanyi, *Z. physik. Chem.*, **B11**, 97 (1930). R. J. Cvetanovic and D. J. LeRoy, *Can. J. Chem.*, **29**, 597 (1951); *J. Chem. Phys.*, **20**, 1016 (1952).

[c] J. F. Reed and B. S. Rabinowitch, *J. Phys. Chem.*, **59**, 261 (1955).

[d] The method consists in allowing alkali metal vapor ($A_2 + M$) to diffuse (with carrier gas, usually) through a nozzle into a halogen or halide. The magnitude of the reaction zone may be observed, since the residual alkali metal atoms will fluoresce strongly when illuminated with resonance radiation, and from that the rate of reaction may be estimated.

to note that all of the data from such systems are consistent with the assumption of steric factors of the order of magnitude of 0.1 to 1.

Also omitted from Table XII.5 are most of the experiments on atoms produced by discharge techniques. Again it is the difficulty in these systems of obtaining reliable, quantitative data which is responsible for the omission.

Within the probable reliability of the data we see that the frequency factors are in good agreement with the theoretical results to be expected for atom-molecule reactions (Table XII.2) and thus provide additional support for the theory.

Tables XII.6 and XII.7 summarize most of the available data on the exchange reactions that involve radicals. Owing to the very large literature on the reactions of methyl radicals, the latter have been separated in Table XII.6, and the remaining reactions have been included in Table XII.7.

The rate data for the reactions of methyl radicals have all been obtained by the method of comparative reactions, and the absolute values presented are based on the adoption of the values of Gomer and Kistiakowsky (Table XII.8) for the rates of recombination of methyl radicals.[a] It can be seen that the bulk of the steric factors for methyl radical exchanges are of the order of magnitude of 10^{-3}. That is about what we might have expected for the reaction of two polyatomic molecules (Table XII.2) of the small mass and moment of inertia of the methyl radical. The individual differences in steric factors show no obvious relations to structure or simple molecular properties. There are some extreme variations such as the values of 3.5×10^{-5} for $CH_3 + CH_2F_2$, as against 1.36×10^{-2} for $CH_3 + CH_3Br$. This 400-fold difference, while not completely outside the range of experimental errors,[b] can be only partially accounted for by the greater loss in the rotational contribution to the partition function of the complex arising from CH_2F_2 as compared with CH_3Br.[c]

Absolute calculations of the frequency factors of bimolecular reactions of the type $H + RH \rightarrow [H \cdot \cdot HR] \rightarrow H_2 + R$ and $CH_3 + RH \rightarrow [CH_3 \cdot \cdot HR] \rightarrow CH_4 + R$ on the basis of available thermodynamic data and an assumed geometry for the transition complex were discussed in Sec. XII.10.

[a] Thus the raw data are given in the form $k_A/k_B^{1/2} = R$, where R is an experimental quantity, k_B is the specific rate constant for the recombination of methyl radicals, and k_A is the specific rate constant for the reaction in question. It then follows that $E_A = E_{obs} + \frac{1}{2}E_B$, where we have assumed $E_B = 0$ in making our calculations of activation energies.

[b] An error of 4 Kcal in the relative activation energies would almost compensate for this difference at 400°K. Secondary reactions could also account for such differences.

[c] That is, the loss in rotational entropy on forming the complex $CH_3 \cdot CH_2F_2$ should be greater than that in forming $CH_3 \cdot CH_3Br$.

TABLE XII.6. BIMOLECULAR EXCHANGE REACTIONS OF METHYL RADICALS[a]

Reactants	Products	Preexponential factor $\log(A/T^{1/2})$, liters/mole-sec	Activation energy, Kcal/mole	Steric factor $\times 10^3$ $P = A/Z'$ $(\sigma, \text{Å})$
$bCH_3 + H_2$	$CH_4 + H$	7.25	10.0	0.95 (3.0)
$bCH_3 + D_2$	$CH_3D + D$	7.61	11.8	2.9 (3.0)
$bCD_3 + H_2$	$CD_3H + H$	7.81	11.1	3.4 (3.0)
$bCD_3 + D_2$	$CH_3D + D$	7.25	10.9	1.3 (3.0)
$bCH_3 + CH_3COCH_3$	$CH_4 + CH_2COCH_3$	7.36	9.7	1.8 (4.0)
$bCD_3 + CD_3COCD_3$	$CD_4 + CD_2COCD_3$	7.57	11.6	3.1 (4.0)
$cCH_3 + C_2H_6$	$CH_4 + C_2H_5$	6.99	10.4	0.72 (4.0)
$cCH_3 + C_2H_4$	$CH_4 + C_2H_3$	6.97	10.0	0.66 (4.0)
$cCH_3 + n\text{-}C_4H_{10}$	$CH_4 + C_4H_9$	6.73	8.3	0.33 (4.5)
$cCH_3 + n\text{-}C_5H_{12}$	$CH_4 + C_5H_{11}$	6.74	8.1	0.28 (5.0)
$cCH_3 + n\text{-}C_6H_{14}$	$CH_4 + C_6H_{13}$	6.83	8.1	0.29 (5.5)
$cCH_3 + \text{neopentane}$	$CH_4 + \text{neopentyl}$	7.02	10.0	0.54 (5.0)
$cCH_3 + \text{diisobutyl}$	$CH_4 + \text{diisobutyl rad}$	6.98	9.5	0.36 (6.0)
$cCH_3 + \text{isobutane}$	$CH_4 + C_4H_9$	6.69	7.6	0.30 (4.5)
$cCH_3 + \text{diisopropyl}$	$CH_4 + \text{diisopropyl rad}$	6.82	7.4 ± 0.5	0.28 (5.5)
$cCH_3 + \text{2-3-4-trimethylpentane}$	$CH_4 + \text{radical}$	7.02	7.9	0.39 (6.0)
$cCH_3 + \text{propylene}$	$CH_4 + \text{radical}$	6.48	7.7	0.22 (4.0)
$cCH_3 + \text{2-butene}$	$CH_4 + \text{radical}$	6.88	7.7	0.48 (4.5)
$cCH_3 + \text{isobutene}$	$CH_4 + \text{radical}$	6.62	7.3	0.26 (4.5)
$cCH_3 + \text{1-butene}$	$CH_4 + \text{radical}$	6.89	7.6	0.48 (4.5)
$cCH_3 + \text{1-pentene}$	$CH_4 + \text{radical}$	6.90	7.6	0.39 (5.0)
$cCH_3 + \text{3-methyl-1-butene}$	$CH_4 + \text{radical}$	6.98	7.4	0.48 (5.0)
$cCH_3 + \text{2,3-dimethyl-1-butene}$	$CH_4 + \text{radical}$	7.33	7.8	0.90 (5.5)
$cCH_3 + \text{cyclopropane}$	$CH_4 + \text{radical}$	6.69	10.3	0.36 (4.0)
$cCH_3 + \text{cyclobutane}$	$CH_4 + \text{radical}$	7.21	9.3	1.0 (4.5)

Reaction				
$^cCH_3$ + cyclopentane	CH₄ + radical	7.09	8.3	0.78	(4.5)
$^cCH_3$ + cyclohexane	CH₄ + radical	7.03	8.3	0.54	(5.0)
$^cCH_3$ + benzene	CH₄ + radical	6.12	9.2	0.07	(5.0)
$^cCH_3$ + toluene	CH₄ + radical	6.83	8.3	0.30	(5.5)
$^cCH_3$ + CH₃OH	CH₄ + radical	6.40	8.2	0.24	(3.5)
$^cCH_3$ + C₂H₅OH	CH₄ + radical	7.18	8.7	1.2	(4.0)
$^cCH_3$ + i-C₃H₇OH	CH₄ + radical	6.71	7.3	0.33	(4.5)
$^cCH_3$ + CH₃NH₂	CH₄ + radical	7.02	8.4	1.0	(3.5)
$^cCH_3$ + (CH₃)₂NH	CH₄ + radical	7.01	7.2	0.78	(4.0)
$^cCH_3$ + (CH₃)₃N	CH₄ + radical	7.56	8.8	2.3	(4.5)
$^cCH_3$ + NH₃	CH₄ + NH₂	7.60	10.0	4.5	(3.0)
$^cCH_3$ + (CH₃)₂O	CH₄ + CH₂OCH₃	7.21	9.5	1.2	(4.0)
$^cCH_3$ + isopropyl ether	CH₄ + radical	6.79	7.3	0.27	(5.5)
$^cCH_3$ + 1-butyne	CH₄ + radical	7.61	9.1	3.1	(4.0)
$^cCH_3$ + 2-butyne	CH₄ + radical	7.36	8.6	1.8	(4.0)
$^dCH_3$ + CH₃F	CH₄ + CH₂F	6.96	8.7	0.88	(3.5)
$^dCH_3$ + CH₂F₂	CH₄ + CHF₂	5.66	6.2	0.035	(4.0)
$^dCH_3$ + CH₃Cl	CH₄ + CH₂Cl	7.40	9.4	2.5	(3.5)
$^dCH_3$ + CH₂Cl₂	CH₄ + CHCl₂	6.66	7.2	0.38	(4.0)
$^dCH_3$ + CHCl₃	CH₄ + CCl₃	6.10	5.8	0.083	(4.5)
$^dCH_3$ + CH₃Br	CH₄ + CH₂Br	8.22	10.1	13.6	(4.0)
$^dCH_3$ + CH₂Br₂	CH₄ + CHBr₂	7.26	8.7	1.2	(4.5)
$^eCH_3$ + CH₃CHO	CH₄ + CH₃CO	7.62	7.5	3.7	(3.7)
$^fCH_3$ + C₂H₄NH	CH₄ + C₂H₄N	5.84	4.4	0.056	(4.0)
$^gCH_3$ + Hg(CH₃)₂	CH₄ + CH₂HgCH₃	6.05	9.0	0.08	(4.5)
		(7.31)	(10.8)	1.5	(4.5)
$^hCH_3$ + ethylene oxide	CH₄ + radical	6.76	9.6	0.44	(4.0)
		(6.83)	(10.8)	(0.51)	(4.0)

TABLE XII.6. BIMOLECULAR EXCHANGE REACTIONS OF METHYL RADICALS[a] (Continued)

[a] All values for rate constants k are calculated from the relation $k = Rk_B^{1/2}$, where R is an observed quantity and k_B is the rate constant for the recombination of methyl radicals. The value used here is from Gomer and Kistiakowsky (see Table XII.5, footnote e). The E listed is in reality $E - \frac{1}{2}E_B$, and we have taken $E_B = 0$. For the reactions of CD_3 we have assumed that $k_B' = (\mu_B/\mu_B')^{1/2}, k_B = 0.91k_B$, or $k_B'^{1/2} = 0.96k^{1/2}$.

[b] E. Whittle and E. W. R. Steacie, $J.$ $Chem.$ $Phys.$, $\mathbf{21}$, 993 (1953). These values differ from those of T. G. Majury and E. W. R. Steacie, $Discussions$ $Faraday$ $Soc.$, $\mathbf{14}$, 45 (1953), and E. Whittle, $ibid.$, 120. They are probably not better than ± 10 per cent for the observed rate constants and ± 0.3 Kcal/mole for E_{act}.

[c] A. F. Trotman-Dickenson and E. W. R. Steacie, $J.$ $Chem.$ $Phys.$, $\mathbf{19}$, 329 (1951).

[d] F. A. Raal and E. W. R. Steacie, $ibid.$, $\mathbf{20}$, 578 (1952).

[e] R. K. Brinton and D. H. Volman, $ibid.$, $\mathbf{20}$, 1053 (1952).

[f] $Ibid.$, $\mathbf{20}$, 25 (1952).

[g] R. Gomer and W. A. Noyes, Jr., $J.$ $Am.$ $Chem.$ $Soc.$, $\mathbf{71}$, 3393 (1949). The values in parentheses have been recalculated from the original data by A. F. Trotman-Dickenson and E. W. R. Steacie, $J.$ $Phys.$ $\&$ $Colloid$ $Chem.$, $\mathbf{55}$, 908 (1951). The reaction mechanism is not yet clear, and so the data are in some dispute.

[h] M. K. Phibbs and B. deB. Darwent, $Can.$ $J.$ $Res.$, $\mathbf{28}$, 395 (1950). Values in parentheses are from R. Gomer and W. A. Noyes, Jr., $J.$ $Am.$ $Chem.$ $Soc.$, $\mathbf{72}$, 101 (1950).

TABLE XII.7. BIMOLECULAR EXCHANGE REACTIONS THAT INVOLVE RADICALS

Reactants and products	Preexponential factor $\log (A/T^{1/2})$ liters/mole-sec	Activation energy, Kcal/mole	Steric factor $\times 10^3$ $P = A/Z'$ (σ, Å)	
$^aC_2H_5 + D_2 \rightarrow C_2H_5D + D$...	7.2	13.3	1	(3.4)
$^bCOCl + Cl_2 \rightarrow COCl_2 + Cl$..	8.0	2.7	13.	(4.0)
$^bCOCl + NOCl \rightarrow$				
$COCl_2 + NO$ or				
$(CO + Cl_2 + NO)$........	9.22	0.8	210	(4.0)
$^cClO + ClO \rightarrow Cl_2 + O_2$......	6.36 (±0.5)	-0.35 ($\pm.65$)	0.5	(4.0)
$^{a,d}C_2H_5 + C_2H_5COC_2H_5 \rightarrow$				
$C_2H_6 + C_2H_4COC_2H_5$......	6.6	7.4	0.3	(5.0)
$^eC_2H_5 + C_2H_5HgC_2H_5 \rightarrow$				
$C_2H_6 + C_2H_4 + Hg + C_2H_5$	6.5	7.4	0.25	(5.0)
$^eC_2H_5 + C_2H_5 \rightarrow C_2H_6 + C_2H_4$	8.24	0.8	30	(4.0)

[a] M. H. J. Wijnen and E. W. R. Steacie, *J. Chem. Phys.*, **20**, 205 (1952). The data given here for ethyl radical reactions are at best tentative, since they came from the interpretation of the photolyses of $(C_2H_5)_2CO$ or $(C_2H_5)_2Hg$, which are not completely elucidated. R. Klein et al., *J. Chem. Phys.*, **30**, 58 (1959), find $E = 7.8$ Kcal/mole in the reaction $D + CH_4 \rightarrow DH + CH_3$.

[b] W. G. Burns and F. S. Dainton, *Trans. Faraday Soc.*, **48**, 39 (1952).

[c] G. Porter and F. J. Wright, *Discussions Faraday Soc.*, **14**, 23 (1953). The uncertainty in the frequency factor is due to the uncertainty of a factor of 3 in the absolute absorption coefficient of ClO radicals. Such a reaction being four-centered is unusual in having no activation energy, and the authors propose a precurser species, the dimer $(ClO)_2$.

[d] L. M. Dorfman and Z. D. Sheldon, *J. Chem. Phys.*, **17**, 511 (1949). K. O. Kutschke et al., *J. Am. Chem. Soc.*, **74**, 714 (1952).

[e] K. J. Ivin and E. W. R. Steacie, *Proc. Roy. Soc. (London)*, **A208**, 25 (1951). K. J. Ivin and M. H. J. Wijnen, *J. Phys. Chem.*, **56**, 967 (1952). The activation energy of 0.8 Kcal/mole is the difference between this reaction and that for the recombination to form butane, which is assigned a value of zero. The steric factor seems unusually high.

14. Bimolecular Association Reactions

The studies of bimolecular association reactions are of special interest because they may be expected to show, at sufficiently low concentrations, the same type of dependence of rate on total concentration as is displayed by unimolecular reactions. Indeed the simplest of such systems, the recombination of atoms at normal gas concentrations, never follow simple second-order kinetics but are rather at the extreme end of the concentration-dependent rate law and their kinetics is found experimentally to follow third-order kinetics. From the discussion of the pressure dependence of unimolecular decompositions (Table XI.2) we would expect the region of total concentration dependence to shift to lower and lower concentrations as the number of atoms in the product molecule increases. This is in qualitative agreement with observation. The association of butadiene molecules shows no deviation from second-order kinetics down to 10 mm Hg pressure (at 200°C), while the rate of recombination of CH_3 radicals already shows some third-order behavior at 10 mm Hg[a] (at 200°C) and the NO_2 recombination is predominantly a third-order reaction at atmospheric pressure (Table XII.8). Because of this it has been necessary to make a somewhat artificial distinction between those bimolecular associations which have been found to follow second-order kinetics over the range of usual laboratory gas concentrations and those which follow third-order kinetics. The data on the former have been summarized in Table XII.8.

Perhaps one of the most striking features of the bimolecular association reactions shown in Table XII.8 is the extreme range of variability found in their steric factors, from about 0.5 for CH_3 radical recombination to about 10^{-6} for the dimerization of cyclopentadienes and other Diels-Alder type additions. In terms of the simple equilibrium theories (Table XII.2) we would expect that the steric factors for the reactions of large molecules would be small. What is perhaps surprising, then, is that the steric factors for the reactions of NO_2, CH_3, C_2F_4, and of butadiene (with cyanogen) should be as large as they are ($> 10^{-3}$).[b] These large values must be attributed to quantum effects (i.e., they cannot be explained on the basis of classical oscillators) which make for an anomalously large entropy for either the critical complex or the transition-state complex.

These effects have already been noted in our discussion in Sec. XII.11 on the calculation of preexponential factors for C_2H_6 decomposition. Thus the large steric factor for the CH_3 recombination can be accepted on the basis of a very loosely bound transition complex $(CH_3 \cdots CH_3)^*$, in which the loss in rotational degrees of freedom of the CH_3 groups has been largely offset by very low frequency vibrations or almost free internal rotations of the transition complex.[c] The same reasonable assumption explains the results for NO_2. These considerations are strengthened when we consider the reverse reactions, the decomposition of N_2O_4 or C_2H_6. The N_2O_4 reaction rate shows a quasi-unimolecular concentration dependence even at 1 atm. For the unimolecular decomposition of N_2O_4 (6 atoms, 12 internal oscillators) we would not expect on the basis of the detailed model to find

[a] G. B. Kistiakowsky and E. K. Roberts, *J. Chem. Phys.*, **21**, 1637 (1953).

[b] A word of caution is in order here. As can be seen from Table XII.8, the accuracy of the data does not usually warrant a specification of the steric factors to better than a factor of 10. The difficulty arises in part from the chemical complexity of the systems (i.e., side reactions) and possibly also from the strongly exothermic nature of the reactions, which tends to cause temperature errors and give a spurious temperature dependence to the activation energy (Sec. XIV.1). This latter has been noted in the case of C_2F_4 (footnote *h*, Table XII.8). See S. W. Benson, *J. Chem. Phys.*, **22**, 46 (1954).

[c] R. A. Marcus, *J. Chem. Phys.*, **20**, 364 (1952), was able to predict the values for recombination by using a detailed (RRK) theory for C_2H_6 with 17 oscillators. He also predicted the range of pressure dependence for the rate.

a quasi-unimolecular character until we were at a characteristic pressure (0°C) of about 1 to 10 mm Hg (Table XI.2). This latter figure, however, was calculated for a mean frequency factor $\bar{\nu} = 10^{13}$ sec^{-1}. However, for quantized oscillators the frequency factor to use is not $\bar{\nu}$ but $\bar{\nu}e^{S*/R}$, where $S*$ is the entropy of activation. Carrington and Davidson (footnote dd, Table XI.4) have calculated that extreme weakening of the doubly degenerate rocking modes of the two NO$_2$ groups in the N$_2$O$_4^*$ critical complex could give this species an excess entropy over N$_2$O$_4$ of 10 to 14 cal/mole-°K, making $e^{S*/R}$ about 150 to 1000 and the frequency factor correspondingly higher[a] than 10^{13} sec^{-1}. An additional factor to consider in this case is the *relatively* low value of the activation energy for the reaction, namely 14.5 Kcal, so that $E*/RT$ at 300°K is about 24. This tends to boost the characteristic pressure by an additional factor of about 10 (Table XI.1).

Similar considerations would apply to the recombination of methyl radicals and the decomposition of C$_2$H$_6$. Here the large value of $S^{\pm}$, the excess entropy of the transition state, tends to boost the characteristic pressure. To account for the values observed, we would have to propose almost completely free rotation of the CH$_3$ groups in the transition state, which is not too unreasonable in view of the large bond energy, 85 Kcal/mole.[b] Kistiakowsky and Roberts found that the rate of recombination of CH$_3$ radicals at 165°C increases about threefold, going from a total pressure of acetone of 1 mm Hg to 10 mm Hg. They further found, rather surprisingly, that acetone was about 40 times more effective than CO$_2$ in deactivation of the complex. These findings explain the rather low values found for the recombination by Ingold and Lossing, who used a mass spectrometer and worked at total pressures (He carrier) in the range of 5 mm Hg. Also explained is the negative temperature coefficient for the recombination found by the latter authors, since the pressure-sensitive regions of the recombination will have a negative temperature coefficient if there is no activation energy for the recombination.

In the light of these findings it may be expected that the reactions of CH$_3$ with NO and O$_2$ have probably also been studied in the third-order region, so that the frequency factors and steric factors in Table XII.8 are probably too low.[c]

The low frequency factors for the Diels-Alder additions reported in Table XII.8 are to be expected in view of the considerable, uncompensated loss in rotational entropy occurring when the critical complex is formed.

---

[a] This is not at all unreasonable in view of the relatively large rotational moments of the NO$_2$ groups.

[b] Note that at the temperatures at which the CH$_3$ radical recombination has been studied, 300 to 500°K, $E*/RT \cong 120$, which tends to lower the characteristic pressure and somewhat counter the entropy effect.

[c] J. G. Calvert, private communication, has evidence for 3 body effects in these reactions.

TABLE XII.8. BIMOLECULAR ASSOCIATION REACTIONS

Reaction	Preexponential factor log $(A/T^{1/2})$, liters/mole-sec	Activation energy, Kcal/mole	Steric factor $A/\bar{Z}'$ (σ, Å)
[a] $CH_3 + CH_3 \rightarrow C_2H_6$	9.32 (9.24)[a']	0 ± 0.7	0.5 (3.5)
[b] $CH_3 + NO \rightarrow CH_2NO$ (or CH_3NOH)	6.2	0	1.5 × 10⁻⁴ (3.5)
[c] $CH_3 + O_2 \rightarrow CH_3O_2$(?)	6.2	0	1.5 × 10⁻⁴ (3.5)
[e] $C_2H_5 + C_2H_5 \rightarrow C_4H_{10}$	8.88 (9.3)[e']	0 (<0.65)	0.13 (4.0)
[g] $NO_2 + NO_2 \rightarrow N_2O_4$	7.78	0	0.010 (4.6)
[h] $2C_2F_4 \rightarrow$ cyclo-C_4F_8	6.58	25.6	7.8 × 10⁻³ (5.0)
[h,i] $2C_2F_3Cl \rightarrow$ cyclo-$C_4F_6Cl_2$	5.91	25.6	1.8 × 10⁻³ (5.0)
[h,i] $2C_2F_3Cl \rightarrow$ cyclo-$C_4F_6Cl_2$	6.30	25.6	4.0 × 10⁻⁴ (5.0)
[h] $C_2F_4 + C_2F_3Cl \rightarrow$ cyclo-C_4F_7Cl	5.80	26.7	4.0 × 10⁻⁵ (5.5)
[j] $C_2H_4 +$ butadiene $\rightarrow$ cyclohexene	5.36	23.1	2.9 × 10⁻⁵ (5.5)
[k] butadiene $\rightarrow$ vinylcyclohexene	(6.46)	(26.0)	(3.6 × 10⁻⁴) (5.5)
[l] Butadiene + $(CN)_2 \rightarrow$ 2-Cyano-3,6-dihydropyridine	7.56	31	2.3 × 10⁻³ (5.5)
[m] 2[2,3-dimethyl butadiene] $\rightarrow$ dimer	5.52	24.7	4.1 × 10⁻⁵ (6.0)
[n] Butadiene + acrolein $\rightarrow$ 1,2,3,6-Tetrahydrobenzaldehyde	4.65	19.3	3.0 × 10⁻⁶ (5.5)
[n] Butadiene + crotonaldehyde $\rightarrow$ 2 methyl-1,2,3,6-tetrahydro-Benzaldehyde	4.49	21.6	1.9 × 10⁻⁶ (5.5)
[m] 2[1,3-pentadiene] $\rightarrow$ dimer	5.92	25.4	1.0 × 10⁻⁷ (6.0)
[m,o] 2-cyclopentadiene $\rightarrow$ dimer	3.39	14.5	3.0 × 10⁻⁶ (6.0)
	(4.47)	(16.3)	(4.5 × 10⁻⁶) (6.0)
[n] Isoprene + acrolein $\rightarrow$ adduct	4.42	18.1	1.5 × 10⁻⁶ (6.0)
[n] Cyclopentadiene + acrolein $\rightarrow$ adduct	4.48	14.8	1.7 × 10⁻⁶ (6.0)
[p] Isobutene + HCl $\rightarrow$ t-C_4H_9Cl	4.6	24.1	3 × 10⁻⁶ (5.0)
[q] Isobutene + HBr $\rightarrow$ t-C_4H_9Br	6.4	22.1	(2 × 10⁻⁴) (5.0)
	(5.7)	(23.8)	(4 × 10⁻⁵) (5.0)
[d] $2CF_3 \rightarrow C_2F_6$	9.05	0	0.15 (4.0)
$CH_3 + C_2H_5 \rightarrow C_3H_8$	9.6	0	0.3 (4.0)

[a] R. Gomer and G. B. Kistiakowsky, *J. Chem. Phys.*, **19**, 85 (1951). Data obtained from sector analysis of photolyses of $Hg(CH_3)_2$ and acetone. K. U. Ingold and F. P. Lossing, *ibid.*, **21**, 1133 (1953), find rate threefold smaller and $E = -2.2 \pm 0.5$ Kcal/mole by mass-spectrographic technique, which must be considered less reliable. R. E. Dodd, *Trans. Faraday Soc.*, **47**, 56 (1951), finds still higher values by sector analysis of photolysis of CH_3CHO. These seem unreasonably high, and the interpretation of the results has been challenged [G. O. Pritchard et al., *J. Chem. Phys.*, **21**, 748 (1953). R. E. Dodd, *ibid.*].

[a'] (See footnote a.) G. B. Kistiakowsky and E. K. Roberts, *ibid.*, **21**, 1637 (1953), give a rate that is 20 per cent less, which is probably more reliable.

[b] R. W. Durham and E. W. R. Steacie, *ibid.*, **20**, 582 (1952), by using Te mirror method in a flow system. Same system gives steric factor for CH_3 recombination of 0.01 (compare above), which is considered a lower limit. The technique cannot be considered very reliable. See also D. M. Miller and E. W. R. Steacie, *ibid.*, **19**, 73 (1951), who found even slower rates by using different techniques.

[c] F. B. Marcotte and W. A. Noyes, Jr., *Discussions Faraday Soc.*, **10**, 236 (1951). Values from photolysis of acetone $+ O_2$ mixtures are tentative pending further study of the oxidative mechanism.

[d] P. B. Ascough, *J. Chem. Phys.*, **24**, 944 (1956).

[e] See Table XII.7, footnote e.

[e'] Values of A. Shepp and K. O. Kutschke, *ibid.*, **26**, 1020 (1957).

[f] C. A. Heller, *ibid.*, **28**, 1255 (1958).

[g] T. Carrington and N. Davidson, *ibid.*, **57**, 418 (1953). The reaction is actually termolecular, and the data given here are for the limiting rates at infinite concentrations of foreign gas calculated from the dissociation rate constant and the known equilibrium constant.

[h] J. R. Lacher et al., *J. Am. Chem. Soc.*, **74**, 1693, (1952). Values for C_2F_4 are in excellent agreement with B. Atkinson and A. B. Trenwith, *J. Chem. Phys.*, **20**, 754 (1952), who also measured the equilibrium. Former authors found considerable self-heating of the reaction due to exothermicity ($\Delta H = -50$ Kcal/mole) and inhibition by C_4F_8. The latter is undoubtedly due to the cooling effect of the compound. The rate constants and activation energies may be in error because of this thermal effect.

[i] The product is the 1,2-dichloro ring adduct with 83 per cent cis and 17 per cent trans isomers.

[j] D. Rowley and H. Steiner, *Discussions Faraday Soc.*, **10**, 198 (1951). Data are from flow system and show scatter of 20 per cent. Calculated heat of reaction at 800°K is 36.3 Kcal, and there is discrepancy between it and the one calculated from difference in activation energies of forward and reverse reactions, which gives 30 Kcal.

[k] G. B. Kistiakowsky and W. W. Ransom, *J. Chem. Phys.*, **7**, 725 (1939), from measurements in a static system. The values in parentheses are of Rowley and Steiner, *loc. cit.*, from measurements at higher temperatures in a flow system. There seems to be a pronounced increase in activation energy with temperature which may be due to the exothermicity of the reaction or side reactions, as proposed by Kistiakowsky and Ransom. They obtain $1.3 \times 10^{11} \exp(-38,000/RT)$ liters/mole-sec for the reaction butadiene $+$ vinylcyclohexene $\to$ trimer. These are crude values and difficult to reconcile with any of the other Diels-Alder condensations.

TABLE XII.8. BIMOLECULAR ASSOCIATION REACTIONS (Continued)

[l] P. J. Hawkins and G. J. Janz, J. Am. Chem. Soc., **74**, 1790 (1952); J. Chem. Soc., 1479, 1485 (1949). Data obtained in flow system are not accurate. There are appreciable side reactions, and H_2 is formed from secondary reaction to give H_2 + 2-cyanopyridine.

[m] J. B. Harkness et al., J. Chem. Phys., **5**, 682 (1937).

[n] G. B. Kistiakowsky and J. B. Lacher, J. Am. Chem. Soc., **58**, 123 (1936). Results are admittedly crude; the rate data show scatter of from 10 to 30 per cent. Same is true of all these Diels-Alder additions.

[o] Results in parentheses are from G. A. Benford et al., Nature, **139**, 669 (1937), which seem more reasonable than those of Harkness et al., loc. cit. See also A. Wasserman, Trans. Faraday Soc., **34**, 128 (1938).

[p] Calculated from the rate of decomposition (Table XI.4) of t-butyl chloride and the equilibrium constant measured by K. E. Howlett, J. Chem. Soc., 1409 (1951). Log K_{eq} (moles/liter) = $-16,900/4.575T$ + 6.33.

[q] Calculated from the rate of decomposition (Table XI.4) of t-butyl bromide and the equilibrium constant measured by G. B. Kistiakowsky and C. H. Stauffer, J. Am. Chem. Soc., **59**, 165 (1937). Log K_{eq} (mole/liter) = $-18,000/4.575T$ + 6.16. Values in parentheses are from the rate constants of K and S. The values for the frequency factor seem discordant with the value of the chloride.

Some attempts have been made both by Rowley and Steiner and Kistia-kowsky and Ransom[a] to calculate this entropy from postulated geometrics and frequencies of the transition complex. In view of the conflicting data reported it is doubtful if such calculations are at present useful.[b] Further experimental work on these systems is needed.

15. Termolecular Reactions: Mechanism

We can define a termolecular reaction as one that requires the partici-pation of three individual particles in a single kinetic process. Thus we might think of the reaction of NO with Cl_2 as proceeding through the simul-taneous interaction of two NO molecules with Cl_2 to form the transient reaction complex $(NO)_2Cl_2$, which then decomposes in a single step into two molecules of NOCl. Or in more general terms, for the reaction of $A + B + C \rightarrow$ products we may write

$$A + B + C \underset{k_2}{\overset{k_1}{\rightleftharpoons}} [ABC]^* \overset{k_3}{\rightarrow} \text{products} \qquad (XII.15.1)$$

An alternative formulation which is in many cases more useful is that termolecular reactions are in reality the result of two successive bimolecular processes involving the participation of a transient complex formed from any pair of the reacting species. Thus for the reaction of $A + B + C$ we can write

$$A + B \underset{k_2'}{\overset{k_1'}{\rightleftharpoons}} AB$$

$$AB + C \overset{k_3'}{\rightarrow} \text{products} \qquad (XII.15.2)$$

or alternatively $A + C \rightleftharpoons AC$ followed by $AC + B \rightarrow$ products, etc., or simultaneous combinations of all such schemes.

If we apply the stationary-state hypothesis to Eq. (XII.15.1), we have for the rate of formation of products

$$\frac{d(\text{products})}{dt} = \frac{k_1(A)(B)(C)}{k_2 + k_3} \qquad (XII.15.3)$$

where the reaction is experimentally third-order and the only restriction on the rate constants is that $(k_2 + k_3)/k_1$ be much greater than the initial concentration of the smallest of A_0, B_0, or C_0.[c]

In similar fashion we find for the alternative scheme [Eq. (XII.15.2)]:

[a] Footnotes j and k, Table XII.8.

[b] It is interesting to note that cyclic complex with a few low frequencies is capable of describing the apparent increase in activation energy with temperature.

[c] See section on stationary-state hypothesis.

$$\frac{d(\text{products})}{dt} = \frac{k_1 k_{3'}(A)(B)(C)}{k_{2'} + k_{3'}(C)} \tag{XII.15.4}$$

$$\xrightarrow{k_{2'} \gg k_{3'}(C_0)} \frac{k_1 k_{3'}}{k_{2'}} (A)(B)(C) \tag{XII.15.5}$$

which is also kinetically third-order if $k_{2'} \gg k_{3'}(C_0)$.[a] This last condition would be generally fulfilled in a reacting gas unless the complex AB was virtually a stable species,[b] and so the two schemes would appear indistinguishable from the point of view of over-all kinetics. The only evidence which could distinguish between them would have to derive from a study of the actual reaction path, namely, evidence which could demonstrate the presence or absence of the intermediates as kinetically important species.[c] It is clear that the same problems will also arise for all higher-order reactions.

There are two distinct types of processes which are kinetically of third order. One of these is the association of atoms or simple molecular species in which a third molecule is required to remove the excess energy of the association. In this case the third body is a true catalyst and acts essentially to transport energy and momentum. The second case is one which involves the chemical reaction of three species.

By using the hard sphere collision model we can compute a collision frequency for three molecules A, B, and C by first computing the stationary concentration of the three possible binary complexes AB, BC, and CA. If we call τ_{AB}, τ_{BC}, and τ_{CA} the mean lifetime of these binary complexes,[d] their stationary concentrations are approximately given by

$$\begin{aligned}
(AB) &= Z'_{AB}\tau_{AB}(A)(B) \\
(BC) &= Z'_{BC}\tau_{BC}(B)(C) \\
(CA) &= Z'_{AC}\tau_{AC}(A)(C)
\end{aligned} \tag{XII.15.6}$$

where Z'_{AB} is the collision frequency of A and B, etc. The over-all rate is then the sum of the rates at which these binary complexes make reactive collisions with the appropriate species

[a] If we included all possible intermediates BC and CA, then Eq. (XII.15.5) would consist of a sum of three kinetically similar terms.

[b] $k_{2'}$ is a unimolecular rate constant with little or no activation energy if AB is a loose complex, while $k_{3'}$ is a bimolecular rate constant which may contain some energy of activation. Thus the frequency factor $A_{2'} \geq 10^{13}$ sec^{-1}, while $A_{3'}(C_0) \leq 10^9$ sec^{-1} if $(C_0) = 1 \times 10^{-2}$ mole/liter [$\sim$0.2 atm(STP)].

[c] In the absence of such evidence the interpretation of the experimental rate constants is necessarily ambiguous. There is an unjustified tendency on the part of many workers to brush aside such questions as academic.

[d] Thus $\tau_{AB}^{-1} = k_{AB}$ = specific rate of decomposition of the complex AB into A and B, etc. See Sec. VII.8F.

$$\frac{d(\text{products})}{dt} = k_{3'}(AB)(C) + k_{3''}(BC)(A) + k_{3'''}(CA)(B)$$

$$= P_C Z'_{AB \cdot C}(AB)(C)e^{-E_C/RT} + P_A Z'_{BC \cdot A}(BC)(A)e^{-E_A/RT}$$
$$+ P_B Z'_{CA \cdot B}(CA)(B)e^{-E_B/RT} \qquad \text{(XII.15.7)}$$

in which the P are steric factors, the Z collision frequencies, and the E activation energies. On substituting from Eq. (XII.15.6) and factoring out the concentrations, we can write for the experimental specific rate constant

$$k_{\text{exp}} = P_C Z'_{AB \cdot C} Z'_{AB} \tau_{AB} e^{-E_C/RT} + P_A Z'_{BC \cdot A} Z'_{BC} \tau_{BC} e^{-E_A/RT}$$
$$+ P_B Z'_{CA \cdot B} Z'_{CA} \tau_{CA} e^{-E_B/RT} \qquad \text{(XII.15.8)}$$

In this last expression, the preexponential factors are all similar in containing a product of two collision frequencies, a steric factor, and a mean lifetime. The latter may be approximated in a number of ways, each of which yields about 10^{-13} sec.[a] Since bimolecular collision frequencies are about 10^{11} liters/mole-sec, this would make $Z'^2 \tau$ about 10^9 liters^2/mole^2-sec. The collision theory thus leads to a frequency of termolecular collisions of about 10^9 liters^2/mole^2-sec, which as we shall see from Table XII.9, is about the order of magnitude observed for the fastest reactions.

From the point of view of the transition-state theory we can write for the specific rate constant of a termolecular process

$$k_{\text{exp}} = \frac{kT}{h} \frac{Q'^{\ddagger}_{ABC}}{Q_A Q_B Q_C} e^{-E_0^*/RT} \qquad \text{(XII.15.9)}$$

while the detailed theory gives[b]

$$k_{\text{exp}} = \bar{\nu} \frac{Q^*_{ABC}}{Q_A Q_B Q_C} e^{-E_0^*/RT} \qquad \text{(XII.15.10)}$$

If we use the crude method of estimating partition functions given in Sec. XII.5, we find an expected value of about 10^{-3} (liter/mole)^2 for the ratio of the partition functions,[c] and by combining this with a mean value of 10^{13} sec^{-1} for either $\bar{\nu}$ or kT/h, we see that the preexponential factors are about 10^{10} liters/mole^2-sec for either of these latter theories, in good agreement with the collision theory and the actual data.

A simple relation between the steric factor of the collision model and the partition functions used in the transition-state theory may be made by employing the relation derived in Eq. (XII.3.14) between frequency

---

[a] Since τ_{AB} is the reciprocal of a unimolecular rate constant, it should be about 10^{-13} sec if we neglect activation energy. Any activation energy term can be absorbed in the experimental term. For alternate derivations see Kassel, op. cit., chap. 4.

[b] See Sec. XII.4 for definitions. We have set the transmission coefficient $\kappa = 1$ in the above.

[c] This would assume a very loosely bound complex. For more tightly bound complexes the values would be lower, corresponding to a negative entropy of activation.

factors and partition functions. By using this relation we can write for the first term in Eq. (XII.15.8)

$$P_C Z'_{AB \cdot C} Z'_{AB} \tau_{AB} = P_C \left(\frac{kT}{h}\right)^2 \left[\frac{q_{tr(ABC)}}{q_{tr(A)} q_{tr(B)} q_{tr(C)}}\right] q'_{rot(AB)} q'_{rot(AB \cdot C)} \tau_{AB} K'_{elec}$$
(XII.15.11)

in which $q'_{rot(AB)}$ refers to the rotational partition function of the complex AB treated as a diatomic molecule and similarly for $q'_{rot(AB \cdot C)}$. But $\tau_{AB} = 1/\nu_{AB}$, where ν_{AB} is the frequency of vibration of group A against group B, so that $kT/h\nu_{AB} \cong q'_v(A \cdot B)$ if $\nu_{AB} \ll kT/h$. If we substitute this last term in Eq. (XII.15.11) and equate to the preexponential term in Eq. (XII.15.9), we find, on solving for the steric factor P_C

$$P_C = K^{\ddagger}_{elec} K^{\ddagger}_{rot} K^{\ddagger}_{vib} [q'_{rot(A \cdot B)} q'_{rot(AB \cdot C)} q'_{vib(A \cdot B)}]^{-1} \quad (XII.15.12)$$

where the K represent ratios of individual partition functions (for the degrees of freedom indicated by the subscripts) of complex ABC to A, B, and C. When two of the species are atoms, $K^{\ddagger}_{rot} \cong q'_{rot(A \cdot B)} q'_{rot(AB \cdot C)}$, while $K^{\ddagger}_{vib}/q'_{vib(A \cdot B)} \cong 1$, so that $P_C \cong K^{\ddagger}_{elec} \leqslant 1$ and the collision theories and equilibrium theories become identical. For A, B, and C complex we will expect considerable contribution at least from the rotational partition functions and will thus expect P_C to be much less than unity.

16. Some Third-order Reactions

There are relatively few systems in either category of termolecular reactions which have been studied in any great detail, and the data for these are presented in Table XII.9.[a] Only three wholly chemical processes are included, and all involve the reaction of NO. The data for the reaction of NO with H_2[b] which has been studied above 1000°K, appear to be third-order, but the mechanism is probably not simple.

Of the three wholly chemical termolecular reactions listed in Table XII.9, only the reaction of NO with O_2 has been studied over an extended range of conditions. All three reactions, however, have preexponential factors of about the same order of magnitude, corresponding to a steric

---

[a] I have omitted many systems for which qualitative data exist, such as the reaction of Na + O_2 [C. E. H. Bawn and A. G. Evans, *Trans. Faraday Soc.*, **33**, 1580 (1937)] to form NaO_2 and the addition of BF_3 to amines [D. Garvin and G. B. Kistiakowsky, *J. Chem. Phys.*, **20**, 105 (1952)]. Also omitted is the growing body of work on the recombination of ions in gas discharges [A. V. Phelps and J. P. Molnar, *Phys. Rev.*, **89**, 1202 (1953), and M. A. Biondi, *ibid.*, **90**, 730 (1953)].

[b] C. N. Hinshelwood and T. E. Green, *J. Chem. Soc.*, **129**, 720 (1926); J. W. Mitchell and C. N. Hinshelwood, *ibid.*, 378 (1936). For mechanism see P. G. Ashmore et al., *Trans. Faraday Soc.*, **52**, 830, 835 (1956).

factor of about 10^{-7}.[a] Gershinowitz and Eyring[b] have shown that the transition-state theory can be made to fit such a frequency factor with reasonable choices for the molecular parameters of the transition complex. On the other hand, either of the two mechanisms that involve the intermediate complexes $(NO)_2$ or $NO \cdot O_2$ will yield a satisfactory explanation of the rate data for the $NO + O_2$ reaction, while $NO \cdot Cl_2$ or $NO \cdot Br_2$ can do likewise for the reaction of NO with Cl_2 and Br_2, respectively. In these cases the observed rate constant may be written as [Eq. (XII.15.5)] $k_{obs} = Kk_3$, in which K is the equilibrium constant for the formation of the intermediate bimolecular complex and k_3 is the bimolecular rate constant for the subsequent reaction of this complex.

Such a formulation also can account for the small, negative activation energy which has been observed for the reaction of NO with O_2. Since k_{obs} is now a combination of an equilibrium constant K with a rate constant k_3, $E_{exp} = \Delta H + E_3$. Since the association equilibrium will be exothermic, ΔH will be negative, and if $|\Delta H| > E_3$, which is positive, E_{exp} can well be negative.[c] The remarkably low activation energies which have been observed for $NO + Cl_2$ and $NO + Br_2$ may very well arise from similar, complex preequilibria. Again, further studies on these systems would be well worthwhile.

The remaining reactions listed in Table XII.9 are all of the type in which the third body acts to remove the excess energy of the association reaction. The experiments from which these data were derived all involve very difficult techniques, and the accuracy of measurement is not great.[d] For the recombinations of Br and I atoms it can be seen that the frequencies are of the expected orders of magnitude, namely, about 10^{10} liters^2/mole^2-sec^{-1}, and the activation energies seem to be near zero. It will, however, be observed that the efficiencies of various third bodies can vary over a considerable range of values. In the case of $I + I$, the efficiency of mesitylene is about 3 times that of the other large molecules and about 100 times that of He. If we visualize the reaction as proceeding via the two possible complexes MI^* or I_2^*, Marshall and Davidson (Table XII.9) have calculated

[a] The frequency of ternary collisions ($2NO + O_2$) can be calculated from Eq. (XII.15.8) to be $1.7 \times 10^8 T$ liters^2/mole^2-sec $= 5.1 \times 10^{10}$ at 300°K, assuming $\sigma_{O_2} = \sigma_{NO} = 3.5$ Å, $\sigma_{NO_3} = \sigma_{N_2O_2} = 4.5$ Å, and $\tau_{NO_2} = \tau_{N_2O_2} = 10^{-13}$ sec.

[b] H. Gershinowitz and H. Eyring, *J. Am. Chem. Soc.*, **57**, 585 (1935).

[c] Kassel, *loc. cit.*, has shown that a dissociation energy of 1.3 Kcal/mole for the $(NO)_2$ dimer would fit the data. The dissociation energy in solution is known to be about 4 Kcal/mole [H. S. Johnston et al., *J. Chem. Phys.*, **19**, 189 (1951)], so that this would be quite reasonable. It seems even more likely, however, that $[O-O \cdots N-O]$ complexes should exist, and it would be worthwhile to study this system further.

[d] Of these, the technique of flash photolysis has been one of the most fruitful. The method involves pulsing a system with a sudden, intense flash sufficient to decompose an appreciable fraction of stable species (for example, I_2 into I atoms) and then following the reverse recombination with a photocell and oscilloscope.

TABLE XII.9. THIRD-ORDER REACTIONS

Reaction	Preexponential factor log A, liters^2/mole^2-sec	Activation energy, Kcal/mole
[a] $2NO + O_2 \rightarrow 2NO_2$	3.02	−1.1
[b] $2NO + Cl_2 \rightarrow 2NOCl$	3.66	3.7
[c] $2NO + Br_2 \rightarrow 2NOBr$	3.5	0 (?)
[d] $2NO_2 + M \rightarrow N_2O_4 + M$	6.5	−2.3 (M≡N₂ at 1 atm. 25°C.)
[e] $H + H + M \rightarrow H_2 + M$	10.0	0 (?) (M≡H₂)
[f] $H + O_2 + M \rightarrow HO_2 + M$	7.1; 8.1	0 (?) (M≡H₂); 0 (?) (M≡O₂); 0 (?)
[g] $Br + Br + M \rightarrow Br_2 + M$	[He] 9.14; [H₂] 9.60; [N₂] 9.64; [CH₄] 9.82; [CO₂] 9.99; [A] 9.4[g']	−1.5
[h-k] $I + I + M \rightarrow I_2 + M$	[He] 9.23[j]; 9.51[h]; 9.50[k]; [A] 9.56[j]; 9.82[h]; 9.64[k]; 9.62[i]; [n-C₅H₁₂] 10.67[j]; 10.82[i]; [Neopentane] 10.76[j]; [Mesitylene] 11.61[j]; [H₂] 9.9[h]; [N₂] 10.1[h]; [CH₄] 10.4[h]; [CO₂] 10.6[h]; [C₆H₆] 11.3[h]; [I₂] 7.8[h]; [n-Butane] 9.36[i]	0 (?)
[l] $O + O_2 + M \rightarrow O_3 + M$	[O₃] 7.78 [CO₂] 7.81; [N₂] 7.42 [He] 7.31; [O₂] 7.39	−4.4; −1.7; −0.6
[m] $O + NO + M(?) \rightarrow NO_2 + M$	[Air] 10.3	0 (?)
[n] $O + NO_2 + M(?) \rightarrow NO_3 + M$	[Air] 11.0	?
[o] $N + N + N_2 \rightarrow 2N_2$	[N₂] 9.8	0

[a] Table XII.4, footnote e. See M. Bodenstein, *Helv. Chim. Acta*, **18**, 743 (1935). The form given fits the data only above 0°C. At much lower temperatures the negative activation energy is about 0.5 Kcal/mole. The negative activation energy can be explained in terms of the preequilibria $NO + O_2 \rightleftharpoons NO_3$ followed by $NO_3 + NO \rightarrow 2NO_2$. The small change in apparent activation energy with temperature is most probably due to the changes in internal energy of the species with temperature.

[b] Table XII.4, footnote f. The above constants are calculated only for data below 40°C, since the radical chain becomes important at high temperatures. $\Delta H° = -9.0$ Kcal/mole NOCl; $\Delta S° = -14.0$ cal/mole-°C. These values can be used to calculate the reverse reaction rate.

[c] M. Trautz and V. P. Dalal, *Z. anorg. Chem.*, **102**, 149 (1918); **110**, 1 (1920). $\Delta H° = -5.50$ Kcal/mole NOBr; $\Delta S° = -14.50$ cal/mole-°C (Br_2 gas at 1 atm). The rates listed are for 273°K. The activation energy is probably small but is not known with any accuracy.

[d] T. Carrington and N. Davidson, *J. Phys. Chem.*, **57**, 418 (1953). Rates calculated from the rate of decomposition of N_2O_4 and the equilibrium data $\Delta H° = -13.87$ Kcal/mole N_2O_4; $\Delta S° = -42.21$ cal/mole-°C. The data are probably not yet close to the lower limit of a true third-order reaction. CO_2 is a more effective third body than N_2.

[e] W. Steiner, *Trans. Faraday Soc.*, **31**, 623 (1935). The data are taken from flow-tube experiments on H atoms produced in a Wood's tube and may be in error owing to wall effects and diffusion.

[f] B. Lewis and G. von Elbe, *J. Chem. Phys.*, **10**, 366 (1942). G. A. Cook and J. R. Bates, *J. Am. Chem. Soc.*, **57**, 1775 (1935), calculate the values shown for $M=O_2$ from experiments on the photooxidation of HI. Neither of these values can be considered very accurate.

[g] E. Rabinowitch and W. C. Wood, *Trans. Faraday Soc.*, **32**, 907 (1936), from direct photometric measurements of Br_2 concentrations during the photostationary state of $Br_2 + h\nu \rightleftharpoons 2Br$. Other measurements by M. Ritchie, *Proc. Roy. Soc. (London)*, **A146**, 828 (1934), and by K. Hilferding and W. Steiner, *Z. physik. Chem.*, **30**, 399 (1935), are based on the inhibition of the photolysis of $H_2 + Br_2$ by foreign gases and are about three- to fourfold lower. Values reported for I and Br recombinations are based on $dX_2/dt = k(M)(X)^2$.

[g'] D. Britton and N. Davidson, *J. Chem. Phys.*, **25**, 810 (1956).

[h] E. Rabinowitch and W. C. Wood, *ibid.*, **4**, 497 (1936). Photostationary state.

[i] R. Marshall and N. Davidson, *ibid.*, **21**, 659 (1953). Direct observation from flash photolysis of I_2 vapor. D. L. Bunker and N. Davidson, *J. Am. Chem. Soc.*, **80**, 5085 (1958).

[j] K. E. Russel and J. Simmons, *Proc. Roy. Soc. (London)*, **A217**, 291 (1953). Flash photolysis. Values by R. L. Strong et al., *J. Chem. Phys.*, **26**, 1287 (1957), are in general agreement on $Br + Br$ and $I + I$.

[k] M. I. Christie et al., *Proc. Roy. Soc. (London)*, **A216**, 152 (1953). Flash photolysis. The high values for benzene and mesitylene suggest that there is appreciable formation of a complex of the form $C_6H_6·I$, etc. The temperature coefficients would be of interest in this regard.

[l] S. W. Benson and A. E. Axworthy, Jr., *J. Chem. Phys.*, **26**, 1718 (1957).

[m] H. W. Ford and N. Endow, *ibid.*, **27**, 1157 (1957).

[n] T. Wentink, Jr., et al., *ibid.*, **29**, 231 (1958). J. T. Herron et al., *ibid.*, **29**, 230 (1958).

from collision theory that the sum of the lifetimes ($\tau_{I_2}^* + \frac{1}{2}\tau_{MI^*}$) is 1.5×10^{-13} sec for M = argon, 1.8×10^{-12} sec for M = pentane, and about 4×10^{-13} sec for M = benzene.[a] These results are in qualitative agreement with what one might expect on the basis of the increasing complexity of the molecules (Sec. XI.4) and the larger number of degrees of freedom through which the energy of the complex can be dissipated. Iodine itself seems to form a metastable species of long life, I_3.

The association of radicals of intermediate complexity may be expected to show pressure dependence under proper conditions. Their lifetime, subject to correction for large entropy changes in the energized state and the contribution of excited states, may be predicted from Eq. (XI.3.4). The pressure range for showing third-order behavior is then predictable from Table XI.2, subject to further correction for the efficiency of deactivation.

The RRK model which forms the basis for these estimates makes some other predictions of interest. Thus we can estimate that, on comparing the efficiencies of two competing associations such as $O + NO$ and $O + O_2$, where the geometry and structure of the reactants are very similar, the more exothermic of the two reactions will be the faster. This rather paradoxical result is a direct consequence of Eq. (XI.3.4), which says that the mean lifetimes of the resulting complexes will be in the ratio $[(E_A^* + skT)/(E_B^* + skT)]^{s-1} \cong (E_A^*/E_B^*)^{s-1}$ for E_A^*, $E_B^* \gg skT$. For the cases above $(E_{NO_2}^*/E_{O_3}^*)^{s-1} \cong 27$ if we use $s = 4$ (3 vibrations + 2 active rotations), while the observation (Table XII.9) is that the $NO + O$ recombination is 240 times faster than $O_2 + O$.[b] The high efficiency of the $NO + O$ recombination relative to $O_2 + O$ is thus a consequence of the fact that the energized NO_2^* has three times as much energy as O_3^* to relocalize before it can split.

The factor of 27 can be increased to 81 if we also admit that NO_2^* with 3 times as many quanta as O_3^* probably also has a statistically threefold larger chance of losing at least one of these quanta in a collision and so being stabilized. This still does not account for the corrected ratio of 480, and the possibility exists that excited electronic states of O_3^* are decreasing its stability further.[c]

The behavior of these two systems can thus be taken as a striking confirmation of the qualitative correctness of the RRK model.

[a] The collision diameters used were $\sigma' = 5.2$ Å, $\sigma_A = 3.6$ Å, and $\sigma_{C_6H_6} = \sigma_{pentane} = 6.2$ Å.

[b] The actual factor to be explained is twice this, or 480, because of the symmetry of the O_2 molecule compared to NO.

[c] Spin restrictions on the combination of triplet O + triplet O_3 would make only one in four orientations permissible, while one in two is acceptable for doublet NO + triplet O. Notice that any excited electronic states formed will have shorter lifetimes than the ground states, since their E^* will be smaller.

17. Rates of Activation

The data on the third-order recombinations of radicals and atoms present us with the possibility of calculating the rates of the inverse process, namely, the rates of bimolecular activation of molecules. To preserve generality, let us consider the association of two active species A and B to form the stable product AB. If the reaction is sufficiently exothermic or the product AB has few internal degrees of freedom, the mechanism of the association is complex and must involve the agency of a third body M. The mechanism may involve either or both of the following paths[a]

$$A + B \underset{2}{\overset{1}{\rightleftharpoons}} AB^* \qquad\qquad A + M \underset{2'}{\overset{1'}{\rightleftharpoons}} AM$$

$$AB^* + M \underset{4}{\overset{3}{\rightleftharpoons}} M + AB \qquad AM + B \underset{4'}{\overset{3'}{\rightleftharpoons}} M + AB$$

$$\text{I} \qquad\qquad\qquad\qquad \text{II}$$

In either case the over-all reaction is given by

$$A + B + M \rightleftharpoons AB + M \qquad K$$

for which we can write the equilibrium constant K if the system proceeds to a final equilibrium state.

If we apply stationary-state kinetics to AB^* and to AM, we find for the rate of formation of AB (by mechanism I):

$$\frac{1}{(M)(A)(B)} \frac{d(AB)}{dt} = \frac{k_1 k_3}{k_2 + k_3(M)} - \frac{k_2 k_4}{k_2 + k_3(M)} \frac{(AB)}{(A)(B)}$$

while from mechanism II: $\qquad\qquad\qquad\qquad$ (XII.17.1)

$$\frac{1}{(M)(A)(B)} \frac{d(AB)}{dt} = \frac{k_{1'} k_{3'}}{k_{2'} + k_{3'}(A)} - \frac{k_{2'} k_{4'}}{k_{2'} + k_{3'}(B)} \frac{(AB)}{(A)(B)}$$

$$\text{(XII.17.2)}$$

When the back reaction is negligible (that is, $K \gg 1$), we can neglect the negative terms in these equations. If in addition we made use of the relations $K = k_1 k_3 / k_2 k_4 = k_{1'} k_{3'} / k_{2'} k_{4'}$, then these expressions reduce to

Mechanism I: $\qquad \dfrac{1}{(M)(A)(B)} \dfrac{d(AB)}{dt} = k_4 K \left[1 + \dfrac{k_3(M)}{k_2} \right]^{-1}$

and $\qquad\qquad\qquad\qquad\qquad\qquad\qquad\qquad\qquad$ (XII.17.3)

Mechanism II: $\qquad \dfrac{1}{(M)(A)(B)} \dfrac{d(AB)}{dt} = k_{4'} K \left[1 + \dfrac{k_{3'}(B)}{k_{2'}} \right]^{-1}$

$$\text{(XII.17.4)}$$

[a] We ignore here for simplicity a third path involving BM and also the different species which are capable of acting as third bodies (for example, AB itself, A, B, etc.).

or in general for the composite mechanism I + II

$$\frac{1}{(M)(A)(B)} \frac{d(AB)}{dt} = K\left\{k_4\left[1 + \frac{k_3(M)}{k_2}\right]^{-1} + k_{4'}\left[1 + \frac{k_{3'}(B)}{k_2}\right]^{-1}\right\}$$
(XII.17.5)

If we further restrict our consideration to the cases in which $k_2 \gg k_3(M)$[a] and $k_{2'} \gg k_{3'}(B)$,[b] this reduces to the very simple form

$$\frac{1}{(M)(A)(B)} \frac{d(AB)}{dt} = k_{exp} = K(k_4 + k_{4'})$$
(XII.17.6)

where k_{exp} is the experimentally determined third-order rate constant.

Since the equilibrium constant K can in principle be determined from thermodynamic data, the experimental rate constant k_{exp} leads to a measure of $k_4 + k_{4'}$. The constant k_4 is a bimolecular rate constant for the rate of activation of AB, while $k_{4'}$ is a bimolecular rate constant for the exchange reaction of M with AB.

When the data on the recombination of atoms are employed (Table XII.9) to calculate $k_4 + k_{4'}$, it is found that the values of $k_4 + k_{4'}$ may be from 100 to 1000 times greater than simple collision frequencies. Thus Marshall and Davidson[c] have calculated that $k_4 + k_{4'}$ is 1.1×10^{13} liters/mole-sec for I_2 in argon, 1.6×10^{14} in neopentane, and 1.8×10^{14} in pentane.[d] This has been commented on by Rice,[e] who pointed out that such anomalously high values may be accounted for in terms of a large positive entropy of activation for the halogens arising from the increase in number of vibrational levels in the critical complex, the increase in electronic states and the increase in the moment of inertia. In the case of H_2 a factor of

[a] When the condition $k_2 \gg k_3(M)$ is not satisfied, then the reaction becomes a quasi-bimolecular reaction and we must make a detailed analysis in which we integrate over all energies of the critical complex AB*. The present restriction implies that we are working at the extreme low-concentration end of the reaction (Sec. XI.4).

[b] Since $k_{2'}$ is a unimolecular rate constant and $k_{3'}$ is a bimolecular rate constant, we may expect the ratio $k_{3'}(B)/k_{2'} \cong 1 \times 10^{11}(B)e^{-E/RT}/1 \times 10^{13}e^{-H/RT} = 1 \times 10^{-2}$ $(B)e^{-(E-H)/RT}$ with (B) in moles/liter, H the bond energy of AM, and E the activation energy of $k_{3'}$. If the bond energy of AM is small (i.e., of the order of kT) then $e^{-(E-H)/RT} \le e$ and our condition will be fulfilled. If, however, H is fairly large and E is near zero, we may find the reverse $k_{2'} \le k_{3'}(B)$, and k_{exp} will show a dependence on B, that is, the reaction becomes less than third order.

[c] Table XII.9, footnote i.

[d] This is based on the assumption that there is no activation energy for k_4 or $k_{4'}$, which is approximately what is observed. If they did have activation energies then the values would be even higher.

[e] O. K. Rice, J. Chem. Phys., 9, 258 (1941). Rice assumes that most decomposition into atoms occurs from species I_2^* with energy within kT of dissociation. This implies that for the reverse reaction of association, deactivation of I_2^* by M goes through small decrements of energy.

about 100 is accounted for[a] principally by the large increase in moment of inertia (r_0 goes from 0.74 to about 4.2 Å) in forming the critical complex. For more complex molecules such as O_3,[b] N_2O_4,[c] N_2O_5,[d] and NO_2Cl,[e] whose decompositions are quasi-unimolecular and which have been studied at or near the "low-pressure" limit where the observed second-order rate constant is a direct measure of the rate of activation, the preexponential factors are roughly 4.6×10^{12}, 2×10^{14}, 2×10^{16}, and 6×10^{13} liters/mole-sec, respectively. These rate constants represent the rates of activation from normal to energized (Sec. XI.4) states, and the large preexponential factors simply reflect the Boltzmann factor[f] $(E/RT)^{n-1}/(n-1)!$ for the excess entropy of the energized state over the ground state.[g] Such reactions represent energy-transfer processes, i.e., the rate at which energy is converted into internal energy, and for polyatomic molecules we can expect the preexponential factors always to be in excess of collision frequencies [Eq. (XI.4.5)]. This is an exception to the usual rule for second-order rate constants.

The increase in value of $k_4 + k_{4'}$ for I_2 in going from argon to the larger hydrocarbons can arise either from the increased entropy of the excited hydrocarbon molecules (since either M or I_2 must have the excitation energy), which is not possible with argon, or it may arise from the bond energy in AM which would tend to lower the activation energy for $k_{4'}$, or finally if there is an increase in entropy on forming AM, this will be reflected in an entropy of activation for $k_{4'}$.[h,i]

It is then interesting to inquire into the reasons that such abnormally large preexponential factors are not observed in bimolecular exchange reactions. The answer must be that in the latter cases there is no comparably large entropy change in the reaction. In such exchanges the species involved are fairly tightly bonded radicals or molecules, in contrast to the

[a] See also *ibid.*, **21**, 750 (1953), and note by G. Careri, *ibid.*, 749, for discussion of H + H.

[b] Table XII.9.

[c] Table XII.8.

[d] Table XI.5.

[e] H. F. Cordes and H. S. Johnston, *J. Am. Chem. Soc.*, **76**, 4264 (1954).

[f] That is, the number of ways in which this energy is distributed over the degrees of freedom of the molecule.

[g] This classically calculated factor is increased by quantum restrictions which assign nearly zero entropy to the vibrationally frozen ground state.

[h] Note that if the activation energy for $k_{4'}$ is less than the heat of the over-all reaction, then the observed recombination coefficient will have a negative activation energy equal to this difference in energies.

[i] This increase in entropy on forming AM is plausible if the bond in AM is very weak (as it usually is), since we are then replacing a tight bond in AB by weak bonds in AM with resulting increase in vibrational entropy and also, very likely, rotational entropy as well.

very weakly bonded species AB* or AM possessing considerable amounts of internal entropy which are formed in the activation processes considered here.[a]

18. Activation Energies for Simple Kinetic Processes

The relation of rate constants to molecular structure is of course the ultimate goal of any molecular theory of kinetics. As we have already seen, the theoretical approach to the problem of correlating preexponential factors with molecular theory has certainly been qualitatively successful, if somewhat lacking from a quantitative point of view. However, the usefulness of even qualitative correlation is considerable in the guide that it provides to the experimentalist in analyzing the details of a complex mechanism. It is unfortunate that at present there is no similarly useful means for correlating activation energies.

With the advent of modern quantum mechanics in 1926, there followed a great activity on the part of physicists and chemists in attempting to calculate the bond energies of molecules. The results have been crude almost to the point of complete discouragement. The difficulty can be readily understood, from the point of view of the kineticist, in terms of the sensitivity of rate constants to activation energy. At 300°K an error of 1.4 Kcal/mole in E_{act} represents a 10-fold error in the rate constant.[b] But 1.4 Kcal is about 0.07 ev, less by at least a factor of 10 than the errors involved in most of the available quantum-mechanical solutions to the Schroedinger equation for molecular systems.

The earliest efforts to apply such methods to the calculation of activation energies were those of London,[c] with results which were extremely crude though promising. Subsequent attempts by Villars,[d] Eyring,[e] and Eyring and Polanyi,[f] and others[g] to improve the accuracy of the method by adopting empirical short cuts have not been singularly successful, and

[a] It is possible that the large steric factor (from 1 to 6) observed for the reaction $Cl + COCl_2 \rightarrow Cl_2 + COCl$ (Table XII.5) is real and arises from a very loosely bound COCl. This is difficult to reconcile with the almost equally high value found for $Cl + COCl \rightarrow CO + Cl_2$, unless we are willing to assume that the excess energy of this latter reaction contributes an excess entropy to the products (i.e., excited species are formed).

[b] This difficulty is much less in the computation of frequency factors which involve the computation of entropies of rotation and vibration. For an error of 10 in k, the error in S_{act} must be 4.6 cal/mole-°K. The approximate methods discussed in this chapter for computing S_{act} are generally about this good because vibrational contributions, the most difficult to assess, are generally quite small.

[c] F. London, Z. physik. Chem., Sommerfeld Festband, 104 (1928); Z. Elektrochem., 35, 552 (1929).

[d] D. S. Villars, J. Am. Chem. Soc., 52, 1733 (1930).

[e] H. Eyring, ibid., 53, 2537 (1931); 54, 3191 (1932); Chem. Revs., 10, 103 (1932).

[f] H. Eyring and M. Polanyi, Z. physik. Chem., B12, 279 (1931).

[g] For a review of this work see Glasstone et al., loc. cit.

further work along these lines will probably have to await improved techniques in performing quantum-mechanical calculations.

Otozai[a] has proposed that the bond length between atoms undergoing chemical change in a molecule is given by the point of inflection in the potential-energy curve for the diatomic molecule of the isolated atoms.[b] This leads, with additional assumptions about the configurations of the complex (not too different from the Eyring method), to values for the activation energies of three- and four-atom systems which are in slightly better agreement with experimental data than the other methods.

Hirschfelder[c] has proposed that for exchange reactions of atoms with molecules of the type $A + BC \rightarrow AB + C$ (exothermic) the activation energy is about 5 per cent of the energy of the bond being broken.[d] For reactions of the type $AB + CD \rightarrow AC + BD$ (exothermic) it is proposed that $E_{act} = 28$ per cent of the sum of the bond energies of the bonds being broken. Neither of these two rules is better than about ±5 Kcal, and they both seem insensitive to changes in homologous series of reactions. Otozai (loc. cit.) has proposed a similar relation $E_{act} = 0.29D_{BC} - 22.2$ for reactions of the first type[e] and $E_{act} = 0.29(D_{AB} + D_{CD})$ for reactions of the second type when $E_{act} \geqslant 22.2$ Kcal. He reports a rule by Uhara for the second type, namely, $E_{act} = 0.65(D_{AB} + D_{CD}) - 0.35(D_{AC} + D_{BD})$, valid for exo- and endothermic reactions. These latter rules seem to be an improvement over the Hirschfelder rules, and from the limited data presented they appear to agree to within about ±3 Kcal with the data. Unfortunately, the occasional large discrepancies that are observed for all of these rules make them generally unreliable except as very rough guides.

Evans and Polanyi[f] have noted that in a homologous series of exothermic abstraction reactions $(A + BC \rightarrow AB + C)$, the change in activation energy is related to the change in heat of reaction by the relation $\Delta E_{act} = \alpha \Delta H_{act}$ where α is a constant for a given series.[g] The more recent and more accurate results on methyl radical abstractions would seem, however, to place limits on the quantitative value of the rule. Voevodskii[h] has pro-

[a] K. Otozai, Sci. Papers of Osaka Univ., **20** (1951); Bull. Chem. Soc. Japan., **24**, 218, 257, 262 (1951).

[b] If the Rydberg equation for the potential energy of a diatomic molecule is used, $U(r) = -D_e[a(r - r_e) + 1]e^{-a(r-r_e)}$, where $r_e =$ the equilibrium distance, and $D_e = D + E_0$, where D is the experimental dissociation energy and E_0 is the zero-point energy of the molecule, then the transition separation is $r^* = r_e + 1/a$, where both r_e and $1/a$ can be computed from spectroscopic data.

[c] J. D. Hirschfelder, J. Chem. Phys., **9**, 645 (1941).

[d] For the reverse reaction the activation energy is higher by the heat of the reaction.

[e] $D_{BC} =$ bond energy of BC.

[f] M. G. Evans and M. Polanyi, Trans. Faraday Soc., **32**, 1933 (1936); **34**, 22 (1938).

[g] H. Steiner and K. H. Watson, Discussions Faraday Soc., **2**, 88 (1947). Bolland, Quart. Revs. (London), **3**, 1 (1949). M. Szwarc, Discussions Faraday Soc., **10**, 143 (1951).

[h] V. V. Voevodskii, Doklady Akad. Nauk. S.S.S.R., **79**, 455 (1951).

posed a set of rules for calculating the bond energies in hydrocarbons and has attempted to calculate the activation energies of abstraction reactions[a] by combining this rule with the rule of Evans and Polanyi. Unfortunately, the data are not as yet sufficiently extensive to estimate the validity of the results.[b]

[a] N. N. Tikhomirova and V. V. Voevodskii, *ibid.*, 993.

[b] The chief stumbling block to most of the attempts at generalization to the present has been precisely the lack of a large body of accurate rate data.

XIII

Some Complex Gas-phase Reactions

1. Complexity of Gas-phase Reactions

Of the hundreds of reactions which have been observed to proceed in the gas phase relatively few, if any, can be described in terms of a single chemical transformation. The large majority proceed through a more or less complex chemical mechanism that involves the formation and destruction of highly reactive free radicals and atoms. Because of the reactivity of these intermediates, they are usually present in extremely low concentrations, and their existence is generally inferred from indirect evidence rather than from direct observation.[a]

This leads, as we shall see, to a situation in which simple measurements of the rate constant, order, and activation energy of a chemical reaction do not provide sufficient data to establish uniquely the detailed mechanism of a chemical reaction. Instead, the task of determining an unambiguous mechanism confronts the experimenter with a problem which requires the application of the utmost ingenuity in devising criteria for testing the validity of the steps in any proposed mechanism.

In the following sections we shall consider in detail some of the better-studied complex reaction systems and some of the methods which have been employed for studying their mechanisms.

Gas-phase reactions will be given an emphasis beyond the other areas of kinetics, precisely because the extensive development of quantitative thermodynamic data and semiquantitative theories makes it possible to interpret the details of these complex reactions of gases to a degree far beyond anything now available for reactions in condensed phases.

[a] There are a few notable exceptions to this statement thanks to the developments of mass-spectrometric and optical techniques (Sec. V.4).

2. The Reaction System $H_2 + Br_2 \rightleftharpoons 2HBr$

In 1906, Bodenstein and Lind[a] showed that the kinetics of the reaction of H_2 and Br_2 (which proceeds at a conveniently measurable rate between 230 and 300°C) could be presented empirically by the equation

$$\frac{d(\text{HBr})}{dt} = \frac{k_t(\text{H}_2)(\text{Br}_2)^{\frac{1}{2}}}{1 + k_i[(\text{HBr})/(\text{Br}_2)]} \qquad (\text{XIII.2.1})$$

Over the temperature range from 25 to 300°C, the inhibition constant k_i has been shown to be less than 1 and to increase only very slightly (from 0.116 at 25°C to 0.122 at 300°C) with temperature.[b] Thus under conditions in which (HBr) < (Br$_2$) the denominator may be ignored and[c]

$$\frac{d(\text{HBr})}{dt} \cong k_t(\text{H}_2)(\text{Br}_2)^{\frac{1}{2}} \qquad (\text{XIII.2.2})$$

The inhibition by HBr and the involved kinetic expression clearly indicate that there is a complex mechanism involved. Solutions to this problem were proposed independently by Christiansen,[d] Herzfeld,[e] and Polanyi,[f] at about the same time. The mechanism which they proposed and which leads to the same kinetic form as Eq. (XIII.2.1) is

$$Br_2 + M \underset{2}{\overset{1}{\rightleftharpoons}} 2Br + M - 46.1 \text{ Kcal}$$

$$Br + H_2 \underset{4}{\overset{3}{\rightleftharpoons}} HBr + H - 16.6 \text{ Kcal} \qquad (\text{XIII.2.3})$$

$$H + Br_2 \overset{5}{\rightarrow} HBr + Br + 41.4 \text{ Kcal}$$

where the heats of reaction listed are for the standard states (25°C, 1 atm) and M may be any of the gaseous species present.

The kinetic equations for the intermediate species in this system are

$$\frac{d(\text{Br})}{dt} = 2k_1(\text{Br}_2)(\text{M}) + k_4(\text{H})(\text{HBr}) + k_5(\text{H})(\text{Br}_2)$$
$$- 2k_2(\text{Br})^2(\text{M}) - k_3(\text{Br})(\text{H}_2) \qquad (\text{XIII.2.4})$$

$$\frac{d(\text{H})}{dt} = k_3(\text{Br})(\text{H}_2) - k_4(\text{H})(\text{HBr}) - k_5(\text{H})(\text{Br}_2)$$

and for the product

$$\frac{d(\text{HBr})}{dt} = k_3(\text{Br})(\text{H}_2) + k_5(\text{H})(\text{Br}_2) - k_4(\text{HBr})(\text{H}) \qquad (\text{XIII.2.5})$$

[a] M. Bodenstein and S. C. Lind, Z. physik. Chem., **57**, 168 (1906).

[b] M. Bodenstein and G. Jung, ibid., **121**, 127 (1926).

[c] Despite its complicated form, Eq. (XIII.2.1) can be integrated exactly and the data employed directly to give the constants k_t and k_i.

[d] J. A. Christiansen, Kgl. Danske Videnskab. Selskab, Mat.-fys., Medd., **1**, 1, 14 (1919).

[e] K. F. Herzfeld, Z. Elektrochem., **25**, 301 (1919); Ann. Physik, **59**, 635 (1919).

[f] M. Polanyi, Z. Elektrochem., **26**, 50 (1920).

If we ignore the pre-stationary-state period, which can be shown to be negligible[a] at all temperatures below 1000°K, then we can make use of the stationary-state hypothesis and set $d(\text{Br})/dt = d(\text{H})/dt = 0$ and solve Eq. (XIII.2.4) for the stationary-state concentrations (H) and (Br).[b]
We find (by setting $K_{1.2} = k_1/k_2$):

$$(\text{Br})_{ss} = \left[\frac{k_1(\text{Br}_2)}{k_2}\right]^{\frac{1}{2}} = K_{1.2}^{\frac{1}{2}}(\text{Br}_2)^{\frac{1}{2}} \qquad (\text{XIII.2.6})$$

$$(\text{H})_{ss} = \frac{k_3(\text{Br})(\text{H}_2)}{k_4(\text{HBr}) + k_5(\text{Br}_2)}$$

$$= \frac{k_3 K_{1.2}^{\frac{1}{2}}(\text{H}_2)(\text{Br}_2)^{\frac{1}{2}}}{k_4(\text{HBr}) + k_5(\text{Br}_2)} \qquad (\text{XIII.2.7})$$

On substitution into Eq. (XIII.2.5) we have

$$\frac{d(\text{HBr})}{dt} = \frac{2k_3 k_5 K_{1.2}^{\frac{1}{2}}(\text{H}_2)(\text{Br}_2)^{\frac{3}{2}}}{k_5(\text{Br}_2) + k_4(\text{HBr})} \qquad (\text{XIII.2.8})$$

or

$$\frac{d(\text{HBr})}{dt} = \frac{2k_3 K_{1.2}^{\frac{1}{2}}(\text{H}_2)(\text{Br}_2)^{\frac{1}{2}}}{1 + k_4(\text{HBr})/k_5(\text{Br}_2)} \qquad (\text{XIII.2.9})$$

which has the same kinetic form as the empirical equation (XIII.2.1). By comparing these two equations we can identify the empirical constants as

$$k_t = 2k_3 K_{1.2}^{\frac{1}{2}} \qquad k_i = \frac{k_4}{k_5} \qquad (\text{XIII.2.10})$$

Since $K_{1.2}$, the equilibrium constant for the dissociation of Br atoms, is known both experimentally[c] and from spectroscopic data,[d] it is possible

[a] F. A. Matsen and J. L. Franklin, *J. Am. Chem. Soc.*, **72**, 3337 (1950). See also S. W. Benson, *J. Chem. Phys.*, **20**, 1605 (1952).

[b] In solving such simultaneous equations it is convenient to use the abbreviated notation of replacing kinetic terms by roman numerals. Thus Eqs. (XIII.2.4) and (XIII.2.5) become

$$\frac{d(\text{Br})}{dt} = 2\text{I} + \text{IV} + \text{V} - 2\text{II} - \text{III} = 0$$

$$\frac{d(\text{H})}{dt} = \text{III} - \text{IV} - \text{V} = 0$$

$$\frac{d(\text{HBr})}{dt} = \text{III} + \text{V} - \text{IV}$$

Then from the second equation III = IV + V, which on substitution into the first equation gives 2I − 2II = 0, or I = II, from which the concentration $(\text{Br})_{ss}$ can be obtained by returning to the original forms. In similar fashion, substitution into the third equation gives $d(\text{HBr})/dt = 2\text{V}$, from which Eq. (XIII.2.8) is obtained. For this and more complicated chains, this type of abbreviated notation is extremely labor-saving.

[c] M. Bodenstein and F. Cramer, *Z. Elektrochem.*, **22**, 327 (1916).

[d] "Selected Values of Chemical Thermodynamic Properties," *Nat'l Bur. Standards Circ.* 500 (1952).

to calculate k_3 from the measured value of k_t. Jost[a] gives it in the form of the collision equation

$$\log k_3 = 12.308 + 0.5 \log T - \frac{17,600}{4.575T} \qquad (XIII.2.11)$$

from which we see that the activation energy $E_3 = 17.6$ Kcal. Since reaction 4 is endothermic by 16.6 Kcal, this implies that $E_4 = E_3 - \Delta E_{34} = 1.0$ Kcal. Moreover, since k_i is essentially independent of temperature, we see that $E_4 = E_5$, so that E_5 is also about 1.0 Kcal.[b]

By using reasonable values for the diameters we can calculate (Table XII.5) the steric factors of reactions 3, 4, and 5 to be 0.12, 0.033, and 0.15, respectively. These seem all quite plausible values for the atom-diatomic molecule reactions.[c]

3. Some Omitted Reactions in the $H_2 + Br_2$ System

We can now ask why it is that some rather obvious reactions have been omitted from the scheme. They are as follows:

$$H_2 + Br_2 \xrightarrow{7} 2HBr + 24.7 \text{ Kcal}$$

$$H_2 + M \underset{9}{\overset{8}{\rightleftharpoons}} 2H + M - 104.2 \text{ Kcal} \qquad (XIII.3.1)$$

$$Br + HBr \xrightarrow{6} H + Br_2 - 41.4 \text{ Kcal}$$

$$M + H + Br \underset{11}{\overset{10}{\rightleftharpoons}} M + HBr + 87.5 \text{ Kcal}$$

To assist in the comparison of the rate constants, the relevant thermal data for this system are compiled in Table XIII.1. The heats of reaction listed in Eq. (XIII.2.3) are obtained from this table.

The reason for omitting reaction 7 is simply one of kinetics. If reaction 7 were important, then there would be added to the expression for $d(HBr)/dt$ a term of the form $k_7(H_2)(Br_2)$. If we are willing to believe the data accurate to ± 2 per cent, we must conclude that if such a reaction

---

[a] W. Jost, Z. physik. Chem., **B3**, 95 (1929).

[b] Note that the equilibrium constant for reactions 3 and 4, namely $K_{3.4}$, may be calculable from spectroscopic and thermal data, so that $k_4 = k_3/K_{3.4}$ may be calculated. This permits us to calculate $k_5 = k_4/k_i$ and also the rate constant for reaction 6 (the reverse of k_5) from $k_6 = k_5/K_{5.6}$, where $K_{5.6}$ is the calculable equilibrium constant. All of these data are given in Table XII.5.

[c] An extensive review of the $H_2 + Br_2$ system is given by E. S. Campbell and R. M. Fristrom, Chem. Revs., **58** (1958).

TABLE XIII.1. THERMAL DATA FOR THE SYSTEM $H_2(g) + Br_2(g) \rightleftharpoons 2HBr(g)$[a]

Species	$\Delta H°$, Kcal/mole	$S°$, cal/mole-°K	C_p°, cal/mole-°K
$H(g)$	52.09	27.39	4.97
$H_2(g)$	0	31.21	6.89
$Br(g)$	26.71	41.81	4.97
$Br_2(g)$	7.34	58.64	8.60
$HBr(g)$	−8.66	47.44	6.96

[a] From *Nat'l Bur. Standards Circ.* 500, 1952. Standard states refer to 25°C and 1 atm pressure (ideal gas).

exists it contributes less than 7 per cent to the over-all reaction path.[a] This would in turn imply that despite its exothermicity, reaction 7 must have an activation energy exceeding 40 Kcal.[b] This seems rather the rule for four-center exchange reactions. When they do occur, they appear to have very high activation energies.

To see why reaction 8 is neglected, we have to compare it with other reactions which consume H_2 molecules and other reactions which produce chain centers. Since the activation energy of reaction 8 is at least 104 Kcal, the heat of reaction, we see that the other radical-producing reaction, 1, which should have a not too much smaller frequency factor and much lower activation energy (that is, ~46 Kcal), is an incomparably faster mechanism for producing radicals.[c] Similarly, reaction 3, which is the alternative mechanism for the consumption of H_2 molecules, has an activation energy of only 17.6 Kcal, so that the relative rates of disappearance of H_2 by these two mechanisms would be

$$\frac{R_8}{R_3} = \frac{k_8(H_2)(M)}{k_3(H_2)(Br)} = \frac{k_8(M)}{k_3(Br)} \qquad (XIII.3.2)$$

or substituting the stationary-state value $(Br)_{ss} = K_{1.2}^{1/2}(Br_2)^{1/2}$ [Eq. (XIII.2.6)]

[a] Neglecting the inhibition by HBr, the over-all reaction would then be $d(HBr)/dt = k_t(H_2)(Br_2)^{1/2}[1 + k_7(Br_2)^{1/2}/k_t]$. If we consider a run in which (Br_2) changes by 50 per cent from its initial value, then $(Br_2)^{1/2}$ changes by only 30 per cent (i.e., from a relative value of 1 to $0.5^{1/2} \cong 0.7$). If now the error in k_t is ±2 per cent, it implies that the variation of the expression in brackets is within ±2 per cent, or under the above conditions, that $k_7\overline{(Br_2)}^{1/2}/k_t \leq 0.02/0.30 = \frac{1}{15} \cong 7$ per cent. Over larger intervals of reaction it would be very difficult to separate out the effect of the inhibition terms, which would no longer be negligible.

[b] The activation energy for k_t is given by [Eq. (XIII.2.10)] $E_t = E_3 + \frac{1}{2}\Delta E_{12} = 40$ Kcal, so that, even if the preexponential factors of k_t and k_7 are comparable, $E_7 > 40$ Kcal in order to give the ratio of $\frac{1}{15}$ for $k_7(Br_2)^{1/2}/k_t$. Further analysis of k_t shows that $E_7 > 50$ Kcal.

[c] At 300°C we can estimate $k_8/k_1 \cong e^{-58,000/RT} \approx 10^{-22}$.

$$\frac{R_8}{R_3} = \frac{k_8}{k_3 K_{1.2}^{\frac{1}{2}}} \frac{(M)}{(Br_2)^{\frac{1}{2}}} \qquad \text{(XIII.3.3)}$$

At 200°C we can calculate from the data of Bodenstein and Cramer,[a] or Table XIII.1, that $K_{1.2}^{\frac{1}{2}} = 3.3 \times 10^{-11}$ (mole/cc)$^{\frac{1}{2}}$, so that in an extreme case when Br_2 is 1 per cent of the total concentration $[(Br_2) = 0.01\ M)]$ and $M = 1$ atm ($\cong 2.6 \times 10^{-5}$ moles/cc at 200°C), then $R_8/R_3 \cong (k_8/k_3) \times 1.5 \times 10^9$. But if k_8 and k_3 have comparable pre-exponential factors, their ratio will be about $e^{-(E_8-E_3)/RT} \cong 10^{-88/2.16} \cong 10^{-41}$, so that $R_8 \ll R_3$.

The importance of reaction 9 may be estimated by comparing its rate with that of the competing reaction which removes H atoms, namely, reaction 5.[b] The ratio of these two steps is

$$\frac{R_9}{R_5} = \frac{2k_9(H)^2(M)}{k_5(H)(Br_2)} = 2\frac{k_9}{k_5}\frac{(M)}{(Br_2)}(H) \qquad \text{(XIII.3.4)}$$

If we substitute from Eq. (XIII.2.7) the stationary concentration of H atoms, we have (neglecting the inhibition by HBr)

$$\frac{R_9}{R_5} = \frac{2k_9(M)}{k_5(Br_2)}\frac{k_3(H_2)}{k_5(Br_2)}(Br) \qquad \text{(XIII.3.5)}$$

To take the most favorable case for reaction 9 when $(M) = (H_2) = 100(Br_2) = 1$ atm and $T = 300°C$, $R_9/R_8 = 2 \times 10^4 k_3 k_9(Br)/k_5^2$. Now $(Br) = K_{1.2}^{\frac{1}{2}}(Br_2)^{\frac{1}{2}} \cong 1.3 \times 10^{-12}$ moles/cc at 300°C, so that $R_9/R_8 = 2.6 \times 10^{-8} k_3 k_9/k_5^2$. If we assume that k_9 goes at every termolecular collision while k_5 has a normal frequency factor, then $k_9 \cong 10^{16}$ (cc^2/mole^2-sec), $k_5 \cong 10^{14}$ cc/mole-sec $\times e^{-E_5/RT}$, and $k_3 \cong 10^{14}e^{-E_3/RT}$ and $R_9/R_8 = 2.6 \times 10^{-6}e^{-(E_3-2E_5)/RT}$, which for $E_3 - 2E_5 = 16$ Kcal is quite small. In fact we see that at 300°C, $E_5 \geqslant 26$ Kcal before R_9 and R_8 become of comparable magnitudes.[c]

[a] Bodenstein and Cramer, loc. cit., give $\frac{1}{2}\log K_{1.2} = -5050/T + 0.375 \log T - 2.05T \times 10^{-4} + 2.4T^2 \times 10^{-8} - 0.683$ when concentration units are mole/cc.

[b] We can see qualitatively that $R_9 \ll R_5$. R_5 has almost no activation energy and a steric factor of about $\frac{1}{4}$, while R_9 at best goes at every termolecular collision. Thus, even at 1 atm, $k_9 M \leq 10^{14}$ cc/mole-sec, while $k_5 \cong 2 \times 10^{12}$ cc/mole-sec. On the other hand $(H)/(Br_2) \ll 10^{-2}$, so that $R_9/R_5 \ll 1$.

[c] It is an interesting result of such calculations that we can show that the stationary-state concentration of H, $(H)_{ss}$, is many powers of 10 higher than the equilibrium concentration $(H)_{eq}$ one would get from $H_2 \rightleftharpoons 2H$. This, as we shall see, is quite common for chain reactions.

In the present case the ratio $(H)_{ss}/(H)_{eq}$ (neglecting HBr inhibition) is given by $(k_3/k_5)[(H_2)/(Br_2)]^{\frac{1}{2}}(K_{1.2}/K_{8.9})^{\frac{1}{2}}$. By using the data of Table XII.5 and thermal data for $K_{8.9}$ from Table XIII.1, we find:

$$\log \frac{(H_{ss})}{(H_{eq})} \cong \frac{1}{2}\log \frac{(H_2)}{(Br_2)} - 0.30 + \frac{2690}{T}$$

which, when $(H_2) = (Br_2)$ and $T = 300°C$, gives $(H)_{ss}/(H)_{eq} = 2.5 \times 10^4$. This is consistent with ignoring reactions 8 and 9, which could be important only when (H) is near $(H)_{eq}$.

The disappearance of H by reaction with Br (reaction 10) can be compared with the competing reaction with Br_2 (reaction 5)

$$\frac{R_{10}}{R_5} = \frac{k_{10}(M)(Br)}{k_5(Br_2)} \tag{XIII.3.6}$$

If we assume that 10 goes on every termolecular collision and set $k_{10} \cong 10^{16}$ cc^2/mole^2-sec, from Table XII.5, $k_5 = 10^{12.83}T^{1/2}10^{-197/T}$, $(Br) = (Br_2)^{1/2}K_{1,2}^{1/2} \cong (Br_2)^{1/2}10^{-5050/T}T^{3/4}10^{-0.8}$ so that

$$\frac{R_{10}}{R_5} \cong 10^{2.4}T^{-1/8}10^{-4850/T}(M)/(Br_2)^{1/2} \tag{XIII.3.7}$$

At a mean temperature $T = 523°K$ and $(M) = 25(Br_2) = 2.4 \times 10^{-5}$ moles/cc ($\sim$1 atm at 250°C) $R_{10}/R_5 \cong 1.3 \times 10^{-9}$, which is certainly negligible.

Finally, reaction 11, the direct decomposition of HBr into atoms, can similarly be shown to be negligible with respect to reaction 1 as a source of Br atoms.[a]

4. The System $H_2 + Br_2$; Photolysis and Wall Effects

In the simple mechanism just considered, the chain reaction of $H_2 + Br_2$ is maintained by an equilibrium supply of Br atoms. While we have given the details for the establishment of this equilibrium [reactions 1 and 2, Eq. (XIII.2.3)] the over-all chain kinetics gives us no information about these details, since only the equilibrium constant for the equilibrium $Br_2 \rightleftharpoons 2Br$ appears in the rate expression. It would be possible to study these steps in detail if we could replace either one of them by a faster kinetic process. Thus if we could substitute for the relatively slow thermal fission of Br_2 molecules a process which produces Br atoms at a greater rate, we would obtain a new rate expression no longer involving the thermal splitting of Br_2. Photochemical activation of Br_2, selective activation by high-energy particles or radiation (e.g., alpha particles, beta particles, gamma rays, and glow discharge) are among the means available for such a process. Alternatively we might find some other fast process for the removal of Br atoms to replace the again relatively slow three-body recombination. That could be done by the introduction of some substance which reacts rapidly with Br atoms such as an olefin[b] or, at low enough temperatures, NO.

[a] This is an unnecessary refinement, because we should normally expect decomposition reactions of HBr to be of importance only when the equilibrium does not proceed appreciably to completion. In the present case the data of Table XIII.1 will show that reverse decomposition of HBr is not important until $T > 1400°K$. For this reason we have also neglected the reverse of reaction 7 [Eq. (XIII.3.1)].

[b] If the reaction is too rapid, as it could be with most olefins, it might be faster than the reaction of $Br + H_2 \rightarrow HBr + H$ and thus utterly suppress that chain. Olefins would not be too interesting a suppressor for the additional reason that they set up their own chains. A free radical trap which added Br directly without creating new radicals would be the ideal reagent.

Of these alternatives, the photochemical reaction has been the most intensively studied.[a] The mechanism is very similar to that for the thermal reaction:

$$Br_2 + I_a \xrightarrow{1'} 2Br$$

$$Br + H_2 \underset{4}{\overset{3}{\rightleftharpoons}} HBr + H$$

$$H + Br_2 \xrightarrow{5} HBr + Br \qquad\qquad (XIII.4.1)$$

$$2Br + M \xrightarrow{2} Br_2 + M$$

$$Br + wall \xrightarrow{12} \tfrac{1}{2}Br_2(wall)$$

In solving this set of equations we can neglect fluorescence and de-activation of excited Br_2 molecules so that the specific rate of reaction $1' = 2I_a$, where I_a is the average number of moles of photons absorbed per cc per second.[b] It is also convenient to assume that a constant fraction of the Br atoms striking the wall is captured to ultimately form Br_2, whereas this accommodation coefficient must certainly depend on the stationary-state concentration of Br atoms, the chemical nature of the wall, the concentrations of other species which may affect the adsorption of Br atoms, etc.

With these assumptions, the stationary-state treatment gives for the stationary-state concentration of H atoms the same result as before [Eq. (XIII.2.7)]:

$$(H)_{ss} = \frac{k_3(H_2)}{k_4(HBr) + k_5(Br_2)}(Br)_{ss} \qquad\qquad (XIII.4.2)$$

while that for (Br) is given by the equation

$$2M_2 k_2(Br)^2 + k_D(Br) - 2I_a = 0 \qquad\qquad (XIII.4.3)$$

or

$$(Br)_{ss} = \frac{(16M_2 k_2 I_a + k_D^2)^{1/2} - k_D}{4M_2 k_2} \qquad\qquad (XIII.4.4)$$

where we have placed the subscript 2 on M to indicate that M_2 is a weighted average of the different species that can aid in the three-body recombination of Br atoms. The rate "constant" k_D is a composite of the mean rate of diffusion of a Br atom to the wall and the accommodation coefficient at the wall. We might expect from simple considerations of diffusion that k_D will have the form[c]

[a] M. Bodenstein and H. Lutkemeyer, Z. physik. Chem., **114**, 208 (1924). W. Jost and G. Jung, ibid., **B3**, 83 (1929). Jung, ibid., 95.

[b] The qualification "average" is important here because it is rare that the exciting light beam is of uniform intensity and that there is no "blind" space in the reaction volume. Also, where there is appreciable absorption from the light beam, the concentration of absorbed photons is always greatest in the front of the reaction cell.

[c] Sec. XIV.6.

$$k_D = \frac{k_D^\circ}{M_D r_0^2} \qquad\qquad (XIII.4.5)$$

where r_0 will be some mean radius of the vessel, M_D will be a weighted concentration[a] of the species responsible for diffusion, k_D° is assumed constant, and convection is neglected.

For vessels of infinite size, the wall effect becomes negligible and $(Br)_{ss} \to (I_a/Mk_2)^{1/2}$, or more precisely, when $k_D^2 < 16M_2 k_2 I_a$, we can expand the expression (XIII.4.4) and we obtain in the region of small wall effects

$$(Br)_{ss} \to \left(\frac{I_a}{Mk_2}\right)^{1/2}\left[1 - \left(\frac{k_D^2}{16Mk_2 I_a}\right)^{1/2}\right] \qquad (XIII.4.6)$$

while in the other extreme when $k_D^2 > 16M_2 k_2 I_a$ and the principal mode of removal of atoms is at the wall, a similar expansion yields

$$(Br)_{ss} \to \frac{2I_a}{k_D} = \frac{2I_a M_D r_0^2}{k_D^\circ} \qquad (XIII.4.7)$$

Since the over-all rate of reaction is proportional to $(Br)_{ss}$,

$$\frac{d(HBr)}{dt} = 2k_5(Br_2)(H)$$

$$= \frac{2k_3(H_2)}{1 + k_4(HBr)/k_5(Br_2)}(Br)_{ss} \qquad (XIII.4.8)$$

we see that the apparent kinetic dependence of the rate on light intensity and total concentration will change from $(I_a/M)^{1/2}$ at small wall effects to $I_a M$ when wall effects predominate.

The condition for small wall effects is $k_D^2 < 16M_2 k_2 I_a$ or, by substituting for k_D from Eq. (XIII.4.5),

$$k_D^{\circ 2} < 16M_D^2 M_2 r_0^2 k_2 I_a \qquad (XIII.4.9)$$

or if we use the approximation $M_D = M_2 = M$

$$(k_D^\circ)^2 < 16M^3 r_0^2 k_2 I_a \qquad (XIII.4.10)$$

We see from this last result that wall effects will be small when the pressure is high, light intensity is high, or the reaction vessel is large.[b]

[a] The concentrations are weighted because the cross section for diffusion depends on the kinetic diameter and mean relative velocity (Sec. VIII.8).

[b] To give a rough numerical example, we can use the Smoluchowsky formula (Sec. XIV.6) for $k_D^\circ = 4\bar{v}L_{12}/3\pi r_0^2$ or $k_D^\circ = 4\bar{v}/3\pi^2 N_{Av}\sqrt{2}\sigma_{12}^2$, where M is in moles/cc, r_0 is in cm, L_{12} is the average mean free path of the radical 1 in species 2, $\bar{v}$ is the mean velocity, and σ_{12} is the kinetic collision diameter. By taking median values, $\sigma_{12}^2 = 2 \times 10^{-15}$ cm^2, $\bar{v} = 4 \times 10^4$ cm/sec, we find $k_D^\circ \cong 3.1 \times 10^{-6}$ moles/cm-sec. If we also take $k_2 = 1 \times 10^{16}$ cc^2/mole^2-sec, $I_a = 10^{-11}$ moles/cc-sec and $r_0 = 3$ cm. Then Eq. (XIII.4.10) can be solved for $M > [(k_D^\circ)^2/(16r_0^2 k_2 I_a)]^{1/3} = 9 \times 10^{-7}$ moles/cc = 16 mm Hg (25°C). If the accommodation coefficient at the wall is smaller than unity, this becomes correspondingly smaller. It may be seen that at total pressures in excess of 100 mm Hg, wall effects are probably small.

The $(Br)_{ss}$ as given by Eq. (XIII.4.4) can be shown to have a maximum value at a concentration given by

$$M_{max} = \left(\frac{k_D^{\circ 2}}{2k_2 I_a}\right)^{\frac{1}{3}}$$ (XIII.4.11)

in deriving which we have assumed $M_2 = M_D = M$. By using the values of the footnote to the preceding paragraph we see that this maximum, which is accompanied by a corresponding maximum in the over-all rate, gives a value for (Br) of

$$(Br)_{ss,max} = I_a^{\frac{2}{3}}\left(\frac{k_D^{\circ}}{2k_2}\right)^{\frac{1}{3}}$$ (XIII.4.12)

and will occur at a pressure in the range of 60 mm Hg (at 25°C).

The effect of a first-order wall recombination for Br atoms is thus to change the kinetic form of the over-all rate expression from a direct dependence on $I_a^{\frac{1}{2}}$ and an inverse dependence on the square root of the total concentration in the region of high pressures [Eq. (XIII.4.6)] to a direct dependence on both absorbed light intensity and total concentration at very low pressures [Eq. (XIII.4.7)].[a]

While not all of these features of the photolysis reaction have been quantitatively verified, the dependence of the rate on $(I_a/M)^{\frac{1}{2}}$ in the high-pressure region has been observed,[b] also in the presence of different inert gases.[c,d] L. Kassel[e] has analyzed the data of Jost and Jung and shown that they are compatible (over a concentration range of 100 fold) with the dual mode of destruction of Br atoms.[f]

That the recombination of Br atoms does obey a two-step mechanism has been directly demonstrated by Rabinowitch and Wood,[g] who have also found the expected maximum in the stationary-state concentration of Br atoms [Eq. (XIII.4.11)] to occur in the range of 200 mm Hg when He is

[a] This treatment ignores the effect of convection which is generated by and proportional in magnitude to density gradients which may be present in the vessel. The velocity of convection will thus be higher in the presence of large density gradients and so be proportional to both pressure and molecular weight at any given temperature. This will tend to make the walls more accessible and so interfere with accurate estimations of recombination coefficients in the high-pressure region.

[b] Footnote [a] page 326.

[c] K. Hilferding and W. Steiner, Z. physik. Chem., 30, 399 (1935).

[d] M. Ritchie, Proc. Roy. Soc. (London), A146, 828 (1934).

[e] L. Kassel, "Kinetics of Homogeneous Gas Reactions," chap. 11, Reinhold Publishing Corporation, New York, 1932.

[f] From Kassel's best values, I have calculated a value for k_D°/r_0^2 of 12×10^{-6} mole/cc-sec. Assuming $r_0 \cong 1$ cm for the vessel, $k_D^{\circ} = 12 \times 10^{-6}$ mole/cm-sec, which is in reasonably close agreement with the value suggested (see footnote page 327) of 3.1×10^{-6} mole/cm-sec.

[g] E. Rabinowitch and W. C. Wood, Trans. Faraday Soc., 32, 907 (1936). See also G. B. Kistiakowsky and J. C. Sternberg, J. Chem. Phys., 21, 2218 (1953), who found maximum at 200 mm Hg in a similar system.

the third body. They suggest that convection may be important in the system.

A final but not unimportant check is provided by the rate constants and temperature coefficients of the photochemical rate constants. From Eqs. (XIII.4.8) and (XIII.4.6) the over-all photochemical rate of formation of HBr in the absence of wall effects may be written as

$$\frac{d(\text{HBr})}{dt} = \frac{k_p(\text{H}_2)(I_a)^{1/2}(\text{M})^{-1/2}}{1 + k_4(\text{HBr})/k_5(\text{Br}_2)} \qquad \text{(XIII.4.13)}$$

where
$$k_p = 2k_3/k_2^{1/2} \qquad \text{(XIII.4.14)}$$

and the activation energy for the quantity k_p is seen to be $E_p = E_3 - \frac{1}{2}E_2$, E_3 and E_2 being the activation energies for k_3 and k_2, respectively. The observed value of $E_p = 17.7$ Kcal, which agrees very well with $E_3 = 17.6$ measured thermally [Eq. (XIII.2.11)].[a] If, on the other hand, the value of k_3 from the thermal reaction is substituted in Eq. (XIII.4.14) and the result solved for k_2, good agreement is obtained with independent, direct measurements of k_2 (Table XII.5). In addition it has been verified that the ratio k_4/k_5 appearing in Eq. (XIII.4.13) is identical with the inhibition constant k_i of the thermal reaction [Eq. (XIII.2.1)].[b]

5. Summary of the Chain Reaction H₂ + Br₂ → 2HBr

There are a number of features of the chain reaction of $\text{H}_2 + \text{Br}_2$ that are worth a retrospective glance. One very striking feature is that while the activation energy for the fission of a Br_2 molecule is 46 Kcal, the activation energy of the over-all reaction is only 40 Kcal. Another point already commented on is that a stationary-state concentration of H atoms is produced in the reaction which is many times greater than that normally in equilibrium with H_2. Finally, it is worth noting that, although the reverse reaction, the decomposition of HBr, is of negligible importance at the temperatures involved, there is a specific and marked inhibition of the rate by HBr.

A survey of the mechanism of the chain is helpful in interpreting these paradoxes:

---

[a] This implies that $E_2 = 0 \pm 0.2$ Kcal. This is indeed the experimental finding from independent measurements on the recombination of Br atoms.

[b] G. K. Rollefson and M. Burton, "Photochemistry and the Mechanism of Chemical Reactions," p. 302, Prentice-Hall, Inc., Englewood Cliffs, N.J., 1939, have warned against the simple algebraic summation of thermal and photochemical rates when both mechanisms are operative. In this latter case $(\text{Br})_{ss} = [K_{1.2}(\text{Br}_2) + I_a/Mk_2]^{1/2} = [(\text{Br})_{ph}^2 + (\text{Br})_{th}^2]^{1/2}$, and in similar fashion the rates must be added vectorially as well:

$$d(\text{HBr})/dt = (R_{ph}^2 + R_{th}^2)^{1/2} = R_{ph}[1 + (R_{th}/R_{ph})^2]^{1/2}$$

When $R_{th} < R_{ph}$, $d(\text{HBr})dt \cong R_{ph} + \frac{1}{2}(R_{th}^2/R_{ph})$, where R_{th} = rate of the thermal reaction [Eq. (XIII.2.9)] and R_{ph} = rate of the photolysis [Eq. (XIII.4.13)].

Radical Supply (Initiation and Termination):

$$Br_2 + M \underset{2}{\overset{1}{\rightleftharpoons}} M + 2Br \qquad E_1 = 46 \text{ Kcal}, \; E_2 = 0$$

Chain Reaction (Propagation):

$$Br + H_2 \underset{4}{\overset{3}{\rightleftharpoons}} HBr + H \qquad E_3 = 17.6 \text{ Kcal}, \; E_4 = 1 \text{ Kcal}$$

$$H + Br_2 \overset{5}{\rightarrow} HBr + Br \qquad E_5 = 1 \text{ Kcal}, \; E_6 = 42 \text{ Kcal}$$

The $Br_2 \rightleftharpoons 2Br$ reaction provides an equilibrium concentration of Br atoms in the system. Once this is achieved, since Br atoms are neither generated nor destroyed by any other appreciably competitive process, the rate processes involved in the equilibrium are of no importance. Thus it is that, aside from a very small induction period, the rate of fission of Br_2 molecules does not enter the picture and only the equilibrium constant is involved. It is not, then, the bond energy of Br_2 but only half the bond energy of Br_2 that is required *for each Br atom entering the chain.* This we will find to be a characteristic of every chain reaction, namely, a reaction which provides free radicals at a relatively small cost *per free radical.*[a]

Once provided with a source of free radicals, to have a rapid chain we need a simple sequence of steps of relatively low activation energy. In the present scheme, the slowest chain step is that of $Br + H_2$; the subsequent chain steps including the inhibition by HBr are all extremely rapid. The result is that as soon as a Br atom has reacted to form HBr + H, the H atom reacts in a very short time to reproduce the Br atom and either a molecule of H_2 if it has reacted with HBr or a molecule of HBr if it has reacted with Br_2. At any instant a certain fraction of the Br atoms which entered the chain will have been replaced by H atoms, but the chain does not deplete the *total number of radicals,* H + Br.

The reason we can have a large supply of H atoms is that it is the system $Br + H_2 \rightleftharpoons HBr + H$ which reaches a state of quasi equilibrium rather than the system $H_2 \rightleftharpoons 2H$.[b] The excess of H beyond the concentration in thermodynamic equilibrium with H_2 is paid for by the free energy liberated by the reacting system. This again we will find to be character-

---

[a] One might argue that, at sufficiently low pressures, the destruction of Br atoms would be principally at the walls and so the stationary concentration of Br atoms would be given by $(Br)_{ss} = 2k_1(M)(Br_2)/k_D$ which, since k_D has no activation energy, would require the full bond energy of 46 Kcal for the Br atom production. However, this argument is misleading because it attributes a non-stationary-state character to the wall, namely, that of acting as a trap for Br atoms without also acting as a third body for their production. In the proper stationary-state model, we must have as many Br atoms leaving the wall as there are striking it, since the thermal equilibrium $Br_2 \rightleftharpoons 2Br$ must be attained throughout the vessel.

[b] Of course as the reaction reaches completion the concentration of both Br and H decreases with it and approaches their real equilibrium values.

istic of chain reactions, namely, that the excess free energy of the reacting system can be used in part to produce superequilibrium concentrations of other substances.

We see that the rate of production of products is determined by two quantities, the first a quasi-thermodynamic quantity, the equilibrium concentration of free radicals, and the second a kinetic quantity, namely, the rate at which each radical can go through a chain cycle. When the cycle is made up of two steps of disproportionate speed as in the present case, it is the slower step (in this case $Br + H_2$) which is of importance in determining the over-all rate. It is this feature which explains in this case the specific inhibition by HBr even though the over-all reaction is essentially not reversible. The slow step in a chain will in general (though not always) be endothermic. This implies that its reverse is exothermic and hence of lower activation energy, and so faster. We can thus always expect inhibition by products in chain reactions except in those cases in which the fast steps are of unusual speed.

We can see that the $H_2 + Br_2$ system is an essentially simple chain in that there is such a separation into slow and fast chain steps. Where this is not the case, then the kinetics become much more complicated and we must deal with two chain carriers whose concentrations are algebraically difficult to separate from each other.

The work done on the system $H_2 + Br_2 \rightleftharpoons 2HBr$ constitutes as nearly consistent a body of kinetic data as we shall find in the entire literature.[a] The thermal and photochemical reactions are in quantitative agreement, and a consistent mechanism for both has been checked from a number of independent points of view. Nevertheless it would be foolhardy indeed to assert that the last word had been spoken on this relatively simple chain system.

6. Two-center Chain Reactions: Chain Lengths and Induction Periods

The $H_2 + Br_2$ reaction provides a classic example of a much more general type of chain reaction, the two-center chain. In the case of $H_2 + Br_2$, the two centers are the H and Br atoms, both of which are required to complete the chain cycle and both of which may be involved in chain termination. A more general type of two-center chain system is the one exemplified by the over-all reaction.[b]

---

[a] The reaction of $Br_2 + D_2 \rightarrow 2DBr$ has been shown by F. Bach et al., *Z. physik. Chem.*, **B27**, 71 (1934), to follow a similar mechanism. The slow chain step is similar, namely, $Br + D_2 \rightarrow DBr + D$ with $E = 19.9$ Kcal, which is 2.2 Kcal higher than for the corresponding reaction with H_2. The inhibition is quantitatively the same.

[b] Specific examples will be the substitution reaction of H_2, Br_2, Cl_2, I_2, etc., with hydrocarbons. The addition reactions of Br_2, etc., to double bonds can also be described by this system.

$$X_2 + RH \rightleftharpoons RX + HX$$

For the chain mechanism (thermal or photochemical), we can write the general scheme:

Initiation:
$$\begin{cases} X_2 + M \xrightarrow{i} 2X + M \\ X_2 + h\nu \xrightarrow{\phi I_a} 2X \end{cases}$$

Chain:
$$\begin{cases} X + RH \underset{2}{\overset{1}{\rightleftharpoons}} R + HX \\ R + X_2 \underset{4}{\overset{3}{\rightleftharpoons}} RX + X \end{cases} \qquad (XIII.6.1)$$

Termination:
$$\begin{cases} X + X + M \xrightarrow{tx} X_2 + M \\ X + R(+M?) \xrightarrow{txR} RX \qquad \text{or disproportionation} \\ R + R(+M?) \xrightarrow{tR} R_2 \qquad \text{or disproportionation} \end{cases}$$

where I_a is the specific rate of light absorption and ϕ is the quantum yield for the production of X atoms on light absorption.

We have now included three possible termination[a] steps among the two chain centers R and X for the most general case. If we apply stationary-state kinetics to this system, we find, assuming long chains,[b]

$$\theta \equiv \frac{(R)_{ss}}{(X)_{ss}} \simeq \frac{k_1(RH)}{k_3(X_2)} \frac{1 + k_4(RX)/k_1(RH)}{1 + k_2(HX)/k_3(X_2)} \qquad (XIII.6.2)$$

with
$$(X)_{ss} = \frac{[k_i(X_2)]^{\frac{1}{2}}[1 + \phi I_a/k_i(X_2)(M)]^{\frac{1}{2}}}{k_{tx}^{\frac{1}{2}}[1 + k_{txR}\theta/k_{tx}(M) + k_{tR}\theta^2/k_{tx}(M)]^{\frac{1}{2}}} \qquad (XIII.6.3)$$

and
$$-\frac{d(X_2)}{dt} \simeq k_3(X_2)(\theta)(X)_{ss} - k_4(RX)(X)_{ss} \qquad (XIII.6.4)$$

where usually the back reaction is unimportant and so only the first term need be considered. In what follows we shall thus neglect reaction 4.

These still-complicated though approximate expressions simplify in three extreme cases, occasioned by one of the three termination processes far exceeding the other two:

Case 1. Termination by $X + X$ (the result here is exact, even for short chains).

[a] There are actually five termination reactions if we include disproportionation, but since this type of reaction has the same kinetic order as the recombination, we will allow both processes to be included under a single rate constant. When R is not too complex a molecule, it is possible that the disproportionation reactions of $R + X$ and $R + R$ may be faster than recombination $X + X$, because the latter may require third bodies. Note that heterogeneous reactions have been omitted.

[b] When the chains are short, meaning for example that $k_3(X_2) \cong k_{tR}(X)$ or $k_{txR}(X)$, then the algebra becomes tediously complicated. Since, however, for most systems of interest, long chains are the rule, the above expressions are quite adequate.

$$-\frac{d(X_2)}{dt} = k_1 \left(\frac{k_i}{k_{tX}}\right)^{\!\frac12} \frac{[1 + \phi I_a/k_i(X_2)(M)]^{\frac12}(RH)(X_2)^{\frac12}}{1 + k_2(HX)/k_3(X_2)} \qquad (XIII.6.5)$$

Case 2. Termination by X + R.

$$-\frac{d(X_2)}{dt} = \left(\frac{k_i k_3 k_1}{k_{tXR}}\right)^{\!\frac12} \frac{[1 + \phi I_a/k_i(X_2)(M)]^{\frac12}(RH)^{\frac12}(M)^{\frac12}(X_2)}{[1 + k_2(HX)/k_3(X_2)]^{\frac12}} \qquad (XIII.6.6)$$

Case 3. Termination by R + R.

$$-\frac{d(X_2)}{dt} = k_3 \left(\frac{k_i}{k_{tR}}\right)^{\!\frac12} \left[1 + \frac{\phi I_a}{k_i(X_2)(M)}\right]^{\frac12}(X_2)^{\frac32}(M)^{\frac12} \qquad (XIII.6.7)$$

We see that, depending on the principal mode of termination, the kinetic order can change considerably and the inhibition by HX can become progressively less and less important as termination involving R becomes more important. In practice it might be difficult to distinguish the successive cases from each other, since the range of concentration of X_2 that is experimentally convenient is not too great. This would be particularly true if the HX inhibition were very large [that is, $k_2(HX) > k_3(X_2)$].

Most halogenation reactions which have been reported are presumed to correspond to case 1. However, a few cases corresponding to case 2 have been reported, and Goldfinger et al.[a] have observed that the photochlorination of C_2Cl_4 changes from case 3 at low temperatures, at which $C_2Cl_5 + C_2Cl_5$ is terminating, to case 2 at higher temperatures, at which C_2Cl_5 becomes less stable and Cl + C_2Cl_5 dominates the termination reaction.

We can define the chain length Φ as the ratio of the number of molecules of product RX formed to the number of radicals produced in the system. In the stationary state, neglecting the back reaction, this becomes

$$\Phi =$$

$$\frac{k_1(RH)[1 + k_2(HX)/k_3(X_2)]^{-1}}{2(M)[k_{tX}k_i(X_2)]^{\frac12}[1 + \phi I_a/k_i(X_2)(H)]^{\frac12}[1 + k_{tXR}\theta/k_{tX}(M) + k_{tR}\theta^2/k_{tX}(M)]^{\frac12}} \qquad (XIII.6.8)$$

If we assume case 1 termination and neglect the inhibition, this simplifies to

$$\Phi_1 = \frac{k_1(RH)}{2(M)[k_i k_{tX}(X_2)]^{\frac12}} \frac{1}{[1 + \phi I_a/k_i(X_2)(M)]^{\frac12}} \qquad (XIII.6.9)$$

for which we can now distinguish the photochemical and thermal chain lengths as

$$\Phi_{1th} = \frac{k_1(RH)}{2(M)[k_i k_{tX}(X_2)]^{\frac12}} \qquad \Phi_{1ph} = \frac{k_1(RH)}{2[\phi I_a k_{tX}(M)]^{\frac12}} \qquad (XIII.6.10)$$

The most important difference between these two cases arises from the

[a] J. Adam et al., *Bull. soc. chim. Belges*, **65**, 533, 549 (1956). J. Adam and P. Goldfinger, *ibid.*, 561 (1956).

fact that in the expressions for the chain lengths only k_1 and k_i have significant activation energies. Thus the activation energy for the thermal chain length is $E_1 - \frac{1}{2}E_i$, which is usually negative because E_i is almost always $> 2E_1$. This means that, as the temperature is increased, the chain length of the thermal reaction decreases, a result which arises from the fact that the rate of chain propagation increases in proportion to the concentration of free radicals, while the rate of chain termination increases as the square of the radical concentration. For the photochemical initiation the activation energy is E_1 and always positive. Note, however, that the rates of both thermal and photochemical reactions do increase with increasing temperature.

The approach to the stationary state in a chain system is not instantaneous but takes a finite time which may be calculated from the kinetic mechanism if the individual rate constants for initiation and termination are known.[a] For the homogeneous chain reaction represented by case 1 Benson[b] has shown how to calculate both the time t_a required to reach any fraction a of the stationary-state concentration of X, and the fraction of reaction F_a occurring in that time. For thermal initiation t_a is given by

$$t_{a,\text{th}} = \frac{1}{4Mk_{tX}K_{\text{eq}}^{1/2}(X_2)^{1/2}} \ln \frac{1+a}{1-a} \qquad (XIII.6.11)$$

where $K_{\text{eq}} = k_i/k_{tX}$, while for photochemical systems, it is

$$t_{a,\text{ph}} = \frac{1}{4(Mk_{tX}I_a\phi)^{1/2}} \ln \frac{1+a}{1-a} \qquad (XIII.6.12)$$

In either case, the fraction of reaction which has occurred is given by $F_a = 1 - (RH)_a/(RH)_0$ where

$$\ln (1 - F_a) = \frac{k_1}{4Mk_{tX}} \ln (1 - a^2) \qquad (XIII.6.13)$$

so that F_a is dependent not on the rate of initiation but only on the relative rates of propagation and termination.

K_{eq} is known for most diatomic molecules (capable of participating in two-center chain reactions) from spectroscopic data, while k_{tX} is known for Br_2 and I_2 for a variety of different third bodies M. This makes it possible to compute, for a number of systems which have been studied, the values of t_a and F_a. Benson and Buss[c] have done so for a number of bromination systems. In addition they have calculated from Eq. (XIII.6.11), with the aid of some simplifying assumptions, "threshold temperatures" for different diatomic species X_2. These are the temperatures at which 90 per cent

---

[a] This time does not depend on the propagation reaction, since the propagation reactions do not alter the total number of radicals but only change their relative proportions.

[b] S. W. Benson, *J. Chem. Phys.*, **20**, 1605 (1952).

[c] S. W. Benson and J. H. Buss, *ibid.*, **27**, 301 (1958).

of the stationary-state concentration of radicals will have been reached within a given time t_a. Table XIII.2 shows these times and temperatures calculated on the assumption that for all species listed M = 1 atm, X_2 = 20 mm Hg, k_{tX} = 3 × 10^{10} liters^2/mole^2-sec. For other conditions or other values of the termination constants, t_a may be computed from Eq. (XIII.6.11).

What is of special interest in Table XIII.2 is that the threshold temperatures are quite high for even such relatively weakly bonded species as I_2 and F_2 and the pre-stationary-state periods are quite long. If we compare them with the thermal systems studied,[a] we come to the conclusion that for many of these systems, either the stationary state has not been attained until relatively late in the experiment or else that, if it has been attained, it must have been by other processes, presumably wall processes. (There is by now a large body of data to indicate that walls act as catalysts to initiate and terminate chains.) In that case, however, it is quite likely that there is also a considerable amount of reaction going on at the walls. It appears that only for the systems H_2 + Br_2 and CH_4 + Br_2 have the thermal reactions been studied under conditions in which homogeneous reactions predominated.

TABLE XIII.2. THRESHOLD TEMPERATURE AND TIMES FOR ATTAINMENT
OF 90 PER CENT OF STATIONARY-STATE CONCENTRATION OF ATOMS
FROM SOME DIATOMIC MOLECULES BY HOMOGENEOUS PROCESSES

Species, X_2:	F_2	Cl_2	Br_2	I_2	H_2	O_2
Time, sec	Temperature to 5% accuracy, °K					
0.1	550	1080	780	560	2330	2710
1.0	455	880	640	470	1860	2150
10.0	390	740	550	410	1560	1600
100.0	340	650	480	350	1340	1540

TABLE XIII.3. TIMES TO REACH 90 PER CENT OF STATIONARY CONCENTRATION
OF ATOMS IN PHOTOLYSIS EXPERIMENTS AT DIFFERENT ABSORBED INTENSITIES

ϕI_a, quanta/cc-sec:	10^7	10^9	10^{11}	10^{13}	10^{15}	10^{17}
Temp, °K	Time, sec					
300	209	20.9	2.1	0.21	0.021	0.0021
500	162	16.2	1.6	0.16	0.016	0.0016

Note: Multiply ϕI_a by 1.6 × 10^{-21} to convert to moles/liter-sec (i.e., for photons, Einsteins/liter-sec).

[a] Benson and Buss, loc. cit.

It is interesting to contrast this with the situation in photolysis experiments, in which the slow thermal initiation is replaced by the much faster photochemical initiation. Table XIII.3 shows "threshold" light intensities required to attain 90 per cent of the stationary-state concentration of atoms at the times indicated. These have been calculated by Benson and Buss (loc. cit.) from Eq. (XIII.6.12), again assuming typical conditions, $M = 1$ atm, and $k_{tX} = 3 \times 10^{10}$ liters^2/mole^2-sec. In contrast to the thermal system it is now seen that in the usual range of light intensities (10^{11} to 10^{13} quanta/cc-sec), the pre-stationary-state times are relatively short and practically independent of temperature.[a]

The times shown in both these tables are in essence the lifetimes of X atoms in the respective systems. Some idea of the importance of the walls may be obtained by comparing the lifetimes with the average times required for atoms to diffuse to the walls. In the photochemical systems usually encountered, the diffusion times (neglecting convection) are of the order of 1 sec. It is clear from Table XIII.3 that, if wall termination is to be neglected, photolysis should be run at high intensities (that is, $> 10^{13}$ quanta/cc-sec), at high total pressure (e.g., by adding inert gas), and in large vessels. This is not always possible, particularly in systems in which chain lengths are long.

Table XIII.4, taken from the work of Benson and Buss (loc. cit.), shows some typical sets of values computed for thermal bromination reactions. Columns headed by R_{tXR}/R_{tX} and R_{tR}/R_{tX} give the ratio of termination expected for R + X and R + R, respectively, compared to X + X. It is seen that, as the RH compound increases in complexity, the bond energy $D(RH)$ decreases and the stationary-state concentration of R increases relative to Br. But concomitantly with this change E decreases also, and so the rate of reaction and the chain length Φ increase as well. Furthermore, termination by R + R and R + X becomes more and more important as compared to X + X. This has not actually been verified experimentally for these systems. However, the data that have been obtained for RH = ethane, isobutane, neopentane, and toluene are not self-consistent and are probably unreliable.

The methods of computation of Φ, t_a, and F_a illustrated in this section are very useful in assessing the validity of a proposed mechanism and the importance in it of heterogeneous steps. With some further assumptions concerning the order of magnitude of the relative rate constants k_{tX}, k_{tR}, and k_{tXR}, it is possible to further check the self-consistency of the scheme and the importance of alternate chain-termination steps, or alternatively,

---

[a] This does not mean that heterogeneous processes are not occurring; on the contrary, termination may still be taking place at the walls. However, it can easily be demonstrated whether or not initiation is taking place at the walls by comparing the "dark" reaction (i.e., reaction with light off) to the "light" reaction.

when these are known to be negligible, to put limits on the magnitudes of other rate constants.[a]

TABLE XIII.4. CHAIN LENGTHS AND RATIOS OF TERMINATION STEPS
FOR SOME BROMINATION REACTIONS, $Br_2 + RH \rightarrow RBr + HBr$

Species	RH	Temp, °K	$[(R)/(Br)]_{ss} = \theta$ Eq. (XIII.6.11)	Chain length[a] Φ_{th}	$\dfrac{R_{tXR}}{R_{tX}}$	$\dfrac{R_{tR}}{R_{tX}}$	Fraction of reaction at $a = 0.9$
H—H	b	570	4×10^{-7}	120	4×10^{-7}	1.6×10^{-13}	8×10^{-6}
CH₃—H	c	570	4.5×10^{-4}	100	2×10^{-2}	9×10^{-6}	6×10^{-7}
BrCH₂—H	d	570	4×10^{-4}	250	2×10^{-2}	7×10^{-6}	5×10^{-6}
C₂H₅—H	e	500	7.5×10^{-3}	3600	0.75	6×10^{-3}	3×10^{-5}
Me₃C—H	f	400	0.57	4×10^{9}	0.38	0.20	0.04
Me₃CCH₂—H	g	470	1.5	2×10^{5}	120	180	2×10^{-4}
ϕCH₂—H	h	440	20	1×10^{7}	2800	5.6×10^{4}	0.005
CCl₃—H	i	440	3.1×10^{-3}	1.5×10^{5}	5×10^{-2}	5×10^{-5}	2×10^{-5}

[a] Computed from Eq. (XIII.6.10) and experimental data.

[b] It is assumed that $k_{tX} = k_{tR} = k_{tRX}$. Computations are made for (M) = (RH) = 1 atm, (X₂) = 20 mm Hg.

[c] The value of G. B. Kistiakowsky and E. K. Roberts, *J. Chem. Phys.*, **21**, 1637 (1953), for CH₃ + CH₃ is used here, and it is assumed that k_{tRX} is equal to it. This is probably an upper limit for k_{tRX}. (RH) = (M) = 390 mm Hg and (Br₂) = 30 mm Hg.

[d] By using the values $k_{tRX} = k_{tR} = k_{tR}$ (CH₃ + CH₃) above. Values computed for (RH) = (M) = 390, (Br₂) = 30 mm Hg.

[e] By using the values $k_{tRX} = k_{tR} = k_{tR}$ (CH₃ + CH₃) above. Values computed for (RH) = (M) = 320, (Br₂) = 40 mm Hg. It is further assumed that $A_1 = A_1(CH_3Br) = 10^{10.7}$ liters/mole-sec and $E_3 = E_2 = 2.9$ Kcal. Note that C₂H₅Br begins to decompose at 500°K. The example is purely hypothetical.

[f] θ calculated by using $k_2/k_3 = 4$, (RH)/(Br₂) = 8 and neglecting inhibition. Assumed also that $k_{tXR} = k_{tR} = k_{tR}(CH_3 + CH_3)$ and (M) = (RH) = 380 mm Hg.

[g] θ calculated by using $k_2/k_3 = 14$, (RH)/(Br₂) = 250 mm Hg/30 mm Hg, and (HBr)/(Br₂) = 1/3. Note that for the thermal runs without added HBr, θ is initially sixfold higher because inhibition has not been neglected.

[h] By using $k_2/k_3 = 2$ (from kinetic data), (RH)/(Br₂) = 200/25, and neglecting inhibition. Assumed also that $k_{tRX} = k_{tR} = k_{tR}(CH_3 + CH_3)$.

[i] By using kinetic data and thermal data; (Br₂) = 47 mm Hg, (RH)/(Br₂) = 16; and neglecting inhibition, which appears unimportant. Same assumption as above for other rate constants.

7. Reactions of H_2 with the Halogens

The kinetics of the thermal and photochemical chain reactions of bromine with saturated, hydrogen-containing compounds (for example, $Br_2 + HR \rightarrow HBr + RBr$), usually has the form[b]

Thermal Reaction:

$$-\frac{d(Br_2)}{dt} = \frac{k_3(HR)(K_{1,2}Br_2)^{\frac{1}{2}}}{1 + k_4(HBr)/k_5(Br_2)} \qquad (XIII.7.1)$$

Photolysis:

$$-\frac{d(Br_2)}{dt} = \frac{k_3(HR)(I_a/k_2M)^{\frac{1}{2}}}{1 + k_4(HBr)/k_5(Br_2)} \qquad (XIII.7.2)$$

Ignoring for the moment the denominators, which account for the in-

[a] Thus if $\theta = (R)/(Br)$ is known to be less than a certain value, this automatically puts limits on the ratio of rate constants, k_1/k_3 [Eq. (XIII.4A.2)].

[b] See Eqs. (XIII.4.13), (XIII.4.14), and (XIII.2.9). We have neglected wall effects in the above equations.

hibition, the numerators in these expressions may be looked upon as a product of two factors, one representing the stationary-state concentration of Br atoms $(K_{1.2}Br_2)^{1/2}$ and $(I_a/k_2M)^{1/2}$, respectively, and the other the net rate at which product HBr is produced per unit concentration of Br atoms. In the chains we have thus far considered, the cycle was only two steps, one slow and one fast.[a] Because of this, only the rate constant k_3 for the slow chain step appears in the over-all rate expression. This is the H atom abstraction: $Br + RH \xrightarrow{3} HBr + R$. For the thermal reaction the over-all chain constant is $k_t = k_3 K_{1.2}^{1/2}$, i.e., a product of an equilibrium constant $K_{1.2}$ and a bimolecular rate constant k_3. Its activation energy, which will be the experimental activation energy of the chain, is given by $E_t = E_3 + \frac{1}{2} \Delta E_{1.2}$.

If we now wish to make comparisons of the chain reactions of Br_2 with those of the other halogens which might be expected to show similar kinetic behavior, we shall need to consider the bond energies of the halogens $\Delta E_{1.2}$ and the activation energies of the atom reactions. In Table XIII.5 are collected some of the relevant thermal data on halogen reactions, while in Table XIII.6 are listed the experimental activation energies for those reactions.

With the exception of F_2, the reaction $H + X_2 \rightarrow HX + X$ is very exothermic, whereas the reaction $X + H_2 \rightarrow HX + H$ is endothermic. In addition, the entropy change accompanying the reaction of H atoms is always more positive than that for the X atom reaction. Consequently, we would expect that the X atoms would be the slow chain carriers, whereas H atoms would react readily and have a lower stationary-state concentration than X. These conditions are the ones required by the $H_2 + Br_2$ chain, so that we might expect to find Cl_2 and I_2 chains following the Br_2 mechanism.

The minimum activation energies for such chains would be $E_t = E_3 + \frac{1}{2} \Delta H_{1.2} \geqslant \Delta H_{3.4} + \frac{1}{2} \Delta H_{1.2}$, or 30.0 Kcal for $Cl_2 + H_2$, 39.7 Kcal for $Br_2 + H_2$, and 50.6 Kcal for $I_2 + H_2$. Since $\Delta S_{1.2}$ is about the same for all the halogens, we might thus expect that these activation energies would account for a major part of the differences that might exist between Cl_2, Br_2, and I_2. This turns out to be the case. For $H_2 + I_2$, the activation energy is sufficiently high that at lower temperatures the bimolecular reaction with an activation energy of about 39 Kcal (Table XII.4) provides the principal reaction path.[b] The $H_2 + Cl_2$ reaction is faster than the

[a] Here we are referring to the mean rate (sec^{-1}) taken by a chain carrier to complete a step in the chain.

[b] S. W. Benson and R. Srinivassan, *J. Chem. Phys.*, **23**, 200 (1955), have shown that at temperatures above 600°K the chain mechanism competes with this direct reaction despite the higher activation energy of the chain. The reason is that $k_t = k_3 K_{1.2}^{1/2}$ and $K_{1.2}^{1/2}$ contributes a relatively large entropy term (Table XIII.5) of about 12.0 cal/mole-°K, which represents a factor of about 420 in the rate constant. This has been verified experimentally by J. H. Sullivan, *J. Chem. Phys.*, **30**, 1292 (1959).

TABLE XIII.5. THERMAL DATA FOR HALOGEN MOLECULE AND ATOM REACTIONS IN GAS PHASE

Reaction no.	Halogen, X:	F		Cl		Br		I	
	Reaction	$\Delta H°$	$\Delta S°$	$\Delta H°$	$\Delta S°$	$\Delta H°$	$\Delta S°$	$\Delta H°$	$\Delta S°$
1, 2	$\frac{1}{2}X_2 \rightleftharpoons X$	19.3	13.6	29.0	12.8	23.0	12.5	17.7	12.0
3, 4	$X + H_2 \rightleftharpoons HX + H$	-30.4	-0.3	1.0	1.4	16.7	1.8	32.9	2.3
3', 4'	$X + CH_4 \rightleftharpoons HX + CH_3$	-32.6	6.1	-1.2	7.7	14.5	8.1	30.7	8.6
	$X_2 + H_2 \rightleftharpoons 2HX$	-128.4	3.1	-44.1	4.7	-24.7	5.0	-3.1	5.1
	$X_2 + CH_4 \rightleftharpoons CH_3X + HX$	-98.1	1.7	-23.8	2.8	-6.6	3.0	12.4	3.4
5, 6	$H + X_2 \rightleftharpoons HX + X$	-98.0	2.8	-45.1	3.3	-41.4	3.2	-36.0	2.8
5', 6'	$CH_3 + X_2 \rightleftharpoons CH_3X + X$	-65.5	2.0	-24.8	1.4	-21.1	-5.1	-20.5	1.1

Note: These data are based on NBS values together with these assumed values: $S°(CH_3) = 47.0$ cal/mole-°K, $\Delta H°_f(CH_3F) = -52$ Kcal/mole. All substances in gas phase. Standard states are 25°C, 1 atm (ideal gases).

$H_2 + Br_2$ reaction. The activation energy of $Cl + H_2 \rightarrow HCl + H$ is about 6 Kcal, so that the over-all chain would have an activation energy

TABLE XIII.6. SOME EXPERIMENTAL ACTIVATION ENERGIES FOR THE REACTIONS OF HALOGEN MOLECULES AND ATOMS IN THE GAS PHASE

Halogen, X:	Cl	Br	I
Reaction	Activation energy		
$X + X + M \rightarrow M + X_2$[a]	0	0	0
$X + H_2 \rightarrow HX + H$	4.6–6.1[b]	17.6	33.4[d]
$H + HX \rightarrow H_2 + X$	3.6–5.1[c]	0.9[c]	0.5[d]
$H + X_2 \rightarrow HX + X$	2–4[e]	0.9[f]	0[d]

[a] There is evidence (Sec. XII.16) that these reactions may have a small negative temperature coefficient, especially when M = benzene or other similarly efficient activator.

[b] Based on the temperature coefficient of the photolysis and the inhibition of HCl formation by NO. The low value 4.6 comes from the low-temperature photolysis.

[c] Computed from the thermal data and the activation energy of the reverse reaction.

[d] J. H. Sullivan, *J. Chem. Phys.*, **30**, 1292 (1959).

[e] Estimated from the effect of O_2 on the photolysis and the small inhibition by HCl.

[f] From the inhibition of HBr.

of 35 Kcal. This is 5 Kcal less than that for $H_2 + Br_2$ and would represent a factor in the rates of 260 at 180°C. On the other hand, in the photochemical reactions only E_3 is involved. Here the activation energies for the photolyses are 6 Kcal for Cl_2, 17.6 Kcal for Br_2, and an estimated 35 Kcal for I_2. The differences are now much larger, and we find that the $H_2 + Cl_2$ chain has chain lengths as much as 10^6 higher than the $H_2 + Br_2$ chain, which is in line with these differences.[a]

The reaction of F_2 with H_2 might be expected to be a rather complicated chain because the F atoms are just about as reactive as the H atoms. In such a case we can no longer neglect chain-ending reactions such as $H + H + M$ and $H + F + M$ compared to $F + F + M$, and the algebraic rate expressions become much more complicated. This reaction is one of the most exothermic known (per unit weight of reactants) and has been used to give intense flames. However, very little is known of the quantitative behavior of the system except the fact that it is very wall-sensitive.[b]

A great deal of work has been done on the thermal and photochemical

[a] W. A. Noyes, Jr., and P. A. Leighton, "Photochemistry of Gases," chap. VI, Reinhold Publishing Corporation, New York, 1941.

[b] M. Bodenstein et al., *Z. anorg. u. allgem. Chem.*, **231**, 24 (1937), found very slow reaction in an Mg vessel at room temperature but appreciable reaction in Pt vessel at −78°C.

reactions[a] of $H_2 + Cl_2$ with not altogether satisfactory results. The system is extraordinarily sensitive to the walls and traces of impurities such as O_2, H_2O, and NCl_3;[b] Johnston and Libby[c] have shown that the exchange of HCl* with Cl_2 is fast and heterogeneous at room temperature in pyrex. It has also been shown[d] that the chain system $H_2 + Cl_2$ cannot reach a stationary state because the chain cycle is too fast for the homogeneous chain-ending process, namely, $2Cl + M \rightarrow Cl_2 + M$. This would unquestionably also be true of the H_2 and F_2 system if it were a chain reaction. At temperatures below 200°K this objection would no longer be valid, and it is interesting to note that the photolysis reaction, $H_2 + Cl_2 \rightarrow 2HCl$, has been found to be well behaved below 172°K and poorly behaved at higher temperatures.[e] At these low temperatures the photolysis rate is given by

$$\frac{d(HCl)}{dt} = k_p I_a^{1/2}(H_2) \tag{XIII.7.3}$$

This shows the expected dependence of the rate on (H_2) and on light intensity for a reaction in which Cl is the slow chain carrier and chain termination is $2Cl + M \rightarrow M + Cl_2$. However, we should expect to find a dependence on total pressure, and this is not observed. At higher temperatures, Potts and Rollefson[f] claim to observe a law of the form

$$\frac{d(HCl)}{dt} = \frac{k_p I_a(H_2)}{1 + k_i(HCl)/(Cl_2)} \tag{XIII.7.4}$$

with an inhibition constant $k_i < 0.1$ and a dependence on the first power of the light intensity presumably indicating first-order disappearance of Cl atoms by diffusion (or convection) to the walls.

This is in more or less reasonable agreement with other work[g,h] in the field, although because of the difficulties already discussed concerning the validity of the stationary-state treatment in this system, all these results must be taken with reservations.

[a] P. G. Ashmore and J. Chanmugam, *Trans. Faraday Soc.*, **49**, 254 (1953). J. C. Morris and R. N. Pease, *J. Am. Chem. Soc.*, **61**, 394, 396 (1939). W. J. Kramers and L. A. Moignard, *Trans. Faraday Soc.*, **45**, 903 (1943). See also Noyes and Leighton, *loc. cit.*

[b] Morris and Pease, *loc. cit.*, give the following values: In photolysis reaction 1 per cent of O_2 at 25°C reduces the rate by a factor of 10^3!! In thermal reaction at 200°C, 1 per cent of O_2 reduces rate only 10 fold. Coating vessel with KCl changes these effects.

[c] W. H. Johnston and W. F. Libby, *J. Am. Chem. Soc.*, **73**, 854 (1951).

[d] S. W. Benson, *J. Chem. Phys.*, **20**, 1605 (1952).

[e] J. C. Potts and G. K. Rollefson, *J. Am. Chem. Soc.*, **57**, 1027 (1935).

[f] *Ibid.*

[g] M. Bodenstein and W. Unger, *Z. physik. Chem.*, **B11**, 253 (1930); M. Bodenstein and E. Winter, *Sitz. ber. preuss. Akad. Wiss., Physik.-math. Kl.*, **1** (1936).

[h] M. Ritchie and R. G. W. Norrish, *Proc. Roy. Soc. (London)*, **A140**, 99, 112, 713 (1933), find $k_i = 1.7$. However, Potts and Rollefson have shown this large value is inconsistent with known thermal data and most likely due to impurities in the HCl.

Craggs[a] and Allmand and Squire[b] worked with very low H_2 concentration and showed that the dependence on light intensity goes from I_a at low Cl_2 concentrations (that is, ~ 0.01 mm Hg) and low light intensities to $I_a^{1/2}$ at high Cl_2 concentrations (that is, ~ 450 mm Hg) and high light intensities. At constant light intensity the rate passes through a maximum value as Cl_2 pressure changes. These characteristics are precisely what we might expect if there is a dual mode of removal of Cl atoms from the system, similar to the case of Br [Eq. (XIII.4.4)].

However, these latter authors have proposed as the basis for the specific effects of Cl_2 as a third body that the unstable intermediate Cl_3 is important in the system. The chain-breaking steps would be then written as[c]

$$Cl + Cl_3 \underset{8}{\overset{7}{\rightleftharpoons}} 2Cl_2 + 58 - D(Cl_3) \qquad\qquad \text{(XIII.7.5)}$$

$$Cl_3 + Cl_3 \overset{2'}{\rightarrow} 3Cl_2 + 58 - 2D(Cl_3) \qquad\qquad \text{(XIII.7.6)}$$

where $D(Cl_3)$ is the heat of the reaction $Cl_3 \rightarrow Cl_2 + Cl$. The authors use only reaction $2'$ as the chain-breaking step which does not seem at all justified, because 7 should compete quite favorably, the ratio of $(Cl)/(Cl_3)$ being $\geqslant 1$ under the conditions of their experiment.

The idea of Cl_3 (or for halogen reactions, X_3 and even HX_2) or HCl_2 as an intermediate in chlorination reactions is indeed a reasonable one and one that has been proposed and looked for in a number of different reactions.[d] It appears feasible from a chemical point of view. From a kinetic

---

[a] H. C. Craggs and A. J. Allmand, *J. Chem. Soc.*, 241 (1936); 1889 (1937).

[b] G. V. V. Squire and A. J. Allmand, *ibid.*, 1869 (1937). G. V. V. Squire et al., *ibid.*, 1878.

[c] The authors use the scheme:

$$Cl_2 + I_a \overset{1'}{\rightarrow} 2Cl \ (2I_a)$$

$$Cl + Cl_2 \underset{8}{\overset{7}{\rightleftharpoons}} Cl_3 \ (K_{7.8})$$

$$Cl + H_2 \underset{4}{\overset{3}{\rightleftharpoons}} HCl + H$$

$$H + Cl_2 \overset{5}{\rightarrow} HCl + Cl$$

$$2Cl_3 \overset{2'}{\rightarrow} 3Cl_2$$

which leads to

$$(Cl)_{ss} = (2I_a + k_8 I_a^{1/2}/k_{2'}^{1/2})/k_i(Cl_2)$$

$$(H)_{ss}/(Cl)_{ss} = \{k_3(H_2)/[k_4(HCl) + k_5(Cl_2)]\}(Cl_3)_{ss} = (I_a/k_{2'})^{1/2}$$

and $\qquad d(HCl)/dt = 2k_5(Cl_2)(H)_{ss} = 2k_3(H_2)(Cl)_{ss}/[1 + k_4(HCl)/k_5(Cl_2)]$

The chlorination reactions of Cl_3 are left out without real justification because, although they are less reactive radicals than Cl, they will be present in proportionately great abundance.

[d] G. K. Rollefson and H. Eyring, *J. Am. Chem. Soc.*, **54**, 170 (1932).

point of view it is precisely what one would propose in the termolecular process $Cl + Cl + Cl_2 \rightarrow 2Cl_2$, where Cl_2 is the third body[a] (Sec. XII.16). If $D(Cl_3)$ is at all appreciable (that is, >3 to 4 RT), we might actually expect Cl_3 to have a reasonable lifetime, especially at low temperatures. To date, however, there has been presented no unambiguous evidence which would give much support to the existence of Cl_3.[b]

8. Reactions of CH_4 with the Halogens

The reactions of the halogens with CH_4 are roughly parallel to those with H_2, which is to be expected in view of the very similar bond strengths involved. The main differences arise from the secondary reactions of the products with the halogens, which add enormously to the complexity of the rates. Of particular interest is the rather large entropy increase associated with the reaction $X + CH_4 \rightarrow CH_3 + HX$ (Table XIII.1) as contrasted with the similar reaction with H_2. There is a correspondingly small entropy change for the reaction $CH_3 + X_2 \rightarrow CH_3X + X$.[c] The result should be a very much diminished, if not negligible, inhibition in these reactions by HX. This seems to be in accord with what little evidence does exist.[d]

9. The Pyrolysis of Hydrocarbons; The Rice-Herzfeld Mechanisms; Decomposition of Butanes

Although there is still considerable controversy about the relative importance of the different possible modes of pyrolysis of saturated hydrocarbons under inhibiting conditions, it is generally agreed that the uninhibited reactions proceed principally by a radical chain mechanism.

The first general mechanism to account for the presumed first-order kinetics of these and other organic pyrolysis reactions was proposed by

[a] The presumed advantage of Cl_3 from a kinetic viewpoint seems largely spurious. The chain-breaking reactions $Cl_3 + Cl \rightarrow 2Cl_2$ or $Cl_3 + Cl_3 \rightarrow 3Cl_2$ are certainly expected to be faster than $Cl + Cl + M \rightarrow M + Cl_2$ in that the former do not require third bodies. However, the process of forming Cl_3, namely, $Cl + Cl_2 \rightarrow Cl_3$, must certainly require a third body, and the over-all advantage of a fast chain-breaking step is completely lost.

[b] It would appear in highest concentrations in Cl_2 vapor at low temperature and high pressure. Thus Cl_2 gas at its boiling point in a photostationary state (i.e., exposed to exciting radiation) would be the most likely place to find Cl_3. For evidence against Cl_3, see G. Chiltz et al., $Bull.$ $soc.$ $chim.$ $Belges$, **67**, 33 (1958).

[c] This may in part arise from an overestimation of the standard entropy of CH_3 by 1 to 2 cal/mole-°K (Table XIII.5).

[d] Of course for $CH_4 + I_2$ the equilibrium is in favor of starting materials, and in fact HI reacts with CH_3I to produce them. In the case of $CH_4 + Br_2$, the equilibrium is not too far over toward $CH_3Br + HBr$ (Table XIII.5).

Rice and Herzfeld.[a] Subsequent work has strengthened their basic premises, which may be summarized as follows:

Initiation:
1. Free radicals are initiated by the splitting of the molecule at its weakest link.

Chain Propagation (First and Second Order):
2. One of these radicals abstracts H from the parent compound to form a small saturated molecule and a new free radical. Both radicals may so react.
3. Free radicals of the type RCH_2—$CH_2\cdot$ can stabilize themselves by splitting off ethylene:[b]

$$RCH_2\text{---}CH_2\cdot \rightarrow R + CH_2\!\!=\!\!CH_2$$

Termination:
4. Chain ending occurs through association or disproportionation of radicals.

These assumptions, together with the relevant thermal data, are of tremendous use in correlating and understanding the extraordinarily complicated product distributions of many pyrolysis reactions.[c]

Thus in the pyrolysis of n-butane at about 550°C, the initial distribution of products is roughly[d] $H_2 = 3$ per cent, $CH_4 = 34$ per cent, $C_2H_4 = 15$ per cent, $C_2H_6 = 14$ per cent, and $C_3H_6 = 34$ per cent. These products can be reasonably well accounted for by the Rice-Herzfeld mechanism:

$$CH_3CH_2CH_2CH_3 \underset{2}{\overset{1}{\rightleftharpoons}} 2C_2H_5 \qquad \text{or } CH_3 + C_3H_7$$

$$C_2H_5 + n\text{-butane} \overset{3}{\rightarrow} C_2H_6 + p\text{-butyl}$$

$$\overset{3'}{\rightarrow} C_2H_6 + sec\text{-butyl}$$

$$C_2H_5 \overset{4}{\rightarrow} C_2H_4 + H$$

$$H + n\text{-butane} \overset{5,5'}{\rightarrow} H_2 + (p\text{- or } sec\text{-}) \text{ butyl}$$

[a] F. O. Rice and K. F. Herzfeld, *J. Am. Chem. Soc.*, **56**, 284 (1934). See also F. O. Rice and K. K. Rice, "The Aliphatic Free Radicals," Johns Hopkins Press, Baltimore, 1935.
[b] Secondary radicals (R—CH_2—$\dot{C}H$—CH_2—R') will split out more complex olefins as will tertiary radicals ($R'R''R'''C\cdot$).
[c] A further hypothesis, the Rice-Teller hypothesis of least motion, is a powerful tool in understanding the chain kinetics. This states that in a metathesis, that path is most likely which involves least concerted motion of nuclei.
[d] E. W. R. Steacie and I. E. Puddington, *Can. J. Res.*, **16B**, 176 (1938). The products show only small variation with temperature, pressure, extent of decomposition (up to about 20 per cent) or presence of inhibitors such as NO (e.g., E. W. R. Steacie and H. O. Folkins, *ibid.*, **18B**, 1 (1940). See also V. A. Crawford and E. W. R. Steacie, *Can. J. Chem.*, **31**, 937 (1953), for further results on both inhibited and uninhibited reactions.

$$(p\text{-butyl}) \ CH_3\text{---}CH_2\text{---}CH_2\text{---}CH_2 \cdot \overset{6}{\underset{7}{\rightleftharpoons}} C_2H_4 + C_2H_5 \qquad (XIII.9.1)$$

$$(sec\text{-butyl}) \ CH_3\text{---}CH_2\text{---}\dot{C}H\text{---}CH_3 \overset{6'}{\underset{7'}{\rightleftharpoons}} C_3H_6 + CH_3$$

$$CH_3 + n\text{-butane} \overset{8,8'}{\rightarrow} (p\text{- or } sec\text{-}) \ butyl + CH_4$$

$$2C_2H_5 \overset{2'}{\rightarrow} C_2H_4 + C_2H_6$$

$$2CH_3 \overset{9}{\rightarrow} C_2H_6$$

$$CH_3 + C_2H_5 \overset{10}{\rightarrow} CH_4 + C_2H_4 \ (\text{or } C_3H_8)$$

We see that there are two alternate chain cycles in the above mechanism, one which proceeds via the primary butyl radical and ethyl radicals (reactions 3 and 6) and whose principal products would be presented by

$$n\text{-}C_4H_{10} \rightarrow C_2H_4 + C_2H_6 \qquad (XIII.9.2)$$

together with some amount of side reaction to produce H_2 from the decomposition of the ethyl radical (reaction 4) followed by H abstraction (reaction 5).

The other chain is produced by the decomposition of the sec-butyl radicals (reaction 6') to give methyl radicals + propylene. The chain cycle consists of reactions 6' and 8' and would be represented by the over-all stoichiometry

$$n\text{-}C_4H_{10} \rightarrow CH_4 + C_3H_6 \qquad (XIII.9.3)$$

with some accompanying ethane production, the relative quantity of which would depend on the speed of this cycle compared to chain-breaking processes[a] involving the CH_3 radical (reactions 9 and 10).

We would expect from independent considerations that the bond strengths of secondary H are less than primary H in hydrocarbons, and this is in accord with the greater yields of sec-butyl radicals which would be required to give the major yields of CH_4 and C_3H_6 which are observed.

The relatively small yield of H via reactions 4 and 5 or 5' is again to be expected[b] from our knowledge of the endothermicity of reaction 4 ($\Delta H_4^\circ \geqslant 38$ Kcal) and the much lower expected activation energies of the competing, ethyl radical reactions (3, 3', 2, 2', 10) for which E should be in the range 0 to 13 Kcal.

[a] A number of such processes involving larger radicals have been omitted from the scheme for purposes of simplification. They are undoubtedly present and important at higher pressures of butane, to judge from the growing importance under these conditions of larger amounts of higher hydrocarbons and polymers.

[b] I have omitted similar reactions such as butyl → butene + H because it is expected that these reactions will be much slower than the competing unimolecular reactions 6 and 6', which are at least 16 Kcal less endothermic and may be expected to have comparable preexponential factors.

When chain lengths are large, the products may be predicted simply by considering only propagation steps. For n-butane this gives:

$$R^{\cdot} + n\text{-butane} \xrightarrow{a} RH + p \text{ or } sec\text{-}C_4H_9$$

$$p\text{-}C_4H_9 \xrightarrow{b} C_2H_4 + C_2H_5$$

$$sec\text{-}C_4H_9 \xrightarrow{b'} C_3H_6 + CH_3$$

We see that when $R^{\cdot} = C_2H_5$ we have a two-step chain $(a + b)$ whose products are $C_2H_4 + C_2H_6(RH)$. When $R^{\cdot} = CH_3$ we also have a two-step chain $(a + b')$ with products $C_3H_6 + CH_4(RH)$.

For purposes of comparison we can consider the pyrolysis of isobutane, which at 575°C decomposes at a slightly slower rate than n-butane[a] but appears to have a slightly smaller temperature coefficient. At about 5 per cent decomposition, the major products of the pyrolysis are $H_2 = C_4H_8 = 30$ per cent, $CH_4 = 16$ per cent, $C_3H_6 = 21$ per cent, $C_2H_4 = C_2H_6 = 1.5$ per cent.[b] In sharp contrast to the n-butane pyrolysis, H_2 and isobutene (C_4H_8) are now major products of the reaction, whereas C_2H_4 and C_2H_6 are almost negligible. CH_4 and C_3H_6 are important products in both systems, although less important for isobutane.

Again the Rice-Herzfeld mechanism seems to give qualitative coherence to the data. A reasonable sequence could be the following:

$$\text{isobutane} \underset{2}{\overset{1}{\rightleftharpoons}} CH_3 + CH_3\text{---}\overset{\cdot}{C}H\text{---}CH_3$$

$$CH_3\text{---}CH\text{---}CH_3 \underset{4}{\overset{3}{\rightleftharpoons}} C_3H_6 + H$$

Chain I:
$$\left\{ \begin{array}{l} H + \text{isobutane} \xrightarrow{5,5'} H_2 + C(CH_3)_3 \quad \text{or } CH_2CH(CH_3)_2 \\[2mm] C(CH_3)_3 \underset{7}{\overset{6}{\rightleftharpoons}} H + CH_2\text{=}C(CH_3)_2 \qquad (XIII.9.4) \end{array} \right.$$

Chain II:
$$\left\{ \begin{array}{l} CH_3 + \text{isobutane} \xrightarrow{8} C(CH_3)_3 + CH_4 \\[2mm] \qquad\qquad\qquad \xrightarrow{8'} CH_2CH(CH_3)_2 + CH_4 \\[2mm] CH_2CH(CH_3)_2 \underset{7'}{\overset{6'}{\rightleftharpoons}} C_3H_6 + CH_3 \end{array} \right.$$

$$2(\text{isobutyl}) \xrightarrow{10} \text{diisobutyl} \quad \text{or disproportionation}$$

Here again there are two radical chains, one provided by CH_3 or by H atoms and going through the t-butyl radical (reactions 5, 8, and 6). The stoichiometry of this chain corresponds to isobutane $\rightarrow H_2 +$ isobutene.

---

[a] E. W. R. Steacie and I. E. Puddington, *Can. J. Res.*, **16B**, 260 (1938).

[b] These yields are quite sensitive to percentage decomposition but relatively insensitive to temperature in the range 520 to 580°C and pressure in the range 20 to 60 cm Hg.

The second chain is again initiated by H or CH_3 but goes through the isobutyl radical, $-CH_2-CH(CH_3)_2$, which can only stabilize itself in a simple way by splitting off CH_3 (reactions 5', 8', and 6'). Its stoichiometry is represented by isobutane $\rightarrow CH_4 + C_3H_6$. From the preponderance of H_2 and C_4H_8 as products of the reaction it would seem that the principal mode of attack of H or CH_3 on isobutane is to abstract the tertiary H atom and produce the t-butyl radical (reactions 5, 8).[a]

There has been a great deal of work done on the kinetics of these pyrolyses,[b,c] with many conflicting and confusing results. Although the uninhibited pyrolyses are described as first order, the evidence for that is very poor. The initial rates fall markedly with decreasing pressure,[d] while the calculated first-order rate constants in any given run will show a decelerating decrease with increasing percentage decomposition to an almost steady value reached near complete decomposition. This decrease is most marked in the first stages of reaction and may amount to 40 to 60 per cent over the first 50 per cent decomposition.

The decrease is attributable in part to the inhibition of the reaction by products (particularly propylene and higher olefins, ethylene is not very effective) and in smaller part to reversible hydrogenation as equilibrium is approached.

Hydrogen has a rather spectacular effect on the rate, accelerating it in some cases by as much as 100 to 200 per cent. This has been attributed to a replacement of the less reactive alkyl radicals CH_3 and C_2H_5 by the more reactive H atoms via the metathesis $R + H_2 \rightarrow RH + H$.

The pyrolyses are somewhat irreproducible in clean, new reaction vessels but seem to achieve stability after the glass surface becomes coated with a carbonaceous deposit, after which the rates seem to be almost independent of surface/volume ratio. This has been reasonably interpreted by Rice and Herzfeld[e] as indicating the making and breaking of chains at the wall. This is also consistent with the fact that the supposedly "inhibited reaction" is much more sensitive to surfaces than is the uninhibited chain reaction. In the case of the inhibited reaction the chains

---

[a] On a purely statistical basis we might expect the abstraction of primary H over tertiary H in the ratio 9:1. The fact that the observed ratio is (judging crudely from the ratio of CH_4/H_2) about 1:2 indicates a total factor of 18 or more in favor of the tertiary H atom. At a temperature of 550°C this would be equivalent to a difference in activation energies for the H abstraction of about 4.8 Kcal favoring the tertiary H. This is in agreement with independent information on the relative strength of C—H bonds: primary > secondary > tertiary. See F. O. Rice and T. A. Vanderslice, *J. Am. Chem. Soc.*, **80**, 291 (1958).

[b] Crawford and Steacie, *loc. cit.*

[c] F. J. Stubbs and C. Hinshelwood, *Proc. Roy. Soc.* (*London*), **A200**, 458 (1950); **A201**, 18 (1950). K. J. Ingold et al., *ibid.*, **A203**, 486 (1950).

[d] Stubbs and Hinshelwood, *op. cit.*, **A201**, 18 (1950), show a 50 per cent increase in rate for n-butane on going from 6 to 50 cm Hg pressure at 530°C.

[e] F. O. Rice and K. F. Herzfeld, *J. Phys. & Colloid Chem.*, **55**, 975 (1951).

presumably start at the walls but may be suppressed in the gas phase as well as at the walls, and so the rate shows an increase with increased surface as well as a sensitivity to the nature of the surface.

The effects of inhibitors such as NO and propylene are rather interesting. There have been a large number of papers on the effects of these inhibitors.[a] Crawford and Steacie, working with NO + n-butane, found a maximal inhibition by NO at about 10 mole per cent. The rates under these conditions were not, however, very reproducible, and contrary to earlier reports gave a different product distribution from the uninhibited reactions.[b]

Comparisons of NO and propylene show that their effects are comparable, the NO being about 10 to 12 times more effective as an inhibitor. Even in the presence of NO, Stubbs et al. found some acceleration of the rate by H_2 which could be interpreted as indicating short chains, although the authors prefer an alternative explanation.

Much has been made of the kinetics of the hydrocarbon pyrolyses despite the fact that they fit no simple kinetic expression and seem very complex. The schemes given in this section for the decompositions of n- and isobutanes [Eqs. (XIII.9.1) and (XIII.9.2)] cannot be solved explicitly for the over-all reaction rates even by using the stationary-state hypothesis.

In the case of the butane pyrolysis, the stationary-state concentration of CH_3 radicals is given to a first approximation (high pressures of n-butane and neglecting inhibition by C_2H_4) by

$$(CH_3)_{ss} = \left[\frac{k_1(n\text{-Bu})}{k_9 + Rk_{10} + R^2(k_2 + k_{2'})} \right]^{1/2} \qquad \text{(XIII.9.5)}$$

$$R = \frac{(C_2H_5)_{ss}}{(CH_3)_{ss}} = \frac{k_8(n\text{-Bu})}{\dfrac{k_4 k_{5'}}{k_5 + k_{5'}} + k_{3'}(n\text{-Bu})} \qquad \text{(XIII.9.6)}$$

and from this we can calculate $d(CH_4)/dt$ as

$$\frac{d(CH_4)}{dt} = (k_8 + k_{8'})(n\text{-Bu})(Me)_{ss} \qquad \text{(XIII.9.7)}$$

We see that, depending on the value of R, the order of the reaction with respect to n-butane is very complicated and may appear to be anything from $\frac{3}{2}$ to $\frac{1}{2}$. The fact that the observed kinetics appear to be close to first order over a limited range of pressures is thus not necessarily a sufficient test of the real kinetics of the system.

---

[a] Stubbs et al., *loc. cit.*, and Steacie et al., *loc. cit.*

[b] For example, the yield of C_2H_6 became extremely small. This is in conflict with the work of Hinshelwood et al. The two groups also differ in their interpretations of the inhibited reactions. See also the discussion by V. V. Voevodskü, *Trans. Faraday Soc.*, **55**, 65 (1959).

10. Pyrolysis of Ethane

Historically, of the various hydrocarbon pyrolyses, that of ethane has been the object of the greatest amount of investigation. As a consequence more is known about it than about the pyrolysis of any other hydrocarbon. Even more important, sufficient thermal and kinetic data are now available so that the rates (and activation energies) of various chain mechanisms can be calculated with about order of magnitude reliability.

In the range 800 to 1000°K ethane decomposes principally into ethylene + H_2 accompanied by smaller amounts of CH_4 and traces of higher hydrocarbons, the latter being principally C_3H_8 and C_3H_6.[a] The over-all reaction is thus

$$C_2H_6 \rightleftharpoons C_2H_4 + H_2 \qquad\qquad (XIII.10.1)$$

together with some

$$C_2H_6 \rightleftharpoons CH_4 + \tfrac{1}{2}C_2H_4 \qquad\qquad (XIII.10.2)$$

and very little

$$C_2H_6 \rightleftharpoons CH_4 + \tfrac{1}{3}C_3H_6 \qquad\qquad (XIII.10.3)$$

A simplified Rice-Herzfeld scheme for this system is

$$C_2H_6 \underset{2}{\overset{1}{\rightleftharpoons}} 2CH_3$$

$$CH_3 + C_2H_6 \underset{4}{\overset{3}{\rightleftharpoons}} CH_4 + C_2H_5$$

Chain:

$$C_2H_5 \underset{6}{\overset{5}{\rightleftharpoons}} C_2H_4 + H$$

$$H + C_2H_6 \underset{8}{\overset{7}{\rightleftharpoons}} H_2 + C_2H_5$$

$$2C_2H_5 \overset{9}{\rightarrow} C_2H_4 + C_2H_6$$

$$\overset{9'}{\rightarrow} C_4H_{10} \qquad\qquad (XIII.10.4)$$

$$CH_3 + C_2H_5 \overset{10}{\rightarrow} CH_4 + C_2H_4$$

$$\overset{10'}{\rightarrow} C_3H_8$$

$$H + CH_4 \overset{11}{\rightarrow} H_2 + CH_3$$

$$H + C_2H_5 \overset{12}{\rightarrow} H_2 + C_2H_4 \text{ (or } C_2H_6)$$

$$\overset{12'}{\rightarrow} 2CH_3$$

[a] A typical set of results at 888°K [L. A. Wall and W. J. Moore, *J. Phys. & Colloid Chem.*, **55**, 965 (1951)] in untreated pyrex gives $H_2 = 15.0$ per cent, $C_2H_4 = 13.1$ per cent, $CH_4 = 3.8$ per cent, and $C_2H_6 = 67.1$ per cent (mole per cent). At this temperature the equilibrium $C_2H_6 \rightleftharpoons C_2H_4 + H_2$ for Moore's system is at about 19 mole per cent C_2H_4, so that the above system is close to equilibrium. (At 900°K, $K_{eq} = 0.05$ atm, so that the back reaction is a serious source of error, particularly at the higher pressures and temperatures.) In the initial stages of the reaction (that is, <10 per cent), the CH_4 amounts to only 2 to 5 per cent of the H_2, which is in good agreement with the earlier work of E. W. R. Steacie and G. Shane, *Can. J. Res.*, **B18**, 351 (1940).

TABLE XIII.7. THERMAL DATA FOR THE SPECIES INVOLVED
IN THE C_2H_6 PYROLYSIS

Substance	ΔH_f°, Kcal/mole	S°, cal/mole-°C	C_p°, cal/mole-°C
H_2	0	31.21	6.89
H	52.10	27.39	4.97
CH_4	−17.89	44.50	8.54
CH_3	32	47 (est)	8.5 (est)
C_2H_4	12.50	52.45	10.41
C_2H_5	26	58 (est)	12.6 (est)
C_2H_6	−20.24	54.85	12.59
C_3H_6	4.86	63.90	15.34
p-C_3H_7	22	67 (est)	17 (est)
sec-C_3H_7	18	66 (est)	17 (est)
C_3H_8	−24.85	64.62	17.66

TABLE XIII.8. SOME EQUILIBRIA IMPORTANT IN THE PYROLYSIS OF C_2H_6

Reaction	ΔH°, Kcal/mole	ΔS°, cal/mole-°C	ΔC_p°, cal/mole-°C	K at 900°K
$C_2H_6 \rightleftharpoons C_2H_4 + H_2$	32.74	28.83	4.70	0.050 atm
$C_2H_6 \rightleftharpoons CH_4 + \frac{1}{2}C_2H_4$	8.60	15.88	1.1	25.8 atm$^{1/2}$
$C_2H_6 \rightleftharpoons CH_4 + \frac{1}{3}C_3H_6$	3.98	10.98	1.0	28.0 atm$^{1/3}$
$C_2H_6 \rightleftharpoons \frac{1}{2}CH_4 + \frac{1}{2}C_3H_8$	−1.11	−0.34	0.5	1.72
$C_2H_6 \rightleftharpoons 2CH_3$	84.2	39	4.4	2.5×10^{-12} atm
$CH_3 + C_2H_6 \rightleftharpoons$ $CH_4 + C_2H_5$	−3.7	0.70	0	11
$C_2H_5 \rightleftharpoons C_2H_4 + H$	38.6	21.8	2.8	4.6×10^{-5} atm
$H + C_2H_6 \rightleftharpoons C_2H_5 + H_2$	−5.9	7.0	1.9	1.4×10^3
$H + CH_4 \rightleftharpoons CH_3 + H_2$	−2.3	6.3	1.9	130
$CH_3 + C_2H_4 \rightleftharpoons p$-$C_3H_7$	−22.5	−32.5	−2.9	0.012 atm^{-1}

Note: The equilibrium constants for the molecular reactions have been obtained from the NBS tables (data at 900°K), whereas the equilibrium constants for the free radical reactions were obtained by extrapolating the data at 300°K. The constants obtained by the two methods are generally within 15 per cent of each other. At 900°K the data are much less sensitive to errors in ΔH_f°. An error in H_f° of ± 2 Kcal will result in an error in K_{eq} of a factor of 3. An error in S_f° of ± 2 cal/mole-°C will result in an error in K_{eq} of a factor of 2.7.

in which we have neglected the reactions that produce higher hydro-carbons.[a] The thermal data which are important in considering this

[a] These are negligible only in the initial stages of the reaction, as will be appreciated from the equilibrium constants and thermal data for the system. At the higher temperatures (800 to 1000°K) the thermodynamic equilibrium favors production of carbon, H_2, and CH_4, and carbonization is quite generally noticed in these reactions during the longer time periods.

scheme are assembled in Table XIII.7. In Table XIII.8 are collected some of the important equilibrium constants for the atomic and molecular reactions in this system, evaluated from the data in Table XIII.7 and high-temperature data of the National Bureau of Standards.

From the equilibrium constants, it is possible to get fairly accurate values (with some simple algebraic approximations) of the equilibrium concentrations of the various species. These are given in Table XIII.9 for two different initial pressures of C_2H_6. One of the most striking fea-

TABLE XIII.9. "EQUILIBRIUM" CONCENTRATIONS OF RADICALS AND MOLECULAR SPECIES PRESENT IN THE PYROLYSIS OF ETHANE AT 900°K

Species	$(C_2H_6)_0$ = 1 atm		$(C_2H_6)_0$ = ⅑ atm		General equation[b] for conc. relative to C_2H_4
	Relative conc.[a]	Partial press., atm	Relative conc.[a]	Partial press., atm	
$C_2H_4 = x$	1.00	0.25	1.00	0.0400	1
C_2H_6	0.076	0.019	0.0216	0.00086	$0.078x^{1/2}G$
CH_4	3.92	0.98	2.78	0.111	$2G$
H_2	0.0152	0.0038	0.038	0.00152	$0.00388x^{-1/2}G$
C_3H_6	0.64	0.16	0.256	0.0102	$1.28x^{1/2}$
C_3H_8	0.00437	0.00109	4.95×10^{-4}	2×10^{-5}	$0.00892xG$
CH_3	8.8×10^{-7}	2.2×10^{-7}	1.2×10^{-6}	4.7×10^{-8}	$4.5 \times 10^{-7}x^{-1/4}G^{1/2}$
C_2H_5/CH_3[c]	0.21	0.46×10^{-7}	0.086	0.40×10^{-8}	$11(C_2H_6/CH_4)$
H/CH_3[c]	3.0×10^{-5}	6.6×10^{-12}	1.05×10^{-4}	5.0×10^{-12}	$0.0077(H_2/CH_4)$

[a] These are ratios with respect to C_2H_4. $(C_2H_6)_0$ = initial pressure of ethane. All other substances assumed absent initially.

[b] These are deduced from the equilibrium constants of Table XIII.8. The quantity $G = 1 + 1.92x^{1/2}$, where $x = C_2H_4$ (atm).

[c] These ratios are no longer referred to C_2H_4.

Note: These concentrations are calculated by assuming the absence of higher hydrocarbons.

tures of these "equilibrium" concentrations is that thermodynamically, the decomposition into $C_2H_4 + H_2$ is relatively unimportant compared to the split into CH_4 + higher olefins, these latter ultimately going to form aromatics and carbon in reactions which are normally quite slow.[a] This implies that in the initial stages of the reaction CH_4 is produced at a much slower rate than is C_2H_4, which is understandable only if the reaction that converts CH_3 to C_2H_5 (reaction 3) is faster than the other reactions that destroy CH_3 (reaction 2), so that CH_3 is kept below its

[a] Wall and Moore, loc. cit., have observed that over longer periods of time H_2 and C_2H_4 go through maxima while CH_4 increases. Packing the vessel and adding NO increase this trend toward true equilibrium and change the product distribution. NO in particular suppresses H_2 and favors CH_4.

thermodynamic equilibrium value and C_2H_5 is kept higher. As we shall see this is in accord with what independent data are available.

If we take what appears to be the simplest set of reactions from the scheme, Eq. (XIII.10.4), they will include

Initiation:
$$C_2H_6 \xrightarrow{1} 2CH_3$$

$$CH_3 + C_2H_6 \xrightarrow{3} CH_4 + C_2H_5$$

Chain:
$$C_2H_5 \underset{6}{\overset{5}{\rightleftharpoons}} C_2H_4 + H \qquad\qquad (XIII.10.5)$$

$$H + C_2H_6 \underset{8}{\overset{7}{\rightleftharpoons}} C_2H_5 + H_2$$

Termination:
$$2C_2H_5 \xrightarrow{9} C_2H_4 + C_2H_6$$

$$\xrightarrow{9'} C_4H_{10}$$

By applying the stationary-state treatment to (H), (CH_3), and (C_2H_5) we find

$$(C_2H_5)_{ss} = \left[\frac{k_1(C_2H_6)}{(k_9 + k_{9'})} \right]^{1/2}$$

$$(CH_3)_{ss} = \frac{2k_1}{k_3} \qquad\qquad (XIII.10.6)$$

$$\frac{(H)_{ss}}{(C_2H_5)_{ss}} = \frac{k_5 + k_8(H_2)}{k_6(C_2H_4) + k_7(C_2H_6)}$$

For the rate of disappearance of ethane we find

$$-\frac{d(C_2H_6)}{dt} = 3k_1(C_2H_6)\left(1 - \frac{1/3}{1 + k_{9'}/k_9}\right)$$

$$+ \frac{k_5(k_1/k_9)^{1/2}}{(1 + k_{9'}/k_9)^{1/2}} \frac{1 - (C_2H_4)(H_2)/K_{eq}(C_2H_6)}{1 + k_6(C_2H_4)/k_7(C_2H_6)} (C_2H_6)^{1/2} \quad (XIII.10.7)$$

in which K_{eq} is the equilibrium constant for the over-all dissociation, $C_2H_6 = C_2H_4 + H_2$, and the term containing K_{eq} represents the correction for the back reaction. In the initial stages of the reaction [i.e., $(C_2H_4) \ll (C_2H_6)$] this reduces to

$$-\frac{d(C_2H_6)}{dt} \to 3k_1(C_2H_6)\left(1 - \frac{1/3}{1 + k_{9'}/k_9}\right) + \frac{k_5(k_1/k_9)^{1/2}}{(1 + k_{9'}/k_9)^{1/2}} (C_2H_6)^{1/2}$$

$$(XIII.10.8)$$

in which the first term represents the destruction of ethane by the initiation and termination steps [Eq. (XIII.10.5)] while the second term represents the added contribution of the chain. When the chain is extremely long, that is, $k_5 \gg 3[k_1k_9(C_2H_6)]^{1/2}$, the over-all reaction will be of $1/2$ order in ethane. If the chain is very short, then the rate will be of first order in ethane. When $k_5 \cong 3[k_1 k_9 (C_2 H_6)]^{1/2}$, the order will be of the complicated form indicated by Eq. (XIII.10.8). It is worth pointing out, however, that the determination of the order of this particular reaction is made extremely complicated by the following difficulties:

1. Since the equilibrium lies at about 20 to 50 per cent decomposition of ethane at 900°K, depending on the pressure, and is lower at lower temperatures, the rate constants calculated from initial pressure changes will be very inaccurate. In addition it will be almost impossible over this limited range of decomposition to establish the order very accurately from points taken during a single run, the difference between half order and first order during such a narrow interval being usually beyond the analytical accuracy.

2. Reaction 5 (and thus 6) may very likely show a quasi-unimolecular behavior because there are so few degrees of freedom in $C_2 H_5$ and the critical ratio E^*/RT for the decomposition at 900°K is so small ($\sim$22). This will have the effect of raising the apparent order of the chain part of the reaction to above $\frac{1}{2}$.

3. The best modern data on the ratio of the rate constants for the ethylene inhibition, k_6/k_7 [Eq. (XIII.10.7)] at 900°K (neglecting possible three-body effects) is 90. Since $E_6 < E_7$, this ratio will increase slowly as the temperature decreases. This implies that the simplified equation (XIII.10.8) is valid only when $(C_2 H_4)/(C_2 H_6) < 1/90$, that is, during the first 1.0 per cent of reaction, an extremely short interval. If, however, the inhibition is included, then the rate expression becomes even more complicated. It would seem on the basis of this analysis that a better approximation during the latter stages of the reaction is to ignore unity in the denominator of Eq. (XIII.10.7), in which case the chain contribution becomes

$$\frac{-d(C_2 H_6)_{chain}}{dt} = \frac{(k_5/k_6)k_7(k_1/k_9)^{1/2}}{(1 + k_{9'}/k_9)} \frac{(C_2 H_6)^{3/2}}{(C_2 H_4)} \left\{ 1 - \frac{(C_2 H_4)(H_2)}{K_{eq}(C_2 H_6)} \right\} \quad \text{(XIII.10.9)}$$

and the order is again complex.

The available evidence[a] on the experimentally observed order is rather confusing and of limited accuracy. All workers have found a falling off of the apparent first-order rate constants with decreasing ethane pressure, the constants being calculated from the initial pressure changes in the system. Sachsse,[b] e.g., found that these first-order rate constants increased by about 50 per cent, going from 30 to 100 mm Hg initial pressure of ethane in the range 850 to 910°K. An attempt by Dintses and Frost[c] to

[a] See E. W. R. Steacie, "Atomic and Free Radical Reactions," 2d ed., vol. 1, sec. IV.9, ACS Monograph No. 125, Reinhold Publishing Corporation, New York, 1954.
[b] H. Sachsse, Z. physik. Chem., **B31**, 87 (1935).
[c] A. I. Dintses and A. V. Frost, Compt. rend. acad. sci. U.R.S.S., **3**, 510 (1934); J. Gen. Chem. U.S.S.R., **3**, 747 (1933).

analyze the rate over the course of a single run led to a mathematical form $-kt = \log(1 - x) + Bx$, where B is a constant. This is neither simple enough nor complex enough to be of much service in interpreting the chain kinetics. It thus appears that the experimental work on the order of the reaction is not of much help in interpreting the chain mechanism. This in turn means that the experimentally measured activation energies are not likely to be too meaningful. Under such conditions it can only be the product distribution and the absolute rates of reaction that can be used with any reliability to check mechanisms.

In the present case the fact that CH_4 is a minor product of the reaction (2 to 10 per cent of H_2) in the initial stages indicates that the rate of the initiation steps shown in Eq. (XIII.10.5) which produce CH_4 must be less than 10 per cent of the rate of the chain which produces H_2. The ratio of the rates of production of H_2 to CH_4, $R(H_2/CH_4)$ can be calculated from the simple scheme Eq. (XIII.10.5) and leads to

$$R(H_2/CH_4) = \frac{d(H_2)}{d(CH_4)} \cong \frac{k_5}{2[k_1 k_{9'}(C_2H_6)]^{1/2}} \qquad \text{(XIII.10.10)}$$

where for simplicity we have set $k_9 = 0$ and ignored the back reaction and

TABLE XIII.10. VALUES OF SOME OF THE RATE CONSTANTS FOR THE REACTIONS OCCURRING IN THE ETHANE PYROLYSIS

Reaction	Arrhenius const[a] A	Activation energy, Kcal/mole	k^a at 900°K	Ref.
$C_2H_6 \xrightarrow{1} 2CH_3$	2×10^{18}	88.0	1.0×10^{-3}	[b]
$CH_3 + C_2H_6 \xrightarrow{3} CH_4 + C_2H_5$	2×10^8	10.4	6×10^5	Table XII.6
$C_2H_5 \xrightarrow{5} C_2H_4 + H$	2.4×10^{13}	40.6	3.3×10^3	[c]
$H + C_2H_6 \xrightarrow{7} C_2H_5 + H_2$	2×10^9	6.4	6×10^7	Table XII.5
$2C_2H_5 \xrightarrow{9} C_2H_4 + C_2H_6$	1.7×10^{10}	0.8	1×10^{10}	Table XII.7
$\xrightarrow{9'} C_4H_{10}$	1.6×10^{10}	0	1.6×10^{10}	[d]
$2CH_3 \xrightarrow{2} C_2H_6$	3.0×10^{10}	0	3×10^{10}	Table XII.8

[a] Units are liters/mole-sec for bimolecular constants and sec⁻¹ for first-order constants.

[b] Calculated from reaction 2 and the equilibrium constant, Table XIII.8.

[c] Calculated from the data of Melville and Robb on the reverse reaction (assuming that $E_6 = 2$ Kcal) and $K_{5.6}$ from Table XIII.8. [H. W. Melville and J. C. Robb, *Proc. Roy. Soc. (London)*, **A196** 494 (1949)]. The frequency factor is 10 fold less than that quoted by Steacie, "Atomic and Free Radical Reactions," 2d ed., ACS monograph No. 125, Reinhold Publishing Co., New York, 1954, which arises in part from the estimate of $K_{5.6}$.

[d] K. J. Ivin and E. W. R. Steacie, *Proc. Roy. Soc. (London)*, **A208,** 25 (1951).

the inhibition due to C_2H_4. In Table XIII.10 are listed some experimental values and some calculated values of the rate constants which can be used to check this ratio. By using this data in Eq. (XIII.10.10) we find that $R(H_2/CH_4) \cong 5$ when $C_2H_6 = 1$ atm and $R \cong 16$ when $C_2H_6 = 0.1$ atm. The importance of the inhibition can be seen at once from Eq. (XIII.10.7). It reduces this ratio R by the factor $[1 + k_6(C_2H_4)/k_7(C_2H_6)]$. At 900°K we find that $k_6/k_7 = 90$, so that if $(C_2H_4)/(C_2H_6) = 0.01$ (that is, 1 per cent of reaction) the inhibition factor is 1.9 and $R(H_2/CH_4)$ is now about 2.6 instead of 5 at $(C_2H_6) = 1$ atm.[a] This is probably an upper limit to the ethylene inhibition because of the possible three-body requirement for the recombination of $H + C_2H_4$, which will tend to reduce the value of k_6 and also make it pressure-dependent. It is clear, however, that the ratios of $(H_2)/(CH_4)$ calculated by the simple mechanism are in the correct range to explain the observed product distribution.

By using the data in Table XIII.10 to calculate the relative rates of disappearance of C_2H_6 we find at 1 atm pressure that $d \ln (C_2H_6)/dt \cong 6 \times 10^{-3}$ sec^{-1} at 900°K and 4×10^{-4} sec^{-1} at 850°K. These values are to be compared with values listed by Steacie (loc. cit.) of about 1.4×10^{-3} sec^{-1} at 900°K and 1.3×10^{-4} sec^{-1} at 850°K. The calculated values appear to be *too high* by about a factor of 4 as compared to the observed values.[b] This is probably better than might have been anticipated in view of the crudity of the assumptions involved in the calculation.

Although the complex form of the rate expression makes it difficult to check an activation energy, it should be noted (neglecting the inhibition and back reaction) that the dominant terms in the rate equation XIII.10.7, have activation energies of 88 and 85 Kcal, respectively. These are both much higher than the best values of the activation energy calculated from observations on the initial rate constants and a supposed first-order mechanism. The latter range from a low value of 69.8[c] to a high of 77 Kcal.[d,e] Despite this high activation energy, the simple mechanism gives a calculated rate constant actually higher than that observed. The reason, of

---

[a] These ratios are not very temperature-sensitive, so that $R(H_2/CH_4)$ will be about the same at 850°K. The inhibition by C_2H_4 is only slightly more effective at lower temperatures because the difference in activation energies of reactions 6 and 7 is only about 4 Kcal.

[b] The calculated rate constants correspond to an apparent activation of about 84 Kcal. However, a ±20 per cent error in the observed rate constants over this temperature range could lead to an average error of ±8 Kcal. Such uncertainties, together with the complexity of the mechanism, make activation energy calculations a very poor criterion of the correctness of the mechanism. The actual spread in absolute rate constants by different laboratories covers a range of about 20 to 80 per cent at any given temperature.

[c] E. W. R. Steacie and G. Shane, *Can. J. Res.*, **B18**, 203 (1940).

[d] R. E. Paul and L. F. Marek, *Ind. Eng. Chem.*, **26**, 454 (1934).

[e] L. Küchler and H. Thiele, *Z. physik. Chem.*, **B42**, 359 (1939).

course, is the abnormally high frequency factor for the unimolecular decomposition of ethane,[a] which compensates for the high activation energy.[b] Küchler and Thiele (loc. cit.) measured the effect of foreign gases upon the initial rates of decomposition of C_2H_6. They found that all gases tested, including H_2, He, A, CO_2, N_2, and CH_4, increased the rate above that for pure C_2H_6. The gases were not too different in their effects except for H_2, which seemed to be a more efficient promoter. On the average, however, a fourfold ratio of added gas/ethane increased the rates of reaction by about 10 to 30 per cent above the high-pressure limiting value obtained in pure ethane.[c] This has been used as an argument in favor of the quasi-unimolecular nature of the initiating step,[d] $M + C_2H_6 \rightarrow 2CH_3 + M$. Since, however, reactions 5 and 6, and particularly 5, are even more likely to show pressure dependence, this evidence is ambiguous as to the origin of the pressure effect.

The inhibition of the ethane decomposition by NO has been studied by a number of workers.[e-g] There seems to be a maximum inhibition of the reaction somewhere in the range 2 to 10 per cent (NO). The effect is least at higher temperatures and also at higher total pressures. In addition the effect seems to be greatest on the initial rate, the inhibited reaction appearing to follow the normal reaction after several per cent decomposition.

One group of workers (Hinshelwood et al.) have interpreted these results as indicating a completely inhibited chain reaction and believe that the residual reaction is the direct unimolecular split $C_2H_6 \rightarrow C_2H_4 + H_2$.

[a] This value rests on the experimental and thermodynamic data for the absolute rate of recombination of CH_3 radicals, thermodynamic data for C_2H_6, CH_4, etc., the bond energy in methane (CH_3—H), and an assumed value for the entropy of CH_3. It is very doubtful if all of these collectively could be sufficiently in error at 900°K to cause an error in k_1 of as much as a factor of 10 to 20. (Note that this would require errors in ΔH of 4 to 5 Kcal and errors in ΔS of 4 to 5 cal/mole-°K.) And observe that the chain varies only as $k_1^{1/2}$.

[b] There have been many dogmatic statements made in the literature about the fact that some proposed mechanism was impossible on the basis of either a too high activation energy or an incorrect order. It is doubtful if the kinetic data always justify such refinement of discrimination.

[c] In pure ethane the initial first-order rate constants fall off with decreasing pressure. Sacchse, loc. cit., found that at about 38 mm Hg the rates were about half their high-pressure limiting values. This is of course quite reasonable in terms of the proposed mechanism.

[d] The recent work on CH_3 recombination (Table XII.10) does indicate some third-body requirement at a total pressure of 20 mm Hg in acetone at 450°K. This pressure region is certain to be higher at higher temperatures because of the negative temperature coefficient for the recombination under such conditions.

[e] L. A. K. Stavely, Proc. Roy. Soc. (London), A162, 557 (1937).

[f] J. E. Hobbs and C. N. Hinshelwood, ibid., A167, 447 (1938).

[g] E. W. R. Steacie and G. Shane, Can. J. Res., B18, 351 (1940).

The chain lengths calculated from the amount of maximal inhibition are about 20 at 50 mm Hg and 6.4 at 500 mm Hg (893°K).[a] Further, there seems to be no change in products with inhibition, a result which is quite striking. However, the interpretation of these inhibited reactions leaves much to be desired.[b] If the proposed initiation is by CH_3 radicals, the inhibition can be interpreted as a stopping of the chain (1) by $CH_3 + NO \rightarrow$ product, in competition with $CH_3 + C_2H_6 \rightarrow CH_4 + C_2H_5$ or (2) by $C_2H_5 + NO \rightarrow$ products. Since, however, the possible products $CH_3 \cdot NO$ or $C_2H_5 \cdot NO$ are not stable at the reaction temperatures and it has been claimed by some that NO does not disappear appreciably, this hardly seems a reasonable solution unless the chain lengths are much larger than supposed.[c]

Rice and Varnerin[d] observed isotopic mixing in the products arising from the decomposition of mixtures of $C_2D_6 + CH_4$, which indicates that a molecular mechanism cannot account for the products of the supposedly maximally inhibited chain reaction. They found that the ratio of $(CH_3D)/(CH_4)$ is directly proportional to the amount of C_2D_6 decomposed in the range 579 to 621°C (initial pressure $\cong$ 100 mm Hg of C_2D_6) and is independent of the amount of NO in the system. Since CH_3D can arise in this system only from a radical exchange mechanism, this seems fairly definite proof of the presence of radicals in what has been considered by the English workers to be the molecular reaction.[e] The same was found for HD and C_2D_5H.

In earlier work by Wall and Moore[f] on the decomposition of mixtures of $C_2H_6 + C_2D_6$ the products were followed by mass-spectrographic analysis. The investigators found some evidence of surface catalysis in packed vessels and found too that NO enhances CH_4 production and—contrary to the findings of other workers—is rapidly consumed. Their results were similar for reactions in quartz, pyrex, and KCl-coated vessels. They also found that the H_2 produced was an almost statistical mixture of H_2, HD, and D_2 at all stages of the NO-inhibited reaction except possibly the very

[a] It is to be noted that this is about what is calculated for the simple mechanism (XIII.10.5).

[b] See Steacie, "Atomic and Free Radical Reactions," op. cit., for a more detailed discussion.

[c] That is, the rate of the initiation is still comparable to the presumed rate of the molecular decomposition and is not then satisfactorily accounted for. If the chains are still larger than is expected, the supposed maximal inhibition is not complete and the interpretation is much more complex.

[d] F. O. Rice and R. E. Varnerin, J. Am. Chem. Soc., 76, 324 (1954).

[e] Under these experimental conditions no isotopic mixing occurs in CH_4-CD_4 or CH_4-D_2 mixtures with or without NO present. Further, CH_4 does not affect the rate of decomposition of C_2D_6.

[f] L. A. Wall and W. J. Moore, J. Phys. & Colloid Chem., 55, 965 (1951).

initial stages, at which the mixing is less than statistical.[a] Similar results were obtained for the isotopic mixing of the product methanes.

While this latter work is open to some criticism on the grounds of too low NO content and too large amounts of over-all decomposition, it must nevertheless be considered as additional evidence against a simple molecular mechanism for the NO-inhibited reaction.

11. Pyrolysis of Ethane: Some Further Steps in the Chain

The enumeration of all the possible reactions involving radicals and molecules in the ethane system would be a tedious task, but one is not really justified in accepting a mechanism for the ethane pyrolysis until such an exhaustive inquiry has been completed. On the other hand, at our present stage of knowledge, the detailed investigation is impracticable if not impossible. The Rice-Herzfeld principle presents about as practical and complete a guide as is at present warranted for the economical discussion of hydrocarbon reactions. However, even this scheme for the ethane pyrolysis [Eq. (XIII.10.4)] has been considerably shortened in the discussion already presented [Eq. (XIII.10.5)], and we may now go back and look at some of the reactions which have been neglected in the latter, simplified chain.

Among the initiation reactions, the following seem of possible importance:

$$C_2H_6 \xrightarrow{13} C_2H_5 + H$$

$$C_2H_6 + C_2H_4 \xrightarrow{14} 2C_2H_5$$

$$CH_4 + C_2H_4 \xrightarrow{15} CH_3 + C_2H_5 \qquad \text{(XIII.11.1)}$$

$$H_2 + C_2H_4 \xrightarrow{16} H + C_2H_5$$

By using the methods employed in the case of the $H_2 + Br_2$ reaction (Sec. XIII.3) we can compare these reactions to the competing reaction 1. By using the available thermodynamic data and making reasonable assumptions about frequency factors it is possible to show that all of these steps are negligibly small, even in the cases of reactions such as 14, 15, and 16, which have appreciably lower activation energies than has reaction 1.

Of the propagation reactions we can consider the following in addition to Eq. (XIII.10.5).

---

[a] They conclude that there is some evidence from this mixing for the presence of a high-activation-energy unimolecular split of $C_2H_6 \rightarrow C_2H_4 + H_2$ at the higher temperatures.

$$CH_4 + C_2H_5 \xrightarrow{4} CH_3 + C_2H_6$$

$$H + CH_4 \xrightarrow{17} H_2 + CH_3$$

$$H + C_2H_4 \xrightarrow{18} H_2 + C_2H_3 \qquad\qquad (XIII.11.2)$$

$$CH_3 + H_2 \xrightarrow{19} CH_4 + H$$

$$CH_3 + C_2H_4 \xrightarrow{20} CH_4 + C_2H_3$$

$$\xrightarrow{21} C_3H_7$$

To understand their role, we must compare these reactions with the reactions in the proposed chain which are competitive. Thus for CH_3 radicals, the competitive chain reaction with C_2H_6 (reaction 3) is to be compared with reactions 19, 20, and 21.

Use of the thermodynamic data and rate constants demonstrates that reaction 19 will become important at 900°K when $(H_2)/(C_2H_6)$ approaches 0.2, but the net effect is merely an increase in chain length arising from the higher H atom concentration.

Reaction 20 has been studied (Table XII.8) and found to have about the same rate as reaction 3, so that C_2H_4 and C_2H_6 are expected to compete about equally for methyl radicals. Reaction 21 is not well known, but from what few data are available[a] it has an activation energy of about 7.0 Kcal and a steric factor that is probably not much lower than that of reaction 20. On this basis we might expect that reaction 21 might be expected to compete about equally with reactions 20 and 3 at 900°K [relative to the ratio $(C_2H_6)/(C_2H_4)$]. However, from the values of the equilibrium constant given in Table XIII.8 we see that, at the usual low values of (C_2H_4) prevailing in the preequilibrium region, the $(p\text{-}C_3H_7)/(CH_3)$ ratio will be less than 10^{-3}. In other words, the $p\text{-}C_3H_7$ radicals produced under these conditions will be very unstable at 900°K, so that unless we go to very high C_2H_4 concentration (i.e., near 1 atm) or much lower temperatures we should not have to concern ourselves with reaction 21.

Reaction 20 should not be important until C_2H_4 has a concentration comparable to that of C_2H_6. In the latter case the C_2H_3 radical must certainly alter the product distribution and the chain becomes much more complex.[b]

[a] L. Mandelcorn and E. W. R. Steacie, *Can. J. Chem.*, **32**, 474 (1954), have made rather crude measurements of reaction 20. They find $k_{20} \cong 4 \times 10^7\, e^{-7000/RT}$ liters/mole-sec with $E_{20} = 7.0 \pm 1.5$ Kcal. This value of E_{20} is in very poor agreement with present thermal data together with the values for E of the reverse reaction.

[b] The C_2H_3 radicals are undoubtedly important in producing higher hydrocarbons such as C_3H_6 by combination with CH_3 or C_4H_8 by combination with C_2H_5. These C_3 and C_4 hydrocarbons do appear in the latter stages of the reaction, and the data may be interpreted as indicating C_2H_4 as a necessary precursor for their production.

Reactions 17 and 18, which involve H atoms, can be compared with the competing reaction 7 of the simple scheme. There are not very reliable data available on reactions 17 and 18 but from what data do appear (Table XII.7), k_7 is expected to be larger than k_{17} or k_{18} and in the initial stage of the reaction only k_7 will be of importance, since in this region $(C_2H_5) \gg (CH_4)$ or (C_2H_4). Once again, it is apparent that, as C_2H_4 and CH_4 build up in the system, the chain will become more complex and reactions 17 and 18 will have to be considered. The same is to be said of reaction 4, which is the reverse of the chain-initiation reaction 3.[a]

Despite the evident complexities of handling additional initiation or propagation reactions, the algebra of these steps is always first-order in radical concentration and the stationary-state equations, although cumbersome, can always be solved explicitly. However, as soon as we introduce more than one termination reaction involving bimolecular participation of two radicals, the equations become nonlinear and explicit solutions are not always possible. Suppose for example that we were to include in the simple scheme Eq. (XIII.10.5), the additional termination reactions

$$C_2H_5 + CH_3 \xrightarrow{10} CH_4 + C_2H_4 \quad \text{or } C_3H_8$$

$$H + CH_3 \xrightarrow{22} CH_4$$

$$H + C_2H_5 \xrightarrow{12} C_2H_6 \quad \text{or } H_2 + C_2H_4 \qquad \text{(XIII.11.3)}$$

$$H + H + M \xrightarrow{23} H_2 + M$$

$$CH_3 + CH_3 \xrightarrow{24} C_2H_6$$

The stationary-state treatment of this more complex chain reaction can be used to derive an expression that governs the total concentration of all radicals in the system, $(H) + (CH_3) + (C_2H_5)$. Since the propagation reactions merely replace one radical by another but do not affect the total concentration of radicals, we can write for the stationary state for all radicals that the sum of all initiation reactions is equal to the sum of all termination reactions. This leads to the equation[b]

$$k_1(C_2H_6) = k_9(C_2H_5)^2 + k_{24}(CH_3)^2 + k_{23}(M)(H)^2 + k_{10}(C_2H_5)(CH_3)$$
$$+ k_{22}(CH_3)(H) + k_{12}(C_2H_5)(H) \quad \text{(XIII.11.4)}$$

This equation, together with the other stationary-state equations, is quadratic in the free radical concentrations and cannot in principle be

---

[a] From the data in Tables XII.6, XIII.8, and XIII.10 we can show that $R(4/5) \cong 2(CH_4)_{atm}$ at 900°K, which is certainly negligible in the initial stages of reaction.

[b] For convenience, the constants are taken to include the two possible reaction paths, combination and disproportionation. Attention must be paid to the definition of the k's to avoid confusion in the use of factors of 2.

solved explicitly.[a] Thus we are forced to consider a more devious method of comparing the relative termination processes and estimating their importance. The first check we can make is to use the results already obtained to calculate the stationary-state concentration of free radicals and see if the results are self-consistent with the assumed chain-ending processes. In Table XIII.11 are given the free radical concentrations computed from Eq. (XIII.10.6) by means of the data of Table XIII.10. We see that, from 850 to 900°K, C_2H_5 is the dominant radical so long as the C_2H_6 pressure is near 1 atm. However, near 0.01 atm, H, CH_3, and C_2H_5 are all nearly of the same order of magnitude, and so the simple chain-ending mechanism, reaction 9, is insufficient at the lower pressures and we must include the other steps of Eq. (XIII.11.4).

TABLE XIII.11. INITIAL FREE RADICAL CONCENTRATIONS
DURING PYROLYSIS OF C_2H_6[a]

Temp, °K	$(CH_3)/(C_2H_5)$	$(H)/(C_2H_5)$	(C_2H_5), moles/liter	$P_{C_2H_6}$, atm
850	0.03	0.0014	6×10^{-9}	1
	0.3	0.14	6×10^{-10}	0.01
900	0.12	0.0041	2.5×10^{-8}	1
	1.2	0.41	2.5×10^{-9}	0.01

[a] Calculated from Eq. (XIII.10.6) and data of Table XIII.10.

It is, however, quite plausible to assume that under these conditions the recombinations of H atoms (reaction 23), H + C_2H_5 (reaction 12), and probably CH_3 + H (reaction 22) are negligible compared to the other processes, since the latter undoubtedly require three-body collisions for stabilization.[b] In this case, H is unimportant in chain-ending steps and Eq. (XIII.11.4) reduces to

[a] If it is assumed that the interradical recombination constants are given by twice the geometric mean of the recombination constants for each free radical, which is not too unreasonable an assumption, then the equation can be linearized and solved explicitly. This implies that $k_{12} = 2[k_9 k_{23}(M)]^{1/2}$, $k_{10} = 2(k_9 k_{24})^{1/2}$, etc., so that Eq. (XIII.11.4) becomes a perfect square which can be extracted to give

$$k_1^{1/2}(C_2H_6)^{1/2} = k_9^{1/2}(C_2H_5) + k_{24}^{1/2}(CH_3) + k_{23}^{1/2}(M)^{1/2}(H)$$

The resultant solutions are still quite complicated but explicit, and we have ignored first-order radical termination processes which may take place on the walls. The factor of 2, above, derives from the symmetry number occurring in the collision frequencies. The approximation is expected to be poor when relating reactions of different order or different M dependence.

[b] Reaction 12, C_2H_5 + H → C_2H_6, will probably also require a third body for stabilization because the excited species formed has at least 12 Kcal energy in excess of that required to dissociate into $2CH_3$ radicals, in which case it does not contribute appreciably to chain termination. This reaction has long been used as the sole chain-breaking step,

$$k_1(C_2H_6) = k_9(C_2H_5)^2 + k_{10}(CH_3)(C_2H_5) + k_{24}(CH_3)^2 \quad (XIII.11.5)$$

and making the not unreasonable assumption that $k_{10} = 2(k_9k_{24})^{1/2}$ we can reduce this to[a]

$$k_1^{1/2}(C_2H_6)^{1/2} = k_9^{1/2}(C_2H_5) + k_{24}^{1/2}(CH_3) \quad (XIII.11.6)$$

which makes possible an explicit algebraic solution for the stationary-state concentration of H, C_2H_5, and CH_3 and thus an explicit equation for the chain reaction. By introducing reactions 10 and 24 into the simple scheme (XIII.10.5) it is found that the stationary-state concentrations of H, CH_3, and C_2H_5 [Eq. (XIII.10.6)] are all reduced by the factor F, where

$$F = 1 + \frac{2}{k_3}\left[\frac{k_1 k_{24}}{(C_2H_6)}\right]^{1/2} \quad (XIII.11.7)$$

and thus the chain rate is reduced by the same factor. At 900°K the data of Table XIII.10 indicate that F is about 1.15 when $C_2H_6 = 1$ atm and 2.5 when $C_2H_6 = 0.01$ atm. F increases rather slowly with temperature, being about 1.05 at 850°K, 1 atm and 1.5 at 950°K, 1 atm. The result is a slight lowering of the order of the chain rate with respect to C_2H_6 and a slight decrease of the over-all activation energy of the reaction.

If we consider in retrospect the work on the pyrolysis of ethane, we are struck by the fact that, while it has produced much controversy and much travail and has been a great stimulus to further work, very little if any quantitative data of interest have come from it. On the contrary, all of the best available data on the steps in the proposed mechanism have come from quite different studies on the behavior of free radicals. And in fact, even at present the best use one can make of the data on this pyrolysis is to check them qualitatively against a proposed mechanism. It is quite doubtful that they can be used to predict individual rate constants with any reliability.

It has been a truism for some time among kineticists that, "If one wishes to understand combustion reactions, one does not study combustions."

---

principally to give a first-order dependence on C_2H_6 and a 70-Kcal activation energy. If it does contribute, it is much more likely to do so via $C_2H_5 + H \rightarrow H_2 + C_2H_4$ and it is unlikely to be the sole important chain-terminating step.

The available evidence on the reaction of $H + C_2H_5$ is provided by the room-temperature data on the Hg-sensitized photodecomposition of C_2H_6 (Steacie, "Atomic and Free Radical Reactions," chap. V.35, op. cit.). The products were 59 mole per cent CH_4, 24 per cent C_3H_8, and 17 per cent C_4H_{10}. The yield of H_2 is almost zero at low pressures and increases to a quantum yield of near 0.5 at 1 atm C_2H_6. CH_4 goes to nearly zero above 200 mm Hg. This indicates that CH_4 must come from the consecutive reactions $H + C_2H_5 \rightleftharpoons C_2H_6^*$, $C_2H_6^* \rightarrow 2CH_3$, $C_2H_6^* + M \rightarrow C_2H_6 + M$. It further indicates that the lifetime of $C_2H_6^*$ with respect to deactivation by collision is about 10^{-9} sec. This would, of course, be even less at 900°K, owing to the higher average energy content of $C_2H_6^*$.

[a] This has been verified by C. A. Heller, J. Chem. Phys., **28**, 1255 (1958).

The work on ethane indicates quite strikingly that the same saying applies to pyrolysis. Our present understanding of pyrolysis reactions, which is not inconsiderable, has for the most part derived, and will continue to derive, from the studies of the thermodynamic and kinetic properties of individual free radicals observed in very carefully selected model systems.

12. Pyrolysis of Di-tertiary Butyl Peroxide; A Thermal Source of Methyl Radicals

Owing mainly to its unique properties as a source of free radicals, the decomposition of dtBP has been the subject of intensive investigation by a number of different laboratories.[a–c] The products of the gas-phase thermal decomposition are principally C_2H_6 + acetone, together with small amounts of methyl ethyl ketone, higher boiling ketones, and methane, the latter being assumed to arise from secondary reactions of the product acetone with CH_3 radicals. The stoichiometry can be represented by

$$(CH_3)_3C\!-\!O\!-\!O\!-\!C(CH_3)_3 \rightarrow 2CH_3COCH_3 + C_2H_6 \quad \sim\!90 \text{ per cent}$$
$$\rightarrow C_2H_5COCH_3 + CH_3COCH_3 + CH_4$$
$$\sim\!10 \text{ per cent} \quad \text{(XIII.12.1)}$$
$$\overset{?}{\rightarrow} (CH_3COCH_2)_2 + 2CH_4$$

The relative amounts of the methane-producing reactions vary somewhat with the reaction conditions. They decrease with decreasing pressure of dtBP, and also with increasing temperature, and increase as the surface/volume ratio increases.[d] There is in addition some difficulty with the stoichiometry of the reaction, the final pressure always being some 4 per cent less than that calculated from the above stoichiometry. It has been suggested that this could arise from the reaction

$$dtBP \rightarrow (CH_3)_3C\!-\!O\!-\!CH_3 + CH_3COCH_3 \quad \text{(XIII.12.2)}$$

but the amount of t-butyl ether formed would have to correspond to about 6 per cent of the acetone produced, and it has not been identified as a product of the gas-phase reaction.[e]

To account for these products, the following scheme has been proposed:

[a] D. Volman and W. M. Graven, *J. Am. Chem. Soc.*, **75**, 3111 (1953).

[b] J. Murawski et al., *J. Chem. Phys.*, **19**, 698 (1951).

[c] E. R. Bell et al., *J. Am. Chem. Soc.*, **72**, 337 (1950); **70**, 88, 95, 1336 (1948).

[d] J. H. Raley et al., *ibid.*, **70**, 88 (1948). N. A. Milas and D. M. Surgenor, *ibid.*, **68**, 205 (1946), found, however, that in the presence of a large amount of glass surface there was no CH_4 or $CH_3COC_2H_5$.

[e] It would of course be very difficult to separate from the larger amounts of acetone and methyl ethyl ketone produced.

$$[(CH_3)_3CO]_2 \xrightarrow{1} 2(CH_3)_3CO$$

$$(CH_3)_3CO \xrightarrow{3} CH_3 + CH_3COCH_3$$

$$CH_3 + CH_3COCH_3 \xrightarrow{4} CH_4 + CH_2COCH_3 \qquad (XIII.12.3)$$

$$CH_3 + CH_2COCH_3 \xrightarrow{5} CH_3CH_2COCH_3$$

$$2CH_3 \xrightarrow{6} C_2H_6$$

From this scheme the stationary-state concentrations of the different radical intermediates are

$$[(CH_3)_3CO]_{ss} = \frac{2k_1(dtBP)}{k_3}$$

$$[CH_2COCH_3]_{ss} = \frac{k_4(CH_3COCH_3)}{k_5} = \frac{k_4}{k_5}(MeAc) \qquad (XIII.12.4)$$

$$[CH_3]_{ss} = \frac{[4k_1k_6(dtBP) + k_4^2(MeAc)^2]^{1/2} - k_4(MeAc)}{2k_6}$$

$$\cong \left[\frac{k_1}{k_6}(dtBP)\right]^{1/2}\left[1 - \frac{k_4(MeAc)}{2k_1^{1/2}(dtBP)^{1/2}}\right]^a \cong \left[\frac{k_1}{k_6}(dtBP)\right]^{1/2}$$

We can see that the total rate of disappearance of peroxide is given by

$$-\frac{d(dtBP)}{dt} = k_1(dtBP) \qquad (XIII.12.5)$$

independent of the relative amounts of secondary reactions.[b] The pressure change in the system is also proportional to the rate of disappearance of peroxide and independent of the particular stoichiometry [Eq. (XIII.12.1)], so that it provides an excellent method for following the reaction and gives directly a value of k_1.

The values reported for k_1 vary from $\log A_1 = 14.60$ (sec^{-1}), $E_1 = 36$ Kcal by Szwarc et al. to $\log A_1 = 16.50$, $E_1 = 39.1$ Kcal by Raley et al.

[a] The equation for CH₃ is $k_6(CH_3)^2 + k_4(CH_3)(MeAc) - k_1(dtBP) = 0$, where the first term represents the rate of formation of C_2H_6 and the second term the rate of formation of CH_4, the last term representing the over-all rate of disappearance of peroxide. Since the yield of CH_4 is never greater than about 10 per cent of the total hydrocarbon, we can neglect the second term to a first approximation and solve the resulting equation for CH₃. On resubstituting this value in the second term of the original equation and solving again for CH₃ we get the second approximation shown above.

[b] There seems to be some evidence for a small amount of chain reaction at lower temperatures from the photolysis of dtBP [L. M. Dorfman and Z. W. Salsburg, *J. Am. Chem. Soc.*, **73**, 255 (1951)].

(*loc. cit.*).[a] If it is assumed that the recombination of t-BuO radicals to reform the peroxide requires no activation energy, then these values are equal to the bond dissociation of the peroxide and in any case provide an upper limit for it. The standard entropy change for reaction 1 may be estimated at about $\Delta S°$, (1 atm) = 40 e.u.,[b] with a probable error of about 3 e.u. (cal/mole-°C). On correction to a standard state of 1 mole/liter this implies that the preexponential factor $A_2{}^c$ is about $10^{-7} \times A_1$ (liters/mole-sec), or in the range 2×10^7 to 2×10^9 liters/mole-sec. If verified these would be among the smallest values observed for the efficiencies of recombination of free radicals in the gas phase. They are not improbable, considering the steric requirements imposed on the recombination by the bulky methyl groups.

If we accept the mean value reported for the activation energies E_1 as the heat of reaction 1, then, together with some thermochemical data on heats of combination and a value of 87.5 Kcal for the heat of dissociation of the tertiary H atom in isobutane, we arrive at the thermal data shown in Table XIII.12.

TABLE XIII.12. SOME CALCULATED THERMODYNAMIC DATA FOR REACTIONS INVOLVING THE t-BUTOXY RADICAL $(CH_3)_3CO$—

Reaction, all gases, $Q + AB \rightarrow A + B$	$\Delta H_f°(AB)$, Kcal	$\Delta H_f°(A)$, Kcal	$\Delta H_f°(B)$, Kcal	Q, Kcal
dtBP → $2(CH_3)_3CO$	-84.7^a	-23.6	—	37.5^b
t-BuOH → $(CH_3)_3CO + H$	-77.1^a	-23.6	52.1	105.6
t-BuOOH → $(CH_3)_3CO + OH$	-53.8^a	-23.6	10.1	40.3
$(CH_3)_3CO → CH_3COCH_3 + CH_3$	-23.6	-51.8	32.0	3.8
$(CH_3)_3CO → (CH_3)_3C + O$	-23.6	3.9	59.2	86.7

[a] Calculated from combustion data of Vaughan et al. [see *Discussions Faraday Soc.*, **10**, 242 (1951), and earlier papers]. Because of uncertainties in the purity, this value and other values listed here are not better than about ±4 Kcal.

[b] Assumed from pyrolysis data, see text.

The qualitative features of the proposed mechanism (XIII.12.3) have been well established. In the presence of good H donors such as ethylene

[a] The latter authors have equated this value to the bond-dissociation energy of the peroxide. To justify this identification they have attempted to compute the bond-dissociation energy of the peroxide from known heat of formation of the peroxide, t-butyl alcohol, isobutane, and the t-butyl radical. Unfortunately, such a calculation can give only the difference in bond-dissociation energies of the peroxide and that for either of the two processes t-BuOH → t-BuO + H or t-BuO → t-Bu + O. Without independent knowledge of one of these last two bond energies, the calculation of the bond energy in the peroxide is not possible.

[b] See Appendix D.

[c] 2 is the reverse of reaction 1.

imine, $\overline{CH_2NHCH_2}$,[a] or CH_3CHO[b] or in the vapor-phase photolysis[c] at lower temperatures, at which the t-butoxy radical may be expected to be more stable, t-butyl alcohol is found as a product. This arises from the reaction

$$(CH_3)_3CO + RH \rightarrow (CH_3)_3COH + R$$

which competes with reaction 3 (XIII.12.3). Seubold, Rust, and Vaughan[d] have shown that dtBP catalyzes the decomposition of the slightly more stable t-butyl hydroperoxide, $(CH_3)_3COOH$, the fast step being the attack of methyl radicals from the dtBP on the H atom in the hydroperoxide. This was verified by the isolation of CH_3D when $(CH_3)_3COOD$ was used with normal dtBP. Further evidence for the t-BuO radical comes from the decomposition of dtBP in various solvents,[e] in which the rate of disappearance of peroxide is first-order and has about the same activation energies and preexponential factors as in the gas phase. The major product in the range 125 to 145°C is now t-BuOH, the ratio t-BuOH/acetone changing from about 19 in tri-n-butylamine to about 4 in the cumene and 0.7 in t-butyl benzene.

The ratio changes with temperature in such a way as to indicate that the differences in activation energies for the competing reactions

$$t\text{-BuO} \xrightarrow{A} \text{acetone} + CH_3$$

$$t\text{-BuO} + \text{solvent} \xrightarrow{B} t\text{-BuOH} + R'$$

have the values $E_B - E_A \cong 13$ Kcal.[f] The interpretation of these results is, however, somewhat complicated by the chain decomposition which is known to set in at higher concentrations of peroxide. Thus in pure liquid dtBP, the principal products are t-BuOH; isobutylene oxide, $(CH_3)_2\overline{CCH_2\text{—}O}$; acetone; and methane, with almost no ethane.

The presence of CH_3 radicals has been shown by the isolation of large amounts of formaldoxime when dtBP is pyrolyzed in the presence of excess NO.[g] This arises from the reaction

$$CH_3 + NO \rightarrow CH_3NO \rightarrow CH_2\text{=}NOH$$

The fact that the rate of decomposition of the peroxide is unaffected by

[a] R. K. Brinton and D. H. Volman, *J. Chem. Phys.*, **20**, 25 (1952).

[b] *Ibid.*, **20**, 1053 (1952).

[c] *Ibid.*, **20**, 1764 (1952). D. H. Volman and W. M. Graven, *J. Am. Chem. Soc.*, **75**, 3111 (1953); L. M. Dorfman and Z. W. Salsburg, *ibid.*, **73**, 255 (1951).

[d] F. H. Seubold, Jr., et al., *J. Am. Chem. Soc.*, **73**, 18 (1951).

[e] J. H. Raley et al., *ibid.*, **70**, 1336 (1948).

[f] This seems extraordinarily high for E_B if E_A has the same value that it has in the gas phase, namely, about 11 Kcal.

[g] Vaughan et al., *Discussions Faraday Soc.*, **10**, 242 (1951).

the added NO indicates that the formaldoxime did not arise from direct reaction of NO with the peroxide. If olefins are added to the decomposing peroxide[a] in a flow system, there are produced higher hydrocarbons whose compositions are indicative of a methyl-initiated polymerization of the olefins. Volman and Graven (loc. cit.) have in fact made use of the radicals from photolyzed dtBP to polymerize butadiene. In the presence of butadiene, acetone production is decreased, indicating that the reaction of t-BuO radical with butadiene to initiate the polymerization is competing successfully with the decomposition into acetone $+ CH_3$. From the temperature dependence of the rate of production of acetone relative to butadiene polymerization they are able to estimate an activation energy of 11.2 ± 2 Kcal for $(CH_3)_3CO \rightarrow CH_3COCH_3 + CH_3$.[b]

From the various estimates of the rate constants and products it is interesting to calculate the stationary-state concentrations of radicals in the gas-phase pyrolysis and consider the implications for some of the competing reactions in the system. By using the relation (XIII.12.4), $(CH_3)_{ss} = [k_1(dtBP)/k_6]^{1/2}$, Vaughan's value for k_1[c] and Table XII.8 for k_6 we find that, at $(dtBP) = 0.01$ atm, (CH_3) varies from 1.3×10^{-6} mm Hg at 100°C to 4×10^{-4} mm Hg at 200°C and 0.017 mm Hg at 300°C.

It is not possible to make similar calculations for $(t$-BuO) or (CH_2COCH_3) because there are no data for k_3 or k_5.[d] However, if it is assumed that k_5 is

[a] F. F. Rust et al., J. Am. Chem. Soc., **70**, 97 (1948). The authors make the interesting comment that there is no reaction of CH_3 with olefins unless large-diameter tubing is used (i.e., no reaction in 15 mm, pyrex; extensive reaction in 70 mm, pyrex). They attribute this to the rapid cleanup of Me radicals by recombination at the walls. This would be rather remarkable if verified, in view of the extremely high efficiency of recombination of CH_3 radicals in the gas phase which has been reported and which would be competitive with wall-phase recombination. This would indicate also that convection is extremely rapid in transporting radicals to the walls.

[b] This seems somewhat inconsistent with the value of 17 ± 3 Kcal estimated from the earlier work on formation of t-BuOH relative to acetone from dtBP photolysis in the presence of ethylene imine by Brinton and Volman, loc. cit.

[c] Although there is a spread in the reported activation energies, the over-all rate constants reported are in fairly good agreement, and since $(CH_3)_{ss}$ depends on the square root of k_1, the values reported above are not very sensitive to the choice made for E_1.

[d] R. E. Varnerin, J. Am. Chem. Soc., **77**, 1426 (1955), has obtained an upper limit on the rate of reaction of $CD_3 +$ acetone $\rightarrow (CH_3)_2(CD_3)CO$ by observing the rate of formation of CH_3COCD_3 relative to H abstraction by CD_3 from acetone, by using CD_3 radicals from the pyrolysis of acetaldehyde, and measuring $(CD_3H)/(CH_3COCD_3)$. Since the rate of reaction of $CD_3 +$ acetone to give CD_3H is known, this gives an upper limit on the addition reaction of $\frac{1}{90}$ the rate of abstraction. If we take the value of Volman and Brinton of 11.2 Kcal for the activation energy of the decomposition of t-BuO, we can calculate a value of 8.7 Kcal for the reverse activation energy of addition of CH_3 to acetone. If we use an estimated $\Delta S° = 39$ e.u. for the reaction, these figures permit us to calculate an upper limit of about 3×10^{13} sec^{-1} for the frequency factor for the decomposition of t-BuO $\rightarrow CH_3 +$ acetone, that for the reverse reaction being about 2×10^6 liters/mole-sec.

about the same as k_6,[a] then the ratio $(CH_2COCH_3)/(acetone)$ varies from about 3×10^{-8} at $100°C$ to 5×10^{-7} at $200°C$ and 3×10^{-6} at $300°C$.

A feature of the reaction which is difficult to understand is the apparent slowness of attack of CH_3 radicals on dtBP itself. This can be contrasted with the assumed reaction of CH_3 with acetone.

$$CH_3 + dtBP \xrightarrow{7} CH_4 + (CH_3)_2(CH_2)COOC(CH_3)_3 \qquad (XIII.12.6)$$

$$CH_3 + CH_3COCH_3 \xrightarrow{4} CH_4 + CH_2COCH_3$$

The ratio of the rates of production of CH_4 from these two sources is given by

$$R\left(\frac{7}{4}\right) = \frac{k_7(dtBP)}{k_4(CH_3COCH_3)} \qquad (XIII.12.7)$$

During the first 20 per cent of the pyrolysis the ratio of $(dtBP)/(CH_3COCH_3)$ has an average value of about 4.5, so that if $R(\frac{7}{4})$ is to be neglected (i.e., less than 0.1) in the chain, it implies, in the range of investigation $k_7 < 45k_4$, a rather startling result when it is considered that the rate of reaction of CH_3 with neopentane is only 3 to 4 times slower (Table XII.6) than k_4.[b]

The result of this inertness of dtBP to attack by CH_3 radicals is a rather marked stability with respect to chain decomposition. This is verified by the lack of effect on the rate of the pyrolysis of the addition of NO or propylene,[c] which are moderately effective as radical traps. Even O_2 in rather large quantities[c] does little to accelerate the rate of decomposition or the production of acetone. However, the addition of HCl to the dtBP leads to a very rapid chain decomposition with the production of iso-

---

[a] L. Mandelcorn and E. W. R. Steacie, *Can. J. Chem.*, **32**, 79 (1954), have shown from material balances and direct observation of the formation of $CH_3CH_2COCH_3$ in the vapor-phase photolysis of acetone that this is in reasonable accord with the data.

[b] The alternative to this conclusion is to assume that the radical produced from reaction 7 is as stable with respect to decomposition as the parent dtBP, so that reaction 7 does not contribute appreciably to chain propagation in the system by the reaction

$$(CH_3)_2(CH_2)COOC(CH_3)_3 \xrightarrow{8} (CH_3)_2\overline{CCH_2O} + OC(CH_3)_3$$

As has been noted, this is a major path for the much faster chain decomposition of pure liquid dtBP. Alternatively, the lack of chains via 7 might be attributed to a rapid recombination of the radical with CH_3 to form t-amyl, t-butyl peroxide:

$$CH_3 + (CH_3)_2(CH_2)COOC(CH_3)_3 \xrightarrow{9} (CH_3)_2(C_2H_5)COOC(CH_3)_3$$

If k_9 is about equal to k_6 and k_8 has the same preexponential factor as k_7, then the relative rates of disappearance of the radical via 8 and 9 is (dtBP = 1 atm, 150°C), $R(8/9) \cong 6 \times 10^3 \times 10^{-E_8/3.9}$, which is negligibly small if $E_8 = E_1$ (39 Kcal) or even half this value.

[c] J. H. Raley et al., *J. Am. Chem. Soc.*, **70**, 2767 (1948).

butylene chlorohydrin, $(CH_3)_2C(OH)CH_2Cl$.[a] This sensitized decomposition can be ascribed to the fairly rapid steps

$$CH_3 + HCl \xrightarrow{10} CH_4 + Cl$$

$$Cl + dtBP \xrightarrow{11} HCl + (CH_3)_2CH_2COOC(CH_3)_3 \quad (XIII.12.8)$$

followed by

$$(CH_3)_2CH_2COOC(CH_3)_3 \xrightarrow{8} (CH_3)_2\overset{\rceil}{COCH_2} + (CH_3)_3CO$$

$$(CH_3)_3CO \xrightarrow{3} CH_3COCH_3 + CH_3$$

$$(CH_3)_2\overset{\rceil}{COCH_2} + HCl \xrightarrow{complex} (CH_3)_2C(OH)CH_2Cl \quad + \text{ isomer}$$

Reaction 10, of CH_3 with HCl, is much faster than the reaction of CH_3 with acetone,[b] so that reaction 10 acts to reduce considerably the stationary-state concentration of CH_3, at the same time replacing it by a very reactive Cl atom. If now the bulk of the CH_3 radicals are routed through reaction 10, followed by a very rapid reaction 11, each split of dtBP (reaction 1) would result in the subsequent production of $2CH_4$, 2 acetone, and 2 dibutylperoxy radicals, or a threefold increase in the rate. A further enhancement will result to the extent that reaction 8 is faster than reaction 1. In the same paper it is noted that HBr has no effect on the rate of decomposition. This is to be expected because of the relative inertness of the Br atom with respect to H abstraction reactions in comparison with CH_3.

It seems clear from all the evidence that the decomposition of dtBP at temperatures above 100°C can be looked upon as producing relatively large and controllable stationary-state concentrations of CH_3 radicals.[c] The original pyrolysis of dtBP will be insensitive to any secondary reactions in the system so long as none of them involves the production of radicals or atoms more active than CH_3. Under these conditions this system provides what is probably one of the most powerful tools yet available for the study of the reactions of the CH_3 radical. As yet it has been used principally for the study of solution reactions. However, with the elucidation of some of the additional steps in the chain, there should be

[a] Ibid.

[b] Steacie, "Atomic and Free Radical Reactions," op. cit. H. O. Pritchard et al., J. Am. Chem. Soc., 77, 2629 (1955), estimate an activation energy $E_{10} \cong 6.5$ Kcal, compared to 9.5 Kcal for E_4, and also a higher preexponential factor.

[c] It can be shown that at these temperatures the reaction of $(CH_3)_3CO$ radicals with CH_3 or stable molecules will usually be much slower than their rate of decomposition to give CH_3 acetone, so that they can usually be neglected in considering secondary reactions.

greater utilization of this system for the study of elementary gas-phase reactions of radicals.[a]

13. Photolysis of Acetone: A Photochemical Source of Methyl Radicals

The body of literature on the photochemical decomposition of acetone is by now quite imposing, and excellent critiques of the work are available.[b–d] This particular photolysis owes its importance to the fact that it has been one of the chief sources of quantitative data on the behavior of methyl radicals. It is therefore of some consequence to examine the principal features of the reaction.

Acetone, like most aliphatic ketones, shows a continuous absorption spectrum through the region 2200 to 3200 Å, with a broad, weak maximum at about 2800 Å.[e] The absorption of light is attended by a chemical reaction whose products are C_2H_6, CO, CH_4, $CH_3COCOCH_3$, $CH_3COC_2H_5$, CH_2CO, and $(CH_3COCH_2)_2$, the latter product never exceeding 3 mole per cent of the sum of $2C_2H_6 + CH_4$. Biacetyl seems to be an important product only at temperatures below 100°C, and its yield is somewhat sensitive to the wavelength of the light used, decreasing with decreasing wavelength. CH_4 yields depend markedly on temperature,[f] light intensity, and acetone pressure. Ethyl methyl ketone is comparable to CH_4 at temperatures below 180°C but[g] the yield decreases with increasing tem-

[a] It is of particular interest to note that these need not be restricted to CH_3 radicals, since the CH_3 radicals produced in the pyrolysis of dtBP can be converted to other radicals by judiciously chosen metathetical reactions. A very common one is the decarbonylation reaction of CH_3 with aldehyde: $CH_3 + RCHO \rightarrow CH_4 + RCO$ followed by the generally rapid $RCO \rightarrow R + CO$.

[b] W. A. Noyes, Jr., and L. M. Dorfman, J. Chem. Phys., 16, 788 (1948).

[c] W. Davis, Jr., Chem. Revs., 40, 201 (1947).

[d] Steacie, "Atomic and Free Radical Reactions," op. cit.

[e] A second, much more intense absorption starts at shorter wavelengths and extends through the vacuum ultraviolet, but little work has been done in this region.

[f] The ratio of yields $(CH_4)/(C_2H_6)$ has been observed by A. F. Trotman-Dickenson and E. W. R. Steacie, J. Chem. Phys., 18, 1097 (1950), to change from about 0.13 at 131°C to 17.5 at 300°C and an acetone pressure of about 24 mm Hg. The ratio is found to vary inversely with the square root of the light intensity and directly as the acetone pressure.

[g] L. Mandelcorn and E. W. R. Steacie, Can. J. Chem., 32, 79 (1954), were able to identify $CH_3COC_2H_5$ mass-spectroscopically and follow its production quantitatively. The same authors, ibid., 32, 331 (1954), also identified C_2H_4 mass-spectroscopically at the higher temperature. They tried unsuccessfully to identify ketene in this high-temperature static system by adding H_2O vapor to the reactants and looking for acetic acid as a product. There is, however, no reason to expect that the vapor-phase reaction, $CH_2CO + H_2O \rightarrow CH_3COOH$, will be significantly fast under these conditions, particularly in competition with other reactions of ketene with free radicals.

perature. Above 200°C ketene becomes an important product,[a] and above 300°C C_2H_4 also begins to appear in increasing quantity. Although bi-acetonyl, $(CH_3COCH_2)_2$, has not been identified in the photolysis, it has been identified in the thermal reaction,[b,c] and the mass balance in the photolysis[d] points to its presence.

The over-all stoichiometry of the photolysis can thus be represented by the following sets of equations:

$$CH_3COCH_3 + h\nu \rightarrow C_2H_6 + CO$$
$$\rightarrow \tfrac{1}{2}(CH_4 + CH_3COC_2H_5 + CO)$$
$$\rightarrow \tfrac{1}{3}[2CH_4 + (CH_3COCH_2)_2 + CO] \qquad (XIII.13.1)$$
$$\rightarrow \tfrac{1}{2}[C_2H_6 + (CH_3CO)_2] \qquad \text{below } 120°C$$
$$\rightarrow CH_4 + CH_2CO \qquad \text{above } 200°C$$
$$\rightarrow \tfrac{1}{2}(2CH_4 + C_2H_4 + 2CO) \qquad \text{above } 300°C$$

Since there is by now convincing evidence that the primary chemical process following the absorption of light is the production of free radicals,[e] the mechanism which has been proposed to account for these products is[f]

$$CH_3COCH_3 + h\nu \xrightarrow{1} CH_3 + COCH_3$$

$$\xrightarrow{2} 2CH_3 + CO$$

$$CH_3 + CH_3COCH_3 \xrightarrow{3} CH_4 + CH_2COCH_3$$

[a] R. C. Ferris and W. S. Haynes, *J. Am. Chem. Soc.*, **72**, 893 (1950), showed that ketene is a major product above 200°C with a maximum yield near 300°C. They used a flow system to prevent decomposition of CH_2CO (which would be expected to be important) and were able to identify it among the trapped products by its rapid reaction with aniline. It would thus appear that CH_2CO may reach a stationary-state concentration in the static systems, and it may be a precursor to the C_2H_4 detected by Mandelcorn and Steacie in the high-temperature photolysis.

[b] F. O. Rice et al., *J. Am. Chem. Soc.*, **56**, 2497 (1934).

[c] J. R. McNesby et al., *ibid.*, **76**, 823, 1416 (1954), found acetonyl acetone, $(CH_3COCH_2)_2$, mass-spectrometrically but were unable to find $CH_3COC_2H_5$ in the pyrolysis at temperatures from 466 to 524°C.

[d] Mandelcorn and Steacie, *loc. cit.*

[e] S. W. Benson and G. S. Forbes, *J. Am. Chem. Soc.*, **65**, 1399 (1943), have shown that the addition of I_2 to acetone suppresses completely the formation of C_2H_6. S. W. Benson and C. S. Falterman, *J. Chem. Phys.*, **20**, 201 (1952), showed further that at 3130 and 2537 Å, the amount of CH_3CD_3 produced from the photolysis of a mixture of CH_3COCH_3 and CD_3COCD_3 was consistent with 100 per cent production of free radicals in the primary process.

[f] We have omitted the decompositions of secondary species and products caused by light absorption. This is a reasonable approximation so long as the absorption coefficients of these species are of the same order of magnitude as acetone itself, or a smaller order of magnitude, since the photolysis reactions seldom proceed to more than a few per cent completion. When, however, reactions are permitted to go to completion or intermediates or products have absorption coefficients much larger than the parent species, there may well be appreciable photochemical decomposition of these products.

$$2CH_3 \xrightarrow{4} C_2H_6$$

$$2COCH_3 \xrightarrow{5} (CH_3CO)_2$$

$$CH_3 + COCH_3 \xrightarrow{6} CH_3COCH_3 \qquad\qquad (XIII.13.2)$$

$$\xrightarrow{6'} CH_4 + COCH_2$$

$$COCH_3 + M \xrightarrow{7} CO + CH_3 + M$$

$$CH_3 + CH_2COCH_3 \xrightarrow{8} CH_3CH_2COCH_3$$

$$CH_2COCH_3 \underset{10}{\overset{9}{\rightleftharpoons}} CH_2CO + CH_3$$

$$2CH_2COCH_3 \xrightarrow{11} (CH_2COCH_3)_2$$

$$COCH_3 + CH_2COCH_3 \xrightarrow{12} (CH_3CO)_2CH_2$$

This is an imposingly complex kinetic scheme that has three radical intermediates and possibly one species that may reach a stationary state, namely, ketene. Because of the second-order character of the reactions proposed for the destruction of radicals,[a] an explicit equation for the stationary-state concentrations is impossible. If we set the rate of reaction $1 = \phi I_a(1 - x)$ and that of reaction $2 = \phi I_a x$, where ϕ = fraction of excited acetone molecules decomposing, I_a = average number of quanta absorbed per cc-sec, x = fraction of excited acetone molecules which split by the second path,[b] then we can equate the ratio of radical production and destruction in the system:

[a] There has been much discussion over the heterogeneous nature of the radical-radical reactions. Originally it was felt that they were all heterogeneous. More recent work has indicated that consistent interpretations of the data are possible if it is assumed that the reactions are homogeneous, and that is the point of view presented here. However, this is certainly one of the weak points in the analysis of the data, and the data cannot be considered to be understood until the questions of heterogeneity have been resolved.

[b] For the process $CH_3COCH_3 \rightarrow CH_3CO + CH_3$, ΔH is estimated at about 80 Kcal, while for $CH_3COCH_3 \rightarrow 2CH_3 + CO$, $\Delta H \cong 90$ Kcal. Since a photon at 3130 Å has 91 Kcal/mole and one at 2537 Å has 112 Kcal/mole, an acetone molecule excited by either of these wavelengths has enough energy to dissociate by either of the two processes. Conversely, if the acetone dissociates by the first process, the acetyl and methyl radicals produced will carry off between them the excess energy of 11 (at 3130 Å) to 32 Kcal (at 2537 Å). These are in this sense "hot" radicals. Since this extra energy can be lost only by collisional deactivation, many collisions may be required to effectively remove it. During this time, which will depend on the collision frequency, the efficiency of a deactivating collision, and the distribution of the extra energy (i.e., vibrational, rotational, translational), the hot radicals have a better than average chance of reacting. Noyes has chosen to lump all of these variables together in the quantity x, which may then be looked upon as the fraction of acetyl radicals formed in the first process that spontaneously decompose before the next collision. The data indicate (Noyes and Dorfman, loc. cit.) that $x = 0.07$ at 3130 Å and 0.22 at 2537 Å. This does not, of course, settle the question of the excess energy possessed by the CH_3 fragments at either wavelength, particularly at 2537 Å.

$$\phi I_a = k_4(CH_3)^2 + k_5(Ac)^2 + k_{11}(CH_2Ac)^2 + (k_6 + k_{6'})(CH_3)(Ac)$$
$$+ k_8(CH_3)(CH_2Ac) + k_{12}(Ac)(CH_2Ac) \quad (XIII.13.3)$$

in which we have abbreviated $COCH_3$ by Ac. If now we make the assumption that $k_6^2 = 4k_3k_5$, $k_8^2 = 4k_4k_{11}$, and $k_{12}^2 = 4k_5k_{11}$,[a] this equation can be simplified by taking the square roots of both sides, giving

$$(\phi I_a)^{1/2} = k_4^{1/2}(CH_3) + k_5^{1/2}(Ac) + k_{11}^{1/2}(CH_2Ac) \quad (XIII.13.4)$$

in which we have made the further approximation of ignoring $k_{6'}$ as a significant termination process.[b] With these approximations, explicit solutions are now possible for the free radical species. Thus we find for the radical concentrations:

$$(CH_3CO)_{ss} = \frac{\phi I_a(1 - x)}{k_7}\left[1 + \frac{2}{k_7}(k_5\phi I_a)^{1/2}\right]^{-1}$$

$$\cong \frac{\phi I_a(1 - x)}{k_7} \quad (XIII.13.5)$$

$$(CH_3)_{ss} \cong$$

$$\left(\frac{\phi I_a}{k_4}\right)^{1/2} \frac{[1 + k_9/2(k_{11}\phi I_a)^{1/2}]}{1 + k_3(MeAc)/2(k_4\phi I_a)^{1/2} + k_9/2(k_{11}\phi I_a)^{1/2} + k_{10}(CH_2CO)/2(k_4\phi I_a)^{1/2}}$$

$$(XIII.13.6)$$

$$(CH_2COCH_3)_{ss} \cong$$

$$\frac{k_3(MeAc)}{2(k_4k_{11})^{1/2}} \frac{1 + k_{10}(CH_2CO)/k_3(MeAc)}{1 + k_3(MeAc)/2(k_4\phi I_a)^{1/2} + k_9/2(k_{11}\phi I_a)^{1/2} + k_{10}(CH_2CO)/2(k_4\phi I_a)^{1/2}}$$

$$(XIII.13.7)$$

Even with the aid of the simplifying assumption made it is evident that these results are still too complex to be very useful in direct interpretations of the data. There is an alternative approach to the problem which has been extremely fruitful. If we consider the original kinetic scheme [Eq. (XIII.13.2)], we see that the only reaction producing C_2H_6 is 4, while only reactions 3 and 6' produce CH_4. The respective rates of production are

$$\frac{d(C_2H_6)}{dt} = k_4(CH_3)^2 \quad (XIII.13.8)$$

$$\frac{d(CH_4)}{dt} = k_3(MeAc)(CH_3) + k_{6'}(Ac)(CH_3)$$

[a] There is as yet very little evidence for or against such an approximation. From a purely collisional point of view it seems quite reasonable, if crude. If not in error by more than an order of magnitude, it will introduce relatively little error into the calculation of stationary-state concentrations. It will introduce much more error into the calculated distribution of products if these are determined principally by termination rather than chain processes.

[b] This will certainly be reasonable in view of the small amounts of ketene produced at low temperatures. It is certainly to be expected that $k_{6'} < k_6$ at low temperatures, while at the higher temperatures (Ac) is so low that it is probably unimportant in chain-termination processes.

If during a run, the light intensity and acetone concentrations are relatively constant, then the stationary-state concentrations of the radicals will also be constant, and consequently Eq. (XIII.13.8) can be integrated to give for the amounts of ethane and methane produced at time t

$$(C_2H_6)_t = k_4 \langle (CH_3) \rangle^2 t \qquad \text{(XIII.13.9)}$$
$$(CH_4)_t = k_3 \langle (MeAc)(CH_3) \rangle t + k_{6'} \langle (Ac) \rangle \langle (Me) \rangle t$$

where the angular brackets denote averages. If now we take the average rates of production R and consider the ratio $R(CH_4)/R^{1/2}(C_2H_6)$, we have

$$\frac{R(CH_4)}{R^{1/2}(C_2H_6)} = \frac{(CH_4)_t/t}{[(C_2H_6)_t/t]^{1/2}}$$
$$= \frac{k_3}{k_4^{1/2}} \langle (MeAc) \rangle + \frac{k_{6'} \langle Ac \rangle}{k_4^{1/2}} \qquad \text{(XIII.13.10)}$$

Thus in a set of photolysis experiments at constant temperature, the ratio of the average rate of formation of CH_4 to the square root of C_2H_6 production should be a linear function of the acetone pressure (ignoring for the moment the term involving $k_{6'}$) and if $k_{6'}$ can be neglected, the ratio should be independent of light intensity and should fall to zero as acetone concentration falls to zero. Work by Dorfman and Noyes (*loc. cit.*) in the range 26 to 123°C has shown that this relation is fairly well obeyed over a 30-fold change in light intensity.[a] Subsequent work by Trotman-Dickenson and Steacie[b] in the temperature range 130 to 300°C confirmed this relationship but indicated that the dependence on acetone concentration decreased in the range below 50 mm Hg. Later work by Linnell and Noyes[c] and a very extensive research by Dodd and Steacie[d] showed the dependence on acetone pressure was more complex than that given by Eq. (XIII.13.10) but could be fitted to a scheme involving the three-body recombination of CH_3 radicals:

$$CH_3 + CH_3 \underset{B}{\overset{A}{\rightleftharpoons}} C_2H_6^*$$

$$C_2H_6^* + M \overset{C}{\rightarrow} C_2H_6 + M^*$$

$$CH_3 + \text{acetone} \overset{3}{\rightarrow} CH_4 + CH_2COCH_3$$

[a] They computed quantum yields by $\Phi(CH_4) = R(CH_4)/I_a$, $\Phi(C_2H_6) = R(C_2H_6)/I_a$, where I_a is the rate of absorption of light (photons/cc-sec). By using these definitions and Eq. (XIII.13.10)

$$\frac{\Phi(CH_4)}{\Phi(C_2H_6)} = \frac{k_3}{k_4^{1/2}} \frac{\langle (MeAc) \rangle}{I_a^{1/2}}$$

[b] A. F. Trotman-Dickenson and E. W. R. Steacie, *J. Chem. Phys.*, **18**, 1097 (1950). Actually, there has never been sufficient work at high pressures of acetone to verify quantitatively the relation (XIII.13.10).

[c] R. H. Linnell and W. A. Noyes, Jr., *J. Am. Chem. Soc.*, **73**, 3986 (1951).

[d] R. E. Dodd and E. W. R. Steacie, *Proc. Roy. Soc. (London)*, **A223**, 283 (1954), have also shown that other gases can be used as third bodies with efficiencies relative to acetone as follows: A $\cong$ 0.03, CO_2 $\cong$ 0.03, C_4F_{10} $\cong$ 0.2. They showed that at 250°C a sevenfold increase in surface has little effect above 1 mm Hg pressure and causes strong enhancement of $R(CH_4)$ below this pressure.

By using this scheme and assuming a stationary-state concentration of activated complexes $C_2H_6^*$ it can be shown that (again ignoring $k_{6'}$):

$$\frac{R(CH_4)}{R^{1/2}(C_2H_6)} = \frac{k_3}{(k_A k_C / k_B)^{1/2}} \langle\langle MeAc\rangle\rangle^{1/2} \left(1 + \frac{k_C}{k_B} \langle\langle MeAc\rangle\rangle\right)^{1/2} \quad \text{(XIII.13.11)}$$

$$\xrightarrow{\langle\langle MeAc\rangle\rangle \to 0} k_3 \left(\frac{k_B}{k_A k_C}\right)^{1/2} \langle\langle MeAc\rangle\rangle^{1/2} \quad \text{(XIII.13.12)}$$

$$\xrightarrow{\langle\langle MeAc\rangle\rangle \to \infty} \frac{k_3}{k_A^{1/2}} \langle\langle MeAc\rangle\rangle \quad \text{(XIII.13.13)}$$

When the third body M = acetone, Dodd and Steacie found that the high-pressure limiting region was approached [Eq. (XIII.13.12)] at pressures of 100 mm Hg[a] while the low-pressure limiting region (XIII.13.13) seemed to be approached below 1 mm Hg.[b]

The work thus seems to indicate that heterogeneous reactions of radicals are not of importance at pressures above 10 mm Hg, and so, as long as the only paths for formation of CH_4 and C_2H_6 are those proposed (reactions 3, 4, and 6), the method of average rates of formation can be used to obtain the ratios of the rate constants. If the mechanism is known, then the photolysis can be carried out in intermittent light and the average lifetime of the methyl radicals can be determined by comparing the experimentally determined rates of formation of ethane at different light frequencies with the curves calculated from the proposed mechanism.[c] In this way Gomer and Kistiakowsky,[d] choosing conditions such that 90 to 96 per cent of the photolysis products were CO + C_2H_6, were able to establish the absolute values of k_4. Under such conditions the $(CH_3)_{ss}$ radical concentration is very closely given by $(\phi I_a / k_4)^{1/2}$ [Eq. (XIII.13.6)] and the small amount of CH_4 production under these conditions is an accurate measure of the average $(CH_3)_{ss}$ if reaction 3 is the only path for the formation of CH_4.[e] This simplifies the experimental work. The work

[a] This indicates at 250°C a minimum lifetime for C_2H_6 complexes of 1.6×10^{-8} sec. The values obtained for k_B/k_C are about 6.3×10^{-5} moles/liter at 147°C and 11×10^{-5} moles/liter at 247°C, indicating an activation energy for k_B of about 2 Kcal. If we put an upper limit of 10^{-3} on the deactivation efficiency of acetone molecules, this gives a maximum lifetime for $C_2H_6^*$ of about 10^{-4} sec, which seems unusually short for a molecule with so many degrees of freedom.

[b] These effects seem to rule out any possibility of observing the influence of the term involving $k_{6'}$ [Eq. (XIII.13.10)], whose reality was suggested to the author by E. W. R. Steacie. Direct observations of ketene production at low temperatures would be needed to say anything about the importance of $k_{6'}$. This might be difficult if it reaches a low stationary-state concentration.

[c] See Noyes and Leighton, "Photochemistry of Gases," *op. cit.*, for a detailed discussion of the sector technique used for such measurements.

[d] R. Gomer and G. B. Kistiakowsky, *J. Chem. Phys.*, **19**, 85 (1951).

[e] Above 100°C the stationary-state concentration of CH_3CO is probably sufficiently small that reactions 6 and 6' are both negligible. This is indicated by the fact that the quantum yield of CO formation is very close to unity above 100°C (i.e., no recombination of $CH_3 + CH_3CO$) and the biacetyl yield falls to zero.

seems to be consistent with the previous work on acetone photolysis, and the values of k_4 are within 10 per cent of each other when the photolysis of $Hg(CH_3)_2$ (which follows a similar scheme) is used as a source of CH_3 radicals. Subsequent work by Roberts and Kistiakowsky[a] has shown excellent agreement with these values and demonstrated the third-order dependence of CH_3 radical recombination in the pressure range 1 to 10 mm Hg. They also showed that CO_2 was only $\frac{1}{40}$ as effective as acetone as a third body for recombination, agreeing with the work of Dodd and Steacie (loc. cit.).[b]

There still remains an inconsistency in the effects of temperature on the ratio $k_3/k_4^{1/2}$. An Arrhenius plot of log $(k_3/k_4^{1/2})$ against $1/T$ gives a value of $E_3 - \frac{1}{2}E_4 = 6.0$ Kcal in the range 25 to 120°C and a value of 9.7 Kcal in the range 120 to 250°C. At pressures such that the three-body recombination of CH_3 is unimportant [Eq. (XIII.13.13)] and $E_4 = 0$, this difference can be taken as the activation energy of reaction 3. Until the question is resolved, the higher-temperature measurements seem more reliable and the value of 9.7 Kcal/mole has been accepted as E_3.[c]

Despite these difficulties of quantitative interpretation, there are sufficient data on the most important steps in the proposed mechanism [Eq. (XIII.13.2)] to indicate that at least qualitatively it is a correct scheme. In Table XIII.13 are collected some estimates of thermodynamic data pertinent to the reaction scheme. The data on reactions 3 and 4 are included in Tables XII.6 and XII.8, respectively. Volman and Graven (loc. cit.) have estimated the activation energy of reaction 7 as 13.5 ± 2 Kcal, which is very close to the heat of reaction estimated at 10 Kcal (Table XIII.13). There are no quantitative data on the value of the frequency factor of 7, but from the low activation energy one might expect that it should show pressure dependence.[d]

One of the very striking features of the acetone photolysis is the absence

[a] G. B. Kistiakowsky and E. K. Roberts, J. Chem. Phys., **21**, 1637 (1953).

[b] Investigations by K. U. Ingold et al., J. Chem. Phys., **21**, 2239 (1953), on the direct mass-spectrometric observation of CH_3 radicals at higher temperatures obtained from pyrolysis of $Hg(CH_3)_2$ and also dtBP confirm the order of magnitude of the recombination constants and also the pressure dependence on the third bodies. K. U. Ingold and F. P. Lossing, J. Chem. Phys., **21**, 368; 1135 (1953), had earlier observed that the radical recombination has a negative temperature coefficient of about 2 Kcal.

[c] There have been attempts to explain this low activation energy at low temperatures in terms of diffusion of radicals to the wall and heterogeneous reaction of CH_3 + acetone, the subsequent photolysis of accumulated biacetyl (which Noyes has calculated to be insufficient), the reaction 6', and possible hot radical effects. None of these has yielded to quantitative analysis, although A. J. Nicholson, J. Am. Chem. Soc., **73**, 3981 (1951), has shown that diffusion of CH_3 radicals to the walls may become important at low intensities, low acetone pressures, or low temperatures. The fact that I_2 does not completely quench formation of CH_4 is an indication that hot radical effects may be important.

[d] This could result in an apparently low frequency factor. Thus if k_7 is in the region of being a bimolecular reaction and has the form $k_7(M)$, we can compute an upper limit for A_7 from the estimated entropy change in the reaction (Table XIII.13) and the

of chains, even at temperatures of 400°C.[a] From the proposed mechanism it is seen [Eq. (XIII.13.2)] that chains are minimized by the fact that ketene, the ultimate product of a chain decomposition, must be a fairly good trap for free radicals. The converse to this is that the acetonyl radical must be fairly stable. Thus if one assigns to Q_1 (Table XIII.13) the bond strength of H in acetone, a value of about 92 Kcal, which seems an upper limit, then the energy required for $CH_2COCH_3 \rightarrow CH_3 + CH_2CO$ is 30 Kcal and the equilibrium constant for this reaction (from the data in Table XIII.13) is about 2×10^{-5} atm at 300°C and 1×10^{-3} atm at 400°C. Thus as soon as ketene builds up to these pressures in the system, the concentration of CH_3 radicals must begin to decline sharply relative to CH_2COCH_3 and the chain which proceeds via CH_3 + acetone → CH_4 + CH_2COCH_3 must slow down.[b] This is in sharp contrast to such closely related photolyses as that of acetaldehyde, which has extremely long chain lengths precisely because the products do not react strongly with the chain-carrying radicals to form more stable species.

The photolysis of acetone seems to be simplest and best understood in the region from 100 to 200°C. In this region the quantum yields for CO are close to unity in agreement with the observation that the CH_3CO radical decomposes into CH_3 + CO very rapidly compared to any second-order reaction it may undergo. The only important products are C_2H_6, CO, CH_4, and small amounts of $CH_3COC_2H_5$. It appears that the acetonyl radical is fairly stable in this region. Under these conditions the system can be used as a good monitor for competing reactions of CH_3 radicals, e.g., hydrogen atom abstraction. If a transparent species RH is introduced into the vessel with acetone, both at known concentrations and the acetone photolyzed, then the important reactions that occur are

$$CH_3COCH_3 + h\nu \xrightarrow{1'} 2CH_3 + CO$$

$$CH_3 + CH_3COCH_3 \xrightarrow{3} CH_4 + CH_2COCH_3$$

$$CH_3 + RH \xrightarrow{3'} CH_4 + R \qquad \text{(XIII.13.14)}$$

$$CH_3 + CH_3 \xrightarrow{4} C_2H_6$$

$$2 \text{ radicals} \xrightarrow{t} \text{stable products}$$

---

assumption that the reverse reaction goes at every termolecular collision with a rate constant of 10^9 liters^2/mole^2-sec. Then at (M) $\cong 4 \times 10^{-3}$ mole/liter, which is a typical pressure, $A_7(M) \leq 2 \times 10^{12}$ sec^{-1}, with $k_7(M) = A_7(M)e^{-E_7/RT}$. T. Iredale and L. E. Lyons, J. Chem. Soc., 588 (1944), have shown the second-order dependence of reaction 7.

[a] E. I. Akeroyd and R. G. W. Norrish, J. Chem. Soc., 890 (1936), found that at 396°C the rate of photolysis was only 50 per cent faster than at 100°C.

[b] J. R. McNesby et al., J. Am. Chem. Soc., 76, 823 (1954), were able to identify biacetonyl, $(CH_2COCH_3)_2$, and acetyl acetone in the pyrolysis of acetone at 466 to 525°C. This indicates the increasing importance of the acetonyl radical, as ketene, which is a major product of the pyrolysis, begins to accumulate in the system.

TABLE XIII.13. THERMODYNAMIC DATA FOR SOME REACTIONS OF INTEREST IN THE PHOTOLYSIS OF ACETONE (ALL GAS PHASE)

Reaction $AB \rightleftharpoons A + B$	AB		A		B		Over-all	
	ΔH_f°	S°	ΔH_f°	S°	ΔH_f°	S°	ΔH°	ΔS°
$CH_3COCH_3 \rightarrow CH_3 + COCH_3$	−51.8	69[a]	32	47*	−4.4[b]	62.5*	79.4	40
$CH_3COCH_3 \rightarrow CH_3COCH_2 + H$	−51.8	69*	$(Q_1 - 105)$	74*	52.0	27.4	Q_1	32
$CH_3COCH_2 \rightarrow CH_3 + COCH_2$	$(Q_1 - 105)$	74*	32	47*	−14.6	58*	$(121 - Q_1)$	31
$(CH_3CO)_2 \rightarrow 2COCH_3$	−78.5[c]	—	−4.4*	62.5*	−26.4	47.3	69.7	
$CH_3CO \rightarrow CH_3 + CO$	−4.4[b]	62.5*	32	47*	52.0	27.4	10.0	32
$CH_3CO \rightarrow CH_2CO + H$	−4.4[b]	62.5	−14.6	58*	52.0	27.4	41.8	23
$CH_3CHO \rightarrow CH_3CO + H$	−39.8	63.5	−4.4[b]	62.5*	52.0	27.4	87.4	26

Note: All values of ΔH° are in Kcal/mole (standard state, 25°C, 1 atm); values of ΔS° are in cal/mole-°C.

[a] Values marked with an * are estimates by the author based on calculations which are considered good to ±2 e.u. (See Appendix D).

[b] Estimated by the author from pyrolysis data, thermal data, and data on the photolysis of biacetyl.

[c] From data on heat of combustion by Parks et al., J. Chem. Phys, 22, 2089 (1954), and estimated heat of vaporization from vapor pressure data.

If again it can be assumed that the only source of CH_4 is reactions 3 and 3' and of ethane, reaction 4, then [Eq. (XIII.13.9)]

$$\frac{R(CH_4)}{R^{1/2}(C_2H_6)} = \frac{k_3}{k_4^{1/2}} \text{ (acetone)} + \frac{k_{3'}}{k_4^{1/2}} \text{ (RH)} \qquad \text{(XIII.13.15)}$$

so that the ratio $R(CH_4)$ to $R^{1/2}(C_2H_6)$ should be a linear function of RH concentration at constant acetone concentration with a slope equal to $k_{3'}/k_4^{1/2}$ and intercept equal to $k_3(\text{acetone})/k_4^{1/2}$. This method has now been used by a large number of authors to obtain relative rate constants for H abstraction reaction of CH_3 radicals from various compounds. Table XII.6 is a compilation of some of these data. It is this type of study which has made the acetone system so important in the study of CH_3 radical reactions[a] and which promises in other systems to give a large body of quantitative data on the kinetic behavior of free radicals.

14. Pyrolysis of Acetaldehyde

Whereas acetone shows little tendency to undergo chain decomposition in photolysis or pyrolysis, acetaldehyde has been found to decompose by a chain mechanism which tends to quite sizable chain lengths as the temperature is raised. As a consequence of this behavior, the decomposition has been found to be remarkably sensitive to the presence of small amounts of substances that can form free radicals more readily than pure acetaldehyde does. A further result of this sensitivity is that the data on the pyrolysis obtained under different conditions or in different laboratories show quite important discrepancies. In compensation for these difficulties the stoichiometry of the pyrolysis seems to be quite simple, the products being $CO + CH_4$, together with very small amounts of C_2H_6 and also some H_2 at temperatures near 500°C.[b,c] These can be represented by[d]

$$CH_3CHO \rightarrow CH_4 + CO$$
$$\rightarrow \tfrac{1}{2}(C_2H_6 + H_2) + CO \qquad \text{(XIII.14.1)}$$

If the order of the reaction is determined from the rate of pressure change in the system during a single run,[e] the points can be fitted very closely to an empirical equation of the type

[a] It is very important if these values are to become established that the effects of hot CH_3 radicals, if any, be investigated. This is particularly true for the substances RH that have relatively low activation energies for H atom abstraction, since hot CH_3 will then require proportionately more "cooling."

[b] M. Letort, *J. chim. phys.*, **34**, 267, 355, 428 (1937).

[c] For a detailed discussion see Steacie, "Atomic and Free Radical Reactions," *op. cit.*

[d] The yields of C_2H_6 and H_2 are less than 1 per cent of the CH_4. While this would make an accurate analysis difficult, it is astonishing that so little work has been done on the quantitative identification of these minor products.

[e] Letort, *loc. cit.*, has shown that this is in agreement with direct chemical analysis.

$$- \frac{d(\text{AcH})}{dt} = k'(\text{AcH})^n \qquad (\text{XIII.14.2})$$

in which $\text{AcH} = CH_3CHO$ and the value of n is very close to 2. However, experiments at different initial pressures of CH_3CHO show that k' is inversely proportional to the square root of the initial CH_3CHO pressure, so that the empirical expression can be reduced to the approximate form

$$- \frac{d(\text{AcH})}{dt} = \frac{k(\text{AcH})^2}{(\text{AcH})_0^{1/2}} \qquad (\text{XIII.14.3})$$

There has been much controversy regarding the order of the reaction (see Steacie, *loc. cit.*) and the importance of a chain mechanism vs. a molecular decomposition of acetaldehyde. The bulk of the present evidence, however, indicates that the chain decomposition is the important path for the reaction,[a,b] and in view of the simplicity of the products a fairly simple Rice-Herzfeld mechanism may be presented.

Initiation: $\qquad\qquad CH_3CHO \xrightarrow{1} CH_3 + CHO$

Chain: $\qquad \begin{cases} CH_3 + CH_3CHO \xrightarrow{2} CH_4 + CH_3CO \\ CH_3CO + M \xrightarrow{3} CH_3 + CO + M \qquad k_3 = k_3'(M) \end{cases}$

$\qquad\qquad\qquad\qquad\qquad\qquad\qquad\qquad\qquad\qquad (\text{XIII.14.4})$

Secondary Chain: $\quad \begin{cases} CHO + M \xrightarrow{4} CO + H + M \qquad k_4 = k_4'(M) \\ H + CH_3CHO \xrightarrow{5} H_2 + CH_3CO \end{cases}$

Termination: $\qquad\qquad 2CH_3 \xrightarrow{6} C_2H_6$

We see that there are four radical species present in the system, two of which are postulated as attacking acetaldehyde. From the observed fact that H_2 and C_2H_6 are extremely minor products we can conclude that the rate of propagation of chains in the system must be very rapid as com-

[a] L. A. Wall and W. J. Moore, *J. Am. Chem. Soc.*, **73**, 2840 (1951); *J. Phys. & Colloid Chem.*, **55**, 965 (1951), working in small pyrex vessels (15 cc) at 500°C and using mixtures of $CH_3CHO + CD_3CDO$, showed that the methane produced had a statistical distribution of light and heavy hydrogens even with added hydroquinone, whereas CH_4 and CD_4, the products expected from a direct molecular split, did not mix under the conditions of the experiment. Hydroquinone was found to inhibit the rate of pyrolysis but not the methane mixing. This is contrary to the earlier result of J. C. Morris, *J. Am. Chem. Soc.*, **63**, 2535 (1941); **66**, 584 (1944). P. D. Zemany and M. Burton, *J. Phys. & Colloid Chem.*, **55**, 949 (1951), agree essentially with Wall and Moore, although their results can be interpreted to indicate a small amount of intramolecular reaction in competition with the chain.

[b] F. O. Rice and R. E. Varnerin, *J. Am. Chem. Soc.*, **76**, 2619 (1954), also studied the mixture $CH_3CHO + CD_3CDO$ and further confirm the absence of molecular split even in the presence of added NO. They find that C_2H_6 and C_2D_6 exert a slight inhibiting effect on the rates of pyrolysis, which is in agreement with the inhibiting effects of foreign gases found by Letort.

pared to the rate of destruction of chain carriers. In fact, since the $(H_2)/(CH_4)$ ratio is always <0.01, we see that, on the average, each radical formed must survive about 100 cycles of the chain before being destroyed.[a]

By applying the stationary-state treatment we find for the stationary-state concentration of the radicals

$$(CH_3)_{ss} = \left[\frac{k_1(AcH)}{k_6}\right]^{1/2} \qquad (H)_{ss} = \frac{k_1}{k_5}$$

$$(CHO)_{ss} = \frac{k_1(AcH)}{k_4}$$

$$\frac{(CH_3CO)_{ss}}{(CH_3)_{ss}} = \frac{k_2}{k_3}(AcH) + \frac{k_1}{k_3}\frac{(AcH)}{(CH_3)} \qquad (XIII.14.5)$$

$$= \frac{k_2}{k_3}(AcH) + \left[\frac{k_1 k_6(AcH)}{k_3^2}\right]^{1/2} \cong \frac{k_2}{k_3}(AcH)$$

and for the rate of disappearnace of CH_3CHO

$$-\frac{d(AcH)}{dt} = 2k_1(AcH) + k_2\left(\frac{k_1}{k_6}\right)^{1/2}(AcH)^{3/2} \qquad (XIII.14.6)$$

where the first term represents the rate of initiation and the second term the chain contribution. Under conditions in which the chain length is great, such as occurs here, we can neglect the first term and write

$$-\frac{d(AcH)}{dt} = k_2\left(\frac{k_1}{k_6}\right)^{1/2}(AcH)^{3/2} \qquad (XIII.14.7)$$

which is almost but not quite in agreement with the empirical relation Eq. (XIII.14.3). Letort and coworkers[b] have proposed that the discrep-

[a] Although each CHO radical produces a molecule of H_2, in doing so it also produces a radical CH_3CO which feeds back into the chain.

[b] Reaction 1 would be replaced by

$$CH_3CHO + M \underset{2'}{\overset{1'}{\rightleftharpoons}} CH_3CHO^* + M$$

$$CH_3CHO^* \overset{D}{\rightarrow} CH_3 + CHO$$

so that in place of k_1 we should write:

$$\frac{k_1' k_D M}{k_2' M + k_D}$$

and a similar scheme for 6:

$$CH_3 + CH_3 \underset{B}{\overset{A}{\rightleftharpoons}} C_2H_6^*$$

$$M + C_2H_6^* \overset{6'}{\rightarrow} C_2H_6 + M$$

would replace k_6 by $k_{6'}k_A M/(k_5 + k_{6'}M)$. If the cross sections for 6', 1', and 2' are different for different species, then the original expressions would be replaced by suitably weighted sums, that is, $k_{1'}M = k_{11'}M_1 + k_{12'}M_2 + \cdots$, $k_{2'}M = k_{21'}M_1 + k_{22'}M_2 + \cdots$,

ancy between these two equations be resolved by making reactions 1 and 6 subject to third-body effects, but it is doubtful that much confidence can be placed in such an interpretation before the other complexities of the reaction such as surface sensitivity and effects of trace impurities (not to mention additional chain steps) have been well investigated.

The experimental activation energy calculated on the basis of a $\frac{3}{2}$-order reaction is found to be 46 Kcal. If we use the simple scheme, this should be equal [Eq. (XIII.14.6)] to $E_2 + \frac{1}{2}(E_1 - E_6)$. The value of E_2 has been measured at 7.5 Kcal (Table XII.6), while $E_6 = 0$. If the bond energy of the H in the HCO radical is assigned the value 15 Kcal, current thermal data give a value of 82.4 Kcal to the enthalpy change of reaction 1, so that if there is no activation energy for the back reaction, this can be taken as the minimum value of E_1 and the calculated value of the chain decomposition is then 48.7 Kcal $- RT/2^a \cong 48$ Kcal at $800°$K, which may be considered to be in excellent agreement with the data.

The empirical rate data of Letort (loc. cit.) can be put in the form (for initial rates) of Eq. (XIII.14.2) with $n = \frac{3}{2}$ and $k' = 7.5 \times 10^{10} e^{-46,000/RT}$ (mole/liter)$^{-\frac{1}{2}}$ sec^{-1}.[b] If we equate this to $k_2(k_1/k_6)^{\frac{1}{2}}$ [Eq. (XIII.14.7)], using the data of Volman and Brinton (Table XII.6) for k_2 and Kistiakowsky

etc. Under these conditions the final chain reaction would have the form (assuming only two foreign bodies M_1 and M_2)

$$-\frac{d(AcH)}{dt} = k_2(AcH)^{\frac{3}{2}} \left(\frac{k_D}{k_A}\right)^{\frac{1}{2}} \left(\frac{k_B + k_{61}'M_1 + k_{62}'M_2}{k_D + k_{21}'M_1 + k_{22}'M_2}\right)^{\frac{1}{2}} \left(\frac{k_{11}'M_1 + k_{12}'M_2}{k_{61}'M_1 + k_{62}'M_2}\right)^{\frac{1}{2}}$$

and if $M_1 = CH_3CHO$, $k_{11}' = k_1$, and $k_{61}' = k_6'$,

$$-\frac{d(AcH)}{dt} = k_2(AcH)^{\frac{3}{2}} \left(\frac{k_1'k_B}{k_6'k_A}\right)^{\frac{1}{2}} \left(\frac{1 + a_1M_1 + a_2M_2}{1 + b_1M_1 + b_2M_2}\right)^{\frac{1}{2}} \left(\frac{1 + c_1 M_2/M_1}{1 + c_6 M_2/M_1}\right)^{\frac{1}{2}}$$

where a_1, a_2, b_1, b_2, c_1, and c_6 are suitably defined ratios of the rate constants. This expression is of course already simplified as compared with the more rigorous expression which we have considered for quasi-unimolecular reactions (Sec. XI.6). K. Bril et al., Bull. soc. chim. Belges, **59**, 263 (1950), have neglected the sum over energies and considered the ratio of the initial rates in the presence of added foreign gases. Under these conditions the ratio of the squares of the initial rates has the approximate form

$$\frac{R_0^2}{R_2^2} = \frac{1 + c_1 M_2/M_1}{1 + c_6 M_2/M_1}$$

where R_0 represents the initial rate in pure acetaldehyde at pressure M_1 and R_2 represents the rate in the presence of added gas 2 at pressure M_2. They find that this last equation fits their results for CO, CH_4, CO_2, H_2, and N_2 with values of c_1 and c_6 always less than 1 but $c_6 > c_1$ for each added gas, so that added gases always have a greater effect on the recombination of CH_3 than on the decomposition of CH_3CHO. The effects are quite small, of the order of 10 to 20 per cent even for large ratios of added gases. In view of the extreme sensitivity of the reaction to surface conditioning and minute traces of O_2, it is doubtful if very much confidence can be placed in the interpretation.

[a] Correction for enthalpies to energies.

[b] A. Boyer et al., J. chim. phys., **49**, 337, 345 (1952), later reported for an experimentally homogeneous and $\frac{3}{2}$-order pyrolysis, the value of

$$k' = 7 \times 10^{10} \exp{(-48,000/RT)}(\text{liters/mole})^{\frac{1}{2}} \text{ sec}^{-1}$$

and Roberts (Table XII.8) for k_6, we find that k_1 is given by (units of $\sec^{-1}$):

$$\log k_1 = 14.30 - \frac{77,000}{2.303RT}$$

If the activation energy is assumed to be 80.8 Kcal at 800°K with the absolute value of the rate constant the same, then this would in turn change the frequency factor to $\log A_1 = 15.34$ instead of the 14.30 indicated.[a]

The extreme sensitivity of CH_3CHO to chain decomposition makes it very susceptible to free radical sensitization. Thus CH_3 radicals from the pyrolysis of azomethane[b] can induce the chain decomposition in CH_3CHO at 300°C, the chain length being as great as 500. The photolysis of azomethane at room temperature can also sensitize the decomposition.[c] Letort[d] showed that CH_3 radicals from dtBP will decompose as many as 50 molecules of CH_3CHO per molecule of dtBP at 160°C.

O_2 is an extremely active catalyst. Letort (loc. cit.) showed that as little as 10^{-3} per cent of O_2 will double the rate of decomposition of CH_3CHO at 477°C, the products being the usual ones. Niclause and Letort[e,f] showed that the decomposition can be induced at temperatures as low as 150°C, the reaction lasting as long as the O_2 supply with the number of CH_3CHO molecules per O_2 molecule reacting being of the order of 100 to 300, going through a minimum at 300°C. At the lower temperatures chain breaking seems to be dominated by wall reactions. NO shows a slight inhibition[g,h] together with pronounced catalysis at higher pressures. Propylene inhibits the rate[i] but not very efficiently.

The decomposition is also sensitized by I_2 at lower temperatures (that is, 250 to 300°C). The mechanism certainly must involve I atoms. Faull and Rollefson[j] showed that the I_2 disappears rapidly in the initial stage of

[a] The entropy change for reaction 1 may be estimated at about 35 e.u., so that the rate constant for the reverse reaction (the recombination of $CH_3 + CHO$ to reform acetaldehyde) would have a preexponential factor of $10^{8.02}$ to $10^{9.06}$ liters/mole-sec, which is smaller than that for CH_3 recombination by at least 1 power of 10.

[b] A. O. Allen and D. V. Sickman, J. Am. Chem. Soc., **56**, 1251, 2031 (1934).

[c] F. E. Blacet and A. Taurog, J. Am. Chem. Soc., **61**, 3024 (1934).

[d] A. Boyer et al., Compt. rend., **231**, 475 (1950).

[e] M. Letort and M. Niclause, Rev. inst. franc. petrole et Ann. combustibles liquides, **4**, 319 (1949).

[f] M. Niclause et al., Compt. rend., **229**, 437 (1949), have shown that samples of CH_3CHO pyrolyzed to 25 to 50 per cent completion will show no more traces of O_2. This seems to be one of the few ways to remove O_2 from CH_3CHO samples in kinetic quantities. The mechanism proposed for the O_2 decomposition is $O_2 + CH_3CHO \rightarrow CH_3CO + HO_2$ (either homogeneous or at lower temperatures, more probably at the walls) followed by $HO_2 + CH_3CHO \rightarrow H_2O_2 + CH_3CO$ and then the usual chain.

[g] J. R. E. Smith and C. N. Hinshelwood, Proc. Roy. Soc. (London), **A180**, 237 (1942).

[h] F. O. Rice and R. E. Vaughan, J. Am. Chem. Soc., **76**, 2629 (1954).

[i] Smith and Hinshelwood, loc. cit.

[j] R. F. Faull and G. K. Rollefson, J. Am. Chem. Soc., **58**, 1755 (1936); **59**, 625 (1937).

the reaction and then is re-formed near the end. They propose a stationary period for I_2 during which the mechanism is

$$I_2 + M \rightleftharpoons 2I + M$$
$$I + CH_3CHO \rightarrow HI + CH_3CO$$
$$CH_3CO \rightarrow CH_3 + CO$$
$$CH_3 + I_2 \rightleftharpoons CH_3I + I$$
$$CH_3 + HI \rightarrow CH_4 + I$$

Note that during this stationary state characterized by absence of I_2 color, most of the iodine is in the form of HI. Their data seem to fit such a scheme, and they have shown that alternate mechanisms involving direct molecular reactions are not reasonable.

The photochemical decomposition of acetaldehyde is enormously more complicated than the high-temperature pyrolysis. Although CO and CH_4 are the main products, H_2, $(CH_3CO)_2$, $(CHO)_2$, HCHO, and C_2H_6 are present in amounts of the order of 1 to 10 per cent of the CO and usually decrease in amount with increasing temperature.[a] The quantum yields are low below 100°C but rapidly climb and reach values reminiscent of the pyrolysis at temperatures of 300°C. There is appreciable evidence accumulating to indicate that photoexcited CH_3CHO molecules can decompose into $CH_4 + CO$ in a single, spontaneous act in competition with the radical split and that the probability of such a split increases with decreasing wavelength. The convincing quality of this evidence is somewhat clouded by the possibility that the effects observed[b] were complicated by excited molecule and hot radical reactions.[c]

At sufficiently high temperatures (i.e., above 200°C) at which HCO and CH_3CO radicals are present in only very small concentrations compared to that of CH_3, the photochemical mechanism is probably similar to the pyrolysis. It is then reasonable to expect that the chain part of the re-

[a] See Steacie, "Atomic and Free Radical Reactions," op. cit., for detailed description.
[b] Thus F. E. Blacet and J. D. Heldman, J. Am. Chem. Soc., **64**, 889 (1942), and also F. E. Blacet and D. E. Loeffler, ibid., **64**, 893 (1942), found that the quantum yield of CH_4 in the photolysis of acetaldehyde-I_2 mixtures went from about 0 at 3130 Å to 0.37 at 2380 Å. Since there is no chain and the quantum yields of CO and CH_3I and CH_4 are all temperature-independent under these conditions, this seems reasonable evidence for a double split of photoexcited acetaldehyde:

$$CH_3CHO^* \begin{array}{c} \nearrow CH_4 + CO \\ \searrow CH_3 + HCO \end{array}$$

the ratio of these two processes being about equal at 2380 Å.

[c] F. E. Blacet and R. K. Brinton, ibid., **72**, 4715 (1950), working with 78 per cent CH_3CDO, obtained results from the photolysis in presence of I_2 that indicate all of the methane came from intramolecular rearrangement. This was based on isotopic analysis of the methane fraction and seems reasonably straightforward. However, other results on the methane composition in the absence of I_2 either point to a hot radical effect on methane production [the corrected $(CH_3D)/(CH_4)$ ratio increasing with decreasing wavelength] or indicate unresolved complexities in the analysis of the data.

action will be given by an expression similar to Eq. (XIII.14.7) for the pyrolysis with k_1AcH replaced by $2\phi I_a$, where ϕ is the fraction of CH_3CHO* decomposing into free radicals and I_a is the specific rate of absorption of light. The form usually observed for the high-temperature photolysis is:

$$\frac{d(CO)}{dt} = k'I_a + k''I_a^{\frac{1}{2}}(CH_3CHO)$$

where the second term represents the chain contribution. The activation energy for the chain photolysis is about 8.6^a to 10.0 Kcal,^{b} which would be ascribed to k_2. While this is somewhat higher than the 7.5 Kcal observed by Brinton and Volman (loc. cit.) by using dtBP as a source of CH_3 radicals, it seems in reasonably good agreement, considering the usual difficulties from wall reactions when chain lengths are so great.

The kinetic scheme for the low-temperature photolysis is almost hopeless at our present state of knowledge of the elementary steps involving CHO and CH_3CO radicals. The scheme is even more complicated than that for the ethane pyrolysis, and as noted earlier, the products are certainly more complicated. It is interesting to note that, where the products are simple because of a long chain, the kinetics become extremely sensitive to walls and impurities. On the other hand, at lower temperatures at which chains are shorter and the reaction is not so sensitive to walls, etc., the chemical complexity of the products becomes important and the investigations just as difficult. With all the work that has been done on CH_3CHO (pyrolysis and photolysis), the elementary mechanism is known with some assurance only at the higher temperatures, and even here the initiation processes are subject to question.^{c} The evidence for three-

---

^{a} D. C. Grahame and G. K. Rollefson, J. Chem. Phys., **8**, 98 (1940).

^{b} J. A. Leermakers, J. Am. Chem. Soc., **56**, 1537 (1934).

^{c} Thermal data do not rule out process A among the following:

$$2CH_3CHO \xrightarrow{A} CH_3CHOH + CH_3CO$$

$$CH_3CHO \xrightarrow{B} CH_3CO + H$$

$$\xrightarrow{C} CH_2CHO + H$$

as contributing appreciably to chain initiation. While reactions B and C have enthalpy changes (estimated at $\Delta H_C^\circ \cong \Delta H_B^\circ \cong 85$ Kcal) only about 3 Kcal higher than the change for $CH_3CHO \xrightarrow{1} CH_3 + CHO$, the entropy changes (estimated at 8 to 9 e.u. lower than the latter) would tend to make them unfavorable. On the other hand reaction A has $\Delta H_A^\circ \approx 55$ Kcal, which is 27 Kcal lower than reaction 1, so that if we assign equal rate constants to the bimolecular reactions which are the reverse of 1 and A, the pre-exponential factors for 1 and A will be in the ratio of $A_1/A_A = \exp(\Delta S_1^\circ - \Delta S_A^\circ)/R$ which on estimating (for standard states of 1 mole/liter) becomes $\exp(28.5 - 4.5)/R = 10^{5.24}$ moles/liter. If E_A is set equal to its minimum value ΔH_A°, the ratio of the rates of production of radicals via 1 and A is $R(1)/R(A) = K_1/K_A(AcH)$, where the K's are the equilibrium constants, and at 800°K, 1 atm pressure of acetaldehyde, $R(1)/R(A) = \frac{1}{2}$, so that even if reaction A had 4 Kcal more of activation energy than estimated it could

body effects on recombination of CH_3 radicals at the higher temperatures of the pyrolysis is of considerable interest theoretically and is qualitatively consistent with the negative temperature coefficient observed for their recombination.[a]

It is still of considerable interest to try to understand the mechanism of catalysis of this reaction by NO and O_2. Studies of these reactions could throw considerable light on the species HNO and HO_2 which have been postulated as important intermediates in NO_2- and O_2-catalyzed oxidation processes and might clarify some of the work done on NO-inhibited reactions of hydrocarbons.

15. Pyrolysis of Dimethyl Ether

The pyrolysis of CH_3OCH_3 has characteristics of the pyrolyses of both CH_3CHO and CH_3COCH_3. It resembles the former in being very susceptible to radical sensitization and showing very long chains and is like the latter in producing fairly large concentrations of an almost stable intermediate, formaldehyde, CH_2O (acetone pyrolysis produces ketene).

The products of the reaction depend on the stage of completion, since the ratio $(CH_2O)/(CH_3OCH_3)$ goes from zero at the beginning of the reaction, when no ether has decomposed, through a maximum and near-stationary value, to as high as 0.25 when the reaction is about half over.[b] The formaldehyde seems to be disappearing at a faster rate than one would calculate from its own observed pyrolysis,[c] so that its disappearance would seem to be a sensitized reaction. Aside from the formaldehyde, the other products in almost equal amounts are CH_4, H_2, and CO, together with about 0.8 per cent of C_2H_6.[d] The stoichiometry of the reaction can thus be represented by

---

still contribute 20 per cent as many radicals as reaction 1 at 800°K and more at lower temperatures. The inclusion of reaction A in the over-all scheme would be, in fact, quite consistent with the observed order being near 1.8 (Letort et al.) during a run, and the inhibiting effects of foreign gases.

[a] At the lower temperatures at which the recombination has been studied, CH_3 radical recombination in the presence of acetone shows three-body effects at pressures of about 10 mm Hg (150°). It is to be expected that the half-life of the $C_2H_6^*$ complex will be shorter at higher temperatures (Sec. XI) and that correspondingly the pressure region of such effects will be higher at the higher temperatures.

[b] P. J. Askey and C. N. Hinshelwood, *Proc. Roy. Soc. (London)*, **A115**, 215 (1927).

[c] C. J. M. Fletcher and G. K. Rollefson, *J. Am. Chem. Soc.*, **58**, 2129 (1936), have estimated that the CH_2O is disappearing in these systems at a rate 15× its own extrapolated rate of pyrolysis. The estimate is somewhat difficult because the pyrolysis of CH_2O is very much complicated by condensation reactions [C. J. M. Fletcher, *Proc. Roy. Soc. (London)*, **A146**, 357 (1934)] and the kinetics are not well established. See also R. Klein et al., *J. Am. Chem. Soc.* **78**, 50 (1956).

[d] E. Leifer and H. C. Urey, *J. Am. Chem. Soc.*, **64**, 994 (1942), followed the reaction products mass-spectrometrically. They identified C_2H_6 both by distillation fractionation

$$CH_3OCH_3 \rightarrow CH_4 + CH_2O$$
$$\rightarrow CH_4 + CO + H_2 \qquad \text{(XIII.15.1)}$$

together with small amounts of

$$CH_3OCH_3 \rightarrow \tfrac{1}{2}(C_2H_6 + H_2) + CO + H_2 \qquad \text{(XIII.15.2)}$$

or for the over-all decomposition with $0 < a < 1,\ 0 < b \ll 1$,

$$CH_3OCH_3 \rightarrow (1 - 2b)CH_4 + aCH_2O + (1 - a)CO$$
$$+ (1 - a + b)H_2 + bC_2H_6 \qquad \text{(XIII.15.3)}$$

Despite the complexity introduced by the intermediate CH_2O, the CH_3OCH_3 pyrolysis can be classified as a well-behaved chain reaction in that it is relatively insensitive to wall effects or traces of O_2.[a] It is also one of the first of the complex organic pyrolyses to lend itself to quantitative treatment via a Rice-Herzfeld mechanism.[b] Benson and Jain[c] have shown that the following scheme is capable of quantitatively accounting for the rate over a range of 90 per cent decomposition and over a considerable range of variation of temperature, pressure, and added gases:

Initiation: $CH_3OCH_3 \xrightarrow{1} CH_3 + OCH_3$

Chain I: $CH_3 + CH_3OCH_3 \xrightarrow{2} CH_4 + CH_2OCH_3$

$CH_2OCH_3 + M \xrightarrow{3} CH_2O + CH_3 + M$

Chain II: $M(?) + CH_3O \underset{5}{\overset{4}{\rightleftharpoons}} CH_2O + H + M$

$H + CH_3OCH_3 \xrightarrow{6} CH_2OCH_3 + H_2$

$H + CH_2O \xrightarrow{7} H_2 + CHO \qquad \text{(XIII.15.4)}$

$CH_3 + CH_2O \xrightarrow{8} CH_4 + CHO$

$M + CHO \xrightarrow{9} M + H + CO$

Chain Transfer: $CH_3 + H_2 \underset{12}{\overset{11}{\rightleftharpoons}} CH_4 + H$

Termination: $CH_3 + CH_3 \xrightarrow{10} C_2H_6$

$CH_3 + CH_2OCH_3 \xrightarrow{10'} CH_3CH_2OCH_3$

$2CH_2OCH_3 \xrightarrow{10''} CH_3OCH_2CH_2OCH_3$

and its mass spectra. It is probable that there may be present very small amounts of other species such as CH_3OH, which is found in the pyrolysis of CH_2O [J. G. Calvert and E. W. R. Steacie, *J. Chem. Phys.*, **19**, 176 (1951)].

[a] S. W. Benson and D. V. S. Jain, *J. Chem. Phys.*, **31**, 1008 (1959).

[b] Benson, *ibid.*, **25**, 27 (1956).

[c] Benson and Jain, *loc. cit.*

If we apply the stationary-state method to the free radical species H, CH_3, CHO, CH_3O, and CH_2OCH_3 and assume that the chains are sufficiently long (that is, >10) that initiation and termination rates may be neglected in comparison to chain steps, we can calculate the following stationary-state relations, where $Me = CH_3$,

$$\varphi = \frac{(H)}{(Me)} = \frac{k_8(CH_2O) + k_{11}(H_2)}{k_6(Me_2O) + k_{12}(CH_4)}$$

$$\theta = \frac{(CH_2OCH_3)}{(CH_3)} = \frac{k_2(Me_2O)}{k_3(M)}\left(1 + \frac{k_6}{k_2}\varphi\right)$$

$$\frac{(MeO)}{(Me)} = \frac{k_5(CH_2O)\varphi}{k_4} \tag{XIII.15.5}$$

$$\frac{(CHO)}{(Me)} = \frac{k_8(CH_2O)}{k_9(M)}\left[1 + \frac{k_7}{k_8}\varphi\right]$$

$$(Me) = \left[\frac{k_1(Me_2O)}{k_{10}}\right]^{\frac{1}{2}}\left[1 + \frac{k_{10'}}{k_{10}}\theta + \frac{k_{10''}}{k_{10}}\theta^2\right]^{-\frac{1}{2}}$$

and $\qquad -\dfrac{d(Me_2O)}{dt} = \dfrac{k_2(k_1/k_{10})^{\frac{1}{2}}(Me_2O)^{\frac{3}{2}}[1 + (k_6/k_2)\varphi]}{[1 + (k_{10'}/k_{10})\theta + (k_{10''}/k_{10})\theta^2]^{\frac{1}{2}}} \qquad$ (XIII.15.6)

If now we further assume that $(k_{10'})^2 = 4k_{10}k_{10''}$ so that the denominator in the last equation can be approximated as a perfect square,

$$-\frac{d(Me_2O)}{dt} \cong \frac{k_2(k_1/k_{10})^{\frac{1}{2}}(Me_2O)^{\frac{3}{2}}[1 + (k_6/k_2)\varphi]}{1 + (k_{10''}/k_{10})^{\frac{1}{2}}[k_2(Me_2O)/k_3(M)][1 + (k_6/k_2)\varphi]}$$
(XIII.15.7)

In the absence of added gases and in the very initial stages of the decomposition $\varphi = 0$, $(Me_2O) = M$ and the equation for the initial rate becomes

$$-\left[\frac{d(Me_2O)}{dt}\right]_0 = \frac{k_2(k_1/k_{10})^{\frac{1}{2}}(Me_2O)^{\frac{3}{2}}}{1 + (k_2/k_3)(k_{10''}/k_{10})^{\frac{1}{2}}} = k_0(Me_2O)^{\frac{3}{2}} \quad (XIII.15.8)$$

that is, a $\frac{3}{2}$-order reaction in ether. This relation has been verified by Benson (loc. cit.) from the data of both of the early workers and was subsequently confirmed by Benson and Jain (loc. cit.) for a 40-fold change in pressure of ether.[a] This further establishes that over the region of interest, reaction 3, the decomposition of $CH_3OCH_2\cdot$ radical behaves like a second-order reaction.[b] This is to be expected because of the low activation energy. At 750°K the critical ratio E^*/RT (Sec. XI.3) is 14. It is also interesting to note that the activation energy E^* of 19 Kcal/mole[b] is

[a] It is worth noting that in the initial stages of the reaction, the products are $CH_2O + CH_4$. This stoichiometry of 2 changes rapidly over the course of the reaction to 3 as $CH_2O \rightarrow CO + H_2$, and observations of pressure changes alone are not sufficient to define the system. In practice CH_2O is measured independently.

[b] The activation energy for reaction 3 was measured as 19 Kcal/mole by R. A. Marcus et al., J. Chem. Phys., 16, 987 (1948), from studies of the Hg-photosensitized decomposition of Me_2O over a considerable range in temperature. They were able to isolate $CH_3OCH_2CH_2OCH_3$ as a major product at temperatures below 200°C.

about 12 Kcal in excess of the heat of the reaction (Table XIII.14), imply-ing a considerable activation energy for the addition of radicals to the double bond in CH_2O.[a]

Over the range 750 to 825°K, the activation energy E_0 corresponding to k_0 [Eq. (XIII.15.8)] is found to be constant and equal to 54.9 Kcal. From Eq. (XIII.15.8) this can be equated to

$$E_0 = E_2 + \tfrac{1}{2}(E_1 - E_{10}) - \frac{x}{1 + x}[E_2 - E_3 + \tfrac{1}{2}(E_{10''} - E_{10})] \quad (XIII.15.9)$$

where $x = k_2 k_{10''}^{\frac{1}{2}}/k_3 k_{10}^{\frac{1}{2}}$. Assuming that $E_{10''} = E_{10} = 0$, using the extrap-olated values of E_2 (Table XII.6) and estimated $E_1 = 81$ Kcal, and as-suming that E_3 is the same 19 Kcal observed by Marcus et al. permits the value of x to be calculated as about 1.2 in this range.[b] An estimate of k_3 can be obtained by making the reasonable assumption that $k_{10} = k_{10''}$. This gives k_3 (780°K) $\cong 5.6 \times 10^5$ liters/mole-sec, and with $E_3 = 19$ Kcal, the Arrhenius factor $\log A_3 = 11.1$.

Similar calculations, using known values for k_2 and k_{10}, give $\log k_1 = 18.1 - 81,000/4.575T$ (units of sec^{-1}), and by estimating the entropy change in reaction 1 at 780°K as 40 e.u. (Table XIII.14) we can calculate for the reverse reaction, the rate of recombination of $CH_3 + CH_3O$, the value $k_{-1} = 1.5 \times 10^{11}$ liters/mole-sec, which is about sevenfold larger than the rate of recombination of CH_3 radicals.[c]

In the presence of added gases, or during a decomposition, the rate of disappearance of ether becomes [Eq. (XIII.15.7)]:

$$S = -\frac{d(Me_2O)^{-\frac{1}{2}}}{dt}$$

$$= \frac{(k_2/2)(k_1/k_{10})^{\frac{1}{2}}[1 + k_8(CH_2O)/k_2F(Me_2O) + k_{11}(H_2)/k_2F(Me_2O)]}{1 + x(Me_2O)/(M)[1 + k_8(CH_2O)/k_2F(Me_2O) + k_{11}(H_2)/k_2F(Me_2O)]}$$

$$(XIII.15.10)$$

where $F = 1 + k_{12}(CH_4)/k_6(Me_2O)$. This, predicts that both H_2 and CH_2O will accelerate the reaction, the amount of acceleration decreasing when CH_4 is added. It further predicts that all added gases (M = Me_2O + $\Sigma k_{3i}M_i$) will exert an accelerating effect on the rate. These predictions have been verified quantitatively for H_2 (and D_2) and CH_2O by observing the initial acceleration produced, and from this, values have been ob-

[a] The addition of CH_3 to C_2H_4 requires about 7 Kcal activation energy. This does not appear to be true of H atoms which require no activation energy.

[b] Note that x changes by less than a factor of 2 from 750 to 825°K. This would be difficult to detect experimentally. Independent observations on added H_2 confirm $x = 1.2$.

[c] This is probably a somewhat high estimate arising from errors in the extrapolation of k_2 or a small error in the estimate of the bond dissociation energy E_1. By definition $k_{-1}/k_{10} = [(1 + x)k_0/k_2]^2 K_{eq}^{-1}$, so that errors in k_2 are magnified.

TABLE XIII.14. THERMODYNAMIC DATA FOR SOME REACTIONS OF INTEREST IN THE PYROLYSIS OF DIMETHYL ETHER (ALL GAS PHASE)

Reaction, AB ⇌ A + B	AB		A		B		Over-all	
	ΔH_f°	S°	ΔH_f°	S°	ΔH_f°	S°	ΔH°	ΔS°
$CH_3OCH_3 \rightarrow CH_3 + CH_3O$	−45.3	63.7	32	47*	2[a]	55*[b]	79	38
$CH_3O \rightarrow CH_2O + H$	2[a]	55*	−27.7	52.3	52	27.4	22	25
$CH_2OH \rightarrow CH_2O + H$	−6[c]	56*	−27.7	52.3	52	27.4	30	24
$HCO \rightarrow CO + H$	11[d]	52*	−26.4	47.3	52	27.4	15	23
$CH_3OCH_2 \rightarrow CH_2O + CH_3$	−3[e]	66*	−27.7	52.3	32	47*	7	33
$CH_3CH_2O \rightarrow CH_2O + CH_3$	−8.1[f]	67.5*	−27.7	52.3	32	47*	12.4	32

Note: Standard state: 1 atm, 25°C; enthalpies in Kcal/mole; entropies in cal/mole-°C (e.u.).

[a] Based on the assumption that for $CH_3OH \rightarrow CH_3O + H$, $\Delta H^\circ = 102$ Kcal.

[b] Entropy values estimated by the author from structural parameters. Probable errors of about ±2 e.u. See Appendix D.

[c] Based on an assumed bond energy H—C in CH_3OH of 94 Kcal.

[d] Based on the bond energy H—C in CH_2O being 90 Kcal. This also seems consistent with independent estimates of the bond energy of the CHO radical as obtained from photolysis measurements.

[e] Based on an assumed bond energy H—C in CH_3OCH_3 of 94 Kcal.

[f] From data of R. E. Rebbert and K. J. Laidler, *J. Chem. Phys.*, **20**, 574 (1952).

390

tained for the ratios k_8/k_2 and k_{11}/k_2 and the energy-transfer coefficients relative to Me_2O.[a]

The results for CH_4 and N_2 indicate minor acceleration compatible with efficiencies relative to Me_2O of about 0.5 and 0.2, respectively. CO shows no acceleration.[b] C_2H_4 shows a strong inhibition of the rate, which is expected on the basis of its ability to scavenge H atoms and CH_3 radicals, replacing them both by the presumably less-active ethyl and C_3H_7 radicals, respectively. Coating the vessel with KCl also cuts the rate[c] by a factor of nearly 2 as might be expected for this material, which is an effective scavenger for H atoms (Sec. XIV.7).

TABLE XIII.15. KINETIC DATA FOR THE PYROLYSIS OF DIMETHYL ETHER

Reaction	Log k (780°K)	Log A	E	Reaction	Log k (780°K)	Log A	E
1	−4.60	18.1	81.0	4	—	—	22.0[a]
−1 (reverse of 1)	11.2	11.2	0	8	5.83	—	—
2	5.83	8.5	9.5	10[b]	10.35	6.4	0
3 M=Me₂O	5.75	11.1	19.0	11	5.61	10.35	10.0
M=H₂	5.15	—	—	11′ (D₂)	5.30	5.61	11.8
M=D₂	5.23	—	—	12	7.39	5.30	8.0
M=CH₂O	5.35	—	—			7.39	
M=CH₄	5.45	—	—				
M=N₂	5.05						

Note: Units for rate constants are in liters/mole-sec; E in Kcal/mole.

[a] From the data of M. K. Phibbs and B. de B. Darwent, *J. Chem. Phys.*, **18**, 495 (1950), on the Hg-sensitized decomposition of CH_3OH, 29 Kcal is found for the activation energy of decomposition of the radical·CH_2OH. This is in excellent agreement with the 30 Kcal estimated for the bond-dissociation energy (Table XIII.14) and indicates no activation energy for the addition of H to CH_2O. Note that the reverse reaction to 4 will probably be the formation of ·CH_2OH rather than its less stable isomer $CH_3O·$.

[b] Values for $k_{10'}$ and $k_{10''}$ are assumed as $2k_{10'} = k_{10'} = 2k_{10}$.

The kinetic data are summarized in Table XIII.15. Together with the thermodynamic data given in Table XIII.14 it is possible to calculate the concentrations[d] of the various radicals present in the reaction. It is also possible to show that none of the other radical termination processes is

[a] For H_2 the ratio $k_{11}/k_2 = 0.60$, while for D_2, $k_{11'}/k_2 = 0.40$, both in reasonable agreement with the values of 0.73 and 0.45 obtained by extrapolation of independent data (Table XII.6) from 180 to 507°C.

[b] These effects are difficult to study accurately because an "inert" gas that has one-third the energy-transfer efficiency of Me_2O will increase the initial rate by only 15 per cent when it is present in equimolar quantity with Me_2O. However, with this much inert gas the precision of the rate (pressure) measurements in terms of ether decreases.

[c] The activation energy is unaffected, however.

[d] At 780°K and 100 mm Hg they are (Me) = 5×10^{-10} mole/liter, (H)/(Me) = 0.002, (CH_3O)/(Me) = 0.05, (CHO)/(Me) = 0.05, and (CH_3OCH_2)/(Me) = 1.4.

important in the system. Chain lengths may be computed; they vary from 21 to 84 at 780°K over the range of 25 mm Hg pressure to 400 mm Hg of ether. Because of the self-acceleration caused by the products, these increase about twofold at 50 per cent decomposition.

A final point of interest is that Me_2O slowly exchanges its H atoms with D_2 during a run.[a] By making reasonable assumptions about the rate constant $CH_3OCH_2 + D_2 \rightarrow CH_3OCH_2D + D$ it is possible to predict the observed rate of exchange within a factor of about 3.

While the pyrolysis is not completely reduced to quantitative terms by these observations, it is more so than any other of the complex organic pyrolyses. Its good behavior would make it well worth further study.

16. Hot Radical Reactions: The Transfer of Energy

The heat exchange in a chemical reaction must appear in the newly formed products if it occurs in one step or in the intermediates as well if it is a complex reaction. Even if the reaction is close to being thermoneutral, it will generally have a slow step which requires at least a moderate activation energy, and for an energy balance, either the products of this step must appear with considerable energy or the products of some subsequent step will. Thus, although the simple bimolecular production of HI from $H_2 + I_2$ has a $\Delta H° = -2.4$ Kcal:

$$H_2(g) + I_2(g) \rightarrow 2HI(g) + 2.4 \text{ Kcal}$$

the activation energy is about 40 Kcal, and so the two HI molecules produced will share between them 42.4 Kcal of energy in excess of the normal thermal distribution. Similarly, in the chain production of HBr, the reaction of H atoms with Br_2, although having an activation energy of only 1 Kcal, is exothermic by about 40 Kcal, and so the products HBr + Br will share between them some 41 Kcal of excess energy:

$$H + Br_2 \rightarrow HBr + Br + 40 \text{ Kcal}$$

In the same reaction the recombination of Br atoms in a termolecular reaction, with for example Br_2 as the third body, will produce two Br_2 molecules sharing an excess of 46 Kcal:

$$Br + Br + Br_2 \rightarrow 2Br_2 + 46 \text{ Kcal}$$

It is of some importance to examine the properties of these excited species and inquire into their fate in the reaction system. Although it is possible to find such excited species in thermal reactions, they can be most conveniently produced with varying amounts of energy in photochemical reactions. In a typical photolysis, the excess energy of the photolysis fragments produced by the primary process will be equal to the energy of the light quantum minus the heat of the reaction. In the photolysis of HI by

[a] Leifer and Urey, loc. cit.

light of wavelength 2537 Å, H and I atoms are produced with an excess energy of about 41 Kcal. At shorter wavelengths the energy will be correspondingly greater, while at longer wavelengths it will be less (e.g., at 1849 Å it will be 82 Kcal and at 3130 Å it will be only 20 Kcal). Following are some examples of step reactions in which the products carry away considerable excitation energy:

$$H + Cl_2 \xrightarrow{1} HCl + Cl + 48 \text{ Kcal}$$

$$I_2 + h\nu(4000 \text{ Å}) \xrightarrow{2} 2I + 35 \text{ Kcal}$$

$$H + O_2 + O_2 \xrightarrow{3} HO_2 + O_2 + 49 \text{ Kcal}$$

$$O + O_3 \xrightarrow{4} 2O_2 + 93 \text{ Kcal}$$

$$HI + \begin{cases} h\nu(3130 \text{ Å}) \xrightarrow{5} H + I + 20 \text{ Kcal} \\ h\nu(2537 \text{ Å}) \xrightarrow{6} H + I + 41 \text{ Kcal} \end{cases}$$ (XIII.16.1)

$$HO + H_2O_2 \xrightarrow{7} H_2O + HO_2 + 32 \text{ Kcal}$$

$$CH_3 + HI \xrightarrow{8} CH_4 + I + 32 \text{ Kcal}$$

$$H + I_2 \xrightarrow{9} HI + I + 36 \text{ Kcal}$$

$$2O + O_2 \xrightarrow{10} 2O_2 + 118 \text{ Kcal}$$

$$2CH_3 + CH_3I \xrightarrow{11} C_2H_6 + CH_3I + 85 \text{ Kcal}$$

$$O_3 + H \xrightarrow{12} HO^* + O_2 + 79 \text{ Kcal}$$

These excited species must carry this excess energy, if they are atoms, as electronic + translational, or if they are not atoms, as rotational + vibrational as well. In all of these cases there are restrictions on the type of energy and the way in which it can be divided between the two products of the reaction. The restrictions are (1) conservation of momentum of the product fragments, which determines the partition of translational energy (i.e., in inverse ratio of the masses),[a] (2) conservation of total angular momentum and also its component along some fixed axis,[b] (3) conservation of total electronic angular momentum,[c] and finally, (4) conser-

---

[a] Thus if the translational energies of the two products are $E_1 = p_1^2/2m_1$ and $E_2 = p_2^2/2m_2$, then since $p_1 = -p_2$ (conservation of linear momentum), $E_1/E_2 = m_2/m_1$.

[b] This is not generally too useful in applications because rotational energies are usually both small and not much involved in these reactions.

[c] For linear molecules this becomes conservation of the component of electronic angular momentum along the molecular axis. This rule is useful principally in cases in which the original collision complex can be considered a linear molecule, in which event it can be used to predict which electronic states of the two product species are compatible with the initial collision complex.

vation of electronic spin, although this latter rule seems of questionable validity for any species containing atoms of atomic number in excess of 10.

If the excess energy is electronic, there is a chance that it may be emitted within a time of 10^{-6} to 10^{-8} sec as radiation if the transition is possible.[a] Otherwise such electronic energy must be degraded by inelastic collisions to other forms of energy, usually vibrational. Generally it is found that the electronic states produced in chemical reactions are metastable,[b] so that we may expect electronically excited states to follow the path of collisional deexcitation. The rare cases of allowed transitions result in what are referred to as chemiluminescent reactions.[c]

For excess energy distributed in the other degrees of freedom, i.e., rotation, translation, and vibration, only collisional degradation is possible. However, it is known that degradation of vibrational energy is a slow process and so a considerable fraction of vibrationally excited species may be expected to persist for some time. On the other hand, rotational and translational energy appear to be exchanged very readily in collisions, so that species thus excited may be expected to be "thermalized"[d] very rapidly.

We can gain some rough idea of the rate of translational degradation of energy from consideration of the hard sphere model for collisions (page 162). In a collision of two hard, smooth spheres of masses m_1 and m_2 and initial energy E_1 and E_2, respectively, the fraction of translational energy E_2 lost by species m_2 (assuming $E_2 > E_1$) is, on the average

$$\left\langle \frac{\Delta E_2}{E_2} \right\rangle = \frac{4\theta}{3(1+\theta)^2}\left[1 + \left(\frac{E_1}{\theta E_2}\right)^{1/2}(\theta-1) - \frac{E_1}{E_2}\right] \qquad (XIII.16.2)$$

or if $E_2 \gg E_1$,[e]

$$\left\langle \frac{\Delta E_2}{E_2} \right\rangle \cong \frac{4\theta}{3(1+\theta)^2}\left[1 + \left(\frac{E_1}{\theta E_2}\right)^{1/2}(\theta-1)\right] \qquad (XIII.16.3)$$

where $\theta = m_1/m_2$. This has a maximum near $\theta = 1$ corresponding to the loss of one-third of the total translational energy per collision. At $\theta = 10$ or 0.1 and $E_2/E_1 = 25$, the average value of $\Delta E_2/E_2$ is about $\frac{1}{6}$, so that the energy exchange is not too sensitive to mass[f] and we should expect to find

---

[a] The lifetime for emission of infrared radiation is of the order of 10^{-2} sec.

[b] That is, they cannot easily radiate their energy.

[c] The reaction $H + O_3 \rightarrow HO + O_2$ has been shown to produce vibrationally excited OH radicals, principally in the ninth vibrational level (64 Kcal of vibrational energy). J. D. McKinley, Jr., et al., *J. Chem. Phys.*, **23**, 784 (1955). This also seems to be true of O abstraction reactions by O atoms. [R. G. W. Norrish et al., *Proc. Roy. Soc. (London)*, 1958 and earlier.] Excited O_2 molecules are formed.

[d] That is, lose their excess energy.

[e] Note that $(E_1/\theta E_2)^{1/2} = v_1/v_2 < 1$. See pages 149, 162.

[f] The number of collisions required to reduce E_2/E_1 from 50 to 9, for example, would be about 5 at $\theta = 1$, about 9 at $\theta = 10$, and about 60 at $\theta = 0.1$. Note that, although

translationally excited species brought very rapidly into thermal equilibrium by collisional exchange of energy. Under what conditions, then, might we expect to find these excited product molecules playing an important kinetic role? The answer is, of course, under conditions in which they may engage in collisions, in which their surplus energy may be in excess of the activation energy required for a chemical change. Thus at least one of the O_2 molecules arising from the reaction of $O + O_3$ [Eqs. (XIII.16.1) to (XIII.16.4)] must have 46.5 Kcal of excess energy, and in a system containing O_3 the bond energy of which is only 24.5 Kcal there would appear to be a fair chance of such an excited O_2 molecule decomposing O_3 and thus giving rise to an energy chain.[a]

We have seen from consideration of the hard sphere model that translational energy is very quickly degraded.[b] Consequently we should not expect to find that translationally excited species will persist long enough to be of importance in a kinetic scheme except for very light masses among extremely heavy masses (i.e., very small θ). The reason for this exception (low θ) can be found in Eq. (XIII.16.3), where it is seen that, as θ becomes very small, the fractional energy transfer becomes small also[c] and at a rate faster than θ.

In the photolysis of HI at shorter wavelengths [Eqs. (XIII.16.1) to (XIII.16.5) and (XIII.16.6)] the excess energy of the radiation appearing in the H atoms may alter their normal probabilities of reacting. There seems to be an increasing body of evidence to indicate that in the shorter-wave photolyses of HI and CH_3I, the H and CH_3 carry off sufficient excess energy to enable them to undergo H-abstraction reactions with abnormally high probability.[d–f] Some of this evidence is semiquantitative in nature,

the equation is nearly symmetrical for values of θ and $1/\theta$ close to 1, at very small θ it shows that it is much more efficient to transfer energy from a light to a heavy mass than vice versa.

[a] This depends, of course, on the relative efficiency for transfer of energy from excited O_2 to the *internal* modes of O_3. In fact such energy chains are not observed in the case of ozone decomposition.

[b] Although the model is idealized, the principles of momentum conservation which were used to calculate the transfer properties apply equally well to more realistic models including quantum-mechanical treatments and the results are the same, the chief differences being in the "effective" molecular diameters and the possibility of "inelastic" collisions.

[c] This has the interesting corollary that a translationally excited species m_2 can transfer only the fraction $m_1/(m_1 + m_2)$ of its energy to a species m_1 in an encounter. Thus very fast-moving but heavy species such as I atoms will be able to use only an insignificant fraction of their translational energy in attacking let us say an H_2 molecule (e.g., about $\frac{1}{65}$). This in turn implies that in a successful encounter $H_2 + I \rightarrow HI + H$, the bulk of the activation energy ($\sim$35 Kcal) must be present in the H_2 molecule prior to the collision as either translational, rotational, or vibrational energy.

[d] R. R. Williams, Jr., and R. A. Ogg, Jr., *J. Chem. Phys.*, **15**, 691, 696 (1947).

[e] R. D. Schultz and H. A. Taylor, *ibid.*, **18**, 194 (1950).

[f] H. A. Schwarz et al., *J. Am. Chem. Soc.*, **74**, 6007 (1952).

and to review it in some detail let us first consider some of the features of the photolysis of the iodides.

17. Hot Radical Reactions: Photolysis of Iodides

Hydrogen iodide has a continuous absorption spectrum in the ultraviolet starting at about 3850 $\text{Å}^{a,b}$ and extending through the far ultraviolet, the absorption coefficient becoming quite large near 2500 Å. The absorption of light leads to the over-all reaction

$$2\text{HI} + h\nu \rightarrow \text{H}_2 + \text{I}_2$$

with a quantum yield Φ_{HI} of 2 over an astonishing range of experimental conditions,[c] and it seems fairly certain that the initial act of absorption is followed by production of a normal H atom and a normal I atom which then participate in the following sequence:

$$\text{HI} + h\nu \xrightarrow{I_a} \text{H} + \text{I} \qquad \Phi = 1$$

$$\text{H} + \text{HI} \xrightarrow{2} \text{H}_2 + \text{I}$$

$$2\text{I} + \text{M} \xrightarrow{3} \text{I}_2 + \text{M}$$

with an inhibitory reaction

$$\text{H} + \text{I}_2 \xrightarrow{4} \text{HI} + \text{I}$$

If we assume stationary states for H and I, we find for the stationary rate of H_2 production:

$$\frac{d(\text{H}_2)}{dt} = \frac{I_a}{1 + k_4(\text{I}_2)/k_2(\text{HI})} = \frac{d(\text{I}_2)}{dt} \qquad (\text{XIII.17.1})$$

where I_a = concentration of photons absorbed per second. For the quantum yield Φ_{H_2} at small conversions in which the ratio $(\text{I}_2)/(\text{HI})$ may be assumed constant

$$\Phi_{\text{H}_2} \equiv \frac{(\text{H}_2)_f - (\text{H}_2)_i}{\int_0^t I_a \, dt} \cong \left[1 + \frac{k_4(\text{I}_2)}{k_2(\text{HI})}\right]^{-1} \qquad (\text{XIII.17.2})$$

so that as I_2 is added to the system one should expect to see Φ_{HI} fall from its maximum value of 2, and similarly, Φ_{H_2} should decrease from its maximum value of 1. Quantitatively we see [Eq. (XIII.17.2)] that the function $1/\Phi_{\text{H}_2}$ should be a linear function of the mole ratio $(\text{I}_2)/(\text{HI})$ with unit intercept and slope $= k_4/k_2$. When the data are treated in this way, it is

[a] C. F. Goodeve and A. W. C. Taylor, *Proc. Roy. Soc.* (*London*), **154**, 181 (1936).

[b] J. Romand, *Compt. rend.*, **227**, 117 (1948).

[c] G. K. Rollefson and M. Burton, "Photochemistry and the Mechanism of Chemical Reactions," Prentice-Hall, Inc., Englewood Cliffs, N.J., 1939.

found that, in pure HI, $k_4/k_2 = 3.8$, independent of temperature, which would imply that reactions 4 and 2 have the same activation energy but that I_2 is 3.8 times more efficient in reacting with H atoms than is HI. This seems an entirely reasonable result. However, when inert diluents such as He, A,[a] cyclohexane,[b] or H_2 are added to the reaction vessel, it is found that the quantum yield Φ_{H_2} decreases, or alternatively, the ratio k_4/k_2 seems to increase and, in addition, show a temperature dependence $(E_4 - E_2 \cong -4.5$ Kcal/mole$)$.

While the experimental data show considerable spread and the precise value of $\Delta E_{4.2}$ is open to question,[c] the interpretation of the results in terms of hot H atoms seems to be the only one reasonably consistent with the data.[d] On this model, H atoms are formed with about 42 Kcal of translational energy which they can lose in consecutive moderating collisions[e] with inert gas. Because of its large mass, HI (and I_2 also) is a very ineffective moderator for hot H atoms. In the absence of moderators, then, the value obtained for k_4/k_2 is that for the reactions of hot H atoms. At very large ratios of inert gas to HI, the value approaches that for the normal, thermal H atoms. By using viscosity diameters, such a model then leads to the conclusion that hot H atoms react with both HI and I_2 on almost every collision, whereas 4.5-Kcal H atoms react about 25 times more efficiently with HI than with I_2.[f] This seems unreasonable,[g] as does the large change in frequency factors ($\sim$100 fold) on going from 42 Kcal H atoms to 4.5 Kcal or thermal H atoms.[h] On the other hand Schwarz et al. claim to have found a very similar change in steric factors for the

[a] Schwarz et al., loc. cit.

[b] R. A. Ogg, Jr., and R. R. Williams, J. Chem. Phys., 13, 586 (1945).

[c] At 114°C the value of k_4/k_2 is 10 at (HI)/(He) = 0.2 and about 4.2 in pure HI. Maximum deviations are about 15 per cent. At higher temperatures the differences are less.

[d] Alternative explanations such as enhanced recombination of H + I helped by increasing M, formation of MH complexes from M + H, or wall reactions of HI do not stand up under closer scrutiny.

[e] Schwarz et al., loc. cit., have shown that the moderating efficiency of H_2, He, A, cyclohexane, and HI are more or less what is to be expected from their reduced masses (vis-à-vis H atoms) [Eq. (XIII.16.3)].

[f] That is, Schwarz et al., loc. cit., obtain by extrapolation to infinite moderator, $k_4/k_2 = 0.04$ exp $(4500/RT)$ which represents, presumably, the reaction rate constants for "thermal" species. There is, however, no theoretical basis for expecting such a low ratio of steric factors (0.04), contrary to the statements by the authors. Transition-state theories would on the contrary predict a ratio of steric factors of about unity. In fact these values would then predict that the frequency factors for the reverse reactions, i.e., $I + H_2 \xrightarrow{2} HI + H$, $I + HI \xrightarrow{4} I_2 + H$ would be in the ratio of 1/200, which would be much lower than any theory would predict.

[g] Ibid.

[h] A value of $E_4 - E_2 \cong 2$ Kcal would bring these frequency factors into a more reasonable range.

analogous reactions of hot H atoms with HBr and Br_2.[a] These systems seem well worth further study.[b]

Another source of evidence for the existence of hot radical reactions is the photolysis of methyl iodide. Alkyl iodides show continuous absorption in the region near 2500 to 2600 Å, with a maximum near 2600 Å. The photochemical evidence is very convincing on the point that in this region the primary process following the absorption of light is the formation of an iodine atom and an alkyl radical. In the case of methyl iodide, the C—I bond energy is about 55 Kcal, so that, if the I atom is in its normal state ($^2P_{3/2}$), there is excess energy of about 57 Kcal shared between the CH_3 and I, while if the I atom is excited ($^2P_{1/2}$), this will be about 35 Kcal of excess energy. Because of the disparity in masses, at least seven-eighths of this excess energy must be given to the CH_3 radical.[c] Consequently if there are any rapid reactions of hot methyl radicals,[d] we may expect to find them here.

The products of the gas-phase photolysis of CH_3I at 25°C are methane,[e−g] ethane, and C_2H_4 in ratios of about 8:1:1. About one-third of the decomposed iodide appeared as CH_2I_2, the remaining iodine being accounted for as I_2 and CHI_3. The quantum yield for CH_3I disappearance is very low, being of the order of 0.05 but increasing toward unity in the presence of 50 per cent or more of HI, HBr, HCl, O_2, and NO.[h] A similar result is obtained from Ag vanes which remove I atoms from the gas phase as an AgI deposit.[i] With rotating Ag vanes, all the iodine is accounted for as AgI, the quantum yield is about 0.5, and $CH_4:C_2H_6:C_2H_4 \cong 1:2:0.7$. This

[a] By using similar techniques they found that the ratio of rate constants $(H + Br_2)/(H + HBr)$ changed from about 0.8 in pure HBr to about 9.5 estimated for infinite ratio of $(H_2)/(HBr)$. These experiments are, however, done at a single temperature.

[b] R. J. Carter et al., *J. Am. Chem. Soc.*, **77**, 6457 (1955), have made further studies on the photolysis of DI by using H_2, C_2H_6, CH_4, and C_5H_{12} as moderators, following ratio of $(D_2)/(HD)$ production vs. $(D_\lambda)/(RH)$, and extrapolating to zero (D_λ). These results confirm qualitatively the hot D atom model. They find, very surprisingly, that H_2 is the poorest moderator for D atoms.

[c] If all the excess were translational, the energy would be shared inversely as the ratio of the masses $127/15 \cong 8/1$. This is a lower limit for the excess energy of CH_3, since it may have vibrational or electronic excitation as well.

[d] Note that the reactions of hot radicals will be picked up only if the steric factors for their reactions are fairly high because deactivation by collision is always a very fast competing process. It is in fact somewhat surprising that hot CH_3 radical reactions are detected at all, since normal CH_3 radicals have steric factors of about 10^{-3} (Table XII.6) for H abstraction reactions.

[e] W. West and L. Schlessinger, *J. Am. Chem. Soc.*, **60**, 961 (1938).

[f] R. Spence and W. Wild, *Proc. Leeds Phil. Lit. Soc.*, *Sci. Sect.*, **3**, 141 (1936).

[g] G. M. Harris and J. E. Willard, *J. Am. Chem. Soc.*, **76**, 4678 (1954).

[h] These increase the quantum yield by reacting with CH_3 radicals, thus preventing the back reaction, $CH_3 + I_2 \rightarrow CH_3I + I$.

[i] R. D. Schultz and H. A. Taylor, *J. Chem. Phys.*, **18**, 194 (1950). It is evident that the low quantum yield for pure CH_3I arises from a back reaction.

evidence is fairly strong in indicating that the reason for the low quantum yield in pure CH_3I arises from the back reaction of $CH_3 + I_2$.

A mechanism which accounts qualitatively for the result is

$$CH_3I + h\nu \xrightarrow{I_a} CH_3 + I$$

$$CH_3 + CH_3I \xrightarrow{2} CH_4 + CH_2I$$

$$CH_3 + I_2 \xrightarrow{3} CH_3I + I$$

$$CH_2I + I_2 \xrightarrow{4} CH_2I_2 + I$$

$$I + I + M \xrightarrow{5} I_2 + M$$

$$CH_3 + CH_3 \xrightarrow{6} C_2H_6$$

$$CH_3 + CH_2I \xrightarrow{7} C_2H_5I^* \xrightarrow{8} C_2H_5 + I$$
$$\xrightarrow{8'} C_2H_4 + HI$$

$$C_2H_5 + I_2 \xrightarrow{9} C_2H_5I + I$$

$$C_2H_5 + C_2H_5 \xrightarrow{10} C_2H_6 + C_2H_4$$
$$\xrightarrow{10'} C_4H_{10}$$

$$CH_3 + HI \xrightarrow{11} CH_4 + I$$

$$C_2H_5I^* + M \xrightarrow{12} C_2H_5I + M$$

$$CH_2I + CH_2I \xrightarrow{13} C_2H_4I_2^* \xrightarrow{14} C_2H_4 + I_2$$
$$+M \xrightarrow{15} C_2H_4I_2$$

While the cracking reactions 8, 8′, and 14 have not been postulated before, they seem to be more reasonable sources of ethylene than the previously postulated secondary photolysis, $CH_2I_2 + h\nu \rightarrow CH_2 + I_2$. The products C_2H_5I, HI, $C_2H_4I_2$, and C_4H_{10} have not been identified as reaction products, but it would be expected that they would be present in small amounts and would be difficult to separate and identify.

However, the interesting aspect of the reaction is the production of CH_4. It was shown by Schultz and Taylor (loc. cit.) and by many workers since[a] that the production of CH_4 is independent of temperature and of added I_2 and is decreased by the addition of "inert" gases such as Ne, N_2, CD_4, He, A, and CO_2. This is precisely the type of evidence which was shown in the case of the photolysis of HI to support a hot radical

[a] R. D. Souffie et al., J. Am. Chem. Soc., 78, 917 (1956). G. M. Harris and J. E. Willard, ibid., 76, 4678 (1954).

mechanism, and indeed, in the present case, one is forced to the same conclusion, i.e., that CH_4 is formed in the photolysis of CH_3I by the reaction of hot CH_3 radicals with CH_3I.[a] Souffie et al. (loc. cit.) estimate that all of the CH_4 is formed by hot radicals and that, further, a considerable fraction of the C_2H_6 similarly arises from a hot radical reaction, presumably of the Walden inversion type, $CH_3 + CH_3I \rightarrow C_2H_6 + I$. Reactions of this type are extremely rare in the free radical chemistry of gases, and in solution reactions seem to involve ionic species. The evidence for such hot C_2H_6 production is that the ratio $(CH_4)/(C_2H_6)$ in the presence of added I_2 has a limiting value of 13.3 independent of temperature and I_2 concentration in the range 28° to 100°C (Souffie et al., loc. cit.).[b] Because of the complexity of the reaction system, the quantitative results on moderator efficiency and extrapolated "normal" thermal CH_3 reactions are poor.

Harris and Willard (loc. cit.) have shown that the hot radical effect is enhanced at 1849 Å, as might be expected for this more energetic radiation. The quantum yield of CH_4 production is raised some 12 fold.[c] It is very likely in all of these cases that the excess energy of the CH_3 is vibrational. However, the evidence is by no means clear at this point.

18. Decomposition of Ozone

One of the simplest chemical systems which is kinetically interesting is the ozone-oxygen system. At temperatures of 70°C and above, pure O_3 and O_3-O_2 mixtures undergo thermal decomposition at rates which are conveniently measurable though tending to explosion at high pressures of O_3. Despite its chemical simplicity, the reaction turns out to be a difficult one to study, being extremely sensitive to catalysis by metals, metal oxides, and trace impurities such as organic matter, peroxides, or oxides of nitrogen. The latter two are particularly difficult to eliminate if the O_2 being ozonized contains traces of N_2 or H_2O.

Early studies on dilute O_3-O_2 mixtures were made by Clement[d] and Chapman and coworkers.[e] The data were not too reproducible but seemed

[a] In the photolysis of acetone-I_2 mixtures, reported by Benson and Forbes, the ratio $(CH_4)/(CO)(\lambda = 2537$ Å$)$ was always about 0.15 to 0.20, indicating either a very low efficiency for reaction 3 relative to CH_3 + acetone or a hot radical effect in this photolysis as well.

[b] In the presence of excess I_2, $\Phi_{I_2} = \Phi_{C_2H_6} = 3.6 \times 10^{-4}$ and $\Phi_{CH_3I} = 0.01$. The latter seems somewhat higher than values reported by other experimenters by a factor of about 2.

[c] A similar result had been obtained by F. P. Hudson et al., J. Chem. Phys., **21**, 1894 (1953).

[d] J. K. Clement, Ann. Physik, IV, 341 (1904).

[e] D. L. Chapman and H. E. Jones, J. Chem. Soc., **97**, 2463 (1910), and earlier papers.

to indicate a second-order dependence of the rate constant on O_3 concentration. Jahn[a] proposed the following mechanism to account for the rate:

$$O_3 \underset{2}{\overset{1}{\rightleftharpoons}} O_2 + O$$

$$O + O_3 \overset{3}{\rightarrow} 2O_2$$

He assumed that O atoms were always in equilibrium with O_2 and O_3, so that $(O) = k_1(O_3)/k_2(O_2)$ with reaction 3 being much slower than reaction 2. Then

$$-\frac{d(O_3)}{dt} = \frac{k_3 k_1 (O_3)^2}{k_2(O_2)} \tag{XIII.18.1}$$

Despite a considerable amount of subsequent work, there has never been any real unanimity of opinion on the correctness of this mechanism or on the homogeneity of the reaction. Glissman and Schumacher[b] in 1933 reported the most extensive series of studies, together with a very complex mechanism involving simultaneous atom and molecular paths and energy chains. They worked with both dilute and relatively concentrated ozone mixtures. Benson and Axworthy[c] have repeated these studies and confirmed the experimental results but have shown that all of the known data, in both dilute and concentrated O_3 systems, can be interpreted in terms of the following modified Jahn mechanism:

$$O_3 + M \underset{1}{\overset{\rightleftharpoons}{{}_2}} O_2 + O + M$$
$$O + O_3 \overset{3}{\rightarrow} 2O_2 \tag{XIII.18.1a}$$

Assuming a stationary state for O atoms leads to

$$(O)_{ss} = \frac{k_1(O_3)(M)}{k_2(O_2)(M) + k_3(O_3)}$$

$$-\frac{d(O_3)}{dt} = \frac{2k_3 k_1 (O_3)^2 (M)}{k_2(O_2)(M) + k_3(O_3)} \tag{XIII.18.2}$$

or, writing

$$k_S = -\frac{1}{(O_3)^2}\frac{d(O_3)}{dt} = \frac{d}{dt}\left(\frac{1}{O_3}\right) = \text{instantaneous second-order rate constant}$$

Equation (XIII.18.2) can be written as

$$\frac{(M)}{k_S(O_3)} = \frac{k_2}{2k_1 k_3}\frac{(O_2)(M)}{(O_3)} + \frac{1}{2k_1} \tag{XIII.18.2a}$$

where (M) is actually the weighted sum of the contributions of different

[a] S. Jahn, Z. anorg. Chem., **48**, 260 (1906).
[b] A. Glissman and H. J. Schumacher, Z. physik. Chem., **B21**, 323 (1933).
[c] S. W. Benson and A. E. Axworthy, Jr., J. Chem. Phys., **26**, 1718 (1957).

gases to reactions 1 and 2 and can be written as $(M) = (O_3) + \Sigma a_i M_i$, where the a_i would represent the efficiencies of the species M_i relative to O_3 in reaction 1.[a] Equation (XIII.18.3) now shows that if the experimentally determined ratio $(M)/k_S(O_3)$ is plotted against $(O_2)(M)/(O_3)$, the result should be a straight line with slope $k_2/2k_1k_3$ and intercept $\frac{1}{2}k_1$.[b] Benson and Axworthy have shown that this is indeed the case for a very large variety of experimental conditions including added foreign gases, different vessels, and a temperature range of 70 to 120°C. The best fit to the data for different foreign gases is given by:

$$(M) = (O_3) + 0.44(O_2) + 0.41(N_2) + 1.06(CO_2) + 0.34(He) \quad (XIII.18.3)$$

where the coefficients are a measure of the efficiency of these various gases (relative to O_3) in transferring energy to and from the internal degrees of freedom of O_3.[c]

Although the data yield only the value of k_1 and the ratio k_1k_3/k_2 [Eq. (XIII.18.2a)], the equilibrium constant $K_{1.2}$ for reactions 1 and 2 can be calculated from thermodynamic and spectroscopic data, which then permits the rate constants k_2 and k_3 to be evaluated. In Arrhenius form, these are

$$k_1 = 4.61 \times 10^{12} \exp\left[-24,000/RT\right] \text{ liters/mole-sec}$$
$$k_2 = 6.00 \times 10^7 \exp\left[600/RT\right] \text{ liters}^2/\text{mole}^2\text{-sec} \quad (XIII.18.4)$$
$$k_3 = 2.96 \times 10^9 \exp\left[-6000/RT\right] \text{ liters/mole-sec}$$
$$K_{1.2} = k_1/k_2 = 7.7 \times 10^4 \exp\left(24,600/RT\right) \text{ moles/liter}$$

Of particular interest in these results are the following points. The frequency factor of reaction 1, a bimolecular reaction, seems to be 20 fold greater than collision frequencies; reaction 2 has a negative activation energy; the activation energy for reaction 1 is less than the bond energy of ozone. It is of further interest to observe that, although reaction 3 is exothermic by 93 Kcal, it nevertheless has an activation energy of 6 Kcal. In addition the O_2 molecules produced by reaction 3 have 99 Kcal of excess energy divided between them, and one might well expect, in view of the low bond energy of O_3 (24.6 Kcal), that an energy chain might be set off by these hot O_2 molecules. The evidence shows, however, that the results are quite consistent with the simple mechanism [Eq. (XIII.18.1a)] and incompatible with the existence of a significant energy chain.

The effect of an energy chain would be incorporated in the following mechanism:

[a] The rate constant 1 now refers to O_3 as the third body M. In the initial stages of the decomposition of pure ozone we can set $(M) \cong (O_3)$. See Eq. (XIII.18.2a).

[b] Alternatively, we can write $k_S(O_3)/(M) = 2k_1 - (k_2/k_3)k_S(O_2)$, which is not quite as convenient for plotting.

[c] The accuracy of these values is ±20 per cent for O_2 and ±10 per cent for the other gases.

$$M + O_3 \underset{2}{\overset{1}{\rightleftharpoons}} M + O_2 + O$$

$$O + O_3 \overset{3}{\rightarrow} 2O_2^*$$ (XIII.18.5)

$$O_2^* + M \overset{4}{\rightarrow} M + O_2 \qquad \text{deactivation}$$

$$O_2^* + O_3 \overset{5}{\rightarrow} 2O_2 + O \qquad \text{energy chain}$$

where, as before, M in reaction 4 would stand for a weighted sum of all possible third bodies including O_3.[a] By using the stationary-state condition for O and O_2^* we now have for the rate of disappearance of O_3:

$$-\frac{d(O_3)}{dt} = \frac{2k_1k_3(M)(O_3)^2}{k_2(M)(O_2) + k_3(O_3)\left[1 - \dfrac{2}{1 + k_4(M)/k_5(O_3)}\right]}$$ (XIII.18.6)

This differs from the previous expression [Eq. (XIII.18.2)] by the bracketed term in the denominator. The energy chain has the effect of increasing the stationary-state concentration of O atoms and thus accelerating the rate. The quantitative effect of this chain should be a function only of the ratio $(M)/(O_3)$. In very dilute O_3, in which reaction 2 tends to become much faster than reaction 3, the mechanism reduces to the simple Jahn mechanism [Eq. (XIII.18.1)] and the effect of energy chains would be negligibly small. In very concentrated O_3, however, and in pure O_3, $k_5(O_3)/k_4(M)$ will have its maximum value, and the chains, if they exist, should become apparent.[b] Now the rates in concentrated O_3 are found to be higher than the rates expected from the non-energy chain mechanism. However, on closer examination it is found that the observed discrepancies do not depend on the ratio $(M)/(O_3)$, as would be expected from Eq. (XIII.18.6), but instead correlate mainly with the over-all rate of decomposition. That is, the faster the decomposition the larger the discrepancy. Benson and Axworthy (*loc. cit.*) have shown that the discrepancies can be reasonably explained in terms of the small temperature gradients which have been observed to exist in these systems and which precede the explosive region, the explosion in fact being caused by thermal self-heating. The energy chains thus turn out to be unimportant.[c]

---

[a] That is, O_3 may act as a reactant, as in reaction 5, or as a moderator, as in reaction 4.

[b] Note that the maximum value of $k_5(O_3)/k_4(M)$ is probably less than 1 when $(M) = (O_3)$ (i.e., pure ozone). A value of 1 would imply equal chances of deactivation or reaction on the collision of $O_3 + O_2^*$.

[c] From the precision of the data, one would estimate that $k_4/k_5 > 15$, which is quite reasonable if the excess energy of O_2^* is translational. One has no way at present of estimating efficiencies of energy transfer of the type vibration-vibration.

However, we can say something about the equilibrium constant in a reaction of the type $O_2 + O_3 \rightleftharpoons O_3^* + O_2$, where O_3 represents an ozone molecule with energy E^* distributed among its internal degrees of freedom. K_{eq} for this reaction is just the number of ways in which E^* can be distributed among the three vibrational degrees of O_3^* (or

The paradoxical values of the rate constants assume more normal proportions if we look upon the decomposition reaction 1 from the point of view of its detailed behavior. Reaction 1 is undoubtedly a unimolecular reaction near its low-pressure limit and can be written as

$$M + O_3 \underset{b}{\overset{a}{\rightleftharpoons}} O_3^* + M$$

$$O_3^* \underset{d}{\overset{c}{\rightleftharpoons}} O_2 + O$$

(XIII.18.7)

where O_3 represents a molecule of critically energized ozone (Sec. XI.5). Under the conditions of the reaction, the concentration of O_3^* never reaches equilibrium with respect to O_3 because $k_c(E)$ is much larger for the higher energy states $(E > E^*)$ than is $k_b(M)$, and so the stationary-state concentration of O_3^* is poor in higher excited states. But this in turn means that the activation energy observed for the split, $O_3 \rightarrow O_2 + O$, will be less than the energy of the reaction.

From the classical theory of unimolecular reactions developed in Sec. XI.4, we see that k_1 can be expressed as

$$\lambda_D Z' \left(\frac{E^*}{RT}\right)^{s-1} \frac{\exp(-E^*/RT)}{(s-1)!}$$

$E_{exp} = E^* - (s - \tfrac{3}{2})RT$. Using the value (NBS thermal data) of $E^* = \Delta E_0^\circ = 24.4$ Kcal and the experimental, $E_{exp} = 24.0$ Kcal, we find that $s \cong 2$ classical degrees of freedom as against three vibrations and two active rotations in O_3. By expressing k_1 in the Arrhenius form this then gives $A_1 = 2.4 \times 10^{12}\lambda_D$ liters/mole-sec (assuming $Z' = 2.0 \times 10^{11}$ liters/mole-sec), which is equal to the observed value for $M = O_2$ with $\lambda_D = 1$.[a] The high frequency factor of reaction 1 thus arises from a high entropy of activation in forming O_3.[b] The negative activation energy for

possibly three vibrations plus two rotations), divided by the number of ways in which this same energy can be divided among the three vibrational degrees of O_3, one vibration of O_2, and two translational degrees of freedom (in the center of mass system of the colliding pair). The equilibrium point lies certainly to the left. This tells us that if the probability of deactivation of O_3^* is λ_2, then λ_1, the chances of activation of O_3 by O_2, is smaller by the equilibrium constant, the collision frequencies being assumed the same.

[a] The maximum value of λ_D is of course 1, since it represents the probability that collision of $O_3^* + O_2$ leads to deactivation of O_3^*. For $s = 3$, $A_1 = 14.4 \times 10^{12}\lambda_D$, and $E^* = 25.1$, while for $s = 4$, $A_1 = 60 \times 10^{12}\lambda_D$, $E^* = 25.9$, giving $\lambda_D = \tfrac{1}{6}$ and $\tfrac{1}{24}$, respectively. Any of these values is in qualitative agreement with the data. However, the classical model used limits the extent to which quantitative agreement may be sought.

[b] Part of the high entropy of O_3 comes from the multiplicity of ways of distributing the energy F into its various degrees of freedom. Another part must come from lengthening of the bonds with consequent increase in the moments of inertia. If we merely

reaction 2 is then a consequency of 1 being at its low-pressure limit. The interpretation to be made is that there is more chance of forming O_3 from $O_2 + O$ if the latter collide with below-average energies, since the complex O_3^* which is formed will have a longer lifetime at lower energies, and thus more time in which to be deactivated. Conversely, the interpretation of the *negative* activation energy for reaction 2 is that the major contribution to the formation of O_3 comes from collisions between O_2 and O which have an average energy 600 calories less than a similar nonreacting pair in the flask.[a]

A further very interesting feature of this reaction system is that the apparent activation energy for the over-all reaction [Eq. (XIII.18.2)] is close to 30 Kcal when $(O_2) \gg (O_3)$, while in almost pure O_3 the initial rate becomes independent of O_2 and the apparent activation energy now appears to be about 24 Kcal. These two extreme concentration regions are characterized also by the fact that O_2 acts as an accelerator in concentrated O_3 and an inhibitor in dilute O_3. In addition, wall recombination of O atoms can be shown to accelerate the rate in dilute O_3 and inhibit the rate in concentrated O_3.[b]

The catalytic decomposition of O_3 is of great interest but also very difficult to study quantitatively. The halogens are known to sensitize the decomposition, but the reaction is very complex,[c,d] leading to the production at not too high temperatures of oxides such as Cl_2O_6, and Br_3O_8.[e] Initiation may be heterogeneous in these systems or, more likely, $O + X_2 \rightarrow XO + X$ following by the chain:

$$X + O_3 \rightarrow XO + O_2$$
$$XO + O_3 \rightarrow XO_2 + O_2$$
$$XO_2 + O_3 \rightarrow XO_3 + O_2 + \cdots \qquad \text{(XIII.18.8)}$$
$$XO_3 + O_3 \rightarrow XO_2 + 2O_2$$

and chain-terminating steps of unelucidated type except for the identified step $2ClO_3 \rightarrow Cl_2O_6$.

count the number of ways of distributing the 24 Kcal among the three vibrations of O_3, it amounts to roughly 65 states (permitting slight discrepancies to be accommodated in rotations and anharmonicities). This would account for 8.3 e.u. and may be taken as a minimum value. A 20 per cent increase in bond lengths would contribute another 3.0 e.u.

[a] When we correct for the temperature coefficient due to a collision term in k_2, this is increased by $\frac{1}{2}RT$.

[b] For further details see papers by A. E. Axworthy, Jr., and S. W. Benson in "Ozone Chemistry and Technology," Am. Chem. Soc. Series, Advances in Chemistry No. 21, 1959.

[c] M. Bodenstein et al., *Z. physik. Chem.*, **B5**, 209 (1929).

[d] K. F. Bonhoeffer, *Z. Physik*, **13**, 94 (1923). B. Lewis and H. J. Schumacher, *Z. physik. Chem.*, **A138**, 462 (1928); **B6**, 423 (1930).

[e] *Ibid.*

In the catalysis by nitric oxides,[a] which are better understood, many of the step reactions being known, initiating steps are the very fast[b]

$$O_3 + NO \rightarrow O_2 + NO_2$$
$$O_3 + NO_2 \rightarrow O_2 + NO_3$$
(XIII.18.9)

These are followed by the well-known synthetic step, $NO_2 + NO_3 \rightarrow N_2O_5$, together with the less well-known step $NO_3 + NO_3 \rightarrow 2NO_2 + O_2$. The decomposition of O_3 catalyzed by N_2O_5 (Schumacher and Sprenger, *loc. cit.*) has the unusual kinetic over-all order of $\frac{2}{3}$ power in N_2O_5 and $\frac{2}{3}$ power in O_3 and is undoubtedly initiated by NO_2 arising from the reversible dissociation of N_2O_5: $N_2O_5 \rightleftharpoons NO_2 + NO_3$. The oxide NO_3 has been identified spectroscopically as a blue gas present in mixtures of N_2O_5 and O_3 emerging from an ozonizer.[c,d]

The catalysis by hydrocarbons and other H donors (for example, H_2O_2) is much less understood but just as efficient. Here again initiation may be heterogeneous although, because of the long chains possible, the homogeneous reaction $O + RH \rightarrow R + OH$ is a quite reasonable source of radicals. This can be followed by the chain originally proposed by Weiss[e] for the solution decomposition

$$HO + O_3 \rightarrow HO_2 + O_2$$
$$HO_2 + O_3 \rightarrow HO + 2O_2$$
(XIII.18.10)

with termination by $2HO_2 \rightarrow H_2O_2 + O_2$.

These catalytic reactions are of great importance in low-temperature oxidation reactions, particularly in atmospheric pollution phenomena (e.g., Los Angeles smog), and their elucidation is required for any elementary understanding of the free radical chemistry of oxygen. The difficulty in studying these systems has been the difficulty of studying ozone itself, plus the evident complexity of intermediates and products in the other system. The results, however, will undoubtedly well repay the efforts.

The photolysis of O_3 has been studied under a variety of conditions,[f–h] but the results are not as complete nor as reliable as for the pyrolysis. In red light,[g] the results seem to fit the mechanism presented except at very

[a] H. J. Schumacher and G. Sprenger, *ibid.*, **A136**, 77 (1928); **B2**, 267 (1929).

[b] H. S. Johnston and D. M. Yost, *J. Chem. Phys.*, **17**, 386 (1949). H. S. Johnston and H. J. Crosby, *ibid.*, **22**, 689 (1954).

[c] Hautefeuille and Chappius, *Compt. rend.*, **92**, 80 (1881); **94**, 112, 1306 (1882). O. Warburg and Leithauser, *Ann. Physik*, **20**, 743 (1906); **23**, 209 (1907); H. J. Schumacher and G. Sprenger, *loc. cit.*

[d] H. J. Jones and O. R. Wulf, *J. Chem. Phys.*, **5**, 873 (1937).

[e] J. Weiss, *Trans. Faraday Soc.*, **31**, 668 (1935). See also M. L. Kilpatrick et al., *J. Am. Chem. Soc.*, **78**, 1784 (1956).

[f] L. J. Heidt and G. S. Forbes, *J. Am. Chem. Soc.*, **56**, 2365 (1934). Heidt, *ibid.*, **57**, 1710 (1935).

[g] G. B. Kistiakowsky, *Z. physik. Chem.*, **117**, 337 (1925). H. J. Schumacher, *J. Am. Chem. Soc.*, **52**, 2377 (1930).

[h] D. H. Volman, *ibid.*, **73**, 1018 (1951).

high $(O_3)/(O_2)$ ratios, at which the quantum yield seems to rise slightly above 2. On the other hand, the rather considerable dark reaction at low temperatures, together with heterogeneous reaction and catalysis, makes these measurements somewhat doubtful. The evidence of Heidt (loc. cit.) on the very high quantum yield (near 6) in relatively concentrated O_3 and at short wavelengths (<2500 Å), if real, may be evidence for a chain carried by electronically excited states of O_2 which may be produced at these short wavelengths. Benson and Axworthy (loc. cit.) have suggested that the sequence is:

$$O_3 + h\nu(\lambda < 3100 \text{ Å}) \rightarrow O_2[^1\Delta_g] + O^*[^1D]$$

Chain: $\quad\quad O^*[^1D] + O_3 \rightarrow O_2^*[^3\Sigma_u^-] + O_2[^3\Sigma_g^-] \quad\quad E_{act} \geqslant 3 \text{ Kcal}$

$\quad\quad\quad O_2^*[^3\Sigma_u^-] + M \rightarrow O^*[^1D] + O[^3P] + M \quad\quad E_{act} = 20 \text{ Kcal}$

together with collisional deactivation of the excited states of O^* and O_2^*. More direct evidence on these excited states and their behavior in such reactions would be desirable.[a]

Volman's study of the photochemical decomposition of mixtures of O_3 and H_2O_2 (loc. cit.) was plagued by a heterogeneous reaction, but qualitatively at least, it seems to confirm the Weiss mechanism (XIII.18.10) with O atom attack on H_2O_2 to give $HO + HO_2$ as the initiating reaction. He then considers that in the photochemical reaction, there is spectral evidence for believing that the optically excited O_3^* has a finite lifetime during which it may be deactivated by collision with an inert gas molecule.[b] From this point of view foreign gases play a dual role, that of deactivation both of O_3^* formed initially on optical excitation, and also the much lower energy O_3^* formed on recombination of $O_2 + O$. Volman uses this argument to reconcile the disparities in data of Glissman and Schumacher on the effect of foreign gases with his own data. The difficulty with such a picture is that it is hard to believe that the lifetime of optically excited O_3^* can be longer than about 10^{-11} sec.[c] Since collisions occur at about 10^{-9}-sec intervals at 1 atm, the importance of primary deactivation of O_3^* by collision cannot exceed $\frac{1}{200}$ of the rate of formation of O_3^* and should be negligible. A more plausible view of the discrepancy is to be

[a] D. H. Volman, J. Chem. Phys., **24**, 122 (1956), has proposed that in the reverse reaction, the synthesis of O_3 from O_2 photosensitized by Hg, the sequence at 2537 Å is $Hg^* + O_2 \rightarrow O_2^* + Hg$ followed by $O_2^* + O_2 \rightarrow O_3 + O$. The O_2^* could be a vibrationally excited ground-state molecule or it could be the electronically excited metastable $^3\Sigma_u^+$ or one of the lower-lying singlet states. He further suggests that at 1849 Å the sequence is $Hg^{**} + O_2 \rightarrow O_2^{**}[^3\Sigma_u^-] + Hg$, followed by predissociation of the excited O_2^{**} into two translationally excited $[^3P]$ O atoms. The evidence is again very indirect. The latter proposal would be in conflict with the chain decomposition proposed above.

[b] It is also possible that in more concentrated O_3 such an excited molecule may cause the decomposition of a second O_3 molecule by $O_3^* + O_3 \rightarrow 2O_3^{**} \rightarrow 2O_2 + 2O$, the energy of O_3^* being more than sufficient for this.

[c] This is what one would expect as a maximum, both in terms of the fuzziness of the optical spectra and on theoretical grounds.

found in the heterogeneous reactions taking place in the system. This is particularly pronounced when H_2O_2 is admitted to the mixture.[a]

19. Oxides of Nitrogen; Decomposition of N_2O_5

The oxides of nitrogen have long been of interest to kineticists both from a theoretical point of view and also from the point of view of their chemical behavior. Greatest interest has been centered about the decomposition of N_2O_5, a first-order reaction originally thought to represent a typical, unimolecular reaction but now a fairly thoroughly investigated, complex decomposition.

The over-all reaction in the gas (or liquid) phase is

$$N_2O_5 \rightarrow N_2O_4 + \tfrac{1}{2}O_2 \qquad\qquad \text{(XIII.19.1)}$$
$$\updownarrow$$
$$2NO_2$$

The reaction proceeds at a convenient rate at room temperature and is experimentally first-order over an extended range of temperatures and pressures.[b,c] The rate constant k_0 (units of sec^{-1}) can be represented by[d]

$$\log k_0 = 13.613 - \frac{24,650}{2.303RT} \qquad\qquad \text{(XIII.19.2)}$$

The reaction can be followed manometrically, since the equilibrium between NO_2 and N_2O_4 is well known and is fast. It can also be followed spectrophotometrically, since the spectra of N_2O_5, N_2O_4, and NO_2 are all quite different, and both ultraviolet[e] and infrared[f] techniques have been successfully employed.

For a long time it was thought that the observed rate was a measure of a slow, initial split of N_2O_5 into intermediates which thereupon reacted rapidly to give products. This implied that the observed rate constant

---

[a] Volman found that the thermal reaction of $O_3 + H_2O_2$ was heterogeneous and strongly catalyzed by Hg vapor and HgO traces. Above 100°C the reaction became more reproducible but the frequency factor and activation energy of the second-order rate constant were both abnormally low, indicative of a surface reaction ($E_{act} = 3.5$ Kcal).

[b] F. Daniels and E. H. Johnston, *J. Am. Chem. Soc.*, **43**, 53 (1921). J. K. Hunt and F. Daniels, *ibid.*, **47**, 1602 (1925). H. C. Ramsperger et al., *Proc. Nat'l Acad. Sci. U.S.*, **16**, 6 (1930).

[c] H. S. Johnston and Yu-sheng Tao, *J. Am. Chem. Soc.*, **73**, 2948 (1951). All of the above studies cover the range 0 to 123°C and 0.04 mm Hg to 1 atm. The above expression fits the entire range with good precision, indicating no variation of activation energy in this range. At lower pressures, the rate constant appears to decrease. We shall discuss this later.

[d] *Ibid.*

[e] J. H. Smith and F. Daniels, *J. Am. Chem. Soc.*, **69**, 1735 (1947). R. L. Mills and H. S. Johnston, *ibid.*, **73**, 938 (1951).

[f] I. C. Hisatsune et al., *ibid.*, **79**, 4648 (1957). See also Ogg et al., *loc. cit.*

was proportional to the rate constant for the suggested initial split. However, the nature of this initial split remained uncertain, none of the proposed reactions being acceptable on energetic grounds, grounds of spin conservation, or on kinetic grounds.[a]

The dilemma posed by this reaction was resolved by Ogg,[b] who proposed the following unusual mechanism:

$$N_2O_5 \underset{2}{\overset{1}{\rightleftharpoons}} NO_2 + NO_3$$

$$NO_2 + NO_3 \overset{3}{\to} NO + O_2 + NO_2 \qquad \text{(XIII.19.3)}$$

$$NO + NO_3 \overset{4}{\to} 2NO_2$$

where it is assumed that $k_2 > k_3$ and $k_4 > k_3$. The unique feature of this mechanism is that reactions 2 and 3 represent competing bimolecular reactions between two identical, simple reactants. This mechanism has been confirmed in quite convincing manner by a number of independent studies.

Assuming stationary states for NO_3 and NO (also an intermediate) gives

$$(NO_3)_{ss} = \frac{k_1(N_2O_5)}{(k_2 + 2k_3)(NO_2)}$$

$$(NO)_{ss} = \frac{k_3}{k_4}(NO_2)$$

and so

$$-\frac{d \ln (N_2O_5)}{dt} = \frac{2k_3 k_1}{k_2 + 2k_3} = k_0 \qquad \text{(XIII.19.4)}$$

which is indeed a first-order equation, but we see that the over-all rate constant k_0 is a composite constant $2k_3 k_1/(k_2 + 2k_3)$. As might be expected for these two competing steps, it turns out that $k_2 \gg k_3$, so that $k_0 \cong 2k_3 k_1/k_2 = 2k_3 K_{1.2}$, the product of a bimolecular rate constant and an equilibrium constant.

The first two steps in this scheme are quite reasonable in terms of spin rules and energies. In addition, NO_3 is a known substance whose spectrum has been observed (Schumacher and Sprenger, *loc. cit.*) and has been found to be a reasonable intermediate in the N_2O_5-catalyzed decomposition of O_3. Further, NO_3 is the observed intermediate in the production of N_2O_5 via the reaction of O_3 with NO_2 (or NO). The mechanism of this very fast reaction is

---

[a] Thus the production of isomeric $N_2O_4 + O$ is energetically impossible, as well as violating spin-conservation rules. The products $N_2O_3 + O_2$ also violate spin rules and in addition would have to be formed via a cyclic complex which would be expected to have a much lower frequency factor. Intermediates which form cannot be permitted to react with N_2O_5, or the rate will no longer be first-order.

[b] R. A. Ogg, Jr., *J. Chem. Phys.*, **15**, 337, 613 (1947).

$$NO_2 + O_3 \xrightarrow{5} NO_3 + O_2$$

(XIII.19.5)

$$NO_3 + NO_2 \xrightarrow{2} N_2O_5$$

Johnston and Yost[a] have studied the reaction $O_3 + NO_2$ and found it to be very fast, second-order mixed, with a rate constant given by

$$-\frac{1}{(NO_2)(O_3)} \frac{d(NO_2)}{dt} = k_5$$

$$= 5.9 \times 10^9 \exp(-7000/RT) \quad \text{liters/mole-sec} \quad \text{(XIII.19.6)}$$

so that reaction 2 does appear to be a reasonable and fast step with probably little or no activation energy.

Additional evidence for the mechanism is provided from two independent sources, the NO-N_2O_5 reaction and the exchange of N-labeled N_2O_5 with NO_2. Ogg and coworkers studied the exchange in CCl_4 solution[b] by using radio N^{13} and in the gas phase[c] by using N^{15}. Since the mechanism for the decomposition predicts a rapid equilibrium between NO_2 and N_2O_5, the rate of exchange of NO_2 with $N_2^{15}O_5$ should provide a direct measure of the rate of primary dissociation, $N_2O_5 \rightarrow NO_2 + NO_3$. As expected, the rate of exchange proved to be much faster than the rate of decomposition of N_2O_5[d] and, in addition, was very sensitive to the pressure of added CO_2.

This effect of CO_2 is at first somewhat surprising because it implies that the rate of dissociation of N_2O_5 is being accelerated by CO_2, which we would expect only if this dissociation were a unimolecular reaction below its high-pressure limit. N_2O_5 has 7 atoms and 15 internal vibrations (or their equivalent) and probably 2 active rotations. For a completely classical molecule (which N_2O_5 is not), the RRK theory (Table XI.2) would predict deviations from the high-pressure limit near 0.1 mm Hg, which is

[a] H. S. Johnston and D. M. Yost, *ibid.*, **17**, 386 (1949). It is of course assumed that this reaction 5 is the slow step in the synthesis of N_2O_5 and may be equated to the experimental rate constant. However, the data cover a fairly wide range of pressures, and it is difficult to see any other mechanism for this very simple reaction. Reaction 2 must of course be fast with respect to reaction 5 for these kinetics to be obeyed, and it appears that reaction 2 must go at about 1 in every 100 collisions between NO_2 and NO_3, which would not be surprising for a recombination reaction of this type. A similar efficiency (1 in 200 collisions) has been found for $2NO_2 \rightarrow N_2O_4$. T. Carrington and N. Davidson, *J. Phys. Chem.*, **57**, 418 (1953).

[b] R. A. Ogg, Jr., *J. Chem. Phys.*, **15**, 613 (1947).

[c] R. A. Ogg, Jr., et al., *ibid.*, **18**, 573 (1950).

[d] The initial rate of appearance of $N^{15}O_2$ is given by $\phi k_1(N_2^{15}O_5)$, where ϕ is the mole fraction of N^{15} in the N_2O_5 (the $N^{15}O_2$ was followed spectrophotometrically in the infrared). At 27°C in the presence of about 500 mm Hg of CO_2, k_1 was about 0.5 sec^{-1} compared to a rate constant of about 4×10^{-5} sec^{-1} for the over-all decomposition. At 50 mm Hg of CO_2, k_1 was about 0.1 sec^{-1}. Such rapid rates are obviously difficult to study with the relatively slowly responding infrared spectrograph, and so the data are only semiquantitative.

about a factor of 500 below the region in which the CO_2 effects appear. This implies that, if we keep a classical model, fewer degrees of freedom are active, or deactivation of $N_2O_5^*$ by collision is very inefficient (e.g., at about 1 per 100 collisions). The alternative to these assumptions is to use the modified Slater theory, in which we assume that energy transfer among the active degrees of freedom is too slow to be important.[a]

The data on the reaction of NO with N_2O_5 provide further evidence concerning the pressure dependence of the unimolecular decomposition of N_2O_5.

Smith and Daniels[b] found that the decomposition of N_2O_5 in excess NO could be described by the over-all equation $NO + N_2O_5 \rightarrow 3NO_2$ and was several orders of magnitude faster than the decomposition of N_2O_5 itself. In addition the reaction appeared to be zero-order in NO and first-order in N_2O_5, indicating that the slow step in the mechanism involved N_2O_5 alone. This is found to be consistent with the Ogg mechanism [Eq. (XIII.19.3)] if we may assume that $k_4(NO) \gg k_3(NO_2)$ in the presence of added NO. Neglecting reaction 3, then, and assuming a stationary state for NO_3 only we find

$$N_2O_5 \underset{2}{\overset{1}{\rightleftharpoons}} NO_2 + NO_3$$

$$NO + NO_3 \overset{4}{\rightarrow} 2NO_2$$

$$(NO_3)_{ss} = \frac{k_1(N_2O_5)}{k_2(NO_2) + k_4(NO)}$$

(XIII.19.7)

$$-\frac{d(NO)}{dt} = -\frac{d(N_2O_5)}{dt} = \frac{k_1(N_2O_5)}{1 + k_2(NO_2)/k_4(NO)} \xrightarrow{(NO_2) < (NO)} k_1(N_2O_5)$$

(XIII.19.8)

which was the mechanism proposed by Smith and Daniels.[c] Now since no

---

[a] The Slater model with no energy transfer would be compatible with such a high-pressure limit. However, it is difficult to believe that transfers are so restricted over lifetimes of the order of 10^6 to 10^7 vibrations. L. S. Kassel, J. Chem. Phys., **21**, 1093 (1953), has shown that the pressure dependence of the rate constant k_1 is compatible with a model having 15 degrees of freedom and an average frequency of 300 cm^{-1} (that is, 25 quanta at 300°K). This, however, seems much too small an average frequency. H. S. Johnston and R. L. Perrine, J. Am. Chem. Soc., **73**, 4782 (1951), have shown that the low-pressure data can be fitted reasonably well with the RRK theory by using 10 oscillators and an average frequency of about 350 cm^{-1}, which is a slight improvement. Both of these calculations suffer from the use of a deactivation probability of unity, and it may be that a lower probability would improve the fit with more reasonable parameters.

[b] J. H. Smith and F. Daniels, J. Am. Chem. Soc., **69**, 1735 (1947).

[c] As we shall see later, because of the pressure dependence, k_1 in Eq. (XIII.19.8) is not the same as the k_1 appearing in Eq. (XIII.19.4). They correspond to the low and high-pressure limits, respectively. The k_1 used here turns out experimentally to vary with total pressure.

inhibition by NO_2 was observed, it was concluded that $k_2(NO_2) \ll k_4(NO)$ (which is certainly true in the beginning of the reaction), and so what is being measured is the rate constant k_1 for the primary unimolecular dissociation of N_2O_5.[a]

Mills and Johnston[b] have studied this reaction over an enormous range of temperatures and total pressures. They confirmed the findings of Smith and Daniels and found also that at low pressures the rate of disappearance of NO becomes dependent on the total pressure of foreign gases in the system. From Eq. (XIII.19.8) we see that this is to be expected because k_1 is the rate constant for a unimolecular decomposition, that of N_2O_5.[c] They were able to develop techniques for studying the reaction over a pressure range of 10^5, which made it possible for them to obtain rate constants near the high- and low-pressure limits of the decomposition.

Mills and Johnston further showed that both this reaction and the decomposition of pure N_2O_5 could be described by a single scheme which incorporated the unimolecular features of the reaction $N_2O_5 \rightarrow NO_2 + NO_3$:

$$N_2O_5 + M \underset{b_i}{\overset{a_i}{\rightleftharpoons}} (N_2O_5)_i^* + M$$

$$(N_2O_5)_i^* \underset{d_i}{\overset{c_i}{\rightleftharpoons}} NO_2 + NO_3$$

$$NO_2 + NO_3 \overset{3}{\rightarrow} NO + O_2 + NO_2 \qquad\qquad \text{(XIII.19.9)}$$

$$NO + NO_3 \overset{4}{\rightarrow} 2NO_2$$

$(N_2O_5)_i^*$ represents a critically energized molecule of N_2O_5 with energy $E_i \geqslant E^*$, where $E^* = \Delta E_{1.2}^\circ$. If we assume that the NO_3 and NO_2 species reach temperature equilibrium and further that NO_3 and each $(N_2O_5)_i^*$ reach stationary states, then

[a] That is, it appears as though NO is a very efficient scavenger for NO_3, much more so than NO_2. This is surprising at first glance, since the latter reaction is a recombination and the former involves atom transfer. A supporting piece of evidence, however, comes from the work of Carrington and Davidson, loc. cit., on the dissociation of N_2O_4. They calculated a relatively low collision efficiency of about $\frac{1}{200}$ to $\frac{1}{1000}$ for the similar recombination of $2NO_2 \rightarrow N_2O_4$ (evaluated for the high-pressure limit). A further resolution of the difficulty is possible if reaction 4 proceeds through formation of a quasi-stable molecule O—N—O—NO_2 which is isomeric with symmetrical N_2O_4.

[b] R. L. Mills and H. S. Johnston, J. Am. Chem. Soc., **73**, 938 (1951). See also H. S. Johnston and R. L. Perrine, ibid., **73**, 4782 (1951).

[c] Mills and Johnston made measurements covering the pressure range 0.07 mm Hg to 10 atm. Over this range of pressures (10^5) the experimentally calculated first-order rate constant changed by a factor of 300. It is surprising that even 10 atm pressure of N_2 did not bring the reaction to better than about 90 per cent of its high-pressure limit. This is the first chemical reaction to be studied near its low- and high-pressure limits.

$$(N_2O_5)_i^* = \frac{(M)(N_2O_5)a_i}{b_i(M) + c_i} + \frac{[\Sigma c_i(N_2O_5)_i^*]d_i/d}{[1 + k_3/d + k_4(NO)/d(NO_2)][b_i(M) + c_i]}$$

$$(NO_3) = \frac{\Sigma c_i(N_2O_5)_i^*}{d(NO_2)[1 + k_3/d + k_4(NO)/d(NO_2)]} \qquad \text{(XIII.19.10)}$$

where[a] $d = \Sigma d_i$ and the sums are over all possible energy states E_i from $E_i = E^*$ to ∞. For the rate of disappearance of N_2O_5 we can write

$$-\frac{d \ln (N_2O_5)}{dt} = \frac{(M)(\Sigma\{a_ic_i/[b_i(M) + c_i]\})[1 + k_4(NO)/k_3(NO_2)]}{1 + k_4(NO)/k_3(NO_2) + (M/k_3)\Sigma\{d_ib_i/[b_i(M) + c_i]\}}$$

(XIII.19.11)

whereupon, setting $r_i = c_i/[b_i(M) + c_i] = 1 - b_i(M)/[b_i(M) + c_i]$ (note that $0 \leqslant r_i \leqslant 1$):

$$-\frac{d \ln (N_2O_5)}{dt} = \frac{(M)[\Sigma a_ir_i][1 + k_4(NO)/k_3(NO_2)]}{1 + k_4(NO)/k_3(NO_2) + (d/k_3)[1 - \Sigma(d_ir_i/d)]}$$

(XIII.19.12)

Now at very high pressures at which for all important states contributing to reaction $b_i(M) \gg c_i$, $r_i \rightarrow c_i/b_i(M) \ll 1$, so that

$$-\frac{d \ln (N_2O_5)}{dt} \xrightarrow{(M)\rightarrow\infty} \frac{[\Sigma(a_ic_i/b_i)][1 + k_4(NO)/k_3(NO_2)]}{1 + k_4(NO)/k_3(NO_3) + d/k_3} \qquad \text{(XIII.19.13)}$$

When there is no initial NO, so that NO can be treated as a stationary intermediate,[b] we find $(NO)_{ss} = k_3(NO_2)/k_4$, and on substitution

$$-\frac{d \ln (N_2O_5)}{dt} \xrightarrow[\substack{(M)\rightarrow\infty}]{(NO)\ll(NO_2)} \frac{2\Sigma(a_ic_i/b_i)}{2 + d/k_3} \qquad \text{(XIII.19.14)}$$

which can be compared with the expression already derived for the decomposition of pure N_2O_5 [Eq. (XIII.19.4)]. We see that $k_1 = \Sigma(a_ic_i/b_i) = K_{1.2}d$ and that $k_2 = d.$[c]

At the other extreme in concentration, at which for all states of im-

[a] Note that d can be identified with k_2 [Eq. (XIII.19.3)].

[b] This is, of course, the minimum value for NO.

[c] Note that $\Sigma (a_ic_i/b_i) = \Sigma K_id_i$, where $K_i = a_ic_i/b_id_i = $ const, is the equilibrium constant for the over-all equilibrium $N_2O_5 \rightleftharpoons NO_3 + NO_2$, which is independent of the energy state of the intermediates i by virtue of our assumption that N_2O_5, NO_2, and NO_3 are at temperature equilibrium in the vessel. This in turn implies that the distribution of NO_2 and NO_3 among the various energy states is independent of the reaction and, similarly, that for N_2O_5 the energy states below ϵ^* are populated according to equilibrium temperature distributions and are not significantly affected by the reaction. These seem to be rather reasonable assumptions and explain the paradox which Mills and Johnston have called attention to in their paper to the effect that the first-order rate constant for decomposition of pure N_2O_5 shows no pressure region of intermediate fall-off.

portance $b_i(M) \ll c_i,$[a] $r_i \to 1$ and Eq. (XIII.19.12) becomes

$$- \frac{d \ln (N_2O_5)}{dt} \xrightarrow{(M) \to 0} (M)\Sigma a_i = (M)a \qquad (XIII.19.15)$$

which is a second-order rate, independent of NO except as NO contributes to the total pressure M. This expression can be compared with the reaction of $NO + N_2O_5$ at very low pressures [Eq. (XIII.19.8)], and we see that k_1 in this latter equation is equal to $(M)\Sigma a_i$ from above and differs from the k_1 of Eq. (XIII.19.4) which, as we saw above, corresponds to the high-pressure limit for N_2O_5 decomposition. $\Sigma a_i = a = k_1^0/(M)$ corresponds to the low-pressure limit, and at intermediate pressures in the NO-N_2O_5 system we can expect k_1 to show a pressure dependence, which is in fact the behavior observed by Johnston et al.

At sufficiently low pressures, N_2O_5, according to Eq. (XIII.19.15), should decompose at the same rate whether or not NO is present. Hodges and Linhorst[b] did actually find that the decomposition of pure N_2O_5 showed a decrease in first-order rate constant in the range 0.05 to 0.004 mm Hg, and they believed that it had become essentially second-order at the latter pressure.[c]

At 27°C their second-order rate constant is 320 liters/mole-sec, which is reasonably close to Mills and Johnston's value of 230 liters/mole-sec for the second-order rate constant in $NO + N_2O_5$ mixtures.[d]

From the observation of Mills and Johnston that NO_2 does not affect the rate constant for $NO + N_2O_5$ at the high-pressure limit one can conclude [Eq. (XIII.19.13)] that $k_4 \geqslant 10d.$[e] This in turn implies that the reaction of $NO + N_2O_5$ at high pressures reduces to

[a] Note that for $E = E^*$, $C(E^*) = 0$, so that there are no pressures low enough for states infinitesimally near E^* to have decomposition rates $C(E^* + \triangle E)$ faster than collision frequencies. However $C(E_i)$ is a monotonically increasing function of E_i (starting from 0 at $E_i = E^*$), while a_i is a monotonically decreasing function of E_i (Sec. XI.3), and so their product must have a maximum for some energy $E > E^*$. It is the states near this maximum which we are considering above and which contribute most to the low-pressure rate.

[b] J. H. Hodges and E. F. Linhorst, *Proc. Nat'l Acad. Sci. U.S.*, **17**, 28 (1931); *J. Am. Chem. Soc.*, **56**, 836 (1934).

[c] There is considerable difficulty with studying reactions at such low pressures because wall collisions at 10^{-3} mm Hg are as frequent as gas-phase collisions in very large vessels.

[d] The work on N_2O_5 in the low-pressure range is certainly unsatisfactory. According to Mills and Johnston's data, the low-pressure region should be observable at about 1 mm Hg pressure of N_2O_5 at 27°C. However, none of the observations shows any significant fall-off before about 0.05 mm Hg. (See L. Kassel, "Kinetics for Homogeneous Gas Reactions," Reinhold Publishing Corporation, New York, 1932, for a discussion of this low-pressure work.)

[e] In the presence of added NO, NO is far in excess of its stationary-state concentration, and so $k_4(NO) \gg k_3(NO_2)$. If k_4 were not larger than d, Eq. (XIII.19.13) would then predict an inhibition term of the form $1 + d(NO_2)/k_3(NO)$, which is evidently not observed and allows the above estimate to be made. It should, however, be noted that

$$-\frac{d \ln (N_2O_5)}{dt} \xrightarrow[\substack{(NO) > (NO_2) \\ (M) \to \infty}]{} \Sigma(a_ic_i/b_i) = k_1^{(\infty)} \qquad (XIII.19.16)$$

Thus the ratio of the high-pressure-limiting first-order rate constants for the systems $NO + N_2O_5$ [Eq. (XIII.19.16)] to pure N_2O_5 [Eq. (XIII.19.14)] should be $1 + d/2k_3$.[a] From the data of Mills and Johnston at high pressures

$$k_1^{(\infty)} \ (sec^{-1}) = 6 \times 10^{14} \exp \left(-\frac{21{,}000 \pm 2000}{RT} \right) \qquad (XIII.19.17)$$

so that with the data on N_2O_5 [Eq. (XIII.19.2)] we find

$$\frac{k_1^{(\infty)}}{k_0} = 1 + \frac{k_2}{2k_3} = 15 \exp (3650/RT) \qquad (XIII.19.18)$$

which at 27°C has the value 7×10^3 and we see that at all temperatures of interest $k_2/k_3 \gg 1$ and we can write this ratio as[b]

$$\frac{k_2}{k_3} = 30 \exp \left(\frac{3650}{RT} \right) \qquad (XIII.19.19)$$

TABLE XIII.16. THERMAL DATA FOR THE OXIDES OF NITROGEN

Substance (gas)	ΔH_f°	S°	C_p	Substance (gas)	ΔH_f°	S°	C_p
O_2	0	49.0	7.0	N_2O	19.5	52.6	9.3
O_3	34.0	56.8	9.1	NO_3	16.5[a]	62[b]	12[b]
N_2	0	45.8	7.0	N_2O_3	20	70[b]	?
NO	21.6	50.3	7.1	N_2O_4	2.3	72.7	18.9
NO_2	8.1	57.5	9.0	N_2O_5	3.6	85[a]	?

[a] From data of N. Davidson and G. Schott, *J. Am. Chem. Soc.*, **80**, 1841 (1958), on the shock-wave dissociation of N_2O_5. They find $\Delta H_{1.2} = 21$ Kcal/mole and $\Delta S_{1.2} = 33.6$ e.u., from which the above have been estimated.

[b] Estimates by the author from known substances of similar structure. All other values from *Natl. Bur. Standards Circ.* 500.

If now we combine the high-pressure data on pure N_2O_5 [Eq. (XIII.19.4)] with the thermal data from Table XIII.16 for the equilibria $N_2O_5 \underset{2}{\overset{1}{\rightleftharpoons}} NO_2 + NO_3$, we find for k_3

---

the system is difficult to study, the data scatter, and there is considerable sensitivity to moisture, surface catalysis, and adsorption effects. The data at the high-pressure limit are for these reasons not too reliable.

[a] There seems to be an error of a factor of 2 in the similar calculation of Mills and Johnston that arises from a confusion in the definition of the rate constants.

[b] J. Jack, *Trans. Faraday Soc.*, **53**, 41 (1957), has obtained rate constants at 25°C for the system N_2O_5 + excess NO + CO_2 which are about threefold smaller than the corresponding rate constants of Mills and Johnston employed in the above calculations. Kassel, *loc. cit.*, has shown that the high-pressure rate data of Mills and Johnston must be in error.

$$k_3 = \frac{k_0}{2K_{1.2}} = 2.3 \times 10^7 \exp\left(-\frac{3650}{RT}\right) \qquad \text{liters/mole-sec} \qquad \text{(XIII.19.20)}$$

and so from the preceding relation

$$k_2 = 6.9 \times 10^8 \text{ liters/mole-sec} \qquad \text{(XIII.19.21)}$$

$$k_1 = 6.0 \times 10^{14} \exp\left(\frac{21,000}{RT}\right) \qquad \text{sec}^{-1}$$

These values indicate a collision efficiency for the recombination of $NO_2 + NO_3$ of about 1 in 300 collisions, which is very close to the value observed for $NO_2 + NO_2$. The frequency factor of reaction 1 is very much higher than the usual factor for unimolecular fissions but lower than the estimated value for the fission of N_2O_4. The frequency factor for k_3 is about 10^4 lower than collision frequencies and is in the expected range of values for atom transfers between large molecules. It is the activation energy for this reaction, $E_3 = 3.65$ Kcal, which seems surprising. From Table XIII.16, one finds that $\Delta H_3 = 5.1$ Kcal:

$$NO_2 + NO_3 \xrightarrow{3} NO + O_2 + NO_2 - 5.1 \text{ Kcal}$$

This implies that either (1) the $\Delta H_f^\circ(NO_3)$ must be lower than the value quoted of 16.5 Kcal or (2) this reaction proceeds by way of the formation of the metastable molecule O—N—O—O whose dissociation into $NO + O_2$ is endothermic by at least 1.5 Kcal.[a] The data are not at present sufficiently accurate to distinguish these possibilities.

Further supporting evidence for the N_2O_5 decomposition mechanism

[a] It seems quite reasonable to suppose that this is the way in which reaction 3 occurs, since there is independent evidence for the metastable complex O—N—O—O. Alternatively, the abstraction could lead to the metastable complex

O
 \
  N—O—O, which
 /
O

could also have a dissociation energy $\geqslant 1.5$ Kcal. There seems to be good evidence for believing that O—N—O—O is the unstable intermediate in the formation of NO_2 from NO, that is,

$$M + NO + O_2 \rightleftharpoons ONOO + M$$
$$ON—O—O + NO \rightleftharpoons 2NO_2$$

R. A. Ogg, Jr., *J. Chem. Phys.*, **21**, 2079 (1953), has shown that NO and N_2O_5 catalyze the exchange of labeled O_2^* with NO_2 (no exchange without catalysis) and has presented very plausible arguments for believing the catalysis to take place via the species O—N—O—O and O_2N—O—O, respectively:

$$(ON + O_2^* \rightleftharpoons) \; O—N—O^*—O^* + O—N—O \rightleftharpoons O—N—O^* + O^*—O—N—O$$

$$(NO_2 + O_2^* \rightleftharpoons) \; \underset{|\atop O}{O—N}—O^*—O^* + \underset{|\atop O}{O—N}—O \rightleftharpoons \underset{|\atop O}{O—N}—O^* + \underset{|\atop O}{O^*—O—N}—O$$

comes from the N_2O_5-catalyzed decomposition of O_3. It has been found[a] that the reaction of O_3 with NO_2 is almost stoichiometric at low O_3 concentrations: $O_3 + 2NO_2 \rightarrow N_2O_5 + O_2$. In the presence of O_3, N_2O_5 undergoes no decomposition, while O_3 goes smoothly to O_2 even at room temperature. The kinetics were measured by Schumacher and Sprenger (*loc. cit.*) and also by Nordberg,[b] both of whom were in excellent agreement in finding the rather unusual rate law

$$-\frac{d(O_3)}{dt} = k_c(N_2O_5)^{2/3}(O_3)^{2/3} \qquad (XIII.19.22)$$

which is well obeyed over a rather large range of concentrations for both O_3 and N_2O_5. The rate constant is calculated to be

$$k_c = 4.2 \times 10^{12} \exp\left(-\frac{20{,}000}{RT}\right) \quad \text{(liters/mole)}^{1/3} \text{ sec}^{-1} \quad (XIII.19.23)$$

and Schumacher and Sprenger proposed the mechanism later adapted by Ogg:

$$N_2O_5 \underset{2}{\overset{1}{\rightleftharpoons}} NO_2 + NO_3$$

$$NO_2 + O_3 \overset{7}{\rightarrow} NO_3 + O_2$$

$$2NO_3 \overset{8}{\rightarrow} NO_2 + O_2 + NO_2$$

By using stationary states for NO_2 and NO_3 we find

$$-\frac{d(O_3)}{dt} = (2K_{1.2}^2 k_7^2 k_8)^{1/3}(N_2O_5)^{2/3}(O_3)^{2/3} \qquad (XIII.19.24)$$

in agreement with the empirical rate law. Since $K_{1.2}$ is available from thermal data[c] and $k_7 = 5.9 \times 10^9 \exp(-7000/RT)$,[d] we can calculate the unknown constant k_8 which turns out to be

$$k_5 = 1.4 \times 10^6 \exp\left(-\frac{4000}{RT}\right) \quad \text{liters/mole-sec} \quad (XIII.19.25)$$

While the frequency factor in this constant is about 10^3 lower than that for a group of similar abstraction reactions, so that the value might be questioned, it is not too unreasonably far from what one might expect in view of the magnification of errors resulting from the calculation.[e] While the

[a] H. J. Schumacher and G. Sprenger, *Z. physik. Chem.*, **A140**, 281 (1929); **B2**, 267 (1929).

[b] Nordberg, *Science*, **70**, 580 (1929).

[c] Table XIII.16.

[d] H. S. Johnston and D. M. Yost, *J. Chem. Phys.*, **17**, 386 (1949).

[e] From Eqs. (XIII.19.22) and (XIII.19.24) $k_8 = k_c^3/2K_{1.2}^2 k_7^2$. Neither $K_{1.2}$, k_7, nor indeed k_c is known with sufficient accuracy to make the separation into frequency factors and activation energies much better than this. Thus if $E_8 = 6.8$ Kcal, A_8 would be boosted to 1.4×10^8 liters/mole-sec, which would be more reasonable.

data are not sufficiently precise to exclude small contributions from the reaction $NO_3 + O_3 \xrightarrow{9} NO_2 + 2O_2$, the contribution must be sufficiently small that one can conclude that its activation energy is probably in excess of 9 Kcal, a value not unreasonable considering the value of 7 Kcal for reaction 7.

Further work has been reported by Johnston[a] on the low-pressure rate in the presence of different inert gases. It is found that the collisional efficiencies of various gases in activating N_2O_5 at low pressures (reaction 1) relative to N_2O_5 itself are N_2O_5, 1; Ar, 0.135; He, 0.124; Ne, 0.090; Kr, 0.159; Xe, 0.147; N_2, 0.234; NO, 0.300; CO_2, 0.400; SF_6, 0.32; and CCl_4, 0.551. These values are more or less what might be expected on the basis of molecular complexity.

The low-pressure rate has never been adequately accounted for because of the difficulties that arise from surface contamination and heterogeneity. While the rate constant is not in too serious error, the activation energy has great uncertainty associated with it. Perrine and Johnston (loc. cit.) report for the low-pressure limit

$$k_1^{(0)} = \sum a_i = 1.28 \times 10^{16} \exp\left(-\frac{19,300}{RT}\right) \qquad \text{liters/mole-sec}$$

$$\text{(XIII.19.26)}$$

As expected from the theory of unimolecular reactions, $k_1^{(0)}$ has an activation energy less than the measured bond energy of 21 Kcal[b] and the frequency factor is abnormally high for a bimolecular reaction. In actual fact, the frequency factor is the highest ever reported for a bimolecular reaction and if verified would imply free rotation of the O_2N—O—NO_2 groups in the activated complex $[N_2O_5]^*$ about the central O atom.[c] From these values, and the RRK theory, Johnston has calculated that the mean life of the $[N_2O_5]^*$ activated complex is of the order of 10^{-7} sec. If the probability of deactivation is less than unity on collision, this time will become proportionately larger.

It appears then that, although the last word on this subject is far from being said, the N_2O_5 decomposition is one of the best understood of our complex reactions, and the mechanism for it is probably as well founded as any of the kinetic mechanisms that have been presented up until now.

[a] H. S. Johnston, J. Am. Chem. Soc., **75**, 1567 (1953). D. J. Wilson and H. S. Johnston, ibid., **75**, 5763 (1953).

[b] In consequence we anticipate that the reverse reaction $NO_3 + NO_2 + M \rightarrow N_2O_5 + M$ has a negative activation energy of 1.7 Kcal at these pressures (i.e., less than 0.1 mm Hg total).

[c] Note that this is difficult to account for if N_2O_5 has as much entropy as is implied in Table XIII.16.

20. Pyrolysis of Nitrites and Nitrates

The alkyl nitrate and nitrite esters contain four different elements and as such may be expected to show a very complex kinetic behavior. This is certainly the case with the nitrate esters, the pyrolysis of which is accompanied by varying amounts of oxidation. For the nitrites, however, the pyrolysis seems to be less complex. The principal reaction products from the pyrolysis of C_2H_5ONO are $CH_3CHO + NO$ together with lesser amounts of $N_2O + H_2O$ and much smaller amounts of C_2H_5OH, HCN,[a] CO,[b,c] and CH_2O.

The decomposition of t-butyl nitrite leads to acetone and formaldoxime, CH_2NOH, as principal products, together with smaller amounts of $HCN + H_2O$ and $NO + t$-butyl alcohol.[d] The other nitrites give similar products, and the free radical mechanism originally proposed by Rice and Rodowskas[e] and amplified by Levy (loc. cit.) seems to give a reasonable account of the observations. The former workers demonstrated the presence of CH_3 radicals in decomposing C_2H_5ONO at 450°C by the mirror-removal technique and proposed a bond energy of about 35 Kcal for what they thought must be the radical initiation process, $C_2H_5ONO \rightarrow C_2H_5O + NO$. This is a quite reasonable value for the RO—NO bond energy and is in good accord with independent estimates of that bond energy. The mechanism for the pyrolysis of C_2H_5ONO as modified by Levy is

$$C_2H_5ONO \underset{2}{\overset{1}{\rightleftharpoons}} C_2H_5O + NO$$

$$C_2H_5O \overset{3}{\rightarrow} CH_3 + CH_2O$$

$$C_2H_5O + NO \overset{4}{\rightarrow} CH_3CHO + HNO \qquad\qquad (XIII.20.1)$$

$$C_2H_5O + HNO \overset{5}{\rightarrow} C_2H_5OH + NO$$

$$2NOH \overset{6}{\rightarrow} N_2O + H_2O$$

$$CH_3 + NO \overset{7}{\rightarrow} CH_3NO \quad \rightarrow CH_2{=}NOH \rightarrow HCN + H_2O$$

On applying stationary states for HNO, CH_3, and C_2H_5O, we find (neglecting reaction 5 compared to reactions 6 and 4)[f]

[a] J. B. Levy, J. Am. Chem. Soc., **78**, 1780 (1956); **76**, 3790 (1954).

[b] E. W. R. Steacie and G. T. Shaw, J. Chem. Phys., **2**, 345 (1934).

[c] See also the comments by M. Szwarc, Chem. Revs., **47**, 75 (1950), on the analyses.

[d] J. B. Levy, Ind. Eng. Chem., **48**, 762 (1956).

[e] F. O. Rice and E. L. Rodowskas, J. Am. Chem. Soc., **57**, 350 (1935).

[f] This seems reasonable as a first approximation, since C_2H_5OH is a minor product compared to N_2O and CH_3CHO at 180°C and 50 mm pressure (Levy, loc. cit.).

$$\frac{(Me)}{(EtO)} = \frac{k_3}{k_7(NO)}; \quad \frac{(HNO)}{(EtO)} = \left[\frac{k_4(NO)}{2k_6}\right]^{1/2}(EtO)^{-1/2}$$

$$(EtO) = \frac{k_1(EtONO)}{(k_2 + k_4)(NO) + k_3}$$

(XIII.20.2)

which gives for the disappearance of the nitrite

$$-\frac{d\ln(EtONO)}{dt} = k_1\frac{k_4(NO) + k_3}{(k_2 + k_4)(NO) + k_3} \qquad \text{(XIII.20.3)}$$

Since it is found that adding NO has little if any effect on the rate of loss of EtONO[a] and initially CH_3CHO accounts for more than 80 per cent of the carbon balance, then $k_4(NO) \gg k_3$,[b] so that the rate expression becomes in the presence of added NO

$$-\frac{d\ln(EtONO)}{dt} = \frac{k_1k_4}{k_2 + k_4} \qquad \text{(XIII.20.4)}$$

This is exactly the behavior found for the decomposition in the presence of added NO, a first-order rate of disappearance of the ester. It is found that the empirical rate constant k_0 is given by

$$k_0 = 6.1 \times 10^{13} \exp\left(-\frac{37,500 \pm 600}{RT}\right) \quad \text{sec}^{-1} \quad \text{(XIII.20.5)}$$

which can be equated to $k_1k_4/k_2 + k_4$ of Eq. (XIII.20.4). Since both k_2 and k_4 are expected to have close to zero activation energy, the value 37.5 Kcal is probably equal to E_1, the bond dissociation energy of C_2H_5O—NO, in good agreement with independent estimates and the results of Steacie and Shaw for pure C_2H_5ONO.

Added acetaldehyde is found to accelerate the rate, and in excess it completely inhibits N_2O. This can be explained by adding the following steps to the scheme:

$$C_2H_5O + CH_3CHO \xrightarrow{8} C_2H_5OH + CH_3CO$$

$$CH_3CO + M \xrightarrow{9} CO + CH_3 + M \qquad \text{(XIII.20.6)}$$

$$CH_3CO + NO \xrightarrow{10} CH_3CO(NO)$$

CH_3CHO would thus compete successfully with NO (reactions 2 and 4) for the C_2H_5O reaction and so accelerate the rate by inhibiting the back reaction. In addition, by replacing the C_2H_5O by CH_3CO (and CH_3), the CH_3CHO can act as a scavenger for NO, which could account for the disappearance of N_2O. If correct, this scheme implies that C_2H_5O can ab-

---

[a] It does have an effect on the products and a slight inhibiting effect on the rate. With excess NO, the acetaldehyde yield → 100 per cent and N_2O → 60 per cent.

[b] This is to be expected if k_4 has a high steric factor ($\sim$0.1) and zero activation energy, both of which are to be expected for disproportionation reactions of radicals which are strongly exothermic as this is expected to be. It is expected that $\Gamma_3 \geqslant 11$ Kcal.

stract H from CH_3CHO at a faster rate than it can from the parent ester via the reaction originally proposed for the decomposition:

$$C_2H_5O + C_2H_5ONO \xrightarrow{11} C_2H_5OH + CH_3CHONO$$

followed by

$$CH_3CHONO \xrightarrow{12} CH_3CHO + NO$$

Reactions 11 and 12, if important, would give a maximum yield for CH_3CHO of 50 per cent. The fact that CH_3CHO is in excess of 50 per cent and that it accelerates the reaction implies that reaction 11 has an activation energy at least 2 Kcal greater than E_8, or else a steric factor at least tenfold smaller. This latter alternative seems much more likely. Levy[a] has shown that EtO radicals produced by the pyrolysis of diethyl peroxide, EtO—OEt, in the presence of EtONO do not attack EtONO significantly at 181°C. In the presence of excess NO, about 60 to 70 per cent yields of EtONO are produced from EtO—OEt even with EtONO present, thus demonstrating that reaction 2 is fast compared with reaction 11.[b]

Two very important aspects of the decomposition are still not very clear. The first is the precise mechanism of reaction 6. It seems very unlikely that $N_2O + H_2O$ are formed directly from two H—N=O molecules colliding in the gas phase.[c] In addition, the proposed scheme does not satisfactorily account for the limiting N_2O yield of about 50 per cent observed with excess NO. The same difficulty occurs in understanding the isomerization of CH_3—NO to CH_2=NOH and its subsequent dehydration to HCN and H_2O. It is likely that both of these latter reactions are heterogeneous.[d]

The decomposition of MeONO[e] leads to $MeOH + CH_2O + NO$ as major products with small amounts of CO. N_2O was not looked for. The mechanism is undoubtedly similar to the one discussed for EtONO, with the scheme initiated by $CH_3ONO \rightarrow CH_3O + NO$. The CO undoubtedly arises from the CH_3O attack on CH_2O to give $CH_3OH + CHO$ with $CHO + R \rightarrow CO + RH$.

The photolysis of CH_3ONO[f] can be rationalized in terms of the thermal

[a] J. B. Levy, J. Am. Chem. Soc., **75**, 180 (1953).

[b] Levy has suggested that another source of C_2H_5OH could be the disproportionation reaction $2EtO \rightarrow EtOH + CH_3CHO$. This is quite possible in the diethyl peroxide pyrolysis, in which the stationary concentrations of EtO are quite high, but not likely in the EtONO pyrolysis, in which NO acts as such an effective suppressor of EtO.

[c] It is far more likely that such a reaction occurs at the vessel surface, particularly in view of the possible catalytic effect of H_2O, which is also present as a product of the reaction.

[d] Heterogeneous reactions certainly occur, since the exchange of NO* with EtONO was found by Levy to give very erratic kinetics. Also the reaction of $EtOH + 2NO_2 \rightarrow EtONO + HNO_3$ is observed to be a very fast heterogeneous reaction in these systems.

[e] E. W. R. Steacie and G. T. Shaw, Proc. Roy. Soc. (London), **A146**, 388 (1934).

[f] J. A. Gray and D. W. G. Style, Trans. Faraday Soc., **48**, 1137 (1952).

decomposition. N_2O is an important product which decreases with rising temperature. A mechanism similar to that for the EtONO would be

$$CH_3ONO + h\nu \xrightarrow{1} CH_3O + NO$$

$$NO + CH_3O \xrightarrow{2} CH_3ONO$$

$$\xrightarrow{4} CH_2O + HNO$$

$$CH_3O + CH_2O \xrightarrow{5} CH_3OH + CHO$$

$$CHO + NO \xrightarrow{8} CO + HNO$$

$$2HNO \xrightarrow{6} H_2O + N_2O$$

For isopropyl nitrite (Levy, *loc. cit.*) the products are acetone + NO together with isopropyl alcohol and CH_3CHO. Minor products are HCN and H_2O.

The decomposition of t-BuONO differs from the other esters in that it has no α H, and so the attack of NO on the t-BuO radical is ruled out. In agreement with this, the rate is found to be inhibited by NO (Levy), the mechanism being

$$t\text{-BuONO} \underset{2}{\overset{1}{\rightleftharpoons}} t\text{-BuO} + NO$$

$$t\text{-BuO} \xrightarrow{3} CH_3COCH_3 + CH_3$$

$$CH_3 + NO \xrightarrow{7} CH_3NO \quad \rightarrow CH_2NOH \rightarrow HCN + H_2O$$

$$t\text{-BuO} + CH_3COCH_3 \xrightarrow{13} t\text{-BuOH} + CH_2COCH_3$$

$$CH_3 + CH_2COCH_3 \xrightarrow{14} C_2H_5COCH_3$$

This leads to the rate law (neglecting reaction 13)[a]

$$-\frac{d \ln (t\text{-BuONO})}{dt} = \frac{k_1 k_3}{k_2(NO) + k_3} \qquad \text{(XIII.20.7)}$$

in good agreement with experiment.

An unexplained feature of the reaction was the observation that added NO inhibits HCN and gives rise to CH_3NO_2. Separate experiments showed that NO reacted with nitrosomethane to give nitromethane (Levy, *loc. cit.*). This is certainly an unexpected result.

Because of the complexity of their pyrolyses, no attempt will be made to discuss in full the decomposition of the nitrate esters. The initial step in the decomposition is probably the split, $RONO_2 \rightarrow RO + NO_2$, the

---

[a] It was found that t-BuOH appeared only in the latter stages of the reaction. The initial acetone yields in the absence of NO were very close to 100 per cent.

energy for which is in the neighborhood of 37 Kcal for $EtONO_2$.[a] Initially the principal product is EtONO, together with very small amounts of $MeONO$, $MeNO_2$, and NO. These account very well for the nitrogen balance for over 90 per cent of the reaction. The carbon and oxygen balance is more difficult because of the production of CO, CO_2, H_2O, and other intermediate oxidation products. The main course of the reaction appears to follow some scheme as

$$C_2H_5ONO_2 \underset{2}{\overset{1}{\rightleftharpoons}} C_2H_5O + NO_2$$

$$C_2H_5O \overset{3}{\rightarrow} CH_3 + CH_2O$$

$$CH_3 + NO_2 \overset{4}{\rightarrow} CH_3O + NO$$

$$\overset{5}{\rightarrow} CH_3NO_2$$

$$C_2H_5O + NO_2 \overset{6'}{\rightarrow} CH_3CHO + HONO$$

$$[2HONO \overset{7}{\rightarrow} H_2O + NO + NO_2]$$

$$[CH_3CHO + NO_2 \overset{8}{\rightarrow} CO + H_2O + NO + ?]$$

$$C_2H_5O + NO \overset{2'}{\rightarrow} C_2H_5ONO \quad \text{or} \quad CH_3CHO + HNO$$

It is observed that NO_2 inhibits the decomposition. This implies that reaction 6' must be considerably slower than reaction 2, which is in marked contrast to the behavior of NO with C_2H_5O and is difficult to understand. In the presence of excess NO one can derive a rate law of the form

$$-\frac{d \ln (EtONO_2)}{dt} = \frac{k_1}{1 + k_2(NO_2)/k_5(NO)} \qquad (XIII.20.8)$$

which seems to fit well the data on the decomposition in excess NO. By extrapolation of the apparent rate constant to zero ratio of $(NO_2)/(NO)$, Levy has estimated k_1 as (sec^{-1})

$$\log k_1 = 16.85 - \frac{41,230}{4.575T} \qquad (XIII.20.9)$$

which is a high but not unreasonable value for the frequency factor A_1.[b] The activation energy, however, is 5 Kcal higher than one would have predicted from the bond dissociation energy, and the result must be considered subject to considerable uncertainty.

The decomposition of HNO_3 provides an interesting comparison with

[a] J. B. Levy, *J. Am. Chem. Soc.*, **76**, 3254, 3790 (1954).

[b] The entropy change in reaction 1 is about 40 cal/mole-°K, so that the value of A_2 can be established at about 3×10^9 liters/mole-sec. That represents a collision efficiency of about $\frac{1}{100}$ for reaction 2, a quite reasonable number compared to $\frac{1}{200}$ for $2NO_2 \rightarrow N_2O_4$, and about the same for $NO_2 + NO_3 \rightarrow N_2O_5$ (see preceding section).

the previous studies. The work of Johnston et al.[a,b] has shown that, at 400°C and higher, the pyrolysis of $HNO_3 \rightarrow \frac{1}{2}[H_2O + 2NO_2 + \frac{1}{2}O_2]$, $NO_2 \rightarrow NO + \frac{1}{2}O_2$ is a homogeneous, chain decomposition[c] for which a reasonable mechanism is

$$M + HONO_2 \underset{2}{\overset{1}{\rightleftharpoons}} M + HO + NO_2$$

$$HO + HNO_3 \overset{3}{\rightarrow} H_2O + NO_3$$

$$NO_3 + NO \overset{4}{\rightarrow} 2NO_2$$

$$NO_3 + NO_2 \overset{5}{\rightarrow} NO + NO_2 + O_2 \qquad\qquad (XIII.20.10)$$

$$HO + NO + M \overset{6}{\rightarrow} HONO + M$$

$$HO + HONO \overset{7}{\rightarrow} H_2O + NO_2$$

$$[2HONO \overset{8}{\rightarrow} H_2O + NO + NO_2] \qquad \text{fast}$$

Many of these reactions have been studied before in the section on N_2O_5 and so will not be discussed again here. In excess NO, the rate becomes nearly first-order over most of the decomposition with a rate constant which is itself a function of the total pressure. NO_2 is an inhibitor for the decomposition, and in consequence the reaction in the absence of added NO shows a steady fall in apparent first-order rate constant with continuing decomposition. In this respect the nitrates and nitrites all seem to have in common the feature that the pyrolysis products inhibit the rate of decomposition. This is to be expected in systems decomposing via radical mechanisms when the products of the reaction include such efficient radical traps as NO and NO_2. It is unfortunate that quantitative data on these systems are at present so sparse and in many cases disparate. This is to be expected for systems that are so complex and show such sensitivity to surface reactions. The free radical chemistry of these systems is, however, a very interesting and important one, and efforts to elucidate it will eventually turn out to be quite rewarding.

[a] H. S. Johnston et al., *J. Phys. Chem.*, **57**, 390 (1953).

[b] H. S. Johnston et al., *J. Am. Chem. Soc.*, **73**, 2319 (1951).

[c] Below 300°C Johnston found the rate to be principally heterogeneous. This accounts in part for his disagreement with the findings of C. Fréjacques, *Compt. rend.*, **232**, 2206 (1951), who studied the reaction through the heterogeneous region. The main discrepancy, however, is only apparent and arises from the quasi-unimolecular behavior of reaction 1.

XIV

The Kinetic Behavior
of Non-stationary-state Systems

1. The Approach to the Stationary State in a Reacting System

A stationary state is by definition one whose description does not change with time. According to this definition it becomes impossible for a closed system which is undergoing chemical reaction to ever achieve a stationary state, because the composition of such a system is constantly changing with time. In this sense it is only open systems (i.e., systems in communication with "infinite" external reservoirs of mass, heat, etc.) which can achieve truly stationary states. However, we have used the description *stationary state* in discussing sequences of consecutive reactions occurring in closed systems. By this we have really meant a *quasi-stationary* state,[a] i.e., a state whose instantaneous description differs from the truly stationary state at that moment by quantities which are negligibly small. And by the word *description* we mean *all* of those parameters (e.g., temperature, pressure, composition, and vessel size) whose specification is required to reproduce a second system that has the same kinetic properties.

Let us examine somewhat generally the behavior of a typical system undergoing reaction at constant volume in a flask whose walls are thermostatted. If initially the reaction mixture was transferred to the flask from some vessel at a lower temperature, there is a finite interval (for gases of the order of 10 to 30 sec) during which the temperature of the mixture reaches that of the flask walls. The processes responsible for this are conduction (both through the gas and across the walls to the thermostat) and convection. The convection arises from the original motion of the gas in

---

[a] The author at this point makes a belated apology to the reader who may have been confused by this somewhat loose nomenclature. It is, unfortunately, common in the literature of the United States and England.

entering the flask and the force of gravity acting on regions whose densities are different (e.g., regions of different temperature).

However, because the system is reacting, the equilibration of temperature never takes place; instead, since every chemical reaction will in general absorb or produce heat, the mixture approaches a temperature distribution with a mean temperature very close to that of the flask walls. If the reaction is exothermic, the gas always remains somewhat hotter than the walls; if endothermic, somewhat colder. We shall discuss these temperature gradients in detail in the next section.

For "slow" reactions the temperature gradients can be shown to be unimportant, but for exothermic reactions they constitute an autocatalytic component of the system which can cause the reaction rate to accelerate very rapidly to what can be described as an explosion. If we consider any given volume element in a *normally* reacting exothermic system, it will have reached a quasi-stationary state of temperature balance when the heat liberated by the developing reaction is balanced by the loss of heat from that volume element due to conduction, convection, and diffusion. When the latter processes are not capable of dissipating the heat of the reaction fast enough, the rate of heat evolution increases and an unstable situation arises, in which the increasing speed of the reaction is limited only by the local supply of reactants. Such instabilities are called thermal explosions, and exothermic reactions can always be brought to such a point by raising the temperature of the system to what can be called a critical explosion temperature.

A very similar picture will also be developed for the establishment of a quasi-stationary state with respect to the concentrations of metastable intermediates in a complex sequence of consecutive reactions. In a system such as $H_2 + Cl_2 \rightarrow 2HCl$, which proceeds via an atomic chain

$$Cl + H_2 \rightarrow HCl + H$$
$$H + Cl_2 \rightarrow HCl + Cl$$

there is some mechanism which generates atoms (e.g., light, heat, or chemical sensitizer) at a nearly uniform rate. The concentration of atoms will reach a quasi-stationary state only if there is a reasonably fast mechanism for atom destruction, i.e., the termination mechanism. If for some reason termination is too slow, the concentration of radicals can grow essentially without limit and the rate of the over-all reaction will grow with it in autocatalytic fashion to the point of explosion. Such instabilities are referred to as chemical or chain explosions as distinguished from the preceding case of thermal explosion, although in practice chain explosions will occur only in exothermic reaction systems and may release sufficient heat to produce a thermal explosion. The two are quite difficult to distinguish.

There is an additional feature of some chemical chains which makes them potentially unstable and liable to chemical explosion. This is the phenomenon known as *branching*. In the reaction of $H_2 + O_2$ to produce water, the system may be described in terms of a rather unusual chain which includes the reaction $H + O_2 \rightarrow HO + O$. Such a reaction in which one chain center, H, produces two new chain centers, $HO + O$, is called a branching reaction. If not balanced by efficient termination processes, the chain centers (in this case atoms and free radicals) may multiply geometrically and produce an explosion.

These autocatalytic instabilities in reaction system which we have described are characterized by a rate of reaction which starts off from some very low value, increases during the entire course of the reaction (which is usually of the order of magnitude of seconds or less) and reaches a maximum only because of the depletion of reactants. This behavior differs from a "normal," slow reaction in that the half-life of the reaction is remarkably short and of the order of magnitude of the time required to develop the quasi-stationary concentrations of intermediates or temperature.[a]

In the following sections we shall consider further the behavior of reaction systems that approach a quasi-stationary state and the explosive instabilities which may exist in such systems.

2. Temperature Gradients in Reacting Systems

For a system in which reaction is accompanied by heat changes, the temperature distribution at any time $T(x,y,z,t)$ may be described by a rather complex second-order differential equation which can be considerably simplified if we neglect convection

$$\frac{\partial T}{\partial t} = \frac{K}{\rho c_v} \nabla^2 T + \frac{RH}{\rho c_v} \qquad \text{(XIV.2.1)}$$

Here K is the coefficient of thermal conductivity, ρ the density, and c_v the specific heat of the medium, R is the specific reaction rate, and H is the molar heat of reaction.[b] The quantity $K/\rho c_v$ is called the thermal dif-

---

[a] Many authors have attempted to draw more fundamental distinctions between normal reactions and explosions without real success. What is used to justify such distinctions is the apparently discontinuous transition between states of normal quasi-stationary reaction and explosion. As we shall see later, such sharp transitions are secondary phenomena associated with the reaction system and not basic properties of the chain.

[b] It is further assumed that K, ρ, and c_v do not change appreciably with time or position in the vessel. If concentration gradients and density gradients are small, this will be a good approximation. When such gradients are not small, the equation must be written

$$\frac{\partial T}{\partial t} = \frac{1}{\rho c_v} \nabla \cdot (K \nabla T) + \frac{RH}{\rho c_v}$$

$$= \frac{1}{\rho c_v} \nabla(K) \cdot \nabla T + \frac{K}{\rho c_v} \nabla^2 T + \frac{RH}{\rho c_v}$$

fusivity of the system and, in analogy with the coefficient of diffusion of matter, has the dimensions of cm^2/sec.[a]

If we consider spherical vessels of radius r_0 such that T will depend only on the distance from the center r, then $T = T(r,t)$, and the Laplacian[b] $\nabla^2 T$ can be written in spherical coordinates

$$\nabla^2 T = \frac{1}{r^2}\frac{\partial}{\partial r}\left(r^2\frac{\partial T}{\partial r}\right) = \frac{2}{r}\frac{\partial T}{\partial r} + \frac{\partial^2 T}{\partial r^2} = \frac{1}{r}\frac{\partial^2}{\partial r^2}(rT) \quad \text{(XIV.2.2)}$$

Such equations are most conveniently solved when the variables are transformed to dimensionless form. A convenient set of dimensionless variables for the system is

$$\theta = \frac{T}{T_0} - 1 \qquad x = \frac{r}{r_0} \qquad \tau = \left(\frac{K}{r_0^2 \rho c_v}\right) t \qquad \text{(XIV.2.3)}$$

where T_0 is the temperature at the walls of the flask (at $r = r_0$) which is assumed to be constant. With these variables our equation can now be written as

$$x\frac{\partial \theta}{\partial \tau} = \frac{\partial^2 (x\theta)}{\partial x^2} + \lambda x \qquad \text{(XIV.2.4)}$$

where $\lambda = RHr_0^2/KT_0$. If we further substitute $\phi = \theta x$, then

$$\frac{\partial \phi}{\partial \tau} = \frac{\partial^2 \phi}{\partial x^2} + \lambda x \qquad \text{(XIV.2.5)}$$

This last equation is a well-known second-order linear partial differential equation.[c] The precise solution is determined by the boundary conditions that $T(r_0,t) = T_0$ (a constant), or equivalently, $\phi(1,\tau) = 0$ and the solution can be written as

$$\theta(x,t) = \frac{T(x,t) - T_0}{T_0}$$

$$= \frac{\lambda}{6}\left[1 - x^2 + \frac{12}{\pi^3 x}\sum_1^\infty \frac{(-1)^n}{n^3}(\sin \pi n x)e^{-n^2\pi^2\tau}\right]$$

$$\text{(XIV.2.6)}$$

The temperature at the center of the flask $T(0,t)$ is given by[d]

[a] The dissipation (or "diffusion") of momentum and matter is also described by equations similar to (XIV.2.1). For momentum, the diffusivity is given by the ratio η/ρ, called the kinematic viscosity, where η is the coefficient of viscosity. For gases at STP all those coefficients have about the same order of magnitude, namely, between 0.1 and 1 cm^2/sec.

[b] In cartesian coordinates

$$\nabla^2 T \equiv \frac{\partial^2 T}{\partial x^2} + \frac{\partial^2 T}{\partial y^2} + \frac{\partial^2 T}{\partial z^2}$$

[c] H. Bateman, "Partial Differential Equations of Mathematical Physics," p. 353, Dover Publications, New York, 1944.

[d] This is obtained from the preceding equation by expanding $\sin(\pi n x)$ in a power series and taking only the first term, that is, $\sin(\pi n x) \to \pi n x$ as $x \to 0$.

$$\theta(0,t) = \frac{T(0,t) - T_0}{T_0}$$

$$= \frac{\lambda}{6}\left[1 + \frac{12}{\pi^2}\sum_1^\infty \frac{(-1)^n}{n^2} e^{-n^2\pi^2\tau}\right] \qquad (XIV.2.6a)$$

Since this last equation gives the maximum temperature difference in the vessel, we are able to consider it rather than the general equation (XIV.2.6).

As τ or t becomes very large, the terms in the sum become zero and we reach a steady state in which the temperature gradient in the vessel is parabolic [Eq. (XIV.2.6)], while from Eq. (XIV.2.6a) the maximum temperature difference is given by

$$\frac{T(0,\infty) - T_0}{T_0} = \frac{RHr_0^2}{6KT_0} = \frac{\lambda}{6} \qquad (XIV.2.7)$$

The mean time needed to achieve this steady state is given approximately by the condition that the first term in the series falls to $1/e$ of its initial value,[a] namely, when $\tau = 1/\pi^2 (n = 1)$ or

$$t = \frac{r_0^2 \rho c_v}{\pi^2 K} \qquad (XIV.2.8)$$

For ideal gases this becomes

$$t = \frac{r_0^2 P C_V}{\pi^2 R T K} \qquad (XIV.2.9)$$

which is independent of molecular weight.

For most liquids the product ρc_v lies between 0.3 and 1.0 cal/cc-°C, while K varies between 0.5 and 1.5×10^{-3} cal/cm-sec-°C. If we choose mean values of 0.6 cal/cc-° and 1.0×10^{-3}, we find that the stationary state is achieved in liquid systems in a time given by $t = 60r_0^2 \text{ sec} = r_0^2$ min when r_0 is in cm. Thus for a 500-cc vessel ($r_0 \cong 5$ cm) the mean time to reach the stationary state is about 25 min, so that it may be expected that convection will play a much more important role than conduction in a liquid system.

Now consider the maximum temperature difference at the stationary state for liquids [Eq. (XIV.2.7)] and assume for convenience that we have a reaction of half-life = 1 hr ($k = 0.69/t_{1/2} \cong 2 \times 10^{-4} \text{ sec}^{-1}$) and a reactive concentration of 0.05 mole/liter such that the average specific rate of reaction R will be about 1×10^{-8} mole/cc-sec. If we further assume a heat of reaction $H = \pm 10$ Kcal/mole and take $K = 1.0 \times 10^{-3}$, we find for $r_0 = 5$ cm

[a] Note that the terms in the series fall off very rapidly because of the appearance of n^2 in the denominator of each coefficient and also in the exponent. Thus when the first exponent is unity, the value of the second term is $\frac{1}{4}e^{-4} = 0.0046$, or only 1.3 per cent of the first term. After this time a good approximation to the general solution is to be had by taking only the first term in the series.

$$T(0,\infty) - T_0 = \pm 0.42°C$$

so that, except for much larger heats or faster rates of reactions, these effects will be rather small.

On the other hand, if we make the same calculations for gases, we find for the mean life[a] at STP with $C_V = 6$ cal/mole-°K

$$t \cong 0.4 r_0^2$$

or $\qquad t = 10$ sec $\qquad$ for $r_0 = 5$ cm

so that stationary states are very quickly established in gas systems.[b] The maximum temperature differences, however, are larger as we can see if we take a 500-cc vessel at STP (~ 0.045 mole/liter) in which the half-life is again 1 hr and the heat of reaction ± 10 Kcal/mole. The product RH is then 9.0×10^{-5} cal/cc-sec and the maximum temperature difference is

$$T(0,\infty) - T(r_0) = \pm 5.3°C$$

so that, unless there is considerable convection, quite appreciable temperature gradients will be established in a reacting gas system.[c] If we take the average temperature over the vessel, we find

$$\left\langle \frac{T - T_0}{T_0} \right\rangle_{av} = 0.4 \left[\frac{T(0,\infty) - T_0}{T_0} \right] = \frac{RHr_0^2}{15KT_0} \qquad (XIV.2.10)$$

[a] K is about 7×10^{-5} cal/cm-°C-sec for gases, except for H_2 and He, which are higher by a factor of about 5.

[b] Note that at pressures of the order of 10 mm Hg these times are correspondingly shortened to the order of 0.1 sec.

[c] For exothermic reactions the temperature will be higher in the flask than at the walls, while the opposite will be true for endothermic reactions. Many pyrolytic reactions show a fall in rate in the neighborhood of 100 to 400 mm Hg. While these are complex chain reactions and the fall in rate may for the most part be due to increased chain breaking at the walls, the existence of a temperature gradient may also be of significance in regard to these pressure effects. Thus, in these pyrolytic reactions, the magnitude of the temperature gradient will be proportional to the rate at which the reaction occurs, which will in turn vary directly with the amount of gas in the system. At high pressures the gradients will be larger than they will be at low pressures. For endothermic reactions such as the pyrolytic decompositions, the average temperature in the flask will decrease as the pressure is raised and the rate constant will appear to decrease with increasing pressure if no convection takes place, an effect opposite to that observed.

If, on the other hand, convection does take place, then it will be the time of establishment of gradients compared to the rate of convection that will govern any observable effects, and here the result is quite the opposite because gradients will be very quickly established at low densities [Eq. (XIV.2.8)]. Despite the fact that the gradients are proportionately lower, the decreased relative efficiency of convection may predominate, and so the rate constants may appear to fall at lower pressure. In any case it can be seen that the time lag in establishing temperature equilibrium together with the above-mentioned gradients will make initial rate data very suspect in gas systems.

which can be used for making corrections to the apparent reaction temperature.[a]

It has been shown by Benson[b] that the existence of laminar convection does not change the average temperature appreciably, so that Eq. (XIV.2.9) is probably a good approximation for calculating temperature gradients of the order of 1 to 2 per cent.[c] Although direct measurement of these gradients is difficult, the rate of the reaction can be used as a "thermometer" if the kinetics are understood. In this way Benson and Axworthy[d] have shown that the peculiarly high reaction rates observed in the decomposition of ozone at high ozone pressures arise from small temperature gradients of the order predicted by Eq. (XIV.2.10). Using the data of Allen and Rice,[e] Benson[f] has shown that the observed, apparent increase in specific rate constant in the thermal decomposition of azomethane near its explosion limit is predicted to within 7 per cent by Eq. (XIV.2.10).[g] In the very rapid endothermic decomposition of HNO_3 at high temperature to give $H_2O + NO_2 + O_2$, Johnston[h] has observed a considerable lowering of the apparent rate constant which can be ascribed to the self-cooling of the reaction system.

3. Thermal Explosions

The temperature distribution in a reacting mixture is stabilized when the rate of loss of heat by conduction or convection from any volume element is equal to that produced by the reaction itself in that volume element. In the case that the rate of heat loss cannot compensate for the rate of heat production, a stationary or quasi-stationary temperature distribution is impossible and the temperature of the reaction mixture increases exponentially, causing the reaction rate to do likewise, and a thermal explosion results. This is illustrated in Fig. XIV.1, which follows

[a] For temperature gradients of this order of magnitude (that is, 2 to 5°C), it is difficult to make direct temperature measurements in the gas since temperature losses to the walls along the leads of any thermometers are orders of magnitude higher than the rate of conduction of heat to the thermometer by the gas.

[b] S. W. Benson, J. Chem. Phys., **22**, 46 (1954).

[c] As the vessel size increases beyond 1 liter or the kinetic viscosity η/ρ becomes very small (that is, <0.1 cm²/sec), turbulence sets in, the heat transfer improves, and the gradients are diminished (see Benson, loc. cit.).

[d] S. W. Benson and A. E. Axworthy, Jr., J. Chem. Phys., **27**, 1718 (1957).

[e] A. O. Allen and O. K. Rice, J. Am. Chem. Soc., **57**, 313 (1935). See also Rice et al., ibid., **57**, 2212 (1935).

[f] Benson, loc. cit.

[g] The agreement here is probably misleading because the different constants going into the equation are not known to better than 5 to 10 per cent.

[h] H. S. Johnston, private communication.

the treatment given originally by Semenoff.[a] The rate of heat evolution as a function of temperature is shown by the family of exponential curves labeled C_1, C_2, and C_3, each corresponding to different concentrations $(C_1 > C_2 > C_3)$ of reactants. These curves are convex to the temperature

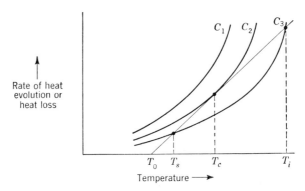

Fig. XIV.1. Heat evolution (curves C_1, C_2, C_3) and heat losses (straight line) as functions of temperature.

axis[b] because of the exponential dependence of the reaction rate constant on temperature. Analytically the rate of heat production by reaction is given by $Q_R = Vf(c)A \exp(-E/RT)H$, where $f(c)$ represents the concentration dependence of the rate,[c] A is the Arrhenius factor, E is the activation energy, H is the molar heat of reaction, and V is the volume of the vessel.

If we consider the entire vessel as our volume element, we may as a first approximation assume that the rate of loss of heat by the gas is proportional to the average temperature difference between the gas and the vessel walls. This corresponds to the straight line in Fig. XIV.1. At low concentrations of reactants (curve C_3), the temperature will mount from T_0, the flask temperature, when the reaction mixture is placed in the flask. As the temperature mounts, the rate of heat loss (although initially small) increases at a faster rate than the rate of heat production until, at some temperature T_s, the two heat terms are just equal to each other and the temperature becomes stationary. Note that this is also a stable temperature distribution, since for any slight displacements in temperature the system will tend to return to T_s. Stable distributions of this type were treated in the last section, and the mean gradients established, $T_s - T_0$, are usually quite small compared to T_0 (about 1 to 2 per cent).

[a] N. Semenoff, "Chemical Kinetics and Chain Reactions," Clarendon Press, Oxford, 1935.

[b] Arrhenius rate constants show a point of inflection at $T = E/2R$. Above this temperature the curves are, of course, concave to the T axis. However, for any reaction of possible interest here, this would correspond to temperatures of $10,000°K$ or more.

[c] That is, for first-order reactions $f(c) = c$, for second-order $f(c) = c^2$, etc.

The curves intersect at two points T_s and T_i. At the upper temperature T_i the distribution is unstable because any fluctuation which reduces the temperature below T_i will bring the system to a point at which the rate of cooling Q_c exceeds the rate of heating Q_R and the temperature will descend rapidly to T_s. Note, however, that any fluctuation which brings the temperature above T_i will do the opposite; i.e., heating exceeds cooling in this region and the system will tend to thermal explosion. T_i is evidently a "kindling" or "ignition" temperature for the system since any process which brings the reaction temperature above T_i, even though the wall temperature remains at T_0, will cause thermal explosion.

It may appear from this that any exothermic reaction system, no matter how dilute, should have an ignition temperature. This is not necessarily the case since the rate of temperature rise will depend on the heat capacity of the system and in very dilute systems an appreciable fraction of the reaction may be over before the ignition limit can be reached.

For a reaction mixture of sufficiently high concentration of reactants (curve C_1, Fig. XIV.1) the rate of heat production may always exceed heat losses (for a given vessel and wall temperature). Such mixtures will always explode, no stable temperature distribution being possible. For any given size vessel and initial T_0 there is a minimum concentration of reactants for which thermal explosion can take place. This is illustrated by curve C_2 of Fig. XIV.1, which is tangent to the rate of cooling curve at temperature T_c. Higher concentrations are always explosive, while lower concentrations are always stable although possessing an ignition temperature.

By using the simple model already proposed we can equate the rates of heat production and heat losses at T_c and also the derivatives of the two curves to solve for T_c and the critical concentration condition $f(C_c)$ for any fixed initial conditions. If we let h be the coefficient of heat transfer between the vessel walls and the gas and let S be the surface area of the vessel, then the rate of heat loss at $T = T_c$ is $Q_c = Sh(T_c - T_0)$, so that our two conditions may be written as

$$Q_R = HVf(c)Ae^{-E/RT_c} = Q_c = Sh(T_c - T_0)$$

(XIV.3.1)

At T_c:
$$\left(\frac{\partial Q_R}{\partial T}\right) = \frac{E}{RT_c^2} Q_R = \left(\frac{\partial Q_c}{\partial T}\right) = Sh$$

(XIV.3.2)

Dividing these two equations to eliminate Sh and Q_R and then rearranging terms gives

$$\frac{RT_c^2}{E} = T_c - T_0$$

(XIV.3.3)

which can be rewritten in terms of the dimensionless variable $\theta_c = (T_c - T_0)/T_0$ as

$$\theta_c = \left(\frac{RT_0}{E}\right)(1 + \theta_c)^2$$

(XIV.3.4)

This equation will have two roots, one of the order of RT_0/E and the other of the order E/RT_0. Now since E/RT_0 is usually quite large for most reactions (i.e., of the order of 35) we see that the first root corresponds to small values of θ_c and the second to absurdly high temperatures of the order of $T_c \cong 36T_0$, so that it is only the first that is of physical interest, namely, $\theta_c \cong RT_0/E$ or $T_c = T_0 + RT_0^2/E.$[a]

If we substitute this result in Eq. (XIV.3.1), we can solve for the critical concentration explosion limits

$$f(C_c) = \frac{ShRT_0^2}{VHAE} e^{E/RT_c} \tag{XIV.3.5}$$

or, on approximating $1/T_c$ in the exponential by $(1 - \theta)/T_0$, we have

$$f(C_c) = \frac{ShRT_0^2}{VHAEe} e^{E/RT_0} \tag{XIV.3.6}$$

In the more customary form, by taking logarithms of both sides and setting $B = ShR/VHAEe$,

$$\ln\left[\frac{f(C_c)}{T_0^2}\right] = \frac{E}{RT_0} + \ln B \tag{XIV.3.7}$$

For the simple case of an nth-order reaction, $f(C_c) = C_c^n$ and Eq. (XIV.3.7) takes the form

$$\ln\left[\frac{C_c^n}{T_0^2}\right] = \frac{E}{RT_0} + \ln B \tag{XIV.3.8}$$

or

$$\ln C_c = \frac{E}{nRT_0} + \frac{1}{n}\ln(BT_0^2) \tag{XIV.3.9}$$

In practice one measures explosion limits by permitting a reaction mixture (of fixed composition) to enter a flask at a previously determined higher temperature and observing the minimum pressure at which explosion takes place. If the rate law for the reaction is known (actually this is seldom true) then any of the above equations can be tested. In this way Sagulin[b] showed that the critical explosion pressures for $H_2 + Cl_2$ mixtures followed an equation of the form

$$\log\frac{P_c}{T} = \frac{2700}{T} - 6.3$$

where T is in °K, P_c is in cm Hg, and the mixture is equimolar in H_2 and Cl_2. Rice and coworkers have similarly shown that the explosion of azo-

---

[a] For E/RT_0 in the usual range near 35, this gives $T_c - T_0 = T_0/35$, or about 10°C if T_0 is about 350°K. Thus these critical explosion temperatures are not much higher than those present in the gradients (T_s, Fig. XIV.1) which are normally set up in exothermic reactions.

[b] A. Sagulin, Z. physik. Chem., **B1**, 275 (1928).

methane,[a] ethyl azide,[b] and probably methyl nitrate[c] can be fitted to these Semenoff type equations. Frank-Kamenetskii[d] has further shown that the explosions of H_2S and N_2O and C_2H_2[e] can also be fitted to such equations and with the use of reasonable values for the parameters, values of B can be calculated in fair agreement with experiment. Unfortunately, most explosions, thermal or otherwise, fit equations of this form, and in most cases the kinetics of the reaction are not well enough known to attempt quantitative interpretations of the parameters. In addition, the coefficient of heat transfer can be sharply dependent on the composition of the mixture, and this will influence the pressure limits when the composition is changed.[f] Thus in Sagulin's work, it was found that at fixed temperature a mixture containing about $\frac{2}{3}Cl_2 - \frac{1}{3}H_2$ had the lowest explosion pressure. From the presumed kinetics of the reaction [i.e., velocity $= k_0(Cl_2)^{1/2}(H_2)$] one would expect this limit to lie closer to $\frac{1}{3}Cl_2 - \frac{2}{3}H_2$. Here it is undoubtedly the high heat-transfer coefficient of H_2 which shifts the expected values.[g]

Many authors have attempted a more precise model for thermal explosions by considering the temperature distribution through the mixture rather than assuming some average temperature. Such an approach starts with the general equation in three dimensions [Eq. (XIV.2.1)]:

$$\frac{\partial T}{\partial t} = \frac{K}{\rho c_v} \nabla^2 T + \frac{f(c)HAe^{-E/RT}}{\rho c_v} \qquad \text{(XIV.3.10)}$$

At the stationary state $\partial T/\partial t = 0$, so that the equation for the stationary distribution becomes

$$\nabla^2 T = \frac{-f(c)HAe^{-E/RT}}{K} \qquad \text{(XIV.3.11)}$$

Both Frank-Kamenetskii and Rice[h] have examined this equation to see under what conditions no stationary solutions are possible, and they find that for spherical vessels there is a limit (which then defines the explosion limit) given by

[a] A. O. Allen and O. K. Rice, *J. Am. Chem. Soc.*, **57**, 310 (1935).

[b] Rice et al., *ibid.*, **57**, 2212 (1935).

[c] O. K. Rice, *J. Chem. Phys.*, **8**, 727 (1940). These data fit as well as the others if the higher heat of reaction of methyl nitrate is used.

[d] D. A. Frank-Kamenetskii, "Diffusion and Heat Exchange in Chemical Kinetics," Princeton University Press, Princeton, N.J., 1955. Originally published in Russian in 1947.

[e] Ya. B. Blyumberg and D. A. Frank-Kamenetskii, *Zhur. Fiz. Khim.*, **20**, 1301 (1945).

[f] This is particularly true for mixtures which contain H_2 or He, both of which have abnormally high thermal conductivities.

[g] The surface, nature, and smoothness of the vessel must also enter the picture, since free radical reactions are as sensitive to surface conditions as to the thermal conductance.

[h] Rice, *loc. cit.*, Frank-Kamenetskii, *loc. cit.*

$$\delta_c = 3.32 = \frac{EHr_0^2 f(C_c) A e^{-E/RT_0}}{RT_0^2 K} \qquad (XIV.3.12)$$

This can be compared with the rearranged form of Eq. (XIV.3.6) taken from our simple model for thermal explosion:

$$\delta_c' = \frac{EHr_0^2 f(C_c) A e^{-E/RT_0}}{RT_0^2 K} = \frac{Sr_0}{Ve} \frac{hr_0}{K}$$

which for spherical vessels becomes

$$\delta_c' = \frac{3}{e} \frac{hr_0}{K} \qquad (XIV.3.13)$$

Now the dimensionless ratio hr_0/K is known as the Nusselt number $Nu(r_0)$, and for systems with convection it takes values of about 5 if the flow is not turbulent.[a] (In the absence of convection h, the heat transfer at the walls is determined by the temperature gradient at the walls, which in turn is proportional to K/r_0.) It is interesting to note that the simple model which permits laminar convection gives values of δ_c' of about the order of 6, which is reasonably close to the value of 3.32 calculated for pure conduction.

Chambre[b] has taken advantage of the fact that the maximum temperature rise preceding explosion is very small to approximate $e^{-E/RT}$ by $e^{-E/RT_0} e^y$, where $y = (E/RT_0^2)(T - T_0)$, so that Eq. (XIV.3.11) can be written as the Poisson-Boltzmann equation

$$\nabla^2 y = -\delta e^y \qquad (XIV.3.14)$$

He then solves this for cylindrical and spherical vessels and shows that the critical values of δ are in good agreement with those calculated numerically by Frank-Kamenetskii and by Rice.

From the simple model [Eq. (XIV.3.1)] it is possible to solve for the mean temperature of the reacting mixture at the stationary state T_s and also for the ignition temperature T_i. These two temperatures are determined by the two points of intersection of the heating and cooling curves (Fig. XIV.1). By rewriting Eq. (XIV.3.1) these temperatures are given by the two lowest roots of the transcendental equation

$$\frac{HVf(c)A e^{-E/RT}}{ShT_0} = \frac{T - T_0}{T_0} \qquad (XIV.3.15)$$

By letting $y = (E/RT_0)[(T - T_0)/T_0]$ and approximating $1/T \cong 1/T_0 +$

---

[a] See, for example, E. R. G. Eckert, "Introduction to the Transfer of Heat and Mass," McGraw-Hill Book Company, Inc., New York, 1950. In the absence of convection the Nusselt number $Nu(r_0)$ becomes a constant.

[b] P. L. Chambre, *J. Chem. Phys.*, **20**, 1795 (1952). He finds $\theta = 1.61$ and $\delta_c = 3.32$ for spherical vessels. For an infinitely long cyclinder, Frank-Kamenetskii finds $\delta_c = 2.00$ and $y_c = 1.37$ and the same values as Chambre for a spherical vessel. Our simple treatment [Eq. (XIV.3.4)] gave $y_c = 1$ for spherical vessels.

$(T - T_0)/T_0$ which is valid for small values of $T/T_0 - 1$, this can be written as

$$\beta = \frac{EHVf(c)Ae^{-E/RT_0}}{ShRT_0^2} = ye^{-y} \qquad (XIV.3.16)$$

where we may further assume that the term β is to a first approximation independent of temperature. Now the function ye^{-y} varies from 0 to a maximum value of $1/e$ at $y = 1$ and then back to 0. For conditions not too far from explosion [that is, $\frac{1}{2} < y < \frac{3}{2}$, $y = 1$ corresponds to critical limits, see Eq. (XIV.3.4)] we may expand ye^{-y} as a power series about its maximum. That leads to the equation

$$(y - 1)^2 \cong 2 - 2e\beta$$

so that the two roots of Eq. (XIV.3.15) are

$$y = 1 \pm (2 - 2e\beta)^{1/2} \qquad (XIV.3.17)$$

and

$$\frac{T_i - T_0}{T_0} = \frac{RT_0}{E}[1 + (2 - 2e\beta)]^{1/2} \qquad (XIV.3.18)$$

$$\frac{T_s - T_0}{T_0} = \frac{RT_0}{E}[1 - (2 - 2e\beta)]^{1/2}$$

Note that the maximum value of β for which a solution is possible is $1/e = 0.368$, which must correspond to the critical explosion limit. At values of $\beta \leq \frac{1}{2}e$, the approximations fail.

The simple model leads to the following differential equation for the rate of change of temperature [Eq. (XIV.3.1)]

$$\frac{dT}{dt} = \frac{Q_R - Q_c}{V\rho c_v}$$

$$= \frac{Hf(c)Ae^{-E/RT}}{\rho c_v} - \frac{Sh}{V\rho c_v}(T_c - T_0) \qquad (XIV.3.19)$$

By using the same approximation as before [Eq. (XIV.3.16)] and the dimensionless variables y and $\tau = Sht/V\rho c_v$ we obtain

$$\frac{dy}{d\tau} = \beta e^y - y \qquad (XIV.3.20)$$

Rice (loc. cit.) has shown how to integrate this equation numerically. However, if we take advantage of the fact that y varies from 0 at $t = 0$ to $y = 1$ at the critical explosion temperature [Eq. (XIV.3.4)], we can approximate e^y over the range by the linear function $1 + (e - 1)y$, so that the equation becomes

$$\frac{dy}{d\tau} \cong \beta - (1 + \beta - \beta e)y \qquad (XIV.3.21)$$

which is readily integrated subject to the condition that $y = 0$ at $\tau = 0$, to give

$$\ln \left[1 - \frac{1 + \beta - \beta e}{\beta} y \right] = -\frac{1 + \beta - \beta e}{\beta} \tau \qquad \text{(XIV.3.22)}$$

so that the time to reach any value of y up to the critical is given by

$$\tau = -\frac{1}{k} \ln (1 - ky) \qquad \text{(XIV.3.23)}$$

where $k = 1/\beta + 1 - e$ and at the critical explosion conditions $\beta = 1/e$, so that $k = 1$. The time required to approach the critical explosion limit is termed the induction period. Equation (XIV.3.23) gives no finite solution for this value, but we can calculate the time required to approach within 90 or 99 per cent of the limit. We see that τ is of the order of unity for that degree of approximation, so that $t_c \cong V \rho c_v / Sh$. The observed values of t_c are of the order of 2 to 10 sec, and Rice (loc. cit.) has shown that the dependence of t_c on various parameters is in qualitative accord with the simple model. However, there is considerable doubt about the precise significance of these induction periods in view of the fact that the time for the gas to reach oven temperature is also of about the same order of magnitude.

The most extensive test which has been made of this conduction model for thermal explosion is to be found in the work of Vanpée[a] on the explosion of $CH_2O + O_2$ mixtures. He used a calibrated thread of 10 per cent Rh-Pt alloy of 20 μ diameter (jacketed by a 50-μ quartz sleeve) suspended at the center of a cylindrical vessel to measure directly his reaction temperature during the induction periods preceding explosion. By using He and Ar as additives and vessels of different diameters he was able to verify the dependence of the critical explosion limits on vessel size and on thermal conductivity of the gas mixture. In addition, he was able to check the maximum predicted temperature at the center of the vessel just prior to explosion and also the value of $\delta_c \cong 2$ [Eq. (XIV.3.12)], the critical explosion parameter for cylindrical vessels. Finally, with a high-speed camera, he was able to show directly that the explosions in this system do start at the center, the hottest region,[b] and propagate to the walls.

4. Chemically Sensitized Explosions; Branching Chains

The simple model of a thermal explosion which we examined in the last section was based on the view that, for an exothermic reaction, heat played the role of an autocatalytic product. It is now interesting to ask if it is possible for chemical species produced as intermediates, or even final

[a] M. Vanpée, Bull. soc. chim. Belges, **64**, 235 (1955).

[b] By using a second filament near the wall he was able to check the logarithmic temperature gradient expected for a cylinder. Under the conditions of these experiments the time constant for thermal conduction was about 0.05 sec [Eq. (XIV.2.8)].

products, to play a similar role in provoking an autocatalysis of the reaction which could culminate in explosion.

If we consider the mechanism of any typical chain reaction, we see that the atoms or free radicals produced are, by virtue of the existence of the chain, autocatalytic agents for the reaction. Thus in the chain reaction $H_2 + Cl_2 \rightarrow 2HCl$ both H and Cl are chain carriers, the chain mechanism being

$$Cl + H_2 \underset{4}{\overset{3}{\rightleftharpoons}} HCl + H$$

$$H + Cl_2 \underset{6}{\overset{5}{\rightleftharpoons}} HCl + Cl \qquad\qquad (XIV.4.1)$$

It is in fact because of the autocatalytic character of the H and Cl intermediates that a fast reaction between H_2 and Cl_2 is observed. But such chain reactions are normally not explosions. In what sense, then, can autocatalysis by chain carriers lead to an explosion? What is required, if catalysis by chain carriers is to lead to an explosion, is that the mechanism of the chain cause an increase in the concentration of chain carriers beyond that present in the normal reaction.

Now if we examine again the $H_2 + Cl_2$ chain system [Eq. (XIV.4.1)], we see that this typical two-center chain cannot change the total concentration of chain centers. All the chain does is change the identity of chain carriers (that is, H to Cl or Cl to H) but not their total concentration. Such chain systems cannot thus provide for an increase in the concentration of chain carriers. Explosions, if they occur in such systems, must be thermal in character.

It is, however, possible to induce explosions in these systems by the use of additives which are frequently referred to as sensitizers. Thus Ashmore[a] has shown that the addition of 0.5 mm Hg of NO to 50 mm Hg of an equimolar mixture of $H_2 + Cl_2$ lowers the critical explosion temperature from 400 to 270°C. The explosion in this case is still, however, a thermal explosion, and it has been shown that the lowering of the explosion temperature was produced by an increase in the concentration of Cl atoms, not by a change in the chain mechanism.[b] This increase in concentration of Cl atoms was produced by the replacement of the slow, high-activation-energy initiation reaction, $M + Cl_2 \rightleftharpoons 2Cl + M(E \geq 57$ Kcal), by the much-lower-activation-energy reaction, $NO + Cl_2 \rightleftharpoons NOCl + Cl(E \cong 22$ Kcal).

Chemical sensitization of this type is very common, but it is not the chain autocatalysis we were seeking. In order for autocatalysis by chain

[a] P. G. Ashmore, "Fifth Symposium (International) on Combustion," p. 700, Reinhold Publishing Corporation, New York, 1955. See also P. G. Ashmore and J. Chanmugam, *Trans. Faraday Soc.*, **49**, 265 (1953).

[b] We distinguish here between that part of the total mechanism which has a cyclic or chain character and produces product and that part which does not.

carriers to lead to a continuously accelerated rate of reaction, the mechanism must be such that the chain itself causes an increase in concentration of chain carriers. Such chains are referred to as branching chains, and they seem to be common to systems undergoing oxidation. Thus in the reaction of $O_2 + H_2$ at temperatures above 400°C there is good evidence for a chain mechanism that involves the following branching chain steps

$$H + O_2 \rightleftharpoons OH + O$$
$$O + H_2 \rightleftharpoons OH + H$$
                                    (XIV.4.2)

In both cases the reaction of one chain carrier (H or O) gives rise to two. Assuming that the OH radicals go on to produce more H atoms via OH + $H_2 \rightarrow$ HOH + H, we see that in one complete cycle of the chain, starting with one H atom, we produce 2HOH and a new 2H for a total of 3H. In 30 cycles of such a chain, in the absence of termination reactions, each original H atom would have grown to 3^{30}, or about 10^{15}, H atoms. It is this Malthusian growth of the branching chain which makes it possible for it to provoke an explosion.[a]

There are, however, more severe requirements for the existence of a branching chain explosion than for a thermal explosion. The most important one is that imposed on the continued growth of radical centers by second-order recombination processes between centers. Such processes place an upper limit on the concentration of radicals which can be realized in any branching system, and when the rate constant for recombination is sufficiently high, the character of the explosion may be significantly altered or else the explosion may actually be inhibited.[b] Complexities of this type make it impossible to present a general model for the branching chain explosion without invoking numerous formal abstractions. Instead, let us look at the experimental features of the branching chain explosion and see if we can find which of these are general and which specific to the chain mechanism.

5. The Branching Chain Explosion; Upper and Lower Limits

The model presented for a thermal explosion predicts that for a reaction mixture of fixed composition and fixed initial temperature, there will be a critical pressure above which explosion will occur and below which a "normal" stationary reaction will take place. The relation between the critical pressure and temperature is given by a modified Arrhenius equation with a negative temperature coefficient [Eq. (XIV.3.8)] which is

[a] As we shall see later, the slowest step in such a chain is the reaction H + $O_2 \rightarrow$ OH + O, which takes about 0.01 sec per H atom at 450°C when $O_2 = 80$ mm Hg. The time for 30 cycles would thus be about 0.3 sec.

[b] By way of contrast, note that in a thermal explosion, once the critical explosion limit is passed, the rate of heat evolution by reaction will always remain greater than the rate of heat loss by conduction, and also convection (Fig. XIV.1).

illustrated in Fig. XIV.2a. The two curves AA' and BB' represent critical limits for reactions having high and low activation energies respectively. The hypothetical curve indicated by $BCC'A'$ might represent the case of a single reaction that has a complex mechanism which changes with increas-

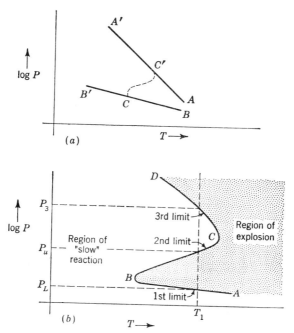

FIG. XIV.2. (a). Relations between pressure and temperature at the critical explosion limits of a thermal explosion. Curve AA': Critical limits for a reaction of high activation energy. Curve BB': Critical limits for a reaction of low activation energy. Curve $BCC'A'$ illustrates hypothetical though unreal case of a reaction which changes its mechanism with increasing pressure from one of a low to one of a high activation energy.

(b). Critical explosion limits for a typical branching chain explosion showing explosion peninsula. Curve $ABCD$ represents explosion limits. The first, second (upper), and third limits are represented by the curves AB, BC, and CD, respectively. Region ABC is called the explosion peninsula.

ing pressure from one of low activation energy to one of high activation energy. Although such a case, as we shall see, would have the features of branching chain explosion, it is difficult to construct a mechanism, aside from a branching chain, that could account satisfactorily for such a transition.[a]

Systems showing the characteristic of a branching chain explosion (e.g., oxidation of P_4, PH_3, NH_3, H_2, etc.) exhibit critical explosion limits of the

---

[a] The limits for a purely thermal explosion depend only on the rate of release of heat. The implication of the curve $BCC'A'$ is that the reaction can actually proceed faster at low pressure than at high!

form illustrated in Fig. XIV.2b. At very low pressures, generally in the range of 0.1 to 10 mm Hg, such systems reach a critical explosion pressure whose limits are characterized by the curve AB (Fig. XIV.2b) and are referred to as the first or low-pressure limit. Below this pressure the reaction is usually negligibly, if not immeasurably, slow. Above this first limit, the explosion persists until a second critical pressure limit or so-called upper limit (curve BC, Fig. XIV.2b) is reached, whereupon the explosion is quenched and a "normal" or measurably slow reaction occurs. With increasing pressure this reaction accelerates until a third limit is reached (curve CD, Fig. XIV.2b) at which explosion once again takes place.

The actual rate of reaction as a function of total pressure at fixed composition and fixed temperature is illustrated in Fig. XIII.3. In the draw-

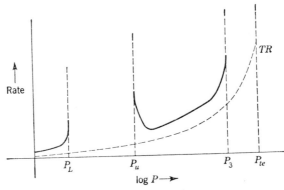

FIG. XIV.3. Over-all rate of reaction of a branching chain reaction as a function of pressure at fixed temperature. P_L is the lower explosion limit and P_u is the upper explosion limit of the explosion peninsula; P_3 corresponds to third explosion limit (Fig. XIV.2b). For purposes of comparison, the dotted curve TR illustrates the rate of a "normal" reaction up to the thermal explosion limit P_{te} (see Fig. XIV.2a).

ing, the curve for the branching chain explosion is shown as discontinuous in the region of the explosion peninsula (i.e., between the first and second limits), whereas for any real reaction the rate reaches some very high but finite value. For purposes of comparison the curve for a normal reaction (TR) below its thermal explosion limit is drawn in the same diagram. Except for the behavior in the vicinity of the explosion peninsula, the curves for the explosion limits are seen to be very similar for the two cases; indeed, it is sometimes very difficult to distinguish between them. There are, however, features of the branching chain explosion which are quite different from the thermal explosion.

It is generally found that the first explosion limit of a branching chain is shifted to lower pressures by the decrease of surface/volume ratio (for example, a system becomes more explosive in larger vessels) or by the addition of an inert gas (for example, N_2 or Ar). While the surface/volume behavior is similar to that found in thermal explosions, the effect of added inert gases is quite different because, in the thermal explosion system, inert gases can only affect the explosion limits if they change the coefficient of heat conductivity or play a role in the reaction mechanism. If they increase the heat conductivity, they can actually raise the explosion limit.[a]

The feature which is unique to the chain-branching system is the paradoxical, upper, or second explosion limit. Here one observes that a reaction proceeding with explosive speed at pressures below the limit is effectively quenched on raising the pressure. In addition, the pressure limit increases if the temperature increases, just opposite to the behavior at the first and third limits. It is the existence of this limit that is the real evidence of the branching chain. It is observed that the limit is much less sensitive to surface-volume effects than is the first limit, while added inert gases always tend here to lower the limit (i.e., quench the explosion).

If we are to construct a kinetic model to explain the preceding behavior, it seems most reasonable to start by saying that in the explosion peninsula, i.e., between the first and second explosion limits, the explosion results from the chain-branching reaction. These reactions are usually first-order in chain centers and, at fixed composition and temperature, first-order in total pressure.[b] Since the explosion is quenched at pressures lower than the first explosion limit, we conclude that, whatever reactions are responsible in the system for removing free radicals, they must be of the first-order, kinetically, in free radicals and less than first-order in total pressure. In view of the sensitivity of the first limit to added inert gases and to chemical composition of the surface, it seems plausible to postulate that chain termination in the low-pressure region is by diffusion of free radicals to the walls of the vessel. When the efficiency of capture of radicals by the walls is very high, such termination will be first-order in free radical concentration and inversely proportional to the pressure of gas through which the radicals must diffuse to reach the walls. When the efficiency is very low, so that absorption of radicals on the proper wall sites is the rate-determin-

[a] Because of the possible chemical complexity of the thermal explosion reaction and because of the fact that all known branching chain reactions are exothermic so that the onset of explosion is also accompanied by considerable heat liberation, none of these tests is decisive in distinguishing the two systems. A good test which is seldom made is the comparison of the effects of two different inert gases of quite different heat conductivities such as He and N_2. In the case of thermal explosions, He addition should always exercise a quenching effect on the explosion relative to N_2. When He exercises an absolute quenching effect (i.e., raises the first explosion pressure limits) this may be taken as a positive indication of thermal explosion or at least the importance of thermal effects if the system is otherwise a chain-branching one. P. Gray and J. C. Lee, *Trans. Faraday Soc.*, **50**, 719 (1954), have shown that the explosion of 1:1 and 1:2 N_2H_4/O_2 mixtures fit the requirements of a thermal explosion; in particular, He raised the pressure limit, while N_2, A, and O_2 lowered the limit by amounts compatible with their thermal conductivities.

[b] For example, in the $H_2 + O_2$ system chain branching by H is given by $k_{bH}(H)(O_2)$ and by O atoms, $k_{bO}(O)(H_2)$ [Eq. (XIV.4.2)].

ing step, the termination reaction is first-order in radical concentration and zero-order in total pressure.[a] In both cases, however, the dependence on pressure is less than first-order and at the first explosion limit chain branching may overtake and exceed chain termination,[b] leading to an exponential growth of the radical concentration and a chain explosion.

If we assume, for concreteness, that chain initiation occurs by chains starting at the surface and diffusing out into the gas where branching takes place and that termination occurs by diffusing back to the surface, then the condition that governs radical concentration may be expressed by the quasi-stationary-state condition

$$k_b(C)P + k_{iw}\frac{S}{V} = \frac{k_{tw}}{P^n}\frac{S}{V}(C) \qquad (XIV.5.1)$$

where k_b represents the second-order specific rate constant for branching of the chain centers C, k_{iw} is the zero-order specific rate constant for initiation of chain centers at the walls, k_{tw} is the diffusion-controlled termination rate constant for capture of radicals at the walls, and P is the total pressure.[c] On solving for $(C)_{ss}$ we find

$$(C)_{ss} = \frac{Sk_{iw}/V}{\dfrac{k_{tw}S}{P^nV} - k_bP} \qquad (XIV.5.2)$$

At very low pressures, $k_{tw}S \gg k_bP^{1+n}V$ and $(C)_{ss} \cong k_{iw}P^n/k_{tw}$. However, as the pressure increases, the branching grows more important relative to wall termination, and as the two terms approach equality, $(C)_{ss} \to \infty$, that is, the reaction proceeds to explode.[d] From Eq. (XIV.5.2) we find for this first explosion limit

[a] Intermediate cases are given by an expression of the form $Sk_sk_D(C)/V[Sk_D + Pk_s]$, where k_D is the coefficient of diffusion per unit pressure P, (C) is the concentration of radicals, k_s is the specific rate of capture at the surface, S is the total surface, and V is the volume.

[b] It is the requirement that chain branching exceed chain termination at the first explosion limit and *lead to explosion* which makes it necessary that the branching reaction be of the same or a higher kinetic order in free radical concentration than the termination reaction. If it were of lower order, the termination reaction would increase in rate relative to branching as the radical concentration grew and quench the reaction.

[c] It might appear that in the interests of microscopic reversibility both k_{iw} and k_{tw} should have the same pressure dependence. This is certainly true, but then the condition for equilibrium requires that (C) be proportional to some power of P [for example, $(OH) = K_{eq}(H_2)^{\frac{1}{2}}(O_2)^{\frac{1}{2}}$], and we have taken the liberty here of throwing all the pressure dependence into the termination reaction.

[d] As pointed out previously, an upper limit is placed upon $(C)_{ss}$ by second-order recombination processes, so that the radical concentration cannot grow without limit. At pressures near the first explosion limit this restriction is unimportant. Thus if we assume that the reaction H + H + M proceeds at every tenth triple collision (that is, $k = 3 \times 10^9$ liters²/mole²-sec) then the lifetime of an H atom at 750°K when M = 8 mm Hg and H = 0.8 mm Hg (!) is about 0.2 sec $[\tau_{\frac{1}{2}} = 1/k(M)(H)]$. This is much slower than the rate of branching in the $H_2 + O_2$ system (see page 457).

$$P_1^{(n+1)} = \frac{k_{tw}}{k_b} \frac{S}{V} \qquad \text{(XIV.5.3)}$$

or for $n = 1$

$$P_1 = \left(\frac{k_{tw}S}{k_b V}\right)^{\frac{1}{2}} \qquad \text{(XIV.5.4)}$$

Since k_{tw} is expected to have a small or zero activation energy, the temperature dependence of P_1 is determined by $1/k_b$ (the reciprocal of the branching rate constant), and P_1 will thus decrease with an increase in temperature, as is observed.

If we apply the same reasoning to the second pressure limit as we did to the first one, we see that, since the explosion is quenched at constant composition and temperature by raising the total pressure, we must postulate a new chain-termination process which is of the same (i.e., first) kinetic order in radical concentration as the branching chain reaction and is at the same time higher than first order in total pressure. Such a reaction cannot be generalized. In the $H_2 + O_2$ system it seems to be satisfied by the chain-transfer process

$$H + O_2 + M \overset{k_{t1}}{\rightarrow} HO_2 + M \qquad \text{(XIV.5.5)}$$

which replaces the fairly active chain carrier H by the relatively inert radical HO_2.[a]

If we add this reaction to our chain-termination system [Eq. (XIV.5.1)] by writing it as a termination step in the form $k_{t1}(C)P^2$, then above the second explosion limit we have

$$(C)_{ss} = \frac{Sk_{iw}/V}{k_{tw}S/P^n V + k_{t1}P^2 - k_bP} \qquad \text{(XIV.5.6)}$$

or, neglecting wall termination,[b]

$$(C)_{ss} \cong \frac{Sk_{iw}/V}{k_{t1}P^2 - k_bP} \qquad \text{(XIV.5.7)}$$

At pressures below the second limit, this higher-order termination process cannot keep pace with chain branching and we have chain explosion. The equation for the second explosion limit is thus given by the equality of these two processes, or

[a] As we shall see later, HO_2 is less active than O atoms with respect to attack on H_2; in addition, it can act to break chains without the requirement of a third body through reactions such as $H + HO_2 \rightarrow H_2 + O_2$, $HO + HO_2 \rightarrow HOH + O_2$; and finally, it can act to circumvent branching by $HO_2 + O \rightarrow OH + O_2$, $H + HO_2 \rightarrow 2OH$.

[b] At regions near the junction of the first and second explosion limits this is a poor approximation and the more exact equation defining both the first and second limits is given by the positive root of the equation $k_{t1}P^{2+n} - k_bP^{1+n}k_{tw}S/V = 0$. At the turning point of the explosion peninsula where the first and second limits merge $k_{t1}P^{2+n} = k_{tw}S/V$. For surfaces of very high efficiency in termination, the approximation is also poor.

$$P_2 = \frac{k_b}{k_{t1}} \qquad\qquad (XIV.5.8)$$

Again, recombination processes k_{t1} such as Eq. (XIV.5.5) are not expected to have activation energies exceeding a few kilocalories, so that P_2 should have the same temperature dependence as the branching chain reaction and show an increase in pressure with increase in temperature, as indeed happens.

It is difficult to postulate without considerable strain a branching process of higher kinetic order in total pressure and first or higher order in radicals which could account for the third explosion limit. Much attention has been given to this problem[a] without a decisive answer appearing. The chief reason lies in the appearance of the thermal explosion limit in this region of pressures above the explosion peninsula. In the absence of specific evidence to the contrary it is usually safer to interpret the third explosion limit as a thermal explosion.

6. Concentration Gradients in Chemical Reactions

It seems to be necessary, in order to account for the influence of the wall on the first explosion limit, to postulate termination of chains on the wall. From other evidence (which we shall discuss later in a few specific cases) on the "slow" rates outside the explosion peninsula it appears equally necessary to postulate chain initiation at the walls. Where surfaces such as the walls play an important role in chemical reactions, we must expect to find that in a certain range of reaction conditions the diffusion to and from such surfaces may exert a limiting effect on the rates of chemical reactions. If the reactions are taking place in the volume of the vessel in competition with reactions at the wall, then we may expect to find concentration gradients within the volume of the vessel.[b]

We can make a crude estimate of the conditions under which diffusion is likely to be important by comparing the time required for diffusion under given circumstances with the time required for a dependent or competing process. Thus from the kinetic theory (Sec. VI.7) of Brownian motion, the time required for a molecule to diffuse a distance x is given approximately by $t_D \cong x^2/D$, where D, the diffusion constant, is inversely proportional to the pressure (that is, $D = D_0/P$). If the dependent process is capture of radicals at the walls, the kinetic theory gives (Sec. VII.8)

[a] D. R. Warren, *Proc. Roy. Soc. (London)*, **A211**, 86, 96 (1952). B. Lewis and G. von Elbe, "Combustions, Flames and Explosions," Academic Press, Inc., New York, 1951. See also N. N. Semenoff, *Doklady Akad. Nauk. S.S.S.R.*, **81**, 645 (1951), and N. S. Akulov, *ibid.*, **83**, 427 (1954).

[b] In systems such as flames or explosions, no surfaces need be present for rather sharp concentration gradients of reactants, intermediates, products, and temperature to exist. Actually in such systems the gradients are so sharp that thermodynamic quantities such as temperature begin to lose their meaning.

$\frac{1}{4}N\bar{c}S$ as the number of wall collisions per second, where $\bar{c}$ is the mean molecular speed, S is the surface, and N the number of molecules per unit volume. The average time for capture of a molecule by collision with the wall is NV, the total number of molecules, divided by the rate of capture, or $t_{cw} = 4V/S\bar{c}\epsilon$, where ϵ represents the probability of being captured on collision with the wall.

For the average free radical in a vessel which is a distance x from the walls, diffusion will be the rate-controlling step in wall termination when $t_D \geqslant t_{cw}$, or since $D = D_0/P$, when

$$P \geqslant \frac{4VD_0}{S\bar{c}\epsilon x^2} \tag{XIV.6.1}$$

or for a spherical vessel of radius r_0, assuming $x \cong r_0$, when

$$P \geqslant \frac{4D_0}{3r_0\epsilon\bar{c}} \tag{XIV.6.2}$$

Now for most gases $D_0/\bar{c}$ is approximately equal to the mean free path and is very close to 10^{-5} cm at STP (Table VIII.3), so that in a 500-cc flask ($r_0 \cong 5$ cm), P must be of the order of $0.002/\epsilon$ mm Hg or higher for diffusion to be important in controlling wall termination. Thus if a radical is captured on every collision ($\epsilon = 1$), diffusion control is important at pressures above 0.002 mm Hg. If, however, $\epsilon = 10^{-4}$, then the range is 20 mm Hg or higher. Below these pressures radicals disappear at the walls, but there are no appreciable gradients present.

From the observation that diffusion-controlled wall termination seems to be the important termination step for most chain explosions at the low-pressure limit, which usually falls between 1 and 10 mm Hg, it appears that for these systems ϵ must be of the order of 10^{-3} or higher.[a]

In the region of concentration in which the contributions of wall reactions are regulated wholly or partially by diffusion, the concentrations of chain centers C will be governed by an equation very similar to the heat flow equation (XIV.2.1)

$$\frac{\partial(C)}{\partial t} = D\nabla^2 C + k_b(C) + k_i \tag{XIV.6.3}$$

where $k_b(C)$ represents the branching reaction, assumed to be first order in radicals C, and k_i represents a constant rate of initiation of radicals in the vessel.[b] This equation must be solved subject to the boundary condition

[a] The efficiency of wall recombination is not necessarily independent of pressure or composition. If the mechanism of wall termination is given by an equation of the form $X_2(g) + W \rightleftharpoons X \cdot W + X(g)$, where W represents a wall site, then an increase in X_2 will change the rates of both forward and reverse reactions.

[b] This would be the case for homogeneous thermal or photochemical initiation in which, over the period of time considered here, the concentration of reactants does not change appreciably. If this is not the case, the diffusion equation above must be solved simultaneously with the kinetic equations. That would be true of flames and explosions.

that the net flux of radicals brought to the walls by diffusion must equal the rate of capture at the wall,[a] that is,

$$-D\left[\frac{\partial(C)}{\partial x}\right]_{x=\text{wall}} = k_{cw}(C)_w = \frac{1}{4}(C)_w \bar{c}\epsilon \qquad \text{(XIV.6.4)}$$

Equation (XIV.6.3) represents, of course, one special case among many possible examples of chain-branching reactions. Variations of this equation may be obtained by adding terms that represent homogeneous first-order termination $k_{t1}(C)$ to Eq. (XIV.6.3) or wall-initiation terms to the auxiliary boundary equation (XIV.6.4). The addition of terms which are second-order in radicals, such as second-order recombination in the gas phase, or second-order branching leads to equations which are nonlinear and which may only be solved either numerically or by approximation.

Equations such as (XIV.6.3) with additional linear terms are well-known equations, and solutions are available for vessels of simple geometry such as long cylinders, flat vessels with linear face dimensions large compared to their separation, and spheres or spherically symmetrical vessels.[b] Such solutions were discussed originally by Bursian and Sorokin,[c] and additional cases have been presented by Lewis and Von Elbe,[d] Semenoff,[e] and Frank-Kamenetskii.[f]

If we make the substitution $N = (C) + k_i/k_b$ and assume that D is constant,[g] then Eq. (XIV.6.3) becomes for a spherical vessel[h] of radius r_0

$$\frac{\partial N}{\partial t} = \frac{D}{r}\frac{\partial^2(Nr)}{\partial r^2} + k_b N \qquad \text{(XIV.6.5)}$$

If we assume that $N(r,t) = F(t)N(r)$, a separation of variables is easily effected by direct substitution:

$$\frac{\partial \ln F(t)}{\partial t} = \frac{D}{rN(r)}\frac{\partial^2[rN(r)]}{\partial r^2} + k_b \qquad \text{(XIV.6.6)}$$

where since the right- and left-hand sides of the equation are functions of different independent variables (that is, r and t), each must be a constant which we shall set equal to A. On substitution we find that $F(t) = Fe^{At}$, F being a constant of integration, while for $N(r)$ we have

$$\frac{\partial^2 Z}{\partial r^2} = -m^2 Z \qquad \text{(XIV.6.7)}$$

---

[a] This is essentially a condition of continuity which avoids the accumulation of radicals on the walls.

[b] Bateman, *op. cit.*

[c] V. Bursian and V. Sorokin, *Z. physik. Chem.*, **B12**, 247 (1931).

[d] G. von Elbe and B. Lewis, *J. Am. Chem. Soc.*, **59**, 970 (1937). See also the prior work of L. S. Kassel and H. H. Storch, *ibid.*, **57**, 672 (1935).

[e] N. N. Semenoff, *Acta Physicochim. U.R.S.S.*, **18**, 93 (1943). (In English.)

[f] Frank-Kamenetskii, *op. cit.*

[g] See the first footnote of Sec. XIV.2; the same restrictions apply here.

[h] Equation (XIV.2.2).

where $Z = rN(r)$ and $m^2 = -(A - k_b)/D$.[a] Now the solution of this last equation gives

$$N(r) = \frac{Z(r)}{r} = \frac{B\sin(mr + \theta)}{r} \qquad \text{(XIV.6.8)}$$

where B and θ are constants of integration. But in order for $N(r)$ to be finite at $r = 0$, the constant $\theta = 0$. The general solution of our differential equation thus has the form

$$N(r,t) = N_m e^{k_b t} e^{-m^2 Dt} \frac{\sin(mr)}{r} \qquad \text{(XIV.6.9)}$$

where we have combined the constants $BF = N_m$.[b]

Now any equation of the form (XIV.6.9) will satisfy the differential equation (XIV.6.5), no matter what the value of m is. Since our original equation was a linear differential equation, any linear combination of solutions will also be a solution, so that the most general solution if m is restricted to discrete values[c] is

$$N(r,t) = e^{k_b t} \sum_m N_m e^{-m^2 Dt} \frac{\sin(mr)}{r} + P(r) \qquad \text{(XIV.6.10)}$$

where $P(r)$ is a solution of the stationary-state equation [see Eq. (XIV.6.5)]

$$\frac{\partial N}{\partial t} = 0 = \frac{D}{r} \frac{\partial^2 (Nr)}{\partial r^2} + k_b N \qquad \text{(XIV.6.11)}$$

Since this equation corresponds to a value for $A = 0$, $P(r)$ must have the value [Eq. (XIV.6.8)]

$$P(r) = \frac{B}{r} \sin \frac{k_b^{1/2} r}{D^{1/2}} \qquad \text{(XIV.6.12)}$$

The fitting of the various constants of integration (that is, B and the N_m) to the boundary conditions [e.g., Eq. (XIV.6.4) and the initial condition that $C(r,0) = 0$ for all r] is a rather tedious exercise in Fourier series[d] the details of which are not very important for our purposes. However, it

---

[a] This restriction of m to real values is necessary only for the stationary solutions of Eq. (XIV.6.5) (i.e., those not leading to explosion). This also implies that $A \leq 0$ for the stationary solutions, so that $|m| \geq (k_b/D)^{1/2}$. Note that for $A > 0$, $F(t)$ will grow without limit. In view of the large gradients in temperature and concentration, the solution of Eq. (XIV.6.5) for the cases leading to explosion are probably not meaningful experimentally. Note that for a thin spherical shell of thickness $2 \, \Delta r \ll r_0$ the solution can be expanded into the solution for the flat vessel.

[b] The subscript N_m is used to indicate that, for any choice of the quantity m, the constant of integration N_m may be different.

[c] It can be shown that so long as r has a finite range of values, the constants m are restricted to a discrete set of values.

[d] Bateman, op. cit., p. 353.

can be shown that the discrete values of m are given by the set of solutions of the transcendental equation[a]

$$\tan(mr_0) = \frac{mr_0}{1 - k_{cw}/D} \qquad \text{(XIV.6.13)}$$

If we let $\phi_i = m_i r_0$ represent the successive roots of this equation, the difference between consecutive values will be of the order of π. Measured in units of $(\pi/r_0)^2$, m^2 thus increases very rapidly. The significance of this is that the exponents in the time-dependent terms of Eq. (XIV.6.10) increase very rapidly, and at times of the order of magnitude of $(m_1^2 D - k_b)^{-1} = (\phi_1^2 D/r_0^2 - k_b)^{-1}$ [where m_1 is the first admissible root of Eq. (XIV.6.13)] the first term is down to $1/e$ of its original value and the subsequent terms will be found to be negligible. This time can be taken as the half-life for the approach to the stationary state, and for reactions near the first explosion limit it is usually found to be of the order of seconds or less.[b] At times greater than that, the solution becomes essentially the stationary-state solution, $N(r,t) \to P(r)$ [Eq. (XIV.6.12)] with the constant B determined by the continuity condition [Eq. (XIV.6.4)][c]

$$B = \frac{k_i k_{cw} r_0}{k_b \left[(k_{cw} - D/r_0) \sin (k_b^{1/2} r_0/D^{1/2}) + (k_b D/r_0)^{1/2} \cos (k_b^{1/2} r_0/D^{1/2}) \right]} \qquad \text{(XIV.6.14)}$$

The solution for the stationary concentration of radicals is given by

$$C(r,\infty) = \frac{B}{r} \sin \left(\frac{k_b^{1/2}}{D^{1/2}} r \right) - \frac{k_i}{k_b} \qquad \text{(XIV.6.15)}$$

Now the function $\sin x/x$ has a maximum value at $x = 0$ and decreases monotonically to 0 at $x = \pi$, after which it becomes negative. The maximum value of the argument of Eq. (XIV.6.15) occurs at $r = r_a$, which puts an upper limit on its value: $k_b r_0^2/D$ must be greater than $[\sin^{-1}(k_i r_0/B k_b)]^2$.

Thus the concentration of free radicals in the vessel at the stationary state has a maximum at the center and decreases smoothly to its minimum

[a] This comes from our boundary condition at the wall, Eq. (XIV.6.4). When $D = k_{cw}$, that is, when every radical striking the wall is captured ($\epsilon = 1$), then $mr_0 =$ an odd multiple of $\pi/2$ or $m = (2n + 1)\pi/2r_0$, with $n = 0, 1, 2, \ldots$. When $k_{cw} \to 0$, that is, no wall termination, the stationary solution becomes impossible. In that case mr_0 equals an integral multiple of π. Thus the admissible solutions for mr_0 lie in the first and third quadrants, the lowest value, m_1, being between 0 and $\pi/2$ if $(k_b/D)^{1/2} < m_1$. If $(k_b/D)^{1/2} > m_1$, the lowest admissible value of m is m_2, which lies between π/r_0 and $3\pi/2r_0$. Note that $(k_b/D)^{1/2}r_0 < \pi$, or else $P(r)$ [Eq. (XIV.6.12)] becomes negative.
[b] Since the criterion for the first explosion limit is that $m_1^2 = k_b/D$, this time is not really well defined.
[c] For consistency, at the stationary state, the rate of loss of radicals at the walls must equal the total rate of radical initiation in the entire volume plus their total rate of branching. These two conditions must be equivalent.

value at the walls.[a] The average value of (C), which is all that is used in computation, may be obtained rigorously from Eq. (XIV.6.15) by averaging over the sphere, or if we approximate this by a parabolic distribution,[a]

$$\overline{C(r,\infty)} \cong \left(\frac{Bk_b^{\frac{1}{2}}}{D^{\frac{1}{2}}} - \frac{k_i}{k_b}\right) - \frac{1}{10}\frac{k_b}{D}r_0^2 \qquad \text{(XIV.6.16)}$$

which can now be used to make corrections for vessel size to steady-state reactions.

Lewis and Von Elbe[b] have considered in detail the special cases of chain breaking at the wall and in the gas phase together with initiation at the walls or in the gas phase. The detailed solutions follow much the same pattern discussed here.

Noyes[c] has made numerical solutions of the case of special interest at higher pressures, that of the second-order recombination of radicals in the gas phase competing with first-order wall recombination. The interesting result here may be stated in terms of the effective thickness r_w over which the wall exerts an appreciable effect. Noyes finds that $r_w \cong 0.9D^{\frac{1}{2}}/(k_i k_{t2})^{\frac{1}{4}}$, where D is the diffusion coefficient, k_i the constant rate of initiation of radicals, and k_{t2} is the rate constant for homogeneous second-order termination of radicals. Within the shell of thickness r_w near the wall, it may be assumed, if the surface is an efficient radical trap, that the concentration of radicals is essentially zero, while outside this shell the concentration of radicals may be assumed to correspond to that which is unperturbed by the walls.

A simple way of arriving at a result which also takes into account the efficiency of the surface is to say that the wall will exert its effect out to a distance r_w, which point is determined by the condition that a radical starting there has an equal chance of being captured on the wall or of undergoing second-order recombination in the gas phase.

The time for diffusion to the wall and capture there is given by

$$t_{tw} = t_D + t_{cw} = \frac{2r_w^2}{D} + \frac{V}{\epsilon Z_w'S}$$

where the symbols have their usual significance and $Z_w'S$ is the frequency of collision with the entire wall. [See Eq. (XIV.6.1) and earlier discussion.[d]] If now the concentration of radicals (C), at this point r_w, is determined by

---

[a] For most practical purposes, the distribution does not differ appreciably from one which is parabolic. That is, we can expand

$$\frac{\sin x}{x} = 1 - \frac{x^2}{6} \qquad \text{so that} \qquad C(r,\infty) \cong \left(\frac{Bk_b^{\frac{1}{2}}}{D^{\frac{1}{2}}} - \frac{k_i}{k_b}\right) - \frac{k_b}{6D}r^2$$

[b] Lewis and Von Elbe, loc. cit.

[c] R. M. Noyes, J. Am. Chem. Soc., **73**, 3039 (1951).

[d] Note the factor of 2 in the expression for t_D. The reason here is that for regions close to the walls only half of the radicals will diffuse toward the walls.

second-order termination, that is, $(C) = (k_i/k_{t2})^{1/2}$, then the mean time for such termination is $t_{t2} = 1/k_{t2}(C) = (k_ik_{t2})^{-1/2}$. If now we equate $t_{tw} = t_{t2}$ and solve for r_w we find

$$r_w = \frac{D^{1/2}}{2^{1/2}(k_ik_{t2})^{1/4}}\left[1 - \frac{V(k_ik_{t2})^{1/2}}{\epsilon Z'_w S}\right]^{1/2} \qquad (XIV.6.17)$$

a result very close to the result of Noyes but which now includes the effect of surface efficiency ϵ.

Gomer[a] has also considered these problems and obtained numerical results for the one-dimensional case of a vessel with faces whose linear dimensions are large compared to their separation. He has considered cases of first- and second-order radical removal both at the walls and in the bulk and for static and flow systems.

7. The Reaction of Hydrogen with Oxygen

Few reactions have been studied as extensively as the classical reaction of hydrogen and oxygen.[b] Because of its relative chemical simplicity, it has served as a prototype and proving ground for theories of branching chain explosions.

At temperatures below about 400°C the reaction is predominantly heterogeneous with a rate which varies considerably depending on the nature of the surface, the previous history, and the extent, that is, the S/V ratio of the reaction vessel. At pressures below 20 mm Hg and in vessels whose walls have been coated by KCl, the rate of reaction is immeasurably slow in this temperature region.

Above 400°C the rate of production of water becomes measurably fast (above 100 mm Hg total pressure), and between about 400 and 600°C the reaction shows all the characteristics of a branching chain explosion replete with three explosion limits. Figure XIV.4, taken from work of Lewis and Von Elbe, illustrates this behavior for a stoichiometric mixture $(2H_2:O_2)$ in a 7.4-cm-diameter pyrex vessel coated with KCl (volume $\cong$ 220 cc).

Below the first explosion limit the rate of reaction seems to be negligibly slow. Above the second explosion limit, studies of the rate of the "normal" reaction have shown considerable irreproducibility among different laboratories and even in the same laboratory. The rates are remarkably sensitive

[a] R. Gomer, *J. Chem. Phys.*, **19**, 284 (1951).

[b] The first monograph that summarized the early work was C. N. Hinshelwood and A. T. Williamson, "The Reaction between Hydrogen and Oxygen," Oxford University Press, New York, 1934. Semenoff, "Chemical Kinetics and Chain Reactions," *op. cit.*, extends this review considerably. G. von Elbe and B. Lewis, *J. Chem. Phys.*, **10**, 366 (1942), discuss the kinetics in great detail and, in their text on explosions (*loc. cit.*), cover the work done through about 1950.

to the surface of the vessel[a] and show an autocatalysis attributable to H_2O formation which has been explained in terms of a decrease in chain-breaking efficiency of the surfaces with increasing vapor pressure of H_2O.[b] For these reasons, the various efforts which have been made to treat the

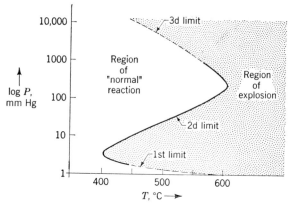

Fig. XIV.4. Explosion limits of a stoichiometric mixture of $H_2 + O_2$ in a KCl-coated spherical vessel (7.4 cm diameter). Explosive region lies in shaded region to right of critical boundary; first limit is somewhat erratic. Dashed parts of curve represent extrapolations.

kinetics in a quantitative manner must be taken with some skepticism. The qualitative and some quantitative features of the kinetic system can be established, however, and the thermodynamics of the free radicals and unstable intermediates which are known to be present are of considerable help in elucidating the kinetic behavior.

Table XIV.1 summarizes the thermodynamic data for the species known to exist in a reacting mixture of $H_2 + O_2$ in the temperature range 400 to 600°C.[c] The simplest kinetic scheme which can be used to account for both the explosion limits and the stationary-state reaction is the following:

[a] Freshly prepared quartz vessels show rates of H_2O formation which may be tenfold or more greater than those in KCl-coated vessels. Even the "same" vessel shows a slow drift with usage, so that the rates may decrease by a factor of 2 over a period of 10 days [see R. R. Baldwin, *Trans. Faraday Soc.*, **52**, 1337, 1344 (1956)].

[b] The kinetic complexity of the system is sufficiently great that the numerous studies which have been made in vessels coated with various salts such as KCl, KOH, H_3BO_3, Na_2WO_4, $BaCl_2$, and Ag all raise as many questions as they attempt to solve. In particular many of these surfaces are appreciably volatile at the temperatures of the experiments, thereby contributing to possible gas-phase reactions, particularly with H_2O. Also, many of the surfaces are unstable and must be renewed. Finally, almost all of them show some amount of steady "aging" and some tendency to erratic behavior. It is very likely that this difficulty of obtaining reproducible results has been one of the attractions of this elusive system.

[c] The only serious difficulties with the use of these data lie in the uncertainties in the H_f° for HO and, particularly, HO_2.

TABLE XIV.1. THERMODYNAMIC DATA FOR SPECIES OF IMPORTANCE
IN THE SYSTEM $2H_2 + O_2 \rightleftharpoons 2H_2O$ (GAS PHASE)

Species	H_f°, Kcal/mole	S_p°, cal/mole-°K	C_p°, cal/mole-°K
H_2	0.00	31.2	4.97
O_2	0.00	49.0	7.02
H_2O	−57.8	45.1	8.03
H_2O_2	−31.8	54[a]	8.5[a]
O_3	34.0	56.8	9.12
H	52.1	27.4	4.97
O	59.2	38.5	5.24
HO	8.0	43.9	7.14
HO_2	12[a]	56[a]	8.5[a]

Data from *Nat'l Bur. Standards Circ.* 500. For substances at 25°C, 1 atm, ideal gases.
[a] Estimates by author. The entropy estimates are not likely to be in error by more than ±2 e.u. The $H_f^\circ(HO_2)$ is uncertain; values estimated in the literature range from 16 to 2 Kcal/mole.

Initiation: $\qquad H_2 + \text{wall} \xrightarrow{k_{iw}} W \cdot H + H$

Propagation: $\qquad H + O_2 \xrightarrow{1} HO + O - 15.3 \text{ Kcal}$

$\qquad\qquad\qquad O + H_2 \xrightarrow{2} HO + H + 1 \text{ Kcal}$

$\qquad\qquad\qquad HO + H_2 \underset{4}{\overset{3}{\rightleftharpoons}} HOH + H + 13.8 \text{ Kcal}$

Chain Transfer (or
Termination): $\quad H + O_2 + M \xrightarrow{5} HO_2 + M + 40 \text{ Kcal}$ $\qquad$ (XIV.7.1)

Termination: $\qquad H + \text{wall} \xrightarrow{k_{vH}} \text{stable species}$

$\qquad\qquad\qquad HO_2 + \text{wall} \xrightarrow{k_{wp}} \text{stable species}$

$\qquad\qquad\qquad HO + \text{wall} \xrightarrow{k_{wOH}} \text{stable species}$

There are a number of features unique to the branching chain which are worth noting. If we consider just the branching part of the chain mechanism, we see that the net reaction (reactions 1 + 2) is $3H_2 + O_2 \rightarrow 2H_2O + 2H$, which is exothermic to the extent of only 5.8 Kcal/mole of H_2O. In fact it may be observed that there is no combination of chain steps which will produce water with more heat output than this figure. The true heat of the over-all molecular reaction is not realized without including termination reactions.[a] This will be a general feature of most branching chain reactions.

[a] The significance of this for the explosion limit is that at the first limit, where termination is principally at the walls, there should be little or no heating of the gas. At the second limit, where termination is principally by reaction 5 in the gas phase, there should be considerable heating of the gas and appreciable temperature gradients should be established in the system.

The stable species referred to in the termination reactions are presumed to arise from reactions of the atoms and free radicals with similar species already adsorbed at the walls.[a] From results of the studies of the low-temperature reaction (400 to 500°C) on silica surfaces which give rates approximately first order in H_2 and zero order in O_2,[b] it seems quite reasonable to suppose that active sites on the walls are covered with O_2[c] and that H_2 can react with them to form a "hydride" (W·H in initiation step) and liberate an H atom into the gas phase. For the explosion limits the initiation reaction is unimportant; however, for the stationary reaction it is an integral part of the rate expression.

From the observation that the first explosion limit is lowered by inert gases such as N_2 and He it is necessary to conclude that, at pressures lower than this limit, termination is predominantly at the surfaces and by a diffusion-controlled mechanism. From the further observation that this limit usually lies in the range 1 to 10 mm Hg for spherical flasks of 5 to 10 cm diameter it is further possible to show, by using our approximate analysis [Eq. (XIV.6.2)], that the efficiency of capture of the wall-terminating radicals is of order of 10^{-4} or larger (i.e., the probability of capture per collision $\epsilon \geqslant 10^{-4}$).

If we apply stationary-state kinetics to the radical concentration, then from the scheme we have[d]

$$\frac{(OH)_{ss}}{(H)_{ss}} = \frac{2k_1(O_2) + k_4(H_2O)}{k_3(H_2) + k_{wOH}} \tag{XIV.7.2}$$

$$\frac{(O)_{ss}}{(H)_{ss}} = \frac{k_1(O_2)}{k_2(H_2)} \tag{XIV.7.3}$$

$$(H)_{ss} =$$

$$\frac{k_{iw}(H_2)}{k_{wH} + k_5(O_2)(M) - [2k_1(O_2) - k_4k_{wOH}(H_2O)/k_3(H_2)]/[1 + k_{wOH}/k_3(H_2)]} \tag{XIV.7.4}$$

[a] Thus if H atoms are adsorbed on active sites on the walls, they can produce H_2 with more H atoms (reverse of reaction 1) or H_2O with OH or $H_2 + O_2$ from HO_2.

[b] C. N. Hinshelwood and C. H. Gibson, Proc. Roy. Soc. (London), **A119**, 591 (1928).

[c] Active sites might be impure metal ions located near the surface, M^{+z} which can reversibly bind O_2, $M_w^{+z} + O_2(g) \rightleftharpoons [M \cdot O_2]_w^{+z}$, as a peroxy or superoxy compound. This would represent a reversible oxidation. The reaction with H_2 could then be $[M \cdot O_2]_w^{+z} + H_2(g) \rightarrow [MO_2H]_w^{+z} + H(g)$. Studies of the exchange of H_2 with D_2 at temperatures near 600°C [G. Boati et al., Nuovo Cimento, **10**, 993 (1953)] show that atoms are initiated at the walls of a silica vessel by O_2 diffusing through the walls of the flask!

[d] We ignore here HO_2, which acts, in the scheme shown, as a terminating species. At higher pressures the reaction $HO_2 + H_2 \rightarrow H_2O_2 + H$ must occur, since H_2O_2 is observed in the system as an unstable intermediate. It is destroyed by radical attack, that is, $R + H_2O_2 \rightarrow RH + HO_2$, so that it cannot be considered to play a very essential role in the mechanism beyond that accorded to HO_2. P. A. Giguère has in fact recently shown that H_2O_2 acts as an inhibitor at the second explosion limit.

This gives for the stationary-state production of H_2O

$$\frac{d\,(H_2O)}{dt} = 2k_1(O_2)(H)_{ss}\frac{1 - k_4k_{wOH}(H_2O)/2k_1k_3(H_2)(O_2)}{1 + k_{wOH}/k_3(H_2)}$$

$$(XIV.7.5)$$

It can be seen that, even for the relatively simple kinetic scheme chosen, the ubiquitous and inconstant surface termination constants are such as to make quantitative analysis of rate data extremely difficult if not impossible.

In the region above the second explosion limit, since all surface reactions are diffusion-controlled and decrease in relative importance with increasing pressure (including k_{iw}), it is likely that the terms in k_{wH} and k_{wOH} are small, and the expression can be somewhat simplified by ignoring them. However, at those pressures, additional termination reactions of HO_2 become important,[a] which indicates that water poisons the active centers and inhibits initiation, or that H_2O has an enormous cross section for the transfer reaction 5 (which is not too likely), or finally that homogeneous termination of H and OH replace wall termination and thus continue the importance of the inhibition reaction 4.[b]

From Eq. (XIV.7.4) for $(H)_{ss}$ we find that no stationary reaction is possible when the denominator vanishes, so that the explosion limits are given by

$$[k_{wH} + k_5(O_2)(M)]\left[1 + \frac{k_{wOH}}{k_3(H_2)}\right] = 2k_1(O_2) - \frac{k_4k_{wOH}(H_2O)}{k_3(H_2)}$$

$$(XIV.7.6)$$

or on rearranging

$$\frac{k_{wH}}{2k_1(O_2)} + \frac{k_5(M)}{2k_1} = \frac{1 - k_4k_{wOH}(H_2O)/2k_1k_3(H_2)(O_2)}{1 + k_{wOH}/k_3(H_2)}$$

$$(XIV.7.7)$$

[a] C. R. Patrick and J. C. Robb, *Trans. Faraday Soc.*, **51**, 1697 (1955).

[b] Lewis and Von Elbe, *loc. cit.*, have shown that, whereas most inert gases will raise the second explosion limit (for example, Ar, He, N_2), indicating the continued importance here of diffusion-controlled termination, both H_2O and CO_2 produce marked lowering of the second limit. It appears almost certain that in both cases the effects must be chemical. With CO_2 a likely reaction is

$$H + CO_2 \xrightarrow{7} HO + CO - 20 \text{ Kcal} \qquad \text{or} \qquad H + CO_2 \rightarrow HO\!-\!\dot{C}\!\!=\!\!0$$

where the unstable $HO\!-\!\dot{C}\!\!=\!\!0$ radical can take part in further chain breaking. As we shall see later, it is the reverse of this first reaction 7 which can account for the enormous acceleration of the $CO + O_2$ reaction by H_2.

The importance of the inhibition reaction 4 must be gauged in terms of the chain step 1 which competes for H atoms. It is seen that 1 is 1.5 Kcal more endothermic than 4, so that, unless there are unusual barriers to it, reaction 4 should begin to compete with reaction 1 as soon as even a few per cent of H_2O have been formed. Another inhibition reaction is $O + H_2O \xrightarrow{8} 2OH - 7.0$ Kcal. We have not considered it here because it competes with reaction 2, which is exothermic and might normally be expected to have a lower activation energy ($E_8 \geqslant 7.0$ Kcal).

At sufficiently low pressures near the first explosion limit where reaction 5 is small compared to wall termination and we may ignore the water inhibition (reaction 4), Eq. (XIV.7.7) becomes

$$\frac{k_{wH}}{2k_1(O_2)} \cong \frac{k_3(H_2)}{k_3(H_2) + k_w(OH)} \tag{XIV.7.8}$$

Baldwin[a] has shown that in KCl-coated vessels varying in diameter from 14 to 51 mm, the first explosion limits can be well correlated with Eq. (XIV.7.8) or (XIV.7.7) (omitting H_2O terms) when the proper pressure dependence of k_{wH} and k_{wOH} are allowed for[b] and reasonable values are assumed for the diffusion coefficients for H and OH.

The form of Eq. (XIV.7.7) is sufficient to fit the second and third explosion limits. However, again the quantitative fitting of the constants is practically meaningless because of the appearance of the quite variable surface terms. For this reason also we shall not attempt to discuss the various numerical results which have been obtained for the different rate constants or the assigned activation energies.[c] Extensive studies of the effects of various surfaces on the different explosion limits have been made by Warren,[d] who was not able to assign a specific chemical role to the behavior of the different surfaces studied[e] or to detect significant differences in effects of the wall on the different radicals present.

A considerable amount of work has been done on the effect of sensitizers on the reaction. NO_2 has been particularly studied[f,g] because of its spec-

[a] R. R. Baldwin, *Trans. Faraday Soc.*, **52**, 1337, 1344 (1956). This work covers a range of temperatures and composition including added N_2 as well.

[b] In the first paper Baldwin solves the diffusion equation for cylindrical and spherical flasks and shows that the diffusion-controlled wall termination constants can be approximated to within 5 per cent by $k_w = (A_0 D/r_0^2)/(1 + A_1 B/P)$, where $B = 4L_0(1 - \epsilon/2)/3r_0$, $A_0 = 5.78$, and $A_1 = 2.75$ for cylinders and $B = L_0(1 - \epsilon)/r_0(1 - L_0/r_0)$, $A_0 = \pi^2$, and $A_1 = 3.1$ for spheres, L_0 being the mean free path at 1 atm, ϵ the probability of sticking to the surface, and D the diffusion constant. [See Eq. (XIV.6.1), which yields similar results.] Baldwin also shows that possible termination of O atoms at the wall cannot be distinguished by the data.

[c] Baldwin, *loc. cit.*, finds values of $E_3 = 9$ Kcal and $E_1 = 15$ Kcal. Since his various assumptions and rate data themselves put a spread of 10 to 20 per cent on the rate constant, the values for E are probably not better than ± 3 Kcal.

[d] D. R. Warren, *Trans. Faraday Soc.*, **53**, 199, 206 (1957).

[e] Boric oxide seems to be an exception in that it has a remarkable accelerating effect on the rate. It reduces the first limit to almost unobservably low pressures and causes the third limit to become very violent. This seems to be related to the known stability of H_2O_2 in B_2O_3, and Warren ascribes the effects to the lack of efficiency of the surface for destroying HO_2. In view of the fast reaction produced at the first limit, it seems more likely that B_2O_3 is an effective catalyst for the production of HO_2 from $H_2 + O_2$.

[f] H. W. Thompson and C. N. Hinshelwood, *Proc. Roy. Soc. (London)*, **A124**, 219 (1929).

[g] F. S. Dainton and R. G. W. Norrish, *ibid.*, **A177**, 393, 421 (1941).

tacular effect in inducing explosion in $H_2 + O_2$ mixtures at points well outside the explosion peninsula. The sensitization is complicated by the fact that ignition is confined to a small but definite range of NO_2 pressures. Below pressures of about 0.05 to 0.1 mm Hg (in $H_2 + O_2$ at about 300 mm Hg) there is no ignition, and that is so also above 5 to 10 mm Hg (same system). These limits of sensitization are affected by vessel diameter, total pressure, and temperature. Figure XIV.5, taken from the work of Dainton and Norrish, shows the limits of the ignition region in very small vessels. Ashmore[a] has shown that, although they vary in their induction periods, the ignition boundaries are much the same for NO_2, NO, NOCl, or mixtures of $NO_2 + NO$.

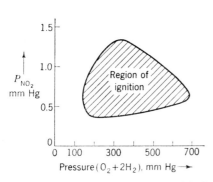

FIG. XIV.5. Variation of NO_2-sensitized ignition limits of $H_2 + O_2$ explosions with composition. Cylindrical vessel was 7 mm in diameter (pyrex, KCl-coated) at 364°C. [From work of F. S. Dainton and R. G. W. Norrish, *Proc. Roy. Soc. (London)*, **A177**, 393, 421 (1941).]

In view of the fact that this sensitization takes place in a region in which chain ending is presumed to be largely by HO_2 and O_3 production,[b] it is likely that these sensitizers act by converting HO_2 into an active radical by reactions such as $R + HO_2 \rightarrow RO + HO$. The suppression of the ignition at higher pressures of sensitizer is then probably due to the increased termination afforded by the very efficient processes.[c]

$$NO + O + M \xrightarrow{9} NO_2 + M$$

$$NO_2 + O \xrightarrow{10} NO + O_2$$

$$NO_2 + O + M \xrightarrow{11} NO_3 + M$$

as well as
$$NO + OH + M \rightarrow HNO_2 + M$$
$$NO_2 + OH + M \rightarrow HNO_3 + M$$

At the temperatures in question the equilibria lie in favor of NO formation from the sensitizers (that is, $NOCl \rightarrow NO + \frac{1}{2}Cl_2$, $NO_2 \rightarrow NO + \frac{1}{2}O_2$)

---

[a] P. G. Ashmore, *Trans. Faraday Soc.*, **51**, 1090 (1955).

[b] O_3 and O are almost certainly in rapid equilibrium at all temperatures and pressures above 10 mm Hg. From the data of Table XIV.1 it can be shown that at about 100 mm Hg pressure of O_2 or higher the $[(O_3)/(O)]_{ss}$ ratio is significantly greater than 1 below 430°C and less than 1 above 430°C.

[c] H. W. Ford et al., Table XII.9, have shown that the ratio of the rate constants at 25°C and 1 atm pressure are $k_9:k_{10}:k_{11} = 1.5:4.0:10$, with reaction 9 occurring at about every triple collision. Thus the activation energy of reaction 10 is, at most, 3 Kcal.

so that their similarity in behavior might be taken to indicate that the active species is NO. This is probably a much oversimplified version of the reaction system.

Sensitization has also been produced by NH_3,[a] C_2N_2,[b] H atoms,[c] O atoms,[d] Cl atoms,[e] and quartz surfaces.[f] Small amounts of iodine show quite spectacular inhibition of the reaction,[g] presumably by virtue of the well-known chain-suppressing behavior of I_2: $R + I_2 \rightarrow RI + I$, the I atoms being quite inert.

8. The Reaction of CO and O_2

Although the system $CO + O_2$ appears to be a simpler one chemically, it is kinetically at least as complex as the $H_2 + O_2$ system and qualitatively shows the same chain-branching features. The existence of a first explosion limit was observed by Sagulin[h] and Garner and Gomm,[i] while the second limit was discovered by Sagulin, Kopp, Kowalsky, and Semenoff.[j] What is particularly striking about the system is the very high explosion temperature. In quartz vessels, the explosion peninsula starts at about 600°C and extends to 800°C at pressures of about 600 mm Hg.

The features of the explosion peninsula are still at issue[k] because of the difficulties associated with minute amounts of H_2O and irreproducible surface effects.[l] In common with most explosions, light is produced; in fact

[a] L. Farkas et al., Z. Elektrochem., **36**, 711 (1930). H. S. Taylor and D. J. Salley, J. Am. Chem. Soc., **55**, 96 (1933).

[b] A. T. Williamson and N. J. T. Pickles, Trans. Faraday Soc., **30**, 926 (1934).

[c] B. Lewis, J. Am. Chem. Soc., **55**, 4001 (1933).

[d] A. Nalbandjan, Acta Physicochim. U.R.S.S., **1**, 305 (1934). F. Haber and F. Oppenheimer, Z. physik. Chem., **B16**, 443 (1932).

[e] R. G. W. Norrish, Proc. Roy. Soc. (London), **A135**, 334 (1931).

[f] F. Haber and H. N. Alyea, Z. physik. Chem., **B10**, 193 (1930).

[g] W. L. Garstang and C. N. Hinshelwood, Proc. Roy. Soc. (London), **A130**, 640 (1931). Organic iodides and HI are also effective; bromides and Br_2 somewhat less so. As little as 0.002 per cent of I_2 can completely eliminate the explosion peninsula.

[h] A. B. Sagulin, Z. physik. Chem., **B1**, 275 (1928).

[i] W. E. Garner and A. S. Gomm, Trans. Faraday Soc., **24**, 470 (1928).

[j] D. Kopp et al., Z. physik. Chem., **B6**, 307 (1930).

[k] A. S. Gordon and R. H. Knipe, J. Phys. Chem., **59**, 1160 (1955), claim that very small amounts of water (e.g., 0.004 per cent) can lower the second explosion limit by 100°C in 10-cm quartz vessels. This has been observed qualitatively by D. E. Hoare and A. D. Walsh, Nature, **170**, 838 (1952), and G. von Elbe et al., "Fifth Symposium (International) on Combustion," p. 610, op. cit. Similar behavior in detonation waves has been reported by G. B. Kistiakowsky et al., J. Chem. Phys., **20**, 994 (1952).

[l] G. Hodman et al., Proc. Roy. Soc. (London), **A137**, 87 (1932), observed that the second limit could not be observed by making up a high-pressure mixture and then reducing the pressure to the second limit. They used instead the method of rapid heating of a known mixture, observing the temperature at which explosion occurs. Gordon and Knipe were able to observe explosion on evacuation by using very dry mixtures. However, they report very erratic results and also relied on the method of heating.

the CO-O_2 flame is one of the most luminous of all known flames.[a] It has been shown by Walsh[b] after a suggestion by Gaydon[c] that the principal emitter is excited CO_2. Added gases such as CO_2 and N_2 raise the second explosion limit, although not markedly. In systems containing small amounts of H_2O they appear to have no effect.

Because of the appearance of a second explosion limit it is necessary to find a branching chain reaction for the system, and two have been proposed. Semenoff (*loc. cit.*) proposed excited CO_2, formed from $O + CO \rightarrow CO_2^*$, as the chain carrier with the branching reaction being

Chain:
$$\begin{cases} CO_2^* + O_2 \xrightarrow{\;1\;} CO_2 + 2O \\ O + CO \underset{3}{\overset{2}{\rightleftharpoons}} CO_2^* \end{cases}$$

Termination:
$$\begin{cases} CO_2^* + M \xrightarrow{\;4\;} CO_2 + M \\ CO_2^* \underset{\text{walls}}{\overset{k_{tw}}{\longrightarrow}} CO_2 \end{cases} \qquad \text{(XIV.8.1)}$$

With such a scheme the explosion limits are given by

$$k_{tw} + k_4(M) = k_1(O_2) \qquad \text{(XIV.8.2)}$$

or

$$M_2 = \frac{k_1(O_2) - k_{tw}}{k_4} \qquad \text{(XIV.8.3)}$$

where M_2 represents the total effective pressure at the second limit.

Gordon and Knipe (*loc. cit.*) present further data in support of such a mechanism, including an additional termination step for O atoms at the walls and a termolecular termination step $CO + O + M \rightarrow CO_2 + M$.[d] Lewis and Von Elbe have objected to such a mechanism on the grounds that $O + CO$ must produce an excited triplet state of CO_2 with a very short lifetime for dissociation. Such an argument is by no means decisive, since a lifetime of even 10^{-11} sec for CO_2^* would be sufficient to account for the branching rates observed.[e]

[a] H. Kondratjewa and V. Kondratjew, *J. Phys. Chem.* (*U.S.S.R.*), **9**, 736, 747 (1937), have shown that the quantum yield is as high as one quantum per 125 molecules CO_2. See also *Acta Physicochim. U.R.S.S.*, **6**, 748 (1937).

[b] A. D. Walsh, *J. Chem. Soc.*, 2266 (1953), showed that the discrete emission in the region 3250 to 6250 Å comes from excited CO_2.

[c] A. G. Gaydon, *Proc. Roy. Soc.* (*London*), **A176**, 505 (1940). A. S. Gordon and R. H. Knipe, *J. Chem. Phys.*, **23**, 2097 (1955), have also found a banded absorption spectrum in the explosion which they ascribe to excited CO. They also ascribe the continuum underneath the banded emission to excited CO.

[d] In principle such a termolecular step is not distinguishable from the two fast consecutive steps, reactions 2 and 4.

[e] The alternative mechanisms would require quite unreasonable reactions to be chain breaking, for example, $O_3 + CO + M \rightarrow O_2 + CO_2 + M$ or $CO_2^* + O_2 + M \rightarrow CO_2 + O_2 + M$. At the temperatures in question the steady-state concentration of O_3 is much less than that of O, so that it seems difficult to invoke O_3 as an important intermediate. Also, the formation of O_3 from $O + O_2 + M$ is quite inefficient (Sec. XIII.18).

Lewis and Von Elbe (*loc. cit.*) have proposed the reaction $O_3 + CO \rightarrow CO_2 + 2O$ as the chain-branching step and $O_3 + CO + M$ as the higher-order chain-breaking step. However, from thermodynamic data (Table XIV.1) above 900°K and 200 mm O_2, $(O_3) \ll (O)$, and in addition it is known that the reaction $O_3 + CO$ is very slow.[a] In their most recent work, Von Elbe, Lewis, and Roth (*loc. cit.*) have produced new data which are compatible with Eq. (XIV.8.3), particularly on the effect of the percentage of O_2 on the second explosion limit,[b] although they still defend a revised version of their original scheme. A detailed discussion of the possible electronic states involved in the $CO + O_2$ reaction has been given by Gordon and Knipe (*loc. cit.*) and by Laidler.[c]

Semenoff (*loc. cit.*) quotes the activation energy for the second limit as 35 Kcal. Because of the catalysis by H_2O and the dependence of the limits on the surface, it is difficult to relate this number to any rate constants. The catalytic effect of H_2O and also of H_2[d] must be ascribed to both surface catalysis and chain branching by the very fast reaction sequence

$$O + \begin{Bmatrix} H_2 \\ or \\ H_2O \end{Bmatrix} \rightarrow \begin{Bmatrix} OH + H \\ or \\ 2OH \end{Bmatrix}$$

$$CO + OH \rightarrow CO_2 + H$$
$$H + O_2 \rightarrow HO + H$$

(XIV.8.4)

Even the HO_2 radical in this system, although it can account for termination of branching, is nevertheless an active radical for the oxidation of CO by the probably fast, exothermic reaction

$$HO_2 + CO \rightarrow CO_2 + HO + 76 \text{ Kcal}$$

9. Ignition and Combustion

In our discussions of branching chain and thermal explosions we examined the critical conditions necessary to establish an explosion in a closed system of finite volume. We saw that, once the critical conditions are at-

[a] D. Garvin, *J. Am. Chem. Soc.*, **76**, 1523 (1954), has shown the reaction to be immeasurably slow at 200°C compared to the reaction of $O + CO$.

[b] To maintain their own mechanism they are forced to add reaction 4 and a concomitant chain-breaking step $CO_2^* + O_2 + M \rightarrow CO_2 + O_2 + M$ which seems very far-fetched.

[c] K. J. Laidler, "The Chemistry of Excited States," Oxford University Press, New York, 1955. It should be noted that the Schumann-Runge O_2 bands which were observed in the CO-O_2 flame have been shown by H. G. Wolfhard and W. G. Parker, *Proc. Phys. Soc.* (*London*), **A65**, 2 (1952), to be thermal in origin.

[d] G. Dixon-Lewis and J. W. Linnett, *Trans. Faraday Soc.*, **49**, 756 (1953), have examined the second limit for the system $H_2 + CO + O_2$ at 510 to 580°C. They find CO increases the second limit of $H_2 + O_2$ to a maximum at 90 per cent CO, whereafter it plunges sharply downward :

tained, the system reacts at an accelerating rate which is limited only by the finite supply of reactants. It is now of interest to discuss the problem of what happens when one element of volume in a system of essentially infinite volume (i.e., an open system) is brought to its explosion or ignition limit.

If the system is one capable of a thermal explosion, there are a number of possible results we may intuitively expect. Within the explosive volume element, the temperature climbs exponentially with the heat output from the reaction. However, as the temperature mounts, the volume element expands to maintain the pressure constant. In so doing it tends to cool itself by two processes. The first is the diminution of the reaction rate brought about by the decrease in concentration (for an ideal gas $c = P/RT$) due to thermal expansion. The second arises from the work done in expanding against the essentially constant pressure of the surrounding system.[a] Both of these cooling processes act to raise the requirements for explosion within our initial volume element [Eq. (XIV.3.1)]. If after making due allowance for this, the reaction element remains above its ignition temperature, then its temperature continues to climb at an exponential rate to an explosion.

Meanwhile the neighboring volume elements will find their temperature increasing because of heat conduction, and since their hot boundary is already at the ignition temperature, they will start to explode. The net result is that as soon as any volume element is brought to a critical ignition limit in the open system, a pressure wave will propagate through the system at the speed of sound followed by a slower temperature wave which propagates through the medium[b] at a rate determined by the rate of heat evolution by the reaction and the thermal conductance of the system. As the wave spreads from the initial source, the reaction rate in the burned gases behind the wave soon reaches a maximum and then decreases to zero. Simultaneously, the temperature reaches a maximum value given by the adiabatic reaction temperature T_b while the density decreases to a minimum.[c] Such a series of events occurring sequentially through a reacting system is described as a combustion wave and, when accompanied by luminosity, as a flame.

From the foregoing description it would appear that the onset of ignition in any volume element of a large enough system is sufficient to es-

---

[a] Since pressure differences are dissipated at essentially the speed of sound and both matter and heat are transferred by much slower diffusion processes, for volume elements whose linear dimensions are very much larger than the mean free path, we can consider their thermal expansion to be adiabatic.

[b] The driving force for this wave is the heat developed by the reaction while the inertial resistance is the reciprocal of the thermal diffusivity $\rho c_p/K$.

[c] For the completely burned gases at final temperature T_b, their molar concentration is given by $n_b = T_u n_u/T_b$, where n_u is the original molar concentration of reactants at T_u (assuming constant pressure).

tablish a self-propagating combustion wave in that system.[a] In closed systems such as spherical flasks or cylindrical tubes, the influence of the fixed boundaries on the developing combustion wave is such that the wave may be dissipated (i.e., quenched) before it has propagated through the system. In an open system it is possible for the wave to reach a constant velocity of propagation from the initial ignition. The common gas flame of a bunsen burner is a somewhat complex example of such a stationary-state combustion wave.

When steady-state flame propagation occurs, the chemical reaction is found to be confined to a very narrow region of thickness δ_f which is of the order of magnitude of $v_f t_f$, where t_f is the half-life of the reaction at the mean flame temperature T_f and v_f is the linear speed of propagation of the flame. For this steady state to exist, the time of diffusion of heat across the flame zone δ_f must be of the same order of magnitude as the half-life of the reaction t_f and so, from the diffusion equation,

$$\delta_f^2 \sim \frac{K}{\rho c_p} t_f \qquad (XIV.9.1)$$

where $K/\rho c_p$ is the thermal diffusivity. On replacing δ_f by $v_f t_f$ we have for the stationary-state flame velocity[b]

$$v_f \cong \left(\frac{K}{\rho c_p} \frac{1}{t_f} \right)^{\frac{1}{2}} \qquad (XIV.9.2)$$

If the velocity of the reaction becomes fast enough and the reaction is sufficiently exothermic, the adiabatic expansion of our reacting zone will occur at a linear rate comparable with the velocity of sound. Under such circumstances a sharp pressure wave begins to be built up ahead of the reaction zone, and it can propagate as a "shock wave" of supersonic velocity in the unburned gases.[c] As the shock front passes through the reaction mixture, it produces adiabatic compression. If the temperature in this adiabatically compressed zone behind the shock wave exceeds the

[a] Note, however, that the ignition temperature increases logarithmically with the surface/volume ratio of the initial element of volume [Eq. (XIV.3.18)], so that there is no unique "ignition temperature" even for a system of fixed composition. This further implies that the ignition energy per unit volume will not be a constant but will increase as the volume of the ignited volume element decreases. The question of whether a minimum size volume element is required in order to propagate a combustion wave has not been settled.

[b] A more precise derivation [Ya. B. Zeldovitch and D. A. Frank-Kamenetskii, *Zhur. Fiz. Khim.*, **12**, 100 (1938)] gives

$$v_f^2 \cong \frac{K}{\rho c_p} \frac{1}{t_b} 2n! \left(\frac{RT_b}{E} \right)^{n+1} \left(\frac{T_b}{T_b - T_0} \right)^{n+1}$$

where t_b is the half-life at the maximum flame temperature T_b and n is the order of the reaction.

[c] The shock wave usually has a pressure gradient across it such that the ratio of pressures $p_1/p_2 \geqslant 2$.

ignition temperature, then a new zone of explosion forms, and from it further shocks originate to sustain and further propagate the original shock wave. Such a wave travels through the gas at supersonic velocities characteristic of shocks, essentially dragging the explosion zone behind it.

While it is observed that the velocity of flame propagation in a combustion wave is of the order of magnitude of 10 to 500 cm/sec, the velocity of detonation waves is of the order of 3×10^5 cm/sec, and the transition between them seems to include velocities not attainable as stationary waves.[a]

Although chain-branching explosions have mechanisms for their kinetic behavior that are quite different from purely thermal explosions, the relations between the explosion limits and temperature are formally similar. However, the mechanism of propagation of a branching chain explosion in the form of a slow combustion wave must be concerned with the diffusion of chain carriers rather than heat. Zeldovitch[b] has shown that as an excellent first approximation, the equations for diffusion of heat and matter can be put into a 1 to 1 correspondence such that the gradients of concentration and temperature will be proportional to each other. Under these conditions the formal equations for combustion wave propagation are the same for the two mechanisms of explosion and, quite independently of the chain mechanism, the gradients of concentration and temperature in a flame will be proportional to each other at all points.[c]

10. Stationary Flames

If we consider a combustion wave as an infinite plane moving through a reaction system, then with respect to the plane itself considered as stationary the unburned gases move toward it at a velocity v_u, while far behind it the burned gases leave with a velocity v_b. The difference in velocities is due to the difference in densities of the burned and unburned gases, ρ_b and ρ_u. The law of conservation of mass requires that the mass flow rate across any surface be constant, so that, if v is the linear gas velocity at any point with reference to the stationary flame front, the mass velocity $\dot{m} = \rho v$ is constant at every point and, in particular, far from the flame front on either side

$$\rho_b v_b = \rho_u v_u = \dot{m} \qquad \text{(XIV.10.1)}$$

[a] For a more complete discussion of combustion waves, detonations, and flame stability the reader is referred to the very detailed exposition in the text by Lewis and Von Elbe, *loc. cit.*, and also the text by Frank-Kamenetskii, *loc. cit.* An excellent qualitative description of the initiation of detonations in gases will be found in an article by G. B. Kistiakowsky, *Ind. Eng. Chem.*, **43**, 2794 (1951).

[b] Ya. B. Zeldovitch and D. A. Frank-Kamenetskii, *Zhur. Fiz. Khim.*, **12**, 100 (1938).

[c] On purely physical grounds this is an expected result because the steepest gradients in temperature must occur where the reaction rate is greatest, which in turn produces the largest concentration of products.

Since the density of the burned gases is less than that of the unburned gases, owing to thermal expansion, the velocity of the burned gases must be greater. Thus there is an increase in momentum of the reaction mixture in passing through the flame zone, and this must be compensated for by a pressure drop across the flame zone. From Newton's equations for the rate of change of momentum[a] we have

$$P_u - P_b = -\rho_u v_u (v_u - v_b) \qquad (XIV.10.2)$$

or on substituting for v_b from Eq. (XIV.10.1)

$$P_u - P_b = \rho_u v_u^2 \left(\frac{\rho_u}{\rho_b} - 1 \right) \qquad (XIV.10.3)$$

For values of these parameters, typical of most steady-state combustion flames [for example, $\rho_u \cong 1.6$ g/liter(STP), $v_u \cong 100$ cm/sec, and $\rho_u/\rho_b \cong 5$], we find $P_u - P_b \cong 0.05$ mm Hg, so that as a reasonable approximation we may neglect this small pressure gradient across the reaction zone of a stationary flame.[b]

Let us consider regions that are far from the flame zone and in which gradients vanish and equilibrium has been achieved. From the first law of thermodynamics we have $\Delta E = \Delta Q - \Delta W$. But the work done by the gas in crossing the flame zone is $P_b V_b - P_u V_u$ per unit mass, where V_b and V_u are the specific volumes. Thus $\Delta E = E_b - E_u = \Delta Q - (P_b V_b - P_u V_u)$ or $(E_b + P_b V_b) - (E_u + P_u V_u) = \Delta H_u = \Delta Q$.

If the combustion can be considered as adiabatic (that is, $\Delta Q = 0$), then it is also isoenthalpic, or $\Delta H = 0$. This is a reasonable assumption for most flames, since radiation is usually quite small compared to combustion. This permits the temperature of the completely burned gases to be determined from the composition and temperature of the reacting mixture and the enthalpy change of the over-all reaction. (Note that we have neglected kinetic energy terms.)

$$\Delta H = H_b - H_u = 0$$
$$= \sum_b (n_i H_i^\circ) - \sum_u (n_i H_i^\circ) + \sum_b \int_{T_a}^{T_b} (n_i M_i c_{pi}) \, dT$$

$$(XIV.10.4)$$

where n_i represents the number of moles of the ith species per unit mass of reaction mixture of molecular weight M_i, molar enthalpy H_i° (at initial temperature T_u), and specific heat at constant pressure c_{pi}. Equation (XIV.10.4), which expresses the law of conservation of energy for the sys-

[a] We are neglecting frictional forces due to viscosity or turbulence, which are quite small.

[b] Unless, of course, we are specifically interested in it, or in the dynamical motion of the gases near the zone. Even for very dense gases in very hot flames, the pressure drop seldom exceeds 1 mm Hg at $P_u = 1$ atm.

tem, can be used to determine the final temperature T_b if the composition of the burned gases is known or is derivable from independent data.[a]

If instead of an element of fixed mass we consider an element in the system of fixed volume and fixed position, then its change in enthalpy $H\rho$ per unit time is given by the equation (neglecting pressure changes)

$$\frac{\partial(H\rho)}{\partial t} = \frac{\partial}{\partial x}\left(\rho \sum H_i n_i \dot{g}_i\right) - \frac{\partial}{\partial x}\left(K \frac{\partial T}{\partial x}\right) \qquad (XIV.10.5)$$

$$= \frac{\partial}{\partial x}\left[\rho \sum H_i n_i(v + v_{iD})\right] - \frac{\partial}{\partial x}\left(K \frac{\partial T}{\partial x}\right) \qquad (XIV.10.6)$$

where $\dot{g}_i$ represents the local space velocity (cm/sec) of the ith species and K is the coefficient of thermal conductivity.[b] The local velocity $\dot{g}_i$ can be split into two parts: v, the average macroscopic, space velocity of the mixture, and v_{iD}, the local space velocity of the ith species relative to the gas. The component v_{iD} is the part of the total velocity of the ith species that arises from the diffusion of the species in the reacting mixture. It is determined in part by the concentration gradient of the species,[c] which is in turn determined by the equation for conservation of species

$$\frac{\partial(n_i\rho)}{\partial t} = \frac{\partial}{\partial x}[\dot{g}_i n_i \rho] - a_i \dot{R} \qquad (XIV.10.7)$$

where a_i are the stoichiometric coefficients[d] in the balanced equation for the over-all reaction and $\dot{R}$ is the local specific rate of reaction, and a complex function of concentration and temperature. In the stationary state $\partial n_i\rho/\partial t = 0$, so that

$$\frac{\partial(g_i n_i \rho)}{\partial x} = a_i \dot{R}_i \qquad (XIV.10.8)$$

The second term on the right-hand side of Eq. (XIV.10.5) represents the net change in heat flux across any volume element that is due to heat

[a] For very hot flames (that is, $T_b > 1500°K$) such computations, which can be made from tables of thermodynamic data, can become extremely laborious, since no explicit equation for T_b can be written. Instead, T_b itself depends on the various equilibria among the product species; at the temperatures concerned, the product species may include appreciable concentrations of free radicals. Such calculations are described by S. R. Brinkley, Jr., in "Combustion Processes," vol. II, chap. C, Princeton University Press, Princeton, N.J., 1956.

[b] We are ignoring radiation losses, viscosity losses, and any possible turbulence. For most flames these are quite fair approximations.

[c] For a more complete discussion of the one-dimensional flame see J. O. Hirschfelder and C. F. Curtiss, "Third Symposium on Combustion and Flame and Explosion Phenomena," The Williams & Wilkins Company, Baltimore, 1949, and also J. O. Hirschfelder et al., "Fourth Symposium (International) on Combustion," The Williams & Wilkins Company, Baltimore, 1953. The local velocity v_{iD} will also depend on gradients of temperature (Sec. VIII.10).

[d] That is, $a_1A_1 + a_2A_2 + \cdots \rightleftharpoons a_nP_n + a_{n+1}P_{n+1} + \cdots$. The specific rate $\dot{R}$ is defined as $|\dot{R}| = (1/a_i)(dn_i/dt)$, which is the same for all species.

conduction. The first term on the right-hand side represents the net change in enthalpy flux across such a volume element that arises from convection v and diffusion v_{iD} of matter across the boundaries of the element [Eq. (XIV.10.6)]. In the stationary state $\partial H\rho/\partial t = 0$, and these two fluxes must be equal to each other. The equation can then be integrated directly to give

$$v\rho \sum H_i n_i \left(1 + \frac{v_{iD}}{v}\right) - K \frac{\partial T}{\partial x} = h \qquad \text{(XIV.10.9)}$$

The constant of integration h in this equation can be determined by reference of these fluxes to either the burned or unburned gases sufficiently far from the flame zone that all gradients are zero, so that $v_{iD} = 0$ and $\partial T/\partial x = 0$. Thus at the two ends, $h = \rho_u v_u \sum_u H_i n_i = \rho_b v_b \sum_b H_i n_i$. With reference to the unburned gases we can thus write Eq. (XIV.10.9) as (note that $\rho v = \dot{m} = $ const):

$$\dot{m} \sum \left[H_i n_i \left(1 + \frac{v_{iD}}{v}\right) - H_i^\circ n_i^\circ \right] = K \frac{\partial T}{\partial x} \qquad \text{(XIV.10.10)}$$

This equation expresses the stationary condition that the two fluxes of enthalpy and heat must be in balance at every point in the flame profile. The summation on the left can be looked upon as a current of excess enthalpy moving toward the flame front with the flame speed $\dot{m}$ and compensated by a heat flux $K(\partial T/\partial x)$ moving in the opposite direction from the hotter to the colder gases. This excess enthalpy comes from the conduction, convection, and diffusion losses of the volume elements in and near the flame zone.[a]

If now we substitute from Eq. (XIV.10.8) into Eq. (XIV.10.6), we have, assuming a stationary state (that is, $\partial H\rho/\partial t = 0$):

$$\frac{\partial}{\partial x}\left(K \frac{\partial T}{\partial x}\right) = \sum \rho n_i \dot{g}_i \frac{\partial H_i}{\partial x} + \dot{R} \sum a_i H_i \qquad \text{(XIV.10.11)}$$

or, assuming K constant, setting $dH_i = M_i c_{pi} dT$, and using total derivatives in place of partials (since x is the only independent variable in the stationary state):

$$K \frac{d^2 T}{dx^2} = \frac{dT}{dx} \sum \rho n_i \dot{g}_i c_{pi} M_i + \dot{R} \Delta H \qquad \text{(XIV.10.12)}$$

$$= \frac{dT}{dx} \sum \rho_i \dot{g}_i c_{pi} - \dot{q} \qquad \text{(XIV.10.13)}$$

$$= \dot{m} \bar{c}_p \frac{dT}{dx} - \dot{q} \qquad \text{(XIV.10.14)}$$

[a] At the regions sufficiently far removed it vanishes along with all gradients. Note that, if we assume the temperature to vary monotonically across the flame, $\partial T/\partial x$ always has the same sign. This implies that the net sum of the diffusion currents v_{iD} (which can be positive or negative) do not change the sign of the excess enthalpy.

where $\Delta H = \Sigma\, a_i H_i$ represents the molar enthalpy change in the reaction (negative for exothermic reactions) while the positive quantity $\dot{q}$ represents the local rate of heat evolution by the reaction and $\bar{c}_p$ is a weight average specific heat.[a] Because of the complex dependence of $\dot{q}$ on concentration and temperature, the formal, explicit integration of Eq. (XIV.10.14) is not possible. However, by means of various approximations it is possible to obtain useful information from this equation.[b]

Figure XIV.6 shows a schematic, not-to-scale representation of the different variables across the profile of a one-dimensional, stationary flame.

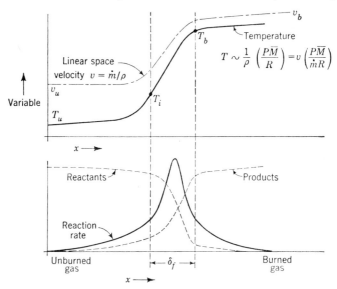

Fig. XIV.6. Schematic representations of the variables in a one-dimensional stationary flame. δ_f represents the thickness of the flame zone in a stationary flame, presumed to be bounded at one end by the ignition temperature T_i and at the other end by lack of reactants. Note that the product of the density and linear space velocity is equal to the constant mass velocity [Eq. (XIV.10.1)]. Also, if the mole change in the combustion can be neglected, then $\rho T = P\overline{M}/R$ const. so that the density is inversely proportional to the absolute temperature and $v/T \cong$ const.

The ignition temperature T_i marks the front of the flame zone, here represented by the thickness δ_f. Up to this point the reaction rate, though relatively fast, is not self-accelerating.[c] Past this point the rate of heat

[a] By definition of $\dot{g}_i$, $\Sigma\rho_i\dot{g}_i c_{pi} = v\Sigma\rho_i c_{pi}(1 + v_{iD}/v) = \rho v\Sigma(\rho_i/\rho)c_{pi}(1 + v_{iD}/v) = \dot{m}\Sigma c_{pi}(\rho_i/\rho)(1 + v_{iD}/v) = \dot{m}\bar{c}_p$, so that $\bar{c}_p$ is a simple weight average specific heat only if the diffusional velocities v_{iD} are small compared to the local space velocity v (which is reasonable) or else their sums cancel.

[b] See Lewis and Von Elbe, loc. cit., for a discussion of some of the qualitative aspects of these approaches.

[c] In comparing Eq. (XIV.10.14) with Eq. (XIV.2.1), which considers heat conduction in a vessel of fixed volume, note that there is an extra term in the former that arises from the cooling of the system due to convection (i.e., the term in $\dot{m}$).

evolution exceeds cooling and the temperature and reaction rate rise at an exponential rate limited only by the finite supply of reactants.

If we assume $\bar{c}_p$ constant, we can integrate Eq. (XIV.10.14) across the flame zone δ_f and obtain

$$\int_{\text{zone}} \dot{q} \, dx = \langle \dot{q}_f \rangle \, \delta_f = -K \frac{dT}{dx}\Big]_{\delta_a}^{\delta_b} + \dot{m}\bar{c}_p \, \Delta T_f \quad \text{(XIV.10.15)}$$

where $\langle \dot{q}_f \rangle = |\Delta H| \cdot \langle R_f \rangle$ represents the average rate of heat evolution in the flame zone and ΔT_f represents the temperature change across the flame zone. If now we assume that the temperature gradient is the same at the two ends of the flame zone,[a] then the conduction term in Eq. (XIV.10.10) vanishes and

$$\dot{m} = \frac{\langle \dot{q}_f \rangle \, \delta_f}{\bar{c}_p \, \Delta T_f} = \delta_f \left[\frac{|\Delta H|}{\bar{c}_p \, \Delta T_f} \right] \langle \dot{R}_f \rangle \quad \text{(XIV.10.16)}$$

$$= \delta_f \left[\frac{|\Delta H|}{\bar{M}\bar{c}_p \, \Delta T_f} \right] \bar{M} \, \langle \dot{R}_f \rangle \quad \text{(XIV.10.17)}$$

where $\bar{M}$ = the mean molecular weight. But in the absence of any net conduction (that is, $dq = 0$, since $\partial T/\partial x$ is the same at the two boundaries of the flame zone), $|\Delta H| = \bar{M}\bar{c}_p \, \Delta T_f$ = the change in temperature of the flame due to reaction, so that

$$\frac{\dot{m}}{\bar{M}} = \bar{\dot{M}} = \delta_f \langle \dot{R}_f \rangle \quad \text{(XIV.10.18)}$$

or the mean molar space velocity $\bar{\dot{M}}$ across the flame zone is equal to the zone thickness times the mean reaction rate, a result which was deduced earlier [Eq. (XIV.9.1)] from simpler arguments.

The integration of Eq. (XIV.10.14) between the unburned gas at $T = T_u$ where all gradients vanish and any arbitrary point near the flame zone leads to

$$K \left(\frac{dT}{dx} \right)_x = \dot{m}\bar{c}_p(T - T_u) - \int \dot{q} \, dx \quad \text{(XIV.10.19)}$$

$$= \dot{m}\bar{c}_p(T - T_u) - |\Delta H| R(x) \quad \text{(XIV.10.20)}$$

where $R(x)$ is the integrated rate along the wave up to the point x.

If we compare this equation with Eq. (XIV.10.10), we see that the terms on the right-hand side of Eq. (XIV.10.20) must correspond to the excess enthalpy of Eq. (XIV.10.10). In particular, if we restrict ourselves to regions in front of the flame zone where no appreciable reaction has occurred [that is, $R(x) = 0$], the excess enthalpy is $\dot{m}\bar{c}_p(T - T_u)$, which corresponds

---

[a] This may not be the case if T_i, the ignition temperature, coincides with the inflection point in the temperature curve where $\partial T/\partial x$ has a maximum. Not much violence is due in this case by moving the boundary toward a more symmetrical position, the distances all being relatively small. Actually, it seems likely that the gradient of temperature is nearly constant across the flame zone.

precisely to the total amount of heat received by the unburned gases at the point (x, T) by conduction from the flame zone.

If we assume that the total rate of heat evolution by reaction of the unburned gases at the leading edge of the flame (temperature $= T_i$) is negligible compared to the excess enthalpy of these gases due to conduction, then we can write Eq. (XIV.10.20) as

$$K \left(\frac{dT}{dx} \right)_{x_i} \cong \dot{m} \bar{c}_p (T_i - T_u) \qquad \text{(XIV.10.21)}$$

If we further assume that the temperature gradient at this edge of the flame zone is the same as the average gradient through the flame, $\Delta T_f / \delta_f = (\partial T / \partial x)_{x_i}$, we can substitute this in the above equation, and on replacing δ_f by its value from Eq. (XIV.10.18),

$$\dot{m} = \left[\frac{K}{\bar{c}_p} \frac{T_i - T_u}{T_f} \overline{M} \langle \dot{R}_f \rangle \right]^{\frac{1}{2}} \qquad \text{(XIV.10.22)}$$

or if we assume that burning is complete in the flame so that $T_f = T_b - T_i$ and substituting $v_u = \dot{m}/\rho_u$, $\bar{n}_u = \overline{M}/\rho_u$:

$$v_u = \left[\frac{K}{\rho_u \bar{c}_p} \frac{T_i - T_u}{T_b - T_i} \bar{n}_u \dot{R}_f \right]^{\frac{1}{2}} \qquad \text{(XIV.10.23)}$$

a result which is again quite close to that deduced earlier for the flame velocity [Eq. (XIV.9.2)][a] by plausibility arguments.

There have been numerous attempts to compare equations such as (XIV.10.23) with experiment[b] and also to obtain other equations which involve explicitly the mechanism of flame propagation by diffusion of free radicals.[c] It is doubtful, however, if the theory as currently formulated [for example, Eq. (XIV.10.23)] can be used for anything other than qualitative or semiquantitative prediction, particularly in view of the complex-

[a] A similar procedure was used by E. Mallard and H. LeChatelier, *Ann. mines*, **8**, series 4, 174 (1883), and subsequently elaborated on by F. Crussard et al. For an excellent critique of these "conduction" theories see W. Jost, "Explosion and Combustion Processes in Gases," translated by H. O. Croft from 1936 edition, McGraw-Hill Book Company, Inc., New York, 1946.

[b] A number of attempts are presented in detail among the papers covered by the "Fourth Symposium (International) on Combustion," *op. cit.*

[c] C. Tanford and R. N. Pease, *J. Chem. Phys.*, **15**, 861 (1947), have developed a model in which propagation occurs by diffusion of radicals. Since wall initiation of chains is not possible in most flames, it seems quite reasonable to expect that initiation by diffusion of chain carriers may be a necessary condition for propagation. It is difficult, however, to see this as a limiting or controlling condition even in fast flames, particularly for branching chain reactions in which, because of the rapid multiplication of centers, even cosmic rays can act as initiations once ignition temperatures have been reached.

ity of real flames and the lack of quantitative knowledge of the kinetics of reactions at the temperatures in question.[a]

Aside from this, the data on burning velocities seem to be in almost quantitative accord with the conduction equation (XIV.10.23) when adapted to flames in finite systems such as cylinders and spheres.[b] The velocity of flame propagation in tubes is complicated by the viscous drag exerted by the walls on the flowing gas, together with the heat losses at the walls. The resulting Poiseuille type of flow tends to make the flame fronts parabolic in these systems.

The stability of open flames attached to flame holders (burners) is also complicated by the hydrodynamics associated with the divergent gas stream from the burner nozzle, heat losses to the burner, and convection of the surrounding air. The stability of such flames relative to their burners and the related phenomena of flash back and blow-off are outside the scope of this book.

11. Shock Waves and Detonations

The pressure drop across a stationary flame front is usually so small that to a first approximation we may consider the flame as isobaric [Eq. (XIV.10.3)]. In addition the kinetic energy associated with the pressure drop may be neglected in comparison with that accompanying heat exchanges. However, for a sufficiently rich reaction mixture and a very exothermic reaction, the rate of linear expansion of the gases in the flame front may begin to approach the speed of sound.[c] Under these conditions the pressure wave from the flame, which is dissipated at the speed of sound, is continuously reinforced by the advancing flame front, and a zone of very sharp pressure, density, and temperature change builds up ahead of the reaction zone and moves into the unburned gases with speeds in excess of sound velocities. Such a phenomenon is referred to as a *shock wave*, and when initiated and accompanied by a combustion it is referred to as a *detonation wave*.

In an explosive mixture a stable detonation wave can be pictured as composed of a shock wave followed by a combustion wave, the two being in symbiotic relationship with each other. By compressing and heating the unburned gases ahead of it, the shock front brings the gases to their

---

[a] It is also unfortunately true that our detailed knowledge of specific heats and thermal conductivities both for multicomponent systems and at the temperatures in question is hardly quantitative.

[b] For a fuller treatment of the stability of conventional burner flames, the reader is referred to the treatises by Lewis and Von Elbe and by Jost, *loc. cit.*, and also the "Fourth Symposium (International) on Combustion," *op. cit.*

[c] To a first approximation, the speed of sound in gases may be taken as the molecular velocity.

ignition point, at which the combustion zone in turn is able to develop heat energy fast enough to maintain the shock wave stationary. If we permit a small zone of no reaction between the back of the shock front and the beginning of the combustion zone, the gases in this zone, although considerably hotter than the unshocked gases, may still be more dense by virtue of their great pressure. Consequently, their local space velocity relative to the shock front is less than that of the unshocked gases before the front. The subsequent chemical reaction, although heating the gases further, still leaves the gases with a greater density and thus lower velocity than the initial, unburned gases. Thus with respect to the detonation front, the burned gases leave with a space velocity less than that of the unburned gases, a situation which is just the reverse of that for an ordinary combustion wave. The profile of a one-dimensional detonation wave is shown schematically in Fig. XIV.7.

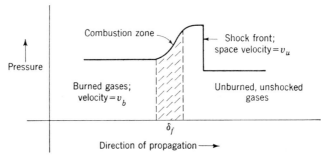

FIG. XIV.7. Pressure profile of a one-dimensional detonation wave. v_u = the velocity of the shock front relative to the unburned gases; $v_b(<v_u)$ is the velocity of the burned gases relative to the unburned gases; δ_f = thickness of combustion zone.

The necessity for a lag between the mechanical shock front and the chemical reaction zone arises from kinetic considerations. In a stationary shock wave traveling through a gas with a supersonic velocity v_u (usually in the range of 10^5 to 10^6 cm/sec), the gradient of density across the shock front is limited by diffusion. The diffusion flux of matter across the shock front of thickness δ_s is given by $D \, \partial\rho/\partial x \cong D \, \Delta\rho/\delta_s$, where D is the average diffusion coefficient in the shock front and $\Delta\rho$ is the change in density. In the stationary state this must be balanced by the mass flow $\rho_u v_u$ into the shock, so that, on solving for δ_s, we find $\delta_s \cong (\Delta\rho/\rho_u)(D/v_u) = (\rho_s/\rho_u - 1) (D/v_u)$. For weak to moderate shocks in which $\rho_s/\rho_u \cong 2$, $D \cong 1$ cm^2/sec and $v_u \cong 10^5$ cm/sec, we find that δ_s is of the order of 10^{-5} cm, which is of the order of magnitude of the mean free path of a molecule at STP.

The time of passage of a molecule through the shock is δ_s/v_u, which is about 10^{-10} sec, that is, the time for about one collision! Since there are no chemical chain reactions which proceed at every collision, this implies that the zone of chemical reaction must be behind the shock front. The time lag between the two aside from possible induction times must be of the order of magnitude of the half-life of the chemical reaction. Since the chemical reaction must keep pace and fuel the advancing shock wave, its half-life must be of the order of magnitude of δ_f/v_{cs}, where δ_f is the thickness of the combustion zone and v_{cs} is of the order of magnitude of the speed of sound in the shocked but unburned gases.

In order to analyze the structure of a detonation wave, we must consider three regions: the unshocked gases, the shocked but unreacted gases, and finally the burned gases behind the reaction zone. From the preceding discussion the structure of the combustion wave behind the shock front should be qualitatively similar to those already considered. The chief difference between ordinary combustion zones and those behind shocks lies in the relatively high temperature and density of the shocked gases being fed into the latter (Figs. XIV.6 and XIV.7). It is therefore of interest to look into the properties of shock waves to see how they may be expected to influence a chemical reaction.

The phenomenon of shock waves in gases seems to have been first predicted by Riemann in 1860 and analyzed quantitatively by Chapman,[a] Jouguet,[b] and then, from a kinetic point of view, Becker,[c] who presented a very picturesque model for the formation of shock waves.

Let us consider in a gas, a very large, flat surface (e.g., a piston) which is uniformly accelerated, in some fixed time t_a, from rest to a final velocity v_b. If we consider the state of the gas in successive increments of time (Fig. XIV.8), we see that each successive increment of motion of the surface imparts to the gas in front of it an excess momentum which is then carried into the gas with molecular velocity, i.e., the speed of sound. However, because of the essentially adiabatic compression occurring in the gas,

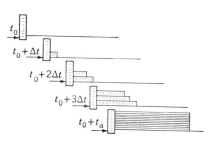

FIG. XIV.8. Becker's model for the formation of a one-dimensional shock wave, showing profile of pressure waves in gas in front of a moving surface accelerating sufficiently rapidly to form a shock wave.

each succeeding momentum wave finds itself moving through a hotter and faster-moving medium and thus with a higher velocity.[d] Consequently, the later waves will overtake the earlier waves at some point in front of the

[a] D. L. Chapman, *Phil. Mag.*, **47**, 90 (1899).

[b] E. Jouguet, *J. math.*, 347 (1905); 6 (1906) (French).

[c] R. Becker, *Z. Physik*, **8**, 321 (1922); *Z. Elektrochem.*, **42**, 457 (1936).

[d] The mean molecular velocity is given by $(8RT/\pi M)^{1/2}$, while the velocity of sound is given by $(\gamma RT/M)^{1/2}$, both increasing as $T^{1/2}$. The velocity also increases because of the increase in space velocity due to the heating.

surface,[a] where they will reinforce each other to form a boundary at which pressure, temperature, and density change almost discontinuously. This is the advancing shock front whose thickness we have seen is of the order of magnitude of a mean free path.

When the surface has attained its terminal velocity v_b, we shall have a stationary shock wave propagating with constant velocity v_u into the gas, followed by a growing column of heated, shocked gas between the surface and the shock front that is moving with the constant velocity v_b of the surface.

The work done on the gas is thus being converted partly into kinetic energy and partly into adiabatic heating. In the stationary state we can calculate the velocity of the shock and the change in state of the gas. To do this let us choose a frame of coordinates moving with the shock front along the x direction. Relative to this front the unshocked gas is moving into the front with a space velocity v_u, density ρ_u, pressure P_u, temperature T_u, and mass velocity $\dot{m} = \rho_u v_u$. Relative to the shock front the shocked gases are moving with a velocity v_s away from the front, so that their velocity relative to the unshocked gases is $v_b = v_s - v_u$ (Fig. XIV.9). The

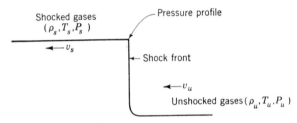

FIG. XIV.9. Pressure profile of a stationary shock wave in a gas.

equation of conservation of mass requires that the mass flow velocity $\dot{m} = \rho v$ be a constant in the stationary state, so that

$$\rho_u v_u = \rho_s v_s = \dot{m} \qquad \text{(XIV.11.1)}$$

while the conservation of momentum gives for the pressure drop across the shock front [Eq. (XIV.10.2)]

$$P_s - P_u = -\rho_s v_s (v_s - v_u)$$

$$= -\rho_s^2 v_s^2 \left(\frac{1}{\rho_s} - \frac{1}{\rho_u} \right) \qquad \text{(XIV.11.2)}$$

From the first law of thermodynamics $\Delta E = \Delta Q - \Delta W - \Delta KE$ where ΔKE is the change in over-all kinetic energy of the system. If we consider a

---

[a] The focusing action of the adiabatic self-heating will bring the pressure wavelets to a focus at a distance in front of the surface which is of the order of magnitude of $t_a v_c$, where v_c is the speed of sound. Unless the final velocity v_b is of the order of magnitude of the speed of sound, the shock will die out.

unit mass of the gas, the total work done by it[a] in crossing the shock boundary $\Delta W = P_s/\rho_s - P_u/\rho_u$. If we neglect radiation and consider conditions sufficiently far from the shock front that conduction is negligibly small, then $\Delta Q = 0$ (i.e., the shock is adiabatic) and $\Delta E = -\Delta W - \Delta KE$. On writing this in terms of a unit mass of gas,

$$E_s + \frac{P_s}{\rho_s} + \frac{v_s^2}{2} = E_u + \frac{P_u}{\rho_u} + \frac{v_u^2}{2} \qquad (XIV.11.3)$$

If we use the preceding equations to eliminate the velocities v_s and v_u from this last equation, we find

$$E_s - E_u = \frac{1}{2}(P_s + P_u)\left(\frac{1}{\rho_u} - \frac{1}{\rho_s}\right) \qquad (XIV.11.4)$$

or in terms of specific volumes $V = 1/\rho$

$$E_s - E_u = \frac{1}{2}(P_s + P_u)(V_u - V_s) \qquad (XIV.11.5)$$

which is known as the Hugoniot equation for shock waves.[b] We can now calculate the temperature of the shocked gases if we know their thermodynamic properties. For perfect gases

$$E_s - E_u = \int_{T_u}^{T_s} c_v \, dT = \bar{c}_v(T_s - T_u)$$

where $\bar{c}_v$ is the average specific heat at constant volume.[c] By using the ideal gas law $P = \rho RT/M$ to eliminate T, this last equation can be written as

$$\frac{P_s}{\rho_s} = \frac{P_u}{\rho_u} + \frac{R}{M\bar{c}_v}(E_s - E_u) \qquad (XIV.11.6)$$

or on substituting for $E_s - E_u$ from Eq. (XIV.11.4) and solving for the ratio ρ_s/ρ_u:

$$\frac{\rho_s}{\rho_u} = \frac{(P_s/P_u)(2M\bar{c}_v/R + 1) + 1}{P_s/P_u + 2M\bar{c}_v/R + 1} \qquad (XIV.11.7)$$

It may be noted that in the limit of very strong shocks, that is, as $P_s/P_u \to \infty$, the density ratio $\rho_s/\rho_u \to 2M\bar{c}_v/R + 1$.[d]

[a] Work is, of course, done on the gas by the moving surface. However, from the point of view of a *fixed* shock front, the gas in moving across the shock front converts part of its kinetic energy into adiabatically compressing itself.

[b] The conservation equations are frequently referred to as the Hugoniot relations.

[c] By choosing volume elements sufficiently far from the shock zone that equilibrium has been established, we avoid the problem of which degrees of freedom of the gas actually respond rapidly enough to contribute effectively to the specific heat. Near the shock front this is a serious problem which can alter significantly the temperature and pressure distribution in that region. Thus there is very likely to be a much higher instantaneous temperature right at the shock front, where vibrational and possible rotational degrees of freedom have not yet achieved their equilibrium importance.

[d] This is not as much of a limit as would appear, since $\bar{c}_v$ increases with increasing temperature when the temperature is sufficiently high. However, for very simple gases such as He $(2\bar{c}_vM/R = 3)$ or N_2 $(2\bar{c}_vM/R = 5)$ which have no low-lying electronic states or rather large dissociation energies, unless $T_s > 4000°K$, the limits are real and are 4 and 6, respectively.

From Eqs. (XIV.11.3) and (XIV.11.4) we find for the velocities

$$v_u = \left(\frac{\rho_s}{\rho_u}\right)^{1/2}\left(\frac{P_s - P_u}{\rho_s - \rho_u}\right)^{1/2} = \left(\frac{\rho_s}{\rho_u}\right)^{1/2}\left(\frac{P_s/P_u - 1}{\rho_s/\rho_u - 1}\right)^{1/2}\left(\frac{P_u}{\rho_u}\right)^{1/2}$$

(XIV.11.8)

$$v_s = \left(\frac{\rho_u}{\rho_s}\right)v_u = \left(\frac{\rho_u}{\rho_s}\right)^{1/2}\left(\frac{P_s/P_u - 1}{\rho_s/\rho_u - 1}\right)^{1/2}\left(\frac{P_u}{\rho_u}\right)^{1/2} \qquad \text{(XIV.11.9)}$$

$$v_b = v_s - v_u = \left(\frac{\rho_u}{\rho_s} - 1\right)v_u \qquad \text{(XIV.11.10)}$$

If we divide v_u and v_s by the respective velocities of sound $(\gamma P/\rho)^{1/2} = v_c$ in the unshocked and shocked gases, we find for the Mach numbers of the gases[a]:

$$\text{Mach}_u = \frac{v_u}{v_{cu}} = \left[\frac{P_s/P_u - 1}{\gamma_u(1 - \rho_u/\rho_s)}\right]^{1/2} \qquad \text{(XIV.11.11)}$$

$$\text{Mach}_s = \frac{v_s}{v_{cs}} = \left[\frac{P_s/P_u - 1}{\gamma_s(\rho_s/\rho_u - 1)}\right]^{1/2} \qquad \text{(XIV.11.12)}$$

Some typical examples of shock properties calculated from the preceding equations are shown in Table XIV.2 for shocks traveling in N_2 at STP, assuming that, in the range of shocks listed, $2\bar{c}_v M/R = 5$. It can be seen that, as the velocity of the moving surface with respect to the gas v_b decreases, the flows at the pressure front decrease toward the speed of sound (Mach 1) and the heating of the gas becomes quite small.[b]

TABLE XIV.2. SHOCK WAVES IN N_2 AT STP

P_s/P_u	ρ_s/ρ_u	$T_s - T_u$, °C	v_u, km/sec	v_s, km/sec	v_b, km/sec	Mach_u	Mach_s
1.1	1.07	8.2	0.354	0.331	0.023	1.047	0.965
2	1.63	63	0.46	0.28	0.18	1.36	0.75
5	2.82	211	0.71	0.252	0.46	2.10	0.56
10	3.81	444	1.00	0.262	0.74	2.96	0.48
50	5.37	2270	2.22	0.413	1.81	6.56	0.40
100	5.67	4530	3.17	0.56	2.61	9.37	0.39

These values are calculated from the equations in the text by assuming a constant molar heat capacity $\bar{C}_v = 2.5R$. (Note the slight minimum in v_s near Mach 2.)

[a] By substituting for $(\rho_s/\rho_u - 1)$ from Eq. (XIV.11.7) it can be shown that $\text{Mach}_u \geqslant 1$, while $\text{Mach}_s \leqslant 1$ if $\rho_s/\rho_u \geqslant 1$, which is actually a consequence of the second law. (See J. O. Hirschfelder et al., "Molecular Theory of Gases and Liquid," Sec. 11.8, John Wiley & Sons, Inc., New York, 1954.) Thus the flow toward the shock is supersonic, whereas the flow away from the shock is always subsonic.

[b] In an ordinary sound wave in which there is a periodic pressure change, the individual compressions can be looked upon as a series of feeble shocks traveling along. The slight heating produced in the sound wave tends to distort the sinusoidal slope of each pulse to a sharp-edged saw-tooth shape with steep gradients at the leading edge.

At very high compression ratios, on the other hand, as ρ_s/ρ_u approaches a limiting or near-limiting value, the temperature climbs precipitously, as does the Mach number. It can be readily shown, by comparing the shock temperatures with those obtained in an adiabatic and isentropic compression ($T^\gamma \propto P^{(\gamma-1)}$) of comparable pressure change that there is much greater heating in the shock wave.[a] It is this ability of a shock wave to produce very large instantaneous temperatures that makes it possible for it to initiate explosions in gas mixtures and support detonation waves. It has been a very useful, semiquantitative tool in investigating the kinetics of very fast reactions at relatively high temperatures.[b]

In a stationary detonation wave, the shock front is followed by a zone of chemical reaction which can be considered as an ordinary stationary-state combustion wave propagating through the denser and hotter gases behind the shock front (Fig. XIV.7). Such a combustion wave is characterized by a pressure decrease and a temperature increase across the flame front. Because of this and because, in the stationary state, the flame front must follow the shock front at a fixed distance, the model of the moving surface is not quite adequate to describe a stationary detonation.[c] A further difference between the two is that, whereas in the mechanical shock the surface velocity v_b was an independent parameter at the disposal of the experimenter, in the detonation the chemical composition of the reacting gases is the "collective" parameter which replaces v_b and is the means by which the experimenter can control the detonation velocity.

If we can be permitted to make the same type of approximations that we made in treating stationary flames, then we can use Eq. (XIV.10.22) to give the mass flow rate of the burned gases with respect to the shock front. This equation, together with the ideal gas law and the conservation conditions (i.e., mass, momentum, energy) for the two zones, completely determines the density and pressure in each of the three regions separated by the zones (i.e., unshocked gases, shocked gases, and burned gases).

[a] The origin of this lies in the extra work done on the gas in the shock process compared to the isentropic process. In the latter, the compression is so slow that the kinetic energy of the gas is negligible. In the shock the kinetic energy of the gas is very large and acts as the source of the additional heating (i.e., from the point of view of the shock front, supersonic gas is slowed in crossing the front to subsonic velocities).

[b] See papers by Davidson et al. on reactions in shock tubes, particularly D. Britten et al., *Discussions Faraday Soc.*, **17**, 58 (1954). In most of this work the shock has been initiated by allowing a gas at high pressure to quickly and adiabatically compress the gas under observation, the latter being originally at much lower pressure.

[c] A closer mechanical analogue would be one in which the surface had holes in it so that the shocked gases could be vented at such a rate that the zone of shocked gas between the moving surface and shock front remained constant in size. It is interesting to note that in such a case, the maximum space velocity of the vented gases relative to the surface would be the local velocity of sound in those gases. As we shall see, this is precisely the result found in detonation waves for the velocity of the burned gases relative to the shock front.

An alternative approach which has been used by most authors since the original work of Chapman and Jouguet is to ignore the region between the shock and combustion zones and just consider the changes across the entire detonation front. The conservation conditions in this case take the form

$$\dot{m} = \rho_b v_b = \rho_u v_u \qquad \text{(XIV.11.13)}$$

$$P_b - P_u = \rho_b v_b (v_u - v_b) \qquad \text{(XIV.11.14)}$$

$$\Delta E = E_b - E_u = \frac{P_u}{\rho_u} - \frac{P_b}{\rho_b} + \frac{v_u^2 - v_b^2}{2} \qquad \text{(XIV.11.15)}$$

where, in the last equation, ΔE now includes changes in composition.[a] That is, $E_b - E_u$ can be written for an ideal gas as $n\,\Delta E_u^\circ + \bar{c}_v(T_b - T_u)$, where n is the number of moles of reaction per gram of mixture, ΔE_u° is the standard internal energy change at temperature T_u, and $\bar{c}_v$ is a mean specific heat for the system.

These equations, together with the ideal gas law, are sufficient to determine P_b, ρ_b, T_b, and v_b in terms of the initial conditions P_u, ρ_u, T_u, and v_u. However, to calculate the detonation velocity $v_u = \rho_b v_b / \rho_u$, an additional equation is needed. Chapman proposed that v_b, the velocity of the burned gases with respect to the detonation front, be taken equal to the local velocity of sound in the burned gases $v_{cb} = (\gamma_b R T_b / M_b)^{1/2}$. It turns out that this choice corresponds to the minimum possible detonation velocity (known as the Chapman-Jouguet condition) and seems to give reasonably good agreement with experiment. However, the arguments for such a velocity, which is also the maximum velocity possible for an expansion wave, are by no means rigorous.[b]

By using the Chapman-Jouguet condition we then have for the detonation velocity $v_u = (v_u - v_b) + v_b$, where $v_u - v_b$ is known as the particle velocity because it corresponds to the velocity of the burned gases with respect to the unburned gases. Alternatively, we have

$$v_u = \frac{\rho_b}{\rho_u} v_b = \frac{\rho_b}{\rho_u} \left(\gamma_b \frac{P_b}{\rho_b} \right)^{1/2} \qquad \text{(XIV.11.16)}$$

The detailed calculations of detonation velocity from thermodynamic data and the equation of state require very laborious, iterated computations following a trial choice of temperature T_b and does not seem to be too sensitive to either the precision of the thermodynamic data or the real

[a] Note that these are the same equations we used earlier for stationary flames [Eqs. (XIV.10.1) to (XIV.10.4)] except for the kinetic-energy terms which we had neglected for flames.

[b] See the discussion in Lewis and Von Elbe on this point. R. L. Scorah, *J. Chem. Phys.*, **3**, 425 (1935), pointed out that such a choice also leads to the maximum rate of degradation of free energy or the minimum rate of production of entropy if the linear equations of irreversible thermodynamics are used. It is doubtful if the present data are sufficiently accurate to discriminate between such a choice and an equation obtained from kinetic considerations such as Eq. (XIV.10.22).

final state including the actual extent of reaction. For this reason, the agreements which have been obtained with experimental data are not easy to evaluate.[a]

There has been a considerable amount of work done in examining the behavior and properties of detonation waves and shock waves. Kistiakowsky and coworkers[b] have examined the thickness of the wave by using Xe as an inert marker and following its density with X-ray absorption. Gilkerson and Davidson[c] used an I_2 marker. The latter workers found a reaction zone behind the shock of about 5 mm thickness. They then showed, by using order of magnitude calculations, that this was sufficiently thick to permit the $H_2 + O_2$ reaction to proceed to completion. They also showed that even a relatively slow reaction like $H_2 + O_2 \rightarrow 2OH$ (assuming reasonable rate constants) was sufficiently fast to initiate radicals and about 10^5 times faster than diffusion in supplying radicals to the reaction zone. Kistiakowsky and Kydd found the same order of magnitude for the zone thickness but found that the lack of exchange of energy between translation and rotation and vibration gave a smaller effective $\bar{c}_v$ for the gases, a smaller initial density, and an apparently faster reaction rate.[d]

12. The Combustion of Hydrocarbons

From the point of view of their technological and economical position in our industrialized society, the oxidation reactions of the hydrocarbons are probably of unrivaled importance. The utilization of partial oxidation reactions in the synthesis of important industrial compounds, while a poor second to the role of oxidation for production of energy, is nevertheless a colossal and growing field. Parallel to the utility of these reactions is their chemical complexity and the multitude of unique phenomena associated with their kinetic behavior.

At relatively low temperature (for example, 100 to 200°C), the gas-phase oxidation of aliphatic hydrocarbons in the absence of catalysts is immeas-

---

[a] The first tests of Eq. (XIV.11.16) were by B. Lewis and J. B. Friauf, *J. Am. Chem. Soc.*, **52**, 3905 (1930), on $H_2 + O_2$ mixtures. D. J. Berets et al., *ibid.*, **72**, 1080 (1950), repeated this work and obtained, with better thermodynamic data, agreement to within about ±3 per cent on detonation velocities.

[b] G. B. Kistiakowsky and P. H. Kydd, *J. Chem. Phys.*, **25**, 824 (1956).

[c] W. R. Gilkerson and N. Davidson, *ibid.*, **23**, 687 (1955).

[d] Because of the slowness of energy transfer, the analysis of even simple shocks in nonreacting mixtures is not easy. D. Horning [see *J. Phys. Chem.*, **61**, 856 (1957) and earlier papers] and coworkers have examined shock fronts by reflected light and shown that, whereas translational equilibrium occurs in a few collisions, rotational equilibrium takes longer and, in a gas like H_2 with very large rotational quanta, as many as 300 collisions may be needed. Vibrational equilibrium is very much longer and depends sharply on the specific gases. These difficulties, together with the apparatus limitations, place rather severe restrictions on the quantitative significance which can be ascribed to the results of kinetic studies in detonations and shocks.

urably slow. With catalysts, the products are usually few in number and can include alcohols, acids, aldehydes, ketones, oxides, and peroxides, with carbon skeletons the same as or only slightly different from the skeleton of the original hydrocarbon. With a proper choice of catalysts, one or more of these products can be made predominant in selected cases.

As the temperature is increased, the homogeneous oxidation reactions[a] occur at faster rates and the products begin to increase in number and also begin to include large amounts of CO and H_2O. Secondary products of importance include species such as CH_2O, H_2O_2, CO_2,[b] CH_4, H_2, and for the higher hydrocarbons, CH_3CHO, etc.

In the neighborhood of 300 to 400°C, the aliphatic hydrocarbons from ethane on up show a rather complex ignition behavior (Fig. XIV.10). At

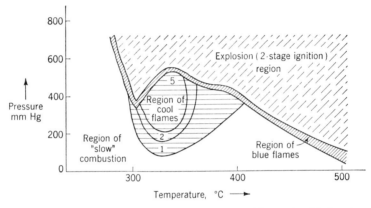

FIG. XIV.10. Ignition limits for equimolar mixtures of C_3H_8 and O_2. Limits after work of D. M. Newitt and L. S. Thornes, *J. Chem. Soc.*, 1656, 1669 (1937). The thin dark-hatched boundary between "slow" combustion and explosion is the region of intense blue flames. The contoured regions designated as "cool" flame regions are numbered after the number of successive cool flames which can be observed in them.

the lower temperatures there is an induction period which in extreme cases can last for many minutes, although it is usually not more than about 1 to 2 min. This is followed by a weak luminosity through the entire reaction flask which can persist to near the end of the reaction. At slightly higher temperatures the luminosity will be followed by a pale-blue flame usually propagating slowly outward from the center of the flask. Sometimes several of these blue flames, or "cool flames," as they are frequently

[a] Surface initiation of radical chains appears to be important over the entire temperature range. However, the bulk of the chain steps are gas-phase.

[b] In the oxidation of hydrocarbons, CO_2 is usually a minor product compared to CO even in cool flames. CO_2 predominates in explosions only. It is also formed in surface-catalyzed decompositions of acids and peroxy acids.

referred to,[a] will follow each other at intervals of a few seconds. At still higher temperatures the "cool flames" are replaced by a more intense luminosity which is followed by an intense blue or yellow flame (depending on temperature and composition) culminating in explosion. The medium-temperature explosions are referred to as two-stage ignitions because of the cool flames which may precede them. None of the hydrocarbons shows the three-limit type of explosion diagram which we have discovered for the cases of H_2 and CO. Both CH_4 and CH_2O are unique in having only a single explosion limit without the cool flame phenomena just described.

Part of the complexity of the hydrocarbon oxidations can be attributed to the fact that free radicals can catalyze the "cracking" reactions of hydrocarbons. Since the oxidation proceeds initially by a free radical process, it provides the possibility of sensitizing the cracking decomposition. With increasing temperature and increasing size of carbon skeleton, these pyrolysis reactions show an increasing importance, since hydrocarbon radicals show an increasing rate of fission with increasing size. In extreme cases, in fuel-rich (i.e., oxygen-poor) mixtures, the O_2 acts as a catalyst to sensitize the pyrolysis of the hydrocarbon.[b] This behavior of O_2 is particularly striking in diffusion flames. Gordon and coworkers[c] have shown that, in the oxygen-poor sections of the flame, the products appearing are those expected for the pyrolysis of the initial reactant. Thus we may expect the combustions to be at least as complicated as the pyrolysis reactions, the latter being already quite complex.

Although there is no direct information on the method of initiation of radicals in the combustion reactions, there seems to be fairly general agreement on the reaction

$$RH + O_2 \xrightarrow{k_i} R + HO_2 \qquad (XIV.12.1)$$

which is likely to have a rather high steric factor (ca. 10^{-1}) and an activation energy $E_i = D(R\!-\!H) - D(H\!-\!O_2)$. If $D(H\!-\!O_2)$ is in the range of 40 to 50 Kcal (Table XIV.1), then E_i will be about 45 Kcal for tertiary H atoms (i.e., isobutane) and aldehydes and probably about 50 Kcal for secondary H atoms and less than both of these if the reaction takes place on surfaces (e.g., with chemisorbed O_2).

[a] The first observations on these cool flames seem to have been made by W. H. Perkin, *J. Chem. Soc.*, **41**, 363 (1882). D. M. Newitt and L. S. Thornes, *ibid.*, 1656, 1669 (1937), give a detailed account of them in $C_3H_8 + O_2$ mixtures. The book by Lewis and Von Elbe discusses them at length.

[b] Most pyrolysis reactions are very sensitive to small traces of O_2 which can sensitize them (see section on hydrocarbon pyrolysis). Interestingly enough, this sensitization diminishes with increasing O_2 content. That is, it appears as if excess O_2 acts as an inhibitor.

[c] S. R. Smith and A. S. Gordon, *J. Phys. Chem.*, **60**, 759, 1059 (1956). Smith et al., *ibid.*, **61**, 553 (1957). The temperatures across the flame are in the range 800 to 1200°C, a range in which even unsensitized pyrolysis reactions will be expected to be fast.

At low temperatures fairly long chains leading to the production of hydroperoxides can be set up:

$$R + O_2 \xrightarrow{1} RO_2$$

$$RO_2 + RH \xrightarrow{2} RO_2H + R \qquad\qquad (XIV.12.2)$$

The chain ending in such cases is by

$$R + R \xrightarrow{k_{tr}} R_2$$

$$R + RO_2 \xrightarrow{k_{trp}} RO_2R \qquad\qquad (XIV.12.3)$$

There is a considerable body of experimental evidence in support of the peroxide chains given above. The most direct evidence comes from the isolation of relatively large yields of peroxides by rapid quenching of hydrocarbon-oxygen reaction mixtures.[a] Newitt and Thornes[b] found up to a 20 per cent yield of peroxides from the 400°C oxidation of propane, while Satterfield and Reid[c] were able to obtain up to 35 per cent (molar basis) yield of H_2O_2 from a 5.5:1 mixture of $C_3H_8 + O_2$ at 425°C. Pahnke et al.,[d] have shown by direct analysis of exhaust from a one-cylinder combustion engine that both H_2O_2 and organic peroxides are important *precombustion* products of the oxidation of alkanes. They found that organic peroxides prevailed at low temperatures, while H_2O_2 dominated at high temperatures (i.e., above 375°C).

This change-over from organic peroxides to H_2O_2 as an important product with increase in temperature is in part attributed to the fact that H_2O_2 is the most stable of the peroxides (most of which decompose rapidly between 100 and 200°C) and to the higher temperature coefficient of the reaction which is presumed to initiate the production of H_2O_2:

$$RCH_2\text{---}\dot{C}HCH_3 + O_2 \xrightarrow{3} RCH{=}CHCH_3 + HO_2$$

followed by $\qquad\qquad\qquad\qquad\qquad\qquad\qquad\qquad\qquad (XIV.12.4)$

$$HO_2 + RCH_2CH_2CH_3 \xrightarrow{4} H_2O_2 + RCH_2\dot{C}HCH_3$$

The mechanism requires that the donor RH be capable of forming a double bond. Note that reactions 3 and 1 are competing bimolecular reactions. It is quite possible that another factor which favors 3 with increasing temperature is the increasing instability of the newly formed hot radical

[a] E. R. Bell et al., *Ind. Eng. Chem.*, **41**, 2597 (1949), were able to prepare large yields of *t*-butyl hydroperoxide by using HBr to sensitize the oxidation of isobutane. In this case, at the lower temperatures the HBr acted as an H donor to the *t*-BuO$_2$ radical. See E. R. Bell et al., *Discussions Faraday Soc.*, **10**, 242 (1951), for a detailed discussion of the gas-phase behavior of peroxy and alkoxy radicals.

[b] D. M. Newitt and L. S. Thornes, *loc. cit.*

[c] C. N. Satterfield and R. C. Reid, "Fifth Symposium (International) on Combustion," p. 511, Reinhold Publishing Corporation, New York, 1955.

[d] A. J. Pahnke et al., *Ind. Eng. Chem.*, **46**, 1024 (1954).

$RO_2 \to R + O_2$. It is estimated that the bond energy $D(R{-}O_2)$ is probably not in excess of about 25 Kcal, so that RO_2 may not be very stable in any case at 400°C.[a]

Peroxides and hydroperoxides are not very stable, the bond dissociation energy of the O—O bond being in the range of from 27 Kcal for the diacyl peroxides to about 36 Kcal for the very stable dtBP. With frequency factors for their unimolecular decompositions in the range of 10^{14} to 10^{16} sec^{-1},[b] these peroxides will have lifetimes of from 10^{-6} to 10^{-3} sec at 400°C and from 10^{-4} to 1 sec at 300°C. Thus we should not expect very high stationary concentrations of peroxides in oxidations in this temperature range.[c] In fact there seems good reason to believe that the onset of the cool flames and ignition in hydrocarbons at temperatures near 300°C is due to this thermal instability of organic peroxides, their subsequent decomposition being equivalent to chain branching:

$$ROOH \to RO + OH \qquad ROOR' \to RO + OR'$$

$$(XIV.12.5)$$

This delayed decomposition of the metastable peroxides has been termed "degenerate chain branching" by Semenoff and has been used by him to account for the ignition limits of hydrocarbon oxidations, in particular for the long induction periods preceding the cool flames.

If the first step in the oxidation of hydrocarbons is the production of peroxy radicals (HO_2, RO_2) and peroxides (ROOH, HOOH, ROOR'), we see that the next stage must involve the decomposition of these intermediates, either by free radical attack on the peroxides[d] or by unimolecular decomposition. An additional mode of destruction is by surface dehydration[e] to form H_2O + aldehyde.

The decomposition of peroxides or peroxy alkyl radicals usually leads

[a] If we assume a unimolecular decomposition for $RO_2 \to R + O_2$ with a frequency factor of 10^{13} sec^{-1}, the half-life at 400°C would be about 10^{-4} to 10^{-5} sec, which would explain quite well the decrease in organic peroxides with increasing temperature.

[b] M. Szwarc, *Discussions Faraday Soc.*, **14**, 125 (1953). M. Szwarc and A. Sehon, *Ann. Rev. Phys. Chem.*, **8** (1957).

[c] There is great difficulty in distinguishing between H_2O_2 and organic peroxides in the reaction products. One very bothersome feature is the fairly rapid reaction of H_2O_2 with aldehydes to form alkoxy peroxides, for example, $RCHO + H_2O_2 \to HOCHROOH$, or with excess RCHO to form $(HOCHR)_2O_2$.

[d] There is considerable evidence for the existence of short chains in the decomposition of dtBP at 160°C and for chain sensitization by more active chain centers such as Cl atoms (Sec. XIII.12).

[e] The sensitivity of peroxides to surface dehydration has caused many investigators to overlook their importance as intermediates. H_2O_2 is very sensitive to catalytic decomposition on glass surfaces, forming $H_2O + O_2$. C. K. McLane, *J. Chem. Phys.*, **17**, 379 (1949), showed that in B_2O_3-coated vessels H_2O_2 is not destroyed. M. D. Scheer, "Fifth Symposium (International) on Combustion," p. 435, *op. cit.*, showed that the kinetics of CH_2O oxidation in vessels of different surface is understandable on the basis of a surface-catalyzed decomposition of peroxyformic acid, $HCO_3H \to H_2O + CO_2$.

to the formation of alkoxy radicals [Eq. (XIV.12.5)] which in turn can produce both aldehydes and alcohols by reactions such as[a]

$$O_2 + R'CH_2O \rightarrow HO_2 + R'CHO$$
$$R + R'CH_2O \rightarrow RH + R'CHO$$
$$2R'CH_2O \rightarrow R'CHO + R'CH_2OH \qquad (XIV.12.6)$$
$$R'CH_2O \rightarrow R' + CH_2O$$

At temperatures above 200°C this last reaction is undoubtedly the principal method for the production of CH_2O in the oxidation of the hydrocarbons, and the sequence of reactions leading to $R'CH_2O$ is typical of the oxidative degradation of hydrocarbon chains.

From this brief account it is seen that the oxidation of hydrocarbons involves the oxidation of all the intermediate oxygenated products as well as their cracking reactions. For this reason it is impossible to give a general scheme for the oxidative mechanisms. Rather, each hydrocarbon must be considered separately with particular attention to the stoichiometry and temperature of the reacting mixture, both of which can profoundly alter the nature of the products and the path of the reaction. Rather than try to summarize all of the work which has been done on the different oxidation mechanisms, let us consider the oxidation of a single one, propane, which although not completely characterized, is sufficiently well understood to serve as a general example of the important processes.

Between 300 and 400°C, the initiation of radicals in mixtures of C_3H_8 and O_2 is probably surface-catalyzed, as has already been discussed. The first important propagation steps will be the attack of these radicals on propane, followed by the formation of peroxides and their decomposition.

$$C_3H_8 + O_2 \rightarrow C_3H_7 + HO_2 \qquad \text{wall?}$$

H_2O_2 *Chain:* $\begin{cases} HO_2 + C_3H_8 \rightarrow H_2O_2 + C_3H_7 \\ C_3H_7 + O_2 \rightarrow C_3H_6 + HO_2 \end{cases}$

Hydroperoxide Chain: $\begin{cases} C_3H_7 + O_2 \rightleftharpoons C_3H_7O_2 \\ C_3H_7O_2 + C_3H_8 \rightarrow C_3H_7O_2H + C_3H_7 \end{cases}$

"Cracking" Reactions: $\begin{cases} p\text{-}C_3H_7 \rightarrow C_2H_4 + CH_3 \\ C_3H_7O_2H \rightarrow C_3H_7O + OH \\ sec\text{-}C_3H_7O \rightarrow CH_3CHO + CH_3 \\ p\text{-}C_3H_7O \rightarrow C_2H_5 + CH_2O \end{cases}$

$$(XIV.12.7)$$

Harris and Egerton[b] found in static sampling experiments on 1:1 C_3H_8/O_2 mixtures at 325°C (1 atm) that in the first 30 per cent of reaction, one mole of C_3H_8 reacting consumed about 1.6 moles of O_2 and produced about 1.8

[a] J. Levy, *J. Am. Chem. Soc.*, **78**, 1780 (1956), has shown that C_2H_5O at temperatures below 200°C is not very active in abstracting H from either C_2H_5ONO or $C_2H_5OOC_2H_5$.
[b] E. J. Harris and A. Egerton, *Chem. Revs.*, **21**, 287 (1937).

moles of H_2O, 0.6 moles of CH_4, 0.53 moles of CO, 0.3 moles of C_3H_6, 0.1 mole each of CH_2O, higher aldehydes, CH_3OH, and CO_2, and about 0.06 moles of H_2. With increasing percentage of reaction, CH_3OH and CO increased appreciably, while C_3H_6 and CH_4 decreased appreciably. No analysis was given for peroxides, which from the work of Newitt and Thornes (*loc. cit.*) are known to be present in large concentrations.

Since propane can be attacked by H-abstracting radicals at either the primary or secondary positions, the initial C_3H_7 radicals consist of both primary and secondary isomers.[a] The early and abundant H_2O formation must certainly arise from the cracking reactions of the peroxides which produce OH radicals, while the aldehydes must come from the cracking of the alkoxy radicals. CH_4 and CO undoubtedly arise from the radical-sensitized decompositions of CH_3CHO, while the sensitized decomposition of CH_2O produces the small amounts of H_2 found, together with more CO.

The blue flames so characteristic of hydrocarbon combustions have been shown to arise from electronically excited states of CH_2O,[b] and the low quantum yields for light production together with the large energy requirement for such activation indicates that a radical-radical reaction is probably the origin of the luminescence. Of the possible reactions responsible, the reactions of CH_3O (formed from CH_3OOH) with H-abstracting radicals are the most likely:[c]

$$CH_3O + \left\{ \begin{array}{c} CH_3 \\ OH \\ C_3H_7 \end{array} \right\} \rightarrow CH_2O^* + \left\{ \begin{array}{c} CH_4 \\ HOH \\ C_3H_8 \end{array} \right\} \qquad (XIV.12.8)$$

For the reactions listed the exothermocities vary from about 98 Kcal for the OH attack to 78 Kcal for the p-C_3H_7 radical attack.

Both CH_2O and CH_3CHO undergo very fast radical-sensitized decompositions such as

[a] Although the primary H atoms are 3 times as abundant, their activation energy for attack seems to be about 3 Kcal higher than for attack on secondary H atoms. This throws the balance over in favor of attack on secondary H atoms, although at 400°C the rates of attack on the two positions should be only 3:1 in favor of the secondaries.

[b] R. A. Day and R. N. Pease, *J. Am. Chem. Soc.*, **62**, 2234 (1940), have estimated quantum yields as low as 10^{-16}, while J. E. C. Topps and D. T. A. Townend, *Trans. Faraday Soc.*, **42**, 345 (1946), have reported much larger yields of about 1 quantum per 10^6 molecules undergoing oxidation.

[c] Since the CH_2O is excited, it is most likely that the CH_2O^* is formed either from a CHO abstraction of H from another radical (such as $2CHO \rightarrow CH_2O + CO$) or else from a structure such as CH_3O in which the bond distances and geometry are significantly different from ground-state CH_2O. In very exothermic dismutation reactions it seems to be the rule that the extra energy will usually appear in the molecule containing the newly formed bond. The abstraction of H from CH_3O, however, could well be an exception. The CHO disproportionation is somewhat lacking in energy (between 51 and 75 Kcal) for the luminescence.

$$R + \begin{Bmatrix} CH_2O \\ CH_3CHO \end{Bmatrix} \rightarrow \begin{Bmatrix} CHO \\ CH_3CO \end{Bmatrix} + RH$$

$$M + CH_3CO \rightarrow CO + CH_3 + M \qquad (XIV.12.9)$$

$$M + CHO \rightarrow CO + H + M$$

while in excess oxygen the acyl radicals can undergo further oxidation to produce peroxy acyl radicals:[a,b]

$$RCO + O_2 \rightleftharpoons RC \overset{\displaystyle O}{\underset{\displaystyle O_2}{\big<}} \qquad (XIV.12.10)$$

which can in turn form peroxy acids by H abstraction

$$R{-}CO_3 + R'H \rightarrow RCO_3H + R' \qquad (XIV.12.11)$$

The subsequent pyrolysis of these acid peroxides will yield CO_2 and possibly acids

$$R{-}C\overset{\displaystyle O}{\underset{\displaystyle OOH}{\big<}} \rightarrow RCO_2 + OH$$

$$RCO_2 + R'H \rightarrow RCO_2H + R' \qquad (XIV.12.12)$$

$$RCO_2 \rightarrow R + CO_2$$

At higher temperatures, the formation of CO_2 is favored and it is not certain that the lifetime of the RCO_2 radicals is great enough above 100°C to form significant quantities of acid $RCOOH$ by H abstraction.[c] There is good evidence that the large amounts of CH_2O and CH_3OH arise from

[a] W. A. Bone and J. B. Gardener, *Proc. Roy. Soc.* (*London*), **A154**, 297 (1936), found up to 20 per cent molar production of performic acid and H_2O_2 in 2:1 CH_2O/O_2 mixtures at 1 atm, 275°C. Formic acid was also produced in about 7 per cent yield.

[b] C. A. McDowell and J. B. Farmer, "Fifth Symposium (International) on Combustion," p. 453, *op. cit.*, have shown the formation of peracetic acid as the principal initial product in the photosensitized and thermal oxidation of acetaldehyde. (See also earlier papers of McDowell and Farmer.) J. Grumer, *ibid.*, p. 447, also showed that, at low O_2 content, C_2H_4, C_3H_6, CO, CH_4, CH_3OH, CH_3CHO, and CH_3CH_2CHO were important products from propane pyrolysis in the range 350 to 475°C. He also found considerable amounts of acetic acid from the oxidation of CH_3CHO in mixtures at 130 to 450°C having about 3 per cent O_2. Such low-O_2 mixtures are, of course, ideal for observing sensitized pyrolysis reactions.

[c] See work by M. Szwarc, *J. Polymer Sci.*, **16**, 367 (1955), on the decomposition of the acyl peroxides in liquid phase. Surface dehydration will produce H_2O and aldehydes from the peroxy acids.

oxidation reactions of CH_3 radicals, the latter being formed in the decomposition reaction of CH_3CO, CH_3CH_2O, p-C_3H_7, and CH_3CO_2 radicals.[a]

The effects of surface on the oxidation reactions have been studied by a number of workers. In general the rates of oxidation tend to decrease with increase of surface/volume ratio. However, Hoare and Walsh[b] have shown that in the "slow" oxidation of CH_4 at 500°C (1 atm) both the rates and products could be profoundly altered by changing either the surface of the vessel or its treatment. They consider these effects as arising from the effectiveness of the surface in destroying peroxy radicals HO_2 and RO_2. Chamberlain, Hoare, and Walsh[c] have used similar arguments to interpret the antiknocking properties of $Pb(C_2H_5)_4$ and PbO in combustion engines. It is also known that KCl surfaces inhibit oxidation rates, explosions, and the formation of peroxides. While the evidence is by no means complete, it seems to be fairly convincing in establishing the surface behavior as fairly specifically directed toward the decomposition of peroxy species.

The cool flame phenomenon seems to be closely tied to the formation of aldehydes and peroxides in oxidation systems. In Fig. XIV.10 is shown a typical example of the explosion limits for a hydrocarbon-oxygen mixture. The explosion region, except for a region of positive slope, resembles the limit curve for a thermal explosion. The transition between "slow" combustion and explosion is characterized by an intense luminous blue flame which appears after a short induction period and is followed by explosion. The induction periods are not more than a few seconds.

The region of cool flames, which also has a region of positive slope (and is in this sense analogous to the second explosion limits for H_2 and CO), has been the subject of much interest. The cool flames are most arresting because of the very long induction periods that precede them. Andrew[d] found that, in n-butane $+ O_2$ mixtures, the induction period decreased ex-

[a] The oxidation of CH_3 radicals formed photochemically has been shown by M. I. Christie, *J. Am. Chem. Soc.*, **76**, 1979 (1954), to give high yields of CH_2O in the range 125 to 175°C, while G. R. Hooey and K. O. Kutschke, *Can. J. Chem.*, **33**, 496 (1955), found that CH_3 from the photolysis of $(CH_3)_2N_2$ formed both CH_3OH and CH_2O in good yields. It seems quite plausible that the common precursor to both CH_3OH and CH_2O in these systems is CH_3O_2H, which can spontaneously decompose into $CH_3O + OH$ or else can be decomposed by radicals $R + CH_3OOH \rightarrow RH + CH_2OOH$, $CH_2OOH \rightarrow CH_2O + OH$. D. M. Newitt and P. Szego, *Proc. Roy. Soc. (London)*, **A147**, 555 (1934), found up to 50 per cent yields of CH_3OH from the oxidation of 90 per cent CH_4, 10 per cent air mixtures at 435°C and 50 atm pressure.

[b] D. E. Hoare and A. D. Walsh, "Fifth Symposium (International) on Combustion," p. 474, *op. cit.*, have shown that, for $CH_4 + O_2$ mixtures at least, the rates can be made to vary by several orders of magnitude, depending on the chemical nature of the surface and its previous conditioning. New vessels may show rates tenfold faster than vessels which have been "aged" and 100 times faster than PbO-coated vessels.

[c] G. H. N. Chamberlain et al., *Discussions Faraday Soc.*, **14**, 89 (1953).

[d] E. A. Andrew, *Acta Physicochim. U.R.S.S.*, **6**, 57 (1937).

ponentially with increasing temperature, falling from 300 sec at 280°C to about 2 sec at 400°C. Conversely, the induction period between the blue flame and the true explosion increased with temperature over the same range from about 0.2 sec to 2.0 sec.[a]

The region of cool flames does coincide with the region of greatest build-up of peroxides and aldehydes in these systems, and while there are some conflicting data, it seems to be generally agreed that small amounts of CH_3CHO[b] or C_2H_5OOH[c] decrease the induction period and sensitize the appearance of explosion.[d] For these and reasons related to the mechanisms of oxidation it has been proposed that the cool flames represent straight chain reaction to produce peroxides and aldehydes and that the induction periods are the times required to build up to the stationary-state radical concentrations. The region of explosion on this basis would be interpreted as one in which the decomposition of the peroxides was sufficiently fast to constitute a chain-branching system. On this basis CH_2O should be an inhibitor of the regular ignition, which is in fact observed at the higher temperatures.

The periodicity of the cool flames and their merging into the region of explosion with increasing pressure are features on which there has been much speculation but as yet little clear-cut evidence. Chamberlain and Walsh[e] have proposed that the catalytic agents responsible for the cool flames are hydroxy alkyl peroxides arising from the condensation of peroxides and aldehydes on surfaces. Frank-Kamenetskii,[f] on the other hand, has made the rather intriguing proposal that the mechanism itself is responsible for the periodicity.[g] This requires that peroxides and aldehydes catalyze each others' production and disappearance in a set of second-order processes such as

$$\frac{d(P)}{dt} = k_1(M)(P) - k_2(P)(A)$$

$$\frac{d(A)}{dt} = k_3(P)(A) - k_4(M)(A)$$

(XIV.12.13)

where P = peroxide, A = aldehyde, and M is hydrocarbon. Although such a scheme satisfies the formal requirements for an oscillating system, it has

[a] For a more complete discussion of cool flames and related phenomena the reader is referred to the work by Von Elbe and Lewis.

[b] B. Aivazov and M. B. Neumann, Z. physik. Chem., **B33**, 349 (1936).

[c] M. Prettre, Ann. office natl. combustibles liquides, **11**, 669 (1936).

[d] E. Blat et al., Acta Physicochim. U.R.S.S., **10**, 273 (1939).

[e] G. H. N. Chamberlain and A. D. Walsh, "Third Symposium on Combustion and Flame and Explosion Phenomena," p. 375, op. cit.

[f] A. A. Frank-Kamenetskii, J. Phys. Chem. (U.S.S.R.), **14**, 30 (1940). See also A. D. Walsh, Trans. Faraday Soc., **43**, 305 (1947).

[g] A. J. Lotka, J. Am. Chem. Soc., **42**, 1595 (1920), was the first to describe a set of kinetic equations for a periodic chemical process. These equations seem to have inspired the suggestion of Frank-Kamenetskii.

difficulties in application to the cool flame phenomena which make it unlikely.[a]

While the kinetic work on hydrocarbon oxidations has been far more quantitative that the preceding discussion would imply, the complexity of the systems, together with their variable sensitivity to vessel surface, makes the quantitative data rather difficult to interpret in terms of elementary reactions, and so we have chosen to omit it.

[a] See Lewis and Von Elbe, *op. cit.*, p. 179.

PART FOUR

Reactions in Condensed Phases

XV

Physical Models for Solution Reactions

1. A Model for Chemical Reactions in Solution

At first glance, the problem of analyzing the rate of chemical reactions in condensed states such as solutions, from a molecular point of view, seems hopelessly complex. The reason is that, in a solution, any particular molecule is at any moment in close contact[a] with a number of nearest neighbors which may vary in number from 4 to 12. The presumably intimate mechanical motions of the atoms and electrons belonging to a molecule (or pair of molecules) undergoing chemical reaction are thus being constantly intruded upon in a random and arbitrary way by an imposingly large number of neighboring molecules. In the case of reactions involving ionic species, the interactions of the neighboring molecules are sufficiently large to become a necessary part of the reaction scheme. In fact, in their absence the reaction may not take place.

Some indication of the importance of the solvent interactions is afforded by the observation that of the thousands of reactions which have been studied in solution, less than 20 have been capable of comparative study in the gas phase. The study of ionic reactions has been almost completely restricted to solutions for reasons which are quite understandable: ionic processes are virtually nil in the gas phase at temperatures below $1000°K$. However, this accounts for most of the solution reactions studied, since, as we shall see, most reactions between polar molecules involve ionic species as intermediates. Thus such common reactions as the hydrolysis of alkyl halides or of esters do not proceed at measurable rates in the gas phase (at least not at temperatures at which other, competing reactions are not dominant). The only large class of reactions which proceed conveniently in both gas and liquid states is the free radical class, and undoubtedly as

[a] By "close contact" we mean that the distances between neighboring molecules are not very much larger than the sum of their "radii."

the type becomes more extensively investigated, kinetic comparisons between the two states will grow.

The molecular description of rate processes in solutions is possible to the extent that a molecular description of the behavior of solutions is available. While no rigorous, molecular theory of solutions has been developed, there are a number of semiempirical models which will provide a framework for discussing the kinetics of solutions. For the purposes of our discussion it is convenient to classify solutions as those in which the average interactions between neighboring molecules are of the order of kT or less (weak interactions) and those in which they are much greater than kT (strong interactions). As we will see later, this corresponds in a crude fashion to the distinction between nonionic and ionic solutions. In the following sections, we shall consider first the behavior of nonionic systems.

2. Collisions in a Solution

The simplest molecular model capable of describing a liquid is the hard sphere molecule with an attractive potential well (Sec. VII.1). There are

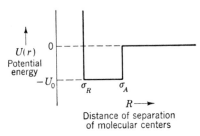

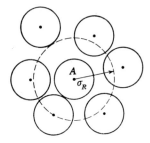

FIG. XV.1. Potential-energy diagram for the hard sphere molecule with rectangular, attractive well.

FIG. XV.2. Cross section through a hypothetical liquid made up of hard sphere molecules with attractive wells showing molecule A surrounded by nearest neighbors. The dotted line represents the distance of closest approach to A, σR.

three parameters of such a model, σ_R, σ_A, and U_0; σ_R is the hard sphere diameter (the distance of closest approach of the centers of two molecules) and σ_A is the distance between centers at which two molecules will experience an infinite attractive force whose potential energy is $U_0 < 0$. The behavior of the potential energy $U(r)$ as a function of distance of separation of a pair of molecules is shown in Fig. XV.1. In Fig. XV.2, we illustrate a cross section through a liquid composed of such molecules.

Because of the close packing in liquids, each molecule in such a system will be surrounded by a number Z of nearest neighbors which will tend to be as large as possible[a] commensurate with geometric packing and the system of intermolecular forces. Depending on the ratio U_0/kT, we will find a more or less tight structure. At very low temperatures (that is, $U_0/kT \gg 1$), e.g., near the freezing point of the liquid, we may expect to find Z very nearly the same for all molecules in the liquid and the average distance between nearest neighbors very close to σ_R. With increasing temperature, fluctuations will become larger and we may find considerable variation in both Z and the average distance between nearest neighbors.

The problem of specifying a collision between two such molecules is complicated by the range of the attractive forces. If we think about it in detail, we will find, as in the case of gases, that for each physical process of interest (i.e., pressure, viscosity, diffusion, etc.), there will be a different definition of a collision. From the point of view of a chemical reaction, no generalization is possible, but we may say for purposes of convenience, as a zero-order approximation, that a pair of molecules will be considered in a state of collision so long as their potential energy of interaction is of the order of magnitude of kT.

With such a definition of a collision and the hard sphere well model we must then conclude that each molecule is in a constant state of collision with *all* of its Z nearest neighbors. If we consider a dilute solution of two species A and B in a third solvent S, then the crowded condition of the system will make it likely that if A and B ever do collide and become nearest neighbors, they may remain so for some time.[b] The mean lifetime of such a collision pair A-B can be estimated crudely by considering the time t_{AB} that it will take for A and B initially r_{AB} apart to diffuse apart to a distance $1.7 r_{AB}$ out of range of each other's attractive forces. This time t_{AB} is of the order of magnitude of[c]

$$ t_{AB} \approx \frac{2 r_{AB}^2}{6 D_{AB}} e^{-w_{ABS}/kT} \qquad (XV.2.1) $$

[a] For spherical force fields as suggested above, the most economical way to fill space is with a hexagonally close-packed structure in which $Z = 12$. Where noncentral forces are present (i.e., rod- or disk-shaped molecules or dipoles), other packing such as cubic packing ($Z = 6$ to 8) or, with strongly directional forces as in ice, tetrahedral packing ($Z = 4$) may be preferred. In mixtures of A + B in which AB forces are much larger than the mean of AA + BB, there may be a tendency to change packing near mole fraction $\frac{1}{2}$.

[b] Such a prolonged collision is referred to as an encounter.

[c] Assuming that we can apply the theory of the random walk to diffusion over distances of the order of a molecular diameter, we find for the mean-square displacement of two noninteracting molecules (Sec. VI.7) A and B during a time t, $\overline{r^2} = 6Dt$. Such a calculation can be justified whether the separation occurs through a continuous sequence of stable configurations or, because of the tightness of the packing and the discrete molecular structure, by an activated jump process. It will only be when we attempt to relate D, the macroscopic diffusion constant, to molecular properties that the mechanism will become of importance.

where $D_{AB} = D_{AS} + D_{BS}$, the sum of the binary diffusion coefficients of A and B with respect to the solvent S, and w_{ABS} is the energy of separation of A and B in the solvent S.[a]

For typical values of $D_{AB} \cong 2 \times 10^{-5}$ cm²/sec and $r_{AB} \cong 4$ Å, we find $t_{AB} \cong 3 \times 10^{-11}$ sec when $w_{ABS} = 0$. This is about 100 times longer than the collision time of about 3×10^{-13} sec estimated for the lifetime of such a pair in the gas phase (Sec. VII.8H) and illustrates an important property of the layer of nearest neighbors, namely, that of impeding access to and from the molecule surrounded by the layer. This long life of molecular pairs in liquids has been described in the literature as due to a "solvent cage," and we shall see later some kinetic consequences of this "caging."

We must note that there is a necessary arbitrariness about the definition of a collision pair, dependent in part on the relative forces between A and B and dependent in part on our model of the liquid state. If, for example, we assume a quasi-lattice model for liquids with short-range but not long-range order, then near neighbors in a solution are restricted in their average separations to what corresponds to lattice positions. For hexagonal close packing of spheres, where the nearest neighbor distance is r_{AB}, the next-nearest neighbor distance is $r_{AB}(\%_3)^{1/2} \cong 1.7 r_{AB}$, and the lattice model implies that intermediate distances between these values are of very low probability. If, however, we assume that the packing forces are not so strong, then there is no strong tendency to lattice-like structure and the probability of positions intermediate between lattice positions is not small.

Assuming for the moment a lattice model for liquid structure, we can calculate the rate at which a molecule A encounters new neighbors in the solution as the rate at which it jumps to a new lattice site times $Z/2$, since it will change half its neighbors at each jump. The frequency is

$$\frac{1}{t(r_{AB})} \cong \frac{6 D_A}{r_{AB}^2} \frac{Z}{2} \qquad (XV.2.2)$$

where D_A is the diffusion constant of A relative to solvent.

The frequency with which A molecules in a solution will encounter B molecules is this frequency multiplied by n_B, the mole fraction of B. For very dilute solutions $n_B \cong N_B/N_S$, the ratio of the molecular densities of B to S molecules.[b] But $1/N_S$, the volume per solvent molecule, can be written as γr_{AB}^3, with γ determined by the packing factor for the lattice. By substituting these relationships in Eq. (XV.2.2) we can write for the encounter frequency of A and B in such a lattice[c]

[a] We are here assuming the range of forces sufficiently short that they are unimportant for distances in excess of $1.7 r_{AB}$. Note that in solution enthalpies and energies are nearly equal.

[b] We have implicitly assumed that A, B, and S have about equal sizes.

[c] E. Rabinowitch, *Trans. Faraday Soc.*, **33**, 1225 (1937), appears to have been the first to use this lattice model to compute collision frequencies in solution.

$$Z'_{AB} = 3Z\gamma r_{AB} D_{AB}$$

$$= 25 r_{AB} D_{AB} \qquad \text{for } Z = 12,\ \gamma = 0.707 \qquad (XV.2.3)$$

where we have replaced D_A by D_{AB} to take into account the relative diffusions of both A and B.

It turns out on further examination that this collision frequency is not significantly altered by using a continuum model for liquid solutions.

In our subsequent considerations of reactions in solution we shall be interested in the rate at which encounters take place. Such a rate will be of interest when the probability of chemical reaction between two species during an encounter is very close to unity. Under these conditions, the rate of the reaction will be limited by the rate of encounters and we describe the reaction as *diffusion-controlled*. Among the reactions in this category are the quenching of fluorescence and the recombination of free radicals.

To calculate the rate of encounters from a somewhat different point of view, let us consider the history of a spherical molecule A, suddenly immersed at random in an ideal solution containing spherical molecules B. By processes of mutual diffusion, A and B will encounter each other in the solution and after a time separate. If we consider just the rate of encounters of A with molecules B which it has not met before, we observe that initially the distribution of B molecules about A is random and consists only of non-encountered species. As time passes, however, A begins to be surrounded by a quasi-stationary gradient of B molecules which it has already encountered once or more than once, plus an equal and opposite[a] gradient of never-encountered B molecules. If we assume these gradients to have a spherical symmetry about A at any time,[b] then, in the stationary state, the flux of "new" B molecules through the solution is constant across any spherical surface S, of radius r, centered on A and is given, from Fick's first law, by

$$D_{AB} S \left(\frac{\partial N_B}{\partial r} \right)_S = 4\pi r_S^2 D_{AB} \left(\frac{\partial N_B}{\partial r} \right)_S = \text{const} \qquad (XV.2.4)$$

This can be solved for the stationary gradient giving

$$\left(\frac{\partial N_B}{\partial r} \right)_S = \frac{a}{r^2} \qquad (XV.2.5)$$

where a is a constant, and on integration we find for N_B

$$N_B = -\frac{a}{r} + \text{const} \qquad (XV.2.6)$$

---

[a] Since the sum of the two types of B, old and new, is constant in the solution, those gradients are equal and opposite.

[b] This is not necessarily correct, since A is moving through the solution.

By assuming that $N_B = N_B^\circ$ at $r = \infty$ and $N_B = 0$ at $r = r_{AB}$,[a] the encounter distance, we can eliminate the constant of integration and, solving for a, we find $a = N_B^\circ r_{AB}$. By substituting this back in Eq. (XV.2.5) and then in Eq. (XV.2.4) we can obtain for the stationary rate of "new" encounters

$$4\pi r_{AB} N_B^\circ D_{AB} \tag{XV.2.7}$$

For the collision frequency of A with B we divide by N_B°:

$$Z'_{AB} = 4\pi r_{AB} D_{AB} \tag{XV.2.8}$$

or in units of liters/mole-sec,

$$Z'_{AB} = \frac{4\pi r_{AB} D_{AB} N_{Av}}{1000} \tag{XV.2.9}$$

If we compare Eq. (XV.2.8) with Eq. (XV.2.3), we see that the latter is about twice as large. This is to be expected because the latter measures the frequency of all A-B encounters, while Eq. (XV.2.8) measures only new encounters. Collins and Kimball[b] have pointed out that in a diffusion-controlled bimolecular reaction between A and B, the initial rate which can be characterized by a random spatial distribution of A and B decays to the lower rate given by Eq. (XV.2.9). The reason for this is that the reaction tends to draw off the A-B pairs in close proximity and leaves a stationary distribution of A-B which approaches that given by the concentration gradient of Eq. (XV.2.6). The relaxation time for such a decay is of the order of[c] $r_{AB}^2/\pi^2 D_{AB}$, which for most molecular systems will be of the order of 10^{-11} sec, or the actual time of an encounter. Noyes[d] has shown that there exist certain experimental systems in which these effects can be observed. We shall say more about them later in our discussion of cage effects in liquids.

For a typical case in which $r_{AB} \cong 4$ Å and $D_{AB} \cong 2 \times 10^{-5}$ cm²/sec we find $Z'_{AB} \cong 6.2 \times 10^9$ liters/mole-sec, a value which is about 30 times smaller than the collision frequencies of about 2×10^{11} liters/mole-sec which are usually found for gases (Sec. VII.8).[e]

The collision frequency Z'_{AB} has a temperature coefficient which can be represented in terms of an activation energy:

[a] This may be recognized as the solution to the problem of the stationary rate of diffusion of particles into an infinite spherical sink of radius r_{AB}. It was first solved by M. V. Smoluchowski, Z. physik. Chem., **92**, 129 (1917), for use in collision problems.

[b] F. C. Collins and G. E. Kimball, J. Colloid. Sci., **4**, 425 (1949).

[c] See Sec. XIV.2 for the similar equation governing heat diffusion.

[d] R. M. Noyes, J. Am. Chem. Soc., **79**, 551 (1957); **78**, 5486 (1956); **77**, 2042 (1955); J. Chem. Phys., **22**, 1349 (1954).

[e] For identical species, A = B, we must divide Eq. (XV.2.9) by 2.

$$E_Z = RT^2 \frac{\partial \ln Z'_{AB}}{\partial T} = RT^2 \frac{\partial \ln r_{AB}}{\partial T} + RT^2 \frac{\partial \ln D_{AB}}{\partial T}$$

$$= RT \frac{\alpha T}{3} + E_D \cong E_D \qquad (XV.2.10)$$

where we have assumed that the average distance between nearest neighbors r_{AB} varies with temperature as one-third the coefficient of cubic expansion $[\alpha \equiv (\partial \ln V / \partial T)]$ and E_D is the activation energy for diffusion. Experimentally it is observed that E_D is about one-third of the energy of vaporization of the solvent and in the range 3 to 5 Kcal/mole for many liquids.[a] Thus, if we write Z'_{AB} in the form of the Arrhenius expression, $A_Z \exp(-E_Z/RT)$, A_Z will be about 5×10^{11} liters/mole-sec for most liquids and E_Z will be about 3 Kcal.

For purposes of calculation, rough estimates of D_{AS} can be made by using the theory of Brownian motion, $D_{AS} = Z_{AS} \lambda_A^2 / 6$, where Z_{AS} is the number of "collisions" per second made by solute molecule A with its near neighbors of solvent S, while λ_A is the mean distance traveled by A during such a collision.[b] But $\lambda_A = \bar{v}_S / Z_{AS}$, where $\bar{v}_S$ is the mean velocity of S. If we further assume that Z_{AS} is proportional to $2r_f / \bar{v}_{AS}$

$$Z_{AS} = \frac{2ar_f}{\bar{v}_{AS}} \qquad (XV.2.11)$$

where $r_f = r_{AS} - (r_A + r_S) = $ the free space between near neighbors and $\bar{v}_{AS}$ is the mean relative velocity of A and S molecules, then

$$D_{AS} = \frac{a}{3} \frac{\bar{v}_S^2}{\bar{v}_{AS}} r_f = \frac{a}{3} \left(\frac{m_A}{m_A + m_S} \right)^{1/2} \bar{v}_S r_f \qquad (XV.2.12)$$

The value of a may be taken as unity, while r_f may be estimated from free volume models for liquids.[c] Values of r_f in the range of 0.3 Å (or about 10 per cent of r_{AS}) seem to fit the data for many liquids to within a factor of about 3.[d]

The collision frequency given by Eq. (XV.2.3) or (XV.2.9) is subject to modification for long-range forces which may exist between A and B. Thus, if A and B are charged ions, then Z'_{AB} will be increased if they have opposite charges and decreased if they are of the same sign (i.e., repel each other).

[a] S. Glasstone et al., "The Theory of Rate Processes," McGraw-Hill Book Company, Inc., New York, 1941.

[b] Such a model is meaningful only for hard sphere molecules. For real force fields, the molecules are always in a state of collision, i.e., interaction which causes momentum exchange.

[c] Glasstone et al., op. cit.

[d] It is clear that the temperature coefficient of D from such a model must come mainly from r_f. The hard sphere, attractive well model does not have a large enough temperature coefficient. This can be improved, however, with the use of a more realistic potential.

The effects in such cases can be considerable, particularly in media of low dielectric constant. Even dipole interactions whose potential varies as $1/r^3$ can give rise to considerably enhanced (or diminished) collision frequencies.[a] The general problem which is involved is that of the rate of diffusion in a force field, and a number of different authors have given solutions to it.[b,c]

Having calculated the mean life [Eq. (XV.2.1)] and the mean rate of formation of a collision pair, we can calculate the equilibrium concentration N_{AB} in the absence of chemical reaction:[d]

$$N_{AB} = N_A N_B Z'_{AB} t_{AB} \qquad (XV.2.13)$$

On substitution for Z'_{AB} and t_{AB} [Eqs. (XV.2.3) and (XV.2.1)]:

$$N_{AB} \cong N_A N_B 8 r_{AB}^3 e^{-W_{ABS}/kT} \qquad (XV.2.14)$$

But the equilibrium constant K_N (in units of molecular concentrations) for the reaction $A + B \rightleftharpoons AB$ is, neglecting activity coefficients:

$$K_N = \frac{N_{AB}}{N_A N_B}$$

$$= 8 r_{AB}^3 e^{-w_{ABS}/kT} \qquad (XV.2.15)$$

or in units of moles/liter, by setting $W_{ABS} = N_{Av} w_{ABS}$,

$$K_c = \frac{8 r_{AB}^3 N_{Av}}{1000} e^{-W_{ABS}/RT} \qquad (XV.2.16)$$

If we write K_c in thermodynamic form as

$$K_c = e^{-\Delta F_c/RT} = e^{\Delta S_c/R} e^{-\Delta H_c/RT} \qquad (XV.2.17)$$

we can identify ΔH_c with W_{ABS}, the energy of formation of a pair[e] while $\Delta S_c = R \ln (8 r_{AB}^3 N_{Av}/1000)$ can be interpreted as the entropy change in

---

[a] As a crude approximation for attractive fields we can say that the effective radius r_{AB} is increased to r'_{AB}, where r'_{AB} is the distance at which the potential energy of the pair AB in the solution is of the order of kT. For repulsive fields the effective radius r_{AB} is diminished by the ratio r_{AB}/r'_{AB}. The more exact treatment for central force fields whose potential is $U(r)$ gives r'_{AB} as

$$r_{AB} \int_1^\infty e U(x)/kT \frac{dx}{x^2} \qquad \text{with } x = r/r_{AB}$$

[b] P. Debye, *Trans. Electrochem. Soc.*, **82**, 265 (1942), considers electrolyte solutions.

[c] S. Chandrasekhar, *Revs. Modern Phys.*, **15**, 1 (1943), has summarized work on the general problem of diffusion and Brownian motion. Also T. R. Waite, *J. Chem. Phys.*, **28**, 103 (1958).

[d] Note that $K_{eq} = k_f/k_r$ and for the reverse reaction $k_r = 1/t_{AB}$. Note also that loose association is only slightly more favored in liquids than in gases, the longer lifetime being offset by the slowness of association.

[e] Note that $\Delta H = \Delta E + \Delta(PV)$. For solutions which are at a constant pressure, generally of 1 atm, $\Delta(PV) = P \Delta V$ is usually quite small compared to RT, so that there is not much error in equating ΔH and ΔE.

shifting a mole of A from a volume of 1 liter to the near-neighbor volume around B of $8r_{AB}^3 N_{Av}/1000$ liters.[a]

A very similar result is obtained if we consider a very dilute solution of both A and B in solvent S, all of which species are made of spherical molecules that have nearly equal volumes.[b] Then the concentration of A-B neighbors in such a solution is given by the product of the concentration of A molecules N_A, the number of near-neighbor sites, Z, the mole fraction of B in the solution n_B, and the Boltzmann function:

$$N_{AB} = ZN_A n_B e^{-W_{ABS}/RT} \qquad (XV.2.18)$$

and since N_A and $N_B \ll N_S$, $n_B = N_B/(N_S + N_A + N_B) \cong N_B/N_S$, hence

$$N_{AB} = \frac{ZN_A N_B}{N_S} e^{-W_{ABS}/RT} \qquad (XV.2.19)$$

so that

$$K_N = \frac{Z}{N_S} e^{-W_{ABS}/RT} \qquad (XV.2.20)$$

If we examine Eq. (XV.2.19) we will note that $1/N_S$ is the volume per solvent molecule, so that Z/N_S is the volume occupied by the layer of neighbor molecules. But from Eq. (XV.2.14) we see that $8r_{AB}^3$ is also approximately the volume occupied by the layer of neighboring molecules, since the volume contained within this sphere contains the central molecule plus half the volume associated with each neighbor.

For typical values such as we have been using, $r_{AB} \cong 4$ Å, $\Delta S_c \cong -R \ln 3.3 = -2.5$ cal/mole-°K,[c] a result which probably underestimates somewhat the entropy decrease in the formation of chemically reacting pairs A-B, since we have not considered the possible loss of internal degrees of freedom of A and B in forming a more closely interacting pair A-B. In addition, the A-B distance in such a pair is probably much closer to bond distances than to the larger Van der Waals or collision distances.

Because of the limitations imposed by activity coefficients and specific interactions, a precise quantitative check of experimental data against the collision formula presented here is not possible. However, the frequency factors of bimolecular reactions which are diffusion-controlled (i.e., those which occur on nearly every collision) such as free radical recombinations,

[a] The significance of neglecting activity coefficients is seen to be the implication that the molecules of A and B, aside from their interaction, find no other differences in their environments on becoming nearest neighbors.

[b] This is a model of a "strictly regular solution" as used by R. Fowler and E. A. Guggenheim, "Statistical Thermodynamics," Cambridge University Press, New York, 1949.

[c] We obtain a similar result from Eq. (XV.2.18), where we see that for nearly close packing with $\overline{Z} = 10$, $\Delta S_c = R \ln Z/C_S$ has the range from 0 for a liquid like benzene with $C_S \cong 10$ moles/liter to -3.3 cal/mole-°K for H_2O with $C_S = 55$ M.

do have rate constants in the range of 10^7 liters/mole-sec[a] (observed in polymerizations) to values of about 8×10^9 liters/mole-sec[b] for the recombination of I atoms in nonpolar solvents. For ionic reactions involving protons, higher values are found; examples are 1.5×10^{11} liters/mole-sec for $H^+ + OH^-$,[c] and 1×10^{11} liters/mole-sec for $H^+ + SO_4^=$.

In Table XV.1 are summarized some data for fast reactions in solution which are presumed to be diffusion-controlled. The range in rate constants of about 10^3 is probably a reflection of the different entropy changes on association. What is of special interest is the low values of the activation energy, which are in the range of diffusion activation energies.

3. Transition-state Model for Solution Reactions

In solution, the intimate contact between solute and solvent molecules, constituting as it does a state of "constant" collision, makes for a rate of energy transfer between solute and solvent as rapid, probably, as that between loosely coupled, normal modes of vibration in a single, large molecule. With the exception of very unusual cases, this will be of the order of magnitude of vibration frequencies (that is, $\sim 10^{13}$ sec^{-1}), which is sufficiently rapid that we may expect to find transition-state complexes in nearly good thermodynamic equilibrium with unreacted species. Under these conditions, we may employ the formalism of any of the transition-state treatments which has been developed earlier.

For an nth-order reaction between species A, B, C, . . . to produce a transition-state complex whose rate of disappearance may be equated to the rate of appearance of products we may write

$$A + B + \cdots \underset{k_2}{\overset{k_1}{\rightleftharpoons}} X \overset{k_r}{\rightarrow} P + Q + \text{products} \qquad (XV.3.1)$$

with $dP/dt = k_r(X)$, where (X) corresponds to the concentration of transition-state complexes for the reaction. It is given by the equilibrium expression:

$$(X) = K_X(A)(B) \cdots \frac{f_A f_B \cdots}{f_X} \qquad (XV.3.2)$$

where K_X is the equilibrium constant $(k_1/k_2 = K_X)$ for the equilibrium

[a] Cheves Walling, "Free Radicals in Solution," John Wiley & Sons, Inc., New York, 1957. The termination rate constants for radical recombinations in polymerizations have activation energies of about 2 to 3 Kcal which compare closely with those expected for diffusion constants.

[b] S. Aditya and J. E. Willard, *J. Am. Chem. Soc.*, **79**, 2680 (1957). They also found that there is an activation energy for the rate constant of 3.2 Kcal/mole which can be compared with the activation energy of 3.3 Kcal/mole for the diffusion of I_2 in CCl_4.

[c] M. Eigen, *Discussions Faraday Soc.*, **17**, 194 (1954). Eigen has pointed out that these higher values are in accord with the corrections for ions of opposite sign and the anomalously high diffusion constant D for protons in water solution.

TABLE XV.1. RATE DATA FOR SOME DIFFUSION-CONTROLLED REACTIONS

Reaction	Solvent	Rate constants, $\times 10^{-9}$, liters/mole-sec	Temp, °C	Activation energy, Kcal/mole
$I + I \rightarrow I_2$[a]	CCl_4	7	23	3.2
	n-Hexane	18	50	—
$2CCl_3 \rightarrow C_2Cl_6$[b]	Cyclohexene + CCl_3Br	0.05	30	<6
	Vinyl acetate + CCl_3Br	0.05	30	<6
β-naphthylamine + $CCl_4 \rightarrow$ fluorescence quenching[c]	Cyclohexane	6	20	2.5
	Isooctane	13	20	1.6
2 polystyryl radicals → stable products[d]	Benzene	0.006	25	1.9
2 polymethyl methacrylate radicals → stable products[e]	Methyl methacrylate	0.024	25	2.8

[a] S. Aditya and J. E. Willard, *J. Am. Chem. Soc.*, **79**, 2680 (1957).

[b] H. W. Melville et al., *Discussions Faraday Soc.*, **14**, 150 (1953).

[c] H. G. Curme and G. K. Rollefson, *J. Am. Chem. Soc.*, **74**, 3766 (1952).

[d] G. M. Burnett, *Trans. Faraday Soc.*, **46**, 772 (1950).

[e] M. S. Matheson et al., *J. Am. Chem. Soc.*, **71**, 497 (1949); **73**, 1700 (1951).

between A, B, . . . and X, and the f_A, . . . are the activity coefficients of the various species.

If the over-all specific rate of appearance of products is given by the expression $dP/dt = k_n(A)(B) \cdots$, where k_n is the experimentally observed nth order specific rate constant, we can write for k_n

$$k_n = k_r K_X \frac{f_A f_B \cdots}{f_X} \qquad (XV.3.3)$$

The product $k_r K_X$ can now be written in terms of the partition functions of A, B, . . . and X, while the activity coefficients f_A, f_B, . . . may be either estimated from various theories of solution or measured experimentally. The value of f_X, like its partition function, can only be assumed from more or less reasonable models for the reaction[a] and theories of the solvation of X in the system.

At sufficiently low concentrations, the activity coefficients become constants independent of composition, and if we choose infinitely dilute solutions as our standard states,[b] we may equate them to unity. Under these conditions, the estimation of k_n a priori is reduced to an estimate of the structural parameters of X required to determine the partition function of X and an estimate of the effect of the solvent on the equilibrium constant K_X.

4. The Effect of Solvent

It is interesting to compare the values of k_n for solution reactions to those obtained for the same reaction in the gas phase. For simplicity, let us assume that the gases and solutions are both sufficiently dilute that we can equate all activity coefficients to unity. Then, from Eq. (XV.3.3):

$$R_n = \frac{k_n(\text{solution})}{k_n(\text{gas})} = \frac{k_r(S)}{k_r(g)} \frac{K_X(S)}{K_X(g)} \qquad (XV.4.1)$$

In the transition-state formalisms k_r is essentially a vibration frequency or else an average vibration frequency (Sec. XII.5), and to a good degree

---

[a] In this sense we never "predict" a rate constant from a transition-state theory; rather, we make an educated guess at a model for the transition state whose properties (within broad limits) appear compatible with the observed rate data. Since the latter is itself based upon an assumed order and mechanism, there may sometimes be a rather tenuous connection between experimental rate observations and transition-state rate constants.

[b] A standard state cannot actually be one of infinite dilution because it must represent a definite thermodynamic state of fixed composition. What we mean by the expression "infinite dilution" as a standard state is really one in which solute-solute interactions are negligible, and the actual choice of a standard state will then be some definite composition in this ideal range. In practice, since molarities are generally employed for solutions, it is usual to select the standard state as "hypothetical 1 molar," that is, a 1 M solution in which solute-solute interactions are zero, hence hypothetical.

of approximation we may expect it to be only slightly affected by solvent interactions, so that $k_r(S) \cong k_r(g)$. The difference in reaction rates thus reduces to a difference in thermodynamic equilibrium constants for the two phases:

$$R_n = \frac{k_n(\text{solution})}{k_n(\text{gas})} \cong \frac{K_X(S)}{K_X(g)} \qquad (XV.4.2)$$

A similar expression can be derived, by the same arguments, for the ratio of rate constants in two different solvents.

The general conclusion from this result is that a reaction will go fastest in the medium that favors the association of reactants. The magnitude to be expected for such an effect can be estimated for nonionic reactions by using a simple model of a solution. If we assume that A, B, . . . and X form ideal solutions with the solvent S which follow Raoult's law over the entire range of compositions, then we can write for the relation between mole fraction x_i of the ith component and its equilibrium vapor pressure above the solution $p_i = x_i p_i^\circ$, where p_i° is the vapor pressure of the pure i species at temperature T. On converting these to concentrations, we have for ideal gases: $p_i = C_{ig}RT$, while for dilute solutions $x_i \cong C_{is}/C_S \cong C_{is}V_S^\circ$ where V_S° is the molar volume of the solvent and $C_S = 1/V_S^\circ$ its concentration (C_{is} refers to solution and C_{ig} to gas).

If now we express the K's in Eq. (XV.4.2) in terms of the equilibrium concentrations in the two phases,

$$R_n = \frac{K_X(S)}{K_X(g)} = \left(\frac{C_{Xs}\, C_{Ag}C_{Bg}\, \cdots}{C_{Xg}\, C_{As}C_{Bs}\, \cdots} \right)_{eq} \qquad (XV.4.3)$$

which becomes, on substituting for the concentrations,

$$R_n = \frac{p_A^\circ p_B^\circ \cdots}{p_X^\circ} \left(\frac{V_S^\circ}{RT} \right)^{n-1} \qquad (XV.4.4)$$

Now, the Clausius-Clapeyron equation,[a] $\ln(p/p_B) = (\Delta H^\circ/RT_B)$ $(1 - T_B/T)$ is a fairly good approximation to the vapor pressure of a pure liquid, and on substitution for p_i°

$$R_n = \exp\left(\frac{-\Delta S_A^\circ - \Delta S_B^\circ + \cdots + \Delta S_X^\circ}{R} \right)$$

$$\exp\left(\frac{\Delta H_A^\circ + \Delta H_B^\circ + \cdots - \Delta H_X^\circ}{RT} \right) \left(\frac{V_S^\circ}{RT} \right)^{n-1} \cdots \qquad (XV.4.5)$$

From the free volume model for a liquid we can write[b]

---

[a] Here ΔH° is the heat of vaporization at the boiling temperature $T_b(P_b = 1 \text{ atm})$. $\Delta S^\circ = \Delta H^\circ/T_b$ is close to 21 for liquid obeying Trouton's rule.

[b] See any text on the liquid state, e.g., J. H. Hildebrand and R. L. Scott, "Solubility of Non-electrolytes," Reinhold Publishing Corporation, New York, 1950. From an operational point of view we can use this equation to define the free volume of a liquid.

$$\Delta S^{\circ} = R \ln \frac{V_g^{\circ}}{V_f^{\circ}} = R \ln \frac{RT_b}{V_f^{\circ} P_b} \qquad (XV.4.6)$$

where $P_b = 1$ atm and V_f° is the free volume per molecule of the liquid at its boiling point. For most liquids, free volumes are about 1 per cent of their molar volumes, within a factor of about 2.[a] On replacing the entropies by the corresponding free volumes, we find

$$R_n = \frac{V_S^{\circ}}{V_{fA}^{\circ}} \frac{V_S^{\circ}}{V_{fB}^{\circ}} \cdots \frac{V_{fX}^{\circ}}{V_S^{\circ}} \exp\left(\frac{\Delta H_{vap}^{\circ}}{RT}\right) \qquad (XV.4.7)$$

where ΔH_{vap}° is the difference in heats of vaporization of X and the reactants. We note that this term will appear as a temperature coefficient and will be responsible for any differences in activation energies of the rate constants for the two phases. Except in cases in which the structure of X is markedly different in terms of charge distribution from the loose complex $(A \cdot B \cdots)$ or in which there are very strong, specific interactions, we may expect to find the energies of vaporization of X nearly equal to $A + B + \cdots$, so that $\Delta H_{vap}^{\circ} = \Delta E_{vap}^{\circ} - (n-1)RT \cong -(n-1)RT$. Alternatively, when the activation energy of a reaction (or the heat of reaction for an equilibrium) is the same in the gas and liquid phase, we are justified in setting $\Delta E_{vap}^{\circ} = 0$ and $\Delta H_{vap}^{\circ} = -(n-1)RT$. In that case, R_n becomes

$$R_n = \frac{V_S^{\circ}}{V_{fA}^{\circ}} \frac{V_S^{\circ}}{V_{fB}^{\circ}} \cdots \frac{V_{fX}^{\circ}}{V_S^{\circ}} e^{-(n-1)} \qquad (XV.4.8)$$

If we assume that the molar volumes of the solvent S and A, B, . . . are about the same and that X is the sum of the molar volumes of A, B, C, . . . , then $V_S^{\circ}/V_{fi}^{\circ} \cong 100$ for $i = A, B, \ldots$, while $V_{fX}^{\circ}/V_S^{\circ} \cong n/100$, then

$$R_n \cong \frac{10^{2n-2}}{ne^{n-1}} \qquad (XV.4.9)$$

For a unimolecular reaction with $n = 1$, $R_n \cong 1$ and the rate constants should be about the same in the gas and solution phases, unless of course there are differences in activation energy between the two phases.[b] The same arguments apply equally well to two different solvents, so that one expects, for ideal solutions, very little difference in rate constants on changing solvent. Although there is little direct experimental evidence

[a] From Eq. (XV.4.6), $V_f^{\circ} = RT_b \exp(-\Delta H_B/RT_b)$ and for a liquid with a Trouton constant of 21, this becomes $V_f^{\circ} \cong 0.6(T_b/273)$ cc. Scott and Hildebrand have shown that for non-hydrogen bonding liquids, a good empirical equation is $\Delta H_B^{\circ}/RT_b = 8.5 + 0.0045T_b = 8.5 + 1.23T_b/273$. By using this latter relation, V_f° (cc) $= 1.27(1 + T_b/273) 10^{-0.53T_b/273} \cong 1.27[1 - 0.23T_b/273 - 0.48(T_b/273)^2]$. In the above calculations, note that the free volumes have been used more as a convenience than with any primary significance in terms of a model of a liquid.

[b] The arguments do not necessarily apply to systems that involve diffusion-controlled reactions, because the rate of diffusion out of or into the solvent cage may become important.

bearing on this point, for the few unimolecular reactions[a] which have been studied, the rate constants are about the same in gas phase as in solution. However, because of the uncertainties in both the mechanism and the activation energies, we shall not attempt to discuss these in detail.

For bimolecular reactions with $n = 2$, $R_n \cong 20$ and we should expect to find these reactions somewhat faster in solution than in gas phase.[b] The same conclusion must also apply to equilibria involving a mole change, $\Delta n = -1$.[c] Again, the experimental evidence for either bimolecular reactions or for equilibria is too sparse or complicated to attempt detailed comparisons. In the main, however, it would appear that rate constants and equilibrium constants are about the same in the two phases, a result at least within order-of-magnitude agreement of the conclusion arrived at here.

If one wished to observe solvent effects in less ambiguous fashion, the best chance would lie in going to higher-order reactions. For a third-order reaction ($n = 3$), $R_n = 450$. Strangely enough, there are no examples of third-order reactions (or equilibria) which have been studied in both gas phase and in solution. The reactions of NO with O_2, Cl_2, and Br_2 are third-order in the gas phase, and it would appear that their study in solution could prove very interesting and profitable.

The effect of solvent can be represented without too much additional difficulty for the case of dilute solutions which are nonideal.[d] For such solutions, Henry's law is obeyed with reasonable accuracy, i.e., the solubility of the vapor is proportional to its concentration in solution. For purposes of comparison with our preceding work we can express the vapor pressure of the solute in terms of the deviations of the system from Raoult's law:[e]

[a] Note that there is a distinction made here between a unimolecular reaction, which implies a single chemical step on the molecular level, and a first-order reaction, whose rate constant may be a complex combination of preequilibria and rate constants.

[b] This is very close to the factor of about 5 estimated in the preceding section from comparisons of the collision rate and lifetime of the collision pairs.

[c] Le Chatelier's principle tells us that the equilibrium $2A \rightleftharpoons A_2$ will be shifted to the right on decreasing the volume available to the system. If the equilibrium is established in the gas phase, the addition of solvent molecules, which are otherwise inert, does precisely this, i.e., they reduce the volume available to the molecules A and A_2. These rough conclusions are somewhat tempered, however, by the fact that the dimer A_2 may lose considerably in rotational entropy on going into solution.

[d] For most solvents, a 0.1 M solution would correspond to a mole fraction of 0.01. So long as we restrict our considerations to solutions not much more concentrated than this, we can neglect the change in activity with concentration.

[e] The emphasis on Raoult's law and the liquid state raises difficulties for solutes which are above their critical temperatures and for which $\Delta H°$, $p°$, and $V°$ appear to have little meaning. From an empirical point of view it appears that the extrapolation of the Clausius-Clapeyron equation to super-critical temperatures gives usable values of $p°$ (see Hildebrand and Scott, loc. cit., chap. 14).

$$p = xp°f°$$ (XV.4.10)

where $f° > 1$ for positive deviations and $f° < 1$ for negative deviations from Raoult's law. If we replace the $p°$ in Eq. (XV.4.4) by $p°f°$ and carry through the same analysis, then we find for R_n [Eq. (XV.4.7)]

$$R_n = \frac{(f°_A f°_B \cdots)}{(f°_X)} \frac{V°_S}{V°_{fA}} \frac{V°_S}{V°_{fB}} \cdots \frac{V°_{fX}}{V°_S} \exp\left(\frac{\Delta H°_{vap}}{RT}\right) \cdots$$ (XV.4.11)

so that the nonideality of the system can be expressed in terms of a ratio of activity coefficients. Since the deviations from Raoult's law are seldom greater than a factor of 2 in the dilute solution range, we see that the nonideality of the system will not cause very serious changes in the ratio R_n over that already considered for ideal systems.

Exceptions to such a conclusion will occur only when there is a marked attraction between solute species or between solvent and one or more solute species. In both these cases, however, the interactions will be manifested as an energy release and, for the solute involved, a very strong, negative deviation from Raoult's law: its energy of vaporization will be greater. Correspondingly, there will be a compensating decrease in the effective free volume of the solute. Since the energy term occurs in the exponent, it will dominate. The net result will be a shift in the equilibrium in favor of the more strongly solvated species.

Such strong solvent effects will manifest themselves in terms of a change in activation energy of the system, partially compensated for by a change in frequency factor. This seems to be a good empirical rule when rates of reaction are strongly influenced by changes in solvent. Marked changes in rate (i.e., of the order of 100 or more) will be accompanied by changes in both activation energy and frequency factor, the former being usually dominant.[a] Exceptions to this rule are provided by systems that involve ions, since the process of ionization will generally involve very large entropy changes on the part of the solvent. A good example of these effects is illustrated in Table XV.2 for the second order (presumably bimolecular) addition of cyclopentadiene to benzoquinone. Despite changes of some 4 Kcal in the energy of activation and a factor of 300 in the Arrhenius A factor, the rate constant does not change by more than a factor of 5.

For the less polar but similar Diels-Alder reaction, the dimerization of cyclopentadiene, which has also been studied in the gas phase, the data compiled in Table XV.3 show that over a very large range of solvents there are small changes in A and E and very little change in k.

---

[a] We can rationalize such behavior for the nonionic systems, i.e., those that involve short-range forces, by observing that $\Delta F = \Delta H - T \Delta S$, so that, if ΔH becomes more negative on changing solvent (i.e., more heat is liberated in the reaction), ΔS tends to decrease, thus compensating the change in ΔF. The reason ΔS decreases is that a more negative ΔH means either more loosely bound reactants or more tightly bound products.

TABLE XV.2. SOLVENT EFFECTS IN THE ADDITION OF CYCLOPENTADIENE
TO BENZOQUINONE[a]

Solvent	Activation energy, Kcal/mole	log A	$-\log k$, at 60°C
C_2H_5OH	12.7	7.0	1.3
C_6H_6	11.5	6.3	1.2
CCl_4	9.2	4.5	1.5
$C_6H_5 \cdot NO_2$	8.8	5.0	0.8

Note: A and k in units of liters/mole-sec.

[a] A. Wasserman, Trans. Faraday Soc., **34**, 128 (1938), and earlier papers. See also R. A. Fairclough and C. N. Hinshelwood, J. Chem. Soc., 236 (1938).

The reaction can be represented by

TABLE XV.3. SOLVENT EFFECTS IN THE DIMERIZATION OF CYCLOPENTADIENE

Medium	Temp range, °C	Log A	E, Kcal/mole	$-\log k$, at 50°C
Gas Phase	79–150	6.1 ± 0.4	16.7 ± 0.6	5.2
C_2H_5OH	0– 55	6.4 ± 0.6	16.4 ± 0.8	4.7
CH_3COOH	25– 70	5.0 ± 0.7	14.7 ± 1.0	5.0
$C_6H_5NO_2$	0– 55	5.5 ± 0.3	15.1 ± 0.4	4.7
CS_2	0– 35	6.2 ± 0.3	16.9 ± 0.5	5.2
C_6H_6	15– 55	6.1 ± 0.4	16.4 ± 0.6	5.0
CCl_4	0– 55	6.7 ± 0.3	17.1 ± 0.4	4.9
Paraffin	−1–172	7.1 ± 0.2	17.4 ± 0.3	4.7
Cyclopentadiene	−2– 35	5.8 ± 0.5	16.2 ± 0.8	5.2

Data compiled by A. Wasserman, Monatsh., **83**, 543 (1952). E, A, and k from Arrhenius equation, with k and A in units of liters/mole-sec.

Note that the errors in A and E are complementary rather than independent. The errors in k are smaller than those in A. The reaction can be represented by

A greater contrast is provided by the first-order rate of decarboxylation of malonic acid. In Table XV.4 are gathered some data from a compilation of Clark (*loc. cit.*) on the effect of different solvents. The data in H_2O of Hall (*loc. cit.*) are included for comparison purposes. Here, over a range

TABLE XV.4. SOLVENT EFFECTS IN DECARBOXYLATION OF MALONIC ACID

Solvent	$\Delta H^{\pm}$, Kcal/mole	$\Delta S^{\pm}$, cal/mole-°K	k (140°C) $\times$ 10³, sec⁻¹
$C_6H_5NH_2$	26.9	−4.5	5.0
m-Cl—$C_6H_4NH_2$	26.6	−6.3	2.6
o-Cl—$C_6H_4NH_2$	26.6	−6.9	2.0
Pyridine	26.0	−4.7	11.1
o-CH_3—$C_6H_4NH_2$	25.7	−7.1	5.6
Glycerine	24.6	−12.2	1.3
o-C_2H_5O—$C_6H_4NH_2$	24.0	−10.7	5.9
4-CH_3-pyridine	23.0	−11.3	15.2
Dimethyl sulfoxide	22.3	−15.0	9.2
3-CH_3-pyridine	21.9	−14.2	15.0
2-CH_3-pyridine	20.9	−17.4	10.8
H_2O (pH $\cong 0.5$)[a]	30.1	0.0	1.3
(pH $\cong 9.0$)	27.8	−10.9	0.082

Data from compilation by L. W. Clark, *J. Phys. Chem.*, **62**, 79 (1958), except for H_2O.
[a] G. A. Hall, Jr., *J. Am. Chem. Soc.*, **72**, 1906 (1950). At pH 0.5 it is the undissociated acid which dissociates, while at pH 9.0 it is the anion of the acid.
Reaction is $CH_2(COOH)_2 \rightarrow CH_3COOH + CO_2$.
$\Delta H^{\pm}$ and $\Delta S^{\pm}$ are calculated from Eq. (XV.5.2) by assuming f_i = constant.

of $\Delta H^{\pm}$ of from 21 to 30 Kcal, the over-all effect on k is not more than a factor of about 10. This reaction probably goes through an ionic intermediate, and it has been proposed that there is specific catalysis by amines acting as bases.[a] In view of the intramolecular H bonding possible, as well as strong specific interaction with solvent, such a simple model hardly seems adequate.

5. Changes in External Variables; Pressure Effects

The transition-state approach permits us to make a separation[b] of the factors constituting an experimental specific rate constant (for an elementary chemical act) into kinetic and thermodynamic factors. Thus, for the transition state $X \rightleftharpoons A + B + C + \cdots$ we can write for the rate constant k_n governing the appearance of products (Sec. XII.4) from X

[a] G. Fraenkel et al., *J. Am. Chem. Soc.*, **76**, 15 (1954). See also review article on decarboxylations by B. R. Brown, *Quart. Revs. (London)*, **5**, 131 (1951).
[b] See, however, Sec. XII.5.

$$k_n = \kappa_X K_X \bar{\nu}_X \cdot \frac{f_A f_B \cdots}{f_X \cdots} \tag{XV.5.1}$$

where K_X is the equilibrium constant that governs the equilibrium between X and the n species A, B, $\ldots$, $\bar{\nu}_X$ is the mean frequency (or probability per unit of time) with which X passes through the configuration that corresponds to the saddle point in the potential-energy diagram, κ_X is the transmission coefficient or probability that an X crossing the saddle point is not reflected back to reactants, and f_X, f_A, $\ldots$ are the activity coefficients of the respective species.

In this last equation the quantities κ_X and $\bar{\nu}_X$ can be described as kinetic, while K_X and the activity coefficients are thermodynamic. In the usual transition-state formalism employed by Eyring et al.[a] it is customary to set $\kappa_X = 1$ and to assume that $\bar{\nu}_X$ corresponds to a normal vibrational frequency whose partition function can be factored out of K_X in classical form $kT/h\bar{\nu}_X$, thus leading to

$$k_n = \frac{kT}{h} K_X^{\ddagger} \frac{f_A f_B \cdots}{f_X} \tag{XV.5.2}$$

where $K_X^{\ddagger} = K_X h\bar{\nu}_X/kT$. These approximations and assumptions have the advantage of eliminating all of the kinetic factors and reducing the theoretical analysis of specific rate constants to an analysis of the thermodynamic factors only.[b] Of the latter, only the thermodynamic properties of X are not accessible to independent measurement but must instead be inferred from the experimental rate data.

If we change any of the external variables governing the system, such as temperature, pressure, etc., then Eq. (XV.5.1) or (XV.5.2) can be used to estimate the effect of such changes on the rate constant[c] so long as the changes in external variables do not alter the mechanism of the reaction. But this last proviso defines a very interesting situation. Since Eq. (XV.5.2) involves only thermodynamic factors, the only external variables that need concern us are the thermodynamic variables of state, i.e., those needed to describe an equilibrium state of a system.

If α is a state variable of the system, then from Eq. (XV.5.2) the effect on k_n of a change in α is given by[d]

[a] Glasstone et al., op. cit.

[b] This is not quite the case, because the resolution of K_X by the method of partition functions still requires the analysis and factoring of $\bar{\nu}_X$.

[c] Section XV.4, in which we treated the effect of solvent changes, can be subsumed under the present generalization if we consider the change from solvent S_1 to S_2 as a change in external variable describing the mole fraction of S_1 and S_2.

[d] In the lack of any detailed knowledge of κ_X and $\bar{\nu}_X$ or of the effect of changes in external variables on them, we shall assume them to be constants for any given transition state, so that for the purposes of the present section, Eqs. (XV.5.1) and (XV.5.2) will yield the same results.

$$\frac{\partial \ln k_n}{\partial \alpha} = \frac{\partial \ln T}{\partial \alpha} + \frac{\partial \ln K_X^{\ddagger}}{\partial \alpha} - \left(\frac{\partial \ln f_X}{\partial \alpha} - \frac{\partial \ln f_A}{\partial \alpha} - \cdots\right)$$

(XV.5.3)

where the usual specification of the independent variables remaining fixed is omitted for simplicity in notation and will be used only where ambiguity could otherwise result.

Since $RT \ln K_X^{\ddagger} = -\Delta \overline{F}_X^{\ddagger}$, the standard, partial molar, free energy change in the reaction $A + B + \cdots \rightleftharpoons X$, we can relate the changes in rate constant k_n with α to corresponding changes in the thermodynamic functions F, H, S, etc. When the infinitely dilute solution is used as the standard state[a] and the range of solute concentrations is less than 0.1 M, then we can neglect the change in activity coefficients because Henry's law is usually well obeyed under these circumstances (except for polyvalent electrolyte systems). Equation (XV.5.3) then takes the simpler form

$$\frac{\partial \ln k_n}{\partial \alpha} = \frac{\partial \ln T}{\partial \alpha} + \frac{\partial \ln K_X^{\ddagger}}{\partial \alpha}$$

(XV.5.4)

For $T = \alpha$, Eq. (XV.5.3) gives us the relation between the experimental activation energy and the thermodynamic enthalpy change plus any changes in activity of the several species. Thus, at constant pressure

$$E_n \equiv RT^2 \left(\frac{\partial \ln k_n}{\partial T}\right)_P$$

$$= RT + \Delta \overline{H}_X^{\ddagger} - RT^2 \left(\frac{\partial \ln f_X}{\partial T} - \frac{\partial \ln f_A}{\partial T} - \cdots\right)$$

(XV.5.5)

since $RT^2 (\partial \ln K_X^{\ddagger}/\partial T)_P = -RT^2 [\partial(\Delta \overline{F}_X^{\ddagger}/RT)/\partial T]_P = \Delta \overline{H}_X^{\ddagger}$.

If we take external pressure as our variable, then ($\alpha \equiv P$)

$$\left(\frac{\partial \ln k_n}{\partial P}\right)_T = \left(\frac{\partial \ln K_X^{\ddagger}}{\partial P}\right)_T - \left[\left(\frac{\partial \ln f_X}{\partial P}\right)_T - \left(\frac{\partial \ln f_A}{\partial P}\right)_T - \cdots\right]$$

(XV.5.6)

Again, by substituting $\ln K_X^{\ddagger} = -\Delta \overline{F}_X^{\ddagger}/RT$ and making use of the thermodynamic relation, $(\partial \overline{F}/\partial P)_T = \overline{V}$, we find[b]

---

[a] See the footnote on page 504 on standard states.

[b] This relation holds true only if we specify concentrations in units of molality or mole fraction, both of which are unaffected by pressure changes. If, however, we use units of molarity (i.e., moles/liter) which are affected by pressure changes (owing to the compressibility of the solution), then we must make the appropriate correction. This can be done by means of the identity $(\partial \overline{F}/\partial P)_{C_i} = (\partial \overline{F}/\partial P)_{x_i} + (\partial \overline{F}/\partial x_i)_P (\partial x_i/\partial P)_{C_i}$, where x_i is mole fraction and C_i is molarity of the ith component. Now, $(\partial x_i/\partial P)_{C_i} = -x_i \beta_S$, where $\beta_S \equiv -(\partial \ln V/\partial P)_T$, the isothermal compressibility of the solution, while in the range of concentrations in which activity coefficients do not change with concentration, $(\partial \overline{F}/\partial x_i)_P = RT/x_i$. Thus we find $\overline{V}_{C_i} = \overline{V}_{x_i} - \beta_S RT$. This leads to the correction: $\Delta \overline{V}_{C_i} = \Delta \overline{V}_{x_i} - \beta_i RT \, \Delta n$, where Δn is the mole change in

$$RT\left(\frac{\partial \ln k_n}{\partial P}\right)_T = -\Delta \overline{V}_{\ddagger}^{X} - RT\left(\frac{\partial \ln f_X}{\partial P} - \frac{\partial \ln f_A}{\partial P} - \cdots\right)$$

$$(XV.5.7)$$

where $\Delta \overline{V}_{\ddagger}^{X}$, the difference in partial molar volumes of X and A, B, ... in their standard states, is usually the only term of importance, the changes in activity coefficients with pressure usually being quite small. For ideal solutions, the activity coefficients are zero and partial molar volumes become identical with molar volumes. Unless there are strong interactions between A, B, ... , $V_X \cong V_A + V_B + \cdots$, so that $\Delta \overline{V}_{\ddagger}^{X} = \Delta V_X \cong 0$ and the rate of the reaction is independent of pressure.

For nonideal solutions in the range of concentrations in which activity coefficients do not change with concentration (e.g., below 0.1 M) we can write with good accuracy

$$RT\left(\frac{\partial \ln k_n}{\partial P}\right)_{T,x_i} = -\Delta \overline{V}_{\ddagger}^{X} \qquad (XV.5.8)$$

$$= RT\left(\frac{\partial \ln k_n}{\partial P}\right)_{T,C_i} + (n-1)RT\beta_S$$

$$(XV.5.9)$$

where, in the last equation, n is the order of the reaction and β_S is the coefficient of compressibility of the solution.

Equations (XV.5.8) and (XV.5.9) predict a change of rate constant with pressure which will depend logarithmically on the partial molar volume change for the transition-state reaction. An exactly similar equation, the Kelvin equation, can be written for the change of equilibrium constant with pressure.[a] Since $\Delta V_{\ddagger}^{X}/RT$ is of the order of magnitude of 10^{-3} atm^{-1} for solution reactions, it is evident that the effect of these pressure changes will be of importance only at pressures in excess of 10^3 atm, and indeed this is verified experimentally.

Despite their small magnitude, study of these pressure effects is both interesting and important for the additional information it gives us about the mechanism of the reaction and the nature of the transition-state complex. If we plot $\ln k_n$ against P, we should get a curve whose slope at any point will be $-\Delta \overline{V}^{\pm}/RT = (\overline{V}_{\ddagger}^{X} - \overline{V}_A - \overline{V}_B - \cdots)/RT$. Since $\overline{V}_i$ can be determined independently for the reactants A, B, etc., it is possible from such a curve to determine $\overline{V}_{\ddagger}^{X}$, the partial molal volume of the transition-state complex, and if data are available over a sufficient range of pressures,

---

the reaction and this last term is not at all negligible. Most authors who have studied rate or equilibria at high pressures have usually chosen to correct their concentrations directly in computing proper rate or equilibrium constants. The present derivation offers an equally useful method.

  [a] Equation (XV.5.8) seems first to have been suggested by J. H. van't Hoff, "Vorlesungen über Theoretische und physikalische Chemie," vol. I, Braunschweig, 1901.

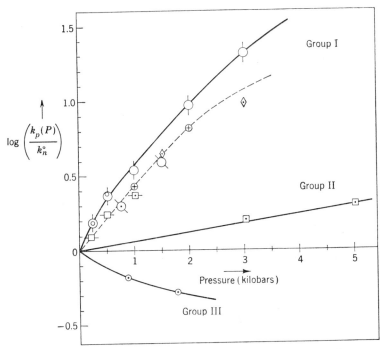

FIG. XV.3. Variation of some rate and equilibrium constants with pressure. ○ = ionization constant for piperidine in methanol at 25°C. [S. D. Hamann and W. Strauss, *Discussions Faraday Soc.*, **22**, 70 (1956).] $C_5H_{11}N + MeOH \rightleftharpoons C_5H_{11}NH^+ + CH_3O^-$. At 45°C the results do not differ significantly. ⊕ = bimolecular rate constant for dimerization of cyclopentadiene at 50°C in monomer as solvent. $2C_5H_6 \rightarrow C_{10}H_{12}$. [D. M. Newitt and A. Wasserman, *Trans. Chem. Soc.*, 735 (1940).] ◇ = bimolecular rate constant for association in methanol at 25°C. [S. D. Hamann and D. R. Teplitzky, *Discussions Faraday Soc.*, **22**, 114 (1956).] $C_6H_5N(CH_3)_2 + CH_3I \rightarrow C_6H_5N(CH_3)_3^+ + I^-$. The same authors have also shown that the apparent first-order rate constant for the solvolysis of allyl bromide in methanol can be superimposed on the same curve (also 25°C). $CH_2 = CHCH_2Br + CH_3OH \rightarrow CH_2 = CHCH_2OCH_3 + H^+ + Br^-$. ⊙ = first-order rate constant for the decomposition of benzoyl peroxide in CCl_4 at 70°C. [A. E. Nicholson and R. G. W. Norrish, *Discussions Faraday Soc.*, **22**, 97 (1956)]. $C_6H_5CO—O—O—CO—C_6H_5 \rightarrow 2C_6H_5COO$.

Note: This reaction is somewhat complicated by the cage effect on the escape of the newly formed free radicals. ⊘ = association constant for $2NO_2 \rightleftharpoons N_2O_4$ in CCl_4 solution at 22°C. [A. H. Ewald, *Discussions Faraday Soc.*, **22**, 138 (1956).] ⊡: $C_2H_5I + C_2H_5O^- \rightarrow C_2H_5OC_2H_5 + I^-$ in C_2H_5OH at 25°C. Second-order rate constant. [R. O. Gibson et al., *Proc. Roy. Soc. (London)*, **A150**, 223 (1935).] ⊟: $H_2O \rightleftharpoons H^+ + OH^-$. [B. B. Owen and S. R. Brinkley, Jr., *Chem. Revs.*, **29**, 461 (1941).] At 25°C, $\Delta \overline{V}_r = -23.4$ cc/mole.

the partial molal compressibility as well.[a] Some typical data are shown in Fig. XV.3.

[a] Over an extended pressure range we can represent $\overline{V}$ by the modified Kirkwood-Tait equation, $\overline{V} = \overline{V}_0/(1 + P/B)^C$, where B and C are positive constants, B in the range of 3000 atm and $C \cong 0.4$. The use of this equation makes it possible to write Eq. (XV.5.8) in the integrated form which may be expanded in a power series in P/B or B/P, de-

The first analysis of the effect of external variables on the rates of reaction seems to have been made by Evans and Polanyi,[a] who used the transition-state model. Soon thereafter, Perrin,[b] by applying the Evans and Polanyi methods to some high-pressure rate data, made a rough classification of pressure effects into three categories.

The first of these was the group of "slow reactions" which had very small Arrhenius A factors. These reactions showed the largest increase in rate with pressure, or alternatively, the most negative values of $\Delta \overline{V}^{\ddagger}$. The second group corresponded to reactions showing a much smaller $\Delta \overline{V}^{\ddagger}$, and Perrin noted that these reactions had Arrhenius A factors which were "normal," that is, in the range 10^{10} liters/mole-sec for bimolecular reactions. The third and not very well-illustrated group were reactions with a negative pressure coefficient,[c] that is, $\Delta \overline{V}^{\ddagger} > 0$. These can be seen plotted together in Fig. XV.3. For purposes of comparison, some equilibrium constants are also plotted.

The effect of pressure on equilibrium constants has been explored in a number of instances, and the Van't Hoff equation (XV.5.8) has been verified from independent studies of the partial molar volumes.[d] This has been reported for the isomerization of cis-dichloroethylene[e] with reasonable accuracy and qualitatively for N_2O_4 dissociation[f] and the ionization of weak electrolytes.[g]

In considering the way in which the volume changes in forming a transition complex, we might, as a first approximation, expect that $\Delta \overline{V}^{\ddagger}$ is some fraction of $\Delta \overline{V}_r$, where $\Delta \overline{V}_r$ is the over-all partial molal change accompany-

---

pending on the pressure range. Over the range 0 to 3000 atm the first two terms will usually suffice, leading to a quadratic dependence of $\ln k_n$ on P.

[a] M. G. Evans and M. Polanyi, *Trans. Faraday Soc.*, **31**, 875 (1935); **32**, 1333 (1936).

[b] Perrin, *ibid.*, **34**, 144 (1938).

[c] Actually this is an almost trivial category, since, when measurable, it might usually be expected that the reverse reactions to groups 1 and 2 would belong to group 3. The only requirement for this is that the over-all partial molal volume change in reaction $\Delta \overline{V}_r$ be less than $\Delta \overline{V}^{\ddagger}$. This condition usually obtains.

[d] A more extensive discussion of pressure effects on chemical systems will be found in the text by S. D. Hamann, "Physico-chemical Effects of Pressure," Butterworth & Co. (Publishers) Ltd., London, 1957.

[e] A. H. Ewald et al., *Trans. Faraday Soc.*, **53**, 991 (1957), measured the I_2-catalyzed isomerization at 180°C from 0 to 3000 atm. For K_{eq} the $\Delta \overline{V}$ observed and calculated agreed to within the accuracy of the data, ± 10 per cent.

[f] A. H. Ewald, *Discussions Faraday Soc.*, **22**, 138 (1956), has commented that the $\Delta \overline{V}$ predicted from bond radii is about 2 cc, whereas $\Delta \overline{V}$ observed is about 23 cc/mole. However, this is a very poor basis for calculation. If we use instead the molar volumes of $N_2O_4 = 62$ cc (at 0°C) and estimate NO_2 by comparison with CO_2 (40 cc at -38°C) and SO_2 (45 cc at -10°C) at about 42 cc/mole, then $\Delta \overline{V}$ becomes 22 cc/mole in good agreement.

[g] B. B. Owen and S. R. Brinkley, Jr., *Chem. Revs.*, **29**, 461 (1941). Also, unpublished work by S. W. Benson.

ing the reaction. Where data are available, this turns out to be a fair approach. Thus, in the dimerization of cyclopentadiene at 40°C,[a] $\Delta \overline{V}^{\ddagger} \cong$ -25 cc/mole (0 to 1000 atm). The over-all change for the reaction can be estimated crudely from the densities of cyclopentadiene (83 cc/mole) and its dimer (135 cc/mole) at 40°C at -31 cc/mole, indicating that in the transition state the structure is already fairly close to that of the tightly bound dimer and not to that of a loosely bound complex.[b]

The ionization of weak electrolytes is generally accompanied by an over-all contraction in volume due to the electrostriction of the solvent produced by the long-range coulombic forces. This is evidenced in a large increase in ionization constant (Fig. XV.3) with increasing pressure. On comparing different solvents we might expect that such an effect will, to a first approximation, be inversely proportional to the compressibility of the solvent, and this seems to be in reasonable agreement with the facts.[c] This electrostrictive effect on the solvent seems to be the principal difference between ionic and nonionic reactions.[d] For the latter class, the $\Delta \overline{V}^{\ddagger}$ can usually be localized in the solute species. Stearn and Eyring[e] have attempted to calculate from models of the transition state the value of $\Delta \overline{V}^{\ddagger}$ for reactions most of which, however, involve ions. Although the agreement is quite good in most cases, it must be looked upon in the instances in which total charge changes as due to a fortuitous cancellation of errors.[f]

Laidler[g] has attempted to predict $\Delta \overline{V}^{\ddagger}$ for reactions that involve changes in total charge on ions by use of an empirical formula for the partial molar

[a] Newitt and Wasserman, *Trans. Chem. Soc.*, 735 (1940).

[b] One would thus predict that for the reverse reaction, dissociation of the dimer, $\Delta \overline{V}_b^{\ddagger} = \Delta \overline{V}_f^{\ddagger} - \Delta \overline{V} = 6$ cc, so that this reaction should be inhibited by pressure increases.

[c] S. W. Benson, unpublished work. Hamann and Strauss, *Discussions Faraday Soc.*, **22**, 70 (1956), have found that in CH_3OH, K_{ion} is more influenced by pressure than in H_2O. This is in accord with the above point of view since $\beta(CH_3OH)/\beta(H_2O) \cong 1.6$ at 0°C.

[d] An admittedly extreme but dramatic example is provided by the study of super-critical solutions. At H_2O densities of about 0.3 g/cc the partial molal volume of NaCl in steam above the critical temperature of water (374°C) can reach values of -5000 cc/mole!! The coefficient of compressibility of steam at this temperature and density is about 20 times greater than for H_2O at 25°C. Data from S. W. Benson et al., *J. Chem. Phys.*, **21**, 2208 (1953).

[e] Glasstone et al., "The Theory of Rate Processes," *op. cit.*

[f] Their approach is to estimate the change in volume by assigning an average cross section to the molecules and an effective bond length in the transition state, thus localizing all volume changes in the solute species. Such an approach, neglecting local packing and long-range electrostriction, cannot account for differences between solvents. Note that when a weak electrolyte ionizes, $MB \rightleftharpoons M^+ + B^-$, there is an expected volume increase due to the increase in number of species which is overshadowed by the electrostriction of the solvent, so that such processes are accompanied by contractions in volume.

[g] K. J. Laidler, *J. chim. Phys.*, 485 (1957); *Discussions Faraday Soc.*, **22**, 88 (1956).

volumes of ions in aqueous solutions. The method, though qualitatively good, is not better quantitatively than about a factor of 2.

For nonionic reactions in which there is bond breaking, one may expect the transition state to have a larger molal volume than the reactant and thus be inhibited by pressure increases. Nicholson and Norrish have verified this for the first-order rate of decomposition of benzoyl peroxide (Fig. XV.3) in CCl_4. From their data, one calculates $\Delta V^{\ddagger} \cong 10$ cc. Considering that the molar volume of the peroxide is about 190 cc, the value of 10 cc could be accounted for by an increase in bond length of about 0.4 Å, a not unreasonable value.[a] For the very similar first-order decompositions of molecules of about the same size, Ewald[b] found $\Delta \overline{V}^{\ddagger} \cong 10$ cc for both the decomposition of 2-2'-azo bisisobutyronitrile and of pentaphenylethane in toluene solution (at 70°C). In both cases, I_2 was used as a scavenger to capture the free radicals formed and permit the reaction to be followed by optical means.

From the meager data that do exist, it appears that in second-order reactions involving the formation of a bond between two species $\Delta \overline{V}^{\ddagger}$ will be about $\frac{2}{3} \Delta \overline{V}_r$ and, in general, of the order of -20 cc. Conversely, for fission reactions we may expect $\Delta \overline{V}^{\ddagger} \cong \frac{1}{3} \Delta \overline{V}_r$, or about 10 cc. In the case of charged species, the electrostrictive effects on the solvent will dominate over these solute effects and we may expect to find $\Delta \overline{V}^{\ddagger}$ negative for the production of charged species. In the cases of group 2 noted by Perrin, in which there was a normal Arrhenius factor and only a small negative $\Delta \overline{V}^{\ddagger}$ (e.g., -4 cc/mole for $OC_2H_5^- + C_2H_5I \rightarrow C_2H_5OC_2H_5 + I^-$), we note that such reactions fit into the bond formation scheme above with possibly somewhat less negative $\Delta \overline{V}^{\ddagger}$ than might be expected in the absence of charge.[c]

[a] Calculated after the method of Stearn and Eyring, loc. cit., where we compute 190 cc/mole = 310 Å/molecule. By assuming a cubic shape for the molecule, we get an edge of 6.8 Å and a cross section of 46 Å². The equivalent bond extension is then $\Delta r = \Delta V^{\ddagger}/46 = 0.4$ Å.

[b] A. H. Ewald, Discussions Faraday Soc., **22**, 138 (1956).

[c] Another example (Perrin et al., loc. cit.) occurs in the basic hydrolysis of chloracetate ion by OH^-. The second-order rate constant is interpreted in terms of ($\Delta \overline{V}^{\ddagger} = -6$ cc/mole in H_2O)

$$\begin{array}{ccc}
\overset{H}{\underset{|}{\overset{\diagdown}{\underset{H}{}}}}\overset{CO_2^-}{\underset{}{\diagup}} & \left[\overset{H}{\underset{|}{\overset{\diagdown}{\underset{H}{}}}}\overset{CO_2^-}{\underset{}{\diagup}} \right] & \overset{H}{\underset{|}{\overset{\diagdown}{\underset{H}{}}}}\overset{CO_2^-}{\underset{}{\diagup}} \\
HO^- + C - Cl \rightleftharpoons & {}^{-\frac{1}{2}}HO\cdots C \cdots Cl^{-\frac{1}{2}} & \rightarrow HO-C + Cl^-
\end{array}$$

Laidler, loc. cit., has attempted to interpret the negative values in these cases as arising from an enhancement of charge in the periphery of the transition species to produce more electrostriction. It appears to the author that this is opposite to the effect which is to be explained, namely, that $\Delta \overline{V}^{\ddagger}$ in this case is not negative enough. Since OH^- is one of the smallest of the univalent negative ions $\overline{V}(Cl^-) - \overline{V}(OH^-) = 23$ cc (a result of its large electrostriction), there are several opposing effects to be considered. One is the diminution in size due to bond formation as already discussed. A second is the in-

6. Interactions between Charged Particles in a Solution

If the model of a solution of uncharged and nonassociated molecules is somewhat crude, it is at least relatively simple and offers a reasonable approximation to the behavior of these solutions. The same, unfortunately, cannot be said of the theory of ionic solutions. The forces responsible for the existence of nonionic liquids can be reasonably described as "short range" in the sense that most of the properties of these liquids can be approximated by neglecting interactions between any but nearest neighbors. Van der Waals forces fall off as $1/r^7$ and the energies of interaction are usually of the order of magnitude of kT for nearest neighbors.[a] The forces between ions, however, fall off as $1/r^2$ and the energies as $1/r$. In addition, the energies are in general large compared to kT. This means that we shall have great difficulty in distinguishing between "independent" molecular units and "molecular complexes" in dealing with ionic solutions.[b] Furthermore, ionic solutions exist at measurable concentrations only because the interactions between ions and solvent molecules are sufficiently large to overcome the interionic forces that would otherwise cause insolubility of the salt. In dealing with ionic systems we are thus involved with what are called "long range" forces between strongly interacting species. To gain some appreciation of the magnitudes of these interactions, let us calculate them by using some overly simple but instructive electrostatic models.

Electrostatic theory tells us that the force between two point charges $z_1 \mathcal{E}$ and $z_2 \mathcal{E}$ at a distance r apart in vacuum is

$$F(r) = \frac{z_1 z_2 \mathcal{E}^2}{r^2} \qquad (XV.6.1)$$

where $\mathcal{E} = 4.80 \times 10^{-10}$ esu, the unit of atomic charge. The potential energy of these two charges is equal to the work required to separate them to ∞:

$$U(r) = \int_r^\infty F(r) \, dr = \frac{z_1 z_2 \mathcal{E}^2}{r} \qquad (XV.6.2)$$

For univalent ions, $|z_1| = |z_2| = 1$, at a distance of 5 Å, $|U(r)| = 66$ Kcal/mole, which is of the order of magnitude of most single bonds. The

creased electrostriction of the transition complex due to its charge of -2, and a third is the compensating effect of its large effective radius. A simple electrostatic theory by itself cannot account for the results of such complex interactions.

[a] For spherically symmetrical particles $F(r) = -cr^{-7}$. The potential $\overline{U}(r) = -cr^{-6}/6$. J. E. Lennard-Jones and A. E. Ingaham, *Proc. Roy. Soc. (London)*, **A107**, 636 (1925), have shown that for face-centered cubic lattices, the contribution of non-nearest neighbors is about 20 per cent of the total binding energy.

[b] This becomes very acute in deciding the status of an ion pair or of the "solvation shell" around an ion.

force $F(r) = 9.2 \times 10^{-5}$ dynes, which would produce on a particle of mass $m = 50/N_{Av} = 8.3 \times 10^{-23}$ g an acceleration of 1.1×10^{18} cm/sec^2, that is, 10^{15} times gravity.[a] For higher charges, both $U(r)$ and $F(r)$ increase proportionally. The magnitudes can become astronomically large for multiply charged ions in closer proximity.

The electrical parameters are equally impressive. The electric field intensity $E(r)$ in the neighborhood of a point charge $z\varepsilon$ is, in vacuo

$$E(r) = \frac{z\varepsilon}{r^2} \cdots \tag{XV.6.3}$$

so that again for a univalent ion at a distance of 5 Å, $E = 5.75 \times 10^7$ volts/cm, a field intensity larger by several powers of 10 than anything yet available in a laboratory.

In a hypothetical, isotropic, structureless, uniform medium of dielectric constant D, the quantities $F(r)$, $U(r)$, and $E(r)$ are all reduced by the factor $1/D$. For water at 25°C with $D = 78.5$, for the preceding cases we find $F(r) = 1.2 \times 10^{-7}$ dynes, $U(r) = 0.83$ Kcal/mole, and $E(r) = 7.2 \times 10^5$ volts/cm, somewhat reduced but still large. In considerations of solvent power it is the ratio $U(r)/kT$ which indicates the extent to which the dielectric medium is important (relative to thermal motion) in shielding charges from each other.

Two equal and opposite charges $\pm\varepsilon$ separated by a distance d constitute an electric dipole of strength $\mu = \varepsilon d$. The potential $U(r)$ at distances $r \gg d$ is given by

$$U(r) = \frac{\mu}{r^2} \cos \theta \tag{XV.6.4}$$

where θ is the angle between the axis of the dipole and the point, distant r from the dipole. The electric field intensity $|E| = (E_r^2 + E_\theta^2)^{1/2}$, where E_r and E_θ are respectively the components along the radius vector r and at right angles to r. They are obtained by differentiating $U(r)$ as follows

$$E_r = -\left(\frac{\partial U}{\partial r}\right)_\theta = \frac{2\mu}{r^3} \cos \theta$$

$$E_\theta = -\frac{1}{r}\left(\frac{\partial U}{\partial \theta}\right)_r = \frac{\mu}{r^3} \sin \theta \tag{XV.6.5}$$

Since most ionizing solvents have dipole moments, we have some interest in calculating the interaction between a point charge ε and a dipole μ. We find for the interaction in the gas phase

$$U(r) = \frac{\varepsilon\mu}{r^2} \cos \theta \tag{XV.6.6}$$

For a water molecule with $\mu = 1.85$ Debye $= 1.85 \times 10^{-18}$ esu and a

[a] Having started from rest such a particle would, in 10^{-13} sec, have moved 0.55 Å and reached a velocity of 1.1×10^5 cm/sec (relative to the other particle).

univalent ion at a separation of 5 Å, $U(r) = 5.1$ Kcal/mole when $\theta = 0.$[a]
At half the distance $U(r) = 20.4$ Kcal/mole and, neglecting water-water
interactions, which are small, we can calculate a value of about 122
Kcal/mole for the hydration of a univalent ion in the gas phase by 6
water molecules at an average distance of 2.5 Å. These interactions are
the same order as the heats of most chemical reactions and would tend to
place such clusters in the category of complex ions rather than loose
aggregates.[b]

The interaction energy of two dipoles can be computed from Eq. (XV.6.5).
If we measure the distance r along a line connecting the dipole centers and
θ is the azimuth angle between each dipole and the line of centers, then the
mutual energy is

$$U(r,\theta,\phi) = \frac{\mu_1 \mu_2}{r^3} (2 \cos \theta_1 \cos \theta_2 - \sin \theta_1 \sin \theta_2 \cos \phi) \cdots$$

$$\text{(XV.6.7)}$$

where ϕ is the angle between two planes, one through each dipole, inter-
secting along the line of centers. The
configuration is shown in Fig. XV.4.

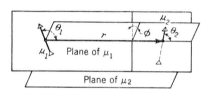

FIG. XV.4. Interactions of two dipoles.
The two planes intersect along r, and each
contains one dipole axis.

The maximum attractive interaction is obtained for the "head-to-
tail" alignment ($\theta_1 = \theta_2 = 0$). For a
pair of water molecules with $\mu_1 = \mu_2$
$= 1.85$ Debyes and $r = 3.1$ Å (the
average distance in liquid water),
$U(r) = 3.2$ Kcal/mole,[c] a rather
appreciable interaction for two
otherwise neutral molecules.[d]

These calculations all neglect significant interactions that arise from
the polarization of electrically interacting molecules. Thus, a neutral

[a] This corresponds to the maximum; when $\theta = \pi$, $U(r) = -5.1$ Kcal/mole, or a
minimum. At the latter position the barrier to rotation is about $5.1/57 = 90$ cal/degree,
so that the bound water molecule is not very free to rotate at room temperature. These
values are all minimum values, since the corrections for polarizability and finite size
of the dipole both tend to increase the interaction.

[b] This neglects the mutual repulsion of the water dipoles which although small per
pair is not negligible because of the large number of repulsions. In considering solvation
in solution we must also reckon with the work required to remove H_2O molecules from
their attractive positions in pure H_2O. This is about 7.5 Kcal/mole H_2O.

[c] If we assume a face-centered cubic lattice for H_2O, this calculation gives about
4.5 Kcal/mole as the dipole contribution to the energy of vaporization of water. On
correcting for finite size of dipole (i.e., assuming $d = 1$ Å), this becomes 6 Kcal.

[d] Equations (XV.6.6 and 7) are accurate only for distances $r \gg d$, where d is the effective length of the dipole. At distances $r \cong d$, there is a correction term in the denominator of amount $1 - d^2/r^2$. Since it is unlikely that r will ever be less than $3d/2$, for
near-neighbor dipoles, this can raise the interaction to a maximum of ⁹⁄₅, the amount
given by the equation.

molecule that has no permanent dipole of its own will, in the presence of an electric field, has induced in it a dipole μ_E. For an isotropic[a] molecule of polarizability α, in a uniform electric field E, the induced dipole μ_E will be opposite to E in direction and of magnitude $\mu_E = \alpha E$. The work done in removing the dipole from the field, which is the interaction energy of the induced dipole with the inducing field, is $U = -\alpha E^2/2$. In the case of the field due to a point charge $z\mathcal{E}$ at a distance r from a spherically symmetrical molecule

$$U(r) = -\frac{\alpha z^2 \mathcal{E}^2}{2r^4} \qquad \text{(XV.6.8)}$$

valid for $r >$ radius of the molecule. Since α is in the range of 1 to 5 cc/mole for many substances, we find by using $\alpha = 3$ cc/mole and $r = 5$ Å that $U(r)$ is about 1.3 Kcal/mole, an interaction which is not small compared to the energies calculated for dipole-dipole interactions.[b] The induced dipole μ_E is about 1 Debye, representing the separation of 1 unit of atomic charge by about 0.2 Å. Although these ion-molecule interactions can have magnitudes as great as or greater than some ion-dipole interactions, it is important to note that they fall off much more rapidly with distance and can no longer be considered long range. It is important also to note that the correction for polarization always represents a net attraction and can amount to several kilocalories for near neighbors.

7. Activity Coefficients in Ionic Solutions

In order to apply the transition-state model to reactions involving charged or polar species, we need to know how the activity coefficients of those species vary with the experimental parameters at our disposal. Because of the long range of electrical interactions, charged particles in a solution will be affected not only by their nearest neighbors but also by ions and molecules at quite considerable distances. Any changes in these non-nearest neighbors will affect the potential energy of the ions and so their activity coefficients. For convenience, we can divide these interactions into two categories: one for the interactions of ions with other ions

---

[a] Most molecules are not isotropic, and in such cases the polarization of the molecule will have components along each of its three principal axes of polarization of magnitude proportional to the field component along those axes. For such cases the reader is referred to more specialized treatises such as J. A. Stratton, "Electromagnetic Theory," McGraw-Hill Book Company, Inc., New York, 1941, or for one that makes less use of vector and tensor notation, J. H. Jeans, "Electricity and Magnetism," 5th ed., Cambridge University Press, New York, 1925.

[b] This model is very crude for small values of r, since it is based on the picture of the polarizable molecule as a conducting spherical shell. A further difficulty also arises from the fact that the polarizability α, which is observed for small field strengths, *may not be* applicable to the enormous fields in the neighborhoods of ions.

in the solution and the second for the interactions with other uncharged molecules, polar or nonpolar as the case may be.

At sufficiently high dilutions the interactions between ions will approach zero and may thus be neglected. At such dilutions the only important interactions experienced by ions will be with the molecules of solvent, and by varying the solvent composition or properties[a] we will observe changes in activity coefficients which are determined by ion-solvent interactions. On the other hand, at high dilution, by changing the concentration of ions, we will observe changes in activity coefficients which arise from ion-ion interactions. Actually, the change in ionic (solute) concentrations will also produce changes in the solvent properties, but at sufficiently high dilutions these will be second-order effects which may be neglected.[b]

The simplest model capable of reproducing the ion-ion interactions at high dilutions is that given by the Debye-Hückel Theory.[c] The model assumes that the solvent can be characterized as a structureless, isotropic medium of dielectric constant D which is identical with its macroscopic dielectric constant. The ions are regarded as point charges imbedded in hard spheres of fixed radius and of dielectric constant 1.

With such a model, the ion-ion interactions are obtained by calculating the most probable distribution of ions around any central ion and then evaluating the energy of the configuration. If $\psi_i(r)$ is the spherically symmetrical potential in the solution at a distance r from a central ion i of charge $z_i\mathcal{E}$,[d] then $\psi_i(r)$ will be made up of two parts: $z_i\mathcal{E}/Dr$, the coulombic field due to the central ion, and an additional part, $\psi_{ai}(r)$, due to the distribution of the other ions in the solution around i. The potentials $\psi_{ai}(r)$ and $\psi_i(r)$ must satisfy Poisson's equation: $\nabla^2\psi_i = -4\pi\rho/D$ at every point r in the solution, $\rho = \rho(r)$ being the charge density at the point r. For a spherically symmetrical potential this can be written as

$$\nabla^2\psi_i(r) = \frac{1}{r^2}\frac{\partial}{\partial r}\left(r^2\frac{\partial\psi_i}{\partial r}\right) = \frac{1}{r}\frac{\partial^2(r\psi_i)}{\partial r^2} = -\frac{4\pi\rho}{D} \qquad (XV.7.1)$$

which must be solved subject to the boundary conditions that $\psi(r) \to 0$ as $r \to \infty$. At the surface of the ion i the electric displacement $DE(r) = -D(\partial\psi_i/\partial r)_{a_i}$ must be continuous, so that $D(\partial\psi_i/\partial r)_{a_i} = -z_i\mathcal{E}/a_i^2$.

The charge density $\rho(r)$ is determined by the Boltzmann equation, which says that the concentration of positive and negative charges in a spherical shell $4\pi r^2\,dr$ at a distance r from the central ion is proportional

[a] This may be done by changing its density via the use of high pressures.

[b] In similar fashion, changes in solvent will also change ion-ion interactions.

[c] For a detailed account of this theory, the reader is referred to treatises on electrolyte solutions such as H. S. Harned and B. B. Owen, "The Physical Chemistry of Electrolytic Solutions," 3d ed., Reinhold Publishing Corporation, New York, 1958, and R. A. Robinson and R. H. Stokes, "Electrolyte Solutions," Butterworth & Co. (Publishers) Ltd., London, 1955.

[d] $\mathcal{E}$ is the unit of atomic charge, 4.80×10^{-10} esu, and z_i will be an integer.

to $n_j \exp(-z_j \mathcal{E} \psi_i / DkT)$, where n_j is the average number of jth type ions in the solution per unit volume of charge $z_j \mathcal{E}$. On summing over all ions, we have

$$\rho(r) = \mathcal{E} \sum_j n_j z_j \exp(-z_j \mathcal{E} \psi_i / DkT) \cdots \qquad (XV.7.2)$$

In order to use this equation, we are forced to make the approximation that $z_j \mathcal{E} \psi_i / DkT \ll 1$ so that the exponential can be expanded in a power series:

$$\rho(r) = \mathcal{E} \left\{ \sum_j n_j z_j \left[1 - \frac{z_j \mathcal{E} \psi_i}{DkT} + \frac{1}{2} \left(\frac{z_j \mathcal{E} \psi_i}{DkT} \right)^2 - \cdots \right] \right\}$$

$$(XV.7.3)$$

Because of the condition of electric neutrality, the first term, $\Sigma n_j z_j$, vanishes and it is the next term which is used:[a]

$$\rho(r) \cong \frac{\mathcal{E}^2 \psi_i}{DkT} \sum n_j z_j^2 = \frac{\varkappa^2 \psi_i}{4\pi} \qquad (XV.7.4)$$

where the parameter $\varkappa$, independent of r, is defined by the last equation. The substitution of this value of $\rho(r)$ into Eq. (XV.7.1) leads to a separable equation whose solution for ψ_i is[b]

$$\psi_i = \frac{z_i \mathcal{E}}{D} \frac{e^{\varkappa a_i}}{1 + \varkappa a_i} \frac{e^{-\varkappa r}}{r} \qquad (XV.7.5)$$

which for small values of $\varkappa r$ can be approximated as

$$\psi_i \cong \frac{z_i \mathcal{E}}{Dr} (1 - \varkappa r) = \frac{z_i \mathcal{E}}{Dr} - \frac{z_i \mathcal{E}}{D(1/\varkappa)} \qquad (XV.7.6)$$

where $\varkappa r < 1$ and $\varkappa a_i < 1$.

From the approximation (XV.7.6) we see that near the ion, at distances $r < 1/\varkappa$, the potential is composed of two parts: the coulombic potential of the central ion $z_i \mathcal{E}/Dr$ and ψ_{ai} the constant coulombic potential of a spherically symmetrical distribution of charge $-z_i \mathcal{E}$ located on the surface of a sphere of radius $1/\varkappa$ that is centered on the ion $z_i \mathcal{E}$. This charge distribution of the neighboring ions is referred to as the "ion atmosphere," while $1/\varkappa$ is called the mean radius of the ion atmosphere.

The electrical work of removing the central ion from its atmosphere, or alternatively, the work of discharging the central ion in its atmosphere (at constant pressure) is equal to $(\partial F_e / \partial n_i) = \mu_{ie}$, the partial change in

[a] For symmetrical electrolytes, note that all odd terms vanish.

[b] There is considerable uncertainty regarding the precise significance to be attached to the parameter a_i which we have been using as the hard sphere radius of the central ion. Customary usage is to regard it as the hard sphere, collision diameter for a pair of oppositely charged ions. In such a case we cannot use the boundary condition for the field intensity E but instead must use the relation that the charge on the atmosphere must be equal and opposite to that on the central ion. As we shall see, a_i is treated as an empirical quantity, and to conform to this practice, we shall drop the subscript i.

Gibbs' free energy due to the electrical interactions of the ion i with its atmosphere. But the latter is given simply by the product of the potential of the atmosphere $\psi_{a_i}^{\circ}$ at the ion i times the charge on i, $\psi_{a_i}^{\circ} z_i \mathcal{E}$. Summing this over all i-type ions in the solution would involve counting the interactions of each i-type ion twice, once at the center of an atmosphere and once in its atmosphere. Thus, the contribution of the ion-ion interactions to the chemical potential is just half of the above figure, or

$$\mu_{ie} = \tfrac{1}{2}\psi_{a_i}^{\circ} z_i \mathcal{E} \cdots \qquad (\text{XV.7.7})$$

From Eq. (XV.7.5) we can calculate the value of $\psi_{a_i}^{\circ} = -z_i \mathcal{E} \varkappa / D(1 + \varkappa a)$, so that

$$\mu_{ie} = -\frac{z_i^2 \mathcal{E}^2}{2D} \frac{\varkappa}{1 + \varkappa a} \qquad (\text{XV.7.8})$$

But since the part of the activity coefficient due to electrical interactions is given by $\mu_{ie}/kT = \ln f_{ie}$, we can write

$$\ln f_{ie} = -\frac{z_i^2 \mathcal{E}^2 \varkappa}{2DkT(1 + \varkappa a)} \cdots \qquad (\text{XV.7.9})$$

By using the definition of ionic strength $\mu = \tfrac{1}{2}\Sigma C_j z_j^2$, together with the relation $n_j = N_{Av} C_j/1000$, where N_{Av} is Avogadro's number, we can express $\varkappa$ [Eq. (XV.7.4)] as

$$\varkappa = \left(\frac{8\pi N_{Av}\mathcal{E}^2}{1000DkT}\right)^{\frac{1}{2}} \mu^{\frac{1}{2}} = A\mu^{\frac{1}{2}} \qquad (\text{XV.7.10})$$

On substitution in Eq. (XV.7.5) we have

$$\ln f_{ie} = -\frac{z_i^2 \mathcal{E}^2 A \mu^{\frac{1}{2}}}{2DkT(1 + Aa\mu^{\frac{1}{2}})} \xrightarrow{\mu \to 0} \qquad (\text{XV.7.11})$$

$$-\frac{z_i^2 \mathcal{E}^2 A \mu^{\frac{1}{2}}}{2DkT} = -z_i^2 A_i C^{\frac{1}{2}} \qquad (\text{XV.7.12})$$

This last equation, valid at extremely low concentrations and lacking the troublesome diameter a, is known as the Debye-Hückel limiting law for ionic activities and has been well verified in aqueous and some non-aqueous systems. The standard state for these electrical activity coefficients is the infinitely dilute solution[a] in which $f_{ie} = 1$.

For a 1:1 electrolyte in water at 25°C, using base 10 logarithms, $A = 0.33$ Å^{-1} M$^{-\frac{1}{2}}$ and $A_i = 0.51$ M$^{-\frac{1}{2}}$. Since a is usually of the order of 3 Å, we see that $Aa\mu^{\frac{1}{2}} = 0.1$ at 0.01 M, and we should not expect to use the limiting law at higher concentrations.[b] Equation (XV.7.11) is reasonably useful up to about 0.1 M if the parameter a is treated as an empirical

---

[a] See footnote on page 504 concerning the standard state.

[b] Since $A \propto CD^{-\frac{1}{2}}$, these considerations must be modified appropriately in considering nonaqueous solvents of different dielectric than water.

quantity.[a] Many equations have been proposed for extending the range of concentration over which the ionic activities can be represented by a simple equation but none of very general scope has been found. Guggenheim[b] has proposed a formula which is reasonably accurate for mixtures of electrolytes up to ionic strengths of 0.1 M. It is

$$\ln f_{ie} = -\frac{z_i^2 A_i \mu^{1/2}}{1 + \mu^{1/2}} + \sum_j B_{ij} C_j \qquad (XV.7.13)$$

where the empirical constants B_{ij} are summed only over the ions of sign opposite to i.[c]

8. The Effect of Ionic Strength on Reactions between Ions

The rate constant for a reaction between two ions A and B:

$$A^{z_A} + B^{z_B} \rightleftharpoons A \cdot B^{z_A + z_B} \rightarrow \text{products}$$

which go through a transition-state complex $X = A \cdot B$ of total charge $z_A + z_B$, is given by Eq. (XV.5.2)

$$\ln k_n = \ln \left(\frac{kT}{h}\right)(K^{\ddagger}) + \ln \frac{f_A f_B}{f_X} + \cdots \qquad (XV.8.1)$$

If we use the limiting law to express the activities and let k_n° represent the rate constant at infinite dilution, we find [Eq. (XV.7.12)]

$$\ln \frac{k_n}{k_n^\circ} = \frac{z_A z_B \mathcal{E}^2 A \mu^{1/2}}{DkT} = 2 z_A z_B A_i \mu^{1/2} \cdots \qquad (XV.8.2)$$

and in H_2O at 25°C by using the known values of A_i

$$\log \frac{k_n}{k_n^\circ} = 1.02 z_A z_B \mu^{1/2} \qquad (XV.8.3)$$

This last equation, derived from Brönsted's[d] and Bjerrum's[e] equations, has been the subject of intensive investigation.[f] It predicts that, at low

[a] For ions of higher charge, the usable concentration range shrinks very rapidly with increasing z.

[b] E. A. Guggenheim, *Phil. Mag.*, **22**, 322 (1936).

[c] The basis for this is the theory of specific ion interaction, proposed by J. N. Brönsted, *J. Am. Chem. Soc.*, **44**, 877 (1922), to the effect that ions will be *uniformly* influenced by ions of the same sign but *specifically* influenced by ions of opposite sign. This is quite reasonable in terms of the above equation, in which the uniform influence is included in the Debye-Hückel terms. The implication is that only ions of opposite sign will approach each other closely enough to produce specific interactions. What the nature of the interactions is, however, is not specified.

[d] J. N. Brönsted, *Z. physik. Chem.*, **102**, 169 (1922); **115**, 337 (1925).

[e] N. Bjerrum, *ibid.*, **108**, 82 (1924).

[f] A very similar expression including a term in the denominator $1 + \varkappa a$ has been obtained by G. Scatchard, *J. Am. Chem. Soc.*, **52**, 52 (1930), from basically the same model.

concentrations, the rate constant for an ion-ion reaction should vary logarithmically with the square root of the ionic strength.[a] In particular, reactions between ions of opposite sign, $z_A z_B < 0$, should slow down with increasing ionic strength, while reactions between ions of like sign should show an increase with increasing ionic strength. Further, the rate of change of the log of the rate constant with $\mu^{1/2}$ should be $1.02 z_A z_B$. Finally, the prediction is that reactions between ions and molecules should not be affected by changing ionic strength. These predictions have been verified nearly quantitatively; Fig. XV.5 shows some results of interest. The difficulty of verifying such a relation quantitatively is manifold. In the first place, the reactions must be measured in the region of low ionic strength if the Debye-Hückel limiting law is to be employed. That means ionic strengths below 0.01 for 1:1 electrolytes and below 0.001 for higher-valence ions. But in that range of ionic strength the total effect sought is about a change of 20 to 50 per cent in the rate constant over the accessible concentration range.

For the higher-valence ions further difficulties are encountered in that the specific ion interaction proposed by Brönsted (loc. cit.) dominates the behavior of the activity, and much better correlation is found between $\log (k/k°)$ and the concentration of polyvalent ions of charge opposite to those entering into the reaction rather than the ionic strength. This has been emphasized by Olson and Simonson,[b] who showed that the reaction $Hg^{++} + CO(NH_3)_5Br^{++}$ (Fig. XV.5) had a rate constant which was dependent on (ClO_4^-) but independent of the positive cation present with it. They showed also that for the reaction $BrCH_2CO_2^- + S_2O_3^= \rightarrow (S_2O_3 \cdot CH_2CO_2)^= + Br^-$, the rate constant depended on (K^+) but not on the negative ion.[c] In both of these cases, at constant concentration of the ion of opposite sign but over a twofold change in ionic strength, the rate constant changed less than 2 per cent, in apparent contradiction to the Brönsted equation.

It has been shown, however, by Scatchard[d] that such ion-ion interaction is to be expected if one uses further terms in the Debye-Hückel equation. If Mayer's[e] somewhat complex theory of electrolytes is applied to polyvalent ions, quantitative agreement is also possible. What is implicit in all of these effects and what is quite reasonable is that polyvalent ions of

[a] In the more general case of reactants, A, B, C, D, . . . , the coefficient of $\mu^{1/2}$ has the form $1.02 (z_A z_B + z_B z_C + z_A z_C + z_A z_D + \cdots) = 1.02 \sum_{i \neq j}' z_i z_j$.

[b] A. R. Olson and T. R. Simonson, J. Chem. Phys., **17**, 1167 (1949). See also comments by M. Kilpatrick, Ann. Rev. Phys. Chem., **2**, 269 (1951).

[c] Negative ions used were NO_3^-, $SO_4^=$, and $Co(CN)_6^{3-}$. In most of these experiments the ionic strength was in the range 0.001 to 0.04 M.

[d] G. Scatchard, Nat'l Bur. Standards Circ., **524**, p. 185, 1953.

[e] J. E. Mayer, J. Chem. Phys., **18**, 1426 (1950).

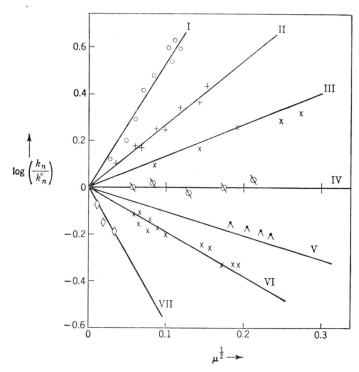

Fig. XV.5. The effect of ionic strength on the rates of some ionic reactions. Reactions:

I: $[Co(NH_3)_5Br]^{++} + Hg^{++} + H_2O \rightarrow [Co(NH_3)_5(H_2O)]^{3+} + (HgBr)^+$
II: $S_2O_8^= + I^- \rightarrow ?[ISO_4^- + SO_4^=] \rightarrow I_3^- + 2SO_4^=$ (not balanced)
III: $[O_2N-N-COOEt]^- + OH^- \rightarrow N_2O + CO_3^= + EtOH$
IV: cane sugar $+ OH^- \rightarrow$ invert sugar   (hydrolysis reaction)
V: $H_2O_2 + H^+ + Br^- \rightarrow H_2O + \frac{1}{2}Br_2$ (not balanced)
VI: $[Co(NH_3)_5Br]^{++} + OH^- \rightarrow [Co(NH_3)_5(OH)]^{++} + Br^-$
VII: $Fe^{++} + Co(C_2O_4)_3^{3-} \rightarrow Fe^{3+} + [Co(C_2O_4)_3]^{4-}$

References for Fig. XV.5.: I, VI. J. N. Brönsted and V. R. Livingston, *J. Am. Chem. Soc.*, **49**, 435 (1927). II. V. K. La Mer, *ibid.*, **51**, 334 (1929). C. V. King and M. B. Jacobs, *ibid.*, **53**, 1704 (1931). III. J. N. Brönsted and A. Delbanco, *Z. anorg. Chem.*, **144**, 248 (1925). IV. S. Arrhenius, *Z. physik. Chem.*, **1**, 111 (1887). V. R. Livingston, *J. Am. Chem. Soc.*, **48**, 53 (1926). VII. J. Barrett and J. H. Baxendale, *Trans. Faraday Soc.*, **52**, 210 (1956). Reactants are shown to indicate the composition of the transition-state complex, and so the equations are not necessarily balanced.

opposite sign will have large enough interactions to form ion pairs or complex ions with each other. Thus, Mg^{++} and $SO_4^=$ at 5 Å separation in H_2O will have a mutual potential energy of 3.3 Kcal/mole [Eq. (XV.6.2)] if we neglect the dielectric saturation of H_2O.[a] This is sufficiently large

---

[a] Since the field strengths in this case are of the order of 1.4×10^6 volts/cm, it is unlikely that H_2O will still have its normal, macroscopic dielectric constant. The evidence available indicates that D begins to decrease significantly at fields in excess of 1×10^5 volts/cm. See discussion in K. J. Laidler and C. Pegis, *Proc. Roy. Soc. (London)*, **A241**, 80 (1957).

compared to kT (0.6 Kcal/mole at 300°K) that the $Mg^{++} \cdot SO_4^-$ ion pair must be considered a species of importance in solutions of $MgSO_4$ at any reasonable concentrations. It is evident that for ions of higher charge such as La^{3+}, ion pairs of this type will be even more important.

Unfortunately, electrolyte theories which include higher terms must introduce additional parameters for ion sizes and repulsive forces. As yet, none of the theories developed appears capable of making a priori predictions of ion-ion interactions for the higher-valence ions or in the more concentrated solutions. In addition, they lead to unwieldy expressions, which makes their application very tedious (Scatchard, *loc. cit.*) and, in view of the assumptions included in these theories, of dubious value.

9. Association of Ions

The Debye-Hückel model for electrolytes must certainly be unreal in its representation of the immediate environment of an ion when the field strengths are so great.[a] The success of the limiting law, at high dilutions, must be ascribed to the fact that in very dilute solutions, changes in concentration do not affect appreciably the immediate neighborhoods of ions. Thus, in 0.001 M NaCl the average distance between ions is about 94 Å, while the ion atmosphere has a radius of 100 Å [Eq. (XV.7.10)]. These are sufficiently large distances that we might expect the model to work well.[b]

However, for higher concentrations the model is no longer valid, and further, the approximations, $\psi_i z_j \varepsilon / DkT < 1$, cannot be valid close to the ion i [Eq. (XV.7.2)]. Bjerrum[c] made the proposal that any pair of ions whose interaction is of the order of $2kT$ or more should be considered as an ion pair, not as independent ions, and that the Debye-Hückel treatment should be reserved only for the free ions separated by distances sufficiently large that their interaction is less than this. If we call this distance r_B and neglect the ion atmosphere[d] around such an ion pair, then for the ion pair z_i and z_j

[a] Additional difficulties beyond that of dielectric saturation already mentioned (page 527) have to do with the enormous electrostrictive pressures in the neighborhoods of ions. These can amount to some 10^5 to 10^7 atm and lead to serious changes in the properties of the solvent (unpublished work by the author). Note that these lead to an increase in dielectric constant due to compression of solvent and tend to cancel the effects of saturation. For some detailed discussion, see paper by H. S. Frank, *J. Chem. Phys.*, **23**, 2023 (1955).

[b] That is, the numbers of pairs of ions at distances less than, say, 20 Å are sufficiently small as not to affect the average behavior represented by the theory.

[c] N. Bjerrum, *Kgl. Danske Videnskab. Selskab, Mat.-fys. Medd.*, **7**, No. 9 (1926); "Selected Papers," p. 108, Einar Munsksgaard, Copenhagen, 1949.

[d] Note that the effect of an ion atmosphere around such a pair is to stabilize the pair.

$$r_{Bij} = \frac{z_i z_j \mathcal{E}^2}{2DkT} \cdots \qquad (XV.9.1)$$

To a first approximation, we may estimate the number of ion pairs $i \cdot j$ by integrating between the distance of closest approach of such a pair a_{ij} and the critical cutoff r_{Bij}:

$$N_{ij} = N_i N_j 4\pi \int_{a_{ij}}^{r_{Bij}} e^{-z_i z_j \mathcal{E}^2/DkTr} r^2 \, dr \cdots \qquad (XV.9.2)$$

On substituting $x = -z_i z_j \mathcal{E}^2/DkTr$ and $b = |z_i z_j| \mathcal{E}^2/DkTa_{ij}$ for r_{Bij} [from Eq. (XV.9.1)]

$$N_{ij} = 4\pi \left(\frac{|z_i z_j| \mathcal{E}^2}{DkT}\right)^3 N_i N_j \int_2^b \frac{e^x}{x^4} \, dx \cdots \qquad (XV.9.3)$$

In terms of molarities, $C_i = 1000 N_i/N_{Av}$. By letting the definite integral be represented by $Q(b)$,

$$C_{ij} = \left[\frac{4\pi Q(b)}{1000}\left(\frac{|z_i z_j| \mathcal{E}^2}{DkT}\right)^3 N_{Av}\right] C_i C_j \cdots \qquad (XV.9.4)$$

We note that the quantity in brackets is the equilibrium constant K_{ij} for the equilibrium $A^{z_i} + B^{z_i} \rightleftharpoons A \cdot B^{z_i + z_j}$. The definite integrals $Q(b)$ have been tabulated as functions of b,[a] so that the K_{ij} is determined by a single parameter.

In terms of fraction of association $\alpha = C_{ij}/C_i$,

$$\alpha = \left[\frac{4\pi Q(b)}{1000}\left(\frac{|z_i z_j| \mathcal{E}^2}{DkT}\right)^3 N_{Av}\right] C_j \cdots \qquad (XV.9.5)$$

A second approximation is to use the Debye-Hückel theory to give the activity of the ions i and j, while a third approximation is to use the Kirkwood or Amis-Jaffe models for ion pairs[b] to evaluate the activity coefficient of the ion pair. Under these conditions we can write:

$$\frac{C_{ij}}{C_i C_j} = K_{ij}\frac{f_i f_j}{f_{ij}} \cdots \qquad (XV.9.6)$$

where K_{ij} can be evaluated from Eq. (XV.9.4).[c]

The Bjerrum treatment suffers from the oversimplifications of the Debye-Hückel model, particularly regarding the correct method of calculating the energy of an ion pair at small distances when molecular structure is surely important. Nevertheless it is certainly a step in the right

---

[a] See Stokes and Robinson, loc. cit. By integration by parts $Q(b)$ can be related to known, tabulated integrals.

[b] See Sec. XV.11.

[c] Since the ionic strength depends on the degree of dissociation, the actual solution of Eq. (XV.9.5) for C_{ij} becomes a matter of successive numerical approximations. One first solves for K_{ij} from Eq. (XV.9.4) and then calculates α, from Eq. (XV.9.5), and the f_i and f_j, by using the revised estimates of ionic strength.

direction, and it has provided a convenient framework for correlation of ion-ion interactions. Fuoss[a] has discussed the problem of the arbitrary choice of the Bjerrum critical distance r_B and has shown that any other choice near $2kT$ for the ion pair interaction distance would serve equally well. He and Kraus[b] have applied the theory to the dissociation of strong electrolytes in media of varying dielectric constant and obtained excellent correlations between conductance and the theory.

Table XV.5 shows the rather dramatic change in K_{eq} for the dissociation of tetrisoamyl ammonium nitrate, $(i\text{-}Am_4N)^+NO_3^-$, with dielectric constant in mixtures of H_2O and dioxane. Although it is possible to get much better agreement with the conductance data by using slightly different values of a_{ij}, the case shown is used to emphasize the essential correctness of the method. Note also that no account has been taken of the preferential solvation of ions by one of the two solvents. The Fuoss and Kraus treatment also gives a simple model for the calculation of ion triplet and quadruplet concentrations.

It can be seen that the added complexity of ion association is likely to make any simple model of ion-ion interactions very difficult to apply without a number of *ad hoc* assumptions concerning ionic radii. This is particularly true for ionic strengths in excess of 0.01 M or for low-dielectric-constant media. However, a further difficulty is raised by the problem of the nature of an ion pair. If we consider the simple case of univalent ions $A^+ + B^-$ forming an ion pair, it is possible to picture the pair as varying in character from one in which the charges remain separated by the sum of the ionic radii of $A^+ + B^-$ to a molecule in which A and B form a covalent bond, not necessarily even polar in character.[c] Nor is it necessarily true that a given species will behave the same in different solvents. If there is a tendency to covalent bond formation, then it is quite possible that the polarity of the A—B bond will depend on the dielectric constant of the solvent.[d] Covalently bound molecules which ionize are considered as "weak" electrolytes, and they are not treated by the methods of Bjerrum, which are meant for "strong" electrolytes. The differences may not always be clear, but the important interactions for the weak electrolyte are with the solvent, and these we shall consider next.

[a] R. M. Fuoss, *Trans. Faraday Soc.*, **30**, 967 (1934).

[b] R. M. Fuoss and C. A. Kraus, *J. Am. Chem. Soc.*, **55**, 21, 1019 (1933).

[c] This problem has been discussed in some detail by Orgel et al. [J. S. Griffith and L. E. Orgel, *Quart. Rev. (London)*, **11**, 381 (1957)] in connection with the type of bonding which exists in the complex ions of metals. It turns out that the description "covalent" or "ionic" is very dependent on the strength of electrical interaction with the complexing group.

[d] An extreme example is the change in N_2O_5 from a "covalently" bound molecule in the gas phase to an ionic lattice of $NO_2^+ \cdot NO_3^-$ on condensation. PCl_5 similarly changes from a nonpolar, covalent molecule in the gas phase (trigonal bipyramid about P) to an ionic lattice of $PCl_4^+ \cdot PCl_6^-$ in the solid.

TABLE XV.5. EFFECT OF DIELECTRIC CONSTANT ON DISSOCIATION OF
TETRISOAMYL AMMONIUM NITRATE (WATER-DIOXANE MIXTURES)

Wt % H_2O	D	$-\text{Log } K$, obs.	$-\text{Log } K$, calc., $a_{ij} = 6.4$ Å	Wt % H_2O	D	$-\text{Log } K$, obs.	$-\text{Log } K$, calc., $a_{ij} = 6.4$ Å
0.60	2.38	15.7	14.7	9.50	5.84	5.8	5.8
1.24	2.56	14.0	13.7	14.95	8.5	4.0	4.0
2.35	2.90	12.0	12.0	20.2	11.9	3.1	3.2
4.01	3.48	9.6	9.9	53.0	38.0	0.60	0.55
6.37	4.42	7.5	7.8				

Data of R. M. Fuoss and C. A. Kraus, *J. Am. Chem. Soc.*, **55**, 21, 1019 (1933). The values of D are somewhat higher than those currently accepted for these mixtures. This would shift a_{ij} but not otherwise affect the general agreement.

Some indication of the importance of specific molecular interactions has been given by Kraus,[a] who has pointed out that the dissociation constant for *o*-chlorophenyl trimethyl ammonium perchlorate [o-$ClC_6H_4N(CH_3)_3$]$^+$ [ClO_4]$^-$ was 44.5×10^{-6} in CH_3CCl_2H and 4.52×10^{-6} in $CH_2Cl \cdot CH_2Cl$, although both these solvents have the same dielectric constant.[b] In Table XV.6 are gathered some additional data, taken from the article of Kraus, which show some inversions in the effect of dielectric constant on dissociation (acetone vs. nitrobenzene) and also the effect of structure on dissociation constant (n-Bu_4N^+ and n-Bu_3NH^+ in nitrobenzene). The comparison of the Li^+ and Na^+ salts in the different solvents is again an indication that ion pairing cannot be characterized adequately by only the two parameters D and a_{ij}.

10. Effect of Solvent on Reactions of Ions

The effect of the solvent on the reactions involving ions has been treated by a number of different authors[c] without great quantitative success. The major factor is undoubtedly the lack of either theoretical or experimental information concerning the behavior of polar molecules in the presence of the enormous fields in the immediate neighborhood of an ion. The problem is to ascertain how the activities of charged particles change with a change in dielectric constant.

Let us assume the same model of the solvent as that used by Debye and Hückel, namely a structureless medium of unvarying dielectric constant D.

[a] C. A. Kraus, *Ann. N.Y. Acad. Sci.*, **51**, 789 (1949).

[b] It has been observed (J. B. Ramsay) that the symmetrical isomer normally exists in the *trans* form but may convert over to the less stable but more polar *gauche* form in the vicinity of an ion.

[c] See discussion in E. S. Amis, "Kinetics of Chemical Change in Solution," The Macmillan Company, New York, 1949.

TABLE XV.6. SOME ANOMALIES IN DISSOCIATION OF PICRATE SALTS
IN SOLVENTS OF DIFFERENT DIELECTRIC CONSTANT

Solvent:	Pyridine, $D = 12.3$	Nitrobenzene, $D = 34.5$	Acetone, $D = 20.5$	Ethylene chloride, $D = 10.2$
Cation	$K \times 10^4$, liters/mole at 25°C			
Li^+	0.81	0.0006	10.3	—
Na^+	0.45	0.28	13.5	—
K^+	1.05	0.86	34.3	—
$n\text{-}Bu_4N^+$	12.8	> 2000	223	—
$(C_2H_5)(CH_3)_3N^+$	8.2	440	—	0.46
$(HOC_2H_4)(CH_3)_3N^+$	9.5	70	—	0.066
$(CH_3OCH_2)(CH_3)_3N^+$	—	240	—	0.264
$n\text{-}Bu_3NH^+$	—	1.90	—	—

Data from C. A. Kraus, *Ann. N.Y. Acad. Sci.*, **51**, 789 (1949).

Then the change in free energy on transferring an ion $z\varepsilon$, of radius r from medium of dielectric constant D_1 to another of constant D_2, should be[a]

$$\Delta F_D = \frac{\varepsilon^2 z^2}{2r} \left(\frac{1}{D_2} - \frac{1}{D_1} \right) \cdots \qquad (\text{XV.10.1})$$

If we use as our reference state a medium of $D = 1$ (vacuo), then

$$\Delta F_D = -\frac{\varepsilon^2 z^2}{2r} \left(1 - \frac{1}{D} \right) \cdots \qquad (\text{XV.10.2})$$

corresponds to the free energy change on transferring a spherical ion from vacuo to solvent D (i.e., the free energy of solvation of an ion).[b] With reference to vacuum, then, as standard state, this solvent interaction contributes to the activity coefficient of an ion an amount

$$\ln f_D = \frac{\Delta F_D}{kT} = -\frac{\varepsilon^2 z^2}{2rkT} \left(1 - \frac{1}{D} \right) \cdots \qquad (\text{XV.10.3})$$

By applying this to the transition-state complex X arising from the reaction $A^{z_A} + B^{z_B} \rightleftharpoons X^{z_A + z_B}$ we can employ Eq. (XV.5.2) to evaluate the

[a] This calculation applies to spherically symmetrical ions. M. Born, *Z. Physik*, **1**, 45 (1920), calculated the work of charging a conducting sphere of radius r, in a medium of dielectric constant D as

$$W = \int_0^{z\varepsilon} \int_r^\infty \frac{q}{Dr^2}\, dq\, dr = \frac{1}{2} \frac{z^2 \varepsilon^2}{Dr}$$

where $q\, dq/Dr^2$ is simply the coulombic force of repulsion of the charge q on the sphere acting on an element of charge dq distant r from the center of the sphere. Such a calculation neglects image forces, important at small distances, but these cancel in comparing two different media.

[b] Note that this is always negative, so that ions are more stable in solvents than in vacuum. For a univalent ion with $r = 2$ Å it amounts to about 150 Kcal/mole for $D > 10$. Compare with Eq. (XV.6.6).

influence of changes in solvent dielectric constant on the reaction rate constant for X:

$$\ln \frac{k_n}{k_n^\infty} = \frac{f_A f_B}{f_X}$$

$$= \frac{\mathcal{E}^2}{2kT}\left(1 - \frac{1}{D}\right)\left[\frac{z_A^2}{r_A} + \frac{z_B^2}{r_B} - \frac{(z_A + z_B)^2}{r_X}\right]$$

(XV.10.4)

where k_n^∞ is the reaction rate constant both at infinite dilution ($\varkappa = 0$) and in vacuum ($D = 1$). If we make the unjustified but simplifying assumption that $r_A = r_B = r_X$, we obtain

$$\ln \frac{k_n}{k_n^\infty} \cong \frac{\mathcal{E}^2 z_A z_B}{r_X kT}\left(1 - \frac{1}{D}\right)$$

(XV.10.5)

On differentiation of k_n with respect to D, we find

$$\frac{\partial \ln k_n}{\partial D} = \frac{\mathcal{E}^2 z_A z_B}{r_X kT D^2}$$

(XV.10.6)

Equation (XV.10.6), first derived by Scatchard,[a] predicts that reactions between ions of like sign will go faster in media of higher dielectric constant, while the reverse will be true for reactions of opposite sign. In particular, if $\ln k_n$ is plotted against $1/D$, Eq. (XV.10.4) predicts a straight line of slope given by

$$-\frac{\mathcal{E}^2}{2kT}\left[\frac{z_A^2}{r_A} + \frac{z_B^2}{r_B} - \frac{(z_A + z_B)^2}{r_X}\right] \xrightarrow{r_A = r_B = r_X} -\frac{\mathcal{E}^2 z_A z_B}{r_X kT}$$

(XV.10.7)

These predictions have been verified qualitatively and semiquantitatively[b] for the displacement reaction

$$S_2O_3^= + BrCH_2COO_2^- \rightarrow S_2O_3CH_2COO_2^- + Br^-$$

in such mixed solvent systems as $C_2H_5OH + H_2O$, urea $+ H_2O$, glycine $+ H_2O$, and sucrose $+ H_2O$.[b] The observed slope is compatible with reasonable values of r_A, r_B, and r_X. For the proton-catalyzed decomposition of the azo dicarbonate ion[c] whose rate-determining step is

$$(OOC—N=N—CO_2)^= + H_3O^+ \rightarrow \text{products } (N_2H_4 + N_2 + CO_3^=)$$

in alkaline mixtures of dioxane and water, it is possible to observe a 400-

---

[a] G. Scatchard, J. Chem. Phys., **7**, 657 (1939). The qualitative predictions are quite reasonable in that they say that increasing dielectric constant will favor dissociation of oppositely charged ions and association of similarly charged ions.

[b] Truly quantitative confirmation is not to be expected, even if the model were more rigorous, the reason being that changes in D are usually achieved by using mixed solvents. In such a situation there is no reason to expect the microscopic behavior to be dictated by the single average macroscopic variable D.

[c] C. V. King and J. J. Josephs, J. Am. Chem. Soc., **66**, 767 (1944).

fold increase in second-order rate constant as the dielectric constant decreases from 79 to about 27. Here r_X [Eq. (XV.10.7)] is about 3.5 Å, a reasonable value.[a] Similar success has been found in interpreting the reaction of $NH_4^+ + CNO^- \rightarrow CO(NH_2)_2$, the classic urea synthesis[b] in mixed solvents. Unfortunately, the rate data cannot distinguish between $NH_4^+ + OCN^-$ or the structurally more probable $NH_3 + HNCO$ as the pair forming the transition complex,[c] since either pair will give the same charge dependence by virtue of the rapid equilibrium between them.

The range of accessible dielectric constant is somewhat limited by the fact that in low-dielectric-constant media, ion pairing begins to become important, and the contribution of ion triplets and quadruplets may also become important. In addition, selective solvation of the ions by the higher-dielectric-constant solvent will here assume its most important aspects.[d] Little information is available on this point, but the kinetic data do show significant deviations from straight lines in the $\ln k_n$ vs. $1/D$ plots at low D.

11. Reactions of Ions with Polar Molecules

Simple electrostatic models can be used to interpret the activity coefficients of polar molecules in terms of just three parameters: a radius, the dipole moment of the solute, and the dielectric constant of the solvent. The continuum model of the solvent can be used to deduce a value for the free energy of solvation of a spherical molecule of radius r containing a point dipole at its center. The value obtained by Kirkwood[e] from electrostatic theory is

$$F_D = \frac{-\mu^2}{r^3} \frac{D-1}{2D+1} \qquad (XV.11.1)$$

This gives for the activity coefficient of the solute, relative to the gas phase $(D = 1)$,

$$\ln f_D = \frac{F_D}{kT} = \frac{-\mu^2}{r^3 kT} \frac{D-1}{2D+1} \qquad (XV.11.2)$$

This expression has been used to correlate activities of polar molecules in mixtures of solvents of low dielectric constant with a success which is

[a] It is interesting to note that the E_{act} for the reaction is virtually unchanged over this range so that the entire effect may be attributed to the entropy of activation.

[b] W. J. Svirbely and S. Petersen, *J. Am. Chem. Soc.*, **65**, 166 (1943). The validity of this work has been questioned [P. A. H. Wyatt and H. L. Kornberg, *Trans. Faraday Soc.*, **48**, 454 (1952)] on the ground that the hydrolysis reaction $CNO^- + 2H_2O \rightarrow HCO_3^- + NH_3$, which is important, has not been taken into account in the product analysis. This can lead to errors in k of up to 40 per cent.

[c] I. Weil and J. C. Morris, *J. Am. Chem. Soc.*, **71**, 1664 (1949). See discussion on pages 561 and 562.

[d] In H_2O-dioxane mixtures, for example, we may expect to find serious changes in solvation in the low H_2O range of composition.

[e] J. G. Kirkwood, *J. Chem. Phys.*, **2**, 351 (1934).

at best qualitative.[a] The reason is that the interactions are now of the order of Van der Waals interactions and it is probably a gross oversimplification to neglect the latter in representing dipole-solvent interactions and to use D to carry the entire weight of the solution behavior. Equation (XV.11.2) has also been used to correlate the effect of solvent on the reactions between polar molecules,[b] but it is questionable if the correlation can really be used to throw much light on either the transition-state complex or on the theory of solutions.[c]

In the case of the reactions of ions with polar molecules, there have been several approaches used. The simplest of these has been to calculate the coulombic energy of a transition complex consisting of a charge $z_A\varepsilon$ and a dipole μ_B.[d] This is given by Eq. (XV.6.6):

$$\Delta F_\mu = \frac{z_A\varepsilon\mu_B}{Dr^2}\cos\theta \qquad (XV.11.3)$$

where we neglect the polarizations of the species. This interaction is a minimum for the head-to-tail orientation ($\theta = 180°$). Now the transition complex has the same charge as the ion A, so that in addition to the dipole energy of Eq. (XV.11.3) there should appear a coulombic term of the Debye-Hückel type due to the charge. This term will, however, be canceled (at $\varkappa = 0$) by the equivalent terms for the ion, so that we can write for the influence of solvent on the rate constant

$$\ln\left(\frac{k}{k^\circ}\right)_{\varkappa=0} = -\frac{z_A\varepsilon\mu_B}{Dr^2kT}\cos\theta \qquad (XV.11.4)$$

Note that this may be positive or negative depending on the value of θ. While in general one may expect θ to be 180°, the value for minimum energy, the geometry of the reaction may dictate other orientations. For reactions of the displacement type, such as the Walden inversion in which a charged group displaces an electronegative group from a polar molecule, e.g.,

θ will be 180° unless the side groups R_1, R_2, R_3 are very polar.

[a] A. R. Martin, *Trans. Faraday Soc.*, **33**, 191 (1937), and earlier papers.
[b] See discussion in K. J. Laidler and H. Eyring, *Ann. N.Y. Acad. Sci.*, **39**, 299 (1940).
[c] It seems to the author that a detailed molecular model for the solution using only near neighbor interactions is no more complex and would probably yield more interesting and useful information. The parameters for such a model would be only the dipole moments and radii of the solute and solvent species.
[d] E. A. Moelwyn-Hughes, "The Kinetics of Reaction in Solution," 2d ed., Oxford University Press, New York, 1947.

Such a model predicts that, where the ion-dipole interaction is attractive ($\theta > 90°$), increasing dielectric constant will slow the reaction, while the opposite will be true for ion-dipole interactions which are repulsive. The real difficulty in applying this model to ion-dipole reactions is that the term given by Eq. (XV.11.4) is of the same order of magnitude as the difference in free energies of hydration of the ion A^{z_A} and the transition complex X^{z_A}. A fairer model for the formation of the transition complex might be to consider it as arising from the displacement of a solvent molecule of dipole μ_S by the solute dipole μ_B. The free energy of formation of the transition state, neglecting changes in dipole-dipole interaction, is now

$$\Delta F_\mu = \frac{z_A \mathcal{E} \mu_B \cos \theta}{D r_B^2} \left(1 - \frac{\mu_S}{\mu_B} \frac{r_B^2}{r_S^2}\right) \qquad (XV.11.5)$$

We see now that the effect of dielectric constant is not separable from the influence of μ_S and r_S, the dipole moment and distance of closest approach of a solvent molecule to the ion. When $\mu_S/r_S^2 = \mu_B/r_B^2$, there should be no effect of changes in solvent, while, depending on which of these is greater, opposite effects may be predicted.[a]

An even more complex but not necessarily more rigorous approach to this problem has been made by Moelwyn-Hughes (loc. cit.), who included polarization interaction and repulsion energy in computing the energy of the ion-dipole complex. Amis and Jaffé[b] have mixed the continuum and molecular approach by using the Onsager model of a dipole imbedded in a solvent to obtain the dipole-solvent interactions. They then use the Poisson equation to obtain the ion-atmosphere effect on the dipole.[c] Both of these approaches lead to very complex equations, and it is doubtful if a meaningful application to experimental data is possible.

The effect of ionic strength on ion-dipole reactions, aside from specific interactions, can be guessed intuitively from considerations of the interactions of the ion atmosphere with the dipole moment of the reactant molecule in the free state and in the complex. For positive interactions of the ion and dipole ($\theta > 90°$) we may expect the ion field to reduce the dipole

[a] The simple model for ion-dipole interactions cited by most writers assumes that the principal contribution to $\Delta F_\mu^{\ddagger}$ is the increase in free energy arising from the distribution of charge over a large ion X in the transition state. On such a basis, all ion-dipole reactions should be accelerated in solvents of decreasing dielectric constant. While such a rule seems to be generally observed for the solvolysis reaction of OH^- with alkyl halides, it is by no means universal.

[b] E. S. Amis and G. Jaffé, J. Chem. Phys., **10**, 598 (1942). See also Amis, "Kinetics of Chemical Change in Solution," op. cit.

[c] The Poisson equation (XV.7.1) is solved in cylindrical coordinates for the atmosphere distribution around a spherically imbedded point dipole. Actually Amis and Jaffé use only one of the special solutions rather than the general solution to the equation. L. C. Bateman et al., J. Chem. Soc., 979 (1940), have shown that the activity coefficient f_D of a dipole will vary as $\log f_D = A\mu/D^2 T^2$ in the limiting form. Here μ is the ionic strength of the solution. This relation has not been verified quantitatively.

field, since the orientation is $\oplus \,(\!\!\!\subset\!\!-\!\!\!-\!\!\!\rightarrow\!\!)$. The result should be a greater stabilization of the free dipole by the ion atmosphere as compared to the complex and thus a decrease of rate with increasing ionic strength. For the negative interaction $\oplus \,(\!\!\!\subset\!\!\!\!-\!\!\!-\!\!\!\rightarrow)$, the opposite should be true.[a]

Unfortunately the data[b] are as yet not sufficiently extensive to make proper tests of these models. The complexity of the model is itself sufficiently great that it is even dubious that any of the correlations which have been made can be given more than qualitative significance.

12. Entropy Changes in Ionic Reactions

In the last few sections we have been using simple electrostatic models to compute the contribution to the free energy changes of the electrical interactions of charged particles. From the relations $\Delta S = -[\partial(\Delta F)/\partial T]_P$, $\Delta H = \Delta F + T\,\Delta S$, $\Delta C_P = [\partial(\Delta H)/\partial T]_P$, and $\Delta V = [\partial(\Delta F)/\partial P]_T$ it is possible from the same models to compute these other thermodynamic properties as well. The two types of interaction of interest are the ion-ion and ion-solvent interactions; of these, the latter is much the larger.

By using the continuum model of the solvent, we have calculated the free energy of separation of a pair of spherical charges $z_A\mathcal{E}$ and $z_B\mathcal{E}$ [Eq. (XV.10.5)] in a dielectric constant D as

$$\Delta F_D = \frac{z_A z_B \mathcal{E}^2}{D r_X} \tag{XV.12.1}$$

where r_X is the equilibrium separation of the charges in a hypothetical ion-pair or dipolar molecule.[c] If we treat r_X as a constant independent of both temperature and dielectric constant, which seems reasonable,[d] then we can write for the other thermodynamic functions

[a] All these considerations are subject to modification by the changes in solvation energy of the displaced solvent molecule and the attached solute molecule.

[b] One difficulty with much of the data is that no consistent attempt has been made to distinguish the effects of ions and ion pairs as reagents. See, for example, J. H. Reinheimer et al., *J. Am. Chem. Soc.*, **80**, 164 (1958).

[c] In principle there is no reason to apply such a treatment to a covalently bound molecule. For such species one expects that the work of ionization should include an extra term that represents the work of forming an ion pair from the covalent molecule. However, it may be expected that this term will be similar in form to the coulombic term, so that the differences may simply involve a constant of proportionality. A more serious objection, which has been raised by K. S. Pitzer, has to do with the neglect of the repulsive-energy terms, in equations such as (XV.12.1), which are responsible for maintaining ion pairs in equilibrium. If these forces change sufficiently rapidly with distance, such as r^{-12}, it can be shown that the repulsive energy is not more than one-twelfth of the coulombic energy. Equally important, however, is the fact that much larger contributions due to mutual polarization and Van der Waals forces of attraction have been neglected.

[d] Unpublished work by the author. See, however, W. Latimer, *J. Chem. Phys.*, **23**, 90 (1955), who explores the possibility of using $\partial r_X/\partial T \neq 0$.

$$\Delta S_D = -\frac{z_A z_B \mathcal{E}^2}{Dr_X T}\left(\frac{\partial \ln D}{\partial \ln T}\right)_P = \frac{\Delta F_D}{T}\left(\frac{\partial \ln D}{\partial \ln T}\right)_P \tag{XV.12.2}$$

$$\Delta H_D = \Delta F_D\left[1 + \left(\frac{\partial \ln D}{\partial \ln T}\right)_P\right] \tag{XV.12.3}$$

$$\Delta C_{PD} = \frac{\Delta F_D}{T}\left\{\left[\frac{\partial^2 \ln D}{\partial(\ln T)^2}\right]_P - \left(\frac{\partial \ln D}{\partial \ln T}\right)_P^2 - \left(\frac{\partial \ln D}{\partial \ln T}\right)_P\right\} \tag{XV.12.4}$$

$$\Delta V_D = -\Delta F_D\left(\frac{\partial \ln D}{\partial P}\right)_T = -\Delta F_D\left(\frac{\partial \ln D}{\partial \ln V}\right)_T\left(\frac{\partial \ln V}{\partial P}\right)_T \tag{XV.12.5}$$

All of these quantities are in absolute magnitude, proportional to ΔF_D and thus inversely proportional to D. We can calculate from Eq. (XV.12.1) that for a pair of charges ± 1 in water at 25°C ($D \cong 79$) with $r_X = 2$ Å, $\Delta F_D \cong 2.1$ Kcal. Since at 25°C $(\partial \ln D/\partial \ln T) \cong -1.4$, $[\partial^2 \ln D/(\partial \ln T)^2] \cong -1.6$,[a] while $(\partial \ln D/\partial \ln V)_T \cong -1.4$ and $\beta = -(\partial \ln V/\partial P)_T = 5 \times 10^{-5}$ atm^{-1}, we find $\Delta S_D \cong -10$ cal/mole-°K, $\Delta H_D \cong -0.8$ Kcal, $\Delta C_{PD} \cong -15$ cal/mole-°K, and $\Delta V_D \cong -6$ cc. The only large group of measurements with which these can be compared are the ionization of weak acids (Table XV.7) for which ΔS is in the range -20 to -30 cal/mole-°K, ΔH is near ± 3 Kcal/mole, ΔC_P ranges from -40 to -60 cal/mole-°K, and ΔV ranges from -10 to -20 cc/mole. The orders of magnitude are certainly in the proper range, showing that, even in these unfavorable cases of covalent compounds, the electrostatic model is not unreasonable.

This continuum model predicts that all of these thermodynamic properties should preserve their sign and increase in magnitude as D is decreased. This is, of course, subject to the behavior of the derivatives $(\partial \ln D/\partial \ln T)$ and $(\partial \ln D/\partial \ln V)$ as D changes. In the case of weak electrolytes this is qualitatively the observed behavior. Some values are given in Table XV.7 for the dissociation of H_2O and acetic acid in water-dioxane mixtures. Since the derivatives of D and V are not the same as the values in pure H_2O, the values cannot be expected to follow D in any simple fashion. However, it is seen that qualitative agreement is obtained. If one uses Eqs. (XV.12.2) to (XV.12.5) and the experimental values of the derivatives for water, it is possible to calculate ΔF_D from the experimental values of each of the thermodynamic quantities in turn. For acetic acid the values are 4.7 Kcal from ΔS, 0.3 Kcal from ΔH, 5.5 Kcal from ΔC_P, and 5 Kcal from ΔV. Except for ΔH, this can be considered quite good correlation compared to $\Delta F_{ion} = 6.5$ Kcal observed directly. For H_2O the values are 4 Kcal from ΔS, 6 Kcal from ΔC_P, 33 Kcal from ΔH, and

---

[a] See compilation of data in Harned and Owen, loc. cit.

7.5 Kcal from ΔV. The comparison with $\Delta F = 19.1$ is much poorer, indicating the crudeness of the continuum model at least as applied to these covalent compounds. We have already seen that for ion pairs the model is much better (Table XV.5).[a]

TABLE XV.7. EFFECT OF DIELECTRIC CONSTANT ON DISSOCIATION
OF WEAK ELECTROLYTES AT 25°C[a]

Electrolyte	Solvent	D	ΔF	ΔH	ΔS	C_P	ΔV
H_2O	H_2O	78.5	19.1	13.5	−18.7	−46.5	−23
	20% dioxane	60.8	19.9	13.5	−21.5	−50.4	—
	45% dioxane	38.7	21.5	13.2	−27.8	−54.0	—
	70% dioxane	17.7	24.4	12.7	−39.2	−49.4	—
$HC_2H_3O_2$	H_2O	78.5	6.5	−0.1	−22.1	−36.5	−15
	20% dioxane	60.8	7.2	−0.05	−24.4	−44.0	—
	45% dioxane	38.7	8.6	−0.4	−30.3	−51.1	—
	70% dioxane	17.7	11.3	−0.6	−40.1	−51.7	—
	82% dioxane	9.6	13.8	−1.3	−50.8	−124.5	—
	10% CH_3OH	74.1	6.7	0.03	−22.3	−43.3	—
	20% CH_3OH	69.2	6.9	0.1	−22.7	−47.1	—
	$HC_2H_3O_2$	6.2	17.2	5.7	−38.6	—	—

[a] Data from compilation of H. S. Harned and B. B. Owen, "The Physical Chemistry of Electrolytic Solutions," 3d ed., Reinhold Publishing Co., New York, 1958, except for pure $HC_2H_3O_2$, which is from W. L. Jolly, *J. Am. Chem. Soc.*, **74**, 6199 (1952). ΔF and ΔH are in Kcal/mole, ΔS and ΔC_P are in cal/mole-°K, and ΔV is in cc/mole.

Since the transition-state model for solution reactions is an equilibrium model, the true testing ground for solution theories will be not kinetic data but rather thermodynamic data. From this point of view, the continuum model of the solvent has had only fair success—for example, in correlating thermodynamic data for ions.[b] Laidler has recently shown that the entropies of aqueous monatomic ions[c] of molecular weight M with reference to $S°(H^+) = -5.5$ e.u. may be represented by the empirical equation:

$$\bar{S}°(M^{\pm z}) = 10.2 + \frac{3}{2} R \ln M - 11.6 \frac{z^2}{r_u} \qquad (XV.12.6)$$

where r_u is Pauling's estimated value for the univalent radius of the ion.

[a] It should also be noted that the characterization of a solvent by its dielectric constant only is always an oversimplification which may ignore quite important nonelectrical interactions, or even quite important electrical interactions such as polarization.
[b] See discussion by K. J. Laidler, *J. Chem. Phys.*, **27**, 1423 (1957), and also by P. C. Scott and Z. Z. Hugus, *ibid.*, 1421 (1957).
[c] K. J. Laidler, *Can. J. Chem.*, **34**, 1107 (1956). This value for $S°(H^+)$ is presumed to be the absolute value.

In a further paper Laidler and Couture[a] have shown that for oxy anions, an equally good relation is[b]

$$\bar{S}^\circ(MO_n^{-z}) = 40.2 + \frac{3}{2} R \ln M - \frac{27.2\, z^2}{0.25\, nr_0} \qquad (XV.12.7)$$

where n is the number of bare O atoms attached to M (for example, 2 in HCO_3^-, 3 in $CO_3^=$) and r_0 = the M-O distance $+ 1.40$ Å, the latter corresponding to the Van der Waals radius of O.

Latimer and Powell[c] have shown that an equally usable expression for monatomic ions with reference to $\bar{S}^\circ(H^+) = 0$ is

$$\bar{S}^\circ(M^{\pm z}) = 37 + \frac{3}{2} R \ln M - \frac{270\, z}{(r_c + x)^2} \qquad (XV.12.8)$$

where r_c is the crystallographic radius but x is 2.00 Å for cations and 1.00 for anions. Powell[d] later showed that an even better fit is given by

$$\bar{S}^\circ(M^{\pm z}) = 47 - 154\, \frac{z}{(r_c + x)^2} \qquad (XV.12.9)$$

and for complex ions, Connick and Powell gave

$$\bar{S}^\circ(MO_n^{-z}) = 43.5 - 46.5(z - 0.28n) \qquad (XV.12.10)$$

where x is 1.3 Å for cations and 0.4 Å for anions.[e] While some theoretical justification has been found for both empirical equations, the Laidler equation corresponds more closely to the continuum model of Born and for this reason may be the more interesting of the two sets. Unfortunately, such correlations do not extend well to other thermodynamic properties without considerable modification to take into account solvent electrostriction[f] and dielectric saturation. The final result is that we still lack a truly quantitative picture of ionic solutions.

The qualitative picture, however, is very important and suggestive. The very large negative magnitudes associated with ΔS, ΔC_P, and ΔV for ionization suggest the long-range effect of the ion-solvent forces, since it is the electrostriction of the solvent which is responsible for these magnitudes (Table XV.7).[g] The $1/r$ dependence of the thermodynamic quantities [Eqs. (XV.12.2) to (XV.12.5)] indicates that 90 per cent of the

---

[a] A. M. Couture and K. J. Laidler, *ibid.*, **35**, 202 (1957).

[b] These relations seem to be good, on the average, to about ± 3.5 e.u., with maximum deviation of the order of about 6 e.u.

[c] R. Powell and W. Latimer, *J. Chem. Phys.*, **19**, 1139 (1951).

[d] R. Powell, *J. Phys. Chem.*, **58**, 528 (1954). R. E. Cornick and R. Powell, *J. Chem. Phys.*, **21**, 2206 (1953).

[e] These differences are ascribed to the different distance of closest approach of water molecules to ions.

[f] See Laidler and Pegis, *loc. cit.*

[g] The values for H_2O are usually smaller (that is, $|\Delta S|$, $|\Delta C_p|$) than for other solvents mainly because of the large D value for H_2O.

TABLE XV.8. CHARGE AND ENTROPY CORRELATIONS FOR SOME IONIC
EQUILIBRIA IN H_2O AT 25°C

Reaction	$\Delta F°$, Kcal/mole	$\Delta H°$, Kcal/mole	$\Delta S°$, cal/mole-°K	Ref.	$\Delta S_D° (r_X = 2 \text{ Å})$, Eq. (XV.12.2)[a]
$H_2O = H^+ + OH^-$	19.1	13.5	−18.7	[b]	−10
$CH_3COOH =$ $H^+ + CH_3COO^-$	6.5	−0.1	−22.1	[b]	−10
$ClCH_2COOH =$ $H^+ + ClCH_2COO^-$ . . .	3.9	−1.2	−17.0	[b]	−10
$HSO_4^- = H^+ + SO_4^=$ . . .	2.7	−5.2	−26.5	[b]	−20
$CH_2(COOH)CO_2^- =$ $CH_2(CO_2)_2^= + H^+$	7.8	−1.2	−29.9	[b]	−20
$H_3BO_3 = H^+ + H_2BO_3^-$ (probably $B(OH)_4^-$) . .	12.6	3.3	−31.1	[b]	−10
$^+NH_3CH_2COOH =$ $H^+ + {}^+NH_3CH_2COO^-$	3.2	1.2	−6.9	[b]	0[c]
$NH_4^+ = H^+ + NH_3$	12.6	12.4	−0.7	[d]	0
$HCNO = H^+ + CNO^-$	5.3	1.6	−12.5	[d]	−10
$CdCl^+ = Cd^{++} + Cl^-$ . . .	2.7	−0.6	−11.	[e]	−20
$CaOH^+ = Ca^{++} + OH^-$	1.9	—	—	[f]	−20
$CaSO_4 \text{ (aq)} =$ $Ca^{++} + SO_4^=$	3.2	—	—	[g]	−40
$Cr(H_2O)_6^{3+} \rightarrow$ $H^+ + Cr(H_2O)_5OH^{++}$	5.2	9.4	14	[h]	+20
$Co(NH_3)_5(H_2O)^{3+} + Cl^-$ $= Co(NH_3)_5Cl^{++}$ $+ H_2O$[i]	−5.4	2.5	26	[d]	30
$CO(NH_2)_2 =$ $NH_4^+ + CNO^-$	6.1	11.1	−16.5	[d]	−10

[a] Note that these estimates are for ion pairs, not covalent species.

[b] Compilation by H. S. Harned and B. B. Owen, "The Physical Chemistry of Electrolytic Solutions," 3d ed., Reinhold Publishing Co., New York, 1958.

[c] The neutral molecule exists in solution as the zwitterion, so that there is a strong ion dipole interaction.

[d] Nat'l Bur. Standards Circ. 500, 1952.

[e] C. E. Vanderzee and H. T. Dawson, Jr., J. Am. Chem. Soc., **75**, 5659 (1953). Values listed are for somewhat uncertain extrapolations to zero ionic strength from a minimum observed value of $\mu = 0.5 M$! This is unfortunately typical of most of the data. At $\mu = 0.5 M$, $\Delta S = -7.2$, $\Delta H = -0.3$, and $\Delta F = 1.9$, so that the extrapolation is not small.

[f] C. W. Davis and B. E. Hoyle, J. Chem. Soc., 233 (1951).

[g] R. P. Bell and J. H. B. George, Trans. Faraday Soc., **49**, 619 (1953).

[h] C. Postmus and E. L. King, J. Phys. Chem., **59**, 1208 (1955). Values given are for $\mu = 0.034 M$.

[i] The free energy data seem at variance with the observation that the reaction in acid is spontaneous to the left!

contribution can be localized in a shell of solvent molecules which is $10r_X$ in radius,[a] not a small distance for concentrated solutions (for example, >0.1 M) but one which would be permissible in the discussion of ion-ion interactions in dilute solutions ($\leqslant 0.01$ M). In no case should we feel justified in using the present continuum model for quantitative discussion of ion-solvent or ion-pair interactions.

TABLE XV.9. SPECIFIC EFFECTS IN THE FORMATION OF ION PAIRS
BY Fe^{3+} IN H_2O^a

Reaction	$\Delta F°$, Kcal/mole	$\Delta H°$, Kcal/mole	$\Delta S°$, cal/mole-°C	$\Delta S_D°$ ($r_X = 2$ Å), Eq. (XV.12.2)	Ref.
$Fe^{3+} + OH^- =$ $Fe(OH)^{++}$..........	-16	-1.2	50	30 (1.2)[b]	c,d
$Fe^{3+} + O_2H^- =$ $Fe(O_2H)^{++}$.........	-12.5	1.8	49	30 (1.2)	e
$Fe^{3+} + F^- = Fe(F)^{++}$..	-4.9	3.3	25 (est.)	30 (1.2)	c
$Fe^{3+} + Cl^- = Fe(Cl)^{++}$	-2.0	8.5	35	30 (1.7)	d,f
$Fe^{3+} + Br^- = Fe(Br)^{++}$	-0.8	6.1	23	30 (2.7)	d,f
$Fe^{3+} + N_3^- = Fe(N_3)^{++}$	-5.7	-4.3	5	30 (12)	c
$Fe^{3+} + C_2O_4^= =$ $Fe(C_2O_4)^+$..........	-13.2	-0.3	45.5	60 (2.7)	c
$Fe^{3+} + CNS^- =$ $Fe(NCS)^{++}$.........	-4.2	0	14	30 (4.0)	f

Data are taken from compilation by N. Uri, *Chem. Revs.*, **50**, 376 (1952).

[a] Because of the difficulty in extrapolating to zero ionic strength and in some cases the dubiousness of the method of analysis, the present figures are of uncertain accuracy.

[b] The values in parentheses are the values of r_X, in Å, required to fit the data by Eq. (XV.12.2).

[c] M. G. Evans and N. Uri, unpublished work.

[d] E. Rabinowitch and W. H. Stockmayer, *J. Am. Chem. Soc.*, **64**, 335 (1942).

[e] M. G. Evans et al., *Trans. Faraday Soc.*, **45**, 230 (1949).

[f] M. W. Lister and D. E. Rivington, *Can. J. Chem.*, **33**, 1572, 1591, 1603 (1955).

A number of authors have attempted to use the continuum model to explain the charge (and solvent) effect on ion-ion reactions. From Eq. (XV.12.2) we can estimate that, on comparing reactions between ions of different charge types in water, we should expect large differences in the entropy of activation $\Delta S^{\ddagger}$ with changes in the value of $z_A z_B$. If, arbitrarily, we use $r_X = 2$ Å in water at 25°C, then we find that $\Delta S_D \cong 10 z_A z_B$ cal/mole-°K.[b] This implies that there should be large changes in entropies

[a] r_X is taken here as the distance of closest approach of ion and solvent molecule.

[b] For reasons which are not obvious, many authors have made extensive generalizations on $\Delta S^{\ddagger}$, assuming r_X = constant for all ions. This is certainly not a good approximation, and such generalizations are suspect.

of activation (of ionic reactions) with charge change. This is, unfortunately, in poor accord with the kinetic data for highly charged ions.[a] The reaction $Fe^{++} + Co(C_2O_4)_3^{3-} \rightarrow Fe^{3+} + Co(C_2O_4)_3^{-4}$ (Table XV.5), which should have an abnormally high A factor and entropy of activation, turns out to have $\Delta S^{\pm} = 0$. Further evidence for specific effects aside from the electrostatic are provided by the pair of similar reactions:[b]

$$S_2O_3^{=} + \begin{cases} BrCH_2CO_2^- \\ ClCH_2CO_2^- \end{cases} \rightarrow (S_2O_3CH_2CO_2)^{=} + \begin{cases} Br^- \\ Cl^- \end{cases}$$

For reasons not at all obvious, the chloride rate constant has an A factor 200 times larger than that for the bromide.

For purposes of comparison we list in Table XV.8 the entropy changes for some ionic equilibria in water at 25°C. Despite the fact that the list includes species of covalent character for which the electrostatic model is a dubious approximation, the over-all entropy change parallels the value predicted from Eq. (XV.12.2). The ΔF° and ΔH° show much less correlation. Some idea of the relative importance of the specific character of the ion pair interaction is provided by the data shown in Table XV.9 on the ion-binding capacities of Fe^{3+}.

13. Cage Effects in Liquid Reactions

Almost any model of a liquid, whether it be a quasi-lattice model or the disordered liquid model adopted here, predicts that once a pair of molecules become near neighbors they will exist in such a relation for a time which is long compared to molecular vibration frequencies. Depending on the precise values of the diffusion constants of the pair and their specific energy of interaction, this time [Eq. (XV.2.1)] was estimated to be in the range of 3×10^{-11} sec, or about 100 times longer than gas-phase collisions. In addition, the slowness of diffusion in liquid systems is such that once a pair (A,B) have separated from an encounter, there is a better than average probability of a reencounter prior to ultimate separation to the average distance of (A,B) pairs in the liquid.

There are a number of cases in which these extended collision times can have kinetic consequences.[c] One of them concerns the primary dissociation of a molecule into reactive fragments, while the second concerns the recombination of active radicals to form an inactive molecule. If we consider first the hypothetical species A-B decomposing in solution to form active species (e.g., radicals) A + B, we can formulate the kinetic scheme by

[a] C. Postmus and E. L. King, *J. Phys. Chem.*, **59**, 1216 (1955).

[b] A. N. Kappana and H. W. Patwardhan, *J. Indian Chem. Soc.*, **9**, 379 (1932).

[c] Among the first to investigate this problem quantitatively were J. Franck and E. Rabinowitch, *Trans. Faraday Soc.*, **30**, 120 (1934). See also E. Rabinowitch and W. C. Wood, *ibid.*, **32**, 1381 (1936).

$$A\text{-}B \underset{2}{\overset{1}{\rightleftharpoons}} [A \cdot B]_C$$

$$[A \cdot B]_C \overset{3}{\rightarrow} A + B \qquad\qquad (XV.13.1)$$

In this scheme $[A \cdot B]_C$ represents two free radicals $A + B$ present as near neighbors in the same solvent cage. Step 1, the dissociation, may occur thermally or photochemically. Step 3 represents the diffusion of A and B from their solvent cage to a distance of separation comparable to the average separation of A-B molecules in the solution.[a]

The stationary concentration of $[A \cdot B]_C$ is given by

$$([A \cdot B]_C) = \frac{k_1(A\text{-}B)}{k_2 + k_3} \qquad\qquad (XV.13.2)$$

while the rate of decomposition of A-B is given by:

$$\frac{-d(A\text{-}B)}{dt} = \frac{k_3}{k_2 + k_3} k_1(A\text{-}B) \qquad\qquad (XV.13.3)$$

This differs from the decomposition in the absence of a solvent cage or diffusion barrier (step 3), such as we might formulate for reaction in the gas phase, by the fraction $k_3/(k_2 + k_3)$. When there is a large steric or energy barrier to recombination, so that $k_2 \ll k_3$, we may expect this cage effect to vanish. When, however, the probability per unit time k_2 for caged radicals A and B to recombine is high (e.g., two I atoms or two CH_3 radicals), then the over-all rate becomes much less:

$$\frac{-d(A\text{-}B)}{dt} \xrightarrow{k_2 \gg k_3} \frac{k_3}{k_2} k_1(A\text{-}B) = \left(\frac{k_1}{k_2}\right) k_3(A\text{-}B) \qquad (XV.13.4)$$

In such a case the caged pair $[A \cdot B]_C$ are in thermodynamic equilibrium with A-B, and the last slow step is the diffusion step 3. Such a reaction can be described as a diffusion-controlled reaction even though it contains a far slower chemical step, reaction 1. The rate will now be sensitive to the relative diffusion constant D_{AB} in the given solvent, and the over-all rate should be lower than the rate in the gas phase.[b] There have been a number of attempts to observe this diffusion control.

[a] This seems like a rather arbitrary definition, and indeed it is. In the present case it provides that the chance of "primary" recombination of an (A,B) pair is no greater than the recombination of A and B with fragments from other A-B molecules. Depending on the competing processes, there will be a different definition of step 3. For a more exact treatment [see R. M. Noyes, *J. Am. Chem. Soc.*, **79**, 551 (1957); **78**, 5486 (1956); **77**, 2042 (1955); *J. Chem. Phys.*, **22**, 1349 (1954)] we must consider the rather high probability of "secondary recombination" for a "nascent" pair A and B which have separated by only a few molecular diameters (i.e., the reverse of 3).

[b] The difficulty with such a conclusion is that there is no a priori way of knowing the value of k_1 in the gas, so that comparisons become somewhat academic.

It has been observed[a] that when acetyl peroxide, $CH_3CO-O-O-COCH_3$, decomposes thermally, the product analysis indicates the initial formation of acetate radicals, CH_3CO_2, which are unstable and subsequently decompose to give $CO_2 + CH_3$. The methyl radicals recombine to form C_2H_6, or in the presence of a hydrogen donor also yield CH_4 in a competing reaction. If a free radical scavenger which reacts efficiently with CH_3 radicals is introduced into the system, it should be possible to inhibit very effectively the formation of C_2H_6. Szwarc[b] has shown that in the presence of small amounts of I_2 the gas-phase decomposition of acetyl peroxide yields no C_2H_6. In the solution phase, however, it is impossible to eliminate C_2H_6 formation. Szwarc has ascribed this to the high efficiency of recombination of CH_3 radicals formed in the same solvent cage. It seems a very direct and elegant demonstration of the "cage effect" and should be worth further study.[c]

Noyes and Lampe[d] have utilized I_2 and allyl iodide, AI, in a solution containing scavenger (dissolved oxygen) which reacts with I atoms and thus competes with recombination. They showed that the quantum yield of I_2 or AI disappearance, which represented attack on the scavenger, increased with decreasing molecular weight of solvent and with increasing temperature in accord with the model for a cage effect. Unfortunately, such results by themselves are not enough to uniquely distinguish the cage effect from other solvent effects on the photolysis.

A simple illustrative example may be taken to indicate the way in which the cage effect can be formulated quantitatively. Let us assume that we are investigating the photolysis of I_2 in the presence of a scavenger S which can react with I atoms to form the relatively inert radical ·SI which does not react with I_2 but which may react with I atoms to form stable SI_2. The kinetic scheme can be represented by

[a] A. Rembaum and M. Szwarc, *J. Am. Chem. Soc.*, **76**, 5975 (1954).

[b] M. Szwarc, *J. Polymer Sci.*, **16**, 367 (1955).

[c] Some of the complexity of such a study is indicated by the findings of J. C. Roy et al., *J. Am. Chem. Soc.*, **78**, 519 (1956), who studied the effect of scavenger iodine on the thermal and photochemical decomposition of azo-bis-isobutyronitrile, $(CN)(CH_3)_2C-N-N-C(CH_3)_2CN$. This is similar in form to the peroxide in that it is presumed to give $2[C(CH_3)_2CN]$ radicals plus N_2 on decomposition. In chlorobenzene as solvent at 80°C with (reactant)/(I_2) ratios covering the range 7 to 1 it was found that a constant amount (60 per cent) of organic iodide, $I-C(CH_3)_2(CN)$, is formed, indicating no "cage effect." It also indicates that about 40 per cent of reaction goes via direct rearrangement! In the photolysis, on the other hand, the scavenging efficiency of I_2 was proportional to $\exp[(I_2)^{1/2}]$, as might be expected for a cage effect. The rate of production of N_2 was not affected by added I_2, so that I_2 did not influence the primary decomposition nor did it set up chains.

[d] F. W. Lampe and R. M. Noyes, *ibid.*, **76**, 2140 (1954). Both I_2 and allyl iodide, AI, are decomposed by light. Radicals disappear by $A + O_2 \rightarrow AO_2$ followed by $AO_2 + I \rightarrow$ products or $2AO_2 \rightarrow$ products. This differs from the example given.

$$I_2 + h\nu \underset{2}{\overset{1}{\rightleftharpoons}} [I \cdot I]_c$$

$$[I \cdot I]_c \underset{4}{\overset{3}{\rightleftharpoons}} 2I$$

$$[I \cdot I]_c + S \overset{5}{\rightarrow} \cdot SI + I$$

$$I + S \overset{6}{-} \cdot SI$$

$$\cdot SI + I \overset{7}{-} SI_2 \qquad\qquad (XV.13.5)$$

In addition, reactions such as $I + [I \cdot I]_c \rightarrow I_2 + I$ and $2[I \cdot I]_c \rightarrow 2I_2$ (or $I_2 + 2I$) may occur, but they can be shown to be minor as compared with those given.[a] If we apply the stationary state to species I, $[I \cdot I]_c$ and $\cdot SI$, we can arrive at the following equation for (I):

$$k_4(I)^2 + k_6(I)(S) \left[1 + \frac{k_3}{k_2 + k_5(S)} \right] - \frac{k_3 I_a}{k_2 + k_5(S)} = 0$$
$$(XV.13.6)$$

When very efficient scavenging is being done, we may neglect reaction 4 and assume that all of the recombination of I_2 is coming from initially caged pairs whose probability of geminal recombination is excessively high even when they have separated by a small distance. A crude analysis of this system has been made by Roy, Hamill, and Williams[b] and a more careful one by Noyes.[c]

It is possible to get a rough result from the theory of the random walk in three dimensions.[d]

The probability that a particle originally at $r = 0$ will be found at time t' to be in a shell of radius r, thickness Δr is[e]

$$P(r,t')r = \frac{4\pi r^2}{(4\pi Dt')^{3/2}} e^{-r^2/4Dt'} \Delta r \qquad\qquad (XV.13.7)$$

If we integrate this over a time t, we will find the total time during the interval t that such a particle will spend in this shell. Multiplying this by k_2[e] gives the probability of geminal recombination of an initial pair during time $t : R(t)$. The average rate is then $R(t)/t = \bar{k}_r$

[a] For simplicity too, we ignore the competing termination step, $2 \cdot SI \rightarrow S + SI_2$.
[b] J. C. Roy et al., *J. Am. Chem. Soc.*, **76**, 3274 (1954).
[c] Noyes, *loc. cit.*
[d] S. Chandrasekhar, *Revs. Modern Phys.*, **15**, 1 (1943). See also Sec. VI.7.
[e] D is the diffusion constant of the two I atoms relative to each other, that is, $D = 2D_I$. r is the near neighbor distance, and $\Delta r \cong r/2$.
[e] k_2 is actually a function of r, since the probability of the recombination must depend on the distance of separation. See Noyes, *loc. cit.*, for a more precise treatment in which primary and secondary recombination are evaluated independently.

$$\bar{k}_r = \frac{R(t)}{t} = \frac{k_2}{t} \int_0^t P(r,t')\Delta r \, dt'$$

$$= \frac{4r^2 k_2 \, \Delta r}{\pi^{\frac{1}{2}}} \int_0^t \frac{e^{-r^2/4Dt'} \, dt'}{(4Dt')^{\frac{3}{2}}}$$

or, on setting $y' = r(4Dt')^{\frac{1}{2}}$,

$$\bar{k}_r = \frac{2k_2 r \, \Delta r}{Dt} \left[1 - \frac{2}{\pi^{\frac{1}{2}}} \int_0^y e^{-y'^2} \, dy' \right] \qquad \text{(XV.13.8)}$$

For small values of y this can be expanded, and taking the first term[a]

$$\bar{k}_r = \frac{2k_2 r \, \Delta r}{Dt} \left[1 - \frac{r}{(\pi Dt)^{\frac{1}{2}}} \right] \qquad \text{(XV.13.9)}$$

If we take for t the mean time of scavenging, then $t = 1/k_5(\text{S})$ and

$$\bar{k}_r = \frac{2k_2 r \, \Delta r \, k_5(\text{S})}{D} \left[1 - r \left(\frac{k_5(\text{S})}{\pi D} \right)^{\frac{1}{2}} \right] \qquad \text{(XV.13.10)}$$

or for the relative rate of recombination to scavenging

$$\frac{\bar{k}_r}{k_5(\text{S})} = \frac{2k_2 r \, \Delta r}{D} \left[1 - \text{r} \left(\frac{k_5(\text{S})}{\pi D} \right)^{\frac{1}{2}} \right] \qquad \text{(XV.13.11)}$$

an equation valid only at relatively high scavenging rates but in reasonable agreement with the data.[b]

[a] In this approximate treatment we have neglected completely the excluded volume effect of the atoms on each other.

[b] Note that the quantum yield is equal to $k_5(\text{S})/[R(t)/t + k_5(\text{S})]$.

XVI

Further Kinetic Aspects
of Solution Reactions

1. Displacement Reactions

If liquid ethyl bromide is shaken with water at 25°C, no appreciable reaction takes place even after several days. The aqueous phase will not show a Br⁻ test with Ag⁺, and the original reactants may be recovered unchanged. With t-butyl bromide, on the other hand, one finds a fairly vigorous reaction with water at 25°C, accompanied by the liberation of heat, to produce t-butyl alcohol and HBr. With benzhydryl bromide, $(C_6H_5)_2CHBr$, the hydrolysis reaction appears to be almost immeasurably fast. Although all of these reactions can be represented stoichiometrically by the same general equation,

$$RBr + HOH \rightarrow ROH + H^+ + Br^- \qquad (XVI.1.1)$$

the rate of reaction is clearly a strong function of the nature of the organic group R. It is, in addition, a strong function of the solvent. It has been found for example that the rate of hydrolysis[a] of t-butyl chloride in acetone-H_2O mixtures increases by 10^4 fold in going from 90 per cent acetone to 90 per cent water. In the gas phase, the reaction does not take place at measurable rates even at 100°C.

Since the C—Br bond dissociation energies are of the order of from 50 to 70 Kcal in both alkyl and aryl bromides,[b] while $D(H—OH) = 118$ Kcal,

---

[a] Very similar results are obtained in CH_3OH-H_2O, dioxane-H_2O, and $C_2H_5OH-H_2O$ mixtures and in pure HCOOH as well. A. Streitwieser, Jr., *Chem. Revs.*, **56**, 571, 620 (1956) (see his Fig. 14), has presented a very comprehensive review of the kinetics of displacement reactions.

[b] Appendix E.

a free radical chain process is not possible for these systems at 25°C.[a] In fact most bond energies are in such a range as to rule out chain reactions of organic compounds via free radicals at temperatures that are below 100°C.[b]

The sensitivity of these reactions to solvent and the failure to react in the gas phase then leave no alternative but a reaction mechanism involving polar or ionic intermediates. Similar conclusions would follow consideration of other reactions of the same class, and as we shall see, this seems to be in accord with experimental findings.

First-order Nucleophilic Substitution Reaction. The most successful description of the mechanism of these reactions is found in the work of Hughes, Ingold, and coworkers,[c] who have proposed two separate categories. The first of these is S_N1, the first-order nucleophilic substitution reaction. This mechanism presupposes that the rate-determining step is the slow ionization of the organic species and that this is followed by a fast reaction with the nucleophilic, or attacking group.[d]

In the case of the hydrolysis of the alkyl or aryl halides in aqueous media this could be represented by

$$R\text{—}X \underset{k_2}{\overset{k_1}{\rightleftharpoons}} R^+ + X^-$$

$$R^+ + H_2O \underset{k_4}{\overset{k_3}{\rightleftharpoons}} \left[R\text{—}O \begin{smallmatrix} H \\ \diagup \\ \diagdown \\ H \end{smallmatrix} \right]^+ \underset{\substack{k_6 \\ (\text{fast})}}{\overset{k_5}{\rightleftharpoons}} ROH + H^+ \qquad (XVI.1.2)$$

where $k_3(H_2O) \gg k_2(X^-)$ and $k_5 \gg k_4$. In essence, every organic carbonium ion R^+ which is formed is attacked by H_2O much faster than it is by X^-. These reactions then have a rate which is simply

$$\frac{-d(RX)}{dt} = k_1(RX) \qquad (XVI.1.3)$$

i.e., the rate of ionization. Experimentally they should be first order in organic halide, zero order in H_2O, and sensitive to the ionizing power of

[a] The minimum activation energy for a chain process would be $\frac{1}{2}D$ (R—Br) $+ E_p$, the activation energy for the propagation step. For this system $E_p > 26$, so that any chain process would have to have an activation energy in excess of 51 Kcal. This would make it prohibitively slow at 25°C.

[b] The exceptions to this are such compounds as peroxides, azo compounds, and systems with redox reagents such as Fe^{3+}, Co^{3+}, etc.

[c] E. D. Hughes et al., *J. Chem Soc.*, 526 (1933). See also the treatise by C. K. Ingold, "Structure and Mechanism in Organic Chemistry," Cornell University Press, Ithaca, N. Y., 1953, for a systematic presentation of these views.

[d] By nucleophilic group is meant an ion or molecule having a closed valence shell which is capable of acting as a Lewis base, i.e., donating electrons. Common nucleophilic agents are the ions Cl^-, Br^-, I^-, OH^-, RO^-, and molecules containing O, N, and S.

the solvent. Since the solvent-ion interaction must be extremely strong in order for ionization to occur, this mechanism can be looked upon as only a first approximation to what is a much more complex situation. It is, however, a very convenient category and one which describes empirically many hydrolysis reactions.

It can be seen that the condition for a first-order reaction depends not only on the ratio of rate constants k_3 and k_2 but also on the concentration of X^- and HOH. It may thus be supposed that, even when $k_3 \gg k_2$, it will still be possible to have X^- sufficiently greater than H_2O to observe the competition of reactions 2 and 3. If we apply the stationary-state treatment to the scheme XVI.1.2, assuming that $k_5 \gg k_4$ and ignoring the back reaction 6,

$$\frac{-d(RX)}{dt} = \frac{k_1(RX)k_3(H_2O)}{k_2(X^-) + k_3(H_2O)} = \frac{k_1(RX)}{k_2(X^-)/k_3(H_2O) + 1}$$

(XVI.1.4)

which, when $k_3(H_2O) \gg k_2(X^-)$, reduces to our previous case. When the converse is true, we have

$$\frac{-d(RX)}{dt} \xrightarrow{k_3(H_2O) \ll k_2(X^-)} \frac{k_3 k_1}{k_2} \frac{(RX)(H_2O)}{(X^-)}$$

(XVI.1.5)

Here the reaction shows mass-law inhibition by one of its products X^-, while the concentration of carbonium ion R^+ is at its equilibrium value. One would expect such limiting behavior only in very exceptional cases, since it would imply reaction in a reasonably strongly ionizing solvent at low water content. Very few cases have been observed to fit this limiting behavior, but one example is provided by the hydrolysis of triphenyl methyl chloride[a] in 85 per cent aqueous acetone. The apparent first-order rate constant for hydrolysis is decreased fourfold by 0.01 M NaCl and remains unchanged on the addition of other salts, such as $NaClO_4$, not having the common Cl^- ion.[b]

The much more common case of hydrolysis reactions, which seem to correspond to the intermediate case of Eq. (XVI.1.4) with $k_2(X^-) \cong k_3(H_2O)$, is difficult to classify with certainty because, with the increase in salt content of the system, there is an increased tendency toward ionization (i.e., increase in $K_{ion} = k_1/k_2$ due to increased ionic strength), which tends to compensate for the mass-law retardation. Because of these ambiguities, other approaches have been employed to throw light on the mechanism. One of these is to study stereochemical changes of RX during reaction, while another is to study competitive reaction of the intermediate R^+. Thus t-butyl chloride in formic acid solution exchanges with radio-

[a] C. G. Swain et al., J. Am. Chem. Soc., **75**, 136 (1953).

[b] The H_2O dependence has not been verified; in general, it is extremely difficult to do so. The interpretation of the data in low dielectric systems is made complex by the problem of ion pairs. Some of these complexities are discussed by F. M. Beringer and E. M. Gindler, J. Am. Chem. Soc., **77**, 3200, 3203 (1955).

active Cl⁻ by a rate law which is not second-order but can be interpreted in terms of Eq. (XVI.1.2). The rate constant k_1 is found from the data to be 3×10^{-4} sec⁻¹ at 15°C.[a] On addition of small amounts of H_2O to formic acid solution of t-butyl chloride, a first-order hydrolysis is observed[b] with $k_1 = 3.7 \times 10^{-4}$ sec⁻¹. The close agreement between these different processes is taken as evidence supporting ionization as the slow, rate-determining step common to both reactions.

An example that illustrates the entire range of behavior is provided by the methanolysis of triphenyl chloride in benzene solutions[c]

$$\phi_3C\text{—}Cl + CH_3OH \rightarrow \phi_3C\text{—}OCH_3 + HCl$$

By using an amine as scavenger to convert the HCl to the quaternary $R_4N^+Cl^-$ salt (in order to avoid catalysis) and radioactive CH_3OH to allow analysis of reactant, Swain and Pegues showed that the reaction was first-order in ϕ_3CCl and first-order in CH_3OH when CH_3OH varied from 2×10^{-6} to 3.5×10^{-3} M. Since this was considerably slower than the exchange of Cl between ϕ_3CCl and radio-$(CH_3)_4N^+Cl^-$, the authors concluded that the mechanism is S_N1 in both cases,[d] the reaction of ϕ_3C^+ with CH_3OH being slower than with Cl⁻. Hence ionization is the slow step in the latter case, while displacement is the slow step in the former. Because of the low dielectric constant, ionization is otherwise negligible and inhibition by Cl⁻ is difficult, if not impossible, to determine. At $(CH_3OH) > 1 \times 10^{-3}$ M, the rate becomes second-order in CH_3OH. This is interpreted as involvement of CH_3OH in the otherwise fast ionization step.[e]

Second-order Nucleophilic Substitution Reaction. The second category proposed by Hughes and Ingold as the antithesis to the S_N1 (slow preionization) mechanism is S_N2, the second-order nucleophilic substitution reaction.

In this case it is presumed that the release of the group X "coincides" with the entering attack of the nucleophilic group. If we generalize the latter as Y, then the reaction is pictured as

$$Y + R\text{—}X \underset{2}{\overset{1}{\rightleftharpoons}} [Y \cdots R \cdots X] \underset{4}{\overset{3}{\rightleftharpoons}} Y\text{—}R + X \qquad (XVI.1.6)$$

where any or all of the groups may be charged. From an experimental point of view the kinetic rate law is second-order mixed, first-order in RX,

[a] W. Koskoski et al., *J. Am. Chem. Soc.*, **63**, 2451 (1941).

[b] L. C. Bateman and E. D. Hughes, *J. Chem. Soc.*, 1187 (1937).

[c] G. C. Swain and E. E. Pegues, *J. Am. Chem. Soc.*, **80**, 814 (1958). An alternative view of this and similar systems has been given by E. D. Hughes et al., *J. Chem. Soc.*, 1206, 1220, 1230, 1238, 1256, 1265, 1279 (1957).

[d] C. G. Swain and M. M. Kreevoy, *J. Am. Chem. Soc.*, **77**, 1122 (1955), found that in benzene the exchange was zero-order in $(CH_3)_4N^+Cl^-$ and first-order in ϕ_3CCl.

[e] This is in contrast to the interpretation of R. F. Hudson and B. Saville, *J. Chem. Soc.*, 4114, 4121, 4130 (1955), on the second-order with respect to EtOH of the reaction, $\phi_3C\text{—}Cl + EtOH \rightarrow \phi_3C\text{—}OEt + HCl$ in CCl_4 solution. These authors attribute the order to the presence of EtOH polymers.

and first-order in the nucleophilic group Y. The classic examples of this S_N2 mechanism are provided by the Walden inversion reactions: the displacement of group X in optically active RX leads to formation of optically active RY, where the asymmetric carbon atom has the inverted configuration.

Holmberg[a] showed that optically active alkyl halides are racemized in solution at rates which are second-order mixed with respect to halide ion and alkyl halide. By using 2-octyl iodide, Hughes and coworkers[b] showed that the rate of exchange with radioactive I$^-$ in acetone solution is precisely equal to the rate of inversion, both rates being mixed second-order. This, of course, is quite reasonable on the assumption that both reactions proceed by S_N2 mechanisms with inversion of configuration:

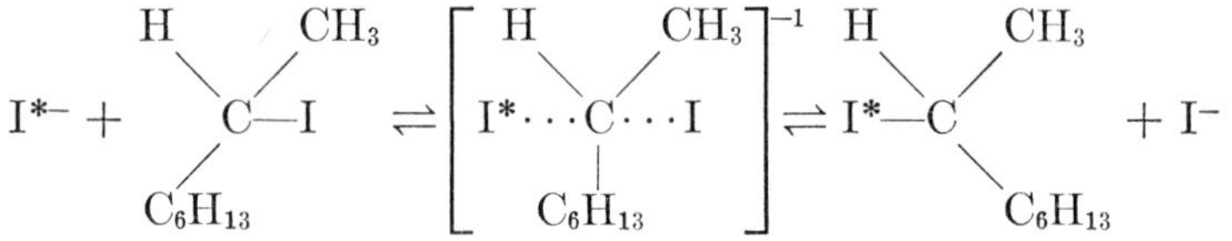

$$I^* + \begin{array}{c} H \quad CH_3 \\ \diagdown \diagup \\ C-I \\ \diagup \\ C_6H_{13} \end{array} \rightleftharpoons \left[\begin{array}{c} H \quad CH_3 \\ \diagdown \diagup \\ I^* \cdots C \cdots I \\ | \\ C_6H_{13} \end{array} \right]^{-1} \begin{array}{c} H \quad CH_3 \\ \diagdown \diagup \\ \rightleftharpoons I^*-C \\ \diagdown \\ C_6H_{13} \end{array} + I^-$$

$$\qquad\qquad (D) \qquad\qquad\qquad (L)$$

where the transition state involves pentavalent carbon. Studies with other halides have confirmed this inversion mechanism for S_N2 reactions, while additional support comes from stereochemical studies of other compounds.[c] Very interesting negative support for the inversion displacement comes from the observation that in halides in which the geometry prevents "back-side" attack on the carbon atom containing the halogen group, there is no S_N2 substitution of halide ion. This was demonstrated in the case of the bicyclic compound, 1-chloroapocamphane,[d] containing a bridgehead chlorine:

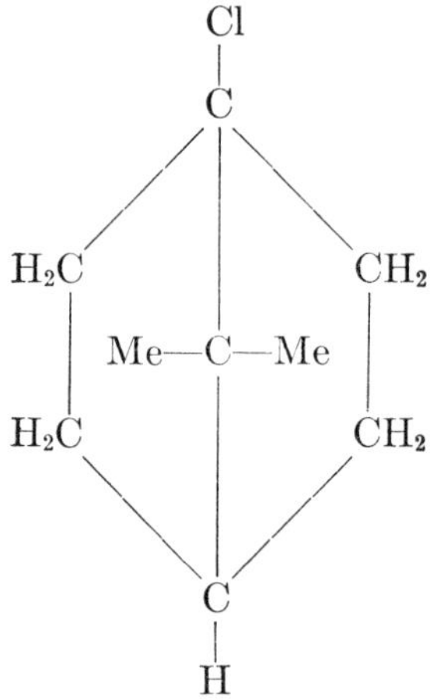

[a] B. Holmberg, J. prakt. Chem., [2]88, 553 (1913).
[b] E. D. Hughes et al., J. Chem. Soc., 1525 (1935).
[c] See treatise by Ingold, loc. cit.
[d] P. D. Bartlett and L. H. Knox, J. Am. Chem. Soc., 61, 3184 (1939).

In this case the structure sterically precludes pentavalent C atom or inversion. However, the Cl displacement does not occur by an S_N1 mechanism either, indicating that the C—Cl bond in this case must also have an extremely low rate of ionization.[a]

The distinction between the S_N1 and S_N2 mechanisms is not necessarily always a sharp one, and if the attacking group Y can facilitate the departure of X, an intermediate case may occur. Such intermediate cases seem to arise in reactions in which the nucleophilic reagent is a solvent molecule, when, unfortunately, kinetic order with respect to solvent is almost impossible to clarify. It is important to note that where the attacking group Y is an ion such as a halide, X^-, or OH^- or RO^- the displacement reaction usually follows fairly clean second-order mixed kinetics.[b] The confusion that arises when Y is a solvent molecule is readily understood when we consider that the mechanism of ionization will involve very strong ion-solvent interactions. In fact ionization is not possible without such interactions.

Reactions when Y is a solvent molecule are referred to as solvolysis reactions of which hydrolysis [Eq. (XVI.1.2)] is a special example involving H_2O. The dual role of solvent as ionizing agent and as nucleophilic agent is what causes the kinetic difficulties. It is the feeling of the author that much of the controversy in the interpretation of solvolysis reactions can be attributed to the fundamental complexity of the molecular systems in contrast to the oversimplification of the models which have been presented to explain them. In fact, until the equilibrium theory of ionic systems shows considerable progress, it is doubtful if the kinetic data will have much "fundamental" interpretation.

It might be supposed that stereochemical evidence would be rather definite in distinguishing S_N1 from S_N2 reaction paths. In S_N1 a stable carbonium ion is formed as an intermediate following ionization [Eq. (XVI.1.2)]. If the carbon center is originally asymmetric and the starting compound, RX, is optically active, then the product, RY, would be racemic.[c] This turns out to be true in many cases. However, when the atom adjacent

---

[a] Some authors have attempted to interpret this as arising from the fact that the carbonium ion must be planar, such structure being precluded by the bicyclic cage. While this may be true, it is certainly not proven by the kinetic data. Steric inhibition of strong solvation is also a possible factor.

[b] The principal exception to this statement involves tertiary carbon, allylic carbon, and benzyl carbon atoms attached to X. In such cases it is often found that the second-order reaction with solvate ion such as RO^- (in ROH solvent) is much slower than apparent first-order reaction with solvent. These systems may be recognized as those giving quite stable carbonium ions, and the solvolysis has been ascribed to an S_N1 mechanism.

[c] The nucleophilic group Y may be expected to attack from either side of the planar carbonium ion.

to this carbon atom contains unsaturated groups,[a] it is frequently found
that the displacement reaction is certainly not S_N2, and yet there is pres-
ervation of optical activity in the products. The explanation offered in
these cases is that the unsaturated group interacts with the carbonium ion
in such a way as to prevent it from becoming symmetrical.

An interesting example of this type has been provided by Cram,[b] who
studied the rates of acetolysis of optically active 3-phenyl-2-butyl tosylate
in glacial acetic acid:[c]

$$
\begin{array}{cc}
\phi \quad H & \phi \quad H \\
| \quad | & | \quad | \\
CH_3-C-C-CH_3 + HOAc \rightarrow CH_3-C-C-CH_3 + HO_3S\phi CH_3 \\
| \quad | & | \quad | \\
H \quad O_3S\phi CH_3 & H \quad OAc
\end{array}
$$

$$(XVI.1.7)$$

The erythro compound, ET, gave the erythro acetate, EA, with almost
quantitative retention of optical activity. On the other hand, the optically
active threo tosylate, TT, gave racemic threo acetate, TA. The explanation for these striking results is given in terms of the different, bridged
carbonium ions formed in each case:

(ET)   (EA)

(TT)   (TA)

If it is assumed that the ionization of the OTs⁻ group is accompanied
by the simultaneous, back-side bridging of the phenyl group, then the carbonium ion formed from the erythro ester, ET, will preserve configuration,
because attachment of an ester group at either of the two bridged C atoms

[a] These can be carbonyl $\left(\begin{array}{c} \diagdown \\ \diagup \end{array} C{=}O \right)$, aryl groups, or even halogen atoms.

[b] D. Cram, *J. Am. Chem. Soc.*, **71**, 3863 (1949); **74**, 2149 (1952).

[c] This is the abbreviation for the ester of paratoluenesulfonic acid, p-CH₃—C₆H₄—
SO₃H.

must yield the same optically active erythro product. The threo compound, however, forms a bridged carbonium ion which has a plane of symmetry such that the configuration is racemic, i.e., the two carbon atoms are mirror images of each other. Addition of ester groups will now produce mirror images by adding at one or the other of the two carbon atoms.

Further supporting evidence is provided by the observation that, after the acetolysis has been allowed to go to 50 per cent completion, the remaining erythro compound, ET, is optically pure, while the remaining threo compound, TT, is racemized.[a] This is considered as evidence also for the intermediate formation of an "intimate" ion pair,[b] since it is found that the reactions are not inhibited by even high concentrations of HOTs.[a] Such a test is not very meaningful, however, since HOTs is extraordinarily little ionized in the low dielectric, acetic acid.

Similar phenomena are found with neighboring ester groups, carboxylate ions, OH, S, N, OR, Cl, Br, I, aryl, alkyl groups, and even H atoms.[c] It is further found that the displacement reactions are significantly enhanced in rate by the participation of some of these neighboring groups in stabilizing the bridged carbonium ion. Winstein and Grunwald[d] have discussed the rate constants of the displacement reactions in terms of the intramolecular structure of the carbonium ion and presented qualitative arguments for anticipating rate enhancement.

The inherent plausibility of metastable bridged carbonium ions as intermediates is supported by two independent types of observation. One is the extensive rearrangements which can occur in allylic systems[e] under conditions in which ionic displacement reactions are possible. A second is the existence of *stable* bridged compounds, including, in the case of boron compounds, pentavalent atoms. Thus diborane and substituted diboranes have stable bridged structures.

[a] D. Cram, *J. Am. Chem. Soc.*, **74**, 2129 (1952).

[b] The two ion pairs which are distinguished are those in which a solvent molecule is interposed between the ions—the "external" ion pair—and the "intimate" ion pair, in which the ions are in contact. The distinction is more readily observed in the case of inorganic complex ions such as $[Cr(H_2O)_5Cl]^{2+}$ and $[Cr(H_2O)_6]^{3+} \cdot Cl^-$, where the solvent shell neighbors are held firmly. The distinction which is more difficult to characterize is that between the original covalent compound and the intimate ion pair.

[c] See review article by Streitwieser, *loc. cit.*

[d] S. Winstein and E. Grunwald, *J. Am. Chem. Soc.*, **70**, 828 (1948).

[e] An extensive review of this field is presented by R. H. deWolfe and W. G. Young, *Chem. Revs.*, **56**, 753 (1956).

Since diborane is isoelectronic with $C_2H_5^+$, the ethyl carbonium ion, and, in general, substituted diboranes are isoelectronic with the corresponding substituted ethyl carbonium ions, it would seem to argue very strongly for the analogous bridged structures for the carbonium ions.[a]

The first-order hydrolysis of neopentyl bromide gives mixtures of *sec*-amyl alcohol and some neopentyl alcohol.[b] While this does not establish the stability of a bridged carbonium ion intermediate, it does indicate the passage through such a state. The reaction is otherwise very difficult to picture. The postulated carbonium ion mechanism is

[a] This also provides a reasonable explanation for the extensive rearrangements which are found to occur in carbonium ions produced in mass spectrometers. See papers by S. Meyerson and P. N. Rylander, *J. Am. Chem. Soc.*, **78**, 5799 (1956); S. Meyerson and H. M. Grubb, *ibid.*, **79**, 842 (1957); and S. Meyerson, *J. Chem. Phys.*, **27**, 1116 (1957).

[b] I. Dostrovsky and E. D. Hughes, *J. Chem. Soc.*, 157, 164, 166, 169, 171 (1946).

In contrast, the reaction in basic alcoholic solution with $Na^+OC_2H_5^-$ gives normal product, ethyl neopentyl ether, with a second-order rate law indicating a direct S_N2 displacement mechanism. Bridged carbonium ions are not always formed as metastable intermediates, other structures being sometimes more stable.[a]

The kinetic identity of the intimate ion pair and the solvent-separated ion pair as distinguishable from the covalent molecule and the normally dissociated ion pairs has been under investigation by Winstein and co-workers.[b] For certain esters in glacial acetic acid, the addition of very small amounts of $LiClO_4$ can produce a large increase in rate (for example, $0.0001\ M$ $LiClO_4$ doubles the rate of acetolysis of cholesteryl tosylate) while much larger increases in $LiClO_4$ concentration produce only a relatively small increase in rate of acetolysis. This is interpreted as involving a stabilization of the solvent-separated ion pair $R^+—S—X^-$ by $Li^+ClO_4^-$ ion pair in the low-dielectric media.[c] The authors represent the kinetic scheme by

$$RX \underset{k_{-1}}{\overset{k_1}{\rightleftharpoons}} [R^+X^-] \underset{\substack{k_{-2}\\(-S)}}{\overset{\substack{k_2\\(+S)}}{\rightleftharpoons}} [R^+—S—X^-] \underset{k_{si}}{\rightarrow} ROS + H^+ + X^-$$

$$\text{Covalent} \quad \substack{\text{Intimate}\\\text{Ion Pair}} \qquad\qquad \Big\updownarrow k_3 \; k_{-3}$$

$$ROS + H^+ \underset{k_s}{\overset{(+SOH)}{\longleftarrow}} R^+ + X^-$$

where S (or SOH) represents a solvent molecule and it is assumed that solvolytic displacement occurs only with the solvent separated ion pair $R^+—S—X^-$ or with the carbonium ion R^+ (although there are cases of reaction with RX and R^+X^- cited by the authors).

The arguments in favor of such a scheme are quite strong but relatively complex and depend on a number of auxiliary data, kinetic, thermo-dynamic, and structural-chemical. The large activity effects on ions and K_{ion} of weak electrolytes in the low-dielectric media have been quantita-tively considered, and it appears that the active species in the system are not ions but rather ion pairs. This makes it likely that the same is true of some of the other work done in glacial acetic acid which has been probably incorrectly interpreted as ionic.

There is one difficulty which always exists in the use of "dry" hydroxylic solvents SOH, and that concerns the equilibrium $2SOH \rightleftharpoons SOS + HOH$. It can be shown from available thermodynamic data that for the simple alcohols and acids the equilibrium constant at 25°C is considerably greater than unity. Thus these systems are metastable with regard to such an

[a] S. Winstein and B. K. Morse, J. Am. Chem. Soc., **74**, 1133 (1952).

[b] See S. Winstein and E. Clippinger, ibid., **78**, 2784 (1956); A. H. Fainberg, ibid., **80**, 459 (1958), and earlier papers, particularly Winstein et al., ibid., **78**, 328 (1956).

[c] In an Am. Chem. Soc. meeting it was suggested that rapid exchange occurs to form $R^+—(S)—ClO_4^-$, which is then capable only of acetolysis. This has since been fairly well established.

equilibrium.[a] Since H_2O has such a profound effect on ionic interactions, slight production of H_2O in these systems is always a point of importance which may affect kinetic and thermodynamic results significantly. It should be explored more fully than has yet been the case.

2. Acid-Base Catalysis

Both acetone and water are strongly polar molecules, the dipole moments being 2.84 and 1.84 Debyes,[b] respectively. Since they mix as liquids in all proportions with minor heat evolution, it is reasonable to expect that there is fairly strong dipole-dipole interaction between their molecules. If pressed for structural details of the liquid, we might guess that in dilute acetone solutions, each acetone molecule may be associated with about 4 H_2O molecules in some such manner as the following:

In view of these interactions it is probably not surprising to find that acetone will slowly exchange its O with[c] H_2O^{18} and its H atoms with[d] D_2O. The reaction with pure H_2O is extraordinarily slow, with a half-life for both these reactions of the order of 1 to 10 weeks. With addition of 10^{-2} M NaOH or HCl, however, the half-life of the D exchange becomes of the order of an hour or so, while the O^{18} exchange is almost too fast to measure.[e] It can be shown that the rate is directly proportional to the stoichiometric concentration of H^+ or OH^- in each case and is essentially independent of the other ion (Cl^- or Na^+).

[a] CH_3OH seems to be particularly difficult to purify. It has been shown by R. B. Porter, *J. Phys. Chem.*, **61**, 1260 (1957), that CH_3OH in pyrex or quartz will always contain small amounts of methyl borate and methyl silicate unless H_2O is present to cause hydrolysis of those esters. "Dry" CH_3OH in pyrex will thus contain $H_2O + (CH_3O)_3B$ in significant amounts.

[b] See compilation by C. P. Smyth, "Dielectric Behavior and Structure," McGraw-Hill Book Company, New York, 1955.

[c] M. Cohn and H. C. Urey, *J. Am. Chem. Soc.*, **60**, 679 (1938).

[d] K. Bonhoeffer and Klar, *Naturwiss.*, **22**, 45 (1934); W. D. Walters and K. Bonhoeffer, *Z. physik. Chem.*, **A182**, 265 (1938).

[e] At 25°C the half-life for exchange of O^{18} between acetone and H_2O^{18} (90 per cent by volume acetone) is about 350 sec for 0.001 M HCl and 10^6 fold longer for 0.001 M NaOH, so that acid catalysis is far more important. Undissociated salicylic acid molecules are about 2000 fold less effective than H^+, while salicylate ions have no measurable effect.

Such observations lead to the designation of reactions as acid- or base-catalyzed. When the catalysis is limited to the species H^+ (or OH^-), the reaction is spoken of as being subject to *specific* H^+ ion (or OH^- ion) catalysis. Many reactions of both organic and inorganic chemistry fit such a designation. However, very early work on such systems[a] soon showed that the catalysis was not limited to H^+ or OH^- but did extend to other species which could be subsumed under the category of what are now called Brönsted acids and bases.

In the Brönsted scheme an acid is any substance HA capable of transferring a proton to a second species called a base B. The stoichiometric equation can be written as

$$HA + B \rightleftharpoons A^- + HB^+ \qquad (XVI.2.1)$$

where A^- is now the conjugate base of HA and HB^+ the conjugate acid of base B. [HA and B can bear charges, so that Eq. (XVI.2.1) is not necessarily a restriction on the nature of HA and B.]

When a reaction is susceptible to catalysis by Brönsted acids and bases, the phenomenon is characterized as *generalized* acid or basic catalysis. One of the earliest reactions discovered to be subject to such generalized acid-base catalysis was the mutarotation of optically active glucose:

α-D-Glucose β-D-Glucose

Because the changes involve the inversion in configuration of H and OH with respect to the pyranose ring and hence a change in optical rotatory power of the sugar, it is easily and conveniently followed in a polarimeter. The reaction has long been a proving ground for the testing of kinetic theories.

In pure water, the mutarotation of glucose, αG, follows a first-order rate law:

$$\frac{-d(\alpha G)}{dt} = k_0(\alpha G) \qquad \text{or} \qquad \frac{-d \ln (\alpha G)}{dt} = k_0 \qquad (XVI.2.2)$$

In the presence of excess H^+ the rate law is found to be

$$\frac{-d \ln (\alpha G)}{dt} = k_0 + k_H(H^+) \qquad (XVI.2.3)$$

---

[a] See the book by R. P. Bell, "Acid-Base Catalysis," Oxford University Press, New York, 1941, for an excellent summary of the subject and its historical background.

where the second-order constant k_H is referred to as the catalytic rate constant of H^+. If the reaction is subject to general acid catalysis by a series of acids HA_j and to general basic catalysis by a series of bases B_n, then, by assigning a specific constant to each acid k_{A_j} and each base k_{B_n}, the overall rate law may be expressed as

$$\frac{-d\ln(\alpha G)}{dt} = k_0 + \sum_j k_{A_j}(HA_j) + \sum_n k_{B_n}(B_n) \qquad (XVI.2.4)$$

The experimental verification of such a complex rate expression is usually carried out by varying the concentrations of a single conjugate acid-base pair at a time. In the mutarotation of glucose in water solution at 18°C, k_0 is found to be 0.0054, while for acetate ion, $k_{Ac^-} = 0.0265$; for phenoxide ion, $k_{\phi O^-} = 4.4$; and for OH^- it is 3800.[a] For catalysis by H^+, k_H is found to be[b] 0.0040 liter/mole-min. It can be seen that the constants cover an exceedingly large range.

What interpretation can be placed on such behavior? In the first place it should be noted that the reactions chosen represent true catalysis in the sense that the catalyst is not consumed in the chemical reaction. Since the behavior common to all of the catalytic species is the ability to transfer protons, it seems fair to assume that there is an interaction between the reacting molecule (or its conjugate acid or base) and the catalytic acids and bases that involves proton transfer. Finally, since the net reaction does not involve protons, we must assume a reverse transfer of protons from this last complex (substrate plus acid or base) to yield final product.

This, in fact, is the interpretation found compatible with the data. In acid-catalyzed reactions, the sequence of reactions usually starts with the transfer of a proton from the acid HA to the substrate molecule M (or its hydrate). In the mutarotation of glucose it is presumed to be:

$$(XVI.2.5)$$

---

[a] Data from compilation of Bell, *loc. cit.* Units are min^{-1} for k_0 and liters/mole-min for other rate constants.

[b] M. Kilpatrick and M. Kilpatrick, *J. Am. Chem. Soc.*, **53**, 3698 (1931).

where we have put in only one of the substituents on the pyranose ring. The slow, and hence rate-determining, step in the reaction is presumed to be reaction 1, the proton transfer from acid HA to the ether linkage in the ring. The resultant changes leading to ring cleavage (3) and open-chain aldehyde (5), are assumed to be much faster. Note that, if reaction 1 were not rate-determining but reaction 3 was instead, the reaction rate would be proportional to MH^+ and independent of the source of the H^+ attached to M. It would in this case show specific H^+ catalysis and not generalized acid catalysis.

In the base-catalyzed reactions, the interpretation involves the transfer of a proton from the substrate MH to the base B as the rate-controlling step. In the case of α or β-D-glucose (now written as G), this could be

$$(XVI.2.6)$$

However, an equally plausible mechanism is just as likely. Since the kinetic measurements merely show a concentration dependence on the product (G)(B) for the base-catalyzed reaction, we cannot distinguish this from a reaction in which the rate-determining step involves the conjugate acid-base pairs, that is, $(G^-)(BH^+)$, where G^- is the conjugate base of glucose. The reasons are evident if we write the acid-base equilibria:

$$G \rightleftharpoons G^- + H^+ \qquad K_G = \frac{(G^-)(H^+)}{(G)}$$

$$BH^+ \rightleftharpoons B + H^+ \qquad K_{BH^+} = \frac{(B)(H^+)}{(BH^+)} = \frac{K_w}{K_B}$$

$$(XVI.2.7)$$

By combining these two equations and rearranging terms we find

$$(G)(B) = \frac{K_{BH^+}}{K_G}(G^-)(BH^+) = \frac{K_w}{K_B K_G}(G^-)(BH^+) \qquad (XVI.2.8)$$

so that the product (G)(B) is always proportional to $(G^-)(BH^+)$, the constant of proportionality being given by $K_w/K_B K_G$, where K_w is the ionization constant for water [that is, $K_w = (H^+)(OH^-)$]. For the catalytic rate term $k_B(G)(B)$ we can equally well write $k_{BH^+}(G^-)(BH^+)$, where from Eq. (XVI.2.8) we see that

$$k_{BH^+} = k_B \frac{K_W}{K_B K_G} \qquad \text{(XVI.2.9)}$$

So long as the equilibria [Eq. (XVI.2.7)] are maintained, no kinetic evidence can distinguish the two rate terms.

When reaction between G^- and BH^+ is rate-determining, the transition state may be pictured as

$$\text{(XVI.2.10)}$$

In this particular transition state, resonance structures which can be written for various double-bond, no-bond interactions will permit charge to be distributed rather widely over the molecule. An analogous picture can be drawn for the corresponding acid-catalyzed transition state that involves GH^+ and A^-, with the rate-determining step involving the transfer of the proton from the extra ring OH^+ group to A^-.

It is to be noted that, to satisfy such an interpretation, any substrate capable of undergoing specific or generalized acid-base catalysis must itself be capable of acting both as acid and as base (i.e., amphiprotic). This seems to be in accord with the facts as known, although not all acid-catalyzed reactions show base catalysis, or vice versa.

It was originally proposed by Lowry,[a] and later argued again by Swain,[b] that since the function of the acid (or base) in these reactions is to facilitate the shift of a proton from one part of the substrate molecule to another (i.e., a prototropic shift), that a mechanism that involves simultaneous removal and addition of a proton might provide the most rapid path. In such case, the reactions should be expected to involve the simultaneous attack of acid and base on the proper parts of the molecule. In the muta-rotation of α-D-glucose the process can be pictured as

[a] T. M. Lowry, J. Chem. Soc., 2554 (1927).
[b] G. D. Swain, J. Am. Chem. Soc., **72**, 4578 (1950). G. D. Swain and J. F. Brown, Jr., ibid., **74**, 2534, 2538 (1952).

$$(XVI.2.11)$$

where for generalized base catalysis the co-attacking acid may be the solvent H_2O, while for generalized acid catalysis the co-attacking base may again be H_2O. This amphiprotic assistance of the solvent could thus account for the observed second-order kinetics.

In support of such an argument Lowry and Faulkner[a] showed that 2,3,4,6-tetramethyl glucose undergoes extremely slow mutarotation in dry pyridine (a moderate base but very feeble acid) and also in cresol (moderate acid, feeble base). In a mixture of the two solvents, however, the reaction is extremely rapid. Swain and Brown, loc. cit., also showed that the same tetramethyl glucose in benzene solution undergoes mutarotation with added phenol and pyridine at a third-order rate, first-order in glucose, phenol, and pyridine. Even more striking is their finding that the amphiprotic species, 2-hydroxy pyridine, attacks the glucose at a rate which is second-order mixed, first-order in glucose, and first-order in catalyst. At a concentration of 0.001 M it is some 7000 times more active a catalyst than a mixture of 0.001 M pyridine + 0.001 M phenol, although it is a 100-fold weaker acid than phenol and a 10,000-fold weaker base than pyridine. The structure of 2-OH-pyridine is admirably suited to undergo the postulated double proton shift with the glucose.[b]

While this termolecular mechanism of acid-base catalysis seems very appealing and demonstrated in a number of cases,[c] it does not necessarily hold in all cases, and in fact Bell and coworkers[d] have shown a number of specific examples in which it does not apply.

[a] T. M. Lowry and I. J. Faulkner, J. Chem. Soc., **127**, 2883 (1925).

[b] 3-OH- and 4-OH-pyridine are structurally less suitable and turn out to be much poorer catalysts.

[c] The generalization of Swain's argument leads to a kinetic equation for the specific rate constant of the form

$$k = \sum_j \alpha_{Aj}(HA_j) \sum_n \alpha_{B_n}(B_n)$$

where the ratio of any two α_{Aj} gives the relative catalytic efficiency of the two respective acids A_j (or bases for α_{B_n}). One of the α's in the equation is superfluous. Note that there is a term corresponding to every possible acid-base pair. The constant k_0 in the former equation (XVI.2.4) is now interpreted as due to solvent, that is, $k_0 = k_0'(H_2O)^2 + k_0''(H_3O)^+(OH^-) = k_0'(H_2O)^2 + k_0''K_w(H_2O)^2$.

[d] R. P. Bell and J. C. Clunie, Proc. Roy. Soc. (London), **A212**, 33 (1952), and R.P. Bell and P. Jones, J. Chem. Soc., 88 (1953). See also Y. Pocker, ibid, 1279 (1957).

3. The Brönsted Relation: Linear Free Energy Relations

The view of generalized acid-base catalysis as a prototropic shift[a] assisted by acids and bases raises, quite naturally, the question of the relationship between the catalytic power of the acid or base and its own ionization constant. It had early been recognized that there is a correlation between the two constants. Taylor[b] proposed the first quantitative relation, that the acid-catalytic constant of an acid k_{HA} was proportional to $K_{HA}^{1/2}$, the square root of its ionization constant. For generalized acid-base catalysis, the Brönsted equation, proposed later,[c] has gained wide empirical use:

$$k_{HA} = G_A (K_{HA})^\alpha \qquad k_B = G_B (K_B)^\beta \qquad \text{(XVI.3.1)}$$

where k_{HA} and k_B are the experimentally observed second-order catalytic constants for acid and base, respectively, and K_{HA} and K_B are the corresponding ionization constants. The exponents α and β are found to be positive fractions $(0 < \alpha, \beta < 1)$ and constant for a given reaction and solvent. The constants G_A (or G_B) will be constant only for structurally similar acids (or bases).

The rationale of essentially empirical relations such as (XVI.3.1) can be demonstrated if we write them in logarithmic form:

$$\ln k_{HA} = \ln G_A + \alpha \ln K_{HA} \qquad \text{(XVI.3.2)}$$

with a similar equation for k_B. Now if we use the transition-state formulation of k_{HA} [Eq. (XV.5.2)] we can write

$$RT \ln k_{HA} = -\Delta \bar{F}_{HA}^{\ddagger} + b \qquad \text{(XVI.3.3)}$$

where $\Delta \bar{F}^{\ddagger}$ is the standard free energy change in forming the transition state from reactants and b is a constant very nearly independent of molecular properties. From thermodynamics we can write $-RT \ln K_{HA} = \Delta \bar{F}_{HA}^{\circ} = $ the standard free energy change in the ionization of HA. If Eq. (XVI.3.2) is correct, it implies a linear relation between $\Delta \bar{F}_{HA}^{\ddagger}$ and $\Delta \bar{F}_{HA}^{\circ}$

$$\Delta \bar{F}_{HA}^{\ddagger} = \alpha \, \Delta \bar{F}_{HA}^{\circ} + \text{const} \qquad \text{(XVI.3.4)}$$

Equation (XVI.3.4) would be expected to hold if the process of ionization were closely parallel to that of formation of the transition state. Since the former involves the transfer from HA of a proton to solvent while the latter involves the *partial* transfer from HA of a proton to reactant, it is certainly not surprising that the free energy changes in the two processes may be related. That the value of the exponent α lies between 0 and 1 is

[a] Note that the further generalization of such a relation to include Lewis acids and bases (i.e., those capable of accepting or donating electrons) would extend the relationship to aprotic solvents. Not very much has been done quantitatively in exploring such systems.

[b] H. S. Taylor, *Z. Elektrochem.*, **20**, 201 (1914).

[c] J. N. Brönsted and K. Pedersen, *Z. physik. Chem.*, **108**, 185 (1924).

expected on the grounds that the transition state represents only a partial transfer of the proton and hence only part of the total free energy change of ionization. However, a precise linear relation would be expected to hold only if there were no specific interactions between the substrate reactant and HA or at least none that were different from the interaction between solvent and HA. That such interactions do exist is indicated by the large deviations which are occasionally observed from the Brönsted relation.

The interpretation given to the constant α is that it measures the "sensitivity" of the reaction (catalysis) to the acidity (or basicity) of the catalyst. In terms of the free energy changes we might say that it is a measure of the amount of the free energy change of ionization that occurs in the formation of the transition state.[a]

The Brönsted relation should not be used as written [Eq. (XVI.3.1)]. Some corrections that arise from symmetry changes in the respective processes and have nothing to do with the intrinsic chemical changes going on in the system must be applied to k_{HA} and K_{HA}. Thus since K and k are expressed in units of moles/liter, we might expect that a dibasic acid in which the two carboxyl groups are widely separated should be twice as effective (per mole) as a monobasic acid such as acetic acid. Conversely, in comparing bases we may expect that formate ion HCO_2^- would be twice as effective statistically in adding a proton as ethoxide ion $C_2H_5O^-$, since the former can add H^+ to either of the two O atoms of the $R-CO_2^-$ ion. These arguments were anticipated by Brönsted and Pedersen, who proposed that Eq. (XVI.3.1) be modified to read

$$\frac{k_{HA}}{p} = G_A \left(K_{HA} \frac{q}{p} \right)^\alpha \qquad \frac{k_B}{q} = G_B \left(K_B \frac{p}{q} \right)^\beta \qquad \text{(XVI.3.5)}$$

where p is the number of equivalent protons on the acid and q is the number of equivalent positions which can add a proton in the conjugate base. For acetic acid $p = 1$ and $q = 2$, while for succinate ion, $HOOC \cdot CH_2 \cdot CH_2 \cdot COO^-$, acting as an acid, $p = 1$ and $q = 4$.

In Fig. XVI.1 are shown plots of log $(K_B p/q)$ vs. k_B/q for the base-catalyzed decomposition of nitramide at 18°C in aqueous solution. The reaction is

$$\begin{array}{c} H \\ \diagdown \\ \\ \diagup \\ H \end{array} N-N \begin{array}{c} O \\ \diagup \\ \\ \diagdown \\ O \end{array} \rightarrow N_2O + H_2O$$

and presumably takes place through the tautomeric azo acid[b]

---

[a] Modified, of course, by any specific interactions between substrate-catalyst-solvent which tends to "spoil" the relation.

[b] Note that this equilibrium is probably a slow, base-catalyzed reaction.

$$
\begin{array}{ccc}
\overset{\displaystyle H}{\underset{\displaystyle H}{\diagdown}}N-N\overset{\displaystyle O}{\underset{\displaystyle O}{\diagup}} & \underset{2}{\overset{1}{\rightleftharpoons}} & H-N=N\overset{\displaystyle OH}{\underset{\displaystyle O}{\diagup}}
\end{array}
$$

The rate-determining step would then be the removal of the proton from this acid

$$
B + H-N=N\overset{\displaystyle OH}{\underset{\displaystyle O}{\diagup}} \xrightarrow[\text{slow}]{3} BH^+ + \left[N=N\overset{\displaystyle OH}{\underset{\displaystyle O}{\diagup}} \right]^- \qquad (XVI.3.6)
$$

$$
\downarrow 4 \; \text{fast}
$$

$$
NNO + OH^-
$$

To account for the observed second-order base catalysis, such a scheme would require that $k_3(B) \ll k_2$. This is quite likely because the azo acid would be isoelectronic with nitric acid and is expected to be a strong acid. The prototropic shift from azo acid back to thermodynamically more stable amine is thus expected to be fast.[a]

Bell (*loc. cit.*) has used four different straight lines with slightly different slopes and intercepts to represent the four classes of bases of Fig. XVI.1. With not appreciably less average deviation, we have chosen to use two parallel lines as shown. For the neutral, B°, and negatively charged bases, B^- and $B^=$, the correlation of Fig. XVI.1 covers a range of 10^5 in k_B and 10^6 in K_{BH}.[b]

Many similar examples of the application of the Brönsted relation may be cited, including the mutarotation of glucose,[c] the iodination of acetone,[d] the bromination of aceto-acetic ester,[e] and the dehydration of 1,1-dihydroxyethane to acetaldehyde[f] and many other closely related compounds (see R. P. Bell, *loc. cit.*). It is usually found that on going from one acid to another structurally very different, the two acids will not fit the same

---

[a] This seems to be quite a general phenomenon. The tautomeric oxyacids of ketones (i.e., enols) and nitro paraffins are all less stable thermodynamically and yet much stronger acids than the corresponding paraffin acids.

[b] Instead of using the base constant K_B we have followed Bell in using the acid constant $K_{BH} = K_W/K_B$. Thus the slope is negative.

[c] F. Westheimer, *J. Org. Chem.*, **2**, 431 (1938). J. N. Brönsted and E. A. Guggenheim, *J. Am. Chem. Soc.*, **49**, 2554 (1927). T. M. Lowry and G. F. Smith, *ibid.*, 2539 (1927).

[d] From data of H. M. Dawson reworked by Bell, *loc. cit.*

[e] K. Pedersen, *J. Phys. Chem.*, **37**, 751 (1933); **38**, 601 (1934).

[f] R. P. Bell and W. C. E. Higginson, *Proc. Roy. Soc.* (*London*), **A197**, 141 (1949). In this most extensively covered case it has been shown that 45 carboxylic acids and phenols fit a Brönsted equation (with $\alpha = \frac{1}{2}$) with a mean deviation of only 0.1 log unit in k_{HA}. The range in pK_{HA} is some 10 units and the maximum deviation is 0.3 log units.

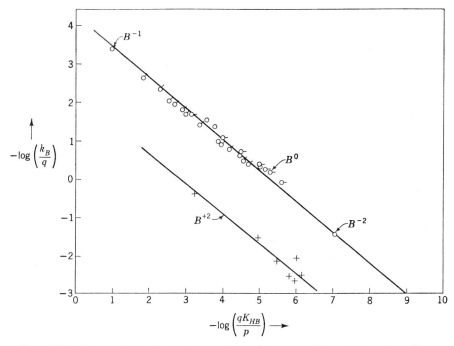

FIG. XVI.1. Brönsted plot for the base-catalyzed decomposition of nitramide. (Data from compilation by R. P. Bell, "Acid-Base Catalysis," Oxford University Press, New York, 1941.) $+$: Doubly charged bases, B^{++}, mostly aquo-ions, $[M(H_2O)_n(OH)]^{++}$. $\circ$: Doubly charged bases, $B^{=}$. $\circ\!\!-$: Uncharged bases, B°; all substituted anilines. $\circ\!\!\!/$: Singly charged bases, B^-; mostly carboxylate ions. B°, B^-, and $B^=$ can all be represented by a single line: $\log k_B/q = -4.40 - 0.84 \log qK_{HB}/p$. B^{++} can be represented by $\log k_B/q = -2.25 - 0.84 \log qK_{HB}/p$.

Brönsted equation. This is not too surprising in view of the specific interactions which may be involved in the catalysis. However, not too much has been done in this regard.

Bell[a] has examined some of the reasons for expecting gross deviations from the Brönsted relation and has suggested that both positive and negative deviations may occur if the transition state represents a structure considerably different in its charge distribution than the base of the conjugate acid (or vice versa).

The cogency of such considerations is, however, subject to question in view of the complex interactions which are involved. If, for example, we wish to make comparisons of the free energy changes in a series of similar reactions and we wish to interpret the results in terms of molecular structure, then we should eliminate all contributions to the ΔF° which are accidental in nature. The simplest of such contributions arises from the symmetry change in a reaction. For any equilibrium

[a] R. P. Bell, *J. Phys. & Colloid Chem.*, **55**, 885 (1951).

$$A + B \rightleftharpoons C + D \qquad (XVI.3.7)$$

we can express the equilibrium constant in terms of individual partition functions (assuming ideal solutions):

$$K_{eq} = \frac{(C)(D)}{(A)(B)} = \frac{Q_C Q_D}{Q_A Q_B} \qquad (XVI.3.8)$$

Since each partition function contains a symmetry number as a simple factor (as was discussed in Chap. IX) we can write $Q_A = Q'_A/\sigma_A$, so that

$$K_{eq} = \frac{Q'_C Q'_D}{Q'_A Q'_B} \cdot \frac{\sigma_A \sigma_B}{\sigma_C \sigma_D} = K_{chem} K_\sigma \qquad (XVI.3.9)$$

where we have defined $K_\sigma = \sigma_A \sigma_B / \sigma_C \sigma_D$. The factor K_{chem} may now be interpreted as the "intrinsic" or "chemical" equilibrium constant from which the accidental symmetry features are removed. One can then argue that, in comparing free energy changes in a series of similar reactions, it is the $K_{chem} = K_{eq}/K_\sigma$ which should be meaningfully compared, not the observed K_{eq}.

By applying such an argument to the Brönsted relation we find the modified expressions

$$\frac{k_A}{K^{\ddagger}_{\sigma A}} = G_A \left(\frac{K_A}{K_{\sigma A}}\right)^\alpha \qquad \frac{k_B}{K^{\ddagger}_B} = G_B \left(\frac{K_B}{K_{\sigma B}}\right)^\beta \qquad (XVI.3.10)$$

where $K^{\ddagger}_\sigma$ represents the symmetry product in going from reactants (acid or base plus substrate) to the transition state. In practice it turns out that this leads to results very similar to the p and q corrections of Brönsted [Eq. (XVI.3.5)], with only occasional disparities arising from ambiguities in the use of the p and q corrections.[a,b]

These symmetry corrections, however, represent only a very first correction. In principle such properties as molecular weight, moment of inertia, and vibration frequency are just as "accidental" and "nonintrinsic" from the point of view of molecular interactions. Nevertheless, they can contribute significantly to free energy changes and should be corrected for if a molecular discussion of such changes is to be rigorous, or at least nonspeculative.

The Brönsted relation represents only one very successful example of what have come to be known as linear free energy relations. It is found that there exist linear relations between the free energy of activation of

---

[a] J. N. Brönsted [*Chem. Revs.*, **5**, 322 (1928)] and also Bell (*loc. cit.*) choose to set $p = q = 1$ for the pair NH_4^+/NH_3. While this may be justified on an empirical basis, it is not justifiable in terms of symmetry. The correction should be $p = 4$, $q = 1$ for the above pair.

[b] S. W. Benson, *J. Am. Chem. Soc.*, **80**, 5151 (1958), has discussed the application of symmetry corrections in general.

many rate constants and the equilibrium constants of closely associated equilibria. We shall examine some of these later.

4. Some Acid-Base-catalyzed Reactions of the Carbonyl Group

The very large dipole moment associated with the carbonyl group, $\diagdown \diagup C=O$, of about 2.50 Debyes makes the group a very attractive reagent for polar and ionic species. The oxygen will be a strong Brönsted base, while the carbon will be a strong Lewis acid. Some consequences of these properties are evidenced in the enhanced water solubility of carbonyl compounds compared to other organic compounds and their strong interactions with water. Thus CH_2O, $HCOCH_3$, and $CO(CH_3)_2$ all interact very strongly with water to form varying amounts of diols[a] according to the equation

$$\begin{matrix} R_1 \\ \diagdown \\ \diagup \\ R_2 \end{matrix} C = O + HOH \rightleftharpoons HO-\begin{matrix} R_1 \\ | \\ C \\ | \\ R_2 \end{matrix}-OH \qquad (XVI.4.1)$$

For CH_2O, K_{eq} (25°C) $\cong 200$ liters/mole in aqueous 0.1 M solution. The ultraviolet absorption spectrum of CH_2O is no longer evident,[b] and the amount of CH_2O left is estimated at less than 0.01 per cent. As we replace H by CH_3, the equilibrium shifts to the left. For CH_3CHO, the equilibrium concentration of diol is about 50 per cent,[c] so that $K \cong 0.02$ liter/mole, while for acetone the diol is estimated at less than 0.01 per cent ($K < 10^{-6}$), although no quantitative data are available. For the related compounds RCOOH, R'COOR no data are available, though polyols may be expected. In the case of CO_2, the equilibrium is found to lie to the left ($K \cong 5 \times 10^{-5}$ liter/mole).[d]

The rate of establishment of the equilibria with water is quite variable, being fast with CH_2O and slow with acetone. However, the rates are very sensitive to catalysis by acids, bases, and even metal ions.[e] In the case of acetaldehyde, the dehydration of the diol in acetone solution is followed

---

[a] These are improperly referred to as hydrates in the literature.

[b] R. Bieber and G. Trümpler, *Helv. Chim. Acta*, **30**, 1860 (1947).

[c] R. P. Bell and J. C. Clunie, *Trans. Faraday Soc.*, **48**, 439 (1952).

[d] D. Berg and A. Patterson, *J. Am. Chem. Soc.*, **75**, 5197 (1953).

[e] The conversion of H_2CO_3 and HCO_3^- to CO_2 is a very important biochemical process which is enzymatically assisted in the lungs.

quite easily by dilatometric methods, and is found subject to generalized acid-base catalysis.[a]

For the acid catalysis, the mechanism is probably

$$
\begin{array}{c}
\underset{\substack{|\\ H}}{\overset{\substack{OH\\ |}}{CH_3-C-OH}} + HA \underset{\substack{2\\ fast}}{\overset{1}{\rightleftharpoons}}
\left[\underset{\substack{|\\ H}}{\overset{\substack{OH\\ |}}{CH_3-C-O}} \underset{\substack{\diagdown\\ H}}{\overset{\diagup H}{\oplus}} \quad A^- \right]
\end{array}
$$

$$
\underset{\substack{|\\ H}}{\overset{\substack{OH\quad H\\ | \quad \diagup}}{CH_3-C-O\oplus}} + A^- \underset{\substack{4\\ slow}}{\overset{3}{\rightleftharpoons}}
\underset{\substack{|\\ H}}{\overset{\substack{O\ominus \quad H\\ | \quad \diagup}}{CH_3-C-O\oplus}} + HA
$$

$$
5 \upharpoonleft\downharpoonright 6 \; fast \qquad\qquad (XVI.4.2)
$$

$$
\underset{\substack{\diagdown\\ H}}{\overset{\substack{O\\ ||}}{CH_3-C}} + \underset{\substack{\diagdown\\ H}}{\overset{\diagup H}{O}}
$$

The rate-determining step is 3, the transfer of the OH proton to the conjugate base A⁻. For the reverse process, the hydration of acetaldehyde, reaction 4 is rate-determining.[b] For the base-catalyzed reaction the proposed mechanism is very similar:

$$
\underset{\substack{|\\ H}}{\overset{\substack{O\quad H\\ || \quad \diagup}}{CH_3-C\cdots O}} + B \underset{3'}{\overset{\substack{slow\\ 4'}}{\rightleftharpoons}}
\underset{\substack{|\\ H}}{\overset{\substack{O^-\\ |}}{CH_3-C-OH}} + HB^+
$$

$$
\underset{\substack{|\\ H}}{\overset{\substack{OH\\ |}}{CH_3-C-OH}} + B \underset{1'}{\overset{2'}{\rightleftharpoons}} \Big\rfloor \; fast \qquad (XVI.4.2a)
$$

The acid-catalyzed exchange of acetone with H_2O^{18} which was discussed in the preceding section must go through an analogous mechanism. The addition of alcohols to aldehydes and ketones to form acetals (and their hydrolysis) is, however, found to be specifically H^+-catalyzed,[c] indicating that in the hydrolysis the slow step is the dissociation of the cation

[a] R. P. Bell and W. C. E. Higginson, *Proc. Roy. Soc. (London)*, **A197**, 141 (1949). The direct hydration reaction has been studied by R. P. Bell et al., *Trans. Faraday Soc.*, **52**, 1093 (1956).

[b] Both hydration and dehydration must show the same rate dependence on acid.

[c] G. W. Meadows and B. de B. Darwent, *Trans. Faraday Soc.*, **48**, 1015 (1952). R. P. Bell and A. D. Norris, *J. Chem. Soc.*, 118 (1941). J. N. Brönsted and W. F. K. Wynne-Jones, *Trans. Faraday Soc.*, **25**, 59 (1929).

$$
\underset{\text{OEt}}{\overset{\text{OEt}}{\text{CH}_3\text{CH}}} \;+\; \text{H}^+ \;\underset{\text{fast}}{\rightleftharpoons}\; \underset{\text{OEt}}{\overset{\overset{+\,\text{H}}{\text{O}}\text{—Et}}{\text{CH}_3\text{CH}}} \;\underset{\text{H}_2\text{O}}{\overset{\text{slow}}{\rightleftharpoons}}\; \text{CH}_3\text{—}\underset{\text{OEt}}{\overset{\overset{+}{\text{O}}}{\text{C}}}\text{—H} \;+\; \text{EtOH}
$$

$$
\underset{\text{H}}{\overset{\text{O}}{\text{CH}_3\text{—C}}} \;+\; \text{HOEt} \;\overset{\text{fast}}{\rightleftharpoons}\; \underset{\text{OEt}}{\overset{\text{OH}}{\text{CH}_3\text{—C—H}}} \;+\; \text{H}^+ \underset{\text{fast}}{\longleftarrow} \qquad\qquad (\text{XVI.4.3})
$$

The acidic properties of the O atom of the carbonyl group is reflected in the tautomeric equilibria of ketones and aldehydes with their corresponding enols:

$$
\underset{\text{R}_2}{\overset{\text{H}}{\text{H—C}}}\text{—}\underset{\text{R}_1}{\overset{\text{O}}{\text{C}}} \;\rightleftharpoons\; \underset{\text{R}_2}{\overset{\text{H}}{}}\text{C}=\text{C}\underset{\text{R}_1}{\overset{\text{OH}}{}} \qquad\qquad (\text{XVI.4.3}a)
$$

Depending on solvent, temperature, and the nature of the groups R_1 and R_2, the equilibrium may lie to either side.[a] Thus it has been found that in pure acetone at 25°C, there is about 2.5×10^{-4} per cent of enol.[b] In contrast, pure acetyl acetone, $CH_3COCH_2COCH_3$, exists as 80 per cent enol, but it is only 15 per cent enol in water[c] solution. The transformation of ketones to enols involves a proton shift, and as we might expect, the rate is susceptible to catalysis by acids and bases.

One of the very first kinetic examples of this catalyzed prototropy was found in the halogenation of acetone. In polar solvents, it is found that the rate of halogenation of acetone is first-order in acetone, zero-order in halogen, X_2, and subject to general acid-base catalysis:[d]

$$
X_2 + CH_3COCH_3 \rightarrow XCH_2COCH_3 + H^+ + X^- \qquad (\text{XVI.4.4})
$$

One of the interesting features of this reaction is that one of the products is H^+. Since the reaction is catalyzed by H^+, the rate accelerates with increasing amounts of reaction, an excellent example of autocatalysis.

The zero order with respect to halogen indicates that the reaction with halogen must occur after the slow, rate-determining step. Considerable evidence points to the slow step as involving the formation of enol in acid solutions, or enolate ion in basic solution:

[a] G. Schwarzenbach and C. Wittwer, *Helv. Chim. Acta*, **30**, 656, 659, 663, 669 (1947).

[b] J. B. Conant and A. F. Thompson, Jr., *J. Am. Chem. Soc.*, **54**, 4039 (1932).

[c] G. Schwarzenbach and E. Felder, *Helv. Chim. Acta*, **27**, 1044 (1944).

[d] A. Lapworth, *J. Chem. Soc.*, **85**, 30 (1904).

Acid Catalysis:

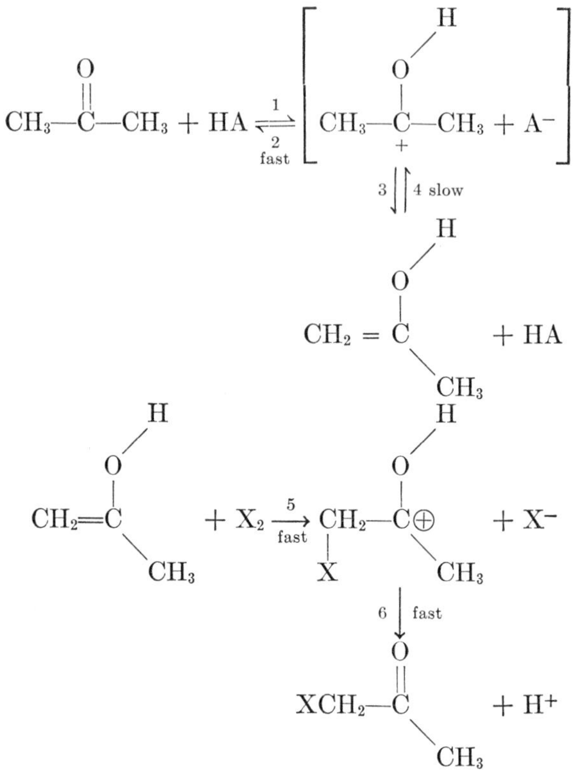

$$CH_3-\overset{\overset{\displaystyle O}{\|}}{C}-CH_3 + HA \underset{\underset{fast}{2}}{\overset{1}{\rightleftharpoons}} \left[CH_3-\overset{\overset{\displaystyle O-H}{|}}{\underset{+}{C}}-CH_3 + A^- \right]$$

$$3 \Big\Updownarrow 4 \text{ slow}$$

$$CH_2 = \overset{\overset{\displaystyle O-H}{|}}{\underset{\displaystyle CH_3}{C}} + HA \qquad \text{(XVI.4.5)}$$

$$CH_2{=}\overset{\overset{\displaystyle O-H}{|}}{\underset{\displaystyle CH_3}{C}} + X_2 \underset{fast}{\overset{5}{\longrightarrow}} CH_2-\overset{\overset{\displaystyle O-H}{|}}{\underset{\displaystyle CH_3}{\underset{|}{\overset{|}{C}\oplus}}} + X^-$$

$$\overset{6}{\underset{fast}{\big\downarrow}}$$

$$XCH_2-\overset{\overset{\displaystyle O}{\|}}{\underset{\displaystyle CH_3}{C}} + H^+$$

Base Catalysis:

$$CH_3-\overset{\overset{\displaystyle O}{\|}}{\underset{\displaystyle CH_3}{C}} + B \underset{\underset{slow}{2}}{\overset{1}{\rightleftharpoons}} \left[CH_3-\overset{\overset{\displaystyle O}{\vdots}}{\underset{\displaystyle CH_2}{C}} \right]^- + HB^+$$

$$\left[CH_3-\overset{\overset{\displaystyle O}{\vdots}}{\underset{\displaystyle CH_2}{C}} \right]^- + X_2 \underset{fast}{\overset{3}{\longrightarrow}} CH_3-\overset{\overset{\displaystyle O}{\diagup\!\!\!\|}}{\underset{\displaystyle CH_2X}{C}} + X^-$$

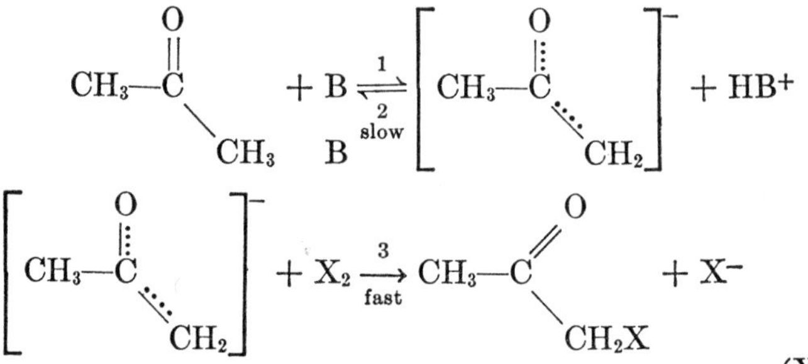

$$\text{(XVI.4.6)}$$

The catalysis has been the subject of careful study by a number of different laboratories,[a] and it is found that the rate of iodination of acetone in aqueous solution of acetate buffer is given by (concentrations in moles/liter)

[a] H. M. Dawson and E. Spivey, *ibid.*, 2180 (1930). R. P. Bell and P. Jones, *ibid.*, 88 (1953).

$$\frac{-d \ln (CH_3COCH_3)}{dt} = k \ (\text{sec}^{-1})$$

$$= 4.63 \times 10^{-10} + 2.73 \times 10^{-5}(H^+) + 0.250(OH^-)$$
$$+ 8.33 \times 10^{-8}(AcOH) + 2.52 \times 10^{-7}(AcO^-)$$
$$+ 3.40 \times 10^{-7}(AcOH)(AcO^-) \qquad (XVI.4.7)$$

Supporting evidence for the mechanism comes from the observation that the bromination and iodination proceed at the same rates. The deuterium exchange is also comparable in absolute rate. Very extensive work with the optically active sec-butyl phenyl ketone,[a] C_2H_5—$CH(CH_3)COC_6H_5$, has shown that the acid-catalyzed iodination, bromination, and inversion have identical rates. The base-catalyzed, OD^-, rates of deuteration and inversion have also been shown to be equal. If the enol and enolate ion can be considered to be planar about the α carbon atom, then these results provide very strong support for the slow enolization step. In fact it is difficult to find any other reasonable interpretation of the data. The enol mechanism is also compatible with the well-known susceptibility of H atoms, in the alpha position to one or more C=O groups, to substitution reactions.

Further kinetic evidence for the importance of enols comes from work on the base-catalyzed condensations of carbonyl-containing compounds. There are a number of such reactions characteristic of aldehydes, ketones, carboxylic acids, esters, amides, etc. Of these the most elementary appear to be the "aldol condensations," which are prototypes of the others. These reactions can be represented by the equation

$$2\ R\!-\!\overset{\displaystyle O}{\overset{\|}{C}}\!-\!CH_3 \rightleftharpoons R\!-\!\underset{\underset{\displaystyle CH_3}{|}}{\overset{\overset{\displaystyle OH}{|}}{C}}\!-\!CH_2\!-\!\overset{\displaystyle O}{\overset{\|}{C}}\!\diagdown_{R} \qquad (XVI.4.8)$$

The reactions (forward and reverse) are acid-base-catalyzed and complicated in basic solutions at higher concentrations of reactants by continued condensation to form polymers of the general formula H—$[CH_2$—$C(R)(OH)]_n$—OH. Depending on R the final diol may revert to the aldehyde or ketone form. In acid solution dehydration of the tertiary alcohol leads to the formation of α-β unsaturated ketones [for example, R—$C(CH_3) = CH$—$CO(R)$], which may condense further.

The equilibrium point depends strongly on R. Thus for acetaldehyde $K_{eq} > 10^4$ liters/mole in basic, aqueous solution at 25°C,[b] while for acetone under the same conditions, $K_{eq} = 0.039$ liter/mole.[c] Because of these

[a] P. D. Bartlett and C. H. Stauffer, J. Am. Chem. Soc., **57**, 2580 (1935). S. K. Hsü and C. L. Wilson, J. Chem. Soc., 623 (1936). Hsü et al., ibid., 78 (1938).

[b] E. H. Usherwood, ibid., **123**, 1717 (1923); 435 (1925).

[c] K. Koelichen, Z. physik. Chem., **33**, 129 (1900). Note that the equilibrium constants have about the same ratio to each other as do the constants for diol formation.

differences, the rate studies for acetone are usually made on the reverse reaction, the catalyzed splitting of diacetone alcohol back into acetone. From the measured equilibrium constant one then gets at the rate of the aldol condensation of acetone. The comparison of the rates of aldol condensation for CH_3CHO and CH_3COCH_3 is quite interesting. In the case of CH_3CHO, the reaction is first-order in acetaldehyde, first-order in base, and seems to show the general properties of general base catalysis.[a] For acetone, on the other hand, the rate is considerably slower; it is second-order in acetone and first-order in OH^- and shows no evidence of generalized base catalysis.[b]

The general scheme for both of these reactions can be represented by

$$
\begin{array}{c}
\underset{\substack{|| \\ R-C \\ \diagdown \\ CH_3}}{O} + B \underset{2}{\overset{1}{\rightleftharpoons}} BH^+ + \underset{\substack{\vdots \\ R-C \\ \diagdown \\ CH_2}}{O^-}
\end{array}
$$

$$
\underset{\substack{\vdots \\ R-C \\ \diagdown \\ CH_2}}{O^-} + \underset{\substack{|| \\ R-C \\ \diagdown \\ CH_3}}{O} \underset{4}{\overset{3}{\rightleftharpoons}} \underset{\substack{|| \\ R-C-CH_2-C-R \\ | \\ CH_3}}{O \qquad O^-} \qquad \text{(XVI.4.9)}
$$

$$
\underset{\substack{|| \\ R-C-CH_2-C-R \\ | \\ CH_3}}{O \qquad O^-} + HOH \underset{\substack{6 \\ \textbf{fast}}}{\overset{5}{\rightleftharpoons}} \underset{\substack{|| \\ R-C-CH_2-C-R \\ | \\ CH_3}}{O \qquad OH} + OH^-
$$

If we assume that steps 5 and 6 are very fast compared to step 4, we can apply a steady-state treatment to the enolate ion and obtain for the over-all rate

$$
\frac{-d(RCOCH_3)}{dt} = \frac{k_1 k_3 (RCOCH_3)^2 (B)}{k_2(BH^+) + k_3(RCOCH_3)} \qquad \text{(XVI.4.10)}
$$

This leads to two possible extreme cases, depending on whether $k_2(BH^+)$ is much greater or much less than $k_3(RCOCH_3)$. When $k_2(BH^+) \gg k_3(RCOCH_3)$:

$$
\frac{-d(RCOCH_3)}{dt} \rightarrow \frac{k_1 k_3}{k_2} (RCOCH_3)^2 \frac{(B)}{(BH^+)}
$$

$$
= \frac{k_1 k_3}{k_2 K_B} (RCOCH_3)^2 (OH^-) \qquad \text{(XVI.4.11)}
$$

where K_B is the dissociation constant for $B + HOH \rightleftharpoons BH^+ + OH^-$. This result corresponds to that observed for acetone, and we see that it

[a] R. P. Bell, private communication, 1958; *J. Chem. Soc.*, 1637 (1937).

[b] V. K. La Mer and M. L. Miller, *J. Am. Chem. Soc.*, **57**, 2674 (1935). C. C. French, *ibid.*, **51**, 3215 (1929); F. H. Westheimer and H. Cohen, *ibid.*, **60**, 90 (1938).

arises from the slowness of the attack of the enolate ion on acetone, relative to its neutralization by BH^+ (or any other weak acid in the system).

If, however, $k_2(BH^+) \ll k_3(RCOCH_3)$, the rate law becomes

$$\frac{-d(RCOCH_3)}{dt} \to k_1(RCOCH_3)(B) \qquad (XVI.4.12)$$

where the rate-determining step is now the ionization of the α H atom. As we may note, this corresponds, very conveniently, to the rate law for the aldol condensation of acetaldehyde.

Thus the gross differences in the rate laws for acetone and acetaldehyde condensation arise not from differences in reaction mechanism, but rather from differences in the relative rates of attack by enolate ion on reactant. In principle, at sufficiently low acetaldehyde concentrations, the rate law for the acetaldehyde should approach that for acetone.

A further evidence for the extreme reactivity of acetaldehyde with its enolate ion is provided by the fact that, when the aldol condensation is run in D_2O, there is no substitution of D atoms on the α carbon positions.[a] This is interpreted as indicating a much more rapid reaction of the enolate ion with acetaldehyde than with D_2O, as might have been already anticipated from the relative slowness of the reaction with BH^+. In contrast, the base-catalyzed exchange of acetone with D_2O[b] is much faster than the rate of dimerization of acetone.

While the foregoing discussion has been directed toward the alkyl carbonyl compounds, it should be pointed out that the groups CH_3, NH_2, and OH are isoelectronic. When attached to carbonyl groups we should expect to find that NH_2 and OH will exhibit similar reaction pathways, and in fact there is a striking parallel between the mechanism outlined for the $COCH_3$ group and the isoelectronic amides and acids.

5. Reactions in Concentrated Solutions; The Acidity Function

In solvents of high dielectric constant such as water, the deviations from ideality caused by ion-ion interactions are reasonably small below concentrations of 0.1 M for 1:1 electrolytes and can be treated adequately by means of the Debye-Hückel theory. For polyvalent electrolytes or for higher concentrations of 1:1 electrolytes, or for either in solvents of lower dielectric constant, the situation is less fortunate. The deviations from ideality can become rather large, and there is no adequate theory for either correlating them or predicting them.

Numerous attempts, both theoretical and empirical, which have been made to treat this problem have met with varying success.[c] The chief

[a] K. F. Bonhoeffer and W. D. Walters, Z. physik. Chem., **A181**, 441 (1938).

[b] K. Bonhoeffer and Klor, Naturwiss., **22**, 45 (1934); W. D. Walters and K. Bonhoeffer, Z. physik. Chem., **A182**, 265 (1938).

[c] See, for example, discussion in treatise by Stokes and Robinson, loc. cit.

difficulty with the more successful efforts has been the use of too many empirical parameters to be of predictive value. One of the more interesting approaches has been made by Hammett and coworkers,[a] who have proposed an empirical function which can be used to measure the acidity (i.e., proton-donating ability) of acid solutions. The rationale of the method may be discussed in terms of the equilibrium between an uncharged base $B°$ and its conjugate acid BH^+ in an aqueous solution of a strong acid. The reaction is

$$B° + H^+ \rightleftharpoons BH^+ \qquad K_B = \frac{1}{K_{BH^+}} \qquad (XVI.5.1)$$

The base constant K_B is expressible in terms of the activities of the various species as

$$K_B = \frac{a_{BH^+}}{a_{B°}a_{H^+}} = \frac{1}{K_{BH^+}} \qquad (XVI.5.2)$$

Or in terms of concentrations and activity coefficients,

$$K_B = \frac{(BH^+)}{(B°)} \frac{f_{BH^+}}{f_{B°}} \frac{1}{(H^+)(f_{H^+})} \qquad (XVI.5.3)$$

If now the base, $B°$, or its conjugate acid, BH^+, is an indicator, or has an appreciable optical absorption coefficient, the ratio $(BH^+)/(B°)$ is experimentally observable. The magnitude of this ratio is a measure of the tendency of the solution to transfer a proton to the neutral base $B°$. On solving for this ratio and expressing the result in logarithmic form we have

$$\log \frac{(BH^+)}{(B°)} = \log K_B + \log \left[\frac{a_{H^+}f_{B°}}{f_{BH^+}} \right] \qquad (XVI.5.4)$$

If we choose infinite dilution[b] and molarity as our standard state, then in very dilute solution as the activity coefficients approach unity, this equation reduces to the simple form[c]

$$\log \frac{(BH^+)}{(B°)} \rightarrow -pK_B + \log (H^+) \qquad (XVI.5.5)$$

What Hammett has proposed is that where the simple concentration (H^+) measures the experimentally disposable protonating power of the dilute solution, the equivalent term in Eq. (XVI.5.4) can be used in concentrated solution to measure the same property. Thus he defines the acidity function of an acid solution with respect to a neutral base as follows:

$$h_0 \equiv a_{H^+} \frac{f_{B°}}{f_{BH^+}} \qquad (XVI.5.6)$$

[a] L. P. Hammett, *Chem. Revs.*, **16**, 67 (1935).

[b] See footnote on page 504.

[c] This might be, for example, 0.01 M $HClO_4$ together with an indicator $B°$ present in concentrations of the order of 1×10^{-4} M or less.

or in logarithmic form

$$H_0 = ph_0 = -\log h_0 = -pK_B - \log \frac{(BH^+)}{(B°)}$$

$$= -\log a_{H^+} \frac{(f_{B°})}{(f_{BH^+})} \qquad (XVI.5.7)$$

Such a function will have a general usefulness only if it can be shown that it is independent of the nature of the base B. This in turn could only be true if the ratio $f_{B°}/f_{BH^+}$ were the same for different bases B°. Although this is not generally the case, and a really broad test of the premise has yet to be provided, it has been shown that for structurally similar classes of bases it does hold. For members of such a class it is then sufficient to measure pK_B in dilute solution and the ratio $(BH^+)/(B°)$ in concentrated solution to obtain the acidity function H_0 for the water-acid system in question. The relation seems to hold quite well for the strong acids and water (for example, H_2SO_4, $HClO_4$, and HNO_3) over an astonishing range of concentrations.[a] For mixtures of water with alcohol or other solvents, however, gross deviations are observed.[b]

Table XVI.1 shows the variation of H_0 with concentration in some H_2O-H_2SO_4 mixtures. What is particularly striking is the fact that, when the molarity changes only 36 fold in going from 5 to 100 per cent H_2SO_4, the protonating power or acidity changes by almost 11 powers of

TABLE XVI.1. ACIDITY FUNCTION IN H_2O–H_2SO_4 MIXTURES AT 20°C

Wt % H_2SO_4	Molarity, H_2SO_4	$-H_0$	Wt % H_2SO_4	Molarity, H_2SO_4	$-H_0$
5	0.50	−0.24	60	8.8	4.46
10	1.05	0.31	70	11.1	5.65
20	2.23	1.01	80	13.5	6.97
30	3.58	1.72	90	16.0	8.27
40	5.10	2.41	95	17.1	8.86
50	6.8	3.38	100	17.9	11.10

Data from review by M. A. Paul and F. A. Long, *Chem. Revs.*, **57**, 1 (1957).

10 in this range, much of the latter coming from the relatively insignificant range, 90 to 100 per cent. One consequence of this enormous acidity of concentrated H_2SO_4 is that substances which may be immeasurably weak bases in water (such as toluene) will become measurably strong bases in

[a] K. N. Bascombe and R. P. Bell, *Discussions Faraday Soc.*, **24**, 158 (1957), have shown that a common H_0-concentration curve can be constructed for these acids and also for concentrated H_3PO_4 and HF solutions by assigning varying hydration numbers to the ionized species and correcting the equation for H_2O activity, that is, $B° + H^+(H_2O)_n \rightleftharpoons BH^+(H_2O)_m + (n - m)H_2O$.

[b] B. Gutbezal and E. Grunwald, *J. Am. Chem. Soc.*, **75**, 559 (1953).

H_2SO_4 and HF.[a] It should be noted that these very large acidities cannot arise from the stoichiometric concentration of H^+ ion in the solution but instead arise from remarkable changes in the activity coefficient ratio $f_{H^+}f_{B^\circ}/f_{BH^+}$. It seems reasonable that much of the increase in this ratio with increasing acid concentration arises from the increase in f_{H^+} due to a loss in H_2O activity.[b]

The acidity function has been used[c] in analyzing the kinetic data on acid-catalyzed reactions in concentrated solutions. Long and Purchase[d] found that the rate of hydrolysis of β–propiolactone in concentrated aqueous solutions of H_2SO_4 and $HClO_4$ is first-order in lactone and proportional to h_0. The over-all reaction is

$$\begin{array}{ccc}
\underset{H_2C-O}{\overset{\displaystyle H_2C-C^{\nearrow O}}{|\quad\quad|}} + HOH \rightleftharpoons \underset{H_2C}{\overset{\displaystyle H_2C-C^{\nearrow O}}{|\quad\quad\searrow_{O-H}}} & & \text{(XVI.5.8)}
\end{array}$$
$$\qquad\qquad\qquad\qquad\qquad\qquad\qquad \searrow_{OH}$$

To account for this dependence on h_0, the following mechanism was proposed:

$$L + H^+ \underset{\underset{fast}{2}}{\overset{1}{\rightleftharpoons}} [L \cdot H]^+$$

where L = lactone.

$$[L \cdot H]^+ \underset{\underset{slow}{4}}{\overset{3}{\rightleftharpoons}} \begin{array}{c} \overset{\displaystyle O}{\underset{\displaystyle H_2C-O}{\overset{\displaystyle H_2C-C^{\nearrow}\oplus}{|}}} \\ \searrow_{H} \end{array} \qquad \text{(XVI.5.9)}$$

[a] E. L. Mackor et al., *Trans. Faraday Soc.*, **54**, 66, 186 (1958), have measured K_B for both substituted and condensed aromatics in liquid HF-BF_3 mixtures by using phase partition to obtain data. H_0 in HF-$NaBF_4$ at 0°C is about −8.6, so the system is surprisingly acidic. Their pK_B values range from 6.3 for toluene to −1.4 for hexamethylbenzene and −6.4 for 9,10-dimethyl anthracene. This last is evidently a reasonably strong base in HF.

[b] In dilute solution H^+ exists as strongly solvated $H_3O^+(H_2O)_n$. If, following Bell, *loc. cit.*, we write Eq. (XVI.5.1) as: $H^+(H_2O)_n + B \rightleftharpoons BH^+(H_2O)_m + (n - m)H_2O$, then a factor a_{H_2O} is contained in f_{H^+} as we have used it. This could well account for much of the enhanced activity if $n - m$ is large.

[c] H_0 and related acidity functions are reviewed by M. A. Paul and F. A. Long, *Chem. Revs.*, **57**, 1 (1957).

[d] F. A. Long and M. Purchase, *J. Am. Chem. Soc.*, **72**, 3267 (1950). Actually, the rate law has an additional term that is first-order in lactone and first-order in H_2O. At concentrations of $HClO_4$ or H_2SO_4 above 3 M this term becomes minor relative to the acid catalysis term.

$$
\begin{array}{ccccccc}
\overset{\oplus}{H_2C}-\overset{}{C}=O & & H_2C-\overset{\overset{\overset{OH_2}{/}}{\oplus}}{C} & & H_2C-\overset{\overset{OH}{/}}{C} & \\
| & +\,H_2O \underset{6}{\overset{5}{\rightleftharpoons}} & | & \underset{8}{\overset{7}{\rightleftharpoons}} & | & +\,H^+ \\
H_2C-O & \text{fast} & H_2C \quad O & \text{fast} & H_2C \quad O & \\
\backslash & & \backslash & & \backslash & \\
H & & OH & & OH &
\end{array}
$$

The rate law for such a mechanism, with $k_2 \gg k_3 \ll k_5(H_2O)$ is given by

$$
\frac{-d(L)}{dt} = k_3([LH]^+) = k_3 K_{1.2}(L) a_{H^+} \cdot \frac{f_L}{f_{[LH]^+}}
$$

$$
= k_3 K_{1.2}(L) h_0 = k_h(L) h_0 \qquad (XVI.5.10)
$$

If h_0 is measured with an indicator B, the only assumption made in Eq. (XVI.5.10) is that $f_L/f_{[LH]^+} = f_B/f_{BH^+}$. In view of the excellent agreement obtained between the observed rate and h_0, it can be looked upon as at least a consistent assumption in this case.[a] Similar relations between h_0 and k_{cat} have been found for the acid-catalyzed hydrolysis of trioxane[b] in H_2SO_4 (0.5 to 8 M), HCl, and mixtures of H_2SO_4 + glacial acetic acid; the hydrolysis of sucrose;[c] the hydrolysis of cyanamide (NH_2—CN) to give NH_4^+ + CO_2[d] in HNO_3; and the decarbonylation of aromatic aldehydes[e] in strong acids (ArCOH → ArH + CO). We may make a tentative generalization and say that, when the transition state represents the decomposition of the conjugate acid of the substrate, the catalytic rate constant will be proportional to (H^+) in dilute acid and h_0 in strong acid. We may expect deviations from such a relation when the assumption about the ratio of activity coefficients $f_{B^\circ}/f_{BH^+} = f_S/f_{SH^+}$ falls down.

When deviations do occur from the h_0 relation and they are not directly traceable to the activity coefficient assumption, then it may be taken as evidence for a mechanism that does not involve simple decomposition of the conjugate acid of the substrate. Such is the case, interestingly enough, in the closely related, acid-catalyzed hydrolysis of γ-butyrolactone, which has been very carefully studied in $HClO_4$ and HCl solutions.[f] To account for the fact that the catalyzed reaction rate was proportional not to h_0 but instead to (H^+), the authors have proposed a mechanism very similar to that for the β-propiolactone [Eq. (XVI.5.9)]:[g]

[a] In order to distinguish between h_0 and simple proton (i.e., specific acid) catalysis, the acid concentrations must be in a range in which h_0 and H^+ are not linearly related.

[b] M. A. Paul, *J. Am. Chem. Soc.*, **72**, 3813 (1950); **74**, 141 (1952).

[c] L. P. Hammett and M. A. Paul, *ibid.*, **56**, 830 (1934).

[d] G. Grube and G. Motz, *Z. physik. Chem.*, **118**, 145 (1925); G. Grube and G. Schmid, *ibid.*, **119**, 29 (1926).

[e] W. M. Schubert and H. K. Latourette, *J. Am. Chem. Soc.*, **74**, 1829 (1952).

[f] F. A. Long et al., *J. Phys. & Colloid Chem.*, **55**, 813, 829 (1951).

[g] Note that H_2O is certainly strongly bound to the acid in LH^+, although we have not written it as such in Eq. (XVI.5.9). Since this latter system follows h_0, we must con-

$$
\begin{array}{c}
\text{CH}_2\!-\!\text{C}\!\!\stackrel{\displaystyle O}{\big\|} \\
\text{CH}_2 \qquad\qquad\quad + \text{H}^+ \underset{\underset{\text{fast}}{2}}{\overset{1}{\rightleftharpoons}} \left[\begin{array}{c} \text{CH}_2\!-\!\text{C}\!\!\stackrel{\displaystyle O}{\big\|} \\ \text{CH}_2 \\ \text{CH}_2\!-\!\text{OH} \end{array}\right]^+ \\
\text{CH}_2\!-\!\text{O}
\end{array}
$$

$$
\left[\begin{array}{c} \text{CH}_2\!-\!\text{CH}_2\!-\!\text{C}\!\!\stackrel{\displaystyle O}{\big\|} \\ \text{CH}_2 \qquad\qquad \text{O} \\ \text{H} \end{array}\right]^+ + \text{H}_2\text{O} \underset{\underset{\text{fast}}{4}}{\overset{3}{\rightleftharpoons}} \left[\begin{array}{c} \text{H} \qquad\quad \text{H} \\ \text{CH}_2 \quad \text{O} \\ \text{CH}_2 \qquad \text{C}\!=\!\text{O} \\ \text{CH}_2\!-\!\text{O} \\ \text{H} \end{array}\right]^+
$$

$$
5 \big\Uparrow\big\Downarrow 6 \quad \text{slow}
$$

$$
\begin{array}{c}
\text{OH} \\
\text{CH}_2\!-\!\text{C} \\
\text{CH}_2 \qquad\qquad \text{O} \quad + \text{H}^+ \underset{\underset{\text{fast}}{8}}{\overset{7}{\rightleftharpoons}} \left[\begin{array}{c} \text{H} \\ \text{O} \\ \text{CH}_2\!-\!\text{C} \qquad \text{H} \\ \text{CH}_2 \qquad\qquad \text{O} \\ \text{CH}_2\!-\!\text{O}\!-\!\text{H} \end{array}\right]^+ \\
\text{CH}_2 \\
\text{OH}
\end{array}
$$

$$(\text{XVI.5.11})$$

The slow step is now the S_N2 displacement of the ether linkage by an H_2O molecule. That such a step occurs in the mechanism has been demonstrated by Olson and Miller[a] by showing that in strong acid (or strong base) solutions in labeled water, H_2O^{18}, there was little or no O^{18} in the resulting OH group of the acid. The rate law for such a mechanism with $k_8 \gg k_5 \ll k_4$ is

$$
\frac{-d(\text{L})}{dt} = k_5(\text{LH}\cdot\text{H}_2\text{O})^+ = k_5 K_{3.4} K_{1.2}(\text{L})(\text{H}_2\text{O})\left(\frac{a_{\text{H}^+}f_{\text{H}_2\text{O}}f_{\text{L}}}{f_{\text{LH}\cdot\text{H}_2\text{O}^+}}\right) \qquad (\text{XVI.5.12})
$$

where $L = $ lactone.

---

clude that whatever hydration exists in LH^+, it is structurally similar to that which occurs in L. Otherwise we should have difficulty in accounting for the h_0 dependence.

[a] A. R. Olson and R. J. Miller, *J. Am. Chem. Soc.*, **60**, 2687 (1938). They also found that in neutral or near-neutral solution, the acid contains the equilibrium concentration of O^{18} in the βOH group. This indicates that the additional term in the rate law that is proportional to H_2O represents an S_N2 displacement by H_2O of the ether linkage attached to the β carbon atom. Actually the lactone used was β-butyro, not β-propio, although the two are similar, as has been shown by T. L. Gresham et al., *ibid.*, **70**, 998, 999, 1001, 1004, 4227 (1948); **71**, 661, 2807 (1949); **72**, 72 (1950). F. A. Long and S. Friedman, *ibid.*, **72**, 3692 (1950), have demonstrated acyl O fission by using O^{18} tracers for the γ-butyro lactone.

Aside from the term for $a_{H_2O} = (H_2O)f_{H_2O}$,[a] this expression differs from the previous one which was proportional to h_0 by a ratio $f_{LH^+}/f_{LH \cdot H_2O^+}$. Over the range 0.4 to 3.3 M HClO$_4$, in which h_0 changes 40 fold, this ratio changes only 6 fold. The fact that the γ-butyro lactone follows (H$^+$) must mean that $a_{H_2O}f_L f_{H^+}/f_{LH \cdot H_2O^+}$ is virtually constant over the entire range of acid concentrations.[b] This is a rather surprising result which has not been pursued[c] further.

It is interesting to note that in the reverse reaction, that of lactone formation from the free [Eq. (XVI.5.11)] γ-OH-butyric acid, the slow step (reaction 6) is now the cyclization of the conjugate acid of the starting reactant. This rate law for this step should follow h_0 and not (H$^+$), and this is indeed what is observed (Long et al., *loc. cit.*).

Hammett and Zucker[d] had proposed that the conformity of an acid-catalyzed reaction to the h_0 or (H$^+$) dependence would be a method of distinguishing between the intervention of a molecule of H$_2$O in formation of the activated state. The studies on the γ-butyrolactone would seem to bear this out, but considerably more work would be required to establish the point.[d]

6. Lewis Acids: Addition to Multiple Bonds

In our discussion of acid catalysis we have up to now restricted our attention to protonic or Brönsted acids. However, the Brönsted acid, defined as a proton donor, calls into existence the Brönsted base, which can act as proton acceptor. But such a base accepts protons by virtue of the fact that it can offer the proton an unshared pair of electrons for bonding, e.g.,

$$H\!-\!\underset{\underset{H}{|}}{\overset{\overset{H}{|}}{N}}\!:\ +\ HA \rightleftharpoons A^- + \left[H\!-\!\underset{\underset{H}{|}}{\overset{\overset{H}{|}}{N}}\!-\!H \right]^+ \tag{XVI.6.1}$$

Lewis[e] saw that such a Brönsted base called for a more generalized acid, namely, one that could accept a pair of electrons. This made it possible to discuss "acid-base" reactions in nonprotonic systems. The reaction of

[a] This term would vanish if we wrote H$^+$ consistently as H$_3$O$^+$ in this last scheme.

[b] Long et al. have pointed out that this factor has a value close to unity over the entire range of conditions studied. The study of the γ-lactone $\rightleftharpoons$ acid system also includes a careful study of salt effects.

[c] A similar result is obtained in the acid-catalyzed enolization of acetophenone, ϕCOCH$_3$, in HClO$_4$ solutions; L. P. Hammett and L. Zucker, *ibid.*, **61**, 2791 (1939).

[d] The present status of the Hammett-Zucker hypothesis is carefully reviewed by F. A. Long and M. A. Paul, *Chem. Revs.*, **57**, 935 (1957), who conclude that, although frequently useful in aqueous systems, it is not well established and is of doubtful help in nonaqueous media.

[e] G. N. Lewis, *J. Franklin Inst.*, **226**, 293 (1938).

BF_3 with tertiary amines or with ethers is an example of a Lewis acid reacting with a Brönsted base:

$$
\begin{matrix}
\text{F} & \text{R} & & \text{F} & \text{R} \\
| & | & & | & | \\
\text{F}\!-\!\text{B} + & :\!\text{N}\!-\!\text{R} & \rightleftharpoons & \text{F}\!-\!\text{B}:\text{N}\!-\!\text{R} \\
| & | & & | & | \\
\text{F} & \text{R} & & \text{F} & \text{R}
\end{matrix}
\qquad\text{(XVI.6.2)}
$$

In aqueous solution, metal ions are Lewis acids and the familiar complex ions such as $Fe(NO)^{2+}$, $Cr(H_2O)_6^{3+}$, and AlF_6^{3-} can be looked on as acid-base complexes. Because of their large valence shells, the nonmetallic atoms beyond the second row in the periodic table (S, P, Cl, Br, I, etc.) can show properties of both Lewis acids and bases. I^- can act as a base with metal ions (Lewis acids) to form quite stable complexes such as HgI_4^-. On the other hand I_2 can act as an acid in its reaction with electrons donors to form complexes of varying stability.[a] With I^- the reaction $I^- + I_2 \rightleftharpoons I_3^-$ proceeds largely to the right in 0.1 M aqueous solution ($K_{eq}^\circ = 140$ liters/mole), $\Delta H^\circ = -4.0$ Kcal.

This generalization of acid-base properties in terms of electron transfer has been used by the English writers[b] as a basis of classification of reagents as either nucleophilic (electron-donating) or electrophilic (electron-accepting). There is then a classification of reactions in terms of these categories as well.

Multiple bonds such as C=C, C=O, C≡C, C≡N, and the conjugated aromatic ring systems, by virtue of their excess of electrons, may act as bases in offering electrons to Lewis acids. On the other hand, by forming negative ions, they may act as Lewis acids to sufficiently strong bases. This behavior is more marked for the polar bonds such as C=O than for the C=C or C≡C bonds. While the basic character of the C=O group is explicable in terms of its highly polar character (dipole moment = 2.5 Debyes), that of the nonpolar C=C is not. Nevertheless there are good data, both thermodynamic and kinetic, for postulating basic character for these nonpolar bonds.

There is direct evidence for the interaction of such acids as Ag^+ and H^+ with double bond systems.[c] In the case of simple olefins, strong acids react in aqueous solution to form alcohols:

[a] Liquid I_2 is actually not too bad a conductor, its specific conductance being 1.7×10^{-4} mho/cm at 140°C. It is also a strong ionizing solvent for salts such as KI. The conductivity is attributed to the reaction, $3I_2 \rightleftharpoons I_3^+ + I_3^-$, for which $K_{eq} = (I_3^+)(I_3^-) \cong 10^{-4}$ moles²/liter².

[b] See text by Ingold, op. cit.

[c] S. Winstein and H. J. Lucas, J. Am. Chem. Soc., **60**, 836 (1938). Lucas et al., ibid., **65**, 227 (1943), have measured the increase in solubility of olefins in aqueous solutions induced by Ag^+ and obtained values for K_{eq} for the formation of Ag^+ olefin complexes.

$$R_1R_2C{=}CR_3R_4 + HOH \overset{H^+}{\rightleftharpoons} R_1R_2CH{-}CR_3R_4OH$$

With the less reactive aromatic compounds, there is complex formation to form the conjugate acid[a]

$$Ar + H^+ \rightleftharpoons ArH^+$$

Evidence for the acidic character of the unsaturated system is demonstrated by their reaction with alkali metals in tetrahydrofuran solutions to form brilliantly colored negative ions:[b]

$$Na + Ar \rightleftharpoons [Na^+ \cdot Ar^-]$$

This acidic behavior is, however, much less prominent than is the basic.

One of the very strong kinetic evidences for such a classification comes from the study of halogen addition to double bonds. The free radical addition of halogens to unsaturated compounds has been extensively studied. In the absence of light, peroxides, or other initiators, the free radical reaction is very slow.[c] However, in glacial acetic acid or more polar solvents, a non-radical mechanism takes place. At low concentration of halogen this reaction is usually second-order, first-order in olefin and first-order in X_2. That the mechanism does not involve direct addition of X_2 across the double bond is indicated by the fact that in solutions of C_2H_4 in H_2O containing Cl^- and NO_3^- in addition to Br_2, the products were $C_2H_4Br_2$, $BrCH_2CH_2Cl$, and $BrCH_2CH_2NO_3$.[d] The implication is clear that the rate-determining step involves olefin and Br_2, and only one bond is made in this step. The final bond is formed subsequently. On the basis of this as well as other kinetic and stereochemical data, Roberts and Kimball[e] proposed that the first step is the formation of a complex between the olefin and Br_2, whose slow ionization was rate-determining:

[a] D. A. McCauley and A. P. Lien, *ibid.*, **74**, 6246 (1952), have measured the base strengths of xylenes in concentrated HF + BF₃ solutions (see also the discussion on page 578).

[b] M. Szwarc et al., *ibid.*, **78**, 2656 (1956), and N. D. Scott, U.S. Patent 2,181,771 (1939), have shown that these aromatic anions can initiate polymerization of olefins by a mechanism of electron transfer to form a radical-anion, followed by dimerization of two radical-anions to form di-anions which can propagate indefinitely:

$$Na \cdot Ar^- + R_1CH{=}CHR_2 \rightarrow Ar + [R_1\overset{\cdot}{C}H{-}\overset{\cdot\cdot}{C}HR_2]^- \cdot Na^+$$

$$2[R_1\overset{\cdot}{C}H{-}\overset{\cdot\cdot}{C}HR_2]^- \cdot Na^+ \rightarrow [R_2\overset{\cdot\cdot}{C}H{-}CHR_1{-}CHR_1{-}\overset{\cdot\cdot}{C}HR_2]^= \cdot (Na^+)_2$$

[c] See the discussion in G. K. Rollefson and M. Burton, "Photochemistry and the Mechanism of Chemical Reactions," Prentice-Hall, Inc., Englewood Cliffs, N.J., 1939.

[d] A. W. Francis, *J. Am. Chem. Soc.*, **47**, 2340 (1925).

[e] I. Roberts and G. E. Kimball, *ibid.*, **59**, 947 (1937).

$$\text{(XVI.6.3)}$$

To account for the production of bromo chlorides and bromo nitrates, it is proposed that the reaction of anions (Br^-, Cl^-, NO_3^-) in step 5 is very fast. Independent evidence for the existence of a metastable, cyclic bromonium ion comes from the observation that the bromine atoms are usually in a threo (trans) orientation in the final product[a] and that in the reverse reaction of acetolysis of dibromides, there is retention of configuration.[b] This would be consistent with the Br-assisted S_N1 ionization of the dibromide. The assisting Br atom would attack the neighboring C atom from the back side to the leaving Br^- ion in a typical Walden inversion. Thus from *trans*-butene-2 we obtain *erythro*-2,3-dibromo butane:

$$\text{(XVI.6.4)}$$

Note, that because of the retention of configuration in the bromonium ion, the same erythro compound is obtained independently of the point of attachment of the entering Br^- ion.

In glacial acetic acid, Ogg and Nozaki[c] found that the second-order rates of bromination of allyl chloride, vinyl bromide, and allyl nitrile gave rate constants which were proportional to the concentrations of LiBr or LiCl in solution, the latter being nearly twice as effective as the former. Although the authors interpreted this as a third-order reaction involving Br^- or Cl^-,

[a] A. McKenzie, *J. Chem. Soc.*, 1196 (1912). E. M. Terry and L. Eichelberger, *J. Am. Chem. Soc.*, **47**, 1067 (1925).

[b] S. Winstein and R. E. Buckles, *ibid.*, **64**, 2780, 2787 (1942). Similar evidence has been found for cyclic iodonium ion [H. J. Lucas and H. K. Garner, *ibid.*, **72**, 2145 (1950)], and for chloronium ion [H. J. Lucas and C. W. Gould, Jr., *ibid.*, **63**, 2541 (1941)].

[c] K. Nozaki and R. L. Ogg, Jr., *ibid.*, **64**, 697, 704, 709 (1942).

it is more probable that they were observing a salt effect[a] on the normal halogenation.

At higher concentrations of halide ($>0.02\ M$) in acetic acid it was found by Walker and Robertson[b] that the rate appears to be third-order over-all and second-order in Br_2. This can be interpreted, in this poorly ionizing solvent, as a Br_2-catalyzed ionization of the olefin-Br_2 complex [step 3, Eq. (XVI.6.3)] to produce the more stable Br_3^- and the cyclic bromonium complex. In similar fashion it is likely that LiBr and LiCl may have catalyzed this step in the work reported by Ogg and Nozaki.[c] In view of the difficulty with salt effects in these low-dielectric solvents, great caution should be exercised in interpreting data obtained with them.

Although the reactions of halogen with aromatics in polar solvents usually lead to a substitution of halogen for H in the aromatic, the first step seems to be the addition of a positive halonium ion to the aromatic and so is similar to the olefin reaction. The mechanism is presumed to be

$$X_2 + ArH \underset{2}{\overset{1}{\rightleftharpoons}} ArHX^+ + X^-$$

$$ArHX^+ \xrightarrow{3} ArX + H^+ \tag{XVI.6.5}$$

with reaction 1 slow and rate-controlling and $k_3 \gg k_2(X^-)$.[d]

Hypobromous acid, HOBr, is found to be a poorer brominating agent than Br_2 at pH 3 (aqueous solutions).[e] However, at higher concentrations of strong acids a third-order rate law is observed:[f]

$$\frac{-d(ArH)}{dt} = k(ArH)(HOBr)(H^+) \tag{XVI.6.6}$$

In this circumstance, the mixture $H^+ + HOBr$ is more effective than Br_2, and this is considered as evidence for the action of $(H_2OBr)^+$ as the effective agent:

$$HOBr + H^+ \rightleftharpoons H_2OBr^+ \tag{XVI.6.7}$$

[a] Note that in acetic acid LiBr and LiCl have dissociation constants of about 10^{-6}, so that (Br^-) and (Cl^-) would be expected to be nearly proportional to $(LiX)^{1/2}$, contrary to the postulated mechanism.

[b] I. K. Walker and P. W. Robertson, *J. Chem. Soc.*, 1515 (1939). P. B. D. de la Mare and P. W. Robertson, *ibid.*, 2838 (1950).

[c] The rate data of Ogg and Nozaki are extremely complex even under their own analysis, and their final rate constants cannot be taken literally. They reported considerable formation of Br^- in the salt-free systems, as well as marked H_2O catalysis. The Br^- ion formation can be interpreted in terms of an acetic acid attack on the bromonium complex, competing with the Br^- (or HBr) present in the solution.

[d] A. E. Bradfield et al., *J. Chem. Soc.*, 1389 (1949).

[e] A. W. Francis, *J. Am. Chem. Soc.*, **47**, 2340 (1925).

[f] E. Shilov and N. Kanyaev, *Compt. rend. acad. sci. U.R.S.S.*, **24**, 890 (1939). W. J. Wilson and F. G. Sopor, *J. Chem. Soc.*, 3376 (1949). D. H. Derbyshire and W. A. Waters, *ibid.*, 564, 574 (1950).

Similar results are obtained[a] with $HOCl + H^+$ and $I_3^- + H^{+}.$[b] In the latter case the rate law for the iodination of aniline is found to be

$$-\frac{d(\phi NH_2)}{dt} = \frac{k(I_3^-)(\phi NH_3)^+}{(I^-)^2} \qquad (XVI.6.8)$$

This is understandable if the mechanism is

$$I_3^- + HOH \underset{2}{\overset{1}{\rightleftharpoons}} H_2OI^+ + 2I^-$$
$$\text{fast}$$

$$H_2OI^+ + \phi NH_2 \underset{\text{slow}}{\overset{3}{\longrightarrow}} [I\cdots\phi NH_2]^+ + H_2O$$
$$\downarrow 4 \text{ fast}$$
$$[I\phi'NH_2] + H^+ \qquad (XVI.6.9)$$

It is not possible to distinguish I^+ from IOH_2^+ as the halogenating agent, but it is doubtful if the difference has much meaning, since I^+ will not exist in the solution except strongly solvated.

A final interesting example of the attack of electrophilic agents on multiple bonds is given by the nitration of aromatics in strong acid solutions. In concentrated H_2SO_4 solutions, HNO_3 is largely dissociated according to

$$HNO_3 + 2H_2SO_4 \rightleftharpoons NO_2^+ + H_3O^+ + 2HSO_4^- \qquad (XVI.6.10)$$

The evidence comes from spectral studies, freezing-point lowerings,[c] and actual isolation of nitronium salts such as $NO_2^+ \cdot ClO_4^-$. In similar fashion it has been shown that the oxides N_2O_5, N_2O_4, and N_2O_3 are completely dissociated to give $NO_2^+ + NO_3^-$, $NO_2^+ + NO_2^-$, and $NO^+ + NO_2^-$ ions, respectively.

The mechanism of nitration supposedly involves the attack of the NO_2^+ ion on the aromatic, followed by the fast release of a proton:

$$NO_2^+ + ArH \underset{\underset{\text{slow}}{2}}{\overset{1}{\rightleftharpoons}} [ArH\cdot NO_2]^+ $$
$$\downarrow 3 \text{ fast}$$
$$ArNO_2 + H^+ \qquad (XVI.6.11)$$

This mechanism agrees with the rate law found by Martinsen:[d]

$$-\frac{d(ArH)}{dt} = k(HNO_3)(ArH) \qquad (XVI.6.12)$$

In concentrated HNO_3 as solvent where $(HNO_3) \gg (ArH)$, the rate becomes simply first-order in ArH.[e] In less acidic solvents, such as nitro-

[a] P. B. D. de la Mare et al., *Research* (*London*), **3**, 192, 242 (1950).

[b] E. Berliner, *J. Am. Chem. Soc.*, **72**, 4003 (1950).

[c] R. J. Gillespie et al., *J. Chem. Soc.*, 2504 (1950). C. K. Ingold et al., *ibid.*, 2576 (1950). D. J. Millen, *ibid.*, 2589, 2600, 2606 (1950).

[d] H. Martinsen, *Z. physik. Chem.*, **50**, 385 (1904); **59**, 605 (1907).

[e] E. D. Hughes et al., *J. Chem. Soc.*, 2400 (1950).

methane or acetic acid, with constant excess of HNO_3 over ArH, the rate for very reactive aromatics[a] becomes zero-order. This is true of benzene, toluene, xylenes, p-chloroanisole, and alkyl benzenes, and all of these compounds nitrate at the same rate. The proposed mechanism now involves bond breaking in nitric acid as the slow step:

$$HA + HNO_3 \underset{2}{\overset{\underset{1}{\text{fast}}}{\rightleftharpoons}} H_2O \cdot NO_2^+ + A^-$$

$$H_2O \cdot NO_2^+ \underset{4}{\overset{3}{\rightleftharpoons}} H_2O + NO_2^+$$

$$NO_2^+ + ArH \underset{6}{\overset{5}{\rightleftharpoons}} [ArH \cdot NO_2]^+$$

$$[ArH \cdot NO_2]^+ \underset{8}{\overset{7}{\rightleftharpoons}} ArNO_2 + H^+ \qquad (XVI.6.13)$$

The acid HA is any strong acid and steps 1 and 2 are considered to be at equilibrium (i.e., faster than subsequent steps) in order to account for the inhibition by added bases A^-. For the reactive ArH which give zero-order rates, step 5 is faster than step 4, so that reaction 3 is rate-determining. With the extremely unreactive ArH such as ethyl benzoate, step 5 is rate-determining, and for ArH of intermediate reactivity a more complex rate law corresponding to competition of 4 and 5 is found. If we apply the steady-state treatment to NO_2^+ and to $[ArH \cdot NO_2]^+$ we find

$$\frac{-d(ArH)}{dt} = \frac{k_3 k_5 k_1 (ArH)(HA)(HNO_3)}{k_2(A^-)[k_5(ArH) + k_4(H_2O)]} \qquad (XVI.6.14)$$

In place of "concentrations" one should write "activities," since these systems are far from ideal. Because of this and other complexities, the complete quantitative confirmation of Eq. (XVI.6.14) in the organic solvents has not been achieved. However, the gross qualitative features are well established.[b]

7. Electron-transfer Reactions: Oxidation and Reduction

The ionic mechanisms which we have been so far discussing in connection with organic reactions have all involved the transfer of atoms or ions in individual steps. Depending on the particular charge of the species transferred, some of the steps may have corresponded, in a formal sense, to oxidation or reduction. None of these reactions, however, corresponded to the simple exchange of charge via electron transfer.

In the field of inorganic chemistry it is customary to speak of oxidation-reduction reactions, for example, $Fe^{++} + Ce^{4+} \rightarrow Fe^{3+} + Ce^{3+}$, as electron

---

[a] G. A. Benford and C. K. Ingold, *ibid.*, 929 (1938). E. D. Hughes and C. K. Ingold, *ibid.*, 2400 (1950).

[b] For detailed discussion see text by Ingold, *op. cit.*

transfer reactions. The very interesting question that arises is whether such reactions actually take place via simple electron transfer or through the agency of atom or ion transfer such as is found in the organic systems.

There is something very appealing about the simple mechanism of electron transfer either directly between species or else indirectly via dissolved, stable electrons in the solvent. The classic example of the blue solutions of the alkali metals in liquid NH_3 certainly indicates the feasibility of having "solvated" electrons in a solution.

The kinetic appeal of an electron transfer between species in solution stems largely from the light mass of the electron and hence the possibility of its penetrating energy barriers via a tunneling mechanism such as has been proposed for the passage of α particles out of a nucleus. Further, its light mass gives the electron an extremely high mobility compared to most molecular species. However, these advantages are considerably offset by the restrictions of the Frank-Condon principle. If we consider, for example, the jump of an electron from Fe^{++} to Ce^{4+} in an aqueous solution, then at thermal electron speeds of about 5×10^6 cm/sec, a distance of 10 Å could be spanned in about 2×10^{-14} sec. Most of the species involved in such a reaction, however (that is, Fe^{++}, Ce^{4+}, H_2O, etc.), have speeds of the order of 3×10^4 cm/sec, so that in such a time they would not move more than about 0.1 Å. But since the existence of such ions in solution is possible only because of the stabilizing solvation of the surrounding water molecules, we see that there would be a considerable solvation barrier to such electron jumps.[a]

Since these heats of hydration are of the order of hundreds of kilocalories for most charge-transfer processes, we see that long electron jumps will not be possible. Instead, ions will have to be very close together at the time of reaction or else there will be too big a solvation barrier to charge transfer.[b]

Because of the enormous changes in solvation energies that accompany charge transfer, we may expect the paths which minimize the necessity for major solvent readjustment to be most favorable. These expectations are usually realized in that solvent, or otherwise inactive, ions in the

[a] That is, the increased solvation around the newly formed Fe^{3+}, arising in part from readjustment of the solvent, which would be required to stabilize this Fe^{3+} is quite considerable. From a Born cycle and the relevant thermal data, this heat of hydration is found equal to $[\Delta H^\circ_{hyd}(H^+) - 326$ Kcal]. If we accept Latimer's value for $\Delta H^\circ_{hyd}(H^+) = -260$ Kcal/mole, this gives the net change in heat of hydration for $Fe^{++} \rightarrow Fe^{3+}$ as -596 Kcal/mole. A similar problem is associated with the shift of H_2O around the newly formed Ce^{3+} ion.

[b] By having the ionic species as close together as possible, we minimize the solvation barrier and confine the solvent readjustment to essentially the first and second solvent shells around the ions.

solution are found to act as very effective catalysts for charge transfer.[a]

An excellent example of these reactions is provided by the studies of radioactive exchange between two species in different oxidation states, involving no net chemical reaction. A very extensively studied system has been the Fe^{++}-Fe^{3+} system in aqueous solution, usually with $HClO_4$ added to maintain constant ionic strength and constant pH and prevent formation of complexes.[b,c] The exchange is relatively fast with a rate law (assuming Fe^{++} is the tagged species):

$$\frac{-d(Fe^{*++})}{dt} = k(Fe^{*++})(Fe^{3+}) \qquad (XVI.7.1)$$

where we have neglected the back reaction. In the presence of added negative ions, there is strong catalysis with a change in rate law indicating first- or second-order dependence on the added ion. The equilibrium constants for the association of Fe^{3+} and most of these ions are known, so that it is possible to write the mechanism as proceeding via a transition state arising from Fe^{++} and an Fe^{3+} complex.[d] Table XVI.2 shows some rate constants obtained in this manner.

TABLE XVI.2. CHARGE-TRANSFER REACTION BETWEEN Fe^{++} AND Fe^{3+} AT 0°C IN THE PRESENCE OF DIFFERENT COMPLEXING IONS

Fe^{3+} (complex)	k, liters/mole-sec	Fe^{3+} (complex)	k, liters/mole-sec
Fe^{3+} (aqua)	0.87	FeF^{++}	9.7
$Fe(OH)^{++}$	1010	FeF_2^+	2.5
$FeCl^{++}$	9.7	FeF_3	0.5
$FeCl_2^+$	15		

Note: $HClO_4$ or salts added to ionic strength of about 0.5 M.

While these data show that negative ions, which probably act as charge-density-reducing species, can accelerate the exchange, they give us no data on the mechanism, that is, whether it occurs by an electron jump or by atom transfer. Some very interesting evidence on this comes from the work of Taube[e] and coworkers on the reaction between $Co(NH_3)_5Cl^{++} + Cr^{++}$ in $HClO_4$ solution to give the Co^{++} and Cr^{3+} species. They find that all

[a] A theory of "pure" electron transfer is described by R. A. Marcus, J. Chem. Phys., **26**, 867, 872 (1957).

[b] R. W. Dodson and J. Silverman, J. Phys. Chem., **56**, 846 (1952).

[c] A. C. Wahl and J. Hudis, J. Am. Chem. Soc., **75**, 4153 (1953).

[d] A word of caution is in order here. The rate law gives us information only on the number of complexing ions in the transition state and not on their mode of attachment. Decisions about the latter must be based on other data.

[e] H. Taube et al., J. Am. Chem. Soc., **75**, 4118 (1953).

the Cr^{3+} formed is in the form of the $CrCl^{++}$ complex and that, when radio Cl^{*-} is used in the Co complex, it all ends up in the $CrCl^{++}$. This is, then, unambiguous evidence for a mechanism of Cl atom transfer via the binuclear transition state:

$$\left[\begin{array}{c} H_3N \qquad NH_3 \quad H_2O \qquad OH_2 \\ \diagdown \quad \diagup \qquad \diagdown \quad \diagup \\ H_3N \text{——} Co \text{——} Cl \text{——} Cr \text{——} OH_2 \\ \diagup \quad \diagdown \qquad \diagup \quad \diagdown \\ H_3N \qquad NH_3 \quad OH_2 \qquad OH_2 \end{array} \right]^{4+} \qquad \text{(XVI.7.2)}$$

A similar result has been reported for the very fast exchange of Cr^{++} with $Cr(H_2O)_5Cl^{++}$ in $1\ M$ $HClO_4$.[a] These same authors also found that Cr^{++} catalyzed the otherwise very slow exchange of Cl^- with $Cr(H_2O)_5Cl^{++}$. The third-order catalytic rate constant in $1\ M$ $HClO_4$ at $0°C$ was 0.5 $liter^2/mole^2$-sec with a transition state proposed as (waters omitted):

$$Cr^{*++} + Cl^- + (CrCl^{++}) \rightleftharpoons [Cr^* \cdots Cl \cdots Cr \cdots Cl]^{3+} \rightleftharpoons$$
$$(Cr^* \text{—} Cl^{++}) + Cr^{++} + Cl^- \qquad \text{(XVI.7.3)}$$

where we have labeled the Cr^{*++} to indicate the shift in charge accompanying Cl exchange.[b] It should be pointed out that such experiments are made possible by the very slow rate at which Cr^{3+} and Co^{3+} will exchange their nearest neighbors in an aqueous solution.

Ball and King[c] have studied the exchange of radio Cr^{51} between Cr^{++} and CrX^{++}, where X is the ion Cl^-, F^-, Br^-, N_3^-, or CnS^-. In all of these exchanges the transition state is shown to contain the ion X^- as a bridge between the two Cr species.[d] At $0°C$, the second-order rate constants (at ionic strength $= 1\ M$, using $HClO_4 + LiClO_4$) are 2.5×10^{-3} liter/mole-sec for CrF^{++}, 11 for $CrCl^{++}$, >60 for $CrBr^{++}$, and >1.2 for CrN_3^{++}. $Cr(NCS)^{++}$, which might be expected to parallel CrN_3^{++}, is estimated to

---

[a] H. Taube and E. L. King, *ibid.*, **76**, 4053 (1954). The bimolecular rate constant is 8 liters/mole-sec at $0°C$. In contrast, R. A. Plane and H. Taube, *J. Phys. Chem.*, **56**, 33 (1952), found for the Cr^{++}-$Cr(H_2O)_6^{3+}$ exchange $k = 4.1 \times 10^{-4}$ liter/mole-sec at $27°C$, with H_2O as the probable bridging group. For the exchange of $Cr(OH)^{++}$ with Cr^{++}, A. Anderson and N. A. Bonner, *J. Am. Chem. Soc.*, **76**, 3826 (1954), report $k = 0.7$ liter/mole-sec. It is quite probable that OH^- is the bridging group, although the authors prefer H atom transfer. This exchange is not catalyzed by Cl^- ion.

[b] There has been considerable speculation on the nature of the transition state in these "atom" transfer, redox reactions. The question involved is whether or not there is a change in the number of ligands attached to the metal ions. J. P. Hunt and H. Taube, *J. Chem. Phys.*, **19**, 602 (1951), have interpreted the slow exchange of H_2O^{18} with $Cr(H_2O)_6^{3+}$ and $Co(NH_3)_5(H_2O)^{++}$ as evidence in favor of a change in ligancy of those ions with a high activation energy requirement. In subsequent work by A. C. Rutenberg and H. Taube, *ibid.*, **20**, 825 (1952), on the Co(III) species, an S_N1 mechanism (i.e., dissociation of H_2O, followed by fast association) is proposed.

[c] D. L. Ball and E. L. King, *J. Am. Chem. Soc.*, **80**, 1091 (1958).

[d] This follows from the facts that the rate is first-order in Cr^{++} and in CrX^{++} and also that there is no decrease in CrX^{++} with exchange.

be 10^6 times slower.[a] From the results of these studies the following sequence can be made of groups in the order of the speed with which they facilitate charge transfer between Cr^{++} and CrX^{++} (all via group transfer): $Br^- > N_3^- > Cl^- > OH^- > F^- > NCS^- > H_2O$ (order for OH^- and H_2O inferred, not demonstrated).

While the evidence seems reasonably clear in the preceding cases that oxidation-reduction takes place via atom or ion transfer between species, there are a number of cases in which rapid reaction takes place without any change in the solvent shells of the redox species. Thus $Fe(CN)_6^{3-}$ and $Fe(CN)_6^{4-}$ undergo rapid exchange of labeled Fe, although neither exchanges CN^- appreciably.[b] The same appears to be true for the exchange between MnO_4^- and $MnO_4^{=}$ [c] and between $Os(bip)_3^{++}$ and $Os(bip)_3^{3+}$, where bip = bipyridyl.[d] Although there have been attempts to interpret some of these reactions as electron transfer reactions proceeding via a tunneling mechanism,[e] the theoretical interpretation seems weak and the experimental evidence is just as readily interpreted in terms of atom (or ion) transfer effected by a bridge involving species in the next nearest solvent layer.[f] The problem is, however, far from settled.

In the case of complex ions such as the oxy-halogen negative ions there is good evidence from O^{18} tracer work that charge transfer from these ions takes place via O atom transfer. Anbar and Taube[g] found that in the reaction $ClO^- + NO_2^- \rightarrow Cl^- + NO_3^-$, the extra O in NO_3^- comes exclu-

[a] The authors ascribe this to the thermodynamically unfavorable $Cr—SCN^{++}$ which would be the product of CNS transfer. For the symmetrical azide, no such objection applies.

[b] R. C. Thompson, *J. Am. Chem. Soc.*, **70**, 1045 (1948). J. W. Cobble and A. W. Adamson, *ibid.*, **72**, 2276 (1950).

[c] H. C. Hornig et al., *ibid.*, **72**, 3808 (1950), and J. C. Sheppard and A. C. Wahl, *ibid.*, **79**, 1020 (1957), suggest on the basis of specific cation effects that cations act as bridges between the two reacting anions. In such case the exchange could take place via metal atom transfer.

[d] F. P. Dwyer and E. C. Gyarfas, *Nature*, **166**, 481 (1950).

[e] B. J. Zwolinski et al., *Chem. Revs.*, **55**, 157 (1955). See also F. S. Dainton, *Chem. Soc.* (*London*), *Spec. Pub. No.* **1**, 18 (1954).

[f] Thus for $Fe(CN)_6^{3-}$ and $Fe(CN)_6^{4-}$ the species could exchange charge via a metal atom or H atom transfer across an outer solvation bridge: $(CN)_5Fe(CN)^{3-}\cdots H\cdots(CN)Fe(CN)_5^{3-}$.

[g] M. Anbar and H. Taube, *J. Am. Chem. Soc.*, **80**, 1073 (1958). They also find that the rate of exchange of O^{18} between ClO^- or BrO^- and H_2O has the form $k(XO^-)/(OH^-)$, implying a transition state $[H_2O\cdots X^+\cdots OH^-]$ in which H_2O displaces OH^-. There is a second path catalyzed by Cl^- for which they suggest the transition state

$$\left[HO\cdots Cl \underset{OH_2}{\overset{Cl}{<}} \right]^-$$

the species $[HOCl\cdots Cl]^-$ being the analogue of a trihalide ion.

sively from originally labeled ClO⁻. They also found that the exchange of Br⁻ with BrO⁻ takes place by O atom transfer between the two species (i.e., a nucleophilic displacement on O (S_N2)).

Tracer work with O^{18} has demonstrated that, in the reaction of $SO_3^=$ with the species XO_3^-, XO_2^-, XO^-, ClO_2, or Cl_2O, there is O atom transfer from oxidizer to $SO_3^=$. The final products in acid solution are $SO_4^=$ and X^-.[a] The same authors also studied the disproportionation of ClO_2 dissolved in alkali:

$$2\ ClO_2 + 2\ OH^- \rightarrow ClO_3^- + ClO_2^- + HOH$$

which has a rate law[b]

$$\frac{-d(ClO_2)}{dt} = k_3(ClO_2)^2(OH^-) \qquad (XVI.7.4)$$

where at 0°C, $k_3 = 230$ liters²/mole² sec. They find that ClO_2^- has the same O^{18} composition as ClO_2, while ClO_3^- has one O atom from the solvent. Such a finding is compatible with a transition state:

$$\left[\begin{array}{c} O \\ | \\ O—Cl\cdots O\cdots H\cdots O—Cl \\ | \\ O \end{array}\right] \rightarrow \left[\begin{array}{c} O \\ | \\ O—Cl—O \\ | \\ O \end{array}\right]^- + H—O—\overset{O}{\underset{}{Cl}} \quad (XVI.7.5)$$

the slow step being H-atom transfer to ClO_2.[c]

A very unusual reaction is $H_2SO_3 + H_2O_2 \rightarrow H_2SO_4 + H_2O$, which has been studied in the pH range 1 to 5.[d] It is found that the $SO_4^=$ contains two labeled O atoms from the $H_2O_2^{18}$, and the authors have proposed a peroxysulfurous acid intermediate, HO—S(O)—O*—O*—H, which undergoes intramolecular rearrangement to HO—S(O)(O*)—O*H (sulfuric acid).

8. Intermediate Valence States in Oxidation-Reduction Reactions

The analogue of free radicals for inorganic ions would be intermediate valence states, and many of the kinetic results on redox reactions have required the postulation of unstable valence states for inorganic ions. One of the classic studies of this type is the slow reaction $2Fe^{3+} + Sn^{++} \rightleftharpoons 2Fe^{++} + Sn^{4+}$. The reaction is very slow in $HClO_4$ solution but is strongly catalyzed by Cl^-.[e] At high Fe^{3+}/Sn^{++} ratios, the rate law has the form

[a] J. Halperin and H. Taube, *ibid.*, **74**, 375 (1952).

[b] W. C. Bray, *Z. anorg. u. allgem. Chem.*, **48**, 217 (1906).

[c] This mechanism would involve a fast preequilibrium $ClO_2 + OH^- \rightleftharpoons (O_2Cl\cdots OH)^-$. The authors suggest a preequilibrium

$$2ClO_2 \rightleftharpoons \overset{O}{\underset{}{O}}—Cl\cdots O\overset{O}{\underset{}{Cl}}$$

followed by OH^- attack on the more substituted Cl atom. The rate data do not distinguish the two mechanisms.

[d] J. Halperin and H. Taube, *J. Am. Chem. Soc.*, **74**, 380 (1952).

[e] W. A. Noyes, *Z. physik. Chem.*, **16**, 576 (1895). M. H. Gorin, *J. Am. Chem. Soc.*, **58**, 1787 (1936). F. R. Duke and R. C. Pinkerton, *ibid.*, **73**, 3045 (1951).

$$\frac{-d(Fe^{3+})}{dt} = k_0(Fe^{3+})(Sn^{++}) \tag{XVI.8.1}$$

where k_0 is strongly dependent on the Cl^- concentration.

Such a rate law cannot be explained without postulating an unstable valence state such as Fe^{4+} or Sn^{3+}. Weiss[a] has proposed a mechanism involving the latter which is in good agreement with the data over a considerable range:

$$Fe^{3+} + SnCl_3^- \underset{2}{\overset{1}{\rightleftharpoons}} Fe^{++} + SnCl_3$$

$$Fe^{3+} + SnCl_3 \overset{3}{\rightarrow} Fe^{++} + SnCl_3^+ \tag{XVI.8.2}$$

By applying the stationary-state treatment to $SnCl_3$ we find for the over-all rate

$$\frac{-d(Fe^{3+})}{dt} = \frac{2k_1k_3(Fe^{3+})^2(SnCl_3^-)}{k_2(Fe^{++}) + k_3(Fe^{3+})} \tag{XVI.8.3}$$

When $k_2(Fe^{++}) \gg k_3(Fe^{3+})$ this reduces to the expression found by Noyes (loc. cit.) on adding Fe^{++} initially:

$$\frac{-d(Fe^{3+})}{dt} \rightarrow \frac{2k_1k_3}{k_2} \frac{(Fe^{3+})^2(SnCl_3^-)}{(Fe^{++})} \tag{XVI.8.4}$$

On the other hand, in very acid solutions (Gorin, loc. cit.) when $k_3(Fe^{3+}) \gg k_2(Fe^{++})$ the rate law becomes

$$\frac{-d(Fe^{3+})}{dt} \rightarrow 2k_1(Fe^{3+})(SnCl_3^-) \tag{XVI.8.5}$$

Duke and Pinkerton (loc. cit.) have shown that the Cl^- ion dependence is not simple but varies in $HClO_4$ media with total Cl^- concentration, being more than third-power in Cl^- at $(Cl^-) > 0.1\ M$. Since both Fe^{3+} and Sn^{++} form Cl^- complexes such as $FeCl^{++}$, $FeCl_2^+$, $SnCl^+$, $SnCl_2$, and $SnCl_3^-$ (for all of which dissociation constants are known in this range) the actual mechanism very likely may be more complex with respect to Cl^- ion than is indicated by the Weiss mechanism.

The closely related exchange of Sn^{++} with Sn^{4+} in 10 M HCl has been shown to follow the rate law (assume Sn^{++} tagged)[b]

$$\frac{-d(Sn^{++})^*}{dt} = k(Sn^*Cl_3^-)(SnCl_6^=) \tag{XVI.8.6}$$

with $k = 4.5 \times 10^7 \exp(-10,800/RT)$ liters/mole-sec and strongly dependent on HCl concentration. The exchange probably gives the Sn^{3+}—Sn^{3+} intermediate via Cl atom transfer, and this latter will rapidly dispropor-

[a] J. Weiss, J. Chem. Soc., 309 (1944).

[b] C. I. Browne et al., J. Am. Chem. Soc., 73, 1946 (1951). The authors were also able to demonstrate the existence of a dimer $Sn_2Cl_{10}^{4-}$ which was in equilibrium with $SnCl_4^=$ and $SnCl_6^=$.

tionate to give exchange. Although it is interesting to discover in such reactions whether two electron changes can take place at once, the present kinetic evidence does not clarify this point.[a]

Intermediate valence states are frequently of importance as catalysts in redox systems. Thus Fe^{3+} will oxidize I^- very slowly in dilute acid solutions to form Fe^{++} and I_3^-.[b] Similarly, the reaction between $Cr_2O_7^=$ (or $HCrO_4^-$) and I^- in acid solutions is extremely slow.[c] If, however, Fe^{++}, I^-, and $Cr_2O_7^=$ are mixed together, a very rapid oxidation of I^- to I_3^- occurs,[d] accompanied by the oxidation of Fe^{++} to Fe^{3+}. In this system Cr must certainly pass through either a 4+ or a 5+ valence state, and current interpretations favor the latter. Further evidence for a Cr(V) state comes from the studies of the exchange between Cr(VI) and Cr(III) in 0.16 M $HClO_4$ solutions.[e] It is found that the rate law is

$$\frac{-d(Cr^{*3+})}{dt} = k(Cr^{3+})^{4/3}(HCrO_4^-)^{2/3} \qquad (XVI.8.7)$$

with the H^+ dependence of k indicating two concurrent transition states differing from each other in composition by two protons. If it is assumed that the equilibrium

$$4H_2O + Cr^{3+} + 2HCrO_4^- \underset{2}{\overset{1}{\rightleftharpoons}} 3H_2CrO_4^- + 4H^+$$

is rapidly established,[f] so that a small amount of Cr(V) is formed in the solution,[g] then a rate-determining step Cr(III) + Cr(V) $\rightleftharpoons$ 2Cr(IV) can account for the observed exchange if the exchange of Cr(V) and Cr(VI) is fast. This last is quite reasonable in view of the rapid exchange of MnO_4^- and $MnO_4^=$ which would be isoelectronic with Cr(V) and Cr(VI). Schematically we could represent the mechanism by[h]

[a] It was thought at one time that changes in valence state could occur with facility only in unit changes. Because of the probability that most of such changes take place by atom transfer, the limitation seems unnecessary. In aqueous systems O atom transfer is equivalent to a two-charge jump, while OH radical or H atom transfer is equivalent to a one-charge jump. From this point of view water may be a good charge transfer medium, providing as it does all types of transfer species.

[b] A. V. Hershey and W. C. Bray, *J. Am. Chem. Soc.*, **58**, 1760 (1936). K. W. Sykes, *J. Chem. Soc.*, 124 (1952). See also Duke and Pinkerton, *loc. cit.*

[c] R. F. Beard and N. W. Taylor, *J. Am. Chem. Soc.*, **51**, 1973 (1929).

[d] C. F. Schönbein, *J. prakt. Chem.*, **75**, 108 (1858). C. Wagner and W. Preiss, *Z. anorg. Chem.*, **168**, 265 (1928). See p. 595.

[e] E. L. King and C. Altman, American Chemical Society Spring Meeting, 1958.

[f] The actual forms for the Cr(VI) and Cr(V) states are not necessary to the kinetic argument except in relation to the pH dependence.

[g] It may appear paradoxical that reactions 1 and 2 which produce the equilibrium concentration of Cr(V) do not cause appreciable exchange. The answer lies in the very small concentration of Cr(V) which is present at equilibrium, so that even when reaction 1 is slow, the small amount of Cr(V) required for equilibrium may be rapidly produced.

[h] Reactions 1 and 2 are not meant to represent an actual reaction path but merely over-all stoichiometry.

$$Cr(III) + 2 Cr(VI) \underset{2}{\overset{1}{\rightleftharpoons}} 3 Cr(V) \qquad \text{fast: } K_{1 \cdot 2} \ll 1$$

Exchange Step:

$$Cr(V) + Cr^*(III) \underset{4}{\overset{3}{\rightleftharpoons}} Cr^*(IV) + Cr(IV) \qquad \text{slow}$$

$$Cr^*(V) + Cr(VI) \underset{6}{\overset{5}{\rightleftharpoons}} Cr^*(VI) + Cr(V) \qquad \text{fast} \qquad (XVI.8.8)$$

Most striking of the catalyzed oxidations are those of $S_2O_8^=$ catalyzed by Ag^+. Cr^{3+},[a] V^{4+},[b] N_2H_4,[c] and Ce^{3+}[d] are all oxidized by $S_2O_8^=$ at identical rates in the presence of Ag^+. In each case the rate law shows first-order dependence on Ag^+, first-order dependence on $S_2O_8^=$, and zero-order dependence on the reducing species. It is proposed that either Ag^{2+} or Ag^{3+}, both of which are present in the solution[e] and in rapid equilibrium with Ag^+ ($2Ag^{++} \rightleftharpoons Ag^{3+} + Ag^+$), is the effective catalyst for the oxidation. The rate-determining step is then the reaction $Ag^+ + S_2O_8^= \rightleftharpoons Ag^{3+} + 2SO_4^=$ (or $Ag^{++} + SO_4^= + SO_4^-$).

Of course, what is interesting but not yet explained in these catalyses is the reason for the speed of the particular path observed and the relative slowness of the uncatalyzed paths. In some cases in which an intermediate valence state is produced from a catalyst species the original state may not be regenerated fast enough compared to continued reaction of the intermediate state. In such cases the "catalyst" is consumed and the system is referred to as an *induced* or *coupled redox system*. An example of such a system is the oxidation of I^- by $HCrO_4^-$ induced by Fe^{++}.[f] Here I^- or Fe^{++} and $HCrO_4^-$ react to form an intermediate state, presumably $Cr(V)$ and I (or Fe^{3+}). In the presence of excess Fe^{++} or I^- the $Cr(V)$ state is further reduced to $Cr(IV)$ or $Cr(III)$. The $Cr(IV)$ goes on to react with another equivalent of I^- to give Cr^{3+} and I. Depending on the relative concentrations, it is found that about two equivalents of I^- and one of Fe^{++} are oxidized[g] for each $Cr(VI)$ reduced to Cr^{3+}. The study of induced reactions is very valuable in giving information on the reactivity and presence of intermediate valence states.

[a] D. M. Yost, *J. Am. Chem. Soc.*, **48**, 152 (1926).

[b] D. M. Yost and W. H. Claussen, *ibid.*, **53**, 3349 (1931).

[c] Yost et al., *ibid.*, **59**, 2129 (1937).

[d] W. H. Cone, *ibid.*, **67**, 78 (1945).

[e] Yost, *op. cit.*

[f] C. Benson, *J. Phys. Chem.*, **7**, 1 (1903). See review by F. H. Westheimer, *Chem. Revs.*, **45**, 419 (1949), for $Cr(VI)$ reactions.

[g] For such reactions an induction factor IF is defined as the number of equivalents of reductant oxidized per equivalent of inducer (here Fe^{++}) oxidized. In essence reducing agent and inducer compete with each other in the further reduction of the intermediate valence states of oxidizing agent.

9. Ion-Radical Reactions; Decomposition of Peroxide

In ionizing solvents, atoms and free radicals are capable of being oxidized or reduced by charge transfer, and their modes of reaction can become very complex. One of the systems in which redox reactions involving atoms and radicals have been extensively studied is that between Fe^{++} and H_2O_2.[a] When equimolar, acidic solutions ($<0.01\ M$) of Fe^{++} and H_2O_2 are mixed together, there is a nearly quantitative reaction

$$H_2O_2 + 2Fe^{++} + 2H^+ \rightarrow 2Fe^{3+} + 2H_2O$$

with a rate law[b]

$$\frac{d(Fe^{3+})}{dt} = k(Fe^{++})(H_2O_2) \qquad (XVI.9.1)$$

The modified Haber-Weiss[c] mechanism which has been proposed to account for this and other observations is

$$Fe^{++} + HOOH \overset{1}{\underset{2}{\rightleftharpoons}} Fe(OH)^{++} + OH$$

$$Fe^{++} + OH \overset{8}{\rightarrow} Fe(OH)^{++} \qquad (XVI.9.2)$$

where in acid solution the $Fe(OH)^{++}$ is rapidly converted to the $Fe^{3+} + H_2O$. Over long periods of time there is a slower evolution of O_2 from the system, indicating a catalyzed decomposition of H_2O_2. This latter reaction is accelerated by increasing concentration of Fe^{3+} or by increasing concentration of H_2O_2. It is also faster in more acid solutions. Tracer studies with O^{18} have demonstrated that both O atoms in the product O_2 come from the same molecule of H_2O_2.[d] These facts would seem to imply that Fe^{3+} acts to oxidize H_2O_2 to $O_2 + H^+$ without rupture of the O—O bond. Since Fe^{2+} acts to reduce H_2O_2 to H_2O, it can be seen that the combined action of the Fe^{++}-Fe^{3+} couple is to catalyze the disproportionation of H_2O_2 to $H_2O + O_2$. A mechanism which seems in reasonable accord with those observations is

Initiation and $\quad \begin{cases} Fe^{++} + HOOH \overset{1}{\underset{2}{\rightleftharpoons}} Fe^{3+} + OH^- + OH \\ \\ Fe^{3+} + HOOH \overset{3}{\underset{4}{\rightleftharpoons}} Fe^{++} + HO_2 + H^+ \end{cases}$

Termination:

---

[a] See N. Uri, *Chem. Revs.*, **50**, 375 (1951), for a review of this and other inorganic reactions involving free radicals.

[b] W. G. Barb et al., *Trans. Faraday Soc.*, **47**, 462 (1951).

[c] F. Haber and J. Weiss, *Naturwiss.*, **20**, 948 (1932). An alternative mechanism involving the production of $Fe^{4+} + 2OH^-$ in step 1 has been put forward by H. Taube and A. E. Cahill, *J. Am. Chem. Soc.*, **74**, 2312 (1952), on the basis of the relative rates of attack on $HOO^{18}H$ and $HOOH$ by Fe^{++}, Cu^+, Sn^{++}, Cr^{++}, and Ti^{3+}. Ti^{3+} has only one upper valence state and behaves slightly differently from the other reductants, which have two or more upper states.

[d] C. A. Bunton and D. R. Llewellyn, *Research (London)*, **5**, 142 (1952).

Chain:
$$HO + HOOH \xrightarrow{5} H_2O + HO_2$$

$$Fe^{3+} + HO_2 \underset{7}{\overset{6}{\rightleftharpoons}} Fe^{++} + O_2 + H^+$$

$$Fe^{++} + HOOH \underset{2}{\overset{1}{\rightleftharpoons}} Fe^{3+} + OH^- + OH$$

Termination: $Fe^{++} + OH \xrightarrow{8} Fe^{3+} + OH^-$ (XVI.9.3)

This scheme has been simplified by ignoring the complexing of Fe^{3+} with OH^-, HO_2^-, or other negative ions which are known to be extensive.[a] In addition it ignores the equilibria[b]

$$
\begin{array}{lll}
H_2O_2 \rightleftharpoons H^+ + HO_2^- & pK_a = 12 & \\
HO_2 \rightleftharpoons H^+ + O_2^- & pK_a \cong 12 & \\
HO \rightleftharpoons H^+ + O^- & pK_a \cong 4 & \text{(XVI.9.4)}
\end{array}
$$

These acid-base equilibria will play an important role in determining the relative importance of the various steps, but as yet there is little clear-cut evidence concerning them. We have omitted the Fe^{4+} oxidation state because the evidence for it is quite tenuous. In addition the chain step of Weiss, $HO_2 + H_2O_2 \rightarrow H_2O + OH + O_2$, is omitted as being structurally unreasonable.[c]

The Fe^{3+} reaction is complicated by the fact that the solution can become supersaturated with respect to O_2 and also by the fact that at the higher pH's, $Fe(OH)_3$ and $Fe(OH)_2$ may precipitate. The production of O_2 via long chains is possible only under conditions such that H_2O_2 is large with respect to both Fe^{++} and Fe^{3+}. Under these conditions we may neglect step 7 and use a stationary-state approximation for Fe^{++} [that is, $d(Fe^{++})/dt = 0$]. Practically all of the H_2O_2 then disappears via the disproportionation chain, sequence 1, 5, and 6 at a rate

$$
\begin{aligned}
\frac{-d(H_2O_2)}{dt} &\cong \frac{2k_1(H_2O_2)(Fe^{++})}{1 + k_2(Fe^{3+})(OH^-)/[k_5(H_2O_2) + k_8(Fe^{++})]} \\
&\cong \frac{2k_1(H_2O_2)(Fe^{++})}{1 + k_2(FeOH)^{++}/k_5(H_2O_2)}
\end{aligned}
\qquad \text{(XVI.9.5)}
$$

Under these same conditions the $(Fe^{3+})/(Fe^{++})$ ratio is nearly stationary and is given approximately by

$$
\left[\frac{(Fe^{3+})}{(Fe^{++})}\right]_{ss}^2 \cong \frac{k_1 k_4}{k_3 k_6}\left[(H^+) + \frac{k_6 k_8 (Fe^{3+})}{k_5 k_4 (H_2O_2)}\right]
\qquad \text{(XVI.9.6)}
$$

It is interesting to see that the Fe^{3+} catalyzed disproportionation has a

---

[a] Uri, *loc. cit.*

[b] The pK values shown for HO_2 and HO are pure guesses based on analogy with H_2O_2 and H_2O (and HF), respectively. The uncertainty is ± 4 pK units or more.

[c] However, W. K. Wilmarth (personal communication) claims that he has found good evidence for this step in the chain reaction of $S_2O_8^{-2}$ with H_2O_2.

rate law quantitatively very similar [Eq. (XVI.9.3)] to that obtained for the stationary-state reduction by Fe^{++} [Eq. (XVI.9.1)]. The chief difference in the two cases is that for the Fe^{3+} catalysis, the Fe^{++} concentration is no longer variable but is essentially fixed by Fe^{3+} and the experimental condition [Eq. (XVI.9.4)].

Other evidence for the existence of radicals in the H_2O_2-Fe^{++}-Fe^{3+} system comes from the observation that alcohols which react only very slowly with peroxide directly are very rapidly oxidized to aldehydes when Fe^{++} is added to the system.[a] The mechanism very likely involves H abstraction from the alcohol by OH radical (or HO_2) followed by oxidation of the radical[b] by Fe^{3+}. For the system Fe^{2+}-H_2O_2-C_2H_5OH it is found[b] that the reaction can be represented by

$$H_2O_2 + C_2H_5OH \rightarrow 2H_2O + CH_3CHO$$

The proposed mechanism for a radical chain is

$$Fe^{++} + H_2O_2 \underset{2}{\overset{1}{\rightleftharpoons}} Fe^{3+} + OH^- + OH$$

$$OH + C_2H_5OH \overset{3}{\rightarrow} CH_3\dot{C}HOH + H_2O$$

$$CH_3\dot{C}HOH + Fe^{3+} \overset{4}{\rightarrow} CH_3CHO + Fe^{++} + H^+ \qquad (XVI.9.7)$$

An alternative reaction for radical attack on peroxide is

$$CH_3\dot{C}HOH + HOOH \overset{5}{\rightarrow} CH_3CH(OH)_2 + OH$$

Although energetically feasible, it seems unlikely in view of the absence of other nucleophilic attacks on peroxide. Further evidence in favor of step 4 rather than 5 comes from the fact that acetone and acetic acid suppress the oxidation of C_2H_5OH and are not themselves appreciably attacked. Since step 4 would not be possible for the $\dot{C}H_2COOH$ or $\dot{C}H_2COCH_3$ radicals while step 5 would, this favors step 4 as the oxidative chain.

A final bit of evidence for radicals in the system is found in the fact that Fenton's reagent ($Fe^{++} + H_2O_2$) will initiate polymerization of vinyl monomers dissolved[c] or suspended[d] in aqueous solution. This arises, presumably, from the addition of OH radicals to monomer:

$$OH + CHX=CHY \rightarrow HOCHX-\dot{C}HY$$
$$HOCHX-\dot{C}HY + CHX=CHY \rightarrow$$
$$HO(CHX-CHY)-CHX-\dot{C}HY \cdots$$

[a] J. H. Merz and W. A. Waters, *J. Chem. Soc.*, 2427 (1949); supplement S15 (1949).

[b] I. M. Kolthoff and A. I. Medalia, *J. Am. Chem. Soc.*, **71**, 3777, 3784, 3789 (1949). The reagent $H_2O_2 + Fe^{++}$ is popularly known as Fenton's reagent.

[c] R. G. R. Bacon, *Trans. Faraday Soc.*, **42**, 140 (1946). L. B. Morgan, *ibid.*, **42**, 169 (1946).

[d] J. H. Baxendale et al., *ibid.*, **42**, 155 (1946). See also F. S. Dainton and M. Tordoff, *ibid.*, **53**, 666 (1957).

A great deal has been done to obtain quantitative estimates or the relevant thermodynamic data for the various steps in the Fe^{++} - Fe^{3+} - H_2O_2 system (see N. Uri, *loc. cit.*), but unfortunately the heat and entropy data for the important radicals OH and HO_2 are subject to such uncertainties as to make these data useless except for the grossest possible analysis.

It is interesting to note that a great deal of evidence has accumulated for believing that the mechanism of the oxidation of Fe^{++} by Br_2,[a] Cl_2,[b] I_2,[c] and O_2[d] is initiated by charge transfer to form the negative ion Br_2^-, etc., $+ Fe^{3+}$, a step very similar to those occurring in the peroxide system. There also seems to be evidence that direct oxidation of Br^-, Cl^-, F^-, and OH^- to the corresponding radicals can be photoinitiated in the presence of Fe^{3+} or Ce^{4+} acting as catalyst.[e]

10. Free Radical Polymerization

Unsaturated organic compounds, i.e., those containing double or triple bonds, are thermodynamically unstable with respect to addition reactions. If the liquid styrene, $C_6H_5CH{=}CH_2$, is left in a bottle at room temperature for any length of time, it will be found to undergo a change to a hard, transparent solid of very high molecular weight. This solid, referred to as polystyrene or styrene polymer, is found to be soluble in solvents such as benzene or toluene, and by various studies it has been shown to consist of a heterodisperse mixture of long chain molecules of the general formula

$$R_1 - \left[\begin{array}{c} \phi \quad H \\ | \quad\ | \\ -C - C - \\ | \quad\ | \\ H \quad H \end{array} \right]_n - R_2$$

R_1 and R_2 may both be H atoms, but the identification is often not certain, and n may vary from a very small integer to 10,000 or more.

This reaction of styrene is very sensitive to changes in experimental conditions. It occurs more slowly at lower temperatures and can be almost permanently inhibited by storage at $-80°C$ (dry-ice temperature). It is accelerated by exposure to sunlight or oxygen and may be inhibited by small amounts of reagents such as quinone, $C_6H_4O_2$, which is frequently

---

[a] P. R. Carter and N. Davidson, *J. Phys. Chem.*, **56**, 877 (1952).

[b] H. Taube, *J. Am. Chem. Soc.*, **68**, 611 (1946), has used Fe^{++} to induce the $Cl_2 + (COOH)_2$ reaction: $Cl_2 + (COOH)_2 \rightarrow 2CO_2 + 2HCl$.

[c] Sykes, *loc. cit.*

[d] R. E. Huffman and N. Davidson, *J. Am. Chem. Soc.*, **78**, 4836 (1956).

[e] E. Rabinowitch, *Revs. Modern Phys.*, **14**, 112 (1942), first interpreted the ultraviolet spectra of the corresponding ion-pair complexes Fe^{3+}—X^- as charge-transfer spectra: Fe^{3+}—$X^- + h\nu \rightarrow [Fe^{++}$—$X]^*$.

used to stabilize liquid styrene. In the presence of strong acids such as H_2SO_4 or mixtures of $HF + BF_3$ the polymerization is found to take place very rapidly even at very low temperatures.[a]

These condensation reactions are not restricted to the $C=C$ double bond systems but seem to be quite general for $C\equiv C$ triple bonds, $C=O$ double bonds, and $C=N$ double bonds. In only some cases, however, is the long-chain product found; quite frequently double molecules or small rings are produced instead.[b] Studies of the details of these polymerizations, as we shall now refer to them, show that there are two mechanisms which can usually be distinguished from one another. One of these is an ionic mechanism which may proceed by the addition of an ion to one end of the double bond and so lead to the formation of an ion which can now add to the double bond of another molecule to start a chain:

$$M^+(\text{or } M^-) + \begin{array}{c} H \\ \diagdown \\ C = C \\ \diagup \quad \diagdown \\ R_1 \qquad R_2 \end{array} \rightarrow M - C - C + (\text{or } -)$$

$$\text{(XVI.10.1)}$$

$$M - \underset{R_1}{\overset{H}{\underset{|}{C}}} - \underset{R_2}{\overset{H}{\underset{|}{C}}} + + \underset{R_1}{\overset{H}{\underset{|}{C}}} = \underset{R_2}{\overset{H}{\underset{|}{C}}} \rightarrow \cdots \rightarrow M \left[- \underset{R_1}{\overset{H}{\underset{|}{C}}} - \underset{R_2}{\overset{H}{\underset{|}{C}}} \right]_n^+$$

Termination in such cases may occur by loss of a proton H^+ to any base in the system, or in the case of anions by the addition of a proton from any acid (or generally, neutralization by Lewis acids or bases).

The second mechanism, which is distinguished from the ionic in that it can occur readily in low-dielectric solvents and also in the gas phase, is the free radical mechanism. Here a free radical R produced in the system thermally or photochemically is added to the double bond in a manner similar to that represented above for ions, and the new radical can now propagate long chains. Termination in such a system usually occurs by mutual combination or disproportionation of two radicals. The kinetic

---

[a] There are a number of excellent treatises on polymerization chemistry and kinetics to which the reader is recommended for further details on this and other aspects. Some of them are (1) P. J. Flory, "Principles of Polymer Chemistry," Cornell University Press, Ithaca, N.Y., 1953; (2) G. M. Burnett, "Mechanism of Polymer Reactions," Interscience Publishers, Inc., New York, 1954; and (3) Cheves Walling, "Free Radicals in Solution," John Wiley & Sons, Inc., New York, 1957.

[b] Thus acetone, acetaldehyde, and esters condense to dimers in the first stage. From aldehydes we also can obtain six-membered rings such as trioxane from CH_2O, paraldehyde from CH_3CHO, etc. Dienes often give dimers reversibly.

scheme for such a free radical polymerization of an unsaturated monomer M can be represented by the following equations:

Initiation: $R \xrightarrow{i} M_1 \cdot$

$$M_1 \cdot + M \xrightarrow{1} M_2 \cdot$$

Propagation: $M_2 \cdot + M \xrightarrow{2} M_3 \cdot$

$$\cdot$$
$$\cdot$$
$$\cdot$$

$$M_n \cdot + M \xrightarrow{n} M_{n+1} \cdot$$

Termination: $M_n \cdot + M_s \cdot \xrightarrow{t} \begin{cases} M_n + M_s \\ \text{or} \\ M_{n+s} \end{cases}$ (XVI.10.2)

We have used the notation of representing radical polymers by $M_n \cdot$ and stable polymer molecules by M_n, the subscript in each case representing the number of monomer units in the chain.[a]

If we apply to the above system of equations the stationary-state hypothesis, then we can write an equation for the rates of initiation and termination of radicals in the system:[b]

$$R_i = R_t = k_t \left(\sum_{n=1}^{\infty} M_n \right)^2 = k_t (M \cdot)^2 \qquad (XVI.10.3)$$

Here we have assumed that the rate of termination of radicals is independent of their degree of polymerization (i.e., of n) and have represented the total concentration of radicals by $(M \cdot) = \sum_n (M_n \cdot)$. The assumption of a universal termination constant for all radicals, independently of molecular weight, is undoubtedly incorrect.[c] However, the kinetics of radical polymerization has not yet attained a stage of quantitative accuracy such that the consequences of this assumption can be explored. To avoid the

[a] The application "chain reaction" is here somewhat confusing. The reaction sequence (XVI.10.2) does not correspond to a chain reaction in the strict sense, since at no point is the original chain carrier reproduced. Only by generalizing the definition of chain carrier in the case of polymers to the point of identifying radical $M_n \cdot$ with $M_{n-1} \cdot$ can we include these polymerizations under chain reactions. Undoubtedly part of the confusion arises from the fact that "chains" are produced as products.

[b] On expanding the sum we would have $k_t M_1^2 + 2k_t M_1 M_2 + k_t M_2^2 + \cdots$. The factor of 2 in the second term arises from two radicals being destroyed per termination act. The first term is half the second merely because of the symmetry factor for collisions between identical species.

[c] S. W. Benson and A. North, *J. Am. Chem. Soc.*, **81**, 1339 (1959).

inordinate mathematical complexity involved in a more exact formulation,[a] we shall proceed with the present scheme.

If we make a similar and perhaps more justifiable assumption that the rate of propagation of the chain is independent of the chain length already achieved, then we can use a single specific, propagation rate constant k_p instead of the infinite set $k_{pn}(n = 1, \ldots, \infty)$ for the separate propagation steps.

This then permits us to write for the rate of disappearance of monomer

$$\frac{-d(\text{M})}{dt} = (\text{M}) \sum_{n=1} k_{pn}(\text{M}_n\cdot) \cong (\text{M})k_p \sum_{n} (\text{M}_n\cdot)$$

$$= (\text{M})k_p(\text{M}\cdot) \qquad\qquad (\text{XVI.10.4})$$

On substituting for $(\text{M}\cdot)$ from Eq. (XVI.10.3) we have

$$\frac{-d(\text{M})}{dt} = k_p \left(\frac{R_i}{k_t}\right)^{\frac{1}{2}} (\text{M}) \qquad\qquad (\text{XVI.10.4}a)$$

where R_i represents the specific rate at which radicals R produced in the system are converted into $\text{M}_1\cdot$. This rate will depend not only on the rate of production of such radicals R but also on the efficiency with which they react with M to initiate chains.

When the reaction is initiated by the photolysis of a chain sensitizer such as acetone or an azo compound, then $R_i = 2\phi I_a$, where I_a is the specific rate of absorption of quanta in the system and ϕ is the fraction of quanta absorbed which result in the initiation of chains.[b] In such case the rate of photopolymerization has the form

$$\frac{-d(\text{M})}{dt} = k_p \left(\frac{2\phi I_a}{k_t}\right)^{\frac{1}{2}}(\text{M}) \qquad\qquad (\text{XVI.10.4}b)$$

The first-order dependence on M and the half-order dependence on light

[a] The more exact formulation of the stationary-state condition would take the form

$$R_i = k_i(\text{M}) = \sum_{n} \sum_{s} k_t(n,s)(\text{M}_n\cdot)(\text{M}_s\cdot)$$

This is frequently, if not justifiably, approximated as a perfect square:

$$k_i(\text{M}) = \left[\sum_{s} k_{ts}^{\frac{1}{2}}(\text{M}_s\cdot)\right]^2$$

[b] Note that ϕ may differ from unity for any of a number of reasons. One is the deactivation of excited molecules by either quenching or fluorescence. A second is the recombination of the two free radicals formed before they initiate chains. Such recombination is favored at low monomer concentration (M) because of the cage effect, i.e., the long time the nascent free radicals will spend as near neighbors because of the solvent cage around them. Finally, in any such system in which they are being continuously initiated, the radicals R can also engage in termination reactions with polymer radicals. In principle such an event should be included in the general rate law.

intensity in photopolymerizations have been verified for a large number of systems over a considerable range of experimental conditions. From the observed rate of such a polymerization, one obtains the empirical rate constant $k_{ph} = k_p(2\phi I_a/k_t)^{1/2}$. It is possible in such experiments to measure I_a, the specific rate of light absorption, but a measure of ϕ is not too simple. One method is to use an initiator such as benzoyl peroxide, ϕCO—OO—$CO\phi$, whose free radicals, phenyl, ϕ, or benzoyl, ϕCOO, can be identified in the final polymer. In principle there should be one benzene ring per chain, and this allows ϕ to be computed.[a] Alternatively, it is possible to measure the number average molecular weight of the final polymer[b] and arrive at the number of chains produced. This also makes it possible to calculate ϕ.

Both of these methods are subject to uncertainty that arises from the manner of termination, i.e., by recombination or by disproportionation. This latter question, if not resolved, can introduce an error of 2 into the estimates of ϕI_a. Assuming that all of these problems have been satisfactorily resolved, the study of photo-stationary-state kinetic systems gives the ratio of rate constants $k_p/k_t^{1/2}$. Within these limitations the results of various laboratories, obtained by the use of different techniques, show reasonably good agreement, usually within 10 to 20 per cent for a given monomer-solvent system.[c]

The use of chemical sensitizers such as benzoyl peroxide, cumene hydroperoxide, or azo-bis-isobutyronitrile, which decompose thermally to give free radicals in a convenient temperature range (i.e., 60°C to 150°C), makes it possible to study polymerizations over an extended temperature range. The form of the rate law with chemical initiations would be given by setting $R_i = 2k_i(In)\phi_i$ in Eq. (XVI.10.4). Here (In) is the initiator concentration, k_i its specific rate constant of decomposition which can usually be measured independently, and ϕ_i is the efficiency with which its radicals initiate chains. The measure of ϕ_i is subject to the difficulties already indicated in connection with the photolysis systems.[d]

---

[a] Such an analysis is complicated by the fact that radicals can also terminate chains, in which case there would be two phenyl groups in such a polymer. This can be shown to be a minor error if the rate of initiation is as fast as the rate of propagation and the chain length is large. A more serious source of error arises from chain transfer reactions, which we shall discuss later. In principle these can be measured and allowed for.

[b] This is usually done by osmotic pressure measurements and is not too simple. The estimate of ϕ is also subject to correction for chain transfer.

[c] A. V. Tobolsky and J. Offenbach, *J. Polymer Sci.*, **16**, 311 (1955), have made careful measurements for styrene, while a similar study of methyl methacrylate, CH_2=$C(CH_3)COOCH_3$, has been made by T. E. Ferrington and A. V. Tobolsky, *J. Colloid Sci.*, **10**, 536 (1956).

[d] A further difficulty which arises in connection with peroxide initiators is that of the radical-induced decomposition of the initiator. This may give chains leading to values of ϕ or ϕ_i in excess of unity.

If we apply the stationary-state hypothesis to each of the radical species, we find for the nth radical, $M_n \cdot$[a]

$$0 = \frac{-d(M_n \cdot)}{dt} = (M)k_p[(M_{n-1} \cdot) - (M_n \cdot)] - (M_n \cdot) \sum_s k_t(M_s \cdot)$$

(XVI.10.5)

By solving for $M_n \cdot$ we have

$$\frac{(M_n \cdot)}{(M_{n-1} \cdot)} = \frac{k_p(M)}{k_p(M) + k_t(M \cdot)} = \frac{1}{1 + k_t(M \cdot)/k_p(M)}$$

(XVI.10.6)

or

$$\frac{(M_n \cdot)}{(M_1 \cdot)} = \left[1 + \frac{k_t(M \cdot)}{k_p(M)}\right]^{1-n}$$

(XVI.10.7)

For the first radical $M_1 \cdot$, the equations give

$$(M_1 \cdot) = \frac{R_i}{k_p(M) + k_t(M \cdot)} = \frac{k_t(M \cdot)^2}{k_p(M) + k_t(M \cdot)}$$

(XVI.10.8)

or

$$\frac{(M_1 \cdot)}{(M \cdot)} = \frac{k_t(M \cdot)/k_p(M)}{1 + k_t(M \cdot)/k_p(M)}$$

(XVI.10.9)

Now the ratio $k_p(M)/k_t(M \cdot) = k_p(M)(M \cdot)/k_t(M \cdot)^2$ can be seen to be the ratio of the rate at which monomer is converted into polymer to the rate of termination of radical chains. If termination occurs by recombination, then this ratio is just one-half the average number of monomer units per final polymer chain, which we may represent by $\bar{n}$, the mean chain length, or mean degree of polymerization. This permits us to write for the stationary radical concentrations

$$\frac{(M_n \cdot)}{(M \cdot)} = \frac{2}{\bar{n}} \left(\frac{1}{1 + 2/\bar{n}}\right)^n \simeq \frac{2e^{-2n/\bar{n}}}{\bar{n}} \qquad \text{for } n = 1, 2, \ldots$$

(XVI.10.10)

which represents a monotonically decreasing, geometrical distribution of radical concentrations.

The rate of production of n-mer, M_n, is given by

$$\frac{d(M_n)}{dt} = \frac{k_t}{2} \sum_{s=1}^{n-1} (M_{n-s} \cdot)(M_s \cdot)$$

(XVI.10.11)

where we have assumed termination by recombination. By substituting from Eq. (XVI.10.10) we find

$$\frac{d(M_n)}{dt} = \frac{2nk_t(M \cdot)^2}{\bar{n}^2} \left(\frac{1}{1 + 2/\bar{n}}\right)^n = 8n \frac{|dM/dt|}{\bar{n}^3} \left(\frac{1}{1 + 2/\bar{n}}\right)^n$$

(XVI.10.12)

so that

$$\frac{-d(M_n)}{d(M)} = \frac{8n}{\bar{n}^3} \left(\frac{1}{1 + 2/\bar{n}}\right)^n \simeq \frac{8n}{\bar{n}^3} e^{-2n/\bar{n}}$$

(XVI.10.13)

[a] Note again the curious cancellation of 2. The rate of loss of $M_n \cdot$ by reaction with $M_n \cdot$ is $2(k_t/2)(M_n \cdot)^2 = k_t(M_n \cdot)^2$. Two radicals $M_n \cdot$ disappear per act, but the rate constant is just half that for $M_n \cdot + M_s \cdot$ ($s \neq n$).

gives the molar rate of appearance of chains of length n per disappearance of 1 mole of monomer. In a stationary-state experiment at low conversion, when R_i, R_t, and (M) may be considered to be nearly constant over the course of the experiment, Eq. (XVI.10.13) also gives the mole fraction of n-mer produced. The weight fraction of n-mer produced is obtained by multiplying by n:

$$\frac{n(M_n)}{(M)} \xrightarrow{\text{low conversion}} \frac{8n^2}{\bar{n}^3}\left(\frac{1}{1+2/\bar{n}}\right)^n \cong \frac{8n^2}{\bar{n}^3}\, e^{-2n/\bar{n}} \quad \text{(XVI.10.14)}$$

If we differentiate this last equation with respect to n at constant $\bar{n}$, we find that the weight fraction has a maximum at $n = \bar{n}$.[a] These results are essentially similar if termination takes place by disproportionation rather than recombination.

The verification of the distribution of n-mer predicted by Eq. (XVI.10.13) or (XVI.10.14) has been made not quantitatively but only qualitatively. This has as much to do with the difficulty of accurately fractionating polymer into different molecular weight ranges as it does with the kinetic difficulties of the experiments that arise from chain transfer.

11. Absolute Rate Constants in Polymerization

It has been pointed out that a study of the stationary rates of polymerization reactions leads to an over-all rate constant which is a composite of initiation, propagation, and termination steps [Eqs. (XVI.10.4) and (XVI.10.5)]. When the initiation rate is assignable, this can finally yield values for the ratio $k_p/k_t^{1/2}$ (page 603). It is of course important to be able to assign specific values to the mean propagation rate constant k_p and the mean termination rate constant k_t. A number of techniques have been developed for doing so.[b] Of them the most extensively used has been the technique of intermittent illumination as applied to photoinitiated reactions. This has been described briefly in Sec. V.4 and will not be enlarged upon here. In principle it depends on the fact that, when a photosensitive polymerization (or chain) system is exposed to light, the rate at which the stationary-state concentration of radicals is approached (or decays) is determined by the second-order rate of termination of radicals. The rate of consumption of monomer, on the other hand, is first-order in radical

---

[a] The weight fraction of this n-mer at maximum is $8/\bar{n}e^2$, so that the yield of $(n\text{-mer})_{max}$ actually decreases very strongly with increasing mean chain length. The half-width of this unsymmetrical distribution is very broad, since at $n = \bar{n}/2$ the weight fraction of $\bar{n}$-mer is $e/4 = 68$ per cent of maximum n-mer.

[b] See discussion by G. M. Burnett and H. W. Melville in "Techniques of Organic Chemistry," vol. 8, "Investigation of Rates and Mechanisms of Reactions," Interscience Publishers, Inc., New York, 1953.

concentration. In consequence of this, the rate of consumption of monomer will be a direct measure of the average radical concentration.

By measuring the "dark" period time which corresponds to the time required for the radicals to decay, one obtains the average radical concentration (from the rate of loss of monomer) and finally the ratio k_t/k_p.[a] Together with the photostationary measurements of $k_t^{1/2}/k_p$, this allows k_t and k_p to be determined separately. This method is useful only in systems in which termination and propagation have different orders with respect to radicals. Although it studies the nonstationary period of the chain reaction, it studies it in a periodic or stationary way.

Other, more direct methods have been devised (Burnett, loc. cit.) for studying the nonstationary period. Most of them involve complex electronic equipment, because the period is usually very short. However, it can be lengthened by choosing sufficiently low radical concentrations,[b] but this means lower rates of polymerization. If sufficiently sensitive means of observing low polymerization rates are available, single radical decay periods can be studied directly. A method for doing so with simple dilatometric equipment has been described by Benson and North.[c]

Because of the inherently large errors involved in the measurements of radical decay rates, the agreements between different laboratories have not usually been better than a factor of 2. Table XVI.3 presents a compilation of some values obtained recently by rotating-sector and other methods. Because of the large errors in determining k_t/k_p, the activation energies listed are probably not better than ± 1 Kcal/mole, with corresponding errors in the Arrhenius A constants.

The Arrhenius A factors for the propagation reactions are low and of the order one would expect from any of the transition-state theories for a bimolecular reaction between two large molecules (Table XII.2). The activation energies E_p for propagation are also low and of the order observed for similar addition reactions in the gas phase of radicals to a double bond. The values of A_t are in the range to be expected for diffusion-controlled reactions (Sec. XV.2) except for vinyl chloride, which must certainly be in error. As pointed out earlier in discussing diffusion-controlled reactions, it is expected that the activation energies will be of the order of a

---

[a] In practice one uses a rotating sector with adjustable slits so that the periods of light and dark may be varied relative to each other. In addition, by varying the speed of rotation the frequency of cycles is varied. For a fixed light/dark interval ratio, there will be a range of frequencies of light interruption during which the mean rate of polymerization changes from a constant high value to a constant, limiting, low value. The frequency in the middle of this range is inversely proportional to the half-life of the radicals and hence to their rate of decay.

[b] The half-life of radical decay is given by $1/k_t(M\cdot)$. Since k_t is of the order of 10^7 liters/mole-sec, when $(M\cdot)$ is 10^{-8} to 10^{-9} M, this period can be varied from 10 to 100 sec.

[c] S. W. Benson and A. North, J. Am. Chem. Soc., **80**, 5625 (1958).

TABLE XVI.3. PROPAGATION AND TERMINATION CONSTANTS FOR SOME
SIMPLE POLYMERIZATIONS

Monomer	Log A_p	E_p, Kcal	Log A_t	E_t, Kcal	$10^{-4}k_t/k_p$, 30°C[a]
Vinyl acetate[b]	7.6	6.3	9.6	3.2	2.0
Methyl methacrylate[c]	6.0	4.7	8.1	1.2	4.3
Methyl acrylate[d]	8.1	7	10.0	5	0.3
Vinyl chloride[e]	6.5	3.7	13.1	4.2	180
Styrene[f]	7.0	7.3	7.8	1.9	5.0

Rate Constants (liters/mole-sec) are in Arrhenius form: $\log k = \log A - E/4.575T$.

[a] The values of k_t reported by British workers are usually twice those given by Americans because of the difference in definition of k_t.

[b] H. Kwart et al., *J. Am. Chem. Soc.*, **72**, 1060 (1950). W. I. Bengough and H. W. Melville, *Proc. Roy. Soc. (London)*, **A225**, 330 (1954); **A230**, 429 (1955).

[c] M. S. Matheson et al., *J. Am. Chem. Soc.*, **71**, 497 (1949).

[d] *Ibid.*, **73**, 5395 (1951).

[e] G. M. Burnett and W. W. Wright, *Proc. Roy. Soc. (London)*, **A221**, 28, 37, 41 (1954). The values for k_t are almost certainly in error here.

[f] M. S. Matheson et al., *J. Am. Chem. Soc.*, **73**, 1700 (1951).

few kilocalories, roughly parallel to the energies observed for viscosity or diffusion in such systems, and that is the range of observed values for E_t.

There is further evidence that radical termination reactions are diffusion-controlled. For many polymers, the rate of polymerization shows a sudden increase when the fraction of polymer produced reaches values near 15 to 30 per cent.[a] In the case of methyl methacrylate, Matheson et al.[b] found that, at 30°C and 15 per cent conversion, k_t has decreased 160 fold, while k_p has not changed appreciably. Vaughan[c] has proposed a simple diffusion model which is in reasonable accord with the data on styrene polymerization at high conversions.

A similar result is observed in the case of polymers, such as the acrylates,[d] which are relatively insoluble in monomer. At very low conversions (as low as 2 per cent for butyl acrylate), polymer can start to precipitate in the form of a colloidal gel. The polymers are usually crosslinked[e] and might be expected to have higher resistance to diffusion of large radicals.

[a] E. Trommsdorf, "Colloquium on High Polymers," Freiburg, 1944, suggested that this increased rate was due to an increase in viscosity with subsequent decrease of k_t. See also data by Matheson, et al., Table XVI.2, footnote d.

[b] M. S. Matheson et al., *J. Am. Chem. Soc.*, **71**, 497 (1949).

[c] M. F. Vaughan, *Trans. Faraday Soc.*, **48**, 576 (1952); *J. Appl. Chem. (London)*, **2**, 422 (1952).

[d] G. V. Shulz and G. Harborth, *Makromol. Chem.*, **1**, 104 (1947).

[e] That is, these polymers are not simple straight chains, but the chains are instead bonded at a number of points to each other. The dienes such as isoprene and butadiene are most prone to crosslinking because of the residual double bonds left in the final polymer: $CH_2\!\!=\!\!CH\!\!-\!\!CH\!\!=\!\!CH_2 \rightarrow R[CH_2\!\!-\!\!CH\!\!=\!\!CH\!\!-\!\!CH_2]_n R$.

Most recently Benson and North[a] have shown that it is possible to use mixed solvents to vary the viscosity over a considerable range and observe a corresponding change in k_t without affecting k_p. Nozaki[b] has shown that, if an aqueous emulsion of vinyl monomer is photolyzed for a sufficiently long period to form stable particles, these latter will contain long-lived polymer radicals which can continue to react with monomer over periods as long as 24 hr or more.[c] Gelled samples of ethylene dimethacrylate show a paramagnetic resonance spectrum indicating 10^{-4} to 10^{-5} M concentration of species with unpaired spins.[d] These samples were quite stable in the absence of O_2.

One of the big difficulties in unraveling the individual rate constants of polymer reactions is caused by the reaction of the polymer with other species in the solution or even by nonadditive reaction with monomer. Of these reactions, those which lead to termination of the radical chain and the production of a new radical are spoken of as chain transfer reactions. A typical example is found in the polymerization of styrene in CCl_4 solution. The final polymer is found to contain chlorine,[e] and the reaction is

$$M_n \cdot + Cl\!-\!CCl_3 \xrightarrow{k_s} M_n\!-\!Cl + \cdot CCl_3$$

The $\cdot CCl_3$ radical usually goes on to start a new chain, so that there is no loss of radicals in the system, nor even any appreciable change in the rate of polymerization. The chief effect of such transfer reactions is found in the mean degree of polymerization. Since each transfer stops one chain and starts a new one, it will lower the mean chain length. Since the mean degree of polymerization $\bar{n}$ is the ratio of number of monomer units polymerized $-d(M)/dt$ to the number of polymer chains formed, we must modify our previous equation (XVI.10.9) to include chain transfer. If we assume termination by recombination

$$\bar{n} = \frac{\text{rate of propagation}}{\text{rate of chain formation}}$$

$$= \frac{k_p(M \cdot)(M)}{\frac{1}{2}k_t(M \cdot)^2 + \sum_s k_s(S)(M \cdot)} \qquad (XVI.11.1)$$

This can be rewritten as

$$\frac{1}{\bar{n}} = \frac{k_t(M \cdot)}{2k_p(M)} + \sum_s \frac{k_s(S)}{k_p(M)} \qquad (XVI.11.2)$$

[a] S. W. Benson and A. North, *J. Am. Chem. Soc.*, **81**, 1339 (1959).

[b] K. Nozaki, U.S. Patent No. 2,666,025 (1954).

[c] Similar results have been observed with gelled samples of polybutadiene by M. S. Kharasch et al., *Ind. Eng. Chem.*, **39**, 830 (1947). These samples were capable of absorbing butadiene vapors and converting them to a polymer that had the appearance of popcorn, hence their name, "popcorn polymer."

[d] G. K. Fraenkel et al., *J. Am. Chem. Soc.*, **76**, 3606 (1954).

[e] R. A. Gregg and F. R. Mayo, *ibid.*, **70**, 2373 (1948).

or substituting from Eq. (XVI.10.3) for $(M \cdot) = (R_i/k_t)^{1/2}$:

$$\frac{(M)}{\bar{n}} = \frac{(R_i k_t)^{1/2}}{2k_p} + \sum \frac{k_s}{k_p}(S) \qquad (XVI.11.3)$$

This predicts that if $(M)/\bar{n}$ is plotted against the concentration of any added material S at a constant rate of initiation R_i, the result should be a straight line of slope k_s/k_p if a single S has been used. The intercept is of course proportional to $R_i^{1/2} k_t^{1/2}/k_p$ in the absence of S.[a] In the study of transfer in the thermal polymerization of styrene,[b] the ratios k_s/k_p at the highest temperature, 100°C, were found to vary from 0.018 for CCl₄, 0.0018 for C_2Cl_4, and 0.00020 for iso-propylbenzene to 1.6×10^{-5} for cyclohexane. Thus transfer is usually a small effect. With very reactive transfer agents, however, such as CBr₄ (in styrene) for which $k_s/k_p = 1.4$ (60°C) or t-butyl mercaptan, $(CH_3)_3CSH$, $k_s/k_p = 3.6$ at 60°C, the polymerization can be completely inhibited and the net reaction becomes simply addition of the transfer agent to the double bond:[c]

$$\text{RCH}{=}\text{CH}_2 + \text{CBr}_4 \rightarrow \underset{\underset{\text{Br}}{|}}{\overset{\overset{\text{H}}{|}}{\text{R}{-}\text{C}}}\underset{\underset{\text{CBr}_3}{|}}{\overset{\overset{\text{H}}{|}}{\text{C}}}{-}\text{H} \qquad (XVI.11.4)$$

These very active transfer agents, particularly the thiols, have become very important industrially in controlling molecular weight.

When there are available sensitive methods for measuring the rate of disappearance of transfer agent, or else its presence in the final polymer, then the equation for the loss of transfer agent S compared to monomer is (for long chains)

$$\frac{-d(M)/dt}{-d(S)/dt} = \frac{k_p(M \cdot)(M)}{k_s(M \cdot)(S)} = \frac{k_p(M)}{k_s(S)} \qquad (XVI.11.5)$$

or

$$\frac{d \ln (S)}{d \ln (M)} = \frac{k_s}{k_p} \qquad (XVI.11.6)$$

so that a plot of ln (S) vs. ln (M) will give a straight line of slope k_s/k_p.

Transfer to solvent or added modifier is easily observed and measured, but transfer to monomer, polymer, or initiator can also occur. If transfer occurs to initiator, the observation can be treated by the method indicated for Eq. (XVI.11.6). The transfer to monomer is, however, a more subtle

---

[a] In thermal polymerizations when R_i is closely second-order in (M) it is more usual to plot $1/\bar{n}$ against (S)/(M) to give a straight line of slope k_s/k_p and intercept equal to $(2k_ik_t)^{1/2}/k_p$. Here $R_i = 2k_i(M)^2$.

[b] R. A. Gregg and F. R. Mayo, J. Am. Chem. Soc., **70**, 2373 (1948); **75**, 3530 (1953).

[c] M. S. Kharasch et al., Science, **102**, 128 (1945), were the first to demonstrate a 1:1 addition of either CCl₄ or CHCl₃ across the double bond of octene-1 to produce CCl_3—CHCl—$(CH_2)_6$—H and CCl_3—$(CH_2)_7$—H, respectively. This has become an important method of synthesis. The reaction is called telomerization.

phenomenon and not so easily measured. If transfer to monomer occurs, we must modify Eq. (XVI.11.2) to

$$\frac{1}{\bar{n}} = \frac{k_t(\text{M}\cdot)}{2k_p(\text{M})} + \frac{k_{sm}}{k_p}$$

$$= \frac{(R_i k_t)^{\frac{1}{2}}}{2k_p}\frac{1}{(\text{M})} + \frac{k_{sm}}{k_p} \qquad (\text{XVI.11.7})$$

This last equation tells us that, in the absence of transfer agents other than M, the reciprocal mean degree of polymerization is a linear function of $R_i^{\frac{1}{2}}/(\text{M})$.[a] R_i can, of course, be varied at constant (M) by using photo-initiation, whence $R_i \propto I_a$, the absorbed light intensity. Alternatively, where chemical initiation is used, $R_i \propto (\text{In})$, the initiator concentration. In this last case, transfer to initiator must be investigated separately in order to distinguish this effect from monomer transfer.

Transfer to polymer may occur by addition of the growing radical end to a double bond of the polymer chain, or it may occur by H atom abstraction from an active H atom of the chain. H atoms in the α position to a double bond (C=C or C=O) or an ether linkage are easily abstractable and thus lend themselves to polymer transfer reactions. The transfer reaction to polymer may be intramolecular or it may be intermolecular. Both have been demonstrated. In the case of intermolecular transfer the demonstration has been made of the production of "graft" polymers. The method is to polymerize a monomer in the presence of "inert" polymer of a different composition.[b] The final product will contain the inert polymer with the new polymer "grafted" onto it.

Transfer to polymer leads to both an increase in molecular weight and to what is termed *branching*. That is, the newly activated polymer chain is no longer linear but has a polymer branch growing from it. Branched polymers have physical properties which can be quite different from linear polymers, and many methods are available for measuring the amount of branching in a given polymer. With sufficient branching, crosslinking to form a three-dimensional network can occur. Such polymers tend to be relatively insoluble and form gels. Dienes are quite susceptible to such gel formation, as are also the acrylates.[c]

Although the process of transfer usually leads to lower degree of polymerization with no change in rate of polymerization, that depends entirely on the reactivity of the new radical produced by transfer. If the latter is

[a] A plot of $1/\bar{n}$ vs. $1/(\text{M})$ at constant R_i will then give a line of intercept equal to k_{sm}/k_p.

[b] R. B. Carlin and N. E. Shakespeare, *J. Am. Chem. Soc.*, **68**, 876 (1946), polymerized *p*-chlorostyrene in the presence of polymethyl methacrylate and succeeded in grafting the polymer of the former into the latter.

[c] See Flory, *op. cit.*

sufficiently unreactive, then the new radicals will not continue polymerization but instead will enter into termination reactions with other polymer radicals:

$$M_n \cdot + SH \overset{k_s}{\rightarrow} M_nH + S \cdot$$

$$M_n \cdot + S \cdot \overset{k_{ts}}{\rightarrow} M_nS$$

Species which can do this are termed retarders or inhibitors. Quinones are good examples of inhibitors. The reaction in this case appears to proceed principally via addition to the O atom to form a semiquinone which then terminates a second radical to form a diether.[a]

$$(XVI.11.8)$$

Sulfur, S_8, O_2, nitrobenzenes, and quinones all show varying degrees of inhibition or retardation, the use of either term depending on the relative inertness of the new radical to further addition of monomer.

A final case of transfer which is of great interest occurs when the transfer species is another monomer, itself capable of polymerization. In such a case, when the reactivities of the two monomers and their respective radicals are comparable, it is possible to produce what is referred to as copolymer: a linear chain that contains varying amounts of the two monomers in a proportion which depends on their concentrations in the initial reaction mixture and their relative rate constants. A great deal of work has been done in the investigation of copolymerizations, and the studies are particularly interesting because they yield values for the relative reactivities of monomers and their radicals. The equation that governs the relative depletion of two monomers M and N is

$$\frac{d \ln (M)}{d \ln (N)} = \frac{r_M(M) + (N)}{(M) + r_N(N)} = \frac{r_M + (N)/(M)}{1 + r_N(N)/(M)} \quad (XVI.11.9)$$

$r_N = k_{MM}/k_{MN}$ is the ratio of rates of reaction of an $M \cdot$ type radical with M and N, respectively, and $r_N = k_{NN}/k_{NM}$ is the similar ratio for $N \cdot$ radicals with N and M, respectively.

12. Structure and Reactivity

An ultimate goal for the science of chemical kinetics is to relate specific rate constants to molecular properties. In principle such a relation is

[a] J. W. Breitenbach and A. J. Renner, *Can. J. Research*, **28B**, 509 (1950). S. G. Cohen, *J. Am. Chem. Soc.*, **67**, 17 (1945); **69**, 1057 (1947).

already contained in the time-dependent Schroedinger equation, together with present theories of electricity and magnetism. Prominent scientists have argued that from such a point of view there can be no new "basic" laws of chemistry, but only secondary relationships at a more complex level of molecular structure. Nature being an open system, it is hardly profitable to discuss the truth or falsity of such a position. Pragmatically speaking, however, it is certain that even with the advent of high-speed computers, the possibility of obtaining solutions to kinetic problems from the Schroedinger equation with even order of magnitude accuracy is quite remote. The reasons for this apparent stalemate lie in the fact that, if we write a rate constant in terms of a free energy of activation, then to predict k to within a factor of 10 we must be able to predict $\Delta F^{\ddagger}$ to within 1.4 Kcal at 300°K. This is 0.06 ev, and even the most optimistic will agree that a priori calculation to within this uncertainty via quantum mechanics would be quite revolutionary.[a] Thus we must take a much less sanguine approach toward such ultimate clarity and for the foreseeable future be content to look for the "secondary" relations between kinetic parameters and molecular properties. There have been a large number of such relations proposed; some have been frankly empirical and some have been placed in the rather peculiar category of "semiempirical."[b] Some of them have been quite useful in correlating data (e.g., the Brönsted-Pedersen relation for generalized acid-base catalysis), and others are more or less qualitative in nature. It is of interest to look for a moment at some of the observations which have been made and which need explaining.

In the field of inorganic chemistry one of the very challenging problems has been the understanding of the stability of ligancy of complex ions. Thus Cr^{3+} and Co^{3+} will usually exchange ligands from their solvent shells, including H_2O, only very slowly. On the other hand, Al^{3+} and Fe^{3+} will exchange complexing ligands such as H_2O and Cl^- very rapidly. As we have already seen, such behavior is closely linked to the problem of the rates of redox reactions and charge transfer. The connection is not unique, however, since such complexes as $Fe(CN)_6^{3-}$ and $Fe(CN)_6^{4-}$, whose ligands are very inert, undergo charge transfer quite readily. Taube[c] has given a qualitative solution to such problems in terms of an orbital model of the valence shell of the ions. What appears to be a more quantitative approach has been recently attempted in terms of the interaction of the electric field

---

[a] Thus the best work on the computation of the bond energy of H_2 has given accuracy of about 0.1 ev. Present methods of computation are in error by about 2 ev when applied to H_2 bond energy and improve to an error of 1 ev only with great labor. See any texts on quantum mechanics.

[b] The implication behind such nomenclature is that there is a formal chain of reasoning proceeding from first principles and leading to the final relation, with one or only a few *ad hoc* assumptions included to maintain the continuity of logic.

[c] H. Taube, *Chem. Revs.*, **50**, 69 (1952).

of the ligands on the relative energies of otherwise equivalent orbitals of the central ion.[a]

In the field of organic chemistry there has developed an enormous empirical art for the relation of chemical reactivity and molecular structure.[b] However, much of this is qualitative and lends itself rather poorly to quantitative prediction. There are, however, a number of attempts which have been fairly successful in correlating large bodies of information. Most of them fall into a common category with the Brönsted-Pedersen relation for generalized acid-base catalysis; they can be classified as linear–free energy relations. Like the Brönsted-Pedersen relation, the basis for these correlations rests on the assumption that, when a given molecule is involved in two similar equilibrium processes, the relative free energy changes in both processes will be affected similarly by changes in structure.

A very striking example of such a relation has been given for aromatic compounds. Mackor et al.[c,d] were able to measure the base strengths K_B of aromatic compounds Ar in concentrated, strong acids (for example, H_2SO_4 or HF):

$$Ar + H^+ \rightleftharpoons ArH^+ \qquad K_B \qquad (XVI.12.1)$$

They also measured the rate of D exchange of such aromatics with the solvent[e] and were able to show that a plot of $\log k_{ex}$[f] vs. $\log K_B$ gave a straight line over a range of many powers of 10 in both constants for a series of substituted aromatics. This is, of course, to be expected, because it is quite likely that D exchange involves formation of the conjugate acid of the hydrocarbons. More striking, however, was their demonstration[d] that a plot of $\log K_B$ vs. $\log k_r$, where k_r is the rate constant for the addition of free radicals to the aromatic,[g] also followed a straight line over 16 units of pK_B and 6 units of pk_r!

In this case the implication of such an agreement is that for the two processes, one rate, k_r, and the other ionization equilibria, K_B, the free energy charges ($\Delta F_r^{\ddagger}$ and ΔF_B°) are similarly affected by changes in structure. This can be stated quantitatively as

[a] See discussion by W. Moffitt and C. J. Ballhausen, *Ann. Rev. Phys. Chem.*, **7**, 107 (1956), for a summary of what has become popularly referred to as "crystal field theory." Applications to inorganic chemistry will be found in F. Basolo and R. G. Pearson, "Mechanisms of Inorganic Reactions," John Wiley & Sons, Inc., New York, 1958.

[b] See for example Ingold's "Structure and Mechanism," *op. cit.*, and E. A. Braude and F. C. Nachod, "Determination of Organic Structures by Physical Methods," Academic Press, Inc., New York, 1955.

[c] E. L. Mackor et al., *Trans. Faraday Soc.*, **53**, 1309 (1957).

[d] Mackor et al., *ibid.*, **54**, 66 (1958).

[e] Mackor et al., *loc. cit.*

[f] k_{ex} is the first-order exchange rate constant.

[g] The radicals in this case were CH_3 radicals as measured by M. Szwarc et al. [M. Levy and M. Szwarc, *J. Am. Chem. Soc.*, **77**, 1949 (1955)] or CCl_3 radicals as measured by E. C. Kooyman and E. Farenhorst, *Trans. Faraday Soc.*, **49**, 58 (1953). Both sets of data gave good lines.

$$F_{r_2}^{\pm} - F_{r_1}^{\pm} = \rho(\Delta F_{B_2}^{\circ} - \Delta F_{B_1}^{\circ})$$ (XVI.12.2)

where the left-hand side represents the difference in free energies of activation of two aromatics, the corresponding term on the right-hand side represents the differences in free energies of base ionization of the same two aromatics, and ρ is a constant. Since the free energies can be related to the corresponding rate or equilibrium constants (Sec. XVI.3) by $\Delta F^{\circ} = -RT \ln K$, etc., we can write this last equation as

$$\log \frac{k_{r_2}}{k_{r_1}} = \rho \log \frac{K_{B_2}}{K_{B_1}}$$ (XVI.12.3)

We might expect that the constant of proportionality ρ will depend on the reactions, solvent, temperature, etc., but not on structure. The effects of structure are represented in the ratios K_{B_2}/K_{B_1} and k_{r_2}/k_{r_1}.

If we take a series of compounds which are related by structural substitutions, such as benzene derivatives, then we can assign the effects of the structural changes to the ratio K_{B_2}/K_{B_1} if we select this property as a standard. Such a system of correlation was first proposed by Hammett[a] and revised and extended by Jaffé[b] to account for the effects of meta and para substituents on the reactivity of benzene derivatives. For convenience, the ionization constant of benzoic acid in aqueous solution at 25°C was chosen as standard and for each meta or para substituent a, a value of σ_a was assigned, where σ_a is equal to log (K_a/K°), K° being the ionization constant for benzoic acid and K_a the ionization constant of the corresponding, substituted benzoic acid.

Jaffé has then shown that for any related rate or equilibrium process, the Hammett equation can be written as

$$\log \frac{k_a}{k^{\circ}} = \rho\sigma_a$$ (XVI.12.4)

or for multiple substitution

$$\log \frac{k_i}{k^{\circ}} = \rho \sum_i \sigma_i$$ (XVI.12.5)

where the effects of multiple substitution in a benzene ring turn out to be additive. These relations have proven very useful in correlating the rates of hydrolysis of aromatic esters in water or alcohol solutions,[c] the ionizations of substituted phenols and thiophenols, etc. (see Jaffé, loc. cit.).

The method does not appear to apply to aliphatic compounds, the acid-base properties of substituents of which are not easily transmitted through the chain, nor does it apply to ortho substituents, because steric effects

[a] L. P. Hammett, Chem. Revs., 17, 125 (1935), Trans. Faraday Soc., 34, 156 (1938).

[b] H. H. Jaffé, Science, 118, 246 (1953); Chem. Revs., 53, 191 (1953).

[c] For each different type of reaction and each change in solvent or temperature, a new value of ρ is needed.

and specific interactions not involving the ring then become important.[a]

Analogues of these relations have been proposed by Winstein and Grunwald[b] to correlate the effect of solvent on S_N1 nucleophilic displacements on carbon. Their equation has the form

$$\log \frac{k_a}{k^\circ} = m\text{Y} \qquad \text{(XVI.12.6)}$$

where k_a and k° are the first-order rate constants for S_N1 solvolysis of a given compound in solvent a and in 80 per cent EtOH—HOH, respectively, $\text{Y} = \log$ of the same ratio for t-butyl chloride in the same solvents, and m is a constant characteristic of the compound but independent of solvent. This equation has been quite successful in the case of a number of solvents.[c]

Similar equations have been proposed by Swain et al.[d] to correlate all nucleophilic displacements on carbon. However, the number of parameters has grown to four, and the usefulness of such a system of bookkeeping is open to question.[e]

It is interesting to note that no such linear correlations have been found between the other thermodynamic functions ΔH° or ΔS° or the corresponding $\Delta H^\ddagger$ and $\Delta S^\ddagger$. This is undoubtedly a reflection of the more erratic behavior of these functions (Table XV.4) and the way in which changes in them tend to cancel each other in yielding ΔF°.[f] However, it makes very suspect many of the so-called basic interpretations of the correlation constants which have been given.

[a] It is interesting to note that the σ constants for different substituent groups are crudely interpretable in terms of their relative affinity for electrons. However, a good many of the interpretations that have been made must be taken with considerable reserve, particularly when the constant ρ is so low that the total range of substituent behavior is confined to 1 pK unit or less.

[b] E. Grunwald and S. Winstein, *J. Am. Chem. Soc.*, **70**, 846 (1948). Winstein et al., *ibid.*, **73**, 2700 (1951).

[c] C. G. Swain and C. B. Scott, *ibid.*, **75**, 141 (1953). C. G. Swain and R. B. Mosely, *ibid.*, **77**, 3727 (1955). Swain et al., *ibid.*, **77**, 3731 (1955). See, however, criticisms by S. Winstein et al., *ibid.*, **79**, 4146 (1957).

[d] *Ibid.*

[e] An extensive review of free energy correlations is given by R. W. Taft, Jr., et al., *Ann. Rev. Phys. Chem.*, **9**, 287 (1958).

[f] J. E. Leffler, *J. Org. Chem.*, **20**, 1202 (1955), discusses the separate H, S, F, relationships. S. Winstein and A. H. Fainberg, *J. Am. Chem. Soc.*, **79**, 5937 (1957), have analyzed the effects of solvent changes on $\Delta F^\ddagger$, $\Delta H^\ddagger$, and $\Delta S^\ddagger$ and have shown them to be quite complex and frequently to depend as much on changes in solvation of reactants as on solvation of the transition state.

XVII

Heterogeneous Reactions

1. Catalysis by Solids

Of all the aspects of chemical kinetics, there is perhaps none which has captured popular fancy so extensively as has the subject of catalysis. The concept of a specially active material which in small amounts can produce gross and specific accelerations of the rates of chemical reactions without itself being consumed in the process has a magical appeal which might be classed by the layman as the chemical analogue of perpetual motion. This appeal is considerably strengthened by the enormous industrial importance which catalysts have had. Some classic examples[a] are the use of iron and the transition metals in the famous Haber synthesis of NH_3 from N_2 and H_2, the use of finely divided platinum in the synthesis of SO_3 (for H_2SO_4) from SO_2 and O_2, the use of silica-alumina gels in the catalytic cracking of petroleum, and more recently, the use of cobalt catalysts in the Fischer-Tropsch synthesis of hydrocarbons from $CO + H_2$.[b]

How does a catalyst affect a chemical reaction? If we maintain the restriction that a catalyst is not appreciably consumed in a chemical reaction, then it can be shown thermodynamically that its role in the reaction cannot be to change the ultimate equilibrium point. Its role is restricted to one of accelerating the rate of approach to equilibrium. However, in most chemical systems, there are many metastable compositions intermediate in free energy between reactants and the state of ultimate equilibrium. We can describe the "specificity" of catalysts in terms of

---

[a] An extensive bibliography of work in industrially important catalysts is given by W. B. Innes in P. H. Emmett (ed.), "Catalysis," vol. 2, chap. 1, Reinhold Publishing Corporation, New York, 1955.

[b] Strong arguments can be made to the effect that World Wars I and II would have been impossible or of much shorter duration without the commercial realizations of the Haber process and the Fischer-Tropsch synthesis, respectively.

their property of accelerating the rate of approach to one of these intermediate states rather than a general, over-all acceleration toward the lowest state.

Because a catalyst affects the rate of reaction and not the ultimate equilibrium, it is not possible to give a general, kinetic description of catalyst behavior. Instead, a proper discussion of catalytic behavior can be made only in terms of mechanism, which is, of course, unique for any given reaction. However, some general classification of catalysts is possible in terms of structure in relation to type of reaction mechanism involved. A useful classification of solids for this purpose is as follows:

Molecular Solids. These are substances such as solid argon or solid CO_2 held together by relatively weak Van der Waals forces or weak polar forces. They are not expected to have significant attractions for gaseous or liquid reactants and thus no important catalytic properties.

Covalent Solids. These are substances such as graphite, diamond, and quartz in which the atoms are bonded to nearest neighbors by covalent linkages forming a macromolecular, two- or three-dimensional network. Atoms at the surfaces and edges of such crystals may be chemically unsaturated and can thus act as centers for initiating free radical or redox reactions.

Ionic Solids. Substances such as NaCl which have a periodic lattice of positive and negative ions. The surfaces and edges of such solids are sites of very intense electric fields, and we may expect that ionic and polar reaction will be facilitated there.

Metals. These are similar to ionic solids, since they contain a lattice of positively charged metal cations. The anions are electrons which differ from anions in salts like NaCl by being mobile. The mobility of the electrons in metals permits metal surfaces to act as very active centers for both free radical and ionic reactions.

Semiconductors. Substances such as ZnO, ZnS, and PbS, while not very good conductors, have electrons which can be thermally excited at very low activation energies (for example, 10 to 20 Kcal) to give electronic conduction.[a] The surfaces and edges of such solids are good centers for redox and possibly free radical reactions.

In considering the possible activity of a given catalyst it is important to note that the mere existence of centers of chemical unsaturation at the surface of the solid is not enough to guarantee catalytic activity. If these centers of activity are sufficiently high in free energy, they will tend to stabilize themselves by forming permanent chemical bonds either with

[a] Doubly charged ions such as $O^=$ and $S^=$ are not stable in vacuo. They are stable in crystals only because of the additional electrostatic attraction furnished by the lattice. At the surface and edges of the crystals a good fraction of the stabilization is lost, and it is likely that $O^=$ ions can then act as effective reducing agents.

the reactants or with any extraneous species which are then equivalent to catalyst poisons.[a] For effective catalytic behavior, unsaturated centers must be capable of forming relatively weak or "labile" bonds with the reactants. This in effect is a valid interpretation of the "catalytic" properties of H_2O (acting as a solvent) in facilitating ionic reactions.

While we have been talking of the surface of a solid catalyst as the site of the catalytic reactions, it is by no means clear that such is always the case. Most crystalline solids are polycrystalline in structure, the interfaces between microcrystals providing many possible sites for catalytic reaction. Amorphous solids such as many metallic hydroxides and oxides may have pores, molecular voids, and irregular surfaces whose accessibility for chemical reaction will depend strongly on the nature of the reactants and the conditions of the experiment.[b]

A further problem arises from the fact that catalysts "work" by forming chemical bonds with the substrate molecules. Thus metals form hydrides with H_2 and H donors such as CH_4 or NH_3, or oxides with O_2 and O donors such as CO_2 or NO_2. It is presumably these surface compounds that play the role of active intermediates in the catalytic reaction. Under slightly different conditions, however, it is possible to extend the catalyst-adsorbate reaction to produce bulk compounds, e.g., hydrides and oxides. In view of this ability of the surface phase to propagate into the bulk in many instances, it is not at all clear that only the "surface" of the solid catalyst is active in the reaction. Such complexities add enormously to the difficulty of interpreting the kinetic data from heterogeneously catalyzed reactions.

2. Initiation of Chains at Solid Surfaces

If we assume for simplicity that the active centers in solid catalysts are confined to their surfaces, then we can picture a number of different pathways for reaction.

[a] There is good evidence now for believing that the poisoning of the silica-alumina gel catalysts used in cracking petroleum represents the neutralization of very strong active acids on the catalyst surface by bases (poisons) such as S^- ion and amines. The acid sites may be H_2O molecules bound to metal ions at the surface. See R. C. Hansford, "Advances in Catalysis," vol. 4, chap. 1, Academic Press, Inc., New York, 1952.

[b] H_2, H_2O, CO, CO_2, NO, and many other gases can be strongly sorbed in such "solids" as glass, quartz, and metals. The quantity of such gases which can be "desorbed" by baking out a glass or metal system in high vacuum is many times greater than that which would correspond to a unimolecular layer adsorbed on the "apparent" geometrical surface. A discussion of such behavior is given by R. M. Barrer, "Diffusion in and Through Solids," Cambridge University Press, New York, 1951. In some zeolites (clays) there are channels of molecular size extending through the silicate structure. These channels are sufficiently small to act as "molecular" sieves, i.e., permit passage of n-butane but not isobutane. A. Wheeler, "Catalysis," vol. 2, chap. 2, op. cit., discusses the quantitative relation between rate of catalysis and surface accessibility.

Surface Initiation or Termination. The surface acts to initiate or terminate radicals or ions which diffuse out into the homogeneous phase, gas or liquid, where a chain reaction takes place. The behavior is certainly characteristic of the activity of glass and quartz surfaces in a great many chain reactions, particularly gas-phase oxidations such as $H_2 + O_2$. When initiation takes place rapidly at the surface, it is likely that termination is also effective there.[a] Such systems can be recognized by the fact that they react at specific rates which are nearly independent of the surface/volume ratio, i.e., zero order with respect to surface or catalyst.

Reaction Entirely at Surface. The rate of reaction is now proportional to the extent of surface, or to the amount of catalyst. Two subcategories of such reactions have been proposed. In the first, all reactions take place between species adsorbed on the surface. These are frequently referred to as Langmuir-Hinshelwood mechanisms. In the second category are those reactions which take place at the surface by reaction between a sorbed species and one from the homogeneous phase. These are sometimes referred to as Rideal mechanisms. The distinctions between the two types are not always clear.

To see how these mechanisms can operate, let us consider a chain reaction in which chain carriers are initiated at the surface and carried into the homogeneous phase where the chain propagates. One of the simplest examples of this type would be a halogenation reaction such as $RH + Br_2 \rightleftharpoons HBr + RBr$, where Br atoms are formed at the surfaces of the solid present.

Ignoring for the moment the question of the homogeneous chain, let us concern ourselves only with the mechanism of Br atom formation at the surface. In the absence of a surface, Br atoms could be formed in the gas phase by the process

$$Br_2 + M \underset{2'}{\overset{1'}{\rightleftharpoons}} 2Br + M \qquad \text{(XVII.2.1)}$$

If we represent by S the unoccupied, active sites on the surface, a competing surface reaction could be

$$S + Br_2 \underset{2}{\overset{1}{\rightleftharpoons}} S \cdot Br + Br$$

$$S \cdot Br \underset{4}{\overset{3}{\rightleftharpoons}} S + Br \qquad \text{(XVII.2.2)}$$

where $S \cdot Br$ represents the surface formation of sorbed Br atoms. If we now apply the steady-state treatment to this system for the minor species $S \cdot Br$, we find for its stationary concentration

$$\frac{(S \cdot Br)}{S^\circ} = \frac{k_1(Br_2) + k_4(Br)}{k_1(Br_2) + k_2(Br) + k_3 + k_4(Br)} \qquad \text{(XVII.2.3)}$$

[a] When the active centers are unchanged by the reaction, the principle of microscopic reversibility makes this a necessary conclusion.

where $S°$ is the number of active sites per unit area initially present. The rate of production of Br atoms is then given by

$$\frac{d(\text{Br})}{dt} = \frac{2k_3k_1S°(\text{Br}_2)[1 - (\text{Br})^2/(\text{Br})_e^2]}{k_1(\text{Br}_2) + k_2(\text{Br}) + k_3 + k_4(\text{Br})} \qquad (\text{XVII.2.4})$$

Here $(\text{Br})_e = K_{eq}^{\frac{1}{2}}(\text{Br}_2)^{\frac{1}{2}}$ is the concentration of Br atoms present at equilibrium. Note that $K_{eq} = k_1k_3/k_2k_4$. This equation simplifies in either of two extreme cases.

Case 1. Sparsely covered surface, $(\text{S·Br}) \ll (\text{S})$. Since at equilibrium $k_3(\text{S·Br}) = k_4(\text{S})(\text{Br})_e$ and $k_1(\text{S})(\text{Br}_2) = k_2(\text{S·Br})(\text{Br})_e$, this implies that $k_3 \gg k_4(\text{Br})_e > k_4(\text{Br})$ and $k_2(\text{Br})_e \gg k_1(\text{Br}_2)$. If we also assume that $k_3 > k_2(\text{Br})_e$, then Eq. (VII.2.4) reduces to

$$\frac{d(\text{Br})}{dt} \to 2k_1S°(\text{Br}_2)\left[1 - \frac{(\text{Br})^2}{(\text{Br})_e^2}\right] \qquad (\text{XVII.2.5})$$

where the negative term in brackets accounts for the back reaction. This case represents surface attack on Br_2 as the rate-limiting case, most of the S·Br formed decomposing rapidly after formation via step 3. The rate constant k_1 can have an activation energy equal to the heat of reaction 1, which in turn is less than the bond dissociation energy of Br_2 by the heat of formation of the S—Br bond. If this latter is in the range of 8 to 15 Kcal, the rate of the surface reaction can be faster than the homogeneous gas-phase reaction 1'. Note that the back reaction will be second-order in Br atoms and will have a negative activation energy equal in magnitude to $D(\text{Br—Br}) - E_1$.

Case 2. Surface nearly saturated, $(\text{S·Br}) \gg (\text{S})$. This implies that $k_3 \ll k_4(\text{Br})_e$ and $k_1(\text{Br}_2) \gg k_2(\text{Br})_e > k_2(\text{Br})$. If we further assume that $k_2(\text{Br}) > k_4(\text{Br})$, so that dissociation of S·Br is the rate-determining step, then

$$\frac{d(\text{Br})}{dt} \to 2k_3S°\left[1 - \frac{(\text{Br})^2}{(\text{Br})_e^2}\right] \qquad (\text{XVII.2.6})$$

The rate is now zero-order in (Br_2) with an activation energy which may be as little as $\frac{1}{2}D(\text{Br—Br})$. The reverse reaction will now be inversely proportional to (Br_2), with a fairly high, negative activation energy and still second-order in Br atoms. An alternative case is possible when $k_4(\text{Br}) > k_1(\text{Br}_2)$. We find

$$\frac{d(\text{Br})}{dt} \to \frac{2k_1k_3}{k_4}S°\frac{(\text{Br}_2)}{(\text{Br})}\left[1 - \frac{(\text{Br})^2}{(\text{Br})_e^2}\right] \qquad (\text{XVII.2.7})$$

When (Br) is near its equilibrium value of $[K_{eq}(\text{Br}_2)]^{\frac{1}{2}}$, this can be written as

$$\frac{d(\text{Br})}{dt} \cong 4k_2K_{eq}^{\frac{1}{2}}S°(\text{Br}_2)^{\frac{1}{2}}\left[1 - \frac{(\text{Br})}{(\text{Br})_e}\right] \qquad (\text{XVII.2.8})$$

where, if the exothermic reaction 2 has a small or zero activation energy, the over-all activation energy which is $E_2 + \frac{1}{2}D(\text{Br}\!-\!\text{Br})$ can be close to $\frac{1}{2}D(\text{Br}\!-\!\text{Br})$. The rate is now $\frac{1}{2}$-order in Br_2, while the reverse reaction Eq. (XVII.2.7) is first-order in Br atoms, with a small, positive activation energy equal to E_2.

This last case is usually observed in the surface-catalyzed recombination of radicals which are homogeneously produced in large concentrations by photolysis or electric discharges. The surface tends to saturate with atoms, and the rate-determining step in their disappearance is reaction 2.

It is important to note that initiation at catalyst surfaces of chain reactions can never give radical concentrations higher than those at equilibrium. In effect the surface can eliminate what is referred to as the induction period of a chain reaction.[a] These can frequently be long enough to completely inhibit the homogeneous reaction even when the chain is fast. The heterogeneous reaction, however, has a considerable handicap compared to the homogeneous reaction owing to the fact that the former is confined to the surface. We can see this very simply by comparing collision frequencies of a molecule with a wall and with another molecule M in the gas phase. This ratio of collision frequencies is (Secs. VIII.6 and VIII.8)

$$\frac{Z_w'}{Z_g'} = \frac{\frac{1}{4}\bar{c}S^\circ/V}{\sqrt{2}\pi M\sigma^2\bar{c}} = \frac{S^\circ/V}{4\pi\sqrt{2}\sigma^2 M} \qquad (\text{XVII.2.9})$$

If we employ average values for molecular collision cross sections of about 20×10^{-16} cm^2 this reduces to (M in atm):

$$\frac{Z_w'}{Z_g'} \simeq \frac{1.2 \times 10^{-6}}{M} \frac{S^\circ}{V} \qquad (\text{XVII.2.10})$$

For the Br_2 reaction [Eqs. (XVII.2.1) and (XVII.2.2)] this unfavorable ratio is further reduced by the fact that the steric factor of reaction $1'$ is about 10^3, while that for reaction 1 will be about 10^{-2}. Thus unless the surface/volume ratio is of the order of 10^{10} cm^{-1} (it is usually near unity for most reaction flasks) or the activation energy at the surface is less by 15 Kcal or more (at 300°K), the surface reaction will not be important.[b] These rule-of-thumb calculations will be found generally applicable in comparing the speeds of homogeneous and heterogeneous reactions.[c]

3. Chemisorption at Solid Surfaces

We shall now consider the case of a catalyzed reaction that takes place exclusively at the surface of the solid catalyst. The over-all reaction can be thought of as proceeding in the following stages:

[a] Sections III.10 and XIII. 6.

[b] Reduced pressures also favor the surface reaction, of course.

[c] Note in this regard that the effective areas of most finely divided commercial catalysts are in the range 10 to 200 m^2/gram, or about 10^6 cm^2/gram. This goes a long way toward overcoming the unfavorable ratio of collision frequencies.

1. Reactants diffuse from a homogeneous phase to the catalyst surface.
2. The reactants are adsorbed on the catalyst surface.
3. Chemical reaction takes place between adsorbed species.
4. Product molecules are desorbed from the catalyst surface.
5. Desorbed product molecules diffuse away from the catalyst surface.

The diffusion of molecules to and from a solid catalyst surface will usually be rapid for gases and slow for liquids. For the latter, depending on the catalyst pore size and accessibility, the over-all rate can become diffusion-controlled (i.e., steps 1 and 5). That is seldom the case for gases. For the moment let us restrict our attention to those catalytic processes whose rates are controlled by steps 2, 3, and 4. Two models for the nature of the sorbed "layer" of reactants at the surface are in common use. In one of them it is assumed that the sorbed layer is loosely bound to the surface and can migrate relatively freely from one surface site to another. The mobile layer can be pictured in the extreme case as representing a "two-dimensional" gas sorbed on the surface. Opposed to this is the model of a strongly bound surface layer such that each sorbed molecule can be thought of as forming a chemical bond with some atom in the catalyst surface. In such a "localized" layer, the migration of reactants may take place slowly either by surface diffusion or by evaporation and reabsorption. The relative slowness of these latter processes can make them rate-controlling.

It is interesting to note that since sorption will be exothermic for most processes, rates of sorption will usually exceed rates of desorption. This means that product molecules in the homogeneous phase will usually be in equilibrium with the sorbed phase. That is not necessarily true for reactants when sorption, being in many cases a chemical reaction with the surface atoms, may have an activation energy and be quite slow.

When the sorbed layer is very loosely bound, as evidenced by the pressure and temperature range in which sorption equilibrium is reached, the process is referred to as "physical adsorption." It is characterized by a rapid and reversible attainment of equilibrium with the gas phase, and the measured heats of adsorption are found to be of the order of magnitude of the heats of liquefaction of the adsorbate. The temperature range in which such sorption occurs lies well below the critical temperature of the adsorbate and is large generally in the vicinity of its boiling point. The forces responsible for physical adsorption are probably the same as those responsible for liquefaction or mixing of two liquids and are subsumed under the heading of Van der Waals forces. The adsorbate can form multimolecular layers on the adsorbent surface at pressures sufficiently close to its vapor pressure at the experimental temperature. At the saturation pressure, the solid surface is simply wetted by liquid.

There has been extensive study of the process of physical adsorption,[a,b] and it has been of great importance in the characterization of catalyst surfaces, particularly in measurements of the "surface area" of catalysts via the measurement of the amount of adsorbate required to form a unimolecular film.[c] However, it is not likely that these weak Van der Waals forces are important in chemical catalysis.

There is a second process referred to as "chemisorption" which is usually distinguishable experimentally from physical adsorption. It is usually a much slower process than physical adsorption and frequently exhibits an increased rate with increasing temperature. Chemisorption is usually irreversible, the desorption process taking place very slowly and requiring higher temperatures. This is a reflection of the much higher heats which are involved in chemisorption; they may be of the order of 10 to 100 Kcal/mole, well in the range of heats of chemical reaction.[d]

The sorption of H_2 on metals which can take place at both high and low temperatures (for example, -180 to $500°C$) is accompanied by dissociation of the H_2 to H atoms and is undoubtedly a chemical process of metal hydride formation on the metal surface. The sorption of O_2 on charcoal, CO, and N_2 on transition metals and of olefins on metals are all accompanied by heat evolutions in the neighborhood of 30 to 100 Kcal[e] and are undoubtedly better viewed as chemical reactions than as loose solvation. For these reasons we shall direct our attention to the process of chemisorption in discussing catalytic reactions.

The formation of a localized monolayer by sorption of reactant molecules A on active sites S can be pictured as a chemical reaction as in the previous section:

[a] S. Brunauer, "The Adsorption of Gases and Vapors: Physical Adsorption," Princeton University Press, Princeton, N.J., 1943.

[b] G. D. Halsey, *Advances in Catalysis*, **4** (1952), presents an excellent critique of the various theories of adsorption in the light of surface heterogeneity. T. L. Hill in the preceding chapter outlines the statistical mechanical and thermodynamics of the different sorption theories.

[c] The BET (Brunauer-Emmett-Teller) theory of multimolecular adsorption has been the pioneering work in this regard.

[d] An example of the range is provided by the sorption of gases on finely divided, crystalline proteins, S. W. Benson and D. A. Ellis, *J. Am. Chem. Soc.*, **70**, 3563 (1948); **72**, 2095 (1950); Benson et al., *ibid.*, **72**, 2102 (1950); S. W. Benson and J. M. Seehof, *ibid.*, **73**, 5053 (1951); **75**, 3925 (1953). "Inert" gases such as N_2, A, and CH_4 in the neighborhood of their respective boiling points are physically adsorbed with heats of sorption of the order of 2 to 3 Kcal. Polar gases such as H_2O and HCl are chemisorbed with heats of sorption for HCl corresponding to the formation of solid hydrochlorides. The polar gases show pronounced physical adsorption in addition to chemisorption at the lower temperatures.

[e] See discussions and papers presented at the Faraday Society, *Discussions Faraday Soc.*, **8** (1950).

$$S + A \underset{2}{\overset{1}{\rightleftharpoons}} A \cdot S \qquad \text{(XVII.3.1)}$$

where $A \cdot S$ represents the sorbed molecule A as a chemical compound. The equilibrium constant for this reaction can be written[a]

$$K_{1.2} = \frac{(A \cdot S)}{(A)(S)} \qquad \text{(XVII.3.2)}$$

If we represent the initial number of sorption sites as $S° = (A \cdot S) + (S) = \text{const}$, we can rewrite the equilibria in terms of the fraction of sites covered, θ:

$$\theta = \frac{(A \cdot S)}{S°} = \frac{(A)K_{1.2}}{1 + (A)K_{1.2}} = \frac{(A)}{K_{2.1} + (A)} \qquad \text{(XVII.3.3)}$$

This expression is known as the Langmuir isotherm.[b] It predicts that, at sufficiently high pressures $[(A) \gg K_{2.1} = 1/K_{1.2}]$, the surface will become saturated with $A(\theta = 1)$, while at low pressures $[(A) \ll K_{2.1}]$, the coverage by A will be proportional to (A). This type of behavior is shown

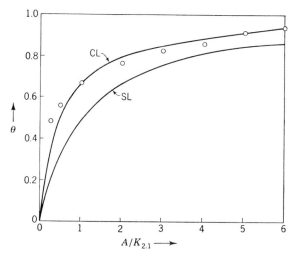

FIG. XVII.1. Various isotherms for gas sorption. Fraction of surface covered θ as function of sorbate concentration (A), measured in units of $K_{2.1}$. Curve SL for simple Langmuir isotherm; homogeneous surface Eq. (XVII.3.3). Curve CL for complex Langmuir isotherm; heterogeneous surface Eq. (XVII.3.5) ($K_M = K_{2.1}$; $K_0 = 0.1 K_M$). o Points for Freundlich isotherm: $\theta = 0.665 \left(\dfrac{A}{K_{2.1}}\right)^{0.194}$. Eq. (XVII.3.6) fitted at $A/K_{2.1} = 1$ and 5.

[a] In principle we should use activities rather than concentrations. Corrections for activities can be made to Eq. (XVII.3.2).

[b] In the original derivation given by Langmuir, the isotherm was derived by considerations of the rates of sorption and desorption at equilibrium. Since the rate constants do not appear in the final expression, this is not the most economical derivation, involving as it does needless assumptions about rates.

in Fig. XVII.1. While some systems show this type of sorption, it is more usual to find exceptions to it. In addition, where data do fit, the interpretation of the constants $S°$ and $K_{1.2}$ in terms of the simple quasi-chemical model presented is not always possible.

Langmuir and subsequent investigators[a] have given reasons for expecting deviations from this simple equation. One argument is that no catalyst surface is homogeneous but on the contrary consists of regions of varying sorptive power. The result of such heterogeneity is that the isotherm may be expected to consist of a series of superposed Langmuir isotherms, each of different numbers of sites $S°$ and each characterized by a different sorption constant $K_{1.2}$. For such a heterogeneous surface we can write an isotherm equation

$$\theta = \frac{\Sigma \theta_i S_i^°}{\Sigma S_i^°} = \frac{(A)}{S°} \sum_i \left[\frac{S_i^°}{(A) + K_{id}} \right] \tag{XVII.3.4}$$

where θ_i is the fractional coverage of the ith surface of $S_i^°$ total sites ($S° = \Sigma S_i^°$) and equilibrium desorption constant $K_{id} = K_{2.1}$. This equation represents a summation of equations of the type (XVII.3.3) for θ_i. Where $S_i^°$ and K_{id} are known functions of i, explicit equations are sometimes possible for θ as a function of (A).

The simplest assumption for $S_i^°$ is that of uniformity. The "uniformly" heterogeneous surface would contain equal numbers of sites for each value of K_{id} over some finite range of values, i.e., from $K_{id} = K_0$ to K_M. Then we can transform our sum (XVII.3.4) into an integral, writing $S_i^° = S° dK/(K_M - K_0)$:

$$\theta = \frac{(A)}{(K_M - K_0)} \int_{K_0}^{K_M} \frac{dK}{(A) + K}$$

$$= \frac{(A)}{(K_M - K_0)} \ln \left[\frac{(A) + K_M}{(A) + K_0} \right] \tag{XVII.3.5}$$

with branches:

$A \ll K_0 < K_M \longrightarrow \dfrac{(A) \ln \left(\dfrac{K_M}{K_0} \right)}{(K_M - K_0)}$

$A \gg K_M \longrightarrow 1 - \dfrac{K_M}{2(A)}$

The isotherms of such a surface approach the Langmuir isotherm characteristic of the weakest binding part of the surface $K_M/2$ at large pressures. At low pressures, $(A) \ll K_0 < K_M$, the sorption takes place nearly linearly on the most strongly binding sites K_0.[b] Over a considerable range of pressures of A between $K_M > (A) > K_0$, the behavior of the

---

[a] For a detailed discussion of these and other isotherms, see S. Brunauer, "Physical Adsorption," Princeton University Press, Princeton, N.J., 1945.

[b] Note that the dissociation constant of the surface-adsorbed atom K_{id} is equivalent [Eq. (XVII.3.3)] in pressure units to the pressure at which one-half of the sites are covered by A.

isotherm will be representable by a fractional power dependence of θ on (A).[a] Such a relation has been used empirically for some time and is known as the Freundlich isotherm:

$$\theta = k(A)^n \qquad (XVII.3.6)$$

where $0 < n < 1$.

The Freundlich isotherm is capable of representing sorption data for a great many substances over a broad range of concentrations, and for purely practical considerations it has been more widely used than the Langmuir equation. However, it suffers from the defect of showing no saturation effect, the sorption increasing indefinitely with increasing adsorbate concentration. In addition, it does not have the neat theoretical interpretation of the Langmuir equation. Thus its use in kinetic studies has been somewhat restricted.

The heat of adsorption may be calculated from a knowledge of the sorption isotherm *at equilibrium* by use of the Clausius-Clapeyron equation.

$$RT^2 \left(\frac{\partial \ln A}{\partial T} \right)_\theta = -\Delta H_S \qquad (XVII.3.7)$$

where A is the pressure of the gas in equilibrium with the surface and the subscript θ is taken to indicate that the derivative is evaluated at constant total sorption. If we apply this to the Langmuir isotherm [Eq. (XVII.3.3)] and to the Freundlich isotherm [Eq. (XVII.3.6)], we find

$$\Delta H_S(\text{Langmuir}) = \Delta H_{1.2}$$

$$\Delta H_S(\text{Freundlich}) = \frac{RT^2}{n} \left(\frac{\partial \ln k}{\partial T} + \frac{\partial \ln n}{\partial T} \ln \frac{\theta}{k} \right) \qquad (XVII.3.8)$$

It is usually found that k decreases, while n increases, with increasing temperature. For the Langmuir isotherm there is no change in heat of sorption with increasing coverage. The Freundlich isotherm has an enthalpy of sorption which increases logarithmically with increasing coverage from $-\infty$ to a finite negative value.[b] Experimentally this is closer to the facts than the Langmuir equation. The complex Langmuir isotherm has a heat of sorption which behaves qualitatively like that of the Freundlich equation but lacks the infinite heat of sorption. It is thus a reasonable model for chemisorption.

The lesson to be learned from any of the isotherm studies is that, if reaction takes place among molecules or species specifically adsorbed on

---

[a] In Fig. XVII.1 the points labeled by ○ are those for a Freundlich isotherm, fitted by $\theta = 0.665$ and $\theta = 0.86$ to the complex Langmuir isotherm. Over the range of concentrations $(A)/K_{2.1}$ from 0.5 to 8, the two isotherms have a maximum deviation of 8 per cent and an average deviation of about 2 per cent, showing the closeness of the fit which is attainable between them. The fit could have been shifted to lower pressures by choosing different parameters.

[b] We have been using θ in the Freundlich isotherm despite the fact that it predicts no surface saturation. Formally this would mean values of $\theta > 1$ are permissible.

the catalyst surface, there may not be a simple relation between the concentrations on the surface and the pressures of reactant species in the gas phase. This makes the molecular interpretation of rate laws for catalytically controlled reactions subject to considerable uncertainty in the absence of precise isotherm data.[a]

4. The Mechanisms of Reactions at Solid Surfaces

The role of a surface in initiating chain carriers which propagate primarily in the homogeneous phase has already been discussed; it is a relatively minor branch of catalytic behavior. Reactions which proceed entirely at the surface of a catalyst appear to be much more common. This latter class is distinguished from the former in showing a direct proportionality between rate of reaction and amount of catalyst, other parameters being held constant. When such a relation does not obtain experimentally, one is faced with either a mixed mechanism or a diffusion-controlled rate, and the mechanistic interpretation of the rate data becomes extremely difficult.

Where the rate is "first-order" in catalyst,[b] an interpretation of the rate data is possible but always subject to uncertainties in our knowledge of the sorption isotherms of all species involved in the mechanism. Let us consider, for example, the decomposition of a single molecular species on a catalyst surface. An example might be the dehydration of alcohols on the surfaces of metal oxides such as alumina gel at 300°C to give an olefin plus water.[c] This endothermic reaction, which is thermodynamically favored in the gas phase by an entropy increase of about 36 e.u., can be written as

$$RCH_2\!-\!CH_2OH \rightarrow RCH\!=\!CH_2 + HOH - 11 \text{ Kcal} \quad (XVII.4.1)$$

The dehydration reaction is usually accompanied by the thermodynamically less favorable side reaction of dehydrogenation:[d]

$$RCH_2\!-\!CH_2OH \rightarrow RCH_2\!-\!CHO + H_2 - 16 \text{ Kcal} \quad (XVII.4.2)$$

At low pressures and temperatures in the range of 200 to 400°C it is found that the rate of dehydration is first-order in alcohol and slower than

---

[a] In practice, wherever possible, the sorption isotherms of the reactant species and products should be measured independently and as close to the experimental conditions of the kinetic runs as is attainable.

[b] Note that this is a necessary but not sufficient condition for a surface reaction because chains starting at the surface and terminating in the gas phase, although not a likely prospect, can also show first-order catalyst dependence.

[c] W. S. Brey, Jr. and K. A. Krieger, *J. Am. Chem. Soc.*, **71**, 3637 (1949).

[d] Transition metals are good selective catalysts for the dehydrogenation. However, in studies of the competitive rates in the cases of C_2H_5OH and $HCOOH$, G. M. Schwab and E. Schwab-Agallides, *ibid.*, **71**, 1806 (1949), have demonstrated that the physical state of the catalyst as well as its chemical structure could affect the relative amounts of the two competing reactions.

the rate of sorption. Such evidence has been interpreted as indicating that a single rate-determining step is taking place on the catalyst surface, and this has been suggested to be the ionization of a sorbed alcohol molecule to give a positive carbonium ion and OH^-, the former then rapidly decomposing to give an adsorbed proton H^+ and the olefin:

$$RCH_2\!-\!CH_2\!-\!OH \xrightarrow[\text{slow}]{1} RCH_2\!-\!CH_2^+ + OH^-$$

$$RCH_2\!-\!CH_2^+ \xrightarrow[\text{fast}]{2} RCH\!=\!CH_2 + H^+$$

$$RCH\!=\!CH_2 \xrightarrow[\text{fast}]{3} RCH\!=\!CH_2(g) \tag{XVII.4.3}$$

$$H^+ + OH^- \xrightarrow[\text{fast}]{4} HOH(g)$$

The over-all rate is then

$$\frac{-d(RC_2H_4OH)}{dt} = k_1(RCH_2CH_2OH)_s \tag{XVII.4.4}$$

If we assume a Langmuir isotherm at low coverage, then the concentration of adsorbed species is proportional to the pressure of alcohol [Eq. (XVII.3.3)],

$$(R'OH)_s \cong K_s S^\circ(R'OH) \tag{XVII.4.5}$$

and

$$\frac{-d(R'OH)}{dt} = k_1 K_s S^\circ(R'OH)$$

where $R'OH$ = alcohol, a result which is consistent with the observed first-order rate law.

It will be noted that the experimental activation energy predicted by this last equation is $E_{exp} = E_1 + \Delta H_S$, where ΔH_S is the heat of sorption (negative) of the alcohol on the surface. If the Langmuir isotherm is obeyed over a broad pressure range, then we should expect that at the higher pressure or on more strongly sorptive catalysts, the surface sites will become saturated and $(R'OH)_s = S^\circ$ = constant. Under these limiting conditions the rate will be zero-order in alcohol and given by[a]

$$\frac{-d(R'OH)}{dt} = k_1 S^\circ \tag{XVII.4.5a}$$

with an activation energy, higher by $|\Delta H_S|$ than the low-pressure rate. At intermediate pressures one may expect a pressure dependence which is apparently fractional-order. For rigorous Langmuir adsorption, the entire range will be given by[b]

[a] G. M. Schwab, *J. Phys. & Colloid Chem.*, **54**, 581 (1950), and earlier papers with coworkers, has shown that the reaction $HCOOH \rightarrow H_2 + CO_2$ is nearly zero-order on a number of metal and alloy surfaces.
[b] *Ibid.*

$$\frac{-d(R'OH)}{dt} = \frac{k_1 K_s S°(A)}{1 + K_s(A)} \qquad (XVII.4.5b)$$

Because of the difficulty of measuring adsorption over the range of reaction conditions, a complete quantitative check of all the terms in Eq. (XVII.4.5b) and their significance has not been made. This inability to explore thoroughly the details of the experimental rate law puts a severe restriction on our ability to explore the details of the surface mechanism. It would be interesting to know if the catalyst surface provides dual sites, both acid and basic, which can interact in a stoichiometric fashion with the ions from the alcohol. Or alternatively, is the function of the surface merely to provide an equivalent, "high-dielectric medium" on which ionization reactions of the adsorbate are facilitated?[a]

An equivalent ionic mechanism can be written for the dehydrogenation reactions

$$RCH_2\text{---}CH_2OH \underset{2}{\overset{1}{\rightleftharpoons}} RCH_2\text{---}CH_2O^- + H^+$$

$$RCH_2\text{---}CH_2O^- \underset{4}{\overset{3}{\rightleftharpoons}} RCH_2\text{---}CHO + H^- \qquad (XVII.4.6)$$

$$H^+ + H^- \underset{6}{\overset{5}{\rightleftharpoons}} H_2(g)$$

where the last two steps may actually occur as one bimolecular step:

$$RCH_2\text{---}CH_2O^- + H^+ \underset{4'}{\overset{3'}{\rightleftharpoons}} H_2(g) + RCH_2\text{---}CHO \qquad (XVII.4.7)$$

In such a case reaction 3 or 3' could be rate-determining and give a rate law first order in adsorbed alcohol. Since, however, the dehydrogenation reactions are facilitated by metal catalysts, particularly the transition metals, it is just as likely that the surface reaction proceeds via free radical species with moderately strong bonding to the metal atoms in the surface. By denoting metal atoms in the surface by M, such a mechanism could be represented by

$$RCH_2CH_2OH(g) + M \rightleftharpoons RCH_2CH_2OH \cdot M$$
$$RCH_2CH_2OH \cdot M + M \rightleftharpoons RCH_2CH_2O \cdot M + M \cdot H$$
$$M + RCH_2CH_2O \cdot M \rightleftharpoons RCH_2CHO \cdot M + M \cdot H$$
$$RCH_2CHO \cdot M \rightleftharpoons RCH_2CHO(g) + M \qquad (XVII.4.8)$$
$$2M \cdot H \rightleftharpoons 2M + H_2$$
$$M \cdot H + RCH_2CH_2O \cdot M \rightleftharpoons RCH_2CHO \cdot M + M + H_2$$

An intensive inquiry into the possibility of such reaction paths has not yet been made.[b]

---

[a] This latter would picture catalysis via ionic mechanisms as taking place in a hypothetical surface phase without regard to specific surface-adsorbate interactions.

[b] There is, however, very good evidence that H_2 chemisorbed on metals is dissociated into atoms or hydride ions; the two are not distinguished by the data. The most direct

5. Inhibition and Competition in Surface Reactions

If we pursue the model of the catalytic surface as a collection of a fixed number of sites S° capable of forming chemical bonds with adsorbed species, then we may expect to find that substances which can compete for the catalytic sites may inhibit the reaction. Furthermore, the effectiveness of the inhibition will depend on the relative pressures of the two adsorbates as well as their sorption constants on the surface. Let us consider as an example the simple Langmuir-type sorption of two sorbates A and B on a surface of S° sites:

$$K_A: \qquad\qquad S + A \rightleftharpoons S \cdot A$$
$$K_B: \qquad\qquad S + B \rightleftharpoons S \cdot B \qquad\qquad \text{(XVII.5.1)}$$

where now at equilibrium

$$(S \cdot A) = K_A(A)(S) \qquad (S \cdot B) = K_B(B)(S) \qquad \text{(XVII.5.2)}$$

with $(S) = S^\circ - (S \cdot A) - (S \cdot B)$. If we eliminate (S) from these equations, we find

$$\frac{(S \cdot A)}{S^\circ} = \theta_A = \frac{K_A(A)}{1 + (A)K_A + (B)K_B}$$

$$\frac{(S \cdot B)}{S^\circ} = \theta_B = \frac{K_B(B)}{1 + (A)K_A + (B)K_B} \qquad \text{(XVII.5.3)}$$

where θ_A and θ_B are respectively the fractions of surface covered by sorbed A and sorbed B.[a] We see that each gas, by occupying surface sites, acts to inhibit the sorption of the other gas.

In a mechanism for which the rate-determining process is the decomposition of sorbed A, B would act as an inhibitor, or if its attachment to the catalyst surface was sufficiently strong, a poison. For the first-order decomposition of sorbed A

$$\frac{-d(A)}{dt} = k_1\theta_A$$

$$= \frac{k_1 K_A(A)}{1 + (A)K_A + (B)K_B} \qquad \text{(XVII.5.4)}$$

A rate law of this type was found by Constable[b] for the dehydrogenation

---

evidence is the fact that the amount of H_2 sorbed is proportional to the square root of the H_2 gas pressure at low coverages or fits a Langmuir-type isotherm at higher coverages of the form $\theta = (H_2)^{1/2}/[K_d + (H_2)^{1/2}]$. This fits the equilibrium $H_2 + 2M \rightleftharpoons 2M \cdot H$. The other evidence comes from the high heats of chemisorption [30 to 60 Kcal; D. D. Eley, *Discussions Faraday Soc.*, **8**, 67 (1950)], which indicates the formation of strong bonds with the surface.

[a] In the sense in which they will be employed, note that $(S \cdot A)$ and $(S \cdot B)$ correspond to total numbers of sorbed species, while θ_A and θ_B refer to their surface concentrations.

[b] F. H. Constable, *Nature*, **117**, 230 (1926); *Proc. Cambridge Phil. Soc.*, **22**, 172 (1926); **23**, 593 (1927). The range of these studies was such that $K_A(A)$ and $K_B(B) \gg 1$.

of ethyl alcohol on copper in the presence of benzene, acetone, or H_2O, all of which act as poisons.

If the rate-determining step in the surface reaction involves the second-order reaction of sorbed A and B, or the transition state involves both sorbed A and B, the rate law will have the form

$$\frac{-d(A)}{dt} = k_1\theta_A\theta_B$$

$$= \frac{k_1 K_A K_B (A)(B)}{[1 + K_A(A) + K_B(B)]^2} \qquad (XVII.5.5)$$

At fixed pressure of A, the rate of reaction will pass through a maximum with increasing pressure of B, the latter being given by

$$(B)_{max} = \frac{1 + K_A(A)}{K_B} \qquad (XVII.5.6)$$

while

$$\left[\frac{d(A)}{dt}\right]_{max} = \frac{K_A(A)}{4[1 + K_A(A)]} \qquad (XVII.5.7)$$

The exchange between D_2 and NH_3 has been found to show such inhibition. The rate law on an activated Fe surface near 150°C can be fitted to the expression[a]

$$\frac{d(NH_2D)}{dt} = \frac{k_1(D_2)^{1/2}(NH_3)}{[1 + k_2(NH_3)]^2} \qquad (XVII.5.8)$$

This has been interpreted by the authors as a competitive adsorption of the two reactants NH_3 and D_2 under conditions such that the NH_3 is always more strongly adsorbed, $K_{NH_3}(NH_3) \gg [K_{D_2}(D_2)]^{1/2}$, and the D_2 is adsorbed as atomic D. The mechanism for such a rate law would be

$$NH_3 + S \overset{1}{\underset{2}{\rightleftharpoons}} NH_3 \cdot S$$

$$D_2 + 2S \overset{3}{\underset{4}{\rightleftharpoons}} 2D \cdot S \qquad (XVII.5.9)$$

$$D \cdot S + NH_3 \cdot S \overset{5}{\underset{6}{\rightleftharpoons}} NH_2D \cdot S + H \cdot S$$

The slow and rate-determining step is presumed to be the displacement reaction 5. From a stereochemical viewpoint this might involve a transition state such as

[a] J. Weber and K. J. Laidler, *J. Chem. Phys.*, **19**, 1089 (1951). See also A. Farkas, *Trans. Faraday Soc.*, **32**, 416 (1936), who reported a different rate law with no maximum and only a first power of the denominator. The data of Weber and Laidler are not precise enough to be certain of the power of the denominator. The maximum they report is not much bigger than their experimental uncertainty.

$$(XVII.5.10)$$

where the exchange is in effect an amide radical, $\cdot NH_2$, transfer[a] and proceeds through an ammonium-type radical.

The very similar rate law suggested by Farkas (*loc. cit.*) would involve noncompetitive sorption by H_2 and NH_3 and would not show the small maximum in the rate with increasing NH_3.[b]

The decomposition of NH_3 on a Pt surface at $1400°K$ has been found to follow a rate law[c]

$$\frac{-d(NH_3)}{dt} = \frac{k(NH_3)}{(H_2)} \qquad (XVII.5.11)$$

The inhibition by the reactant hydrogen can be explained by assuming that H_2, which is dissociatively adsorbed on Pt as atoms, is more strongly bound than NH_3 and that the rate-determining step is the dissociation of NH_3 on the surface:[d]

$$NH_3 \cdot S + S \rightarrow NH_2 \cdot S + H \cdot S$$

[a] An alternative atom abstraction reaction would involve the production of singly attached H_2 or HD molecules and also NH_2 radicals

$$D \cdot S + NH_3 \cdot S \rightleftharpoons NH_2 \cdot S + HD \cdot S$$

This would be followed by the rapid dissociation of $HD \cdot S$ by reaction with S or by the back reaction with $NH_2 \cdot S$. This does not seem as likely a mechanism although compatible with the rate law.

[b] Although the maximum is very weak and not a strong argument against noncompetitive adsorption, a supporting evidence for the latter is the qualitative observation that on the addition of NH_3 to a metal tube which has been exposed to H_2 and "evacuated," additional H_2 is liberated.

[c] G. M. Schwab and H. Schmidt, *Z. physik. Chem.*, **B3**, 337 (1929).

[d] A number of alternative mechanisms are also compatible with this rate law. One would be the decomposition of $NH_2 \cdot S$, assumed to be in equilibrium with $NH_3 \cdot S$ and $H \cdot S$. This could happen in a first-order step or via metathesis with $H \cdot S$:

$$NH_2 \cdot S + H \cdot S \rightarrow NH \cdot S + S + H_2$$

A final path could be the reaction of $H \cdot S$ with $NH \cdot S$ to form N atoms and H_2, the $NH \cdot S$ being assumed to be at equilibrium with NH_3:

$$NH_3(g) + 2S + S' \rightleftharpoons NH \cdot S' + 2H \cdot S$$
$$NH \cdot S' + H \cdot S \rightarrow N \cdot S' + H_2 + S$$

This mechanism will satisfy the rate law if the sites S' which hold N are different from the sites S which hold H atoms.

On Pt^a and crushed, fused quartz,[b] the reaction $CO + \frac{1}{2}O_2 \rightarrow CO_2$ follows the rate law

$$\frac{d(CO_2)}{dt} = k \frac{(O_2)}{(CO)} \qquad (XVII.5.12)$$

This can be interpreted in terms of a rate-controlling reaction between weakly sorbed O_2 and strongly sorbed CO which saturates the surface:

$$O_2 \cdot S + CO \cdot S \rightarrow CO_2 \cdot S + O \cdot S$$

This slow step would be followed by the fast reaction of $O \cdot S$ with $CO \cdot S$ to form more CO_2.

In contrast to this behavior, the catalysis of the reaction by CuO is found to be zero-order in O_2 and first-order in CO^c at about 300°C. This can be interpreted as either the reaction of weakly sorbed CO with strongly sorbed O_2 (noncompetitive sorption) or else a Rideal type mechanism in which the slow step is the sorption of CO on the surface. The latter seems quite reasonable.

At about 0°C, on CuO, the rate law becomes first-order in O_2 and zero-order in CO. The mechanism has been clarified by Garner et al.[d] through surface studies of CuO. These have shown that the CuO surface is covered with a layer of strongly sorbed CO, probably in the form of CO_3^- ions. Even at 0°C this layer can react with CO to liberate CO_2. If it is assumed that $CO_3^- \cdot S$ and $CO \cdot S$ are in fast equilibrium with $O_2 \cdot S$ and the rate-determining step is reaction between $CO \cdot S$ and $CO_2 \cdot S$, the rate law can be explained.

$$2e^- + 2S + O_2(g) \underset{2}{\overset{1}{\rightleftharpoons}} 2[O^- \cdot S] \qquad \text{fast}$$

$$CO \cdot S + 2[O^- \cdot S] \underset{4}{\overset{3}{\rightleftharpoons}} CO_3^- \cdot S + 2S \qquad \text{fast} \qquad (XVII.5.13)$$

$$CO_3^- \cdot S + CO \cdot S \overset{5}{\rightarrow} 2CO_2(g) + 2e^- + 2S \qquad \text{slow}$$

This gives for strong sorption by $CO(\theta_{CO} \cong 1)$:

$$(CO_3^- \cdot S) = K_{3.4} \frac{(CO \cdot S)(O^- \cdot S)^2}{(S)^2} = K_{1.2}K_{3.4}(CO \cdot S)(O_2)$$

where we have assumed a steady state for e^-. Then for the rate

$$\frac{d(CO_2)}{dt} = k_5 \theta_{CO_3^-} \theta_{CO}$$

$$= k_5 K_{1.2}K_{3.4}\theta_{CO}^2(O_2) \qquad (XVII.5.14)$$

$$\cong k_5 K_{1.2}K_{3.4}(O_2)$$

[a] I. Langmuir, *Trans. Faraday Soc.*, **17**, 621 (1922).

[b] M. Bodenstein and F. Othmer, *Z. physik. Chem.*, **53**, 166 (1945).

[c] G. M. Schwab and G. Drikos, *ibid.*, **52**, 234 (1942).

[d] W. E. Garner et al., *Proc. Roy. Soc.* (*London*), **A211**, 472 (1952). Garner et al., *ibid.*, **A197**, 294 (1949); *Discussions Faraday Soc.*, **8**, 246 (1950).

While the data presented seem to accommodate quite readily to inter-pretation in terms of the simple Langmuir isotherm, this is by no means generally the case. More often than not, over a large pressure range, it may be found that the data do not lend themselves readily to a description in terms of a simple order or Langmuir isotherm. In such cases the non-uniformity of surfaces has been invoked together with more complex iso-therms. While this usually permits fitting the data with a simple chemical mechanism, the complexity of the resulting expressions and the large number of adjustable parameters make the kinetic interpretation very unsatisfying.[a] This problem of separating isotherm data from surface kinetics is one of the major obstacles to the elucidation of the mechanisms of catalytic reactions.

6. Some Catalytic Reactions Involving H_2 and Metals

A considerable proportion of the investigations of catalytic reactions involve hydrogen or hydrogen-containing compounds. In all of the cases studied, intermolecular transfer of H atoms takes place and the evidence points to an intermediate chemisorption of H atoms (or ions) on the catalyst surface.

H_2 gas itself is rapidly sorbed by the transition metals and more slowly by metal oxides and elements such as carbon (graphite) and germanium.[b] On the oxides the sorption frequently leads to the formation of hydroxides, and on heating H_2O may be desorbed.[c,d] Some reversible sorption occurs as well, and it has been suggested that this corresponds to a hydride for-mation with the surface metal ions. In the case of metals, H_2 gas is sorbed rapidly even at 78°K with a heat of sorption which may be of the order of 40 Kcal or more, decreasing slowly with increasing coverage until near saturation, when it can approach zero.[e] A considerable account of evi-dence supports the view that the sorption on metals is a direct 1:1 stoichio-

[a] M. Temkin and V. Pyzhev, *Acta Physicochim. U.R.S.S.*, **12**, 327 (1940), have used what is equivalent to a complex Langmuir isotherm to explain the very complex kinetics of NH_3 decomposition on Fe surfaces. A theoretical discussion of nonuniform surfaces is given by G. D. Halsey, *J. Chem. Phys.*, **17**, 758 (1949).

[b] See B. M. W. Trapnell, "Chemisorption" Academic Press, Inc., New York, 1955, for a detailed discussion of chemisorption and catalysis.

[c] D. A. Dowden and W. E. Garner, *J. Chem. Soc.*, 893 (1939).

[d] R. A. Beebe and D. A. Dowden, *J. Am. Chem. Soc.*, **60**, 2912 (1938).

[e] B. M. W. Trapnell, *Proc. Roy. Soc. (London)*, **A206**, 39 (1951). O. Beeck, *Discussions Faraday Soc.*, **8**, 118 (1950). This decrease in heat of sorption is still the subject of controversy. Some authors ascribe it to surface heterogeneity, others to repulsion between adsorbate molecules (or ions). Note that even when $\Delta H_S = 0$, the S—H bond is $\frac{1}{2}D(\text{H—H}) = 52$ Kcal.

metric reaction with the metal ion leading to hydride formation,[a,b]
$$H_2 + 2S\cdot \rightarrow 2H\cdot S.$$

When "clean" tungsten films are exposed to H_2 gas at 78°K, there is a rapid, dissociative sorption leading to saturation of the surface in the neighborhood of 10^{-2} mm Hg pressure of H_2. About the first 75 per cent of the H_2 is sorbed exothermally with $\Delta H_S < -40$ Kcal. We would not expect this tightly sorbed H_2 to be desorbed at these temperatures, and in fact Roberts (loc. cit.) has shown that desorption is not appreciable below 600°K. However, in the range of coverage from 0.80 to 1.0, Trapnell (loc. cit.) showed that ΔH_S approached zero and that this last fraction could be desorbed even at 78°K. The strongly and weakly bound H atoms can exchange with each other on the surface at 200°K[c] but not at 78°K[d] as evidenced by work on H_2-D_2 exchange. Gundry[e] was able to show that on evaporated Ni films which chemisorb N_2 weakly ($\Delta H_S \cong -10$ Kcal/mole) at 78°K, H_2 and N_2 are in competition for sorption sites and a small amount of sorbed N_2 will inhibit the exchange of H_2 and D_2.

These observations have been very useful in clarifying the mechanism of ortho- to para-H conversion and H_2-D_2 exchange, both of which reactions occur readily in contact with W, Ni, Fe, Pt, Pd, and other of the transition metals. The initial rate of conversion of para-H to ortho-H at constant pressure was found to be proportional to the partial pressure[f] of para-H. There now seems to be general agreement that this can be satisfactorily accounted for in terms of the rates of dissociative sorption and desorption of the gas. Thus for the dissociative sorption of H_2,

$$H_2(g) + 2S \underset{2}{\overset{1}{\rightleftharpoons}} 2H\cdot S$$

the initial rate of reaction is (at constant pressure) proportional to the rate of desorption, which is, of course, equal to the rate of sorption

$$R = k_1(H_2)(S)^2 \qquad\qquad (XVII.6.1)$$

By using the experimental isotherm to obtain (S), Trapnell (loc. cit.) has shown that this equation is in accord with the data.[g]

[a] J. K. Roberts, Proc. Roy. Soc. (London), **A152**, 445 (1935).

[b] O. Beeck and A. W. Ritchie, Discussions Faraday Soc., **8**, 159 (1950).

[c] D. D. Eley, Proc. Roy. Soc. (London), **A178**, 452 (1941).

[d] P. M. Gundry, Advances in Catalysis, **9**, 692 (1957).

[e] Ibid.

[f] A. Farkas and L. Farkas, J. Am. Chem. Soc., **60**, 22 (1938). A. Farkas, Trans. Faraday Soc., **32**, 416 (1936). D. D. Eley and E. K. Rideal, Proc. Roy. Soc. (London), **A178**, 429 (1941). The rate has the form $-d(p\text{-}H_2)/dt = k[(p\text{-}H_2) - (p\text{-}H_2)_e]$, where $(p\text{-}H_2)_e$ = the equilibrium concentration of p-H_2.

[g] Note that, if H_2 followed a Langmuir isotherm, $1/(S) = K_S^{1/2}(H_2)^{1/2}$, we would obtain on substitution in Eq. (XVII.6.1) a zero-order rate law, which is in fact observed. The true isotherms, however, follow $1/(S) \cong k_S'(H_2)^{0.2}$, and it is only when allowance is

The data on the H_2-D_2 exchange also follow a similar rate law, and there seems little question but that the exchange occurs between atoms sorbed on the surface. The rates of ortho- to para-H conversion and H_2-D_2 exchange are usually of comparable magnitudes, where the former is not magnetically catalyzed,[a] and have identical activation energies.

The metals which are active in H_2-D_2 exchange are also active in the hydrogenation of unsaturated compounds. The reaction of H_2 with C_2H_4 has been extensively studied as a prototype of this type of reaction. While a great deal has been learned, the mechanism is not yet completely resolved, and from work on higher olefins it appears that each hydrogenation may have features unique to itself.

Cu, Pt, Ni, and W surfaces will sorb C_2H_4 at temperatures from -70 to $200°C$. The sorption may be accompanied by a self-hydrogenation, liberating C_2H_6 and leaving C_2H_2 strongly sorbed on the surface.[b,c] This latter reaction is very slow and tends to poison the catalyst. The stoichiometry of the reaction has been suggested as

$$2S\cdot + C_2H_4(g) \rightleftharpoons S\cdot C_2H_4\cdot S$$
$$2S\cdot + S\cdot C_2H_4\cdot S \rightleftharpoons S\cdot C_2H_2\cdot S + 2H\cdot S \qquad \text{(XVII.6.2)}$$
$$C_2H_4(g) + 2H\cdot S \rightleftharpoons C_2H_6 + 2S\cdot$$

with an over-all stoichiometry:

$$2S\cdot + 2C_2H_4 \rightleftharpoons S\cdot C_2H_2\cdot S + C_2H_6$$

In the presence of D_2 gas there is an accelerated addition reaction to form ethanes and a simultaneous exchange to form substituted deuteroethylenes. The addition usually predominates at low temperatures, while the exchange begins to predominate above $90°C$. One of the interesting observations on the addition at low temperatures is that the initial product of the reaction of $C_2H_4 + D_2$ is C_2H_6. This would seem to indicate a preequilibrium between the pool of D atoms formed from dissociative sorption of D_2 and the H atoms dissociated from sorbed C_2H_4.[d] It is further

---

made for the pressure dependence of $k_1 \propto (H_2)^{-0.6}$ that one finds the zero-order rate. This apparent pressure dependence of the rate of sorption, $k_1(H_2) \propto (H_2)^{0.4}$, arises from the change in activation energy of sorption with increasing coverage.

[a] D. D. Eley, *J. Phys. & Colloid Chem.*, **55**, 1017 (1951).

[b] O. Beeck, *Discussions Faraday Soc.*, **8**, 122 (1950). B. M. W. Trapnell, *Trans. Faraday Soc.*, **48**, 160 (1952). See, however, D. A. Schissler et al., *Advances in Catalysis*, **5**, 37 (1957), who found no C_2H_6 from C_2H_4 at 3 mm Hg pressure and $-78°C$.

[c] The heat of sorption of C_2H_4 is much greater than that of H_2, varying from 138 Kcal/mole on Ta to 102 Kcal/mole on W and 58 Kcal/mole on Ni.

[d] The production of C_2H_6 would then correspond to the product favored by thermodynamic equilibrium:
$$\tfrac{1}{3}C_2H_6 + D_2 \rightleftharpoons \tfrac{1}{3}C_2D_6 + H_2 \qquad K < 1$$

found[a] that the Ni-catalyzed H_2-D_2 exchange is inhibited by C_2H_4 even when the addition reaction is taking place.[b]

Although there has been much investigation of the kinetics of both the exchange and addition reactions, there is no substantial agreement on either order or mechanism. In the usual range of experimental conditions, it is generally found that on Pt and Ni the rate law for hydrogenation appears to be first-order in H_2 and to vary from zero order in C_2H_4 at low temperatures to some fractional order or unity at higher temperatures (0 to 200°C). Such behavior can be interpreted in terms of a transition-state complex on the surface that contains C_2H_4 + 2H atoms. This could be the slow combination of C_2H_5 radicals with H atoms:

$$S \cdot C_2H_4 \cdot S + H \cdot S \underset{2}{\overset{1}{\rightleftharpoons}} S \cdot C_2H_5 + 2S \qquad \text{fast}$$

$$\text{(XVII.6.3)}$$

$$S \cdot C_2H_5 + H \cdot S \overset{3}{\to} C_2H_6(g) + 2S \qquad \text{slow}$$

The rate law for such a mechanism would be

$$\frac{d(C_2H_6)}{dt} = k_3 K_{1.2} \frac{(S \cdot C_2H_4 \cdot S)(H \cdot S)^2}{(S)^2} \qquad \text{(XVII.6.4)}$$

and if both C_2H_4 and H_2 are in equilibrium with the same surface site S, then:

$$(H \cdot S) = K_{H_2}(H_2)(S)^2 \qquad (S \cdot C_2H_4 \cdot S) = K_{C_2H_4}(C_2H_4)(S)^2$$

and

$$\frac{d(C_2H_6)}{dt} = k_3 K_{1.2} K_{H_2} K_{C_2H_4}(H_2)(C_2H_4)(S)^2 \qquad \text{(XVII.6.5)}$$

where if $(H \cdot S) \ll (S \cdot C_2H_4 \cdot S)$, application of the Langmuir isotherm gives

$$S^\circ = (S) + 2(S \cdot C_2H_4 \cdot S)$$
$$= (S) + 2K_{C_2H_4}(C_2H_4)(S)^2 \qquad \text{(XVII.6.6)}$$

At high temperatures, at which the active sites[c] $(S) \gg (S \cdot C_2H_4 \cdot S)$, this leads to first-order dependence of the rate on both (H_2) and (C_2H_4). At low temperatures, at which most active sites are covered by C_2H_4, $(S)/S^\circ \cong 1/[2S^\circ K_{C_2H_4}(C_2H_4)]^{1/2}$ and Eq. (XVII.6.6) gives zero-order dependence on C_2H_4. For such a mechanism to hold, the $S \cdot C_2H_5$ species which is formed must be such that the added H (or D) atom is in some

---

[a] G. H. Twigg and E. K. Rideal, *Proc. Roy. Soc. (London),* **A171,** 55 (1939).

[b] Schissler et al., *loc. cit.,* have shown that a nickel-supported catalyst (kieselguhr support) shows negligible inhibition of the H_2-D_2 exchange by C_2H_4. This may indicate different modes of adsorption for the supported catalyst.

[c] The problem of what constitutes the active surface at any given temperature is a difficult one which cannot be resolved by assumption of the simple Langmuir isotherm. As we have noted in the case of H_2, large variations in heat of sorption will imply severe changes in active surface temperature which may thus give rise to anomalous pressure dependence of the rate of reaction. These important details can be resolved only by direct isotherm measurement.

way more loosely bound than its neighbors. Otherwise, exchange would be always more rapid than addition.

Other mechanisms that have been proposed for the addition reaction differ from the one presented, but it seems fruitless at the present stage of experimental understanding to pursue them in detail.

The order of the D_2 exchange reaction has been reported as first order in D_2 and zero order in C_2H_4, but the data are not good enough to be sure of these orders.[a]

A comparable study of the Ni-catalyzed reactions of isobutene, 1-butene, and cis-butene-2 in the presence of H_2 and D_2 has revealed[b] a very complex series of chemical reactions that take place, including induced isomerization of 1-butene to 2-butene, deuterium exchange, induced cis-trans isomerization of butene-2, and finally, addition to the double bond. Below 200 mm Hg pressure, the rates of exchange, addition, and isomerization are about equal for 1-butene and are about $\frac{1}{2}$ order in olefin and $\frac{1}{2}$ order in H_2. With increasing excess of H_2 this approaches zero order in olefin and $\frac{1}{2}$ order in H_2, while for large excess of 1-butene, all reactions become inhibited (30 to 150°C). Although the authors have attempted to discuss mechanisms in connection with the data, the lack of information on the isotherms makes that of dubious value.

A study of the catalyzed reactions (Pt, Ag, and Pd) of C_6H_6 and D_2 showed that the ratio of exchange to addition (to produce cyclohexanes) is high on Ni and low on Pt.[c] It is found, in contrast to work on simple olefins, that once addition starts, there is essentially no exchange of H by the intermediate cyclohexenes or product cyclohexane. An incomplete study of the concentration dependence of the rate gave a close to zero order in hydrocarbon for both exchange and addition. The exchange was nearly $-\frac{1}{2}$ order in H_2, while addition was between $\frac{1}{2}$ and first order in H_2. From these and other observations the authors have postulated two independent mechanisms for the two reactions. Again, the absence of isotherm data makes the interpretation of doubtful value.[d]

[a] Twigg and Rideal, loc. cit., performed the experiments which have been criticized by K. J. Laidler et al., J. Chem. Phys., **20**, 1331 (1952), who proposed a rate $\frac{1}{2}$ order in D_2. A law $\frac{1}{2}$ order in D_2 and zero order in C_2H_4 can be obtained from Eq. (XIII.6.3) by adding a slow step:

$$[C_2H_4 \cdot D] \cdot S + S \xrightarrow{4} C_2H_3D \cdot S + H \cdot S$$

followed by rapid equilibration of singly bound $C_2H_3D \cdot S$.

[b] T. I. Taylor and V. H. Dibeler, J. Phys. & Colloid Chem., **55**, 1036 (1951). See also C. D. Wagner et al., J. Chem. Phys., **20**, 338 (1952).

[c] J. R. Anderson and C. Kemball, Advances in Catalysis, **9**, 51 (1957).

[d] The authors have proposed that phenyl radical is the intermediate for exchange and that addition proceeds by D_2 molecules (or two D atoms) adding to benzene. It is interesting to note that, if benzene and D_2 are adsorbed on noncompetitive sites, the data are consistent with $\phi H \cdot S + S' \rightleftharpoons \phi \cdot S + H \cdot S'$ as being rate-determining for exchange, while a mechanism similar to that for C_2H_4 [Eq. (XVII.6.3)] could account for

Some very fascinating features of these hydrogenation reactions which we have not discussed involve the relation between catalyst structure in relation to activity and the stereospecificity of catalytic reactions. It is generally found that additions to the double bond[a] or to unsaturated rings[b] usually take place in a specific manner, the cis isomer being favored. Thus substituted acetylenes give[c] cis olefins and substituted cyclohexenes[d] give principally cis adducts. The further elucidation of these aspects of catalysis will undoubtedly be of considerable interest in the chemical understanding of the process.

7. Oxidation of Metals; Rates of Chemisorption

The bulk of evidence which we have discussed so far indicates that the mechanism of catalysis at solid surfaces takes place via the reaction of catalyst atoms (or ions) with the adsorbate to form a monolayer of chemically active intermediates. Since the initial act of chemisorption is a chemical reaction, it is not surprising to find that it may be accompanied by an activation energy of sorption. In general, however, the act of chemisorption is very rapid and occurs at a reasonable proportion of the estimated collisions of the gas molecule with the geometrical surface.[e] Even when we might expect the rates of sorption to decrease as the surface monolayer nears completion, it is often found that the rate is only slightly diminished.[f] This has been interpreted as due to the formation of a loosely held second sorbate layer, formed on top of the monolayer, which is capable of migrating fairly rapidly to uncovered sorption sites.

The rate of chemisorption may be estimated from the kinetic theory formula for wall collisions [Eq. (VII.6.6)]:

$$\frac{dN}{dt} = \alpha Z_w = \tfrac{1}{4} N_g \bar{c} \alpha \qquad (XVII.7.1)$$

where Z_w is the number of collisions per unit area of molecules N at a

---

addition. Such dual-site catalysis has been suggested by A. Amano and G. Parravano, *Advances in Catalysis*, **9**, 716 (1957), to account for benzene hydrogenation on supported metal catalysts.

[a] K. N. Campbell and B. K. Campbell, *Chem. Revs.*, **31**, 77, 145 (1943).

[b] R. P. Linstead et al., *J. Am. Chem. Soc.*, **64**, 1948 (1942).

[c] *Ibid.*

[d] S. Siegel and M. Dunkel, *Advances in Catalysis*, **9**, 15 (1957).

[e] J. L. Morrison and J. K. Roberts, *Proc. Roy. Soc. (London)*, **A173**, 1 (1939), were able to show that O_2 is adsorbed on about $\tfrac{1}{16}$ of all collisions with a tungsten surface. Similar results were obtained by J. A. Becker and C. D. Hartman, *J. Phys. Chem.*, **57**, 153 (1953), for N_2 on W. Such high efficiencies of sorption are not unreasonable if the initial act is charge transfer, i.e., formation of an O_2^- or N_2^- ion on the metal surface. The cases of slow chemisorption requiring activation are relatively few (for example, CH_4 on Ni and Pt) and probably involve bond breaking in the initial step.

[f] *Ibid.*

molecular density in the gas N_g and mean velocity $\bar{c}$ and α is the probability of sticking per collision. By dividing by N_0, the number of active surface sites per unit area, we can write for the rate of increase of surface coverage $\theta = N/N_0$:

$$\frac{d\theta}{dt} = \frac{1}{4}\frac{N_g}{N_0}\bar{c}\alpha \qquad (\text{XVII.7.2})$$

When the sorption takes place only by direct contact with bare sites, α will be proportional to $1 - \theta$ for single-site sorption or to $(1 - \theta)^2$ for double-site sorption (for example, $N_2 + 2S \rightarrow 2N \cdot S$). If in addition there is an activation energy for sorption, then α will have the form

$$\alpha = f(\theta)e^{-Es/RT} \qquad (\text{XVII.7.3})$$

It has been found that most slow chemisorptions have rates which can be described by an equation first suggested by Zeldovitch[a]

$$\frac{d\theta}{dt} = be^{-\beta\theta/RT} \qquad (\text{XVII.7.4})$$

The integrated form ($\theta = 0$ at $t = 0$) can be written as

$$\theta = \frac{RT}{\beta}\left[\ln\left(t + \frac{RT}{\beta b}\right) - \ln\frac{RT}{\beta b}\right] \qquad (\text{XVII.7.5})$$

so that the amount adsorbed, which is proportional to the coverage θ, is a logarithmic function of time. The exponential decrease in rate of sorption with amount sorbed can be explained quite reasonably in terms of an increase in activation energy for chemisorption with increasing coverage.[b] This may arise from interactions between adsorbate molecules[c] which could account for such behavior even on uniform surfaces. Much more likely, however, is the explanation suggested by Halsey[d,e] that such effects arise from nonuniformity of the surface.

Chemisorption is a thermodynamically favorable process, and, more often than not, the chemical reaction is restricted to the surface layer only

[a] Y. Zeldovitch, *Acta Physicochim. U.R.S.S.*, **1**, 449 (1934). Y. Zeldovitch and S. Roginsky, *ibid.*, **1**, 59, 554 (1934). These papers showed that the slow sorption of CO on MnO_2 followed such an exponential law. The equation is nevertheless frequently referred to as the Elovich equation after a paper by S. Y. Elovich and G. M. Zhabrova, *Zhur. Fiz. Khim.*, **13**, 1716, 1775 (1939), in which the same sorption was noted for both H_2 and C_2H_4 on Ni.

[b] An alternative explanation by H. A. Taylor and N. Thon, *J. Am. Chem. Soc.*, **74**, 4169 (1952), suggests that the act of sorption provokes an accompanying decrease in the number of active sites at a rate which is proportional to the product of (S) and $d\theta/dt$. The physical model for such a decay seems quite unreasonable.

[c] M. Boudart, *ibid.*, **74**, 3556 (1952).

[d] G. Halsey, *J. Phys. Chem.*, **55**, 21 (1951), uses what is essentially the uniformly nonuniform surface. See page 625.

[e] A. S. Porter and F. C. Tompkins, *Proc. Roy. Soc. (London)*, **A217**, 529 (1953), have adapted Halsey's treatment for a surface on which surface diffusion is rate-controlling.

because of a large activation energy, which inhibits further attack of the sorbate on the sorbent. This is notably the case for the chemisorption of O_2 gas on metals or metallic oxides. With most metals and with lower-valence metallic oxides such as Cu_2O, the stable state of the system is usually the higher valence oxide. The resistance of bulk metals such as Al and Ni to rapid oxidative attack, even at high temperatures, can be attributed to the protective action of the chemisorbed oxide layer. Such protection occurs when the metal ion in the oxide layer has a larger radius than it does in the metal substrate. At least for these metals, it is found that the oxide film is compact and strongly adherent to the metal substrate.[a]

The continued oxidation of the metal substrate beneath the protective oxide layer must become a diffusion-controlled process for thick enough oxide films in which either metal atoms[b] or oxygen atoms diffuse through the metal oxide layer to the appropriate interface where reaction proceeds. Let us assume a thick enough oxide layer on a plane metal surface where a steady state has been achieved. Then we can write for the rate of formation of metal oxide, MO, per unit area (assuming metal ion diffusion):

$$\frac{d(\text{MO})}{dt} = k_0 D_M \left[\frac{\partial(\text{M})}{\partial x}\right]_{\text{ss}} = \rho_0 \frac{dx}{dt} \qquad (\text{XVII.7.6})$$

where D_M is the diffusion constant of metal M across the metal oxide layer while $[\partial(\text{M})/\partial x]_{\text{ss}}$ is the stationary state concentration gradient of metal atoms across the oxide layer of thickness x and density ρ_0. This constant gradient is simply $(\text{M})_0/x$, where $(\text{M})_0$ is the fixed, surface concentration of metal atoms at the metal–metal oxide interface.

By arranging terms after substitution, we have

$$\frac{dx}{dt} = \frac{k_0 D_M (\text{M})_0}{\rho_0 x} \qquad (\text{XVII.7.7})$$

or on separating variables and integrating ($x = x_0$ at $t = 0$):

$$x^2 - x_0^2 = \left[\frac{2 \, k_0 (\text{M})_0 D_M}{\rho_0}\right] t \qquad (\text{XVII.7.8})$$

This last equation, which predicts a parabolic growth of the oxide film with time, is characteristic of the oxidation of most metals aside from the alkali metals and alkaline earths.[c,d]

A more explicit model for the oxidation process involving the motion of O^- and M^+ ions together with electrons and lattice defects has been formulated by Wagner,[e] and many efforts have been made to relate the

[a] For the alkali metals and alkaline earths (except Be), the oxide layer is porous and nonadhesive and the metal ions have smaller volumes than in the pure metal.

[b] Or more likely, metal ions and electrons, separately.

[c] E. A. Gulbransen and K. F. Andrew, *Trans. Electrochem. Soc.*, **96**, 364 (1949).

[d] C. Wagner and K. Grünewald, *Z. physik. Chem.*, **B40**, 455 (1938).

[e] C. Wagner, *ibid.*, **B21**, 25 (1933); **B32**, 447 (1936). E. A. Gulbransen has summarized the classical picture in terms of diffusion [*Ann. N.Y. Acad. Sci.*, **53**, 830 (1954)].

composite rate constant of Eq. (XVII.7.8) to properties of the various components in the system.[a] The problem is complicated by the space charge effects present in the ionic media, and for very thin oxide films the interfaces will exert electrical effects on each other.[b] The qualitative features of the models seem reasonably well confirmed by experiment, although many unexplained features of the oxidation process persist.[c]

8. Electrode Reactions

If two inert, plane, metal electrodes are placed parallel to each other in a solution that contains an electrolyte and a very small electrical potential of magnitude E is applied across them, a small current I that decreases with time will be observed to flow between them. The current will consist of the motion of positive ions to the cathode and negative ions to the anode. Initially it will obey Ohm's law, $I = E/R$, where the resistance of the solution R is inversely proportional to the mobility of the ions present. However, after a short time, the accumulation around each electrode of ions of opposite charge, together with the depletion of ions of like charge, will produce an opposing potential in the solution, the polarization potential, which will cause the current to fall to zero[d] as an equilibrium state is reached.

As the potential is increased, there is a point at which no equilibrium state is reached, but instead, an appreciable steady current flows which will obey Ohm's law over a reasonable range of applied potential. The potential at which this steady current is observed is called the decomposition potential because it is accompanied by chemical reaction (electrolysis) at the electrode surfaces. These electrode reactions are quite generally the oxidation (anode) and reduction (cathode) of ionic or molecular species present in the solution. If the reactions at the electrodes are reversible, then the decomposition potential E_D is related by the Nernst equation to the free energy changes of the electrode reactions

$$E_D = \frac{-\Delta F}{n\mathfrak{F}} \qquad\qquad (XVII.8.1)$$

In this equation $\mathfrak{F}$ is the Faraday (96,500 coulombs/equivalent) and $\Delta F/n$

---

[a] For an excellent and detailed summary on oxidation of metals, see chap. 14, by T. B. Grimley, of W. E. Garner (ed.), "Chemistry of the Solid State," Academic Press, Inc., New York, 1955.

[b] The theory for thin films has been worked out by N. F. Mott, *J. chim. Phys.*, **44**, 172 (1947); *Trans. Faraday Soc.*, **43**, 429 (1947).

[c] F. S. Stone, *Advances in Catalysis*, **9**, 492 (1957), has made the interesting observation that the large heat of reaction in the oxidation of Ni films can greatly affect the structure of the oxide layers and the rate of oxidation of the metal.

[d] We assume here that there are no disturbances due to gravitational convection.

is the sum of the free energy changes that take place at the electrodes per equivalent of charge transported through the solution.

As an example, the electrolysis of $ZnBr_2$ solution will take place reversibly at a decomposition potential of about 1.3 volts to produce Zn metal at the cathode and liquid bromine, Br_2 (and Br_3^-), at the anode according to the stoichiometric equation

$$Zn^{++} + 2Br^- \rightarrow Zn + Br_2 \qquad n = 2$$

In many cases, the electrode reactions are not reversible and the decomposition potential is observed to be in excess of the thermodynamically calculated value. The excess voltage, referred to as an overvoltage, is found to vary with the nature and surface area (e.g., roughness) of the electrodes, impurities in the solution, and the actual current density passing through the solution. The relation between current density I_D and overvoltage E was investigated by Tafel,[a] who proposed the very successful empirical equation

$$E = a + b \log I_D \qquad (XVII.8.2)$$

where the parameters a and b are characteristic of a given system.[b]

In the case of both reversible and irreversible electrode reactions, methods are now available for studying the steady-state and transient currents, and there has been much progress in the analysis of these currents in terms of the kinetic processes involved.[c]

These reactions at electrode surfaces, to which may be added the processes of the solution of metals, corrosion, etc., are closely related to the catalytic reactions at interfaces which we have already discussed. As in the latter cases, the over-all rate may be separated into a number of distinct steps, any one of which under appropriate circumstances can become rate-controlling. These are

1. Diffusion of reactant species to the electrode
2. Chemisorption of reactant species at the electrode
3. Chemical reactions on the electrode surface
4. Desorption of product species from the electrode
5. Diffusion of product species away from the electrode

[a] J. Tafel, *Z. physik. Chem.*, **50**, 641 (1905).

[b] The value of b is generally in the range of 0.1 to 0.2 volts and can be written in terms of the Nernst coefficient as $RT/\alpha \mathfrak{F}$; at 25°C α will be in the range 0.3 to 0.5. The value of a is much more sensitive than is b to the other parameters of the system.

[c] For a more detailed discussion of these points the reader is referred to G. Kortum and J. O'M. Bockris, "Textbook of Electrochemistry," Elsevier Publishing Company, Amsterdam, 1951, and to I. M. Kolthoff and J. J. Lingane, "Polarography," Interscience Publishers, Inc., New York, 1952. More quantitative treatments will be found in J. O'M. Bockris, "Modern Aspects of Electrochemistry," Butterworth & Co. (Publishers) Ltd., London, 1954, and in P. Delahay, "New Instrumental Methods in Electrochemistry," Interscience Publishers, Inc., New York, 1954.

In the case of very simple chemical changes, such as the reversible oxidation of Fe^{++} to Fe^{3+} at an anode surface, steps 2, 3, and 4 may all merge into the formation of a transition complex with some species on the electrode surface and the transfer of an electron from Fe^{++} to the electrode.[a]

In the case of the deposition of metal ions, such as the electroplating of Zn^{++}, steps 4 and 5 are merely the reverse of 1 and 2. In the case of the deposition of gases (for example, H_2 and O_2) or the more complex reduction of species such as NO_3^-, the surface chemistry may become very involved, with one or more metastable intermediates playing an important role.[b]

Because of the possible complexity of the electrode process no general treatment is possible, but some of the main concepts may be illustrated in terms of a simple model in which we shall *assume* a definite kinetic path. Let us consider an electrolysis system that consists of an inert metal electrode in contact with oxidized and reduced forms of a dissolved ionic species which can react at the electrode according to the stoichiometric equation

$$O^z + ne^- \underset{2}{\overset{1}{\rightleftharpoons}} R^{z-n} \qquad\qquad (XVII.8.3)$$

where z and $z - n$ are the respective charges on O and R.[c]

At equilibrium the ions O^z and R^{z-n}, together with the negative ions and solvent species (H_2O, OH^-, H^+), will reach some equilibrium concentration on the surface of the metal electrode.[d] There will be a potential difference between the metal electrode and the bulk of the solution whose magnitude may be measured relative to some reference electrode such as the standard hydrogen or calomel electrodes. For convenience let us refer our working electrode to the standard hydrogen electrode taken as zero. Its potential is then related to the concentrations (O) and (R) in the solution by the Nernst equation

$$E = E_{ox}^{\circ} - \frac{RT}{n\mathfrak{F}} \ln \frac{(O)}{(R)} \qquad\qquad (XVII.8.4)$$

[a] As in the case of homogeneous reactions, this may be accomplished by atom transfer (for example, H atoms).

[b] See Bockris, *loc. cit.*, for detailed discussion. The deposition of metals from solvated ions is particularly difficult to study because of the speed and sensitivity to electrode surface of the electrode process. For deposition the chemical process on the surface may involve migration of a chemisorbed metal atom (or ion) to lattice sites.

[c] Typical examples might be $Fe^{3+} + e^- \rightleftharpoons Fe^{++}$ or $Fe(CN)_6^{3-} + e^- \rightleftharpoons Fe(CN)_6^{4-}$.

[d] The chemical system at the surface will be quite complex. There will be strongly bound species at various sites on the metal surface. Where these are ionic, the electric field established at the surface will tend to attract ions of opposite charge from the solution. The first layer has been termed the electrode double layer, while the gegenion distribution in the solution is called the diffuse double layer. A theoretical analysis of the double layer has been made by Gouy and Chapman and adapted to kinetic analysis by Stern. For references, and discussion see paper by D. C. Grahame, *J. Chem. Phys.*, **21**, 1054 (1953).

where E°_{ox} is the standard oxidation potential (Latimer-Lewis-Randall convention) of the working half-cell and again for convenience we are using concentrations in place of the more proper activities.

If this cell is now connected to an adjustable voltage (e.g., a battery), the potential E can be changed at will and the reaction system, $O \rightleftharpoons R$, driven in the forward or reverse direction. Let us assume that the potential at E is made more positive, so that the net reaction at the electrode is the reduction of O to R. Eventually there will be reached a new equilibrium state in which the ratio $(O)/(R)$ will be smaller as indicated by Eq. (XVII.8.4):[a]

$$\frac{[(O)/(R)]_f}{[(O)/(R)]_i} = \exp\left(\frac{-n\mathfrak{F}\,\Delta E}{RT}\right) \qquad (XVII.8.5)$$

where ΔE is the change in potential of our working electrode.

Since the increase in potential acts to decrease O and increase R, this may take place by either an increase in the rate of reduction of O or by an accompanying relative decrease in the reverse oxidation of R to O.[b] For convenience it is usual to take both these possibilities into account by associating a fraction α of the potential change ΔE with increase in the rate of reduction and the residue $(1 - \alpha)\,\Delta E$ with the decrease in the rate of oxidation. While in principle α may be positive or negative, it has been found to lie near the value of $\frac{1}{2}$, which can be interpreted as a symmetry in the charge transport with respect to potential in the transition state.[c] That is, half of the change in potential accelerates the reduction, while the other half decelerates the oxidation.

If we assume for purposes of an example that the rate of oxidation is simply proportional to the concentration of R in the vicinity of the electrode, then by considerations of the equilibrium, the rate of reduction is first-order in O and we can write for the rates of change

$$\frac{-d(O)}{dt} = \frac{d(R)}{dt} = k_R(O) - k_O(R) \qquad (XVII.8.6)$$

which is zero at equilibrium, and from the Nernst equation (XVII.8.4)

---

[a] We are assuming for the moment that the changes in our reference half-cell are of negligible importance or else compensated, so that all of the changes take place in the working half-cell.

[b] Since the potential difference between the electrode and the solution, at equilibrium, occurs across the two double layers, near the interface, it is likely that this difference in rates represents a difference in the work of transport of charge across these double layers. Depending on the mechanism it may be associated with the transport of either O or R to the electrode. Note that the actual charge transfer may occur by atom transfer from a solvated O (or R) to species sorbed on the electrode surface.

[c] M. Volmer and T. Erdey-Grúz, Z. physik. Chem., **A159**, 165 (1931). T. Erdey-Grúz and H. Wick, ibid., **A162**, 53 (1932). See also J. A. V. Butler, Trans. Faraday Soc., **28**, 379 (1932), and J. A. V. Butler and G. Armstrong, J. Chem. Soc., 743 (1934).

$$\frac{(O)_e}{(R)_e} = \frac{k_O}{k_R} = \exp\left[\frac{-n\mathfrak{F}(E - E^\circ_{ox})}{RT}\right] \qquad \text{(XVII.8.7)}$$

$$= \frac{k^\circ_O}{k^\circ_R} \exp\left[\frac{-n\mathfrak{F}E}{RT}\right] \qquad \text{(XVII.8.8)}$$

where the standard rate constants k°_O and k°_R are the respective values when $E = 0$ [i.e., when O and R are in their standard states and the system is at equilibrium]. If we now make use of the parameter α, we can separate k_O and k_R:

$$k_O = k^\circ_O \exp\left[\frac{-(1 - \alpha)n\mathfrak{F}E}{RT}\right]$$

$$k_R = k^\circ_R \exp\left[\frac{\alpha n\mathfrak{F}E}{RT}\right] \qquad \text{(XVII.8.9)}$$

The current density which will be observed at our working electrode is equal to the difference in the specific rates of oxidation and reduction

$$\begin{aligned} I &= I_O - I_R \\ &= n\mathfrak{F}[k_R(O) - k_O(R)] \\ &= n\mathfrak{F}k_R(O)\left[1 - \frac{k_O(R)}{k_R(O)}\right] \end{aligned} \qquad \text{(XVII.8.10)}$$

In the equilibrium state, this is, of course, zero.[a] When the potential is increased, then the concentration (O) is depleted near the surface, while (R) is increased. Eventually a quasi-stationary state will be established, one in which the depletion of O at the electrode is balanced by a diffusion of O from the solution, while the excess of R will similarly be balanced by a diffusion of R away from the electrode to the solution.[b] We can write for such a stationary diffusion state:

$$\frac{I}{n\mathfrak{F}} = -k_{DO}[(O) - (O)_B] = k_{DR}[(R) - (R)_B] \qquad \text{(XVII.8.11)}$$

where the mass transfer coefficients k_{DO} and k_{DR} are functions of the diffusion coefficients of O and R, respectively, $(O)_B$ and $(R)_B$ refer to the concentrations in the bulk of the solution, and (O) and (R) refer to the

---

[a] At equilibrium the equal and opposite current densities I_O and I_R are referred to as the exchange currents. They could be measured, in principle, by measuring the rate of exchange of suitably labeled tracer O or R as catalyzed by the surface. For a discussion of such exchange measurements see C. V. King, *Ann. N.Y. Acad. Sci.*, **58**, 910 (1954).

[b] The time required to establish such stationary states will depend on the diffusion coefficients of the ions in the solution and the size of the electrodes. For a small spherical electrode of radius r_0 the time for establishment of a quasi-stationary state will be of the order of $t \cong r_0^2/\pi^2 D$. For $r_0 = 0.1$ cm and $D \cong 10^{-5}$ cm²/sec, t is about 100 sec, so that for large electrodes the times can become quite long. For ions, the diffusion of O and R is not independent of the speed of negative ions in the solution because the condition of electroneutrality requires a coupled diffusion mechanism.

vicinity of the electrode. By eliminating (O) and (R) between Eq. (XVII.8.10) and Eq. (XVII.8.11) we find

$$I = \frac{n\mathfrak{F}k_R(O)_B[1 - k_O(R)_B/k_R(O)_B]}{1 + \dfrac{k_R}{k_{DO}} + \dfrac{k_O}{k_{DR}}} \qquad \text{(XVII.8.12)}$$

By using Eq. (XVII.8.9) this becomes

$$I = \frac{n\mathfrak{F}k_R^\circ(O)_B e^{\alpha n\mathfrak{F}E/RT}\{1 - [k_O^\circ(R)_B/k_R^\circ(O)_B]e^{-n\mathfrak{F}E/RT}\}}{1 + (k_R^\circ/k_{DO})e^{\alpha n\mathfrak{F}E/RT} + (k_D^\circ/k_{DR})e^{-(1-\alpha)n\mathfrak{F}E/RT}} \qquad \text{(XVII.8.13)}$$

In the extreme case of very high overvoltages ($n\mathfrak{F}E \gg RT$), the rate of reaction becomes limited by the diffusion of O to the electrode and the above equation reduces to

$$I \xrightarrow{n\mathfrak{F}E \gg RT} I_D = n\mathfrak{F}k_{DO}(O)_B \qquad \text{(XVII.8.14)}$$

This is a situation which has been studied with microelectrodes[a] and in which the precise solution of the diffusion equation can be easily applied to the data. In the case that the chemical reaction at the electrode is not sufficiently fast compared to the diffusion to the electrode, then Eq. (XVII.8.13) or its appropriate variant can be used to explore the mechanism of the electrode process.

A considerable amount of work has been done on the deposition of H_2 gas at cathodes and on O_2 gas at anodes. The reduction of H^+ ions at a metal cathode to form H_2 gas is no less complex than the process of catalytic hydrogenation. The current-voltage relation is very sensitive to trace impurities[b] in the solution and also to the metal used and its conditioning. The data obtained with smooth Pt and Pd electrodes has been interpreted as due to a slow recombination of sorbed H atoms,[c] while on Hg[d] the mechanism has been presented as a slow H atom transfer from H_3O^+ near the surface to the metal electrode.

The kinetics of O_2 evolution is very complex because of the intermediate formation of the not too stable H_2O_2. On an Hg surface however, the $(O_2)/(H_2O_2)$ equilibrium has been satisfactorily studied, and reproducible data on the EI curve have been obtained.[e] For a detailed discussion of some of the more complex redox systems, the reader is referred to the texts by Bockris.

The solution of a metal by displacement of H^+ from solution as H_2 gas

[a] See text by Kolthoff and Lingane, loc. cit.

[b] A. M. Azzam et al., Trans. Faraday Soc., 46, 918 (1950), showed that even at 10^{-10} moles/liter of "impurities," effects on the EI curves could be detected. This is, of course, reasonable if sorption and desorption on "active" sites of the electrode surface are a rate-limiting process, as they are in the ortho- to para-H conversion.

[c] Text by Bockris, loc. cit.; S. Schuldiner, J. Electrochem. Soc., 99, 488 (1952).

[d] F. P. Bowden and K. E. W. Grew, Discussions Faraday Soc., 1, 86, 91 (1947).

[e] N. E. Yablokova and V. S. Bagotski, Doklady Akad. Nauk S.S.S.R., 85, 599 (1952.)

represents a case in which anodic and cathodic processes take place at the same electrode (these are called polyelectrode processes) and diffusion and chemical processes can both become slow processes. Early work on the solution of Na amalgams[a] in acids and bases indicated that the rate was first order in H^+ and approximately $\frac{1}{2}$ order in the Na concentration. This has been interpreted as a diffusion-controlled process in acid solutions. However, studies by Kilpatrick et al.[b] on the rate of solution of metals in various acids and solvents have shown that the rate is proportional to the concentration of undissociated acid and that the relative rate constants with different acids could be correlated by a Brönsted plot. They have interpreted all this as evidence for a specific acid-catalyzed transfer of a proton from the undissociated acid molecule to the metal surface[c] as a rate-controlling step.

The rate of solution of salts such as NaCl has been interpreted as a diffusion-controlled reaction in stirred systems, the diffusion taking place across a boundary layer of saturated salt solution on the surface of the salt crystals. While this is probably a correct picture for simple salts such as the alkali metal halides, it may not necessarily be the case for the salts of the higher-valence transition metals, where the slowness of ligancy change may be rate-controlling. Thus anhydrous $CrCl_3$ is only very slowly soluble in water, the rate being independent of the rate of stirring. It is found that small amounts of Cr^{++} in the solution have an enormous effect on the rate, the mechanism presumably involving charge transfer between Cr^{++} in solution and Cr^{3+} in the solid.[d] It would seem that such systems would be well worth further investigation.

9. Phase Formation

Physical changes of state are observable under suitable conditions as well-defined phenomena. However the very frequent occurrence of superheating and supercooling in liquids, supersaturation of vapors (e.g., in closed chambers), and the persistence of metastable solids (e.g., monoclinic sulfur at 0°C) show that these phase changes can be at times exceedingly

[a] J. N. Brönsted and N. L. R. Kane, J. Am. Chem. Soc., **53**, 3624 (1931). The study of metal corrosion is, of course, a closely related field.

[b] M. Kilpatrick and J. H. Rushton, J. Phys. Chem., **38**, 269 (1934). M. Sclar and M. Kilpatrick, J. Am. Chem. Soc., **59**, 584 (1937). F. A. Fletcher and M. Kilpatrick, J. Phys. Chem., **42**, 113 (1938). W. G. Dunning and M. Kilpatrick, ibid., 215 (1938).

[c] Their amalgam studies showed that the rate for very strong acids and some of the weak acids, under conditions in which the H_2O reaction with the amalgam can be neglected, is zero-order in metal concentration. The water reaction with Na-Hg is $\frac{1}{2}$ order in metal but independent of acid.

[d] Interpretation by R. A. Ogg, Jr., private communication.

slow.[a] While it seems to have been generally appreciated that the mechanism of phase transition had to take place through the growth of small nuclei of the new phase which were somehow created in the old phase, it was Gibbs[b] who was among the first to point out that, for it to grow, a small nucleus would have to exceed a certain critical size.

Let us consider as an example the case of a saturated vapor which has been suddenly and adiabatically compressed to a vapor pressure P which is in excess of its equilibrium vapor pressure P_0 at the final temperature T. In order for liquid to form, it must grow by the growth of small droplets. If, however, we consider a very small droplet of the liquid phase present in the vapor, it will have an excess free energy, compared to bulk liquid, that is due to its extra surface. The magnitude of the excess surface energy is $4\pi r^2\sigma$, where σ is the surface tension and r is the radius of the drop. In order for the drop and vapor to be in equilibrium, the vapor pressure P must exceed the saturation vapor pressure P_0 by an amount which can be calculated from the Gibbs-Kelvin equation:[c]

$$\ln \frac{P}{P_0} = \frac{2\sigma v_M}{rRT} \tag{XVII.9.1}$$

where v_M is the molar volume of the liquid.

For any supersaturation ratio P/P_0 this equation gives the radius of a critical size drop whose vapor pressure corresponds to P. Drops of smaller radius will have a larger vapor pressure and tend to evaporate, while larger drops will have smaller vapor pressures and will tend to grow in size indefinitely.[d]

[a] Associated phenomena such as the apparent inability of liquids to support tension and the related process of cavitation produced in liquids under high shear rates have been shown by E. N. Harvey et al., *J. Am. Chem. Soc.*, **67**, 156 (1945), to be due to microscopic bubbles of gas entrained on surfaces. In the absence of such nuclei L. J. Briggs, *J. Chem. Phys.*, **19**, 970 (1952), has observed reproducible tensile strengths for H_2O and organic liquids of the order of 100 to 300 atm, which is what might be expected on theoretical grounds.

[b] J. W. Gibbs, "Collected Works," vol. 2, Yale University Press, New Haven, Conn., 1948.

[c] This equation can be derived from the condition that at constant temperature the total free energy of the system (vapor + drop), $F = n_v\mu_v + n_l\mu_l + 4\pi r^2\sigma$, must be a minimum with respect to exchange of particles between vapor and drop. Thus $dF = \mu_v\,dn_v + \mu_l\,dn_l + 8\pi r\sigma\,dr = 0$ and together with $-dn_v = dn_l = 4\pi r^2\,dr/v_M$ this leads to the condition that the difference in chemical potentials, $\mu_v - \mu_l = 2\sigma v_M/r$. For an ideal gas this can also be written as $\mu_v - \mu_l = -RT\ln(P/P_0)$.

[d] A similar argument can be made to show that there will be a critical size bubble of vapor in a liquid which has been superheated or put under tension. The pressure P in such a critical size bubble of vapor must exceed the external pressure on the system P_0 by an amount given by $P = P_0 + 2\sigma/r$. P_0 is determined by the external pressure (or tension) applied to the liquid, while P is determined by the vapor pressure of the liquid and for rough calculations may be approximated by the Clausius-Clapeyron equation.

The model which has been successful in explaining condensation from a supersaturated vapor assumes that in the saturated vapor there is an equilibrium distribution of small droplets whose concentrations can be calculated from the equilibrium constant K_n of the equation

$$nN \rightleftharpoons N_n \qquad (XVII.9.2)$$

where N_n represents a drop (or cluster) that contains n molecules of the vapor. On increasing the pressure from P_0 to P, there is a change to a new equilibrium distribution of droplets characteristic of the new pressure. However, at this new pressure all droplets in excess of the critical size will grow indefinitely, because they are metastable, and this will lead to condensation.

As a first approximation we can assume that the rate of condensation R_c will be given by the concentration of critical size drops N_c, multiplied by the rate Z_c at which vapor molecules condense on their surfaces S_c

$$\frac{-d(N)}{dt} = R_c = N_c S_c Z_c \alpha \qquad (XVII.9.3)$$

$$= K_c(N)^c (4\pi r_c^2)\alpha \left(\frac{RT}{2\pi M}\right)^{\frac{1}{2}} \qquad (XVII.9.4)$$

where K_c is the equilibrium constant for the formation of a critical size cluster containing c molecules of radius r and α is the probability that a molecule of vapor N will stick to such a cluster on collision.

Now the change in free energy on forming a single cluster of n molecules from the vapor can be expressed as

$$\frac{\Delta F_n}{N_{Av}} = \frac{n}{N_{Av}}(\mu_l - \mu_v) + 4\pi r_n^2 \sigma$$

$$= \frac{4}{3}\frac{\pi r^3}{v_M}(\mu_l - \mu_v) + 4\pi r_n^2 \sigma \qquad (XVII.9.5)$$

where μ_l and μ_v are the chemical potentials of liquid and vapor, respectively, and v_M is the molar volume of the liquid.[a] For the critical size cluster N_c which is at equilibrium with the supersaturated vapor, we can use the Gibbs-Kelvin relation,[b] $\mu_l - \mu_v = -2\sigma v_M/r$ in Eq. XVII.9.5 to get

$$\Delta F_c = \frac{4}{3}\pi r_c^2 \sigma N_{Av} \qquad (XVII.9.6)$$

and on substitution from Eq. XVII.9.1 for r

[a] In this crude model we are assuming that it is meaningful to use the macroscopic quantities σ and v_M in discussing the properties of a cluster of molecules which may contain as few as 50 molecules. Further refinement of such a model is hardly necessary at the present stage of experimental development.

[b] Note that this corresponds to the condition $\partial(\Delta F_n)/\partial n = 0$ given in Eq. (XVII. 9.5).

$$\Delta F_c = \frac{16\pi\sigma^3 v_M^2 N_{\text{Av}}}{3R\,T\,(\ln P/P_0)^2} \qquad \text{(XVII.9.7)}$$

Since $RT \ln K_c = -\Delta F_c$, we can rewrite Eq. (XVII.9.4) as[a] ($N_1^\circ \cong PV/kT$):

$$\frac{-d \ln (N)}{dt} = \frac{64\pi^2\alpha\sigma^\circ v_M^2}{R^2 T^2 (\ln P/P_0)^2} \left(\frac{RT}{2\pi M}\right)^{1/2} N_1^\circ \exp\left[\frac{-16\pi\sigma^3 v_M^2 N_{\text{Av}}}{3R^3 T^3 (\ln P/P_0)^2}\right]$$

$$\text{(XVII.9.8)}$$

The interesting feature of this last formula is that the rate of condensation depends exponentially on the supersaturation ratio P/P_0. Thus for H_2O at 25°C, the exponent in the equation is -200 when P/P_0 is 2.8. However, when $P/P_0 = 7.2$, the exponent has reduced to -50, i.e., the rate of condensation has increased by 10^{150}![b] It can be seen that the rate of condensation may be negligibly small at one value of P/P_0 and become immeasurably fast at some slightly higher value. This is in accord with observations which have been made. From the very nature of the process it can be seen that quantitative verification of the equation will be extremely difficult. However, qualitative agreement has been obtained.[c]

The problem of crystal growth from supercooled liquids has been formulated in terms of a similar model based on the interfacial tension of microcrystals in the solution. A number of experimental studies which have been made have given further support to the qualitative concepts of the model.[d] The time lag in nucleation required for the distribution of nuclei to change from the equilibrium value at saturation to the stationary concentration at supersaturation has been discussed in some detail by Kantrowitz.[e]

Depending on the mutual solubilities of reactants and products, chemical and physical reactions in (or between) solids may involve phase changes. In most reactions of solids, the process of diffusion is sufficiently slow that it becomes a rate-controlling process and nucleation is relatively unimportant. This has been found to be the case in the reaction of

[a] By choosing normal liquid in equilibrium with vapor as the standard state for measuring free energies, the factor $(N)^c$ in Eq. (XVII.9.4) becomes unity.

[b] For water at 25°C with $\sigma = 80$ dynes/cm and $v_M = 18$ cc/mole, $r_c \cong 12$ Å at $P/P_0 = 2.8$ and $c \cong 160$ molecules. Because of the small variation of σv_M among liquids, these numbers will not be very different for other liquids.

[c] M. Volmer and A. Weber, Z. physik. Chem., **119**, 277 (1925), were the first to develop this model. It was given a more kinetic turn by R. Becker and W. Döring, Ann. Physik, **24**, 719 (1935). Further refinements are discussed by M. Volmer, "Kinetik der Phasenbildung," J. W. Edwards, Publisher, Inc., Ann Arbor, Mich., 1948; by H. N. V. Temperley, "Changes of State," Interscience Publishers, Inc., New York, 1956; by J. Frenkel, "Kinetic Theory of Liquids," Oxford University Press, New York, 1946; and by W. G. Dunning in W. E. Garner (ed.), "Chemistry of the Solid State," Academic Press, Inc., New York, 1955.

[d] See references in preceding footnote for a bibliography of experimental studies.

[e] A. Kantrowitz, J. Chem. Phys., **19**, 1097 (1951).

CoO + ZnO to form mixed oxides[a] and in the reactions of KCl + CsBr.[b] The reaction of solid Ag and sulfur, using particles of known size, has been shown to fit the equation for diffusion into a spherical sink with very reasonable values for the various diffusion parameters.[c]

In a number of other cases, however, notably in the exothermic decompositions of solids which can become explosive[d] and in the endothermic transformation of hydrates of salts and of carbonates (to oxides),[e] the slow processes seem to be nucleation. The rate laws for such nucleation-controlled processes[f] can be very complex, and the rate studies are difficult to make and to reproduce. In many of these cases small amounts of impurities play an important role in governing the development and growth of nuclei, and there are a number of instances of small amounts of water vapor having a significant catalytic effect.[g]

Such effects make quantitative studies of these reactions very difficult but by no means less interesting. Another difficulty associated with all studies of the reactions of solids is the dependence of the reaction rate on the previous history of the solid. A major contribution to such erratic behavior is the property of solids of being able to exist for long periods of time in metastable states of physical stress. This introduces into the description of solids an additional set of thermodynamic variables which are not necessarily at the disposal of the experimenter or even observable.

10. Enzyme-catalyzed Reactions

Biological processes at the level of the single cell or at the level of the more complex, multicellular forms of life constitute some of the most intricate and challenging problems of chemistry and chemical kinetics. From the enormous amount of work that has been done on elucidating the elementary kinetic pathways in biological processes, some few generalizations can be made. One of these is that most discrete structural steps[h] in biochemical processes are catalyzed by large molecules called enzymes.

[a] J. A. Hedvall, Z. anorg. Chem., **86**, 201, 296 (1914); **92**, 301, 369, 381 (1915).

[b] H. L. Link and L. J. Wood, J. Am. Chem. Soc., **60**, 2320 (1938). H. F. Mason, J. Phys. Chem., **61**, 796 (1957), has discussed the diffusion processes for the solid alkali metal halides.

[c] W. P. Rieman, J. Phys. Chem., **61**, 813 (1957).

[d] See articles by W. E. Garner, P. W. M. Jacobs and F. C. Tompkins, and C. E. H. Bawn and A. R. Ubbelhode in Dunning, "Chemistry of the Solid State," op. cit. A very interesting symposium on molecular mechanism of rate processes in solids will be found in the Discussions Faraday Soc., **23** (1957).

[e] J. H. DeBoer, Discussions Faraday Soc., **23**, 171 (1957).

[f] Garner, Jacobs, Tompkins, Bawn, and Ubbelhode, loc. cit.

[g] S. W. Benson and R. L. Richardson, J. Am. Chem. Soc., **77**, 4206 (1955), have noted that the reaction $Na_2SO_4(s) + HCl(g) \rightarrow NaCl \cdot NaHSO_4(s)$ is extremely slow at 25°C but goes fairly rapidly in the presence of small partial pressures of H_2O even under conditions such that $Na_2SO_4 \cdot 10H_2O$ is unstable.

[h] Exclusive of simple ionization.

These enzymes can be generally classified as molecules that are composed of a proteinlike section which accounts for most of the molecular weight, and usually one "prosthetic" grouping of smaller molecular weight which is presumed to be the site of the catalytic activity. The porphyrin nucleus containing a chelated metal ion (Fe^{++}, Co^{++}, Cu^{++}, etc.) is found to be a common part of prosthetic units in plant and animal systems. Thus animal hemoglobin contains such a unit with Fe^{++} (the heme) attached to the protein moiety (globin). It is similar in structure to the Mg^{++}-containing prosthetic unit in the chlorophyll of plants and one-celled animals. While the molecular weights of the proteins are usually in the range of 30,000 to 80,000, they may be lower or very much higher. The specificity of enzymes is usually indicated by the very particular reaction they catalyze. Thus the isolatable material fumarase catalyzes the equilibrium between malic and fumaric acid:[a]

$$
\begin{array}{c}
\text{H} \\
| \\
\text{H—C—COOH} \\
| \\
\text{HOOC—C—H} \\
| \\
\text{OH}
\end{array}
\rightleftharpoons \text{HOH} +
\begin{array}{c}
\text{H—C—COOH} \\
\| \\
\text{HOOC—C—H}
\end{array}
$$

Urease catalyzes the hydrolysis of urea to ammonia, while catalase is capable of rapidly decomposing H_2O_2 to $H_2O + O_2$.[b] In a number of cases, enzyme catalysis requires the participation of a smaller molecule usually referred to as a coenzyme. Many of the vitamins and simple nucleotides such as adenosine triphosphate (ATP) have been shown to act as coenzymes.

The kinetic role of enzymes was first given a general formulation by Michaelis and Menten,[c] They proposed that the molecule undergoing reaction (substrate S) is adsorbed reversibly on a specific site E of the enzyme to form a stable enzyme-substrate $S \cdot E$ complex whose subsequent decomposition into products is rate-controlling. This scheme, which resembles that suggested by Langmuir for surface catalysis, can be represented by

$$S + E \underset{2}{\overset{1}{\rightleftharpoons}} S \cdot E$$

$$S \cdot E \overset{3}{\rightarrow} E + \text{products}$$

(XVII.10.1)

[a] R. A. Alberty and G. G. Hammes, *J. Phys. Chem.*, **62**, 154 (1958), have shown that the very fast specific rate constants observed for this reversible reaction are compatible with a diffusion-controlled rate such that every collision of a fumarate ion with an active site leads to reaction. This is apparently true also for the combination of NO with iron hemoglobin and H_2O_2 with yeast peroxidase.

[b] A very broad survey of the respiratory enzymes will be found in H. A. Lardy (ed.), "Respiratory Enzymes," Burgess Publishing Co., Minneapolis, 1949.

[c] L. Michaelis and M. Menten, *Biochem. Z.*, **49**, 333 (1913).

If the total initial concentration of enzyme is $(E_0) = (E) + (S \cdot E)$ and we assume that $(S \cdot E)$ reaches a stationary state, the latter is given by

$$(S \cdot E)_{ss} = \frac{k_1(S)(E)}{k_2 + k_3}$$

$$= \frac{k_1(S)(E_0)}{k_1(S) + k_2 + k_3}$$

(XVII.10.2)

where $(S \cdot E)_{ss}/(E_0)$, in analogy to the Langmuir scheme, can be looked upon as θ, the fraction of occupied sites.

The stationary rate of reaction is then given by:

$$R = -\frac{d(S)}{dt} = \frac{k_1 k_3 (S)(E_0)}{k_1(S) + k_2 + k_3} = \frac{k_3(S)(E_0)}{(S) + K_M}$$

(XVII.10.3)

where $K_M = (k_2 + k_3)/k_1$.

By inverting both sides, this can also be written as

$$\frac{1}{R} = \frac{1}{k_3(E_0)} + \frac{K_M}{k_3(E_0)} \frac{1}{(S)}$$

(XVII.10.4)

If the data are plotted in the form of (S) against t and a smooth curve drawn through the individual points, then the slope at any point will be R and a replot of $1/R$ against $1/(S)$ should give a straight line of intercept $1/k_3(E_0)$ and slope $K_M/k_3(E_0)$. If the constant (E_0) is known, then it is possible from such data to obtain the two constants K_M and k_3.[a]

Equation (XVII.10.3) can be integrated directly to give

$$(S_0) - (S) + K_M \ln \frac{(S_0)}{(S)} = k_3(E_0)t$$

(XVII.10.5)

where $(S) = (S_0)$ at $t = 0$. Approximate forms of this last equation, valid during the early stages of the reaction, e.g., when $[(S_0) - (S)]/(S_0) \ll 1$ can be obtained as

$$\ln \frac{(S_0)}{(S)} \cong \frac{k_3(E_0)t}{S_0 + K_M}$$

(XVII.10.6)

or with nearly equal precision, during this same period:

$$(S_0) - (S) = \frac{k_3(S_0)(E_0)t}{(S_0) + K_M}$$

(XVII.10.7)

The appropriateness of the stationary-state assumption is open to question in such a system when the amount of enzyme-substrate complex is large compared to free enzyme and (E_0) is not very much smaller than (S) (Sec. III.9). When, however, $(S) \gg (E_0)$ or $(E \cdot S) \ll (E_0)$ then the stationary-state hypothesis is valid for amounts of reactiᵣ $(S_0) - (S) > (E \cdot S)$.

The phenomenon of competitive inhibition exhibited by heterogeneᴄ

[a] It is not possible from such data alone to obtain the individual constants k_1 and k_2 or even the ratio $k_1/k_2 = K_{1,2}$.

catalysts is also demonstrated by enzymes. A competitive inhibitor S′ is assumed to be strongly sorbed at the same reaction sites E as the substrate S and thus blocks the reaction by cutting down the effective number of active sites. If we add to the scheme [Eq. (XVII.10.1)] the inhibition step

$$S' + E \underset{i'}{\overset{i}{\rightleftharpoons}} S' \cdot E \qquad K_i = \frac{k_i}{k_{i'}} \qquad \text{(XVII.10.8)}$$

and apply the stationary-state hypothesis to $(S' \cdot E)$ and $(S \cdot E)$, then the stationary concentrations are

$$\theta = \frac{(E)}{(E_0)} = \frac{1}{1 + K_i(S') + (S)/K_M}$$

$$(S' \cdot E)_{ss} = K_i(S')(E) = K_i(S')(E_0)(\theta) \qquad \text{(XVII.10.9)}$$

$$(S \cdot E)_{ss} = \frac{(S)(E)}{K_M} = \frac{(S)(E_0)(\theta)}{K_M}$$

This gives for the stationary rate of reaction

$$R = -\frac{d(S)}{dt} = k_3(S \cdot E)_{ss} = \frac{k_3(S)(E_0)}{K_M + K_iK_M(S') + (S)} \qquad \text{(XVII.10.10)}$$

Rewriting this as

$$\frac{1}{R} = \frac{K_M[1 + K_i(S')]}{k_3(E_0)(S)} + \frac{1}{k_3(E_0)} \qquad \text{(XVII.10.11)}$$

we see that a plot of $1/R$ against (S) at fixed values of (E_0) and (S') should yield a straight line of intercept $1/k_3(E_0)$ and slope $K_M[1 + K_i(S')]/k_3(E_0)$ from which a value of $K_M[1 + K_i(S')]$ can be obtained. A plot of this quantity against (S') obtained from experiments with different added concentrations of inhibitor should then give a straight line of slope K_iK_M and an intercept [at $(S') = 0$] of K_M. The kinetics of many inhibited enzyme reactions have been shown to be amenable to such treatment.[a]

There may be observed a noncompetitive type of inhibition in which it is assumed that the inhibitor S′ operates either by being sorbed on a site adjacent to the substrate, where it slows down the rate-controlling step 3, or inhibiting the sorption of a coenzyme or other species on this adjacent site. The rate expressions in such a case have a form different from that just given for competitive inhibitions.[a,b]

The reality of an enzyme-substrate complex was first demonstrated by Stern,[c] who showed that the brown color of a catalase solution changed to

[a] A more extensive discussion will be found in the article by F. M. Huennekeus in A. Weissberger (ed.), "Technique of Organic Chemistry," vol. 13, Interscience Publishers, Inc., New York, 1953.

[b] E. R. Ebersole et al., Arch. Biochem., **3**, 399 (1944).

[c] K. G. Stern, J. Biol. Chem., **114**, 473 (1936). D. Keilin and T. Mann, Proc. Roy. Soc. (London), **B122**, 119 (1939), observed a similar change with H_2O_2. H. Theorell, Enzymologia, **10**, 250 (1941), was able to show that a green complex precedes the red

red on addition of C_2H_5OOH. By using spectrophotometric techniques to observe complex directly, Chance[a] was able to study the very rapid rate of complex formation between H_2O_2 and catalase and also the subsequent fast bimolecular step

$$E \cdot H_2O_2 + H_2O_2 \rightarrow E + 2H_2O + O_2$$

in which O_2 is produced. The rates of competing reactions of the complex with other H donors such as alcohols, which were oxidized to aldehydes, were also measured.

When accurate data can be obtained over a range of both concentrations and temperatures,[b] it is possible from the Michaelis-Menton model to obtain data on the first-order rate constant k_3 and the constant $K_M = (k_2 + k_3)/k_1$ and their apparent activation energies E_3 and E_M.[c] Unfortunately, most of the values quoted in the literature for the activation energies of enzyme-catalyzed reactions are derived from the use of overly simple first-order equations to describe the reaction. Consequently these values are a composite of K_M, k_3, and the other constants in the Michaelis-Menton equation and cannot be used for interpretive purposes. Where the constants have been separated[d] it is found that the values of E_3 are low and of order of magnitude of 5 to 15 Kcal/mole. It is of interest to note that enzyme preparations from different biological sources, which may show different specific activity for a given reaction, have very nearly the same temperature coefficient for their specific rate constants.[e]

Proteins are composed of amino acids whose side chains may contain acid or basic groups, and many proteins have concentrations of such groups in the range of 1 millimole per gram of protein. In addition, the peptide

complex. R. C. Jarnagin and J. H. Wang, *J. Am. Chem. Soc.*, **80**, 786 (1958), have shown by tracer studies that the reaction of the complex with H_2O_2 to produce O_2 leaves the O—O bond intact.

[a] B. Chance in "Technique of Organic Chemistry," vol. 8, *op. cit.* The value obtained for the second-order rate constants for the formation of catalase-H_2O_2 complex is about 1×10^7 liters/mole-sec. This is sufficiently close to the estimated value of 10^9 expected for a diffusion-controlled reaction to indicate that almost every encounter of H_2O_2 with an active site is effective in complex formation. The rate constant for complex + $H_2O_2 \rightarrow O_2$ is also of the same order of magnitude, which seems surprisingly high for a reaction which involves bond breaking. Spectrophotometric techniques have been used by B. L. Vallee et al., *J. Am. Chem. Soc.*, **80**, 397 (1958), to demonstrate reversible complexing of inhibitors to zinc-containing enzymes.

[b] It is not always possible to obtain data over too broad a temperature range because the rates may be too slow at low temperatures, while protein denaturation can become a limiting factor at temperatures in excess of 40 or 55°C.

[c] Note that $E_M = [(k_2E_2 + k_3E_3)/(k_2 + k_3)] - E_1$, which, depending on the relative magnitudes of k_2 and k_3, can take on values over the range $E_2 - E_1$ to $E_3 - E_1$. Thus E_M may show strong temperature dependence.

[d] J. P. Hoare and K. J. Laidler, *J. Am. Chem. Soc.*, **71**, 2699 (1949); **72**, 2487 (1950).

[e] I. W. Sizer, *Advances in Enzymol.*, **3** (1943).

linkages in the protein skeleton will be quite polar and capable of acting as weak acids and bases.[a] As a result, the properties of proteins are very sensitive to pH and there is usually a corresponding sensitivity of enzyme activity to pH.[b] The pH dependence of the Michaelis-Menton constants k_3 and K_M can be quite complex and is not too easy to study. One of the reasons for this is that the pH study requires a buffer solution, and it is not infrequent to find that there are specific interactions between the buffer components (notably $HPO_4^{=}$) and the enzyme. In addition there are effects of ionic strength on the protein activity and substrate activity which further complicates the interpretation of buffer behavior.[c] The requirements of buffers make it difficult, if not impossible, to make the usual extrapolations of ionic strength to infinite dilution, and this makes theoretical interpretation of the data very difficult.[d]

All of the preceding remarks apply equally well to the studies which have been made of the effects of high pressures on the rate constants of enzyme reactions[e] and also protein reactions.[f] The values of $\Delta V_3^{\ddagger}$ and $\Delta V_M^{\ddagger}$ which can be obtained from the pressure coefficients of the rate constants cannot be naively interpreted in terms of simple volume changes of the enzyme or protein without a careful assessment of the other parameters of the system and their changes with pressure.[g]

Finally, it should be remarked that proteins and enzymes, being macromolecular, may not necessarily represent identical structures; instead they may be composed of a range of closely related but slightly different molecules. In such a case we might expect to find that they would have to be characterized by additional parameters and that their sorption isotherms

[a] S. W. Benson and J. M. Seehof, *J. Am. Chem. Soc.*, **77**, 2579 (1955), and earlier papers.

[b] This is further complicated by the acid and base groups which may be attached to the prosthetic group of the enzyme.

[c] This has not always been appreciated. In all of the equations used in this section, concentrations should be replaced by activities. When substrate concentrations have been varied over large ranges, activity corrections may become important. Thus in the urease-urea reaction, G. B. Kistiakowsky and A. J. Rosenberg, *J. Am. Chem. Soc.*, **74**, 5020 (1952), have studied the rate over a urea range of 0.0003 to 2.0 M. The fall-off in rate at the high urea concentrations which has been observed is probably attributable to changes in activity of species in the system rather than to changes in mechanism.

[d] Perhaps one of the most extensive investigations to date of all of these effects is to be found in the work of C. Niemann and collaborators, *J. Am. Chem. Soc.*, **80**, 1457, 1465, 1469, 1473, 1481 (1958), who studied the α-chymotrypsin–catalyzed hydrolysis of various esters in aqueous solution.

[e] F. H. Johnson and I. Lewin, *J. Cellular Comp. Physiol.*, **28**, 1, 23, 47, 77 (1946), and earlier papers.

[f] D. H. Campbell and F. H. Johnston, *J. Am. Chem. Soc.*, **68**, 725 (1946).

[g] Thus ionization of various groups is usually accompanied by a decrease in partial molar volume due to electrostriction of the solvent. Changes in ionization with pressure may easily obscure other apparent pressure effects on the rate constant.

might resemble the Freundlich or the complex Langmuir isotherm. Such fine distinctions are rendered somewhat obscure by the difficulty associated with obtaining enzymes of reproducible and constant specific activity. This is in part attributable, as in the case of solid catalysts, to changes in the active sites or their number, perhaps caused by trace amounts of ions which are strongly sorbed and can act as poisons.

APPENDIX A

Some Atomic Constants[a]

Name	Symbol	Value and unit
Electronic charge...........	e	4.803×10^{-10} esu
Electronic mass............	m_e	9.108×10^{-28} g
Avogadro's number.........	N_{Av}	6.025×10^{23} molecules/mole
Velocity of light...........	c	2.998×10^{10} cm/sec
Planck's constant...........	h	6.625×10^{-27} erg-sec
	$\hbar$	$h/2\pi = 1.0545 \times 10^{-27}$ erg-sec
Gas constant...............	R	8.317×10^7 ergs/mole-°K
Faraday...................	F	96,493 abs coulombs/equivalent
Boltzmann constant........	k_B	$R/N_{Av} = 1.380 \times 10^{-16}$ erg/molecule-°K
First Bohr radius..........	a_0	$\hbar^2/m_e e^2 = 5.292 \times 10^{-9}$ cm
Rydberg constant for H.....	R_H	109,678 cm^{-1}
Rydberg constant for ∞ mass	R_∞	109,737 cm^{-1}
Bohr magneton............	μ_0	$\hbar e/2m_e c = 0.9273 \times 10^{-20}$ erg/gauss
Loschmidt number.........	N_0	2.687×10^{19} molecules/cc (STP) (perfect gas)
Molar volume at STP.......	V_0	22,414 cc (STP)/mole (perfect gas)
Magnetic moment of electron	μ_e	0.9284×10^{-20} erg/gauss
Magnetic moment of proton	μ_p	1.410×10^{-23} erg/gauss

[a] Values taken from "American Institute of Physics Handbook," sec. 7-3, McGraw-Hill Book Company, Inc., New York, 1957.

APPENDIX B

Some Useful Mathematical Relations

Series Expansions

Binomial

$$(x \pm y)^n = x^n \pm nx^{n-1}y + \frac{n(n-1)}{2}x^{n-2}y^2 + \cdots + \frac{n!}{k!(n-k)!}x^{n-k}(\pm y)^k + \cdots$$

Taylor's Series

$$f(x_0 + x) = f(x_0) + xf'(x_0) + \frac{x^2}{2}f''(x_0) + \cdots + \frac{x^n}{n!}f^n(x) + \cdots$$

Exponential

$$e^x = 1 + x + \frac{x^2}{2} + \cdots + \frac{x^n}{n!} + \cdots$$

Logarithmic

$$\ln(1 + x) = x - \frac{x^2}{2} + \frac{x^3}{3} - \cdots - \frac{(-x)^n}{n} + \cdots \qquad -1 < x < 1$$

Trigonometric

$$\sin x = x - \frac{x^3}{3!} + \frac{x^5}{5!} - \frac{x^7}{7!} + \cdots$$

$$\cos x = 1 - \frac{x^2}{2!} + \frac{x^4}{4!} - \frac{x^6}{6!} + \cdots$$

$$\tan x = x + \frac{x^3}{3} + \frac{2x^5}{15} + \frac{17x^7}{315} + \cdots$$

Sums

$$\sum_1^n k = 1 + 2 + 3 + \cdots + n = \frac{n(n-1)}{2} \qquad \sum_1^n k^3 = \frac{n^2(n+1)^2}{4}$$

$$\sum_1^n k^2 = \frac{n(n+1)(2n+1)}{6} \qquad \sum_1^\infty \left(\frac{1}{k^2}\right) = \frac{\pi^2}{6}$$

Integrals

Gamma Function

$$\Gamma(n) = \int_0^\infty x^{n-1}e^{-x}\,dx = (n-1)\Gamma(n-1)$$

For $n = integer$

$$\Gamma(n) = (n-1)! \qquad \Gamma(1) = 1$$

$$\Gamma(\tfrac{1}{2}) = \int_0^\infty x^{-\frac{1}{2}}e^{-x}\,dx = 2\int_0^\infty e^{-y^2}\,dy = \sqrt{\pi}$$

$$\Gamma(n+\tfrac{1}{2}) = (n-\tfrac{1}{2})\,\Gamma(n-\tfrac{1}{2}) = (n-\tfrac{1}{2})(n-\tfrac{3}{2})\cdots\tfrac{1}{2}\sqrt{\pi}$$

$$\int x^m e^{ax}\,dx = \frac{x^m e^{ax}}{a} - \frac{m}{a}\int x^{m-1}e^{ax}\,dx = \frac{m!\,e^{ax}}{a^{m+1}}\left[\frac{(ax)^m}{m!} - \frac{(ax)^{m-1}}{(m-1)!} + \cdots + (-1)^m\right]$$

$$\text{for } m = integer$$

$$\int x^m e^{-ax^2}\,dx = \frac{1}{2a^{(m+1)/2}}\int y^{(m-1)/2}e^{-y}\,dy$$

which can now be treated like preceding case

$$\text{Also} \qquad = -\frac{x^{m-1}e^{-ax^2}}{2a} + \frac{(m-1)}{2a}\int x^{m-2}e^{-ax^2}\,dx$$

$$\int_0^\infty x^m e^{-ax^2}\,dx = \frac{1}{2a^{(m+1)/2}}\int_0^\infty y^{(m-1)/2}e^{-y}\,dy = \frac{\Gamma[(m+1)/2)]}{2a^{(m+1)/2}}$$

$$\int_0^\infty x^m e^{-ax}\,dx = \frac{\Gamma(m-1)}{a^{m+1}}$$

APPENDIX C

Some Thermodynamic Data for Atoms, Molecules, and Free Radicals in the Gas Phase

Atoms

Atom or species	C_p°	S°	ΔH_f°	$D^{\circ\ a}$
H.	4.97	27.4	52.1	
D.	4.97	29.5	53.0	
O.	5.24	38.5	59.2	
F.	5.44	37.9	18.3	
Cl.	5.22	39.5	29.0	
Br.	4.97	41.8	26.7	
I.	4.97	43.2	25.5	
S.	5.66	40.1	56.3	
N.	4.97	36.6	112.8	
C.	4.98	37.8	171.7	
Hg.	4.97	41.8	14.5	
Li.	4.97	33.1	37.1	
Na.	4.97	36.7	26.0	
K.	4.97	39.3	21.5	

Diatomic Molecules and Radicals

	C_p°	S°	ΔH_f°	$D^{\circ\ a}$
H_2.	7.0	31.2	0.0	104.2
D_2.	7.0	34.6	0.0	106.0
O_2.	7.0	49.0	0.0	118.3
F_2.	7.5	48.6	0.0	36.6
Cl_2.	8.1	53.3	0.0	58.0
Br_2.	8.6	58.6	7.3	46.1

Diatomic Molecules and Radicals (*Continued*)

Atom or species	C_p°	S°	ΔH_f°	D° [a]
I₂.....................	8.8	62.3	14.9	36.2
S₂.....................	—	—	29.9	82.7
N₂.....................	7.0	45.8	0.0	225.6
Li₂....................	8.5	47.1	47.6	26.6
Na₂....................	8.8	55.0	34.0	18.0
K₂.....................	8.8	60.0	30.8	12.2
HF.....................	7.0	41.5	−64.2	134.6
HCl....................	7.0	44.6	−22.1	103.2
HBr....................	7.0	47.4	−8.7	87.5
HI.....................	7.0	49.3	6.2	71.4
HO.....................	7.1	43.9	8.0	103.3
DO.....................	7.1	45.4	—	—
HS.....................	7.1	46.7	—	—
NO.....................	7.1	50.3	21.6	150.1
CO.....................	7.0	47.3	−26.4	257.3
SO.....................	7.0	53.0	−15.4	130.9
ClO....................	7.0	—	25	63

Species with Three and Four Atoms

Atom or species	C_p°	S°	ΔH_f°	D° [a]
H₂O....................	8.0	45.1	−57.8	117.9
D₂O....................	8.2	47.4	−59.6	—
H₂S....................	8.1	49.2	−4.8	—
CH₂....................	(8.1)[c]	—	(78)[c]	—
NH₂....................	(8.1)[c]	—	(101)[c]	—
N₂O....................	9.3	52.6	19.5	39.7 (N₂—O); 114.9 (N—NO)[b]
NO₂....................	8.7	57.5	8.1	72.7
CO₂....................	8.9	51.1	−94.1	126.8
O₃.....................	9.1	56.8	34.0	25.2
Cl₂O...................	10.9	63.7	18.2	36
F₂O....................	10.4	59.0	7.6	—
ClO₂...................	10.0	60.6	24.7	60
SO₂....................	9.5	59.4	−71.0	114.8
CHO....................	—	—	(11)[c]	(15)[c]
HCN....................	8.6	48.2	31.2	—
NH₃....................	8.5	46.0	−11.0	(108)[c]
CH₃....................	(8.5)[c]	(47.0)[c]	32.0	(98)[c]
C₂H₂...................	10.5	48.0	54.2	—
H₂O₂...................	10.4	55.5	−32.5	48.5
CH₂O...................	8.5	52.3	−27.7	(90)[c]
SO₃....................	12.1	61.2	−94.5	82.7

Hydrocarbons

Atom or species	C_p°	S°	ΔH_f°	
C₂H₄...................	10.4	52.5	12.5	
C₂H₆...................	12.6	54.9	−20.2	
n-C₄H₁₀...............	23.3	74.1	−30.1	

Hydrocarbons *(Continued)*

Atom or species	C_p°	S°	ΔH_f°	$D^{\circ\ a}$
i-C_4H_{10}...............	23.1	70.4	-32.1	
Neopentane.............	29.1	73.2	-39.7	
$CH_3CH{=}CH_2$...........	15.3	63.8	4.9	
1-butene...............	20.5	73.0	0.0	
2-butene-cis............	18.9	71.9	-1.7	
-trans...........	21.0	70.9	-2.7	
CH_3-$C{\equiv}CH$	14.5	59.3	44.3	
Cyclopropane...........	13.3	56.8	12.7	
Cyclobutane............	17.3	63.4	6.3	
Cyclopentane...........	19.8	70.0	-18.5	
Benzene...............	19.5	64.3	19.8	
Toluene...............	24.8	76.4	12.0	
Styrene................	29.2	82.5	35.2	

Miscellaneous Compounds

	C_p°	S°	ΔH_f°	$D^{\circ\ a}$
CH_3OH.................	10.5	57.3	-48.1	
C_2H_5OH................	15.6	67.3	-56.2	
CH_3OCH_3..............	15.7	63.7	-45.3	
CH_3ONO...............	—	$(68)^c$	-14.9	
CH_3ONO_2..............	—	76	-29.0	
CH_3Cl.................	9.7	55.8	-19.6	
C_2H_5Cl................	15.0	66.2	-25.7	
t-C_4H_9Cl...............	27.3	77.1	-42.8	
C_2H_5Br................	$(15.3)^c$	$(68.7)^c$	-15.3	
$HCOOH$................	9.1	60.1	$(-88)^c$	
CH_3COOH.............	—	70.1	-103.7	
CH_3NH_2...............	12.9	57.7	-6.7	
$C_2H_5NH_2$..............	16.7	$(68.5)^c$	-11.6	
C_2H_5O—OC_2H_5..........	$(29.0)^c$	$(94.3)^c$	-47.8	
CH_3CO—O—O—$COCH_3$...	—	—	$(-119)^d$	
CH_3CHO...............	13.1	63.2	-39.8	
CH_3COCH_3.............	17.9	70.5	-51.7	
CH_3COOCH_3...........	—	$(80.1)^c$	-99.2	
$CH_2{=}C{=}O$.............	11.4	57.1	-14.6	
CH_3SH................	12.1	60.9	(-4.0)	
CH_3SCH_3..............	17.7	68.3	-9.0	

Notes:

1. Standard state: ideal gas phase, 25°C, 1 atm pressure. $\Delta H_f^\circ = 0$ for the elements at 25°C, 1 atm.

2. Principal source of data: *Nat'l Bur. Standards Circ.* 500, 1952, and recent literature. See S. W. Benson and J. H. Buss, *J. Chem. Phys.*, **29** (1958), for details of source references.

3. Units are Kcal/mole for ΔH_f° and cal/mole-°K for C_p° and S°.

^{a} Standard bond dissociation energy of the species X—Y is defined here as the standard enthalpy change for the reaction X—Y $\rightleftharpoons$ X + Y; $D^\circ(XY) = \Delta H_f^\circ(X) + \Delta H_f^\circ(Y) - \Delta H_f^\circ(X{-}Y)$.

^{b} Parentheses show bond broken in case of unsymmetrical species.

^{c} Estimate by author.

^{d} Value uncertain by ± 4 Kcal.

APPENDIX D

The Estimation of Thermodynamic Properties of Simple Molecules in the Ideal Gas State

In Tables D.1 to D.3 are tabulated values to be used in estimating C_p°, S°, and H_f° for gas-phase molecules. The tables are compiled on the basis of the assumption that these thermodynamic properties may be con-

TABLE D.1. PARTIAL ATOMIC CONTRIBUTIONS FOR THE ESTIMATION OF C_p° AND S° FOR SPECIES IN THE GAS PHASE (25°C, 1 ATM)

Atom	C_p°	$S^{\circ\ a}$
H	0.85	21.0
D	1.20	21.7
C	3.75	$(-32.6)_4$, $(-13.5)_3$, $(5.2)_2$, $(22.0)_1$
N	3.40	$(-12.1)_3$, $(5.8)_2$, $(22.9)_1$
O	3.40	$(8.8)_2$, $(25.5)_1$
F	2.40	25.5
Cl	3.70	$(28.4)_1$, $(10.0)_2$
Br	4.20	31.3
I	4.60	33.3
Si	5.90	$(-29.3)_4$
P	—	$(-9.5)_3$
S	4.70	$(12.8)_2$, $(-11.0)_3$, $(-33.5)_4$, $(27.0)_1$

Note: See Appendix C, Notes, for units, standard states, and sources of data.

[a] Subscript designates ligancy of atom, i.e., number of atoms bonded to central atom. The entropy contribution must be corrected by the addition of any electronic entropy, $R \ln q_n$, where q_n is the electronic partition function. The quantity $R \ln \sigma$ must also be subtracted from the total entropy to correct for symmetry. σ is the symmetry number of the final species. Values taken from S. W. Benson and J. H. Buss, *J. Chem. Phys.*, **29** (1958). These quantities are not to be used for cyclic structures such as benzene compounds. Estimates of C_p° and S° are good to about ± 2 cal/mole-°K for most species but may be poorer for heavily substituted species such as neopentane. They may also be poorer for very simple H-containing species such as NH_3 and CH_4.

sidered as made up of individual, additive contributions of the parts of the molecules. Three different sets of tables represent successive degrees of approximation and reliability starting with the assumption of additivity of atomic properties, next the additivity of bond properties and finally the additivity of group properties. The reliability of each table and its method of use are included.

Examples of Use of Partial Atomic Contribution to C_p° and S°

1. For CH_2Cl_2:

$C_p^\circ = 3.75 + 2(0.85) + 2(3.70) = 12.85$
$C_{p,obs}^\circ = 12.2$
$S^\circ = -32.6 + 2(21.0) + 2(28.4) - R \ln 2 = 64.8$
$S_{obs}^\circ = 64.6$

TABLE D.2. PARTIAL BOND CONTRIBUTIONS FOR THE ESTIMATION OF C_p°, S°, AND ΔH_f° OF GAS-PHASE SPECIES AT 25°C, 1 ATM

Bond	C_p°	S°	ΔH_f°	Bond	C_p°	S°	ΔH_f°
C—H	1.74	12.90	−3.83	S—S	5.4	11.6	—
C—D	2.06	13.60	−4.73	C_v—C^a	2.6	−14.3	6.7
C—C	1.98	−16.40	2.73	C_v—H	2.6	13.8	3.2
C—F	3.34	16.90	—	C_v—F	4.6	18.6	—
C—Cl	4.64	19.70	−7.4	C_v—Cl	5.7	21.2	−0.7
C—Br	5.14	22.65	2.2	C_v—Br	6.3	24.1	9.7
C—I	5.54	24.65	15.0	C_v—I	6.7	26.1	—
C—O	2.7	−4.0	−12.0	>CO—H^b	4.2	26.8	−13.9
O—H	2.7	24.0	−27.0	>CO—C	3.7	−0.6	−14.4
O—D	3.1	24.8	−27.9	>CO—O	2.2	9.8	−50.5
O—O	4.9	9.1	21.5	>CO—F	5.7	31.6	—
O—Cl	5.5	32.5	9.1	>CO—Cl	7.2	35.2	−27.0
C—N	2.1	−12.8	9.3	ϕ—H^c	3.0	11.7	3.25
N—H	2.3	17.7	−2.6	ϕ—C^c	4.5	−17.4	7.25
C—S	3.4	−1.5	6.7	(NO_2)—O^c	—	43.1	−3.0
S—H	3.2	27.0	−0.8	(NO)—O^c	—	35.5	+9.0

Notes: See Appendix C, Notes, for units, standard states, and sources of data. See Table D.1 for corrections to entropy for symmetry and electronic contributions. C_p° and S° estimated from rule of additivity of bond contributions are good to about ±1 cal/mole-°K but may be poorer for heavily branched compounds. The values of ΔH_f° are usually within ±2 Kcal/mole but may be poorer for heavily branched species. Peroxide values are not certain by much larger amounts.

All substances in ideal gas state.

[a] C_v represents the vinyl group carbon atom. The vinyl group is here considered a tetravalent unit.

[b] >CO— represents the bond to carbonyl carbon, the latter being considered a bivalent unit.

[c] NO and NO_2 are here considered as univalent, terminal groups, while the phenyl group ϕ, C_6H_5, is considered as a hexavalent unit.

TABLE D.3. PARTIAL GROUP CONTRIBUTIONS FOR THE ESTIMATION OF C_p°, S°, AND ΔH_f° OF GAS-PHASE SPECIES (25°C AND 1 ATM)

Group	C_p°	S°	ΔH_f°
Hydrocarbons			
C-(H)₃(C)	6.20	30.41	−10.08
C-(H)₂(C)₂	5.45	9.42	−4.95
C-(H)(C)₃	4.47	−12.07	−1.48 (−1.90)[a]
C-(C)₄	4.35	−35.10	1.95 (0.50)[a]
Correction for each gauche[a] configuration of large groups	—	—	0.70[a]
C_d-(C_d)(H)₂	5.20	27.6	6.25
[C_d-(C_d)(C)(H)] + [C-(C_d)(H)₃]_av	10.00	38.5	−1.20
Correction for cis/trans isomers	∓1.00[b]	±0.6[b]	±0.50[b]
[C_d-(C_d)(C)₂] + 2[C-(C_d)(H)₃]	16.10	48.7	−9.85
[C-(C_d)(C)(H)₂] − [C-(C_d)(H)₃]	−0.8	−20.9	5.2
[C-(C_d)(C)₂(H)] − [C-(C_d)(H)₃]	1.0	−42.9	8.2
[C-(C_d)(C)₃] − [C-(C_d)(H)₃]	−3.5	−66.5	10.9
C_t-(C_t)(H)	5.3	24.7	27.1
[C_t-(C_t)(C)] + [C-(C_t)(H)₃]	9.3	36.7	17.5
[C-(C_t)(C)(H)₂] − [C-(C_t)(H)₃]	−1.1	−20.2	5.2
[C-(C_t)(C)₂(H)] − [C-(C_t)(H)₃]	−2.0	−41.6	8.4
C_B-(C_B)₂(H)	3.25	11.55	3.30
[C_B-(C_B)₂(C)] + [C-(C_B)(H)₃]	9.25 (8.75)[c]	22.0 (22.5)[c]	−4.20 (−4.50)[c]
Correction for each ortho configuration of large groups[c]	1.2	−1.3	0.3
[C-(C_B)(C)(H)₂] − [C-(C_B)(H)₃]	−0.8 (−0.4)	−20.4 (−20.8)	4.9 (5.2)
[C-(C_B)(C)₂(H)] − [C-(C_B)(H)₃]	−0.8 (−0.4)	−42.2	9.0
[C_B-(C_B)₂(C_d)] + [C_d-(C_B)(C_d)(H)]	7.7	−1.5	12.5
Halogen and Other Compounds			
C-(Cl)(H)₂(C)	8.8	37.8	−15.7
C-(Br)(H)₂(C)	9.1	40.5	−5.2
C-(I)(H)₂(C)	—	—	8.8
C-(Cl)(H)(C)₂	8.5	17.7	−14.4
C-(Br)(H)(C)₂	—	(20.3)[d]	−2.3
C-(Cl)(C)₃	8.7	−7.6	−12.7
C-(Br)(C)₃	9.3	−4.2	−0.4
C-(Cl)₂(H)(C)	12.2	44.2	—
C-(Cl)₃(C)	15.8	50.1	—
C-(F)₃(C)	12.5	42.0	—
C-(Br)(Cl)(H)(C)	12.5	47.0	—
C-(F)₂(Cl)(C)	13.6	43.0	—
C-(F)(Cl)₂(C)	15.0	45.6	—
C_d-(C_d)(Cl)(H)[e]	7.9	35.3	2.7
C_d-(C_d)(Cl)₂	11.5	42.3	—
C_d-(C_d)(Br)(H)[e]	8.1	38.1	12.4
C_d-(C_d)(I)(H)[e]	8.6	40.5	—
C_d-(C_d)(F)₂	9.6	37.2	—
C_d-(C_d)(Br)₂	11.6	46.9	—
C_d-(C_d)(F)(Cl)	10.4	39.8	—
[O-(C)(H)] + [C-(O)(H)₃]	10.5	59.5	−48.1
[O-(C)(C)] + 2[C-(O)(H)₃]	15.7	69.4	−45.3
[C-(O)(C)(H)₂] − [C-(O)(H)₃]	−1.1	−20.4	2.0
[C-(O)(C)₂(H)] − [C-(O)(H)₃]	(−1.2)[d]	−44.3	3.0
[C-(O)(C)₃] − [C-(O)(H)₃]	—	—	3.3
O-(O)(H)	5.2	28.5	−16.3
O-(O)(C) + C-(O)(H)₃	(9.3)[d]	(40.0)[d]	−15.8
[CO-(C)(H)] + [C-(CO)(H)₃]	13.1	65.3	−39.6
[CO-(C)₂] + 2[C-(CO)(H)₃]	17.9	76.2	−51.7
[C-(CO)(C)(H)₂] − [C-(CO)(H)₃]	(−0.8)[d]	(−20.5)[d]	3.3

TABLE D.3 (Continued)

Group	$C_p°$	$S°$	$\Delta H_f°$
Miscellaneous Compounds (Continued)			
[CO-(O)(H)] + [O-(CO)(H)]................	9.1	60.1	−89.6
[CO-(O)(C)] + [O-(CO)(H)] + [C-(CO)(H)₃]...	—	72.3	−104.0
[O-(CO)(C)] − [O-(CO)(H)] + [C-(O)(H)₃]....	—	12.2	5.5
[N-(C)(H)₂] + [C-(N)(H)₃]..................	12.9	59.9	−6.7
[N-(C)₂(H)] + 2[C(N)(H)₃].................	16.6	69.7	−6.6
[C-(C)(N)(H)₂] − [C-(N)(H)₃]..............	−2.4^f	(−20.7)^d	5.2
[N-(C)₃] + 3[C-N(H)₃]....................	—	77.7	−6.6
[S-(C)(H)] + [C-(S)(H)₃]...................	12.12	63.1	−5.5
[S-(C)₂] + 2[C-(S)(H)₃]..................	17.7	74.0	−9.0
[S-(S)(C)] + [C-(S)(H)₃].................	11.0	43.1	—
[C-(S)(H)₂] − [C-(S)(H)₃]................	−1.1	−20.5	4.6
[C-(S)(C)₂(H)] − [C-(S)(H)₃]............	−1.6	−42.2	7.7
[C-(S)(C)₃] − [C-(S)(H)₃]................	−1.8	−64.8	9.3
C-(C)(H)₂(NO₃)...........................	—	58.3	−26.7
C-(C)₂(H)(NO₃)...........................	—	(36.5)^d	−25.4
C(C)(H)₂(CN)	11.0	39.6	—

Notes: See Appendix C, Notes, for units and standard states. See Table D.1 for corrections to entropy for symmetry and electronic contributions. Values of $C_p°$ and $S°$ estimated from these tables are usually good to ±0.3 cal/mole-°K but may be as poor as ±1.5 cal/mole-°K for very heavily substituted compounds. Values of $\Delta H_f°$ are usually within ±0.4 Kcal/mole but for heavily substituted compounds may be in error by as much as ±3 Kcal/mole.

C_d represents a double-bonded C atom, C_t a triple-bonded C atom, and C_B a C atom in a benzene ring.

^{a} Values in parentheses are to be used when corrections for gauche configurations are made. The correction applies to each such configuration. Thus 2-methyl butane has one gauche correction, 2,3-dimethyl butane has two, and 2,2,3-trimethyl pentane has five.

^{b} The upper sign represents correction for cis isomer; lower sign is that for trans isomer. In molecules such as 2-methyl butene-2 in which two methyl groups are cis to each other a cis correction should be applied. Cis correction should be applied for every pair of large groups that are on the same side of a double bond (e.g., two corrections in tetramethylethylene).

^{c} Correction is applied to each independent pair of large groups that are ortho to each other, in which case the values in parentheses should be used as well. For example, two corrections are made in 1,2,3-trimethyl benzene and six corrections in hexamethyl benzene.

^{d} Estimate by author.

^{e} These are averages for cis-trans isomers.

^{f} Although obtained from the data on $C_2H_5NH_2$, this value seems too low by about 1.5 units.

2. For CH_3OCH_3:

$C_p° = 2(3.75) + 6(0.85) + 3.40 = 16.0$
$C_{p,obs}° = 15.7$
$S° = 2(-32.6) + 6(21.0) + 8.8 - R \ln 18 = 63.9$
$S_{obs}° = 63.7$

3. For NO_2:

$C_p° = 3.40 + 2(3.40) = 10.2$
$C_{p,obs}° = 9.1$
$S° = 5.8 + 2(25.5) + R \ln 2 \text{ (electronic)} - R \ln 2 = 56.8$
$S_{obs}° = 57.5$

Examples of Use of Partial Bond Contribution

1. For C_3H_8:

$C_p° = 2(1.98) + 8(1.74) = 17.88 \qquad C_{p,obs}° = 17.6$
$S° = 2(-16.40) + 8(12.90) - R \ln 18 = 64.7 \qquad S_{obs}° = 64.5$
$\Delta H_f° = 2(2.73) + 8(-3.83) = -25.18 \qquad \Delta H_{f,obs}° = -24.8$

2. For ethyl benzene, ϕC_2H_5:

$C_p^\circ = 5(3.0) + 4.5 + 1.98 + 5(1.74) = 30.2 \qquad C_{p,\text{obs}}^\circ = 29.7$

$S^\circ = 5(11.7) + 1(-17.4) + 1(-16.4) + 5(12.90) - R \ln 6 = 85.6$

$S_{\text{obs}}^\circ = 85.0$

$\Delta H_f^\circ = 5(3.25) + 7.25 + 2.73 - 5(3.83) = 7.08$

$\Delta H_{f,\text{obs}}^\circ = 7.1$

Examples of Use of Partial Group Contribution

1. S° for $CH_3\!-\!CH\!=\!C(CH_3)CH_2CH_3$ (3-methyl pentene-2):

The compound has the following groups:

$2[C\!-\!(C_d)(H)_3] + [C_d\!-\!(C_d)(H)] + [(C_d)\!-\!(C)_2] + [C\!-\!(C)(H)_3] + [C\!-\!(C_d)(C)(H)_2]$

From Table D.3 we find the following contributions:

$[C_d\!-\!(C_d)(H)(C)] + [C\!-\!(C_d)(H)_3] = \quad 38.5$

$[C_d\!-\!(C_d)(C)_2] + 2[C\!-\!(C_d)(H)_3] = \quad 48.7$

$[C\!-\!(C_d)(C)(H)_2] = [C\!-\!(C_d)(H)_3] = \qquad\qquad -20.9$

$[C\!-\!(C)(H)_3] = \quad 30.4$

Correction for bulky cis groups $\quad = \quad \underline{\quad 0.6 \quad}$

$\qquad\qquad\qquad\qquad\qquad\qquad\qquad 118.2 \quad -20.9 = \quad 97.3$

Symmetry contribution

(3 methyl groups), $-R \ln \sigma \qquad = \qquad -R \ln 27 = \underline{-6.5}$

$\qquad\qquad\qquad\qquad\qquad\qquad\qquad\qquad\qquad S^\circ = \quad 90.8 \text{ cal/mole-}^\circ K$

2. S° for $CCl_3 \cdot CH_2CO \cdot OCH(CH_3)CH_2Br$. Working from the carbonyl group we have the following contributions:

$[CO\!-\!(C)(O)] + [O\!-\!(CO)(H)] + [C\!-\!(CO)(H)_3] = \quad 72.3$

$[O\!-\!(CO)(C)] - [O\!-\!(CO)(H)] + [C\!-\!(O)(H)_3] = \quad 12.2$

$[C\!-\!(CO)(C)(H)_2] - [C\!-\!(CO)(H)_3] \qquad\qquad = \qquad -20.5$

$[C\!-\!(O)(C)_2(H)] - [C\!-\!(O)(H)_3] \qquad\qquad = \qquad -44.3$

$[C\!-\!(Cl)_3(C)] \qquad\qquad\qquad\qquad\qquad\qquad = \quad 50.1$

$[C\!-\!(H)_3(C)] \qquad\qquad\qquad\qquad\qquad\qquad = \quad 30.4$

$[C\!-\!(Br)(C)(H)_2] \qquad\qquad\qquad\qquad\qquad = \quad \underline{40.5}$

Total $\qquad\qquad\qquad\qquad\qquad\qquad\qquad = 205.5 \quad -64.8 = 140.7$

Symmetry contribution, $-R \ln \sigma \qquad\qquad = \qquad -R \ln 9 = -4.4$

Optical activity (entropy of mixing) $\qquad = \qquad R \ln 2 = \underline{\quad 1.4}$

$\qquad\qquad\qquad\qquad\qquad\qquad\qquad\qquad\qquad S^\circ = 137.7 \text{ cal/mole-}^\circ K$

Approximate Bond Dissociation Energies of Some Organic Compounds (R-X) in the Gas Phase (25°C, 1 atm)

X:	H	Cl	Br	I	OH	CH_3	C_2H_5
R	Dissociation energies						
H	104	103	88	71	118	102	98
CH_3	102	81	68	54	88	84	83
C_2H_5	98	81	68	53	90	83	82
$(CH_3)_2CH$	94	81	67	51	90	81	80
$(CH_3)_3C$	90	78	63	47	89	78	77
C_6H_5	105	87	73	61	96	95	94
$C_6H_5CH_2$	84	66	51	38	77	69	68
$CH_2{=}CH{-}CH_2$	80	70	53	41	76	65	64
CH_3CO	89	79	—	—	110	81	83
C_2H_5O	102	—	—	—	42	80	82

Note: These are values in Kcal/mole for the standard enthalpy change at 25°C, 1 atm. The values for the diatomic species are accurate to within 0.5 Kcal. For all other species the expected reliability is ± 2 Kcal and may be as large as ± 4 Kcal. Where heats of formation of radicals are known, the method of group contributions may be used to deduce values for other radicals related by the additions of known groups. Thus

$$\Delta H_f^{\circ}(CH_3{-}\dot{C}H_2) = 26 \text{ Kcal/mole}$$

from which we can deduce a value of 36 Kcal/mole for the radical group $\dot{C}{-}(C)(H)_2$. This can now be used together with the group values in Table D.3 to deduce values for related radicals.

As examples we can estimate the ΔH_f° for the following radicals: $CH_3CH_2\dot{C}H_2 = 21$, $ClCH_2\dot{C}H_2 = 20$, $HOCH_2CH_2\dot{C}H_2 = -15$, etc. Because of the lack of sufficient experimental data, it is not yet possible to estimate the reliability of these group additivity rules for free radicals.

Estimate of C_p° and S° for free radicals may be made to within ± 2 cal/mole-°K by taking the values for closely related compounds and making corrections for symmetry and electronic states (Table D.1) to S°. Thus C_p° and S° for NH_3 and $\dot{C}H_3$ are probably very close to each other. In similar fashion the values of C_p° and S° (properly corrected) are probably also within ± 2 cal/mole-°K for the radicals $H\text{—}C\equiv\dot{C}$, $H\text{—}\dot{C}\text{=}O$, and $H\text{—}O\text{—}\dot{O}$ and the molecules HCN and HNO (a correction of about 3 cal/mole-°K may be made to S° for nonlinearity). The methods of partial bond and atom contributions may also be used to estimate C_p° and S° for radicals to about the same reliability as indicated for the molecular species.

Problems

1. The specific rate constant for a $3/2$-order reaction of a single reactant has the value 3.0×10^{-5} liter$^{1/2}$/mole$^{1/2}$-sec at 30°C. Calculate the value of the rate constant at 30°C in the following units: (a) cc$^{1/2}$/mole$^{1/2}$-min; (b) cc$^{1/2}$/molecule$^{1/2}$-sec.

If the initial concentration of reacting material is 0.042 mole/liter, calculate the initial rate of disappearance and the half-life in seconds of the material. Calculate also the time for the concentration to fall to 0.002 mole/liter. All at 30°C.

2. A pure gas X is found to decompose into products by a reaction which can be described by a simple stoichiometric equation. When pure X is introduced into a flask at 150°C, the following series of pressure readings is observed:

Total pressure (mm Hg)	150.3	157.5	167.5	193.1	249.8
time (min)	0	90	210	516	∞

Compute from these data: (a) the stoichiometry of the reaction, (b) the order of the reaction, (c) the rate constants in units of mm Hg and minutes, and (d) the rate constant in units of moles/liter and seconds.

3. A second-order reaction (single component) is found to be 75 per cent completed in 92 min when the initial concentration of reactant is 0.24 M. How long will it take for the concentration to reach 0.16 M under the same conditions?

4. The reaction $NO + \frac{1}{2}O_2 \rightarrow NO_2$ is found to be first-order in O_2 and second-order in NO. Show that, when the NO is present in very large excess, the rate of formation of NO_2 is closely represented by an equation which is first-order in O_2. What is the limiting approximate form of the rate law for NO_2 when O_2 is present in large excess?

5. The reversible dimerization of C_2F_4 proceeds in the gas phase between 300 and 450°C to form cyclo-C_4F_8. The forward reaction is second-order in C_2F_4 and the reverse is first-order in C_4F_8:

$$2C_2F_4 \underset{k_2}{\overset{k_1}{\rightleftharpoons}} C_4F_8$$

(a) Derive the differential rate expression for C_2F_4 in terms of the initial and instantaneous concentrations and k_1 and k_2. (b) Obtain the integrated expression for C_2F_4 as a function of t.

6. The N_2O_5-catalyzed gas-phase decomposition of O_3 to produce O_2 is presumed to go through the following sequence of reactions:

$$N_2O_5 \overset{1}{\underset{2}{\rightleftharpoons}} NO_2 + NO_3 \qquad K_{1,2} \ll 1$$

Chain:
$$\begin{cases} NO_2 + O_3 \overset{3}{\to} NO_3 + O_2 \\ \\ 2NO_3 \overset{4}{\to} 2NO_2 + O_2 \end{cases}$$

(a) Assume that reaction 1 is first-order and all other steps are second-order and write down the rate equations for all species. (b) By employing the stationary-state hypothesis for the intermediates NO_2 and NO_3 calculate their stationary-state concentrations and also the expression for the rate of reaction of O_3 in terms of the concentrations of O_3 and N_2O_5. Note that N_2O_5 is not consumed in the reaction. (c) Calculate the rate law for O_3 disappearance if step 4 is replaced by the second-order step: $O_3 + NO_3 \overset{5}{\to} 2O_2 + NO_2$. Do you think it possible to distinguish these rate laws experimentally?

7. The stoichiometric reaction $2A + B \to 2D$ is found to proceed via the formation of an intermediate C which is not necessarily present in small concentration. The mechanism is

$$A + B \overset{1}{\to} C \qquad C + A \overset{2}{\to} 2D$$

where both steps are second-order mixed. By using the technique of eliminating time as a variable (Sec. III.7A) obtain an equation for C as a function of B, assuming $C = 0$ at $t = 0$. Obtain also the equation for dB/dt in terms of A_0, B, B_0, and k_1 and k_2.

8. For the reaction of Prob. 7 compute the stationary-state concentration of C. Is this also its maximum concentration? Explain. When $A_0 = B_0$ at $t = 0$ and $k_2 = 25k_1$, calculate the amount of B used up when $C = 95$ per cent of its stationary-state concentration. How much D will have been formed at this time? If $A_0 = 0.02\ M$ and $k_1 = 3.0 \times 10^{-4}$ liter/mole-sec, calculate approximately how long this will take.

9. At 60°C the reaction of a pure gas A is found to obey a rate law $-dA/dt = kA^3$ with $k = 4.2 \times 10^{-6}$ when A is measured in mm Hg pressure. At 100°C $k = 1.4 \times 10^{-4}$ in the same units. (a) Calculate the activation energy for the reaction and the Arrhenius A factor. (b) Compute the values of k at both 60 and 100°C when A is measured in units of moles/liter. (c) Calculate the activation energy and the Arrhenius A factor for the rate constant measured in the units of moles/liter and seconds. Are these the same values as those obtained in (a)? Is there a true activation energy for the reaction? Explain.

10. At 400°C the decomposition of NO_2 into $NO + O_2$ proceeds to completion with a second-order rate constant which can be represented by $\log k = 8.80 - 25{,}600/4.575T$ with k in units of liters/mole-sec.

If 200 mm Hg pressure of NO_2 gas are introduced into a flask at 400°C, calculate how long it will take for the pressure to reach 240 mm Hg.

11. A reversible reaction is found to follow the stoichiometry:

$$2A + B \overset{1}{\underset{2}{\rightleftharpoons}} 2C + 20\ \text{Kcal} \qquad \text{exothermic}$$

At 100°C the forward reaction 1 is found to follow the rate law

$$\frac{-dA}{dt} = k_1A^2 \qquad \log k_1 = 10.42 - \frac{21{,}600}{4.575T}$$

in units of liters/mole-sec for k_1. (a) What is the rate law for the back reaction 2? Is this rate law uniquely defined? Explain. (b) What is the activation energy for the back reaction? (c) At 100°C, $K_{eq} = 6.2 \times 10^3\ \text{atm}^{-1}$. Compute the value of k_2 both at 100 and at 200°C. (d) Calculate the entropy change in the reaction. (e) The empirical rate law for the forward reaction is zero order in B. Can this be a valid expression for all *possible* concentrations of A and B? Explain.

12. The hydrolysis of an alkyl halide RCl in an alkaline buffer solution at 40°C is followed by withdrawing small aliquots from the reaction mixture at intervals, quenching the sample in acetic acid, and then titrating the mixture for Cl^- ion. The over-all reaction is $RCl + HOH \rightarrow ROH + H^+ + Cl^-$ and the rate is first-order in RCl with an activation energy of about 18 Kcal/mole and a half-life of 3000 sec. (a) If the analytical measurement of Cl^- is good to ± 0.3 per cent and the temperature control to $\pm 0.1°C$ and the sampling time is about ± 2 sec, calculate the expected accuracy in determining k from a rate study in which samples are taken at 1000, 2000, 3000, and 4000 sec and the value of RCl initially was 0.1885 M. (b) If the same procedure is followed at 30°C to obtain a second value of k_1, what is the expected precision of the activation energy determined from these two values of k_1?

13. The isomerization reaction $A \rightleftharpoons B$ is found to go to completion at 65°C. The reaction is found to be of $\frac{4}{3}$ order in A on the basis of two measurements of the half-life determined on initial concentrations of A differing by a factor of 2. The analytical accuracy is ± 2 per cent, the error due to time is negligible, and the temperature control is $\pm 0.1°C$, with $E_{act} = 20$ Kcal. (a) What is the expected precision of measurement of the half-life? (b) Do you believe that the precision of these measurements is sufficient to tell whether the rate is $\frac{3}{2}$ order or $\frac{4}{3}$ order? Explain.

14. The rate of decomposition $A \rightarrow$ products is found to follow the rate law: $-dA/dt = kA^2$ and the following data are obtained in a single experiment

t, ±1 sec	A, mg/cc, ±1%	t, ±1 sec	A, mg/cc, ±1%
0	16.4	600	12.8
200	15.0	800	11.5
400	13.4		

The expected errors of time and analysis are as indicated. Calculate the best second-order rate constant from these data by using pairs of points and weighting each result according to its expected precision. Of these pairs, how many may be omitted because of the inherently low precision?

15. The photosensitized reaction of Br_2 with H_2 at 150°C has been shown to be a chain reaction with the following mechanism in large excess of H_2:

$$\begin{array}{ll} & (Rate) \\ Br_2 + h\nu \rightarrow 2Br & 2I_a \\ \\ Br + H_2 \underset{2}{\overset{1}{\rightleftharpoons}} HBr + H & \text{second order mixed} \\ \\ H + Br_2 \overset{3}{\rightarrow} HBr + Br & \text{second order mixed} \\ \\ Br + Br + H_2 \overset{4}{\rightarrow} Br_2 + H_2 & \text{third order mixed} \end{array}$$

(a) Using the stationary-state hypothesis for Br and H atoms, calculate their stationary-state concentrations in terms of the other concentrations, rate constants, and the rate of absorption of photons I_a. (b) Calculate the differential expression for the stationary rate of production of HBr. (c) Calculate the expression for the quantum yield of HBr production. Under what conditions will the quantum yield be highest? Does this make physical sense? Explain.

16. Two points x_1 and x_2 are selected at random from the x axis between the limits 0 and d. (a) What are the distribution functions for x_1 and x_2? (b) Calculate the average values of x_1 and x_2. (c) Calculate the dispersion of x_1. (d) Calculate $\overline{(x_1 - x_2)}$ and $\overline{(x_1 - x_2)^2}$. (e) If $x_1 \leqslant x_2$, what is the distribution function for x_1? Calculate $\overline{x_1}$ and $\overline{(x_2 - x_1)^2}$ under these conditions.

17. N identical molecules are distributed at random in a flask of volume V. Assume they behave as an ideal gas. (a) What is the probability of finding any given molecule in a small volume element v? (b) Derive an expression for the probability distribution $P(n,v)$ for finding *exactly* n molecules in v. (c) By using Sterling's approximation for $x!$, derive the limiting value of $P(n,v)$ when $v \ll V$. Note, for small x, $\ln (1 + x) = x - x^2/2$.

18. Prove that, if $y = a_1 x_1 + a_2 x_2$, where x_1 and x_2 are independently distributed variables, then $\overline{y} = a_1 \overline{x}_1 + a_2 \overline{x}_2$ and $\sigma_y^2 = a_1^2 \sigma_{x_1}^2 + a_2^2 \sigma_{x_2}^2$.

19. Two spheres of radius r_1 and r_2 are separated by a distance r_0. R_{12} represents the distance between two points R_1 and R_2, each one chosen at random on the surface of each sphere respectively. Calculate the following average quantities: (a) $\overline{R}_{12}$, (b) $\overline{R_{12}^2}^{1/2}$, and (c) $\overline{(1/R_{12})}$.

20. Criticize the following argument: When asked to predict the toss of a coin, most people will call "heads" 60 per cent of the time. Since with a fair coin, heads will appear only 50 per cent of the time, the best winning strategy in coin-tossing games is to let the other person call the toss.

21. The Lennard-Jones potential for the energy of interaction between two gas-phase molecules is usually written as $E(r) = A/r^{12} - B/r^6$, where r is the distance between the centers of the assumed spherical molecules. (a) Calculate the value of r_0, the separation of minimum energy, and E_0, the work required to separate the two species from this point to infinity. (b) Sketch qualitatively the shape of the curve $E(r)$ as a function of r and also the function $F(r)$, the force between the two molecules.

22. Calculate the half-life (in seconds) for the pressure to fall in a 1-liter flask containing Br_2 gas at 25°C and 1.0 mm Hg pressure when there is a hole 0.002 cm in diameter in a thin wall of the flask connected to a vacuum pump.

23. Calculate the rate at which molecules in a 2-dimensional container will make collisions with the container walls (per unit length).

24. At 0°C the coefficient of viscosity of $N_2 = 1.66 \times 10^{-4}$ poise. From this compute the collision diameter of N_2 and compare it with the diameter of N_2 computed from (a) the volume of solid N_2 assuming hexagonal close packing (i.e., 12 nearest neighbors), (b) Van der Waals constant b, computed from the critical volume of N_2 gas. (c) Compute the coefficient of self-diffusion of N_2 gas at STP.

25. (a) Calculate the number of binary collisions/cm^3-sec in O_2 gas at 25°C, 1 atm pressure, assuming a collision diameter of 3.0 Å. (b) If the duration of a collision between O_2 molecules is 10^{-13} sec, compute the concentration of O_4 complexes in O_2 gas at 25°C and 1 atm pressure. (c) By assuming the diameter of an O_4 molecule is 4.0 Å, compute from your answers in (a) and (b) the number of triple collisions between O_2 molecules per cc per second at 25°C and 1 atm pressure.

26. By using the following molecular data compute the equilibrium constant for the following reaction at 400°C: $H + HBr \rightleftharpoons H_2 + Br$. $r_0(HBr) = 1.41$ Å, $r_0(H_2) = 0.742$ Å, $E_0^\circ = 16.7$ Kcal, $\nu_0(HBr) = 2650$ cm^{-1}, $\nu_0(H_2) = 4395$ cm^{-1}, $g_{elec}(H) = 2$, and g_{elec} (Br) = 4.

27. Given the mass of the earth as 6.0×10^{27} g, the radius as 6.5×10^8 cm, and the gravitational constant $G = 6.67 \times 10^{-8}$ (cgs units). (a) Calculate the kinetic energy required by a particle of mass m to escape from the earth's surface. (b) At 300°K what fraction of the molecules of H_2, H_2O, and O_2 would have velocities (in one direction only) in excess of escape velocity? (c) By assuming that the earth's atmosphere may be idealized by a spherical shell of thickness 1×10^7 cm with a pressure that varies uniformly from 1 atm down to 1×10^{-6} atm at the top surface, calculate the half-life of

H_2, O_2, and H_2O in the earth's atmosphere at 300°K. (*Note:* Assume escape takes place only from the very top of the atmosphere and neglect collisions.)

28. (*a*) What is the mean free path of an O_2 molecule in a 50 mole per cent mixture of $H_2 + O_2$ at STP? (*b*) What is the mean free path of the O_2 molecules with respect to an H_2 molecule? (*c*) What fraction of the O_2 molecules will go 1 cm or more before striking an H_2 molecule?

29. How long will it take a molecule of Cl_2 (at 10 mm Hg pressure, 30°C) at the center of a spherical flask of radius r_0 to diffuse to the walls? How many collisions will it have made in this time with other Cl_2 molecules? [*Note:* Use the distribution function for the random walk to compute mean displacement, Eq. (VI.7.8).]

30. At what pressure in a 1-liter spherical flask at 30°C will the rate of collision of N_2 molecules with each other be equal to the rate of collisions with the wall? At this pressure what is the ratio of flask surface to N_2 collision surface?

31. (*a*) Calculate the probability that in a dilute gas of hard sphere molecules, a single molecule of translational energy $10RT$ will emerge from its next collision with energy in excess of $10RT$. (*b*) Calculate the average energy this molecule will have after its next collision.

32. One method of measuring the vapor pressures of solids or liquids having very low vapor pressures is to measure the loss in weight of the material when placed in an evacuated container at fixed temperature having an effusive leak to a vacuum chamber. The device is called a Knudsen cell, and its use rests on the principle that an equilibrium vapor pressure will be established in the chamber. (*a*) The area of the leak in a Knudsen cell is 0.0050 cm². It is found that 18.2 mg of a material having a molecular weight of 320 g/mole is lost from the cell in 3 hr and 45 min at a temperature of 500°C. Calculate the vapor pressure of the material at this temperature. (*b*) Is it true that the observed vapor pressure P_{obs} in the cell will be the equilibrium vapor pressure P_e which would exist in the absence of a leak? Show that the ratio $P_{obs}/P_e = (1 + A_h/aA_c)^{-1}$, where A_h/A_c is the ratio of the area of the leak to that of the cell and a is the probability that a vapor molecule striking the inner surface of the cell will stick to it. (*c*) If there exists a reversible vapor equilibrium of the form $2A \rightleftharpoons A_2$, can the Knudsen cell data still be used to determine vapor pressures? Explain.

33. C_2H_5 radicals and H atoms recombine to form an excited state $C_2H_6^*$ of ethane, the bond energy being 98 Kcal at 300°K. If the bond dissociation energy of the process $C_2H_6 \rightarrow 2CH_3$ is 84 Kcal at 300°K, calculate, by using the RRK theory and any reasonable assumptions about A and n, the mean lifetime at 300 and 600°K for decomposition of $C_2H_6^*$ into $2CH_3$. If A is 10^{17} sec⁻¹ and deactivation occurs only at 1 in 10 collisions, calculate the pressure at 300°K at which CH_3 production is inhibited by a factor of 2 from the low-pressure limit.

34. $CH_2CO \rightarrow CH_2 + CO - 70$ Kcal. If the RRK A factor for the decomposition of CH_2CO is 10^{16} sec⁻¹ and arises from an entropy of activation in the transition state, calculate the pressure range at 850°K at which the decomposition may be expected to be about one-half the limiting high-pressure rate. Make any reasonable assumptions about n and deactivation. If the high A factor is associated not with the transition state alone but also with the energized state, how does that affect your answer?

35. The reaction $CH_3Cl \rightleftharpoons CH_3 + Cl - 81$ Kcal ($\Delta H°$) may proceed through an upper excited electronic state $[CH_3Cl]_e$ whose dissociation energy is only 18 Kcal to give the same $CH_3 + Cl$. (*a*) How would this affect the rate of decomposition of CH_3Cl? Consider in your answer the pressure dependence. (*b*) How would this affect the rate of recombination at 500°K of $CH_3 + Cl$? Discuss the pressure dependence of the rate.

36. Energy transfer and metathesis are both second-order reactions. The former can have preexponential factors much in excess of collision frequencies, while that of the latter never exceeds collision frequencies. Explain in qualitative terms the reasons for this.

37. Explain qualitatively why the recombination of $CH_3 + O_2$ is more efficient at low temperatures than at high temperatures.

38. The molecule X can decompose unimolecularly in either of two ways: $X \xrightarrow{1} A + B - 60$ Kcal, or $X \xrightarrow{2} C + D - 50$ Kcal. (a) Show that the existence of the low-energy path, reaction 2, makes the high-energy reaction 1 less likely to happen. (b) By using the RRK model, derive a general expression for the rates of the two reactions.

39. The gas-phase decomposition of cyclopropane, C, to propylene, P, has been proposed as proceeding through a trimethylene biradical, T.

Path:
$$\begin{cases} C \underset{2}{\overset{1}{\rightleftharpoons}} T - 54 \text{ Kcal} \qquad \Delta S° \cong 10 \text{ cal/mole-°K} \\ \\ T \xrightarrow{3} P + 61.5 \text{ Kcal} \qquad \Delta S° \cong 2 \text{ cal/mole-°K} \end{cases}$$

Over-all: $C \rightleftharpoons P + 7.5$ Kcal $\Delta S° \cong 8$ cal/mole-°K

(a) By neglecting pressure effects, calculate the rate of isomerization in terms of k_1, k_2, and k_3. (b) If the activation energy of k_3 is 11 Kcal/mole while E_2 is only 4 Kcal/mole, calculate the over-all activation energy for the isomerization. What is the estimated mean lifetime of a molecule of T in the system, assuming that $A_2 = 1 \times 10^{14}$ sec^{-1}? (c) If the preexponential factor $A_3 = 2 \times 10^{13}$ sec^{-1}, compute the over-all, first-order rate constant for the decomposition.

40. By assuming that the rate constant for the reaction of a trimethylene free radical with RH is 1×10^{-8} exp $(-10,000/RT)$ liter/mole-sec (Table XII.6), calculate the pressure of RH (at 800°K) necessary for H abstraction to compete equally with reaction 3 in the preceding problem. Repeat for reaction 2. Suppose that NO combines with free radicals at 1 in 100 collisions. What pressure of NO at 800°K is needed to compete equally with reaction 3?

41. The reaction $CH_3 + C_2H_5 \rightarrow C_3H_8$ has a rate constant of 4×10^{10} liters/mole-sec ($E_{act} = 0$) at 400°K. By using any reasonable estimates of the entropies of CH_3 and C_2H_5 together with known values of C_3H_8, calculate the frequency factor of the first-order rate constant of the reverse reaction at 300°K.

42. Making any reasonable assumptions about the geometry and force constants in the transition state calculate a value for the frequency factor of the metathesis reaction $Cl + NOCl \rightarrow NO + Cl_2$. *Note:* See paper by Pitzer, Johnston, et al. for one method. Refer to Herzberg for possible values of force constants and geometry.

43. Assuming that direct bimolecular reaction of $H_2 + Br_2$ has a rate constant given by $\log k = 10.0 - 50,000/4.575T$ in units of liters/mole-sec, calculate at 1000°K the pressure of Br_2 at which its ratio to the normal chain rate would amount to 0.10.

44. The recombination of Br atoms in the presence of N_2 at 1000°K has a third-order constant equal to 2×10^9 liters^2/mole^2-sec. Using the thermodynamic data for the dissociation of Br_2, calculate at 1000°K the specific second-order rate constant for the rate of dissociation of Br_2 by N_2. What is the frequency factor for the rate constant? How does it compare with collision frequencies?

45. At 1000°K NOCl rapidly decomposes into $NO + Cl_2$. The bond dissociation energy of NOCl is 36 Kcal. Show quantitatively that the rate at which Cl atoms are generated from Cl_2 molecules can be enormously accelerated at 1000°K by the presence of small amounts of NO. Make any assumptions that seem reasonable to you but state them explicitly and also state your reasons for choosing them.

46. At 200°C, neither $COBr_2$ nor HCOBr is a stable compound. Propose a radical chain to account for the thermal bromination of CH_2O to produce CO + HBr with the initiation step being the thermal dissociation of Br_2. Neglect wall reactions. Compute expressions for the stationary-state concentrations of all radical species and the rate expression for $-d(Br_2)/dt$. Try to justify all steps included in the scheme.

47. The decomposition of NO_2 to give $2NO + O_2$ has been found at 500°C to follow a second-order rate law with log $k_2 = 8.80 - 25,600/4.575T$ (liters/mole-sec). (a) If you wish to represent this rate constant over the temperature range 400 to 600°C by the collision equation $k = ZT^{1/2} \exp{(-E/RT)}$, calculate the values to be used for Z and E. (b) If the data are presented in terms of the transition-state theory $k = (RT/N_{Av}h) \exp{[\Delta S^{\ddagger}/R - \Delta H^{\ddagger}/RT]}$, calculate $\Delta S^{\ddagger}$ and $\Delta H^{\ddagger}$. (c) If the data are presented in terms of the modified Arrhenius equation $k = AT^3 \exp{(-E/RT)}$ with A in units of mm Hg, calculate A and E.

48. If $k(r)$ represents the probability that molecules A and B, separated by a distance r in a solution, will react and $P(r)$ represents the steady-state probability distribution for finding a pair of molecules A and B distant r apart, show that $P(r)$ is close to the random value in a nonreacting solution only when $k(r)$ is very small for small r. Show that when the probability of reaction of A and B on an encounter is very high, $P(r)$ has very strong deviations from equilibrium (i.e., there is very much less chance of finding A near B).

49. Describe qualitatively the behavior of the temperature and the density in a shock wave passing through a gas such as CH_4 in which the equilibration of rotational and vibrational energies with translational energy is slow. What effect does the chemical dissociation into CH_3 and H have on the shock?

50. A solution contains reactant A which decomposes at a first-order rate, $-dA/dt = k_A A$, to give products B + C. It is flowed at constant velocity, V cc/sec, through a cylindrical tube of length X and radius r_0, immersed in a thermostat. (a) Assuming that no reaction occurs in the connecting leads to the reactor tube, calculate the concentration of A as a function of position in the tube. (Neglect mixing in the flow stream.) (b) What is the effect on your calculations of assuming nonturbulent Poiseuille streamline flow through the tube with a parabolic distribution of velocities from the center of the tube to its edges? With this type of flow show how to calculate the average concentration of A in the exit stream.

51. It is found that the kinetics of the redox reaction, $Hg_2^{++} + Tl^{3+} = 2Hg^{++} + Tl^{+}$, follows the rate law

$$-\frac{d(Tl^{3+})}{dt} = \frac{k(Hg_2^{++})(Tl^{3+})}{(Hg^{++})}$$

at constant acidity and constant ionic strength. (a) Devise a mechanism to account for this rate law. (b) It is found that the observed rate constant k decreases as the acidity of the solution is decreased. How can you account for this?

52. The acid-catalyzed bromination of an ester RCH_2COOR' is found to follow the Brönsted-Pedersen law with an exponent $\alpha = \frac{1}{2}$. If the monosuccinate acid ion is found to have a specific rate constant of 3.6×10^{-4} liter/mole-sec, calculate the specific rate constant you would expect for formic acid. K_{ion}(formic acid) $= 2.1 \times 10^{-4}$; K_{ion} (monosuccinate ion) $= 2.4 \times 10^{-6}$.

53. (*Special assignment.*) Browse through the journals of the past two years and select an experimental article on kinetics which you believe to be interesting and of some, although not necessarily great, importance.

Prepare a detailed review of this article in which you first outline briefly the study, experimental range of parameters, and results. Make some spot checks of the author's data to satisfy yourself that he has not been careless in using them. Then comment on the method of treatment of the data, on the errors of analysis, and the expected errors in the results reported. Comment on the validity of the proposed rate law. If a mechanism has been proposed, check its consistency and, to the best of your own ability, comment on its uniqueness and also on alternative mechanisms. How might some of these be tested? Where kinetic models or theories have been employed, check to see that they have been used correctly.

Make a summary statement in which you point out the weaknesses and omissions of and the critical experiments suggested by, the article and in which you decide whether or not the article represents a proper contribution to the literature (i.e., if you were a panel member of an agency allotting research funds, would you vote to support the work?).

Note to the Student: Your report will be judged in part on your choice of article. Avoid articles which are not sufficiently detailed or which do not contain enough experimental work to merit discussion. Articles which are patently in error or arrive at incorrect results may nevertheless be worthy of selection.

Note to the Instructor: If this special assignment is to be made, it is worthwhile to inform the class about it as early as possible in order to give them time to browse properly. In the author's course the problem has been used in place of any other homework assignments during the last month and has counted significantly in the final grade.

Name Index

Adam, J., 333
Adams, G. K., 263
Adamson, A. W., 591
Adirovich, E. I., 55
Aditya, S., 503
Aivazov, B., 488
Akeroyd, E. I., 377
Alberty, R. A., 653
Allen, A. O., 383, 431, 435
Allmand, A. J., 342
Altar, W., 285
Alter, H. W., 257
Altman, C., 594
Alyea, H. N., 459
Amis, E. S., 531, 536
Anbar, M., 591
Anderson, A., 590
Anderson, J. R., 638
Anderson, W. F., 255
Andrew, E. A., 487
Andrew, K. F., 641
Armstrong, G., 465
Arnell, A. R., 263
Arrhenius, S., 527
Ascough, P. B., 303
Ashmore, P. G., 291, 293, 341, 439, 458
Askey, P. J., 77, 386
Atkinson, B., 259, 303
Axworthy, A. E., Jr., 225, 239, 269, 293, 311, 401, 405, 431
Azzam, A. M., 647

Bach, F., 99, 331
Bacon, R. G. R., 598
Bagotski, V. S., 647
Balandin, A. A., 42, 55
Baldwin, R. R., 453, 457
Ball, D. L., 590
Ballhausen, C. J., 613
Barb, W. G., 596
Barnard, J. A., 259
Barrer, R. M., 618

Barrett, J., 527
Bartlett, P. D., 105, 114, 552, 573
Barton, D. H. R., 259, 265
Bascombe, K. N., 577
Basolo, F., 613
Bateman, H., 428, 448, 449
Bateman, L. C., 536, 551
Bates, J. R., 311
Bauer, S. H., 65
Bawn, C. E. H., 263, 308, 652
Baxendale, J. H., 527, 598
Beard, R. F., 594
Becker, J. A., 639
Becker, R., 473, 651
Beebe, R. A., 634
Beeck, O., 634–636
Bell, E. R., 363, 482
Bell, R. P., 541, 559, 560, 563, 566–570, 572, 574, 577, 578
Bellenot, H., 114
Benford, G. A., 304, 587
Bengorath, W. I., 607
Benington, F., 105
Benson, C., 595
Benson, S. W., 38, 46, 54, 78, 79, 108, 225, 239, 262, 269, 289, 293, 300, 311, 334–336, 338, 371, 387, 388, 400, 401, 405, 431, 515, 516, 568, 601, 606, 608, 623, 652, 657, 664, 665
Berets, D. J., 479
Berg, D., 569
Beringer, F. M., 550
Berlie, M. R., 115, 293
Berliner, E., 586
Berthoud, A., 114
Bieber, R., 569
Biondi, M. A., 308
Bjerrum, N., 525, 528
Blacet, F. E., 104, 383, 384
Blades, A. T., 259, 260
Blagg, J. C. L., 291
Blat, E., 488
Blyumberg, Ya. B., 435

681

Daniels, F., 16, 108, 260, 261, 263, 265, 408, 411
Darwent, B. deB., 293, 298, 570
Davidson, N., 260, 265, 293, 301, 303, 311, 410, 412, 415, 477, 479, 599
Davies, G., 20
Davis, C. W., 541
Davis, T. W., 102
Davis, W., Jr., 370
Dawson, H. M., 566, 572
Dawson, H. T., 541
Day, R. A., 485
Debye, P., 500, 531
Delahay, P., 643
De la Mare, P. B. D., 585, 586
Delbanco, A., 527
Denbigh, K. G., 62
Derbyshire, D. H., 585
De Wolfe, R. H., 555
Dibeler, V. H., 638
Dickel, G., 188
Dickinson, R. G., 114
Dintses, A. I., 353
Dixon-Lewis, G., 461
Dodd, R. E., 303, 374, 375
Dodgen, H., 265
Dodson, R. W., 589
Dole, M., 190
Dootson, F. W., 188
Dorfman, L. M., 299, 366, 370, 372, 374
Döring, W., 651
Dorman, L. M., 364
Dostrovsky, I., 556
Doumani, T. F., 105
Dowden, D. A., 634
Dribos, G., 633
Dubois, J. T., 221
Duffield, R. B., 85
Duke, F. R., 592, 594
Duncan, N. E., 260
Dunkel, M., 639
Dunning, W. G., 648, 651, 652
Durham, R. W., 303
Dushman, S., 160, 161, 175, 176, 181
Dux, W., 111
Dwyer, F. P., 591

Ebersole, E. R., 655
Eckert, E. R. G., 436
Egerton, A., 263, 484
Eggert, H., 107
Eichelberger, L., 584
Eigen, M., 502
Einstein, A., 65
Eley, D. D., 630, 635, 636
Eliashevich, M., 233
Ellis, D. A., 623
Elovitch, S. Y., 640
Eltenton, G. C., 103
Emde, F., 143
Emmett, P. H., 616
Endow, N., 311

Erdey-Gruz, T., 465
Erofeev, B. V., 42
Esson, W., 42
Eucken, A., 179
Evans, M. G., 247, 308, 317, 318, 515, 542
Ewald, A. H., 514, 515, 517
Eyring, H., 42, 247, 256, 283, 285, 309, 316, 342, 517, 535

Fainberg, A. H., 557, 615
Fairclough, R. A., 509
Falterman, C. W., 108, 371
Farenhorst, E., 613
Farkas, A., 16, 25, 58, 106, 292, 631, 635
Farkas, L., 16, 25, 106, 292, 293, 459, 635
Farmer, J. B., 486
Faulkner, I. J., 563
Faull, R. F., 383
Fawcett, E. W., 31
Felder, E., 571
Feller, W., 128, 129
Ferrington, T. E., 603
Ferris, R. C., 371
Fletcher, C. J. M., 386
Fletcher, F. A., 648
Flory, P. J., 600, 610
Foerster, F., 17
Folkins, H. O., 344
Forbes, G. S., 371, 400, 406
Ford, H. W., 311, 458
Fowler, L., 293
Fowler, R. H., 190, 272, 501
Fraenkel, G. K., 510, 608
Francis, A. W., 583, 585
Franck, J., 544
Frank, H. S., 528
Frank, J., 110
Frank-Kamenetskii, D. A., 54, 435, 436, 448, 463, 464, 488
Franklin, J. F., 42
Freiling, E. C., 291
Frejacques, C., 248, 424
French, C. C., 574
Frenkel, J., 651
Friauf, J. B., 479
Friedman, S., 580
Fries, H., 114
Frisch, P., 260
Fristrom, R. M., 322
Frost, A. V., 353
Fugassy, P., 259
Fuguitt, R. E., 255
Fuoss, R. M., 55, 530, 531
Furry, W. H., 188
Fürth, R., 187

Gardener, J. B., 486
Garland, C. W., 200
Garner, H. K., 584
Garner, W. E., 459, 633, 634, 642, 651, 652

Subject Index